Martin Tomlinson · Cen Jung Tjhai
Marcel A. Ambroze · Mohammed Ahmed
Mubarak Jibril

Error-Correction Coding and Decoding

Bounds, Codes, Decoders, Analysis and Applications

Martin Tomlinson
School of Computing, Electronics and Mathematics
Plymouth University
Plymouth, Devon
UK

Cen Jung Tjhai
PQ Solutions Limited
London
UK

Marcel A. Ambroze
School of Computing, Electronics and Mathematics
Plymouth University
Plymouth, Devon
UK

Mohammed Ahmed
School of Computing, Electronics and Mathematics
Plymouth University
Plymouth, Devon
UK

Mubarak Jibril
Satellite Applications and Development
Nigeria Communications Satellite Limited
Abuja
Nigeria

DOI 10.1007/978-3-319-51103-0

Library of Congress Control Number: 2016963415

This book is dedicated to our families and loved ones.

Preface

The research work described in this book is some of the works carried out by the authors whilst working in the Coding Group at the University of Plymouth, U.K. The Coding Group consists of enthusiastic research students, research and teaching staff members providing a very stimulating environment to work. Also being driven by academic research, a significant number of studies were driven by the communications industry with their many varying applications and requirements of error-correcting codes. This partly explains the variety of topics covered in this book.

Plymouth, UK	Martin Tomlinson
London, UK	Cen Jung Tjhai
Plymouth, UK	Marcel A. Ambroze
Plymouth, UK	Mohammed Ahmed
Abuja, Nigeria	Mubarak Jibril

Acknowledgements

We would like to thank all of our past and present research students, our friends and fellow researchers around the world who have helped our understanding of this fascinating and sometimes tricky subject. Special thanks go to our research collaborators Des Taylor, Philippa Martin, Shu Lin, Marco Ferrari, Patrick Perry, Mark Fossorier, Martin Bossert, Eirik Rosnes, Sergey Bezzateev, Markus Grassl, Francisco Cercas and Carlos Salema. Thanks also go to Dan Costello, Bob McEliece, Dick Blahut, David Forney, Ralf Johannason, Bahram Honary, Jim Massey and Paddy Farrell for interesting and informed discussions. We would also like to thank Licha Mued for spending long hours editing the manuscript.

Contents

Acronyms

AG	Algebraic Geometry
ANSI	American National Standards Institute
ARQ	Automatic Repeat Request
AWGN	Additive White Gaussian Noise
BCH	Bose–Chaudhuri–Hocquenghem
BCJR	Bahl–Cocke–Jelinek–Raviv
BDD	Bounded Distance Decoding
BEC	Binary Erasure Channel
BER	Bit Error Rate
BP	Belief Propagation
BSC	Binary Symmetric Channel
CRC	Cyclic Redundancy Check
dB	Decibel
DFT	Discrete Fourier Transform
DRP	Dithered Relative Prime
DVB	Digital Video Broadcasting
EG	Euclidean Geometry
FEC	Forward Error Correction
FER	Frame Error Rate
FSD	Formally Self-Dual
GF	Galois Field
HARQ	Hybrid Automatic Repeat Request
IR	Incremental Redundancy
IRA	Irregular Repeat Accumulate
LDPC	Low-Density Parity-Check Codes
MDS	Maximum Distance Separable
ML	Maximum Likelihood
MRRW	McEliece–Rodemich–Rumsey–Welch
MS	Mattson–Solomon
NP	Nondeterministic Polynomial

PDF	Probability Density Function
PEG	Progressive Edge Growth
PIN	Personal Identification Number
QAM	Quadrature Amplitude Modulation
QR	Quadratic Residue
RS	Reed–Solomon
SDD	Soft-Decision Decoding
SNR	Signal-to-Noise Ratio
WD	Weight Distribution

Part I
Theoretical Performance of Error-Correcting Codes

This part of the book deals with the theoretical performance of error-correcting codes. Upper and lower bounds are given for the achievable performance of error-correcting codes for the additive white Gaussian noise (AWGN) channel. Also given are bounds on constructions of error-correcting codes in terms of normalised minimum distance and code rate. Differences between ideal soft decision decoding and hard decision decoding are also explored. The results from the numerical evaluation of several different code examples are compared to the theoretical bounds with some interesting conclusions.

Chapter 1
Bounds on Error-Correction Coding Performance

1.1 Gallager's Coding Theorem

The sphere packing bound by Shannon [18] provides a lower bound to the frame error rate (FER) achievable by an (n, k, d) code but is not directly applicable to binary codes. Gallager [4] presented his coding theorem for the average FER for the ensemble of all random binary (n, k, d) codes. There are 2^n possible binary combinations for each codeword which in terms of the n-dimensional signal space hypercube corresponds to one vertex taken from 2^n possible vertices. There are 2^k codewords, and therefore 2^{nk} different possible random codes. The receiver is considered to be composed of 2^k matched filters, one for each codeword and a decoder error occurs if any of the matched filter receivers has a larger output than the matched filter receiver corresponding to the transmitted codeword. Consider this matched filter receiver and another different matched filter receiver, and assume that the two codewords differ in d bit positions. The Hamming distancc bctween the two codewords is d. The energy per transmitted bit is $E_s = \frac{k}{n} E_b$, where E_b is the energy per information bit. The noise variance per matched filtered received bit, $\sigma^2 = \frac{N_0}{2}$, where N_0 is the single sided noise spectral density. In the absence of noise, the output of the matched filter receiver for the transmitted codeword is $n\sqrt{E_s}$ and the output of the other codeword matched filter receiver is $(n - 2d)\sqrt{E_s}$. The noise voltage at the output of the matched filter receiver for the transmitted codeword is denoted as $n_c - n_1$, and the noise voltage at the output of the other matched filter receiver will be $n_c + n_1$. The common noise voltage n_c arises from correlation of the bits common to both codewords with the received noise and the noise voltages $-n_1$ and n_1 arise, respectively, from correlation of the other d bits with the received noise. A decoder error occurs if

$$(n - 2d)\sqrt{E_s} + n_c + n_1 > n\sqrt{E_s} + n_c - n_1 \tag{1.1}$$

that is, a decoder error occurs when $2n_1 > 2d\sqrt{E_s}$.

M. Tomlinson et al., *Error-Correction Coding and Decoding*,
Signals and Communication Technology,
DOI 10.1007/978-3-319-51103-0_1

The average noise power associated with n_1 is $\mathrm{d}\sigma^2 = \mathrm{d}\frac{N_0}{2}$ and as the noise is Gaussian distributed, the probability of decoder error, p_d, is given by

$$p_d = \frac{1}{\sqrt{\pi d N_0}} \int_{\mathrm{d}\sqrt{E_s}}^{\infty} \mathrm{e}^{\frac{-x^2}{dN_0}} \,\mathrm{d}x \tag{1.2}$$

This may be expressed in terms of the complementary error function (erfc)

$$\mathrm{erfc}(y) = 2\frac{1}{\sqrt{2\pi}} \int_{y}^{\infty} \mathrm{e}^{\frac{-x^2}{2}} \,\mathrm{d}x \tag{1.3}$$

and

$$p_d = \frac{1}{2}\mathrm{erfc}\left(\sqrt{d\frac{k}{n}\frac{E_b}{N_0}}\right) \tag{1.4}$$

Each of the other $2^k - 2$ codewords may also cause a decoder error but the weight distribution of the code $\mathscr{C}_i$ is usually unknown. However by averaging over all possible random codes, knowledge of the weight distribution of a particular code is not required. The probability of two codewords of a randomly chosen code $\mathscr{C}_i$, differing in d bit positions, $p(d|\mathscr{C}_i)$ is given by the binomial distribution

$$p(d|\mathscr{C}_i) = \frac{\binom{n}{d}}{2^n}, \tag{1.5}$$

where $\binom{a}{b} = \frac{a!}{(a-b)!b!}$. A given linear code $\mathscr{C}_i$ cannot have codewords of arbitrary weight, because the sum of a subset of codewords is also a codeword. However, for non linear codes, p_d may be averaged over all of the codes without this constraint. Thus, we have

$$\overline{p_C} = \sum_{i=1}^{2^{n2^k}} p(d|\mathscr{C}_i)p(\mathscr{C}_i) < \frac{1}{2^{n2^k}} \sum_{d=0}^{n} \sum_{i=1}^{2^{n2^k}} \frac{\binom{n}{d}}{2^{n+1}} \mathrm{erfc}\left(\sqrt{d\frac{k}{n}\frac{E_b}{N_0}}\right) \tag{1.6}$$

Rearranging the order of summation

$$\overline{p_C} < \frac{1}{2^{n2^k}} \sum_{i=1}^{2^{n2^k}} \sum_{d=0}^{n} \frac{\binom{n}{d}}{2^{n+1}} \mathrm{erfc}\left(\sqrt{d\frac{k}{n}\frac{E_b}{N_0}}\right) \tag{1.7}$$

and

$$\overline{p_C} < \frac{1}{2^{n+1}} \sum_{d=0}^{n} \binom{n}{d} \mathrm{erfc}\left(\sqrt{d\frac{k}{n}\frac{E_b}{N_0}}\right). \tag{1.8}$$

Remembering that any of the $2^k - 1$ matched filters may cause a decoder error, the overall probability of decoder error averaged over all possible binary codes $\overline{p_{\text{overall}}}$, is

$$\overline{p_{\text{overall}}} = 1 - (1 - \overline{p_C})^{2^k - 1} < 2^k \overline{p_C} \tag{1.9}$$

and

$$\overline{p_{\text{overall}}} < \frac{2^k}{2^{n+1}} \sum_{d=0}^{n} \binom{n}{d} \text{erfc}\left(\sqrt{d \frac{k}{n} \frac{E_b}{N_0}}\right). \tag{1.10}$$

An analytic solution may be obtained by observing that $\frac{1}{2}\text{erfc}(y)$ is upper bounded by e^{-y^2} and therefore,

$$\overline{p_{\text{overall}}} < \frac{2^k}{2^n} \sum_{d=0}^{n} \binom{n}{d} \text{e}^{-d \frac{k}{n} \frac{E_b}{N_0}} \tag{1.11}$$

and as observed in [21],

$$\left(1 + \text{e}^{-\frac{k}{n} \frac{E_b}{N_0}}\right)^n = \sum_{d=0}^{n} \binom{n}{d} \text{e}^{-d \frac{k}{n} \frac{E_b}{N_0}} \tag{1.12}$$

and

$$\overline{p_C} < \frac{1}{2^n} \left(1 + \text{e}^{-\frac{k}{n} \frac{E_b}{N_0}}\right)^n \tag{1.13}$$

$$\overline{p_{\text{overall}}} < \frac{2^k}{2^n} \left(1 + \text{e}^{-\frac{k}{n} \frac{E_b}{N_0}}\right)^n \tag{1.14}$$

Traditionally, a cut-off rate R_0 is defined after observing that

$$\frac{2^k}{2^n} \left(1 + \text{e}^{-\frac{k}{n} \frac{E_b}{N_0}}\right)^n = 2^k \left(\frac{1 + \text{e}^{-\frac{k}{n} \frac{E_b}{N_0}}}{2}\right)^n \tag{1.15}$$

with

$$2^{R_0} = \frac{2}{1 + \text{e}^{-\frac{k}{n} \frac{E_b}{N_0}}} \tag{1.16}$$

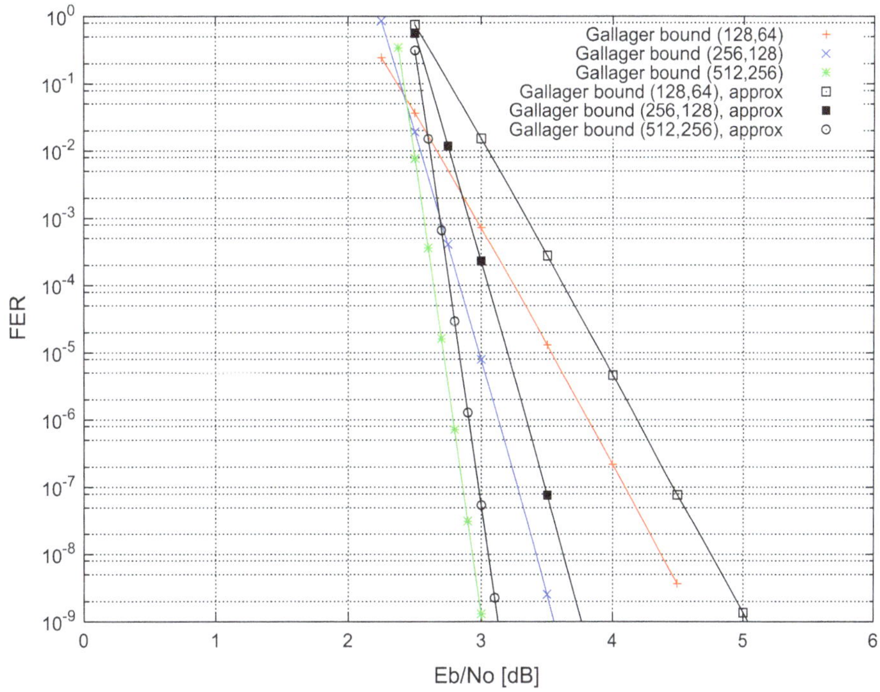

Fig. 1.1 Approximate and exact Gallager bounds for $(128, 2^{64})$, $(256, 2^{128})$ and $(512, 2^{256})$ nonlinear binary codes

then

$$\overline{p_{\text{overall}}} < 2^k 2^{-nR_0} = 2^{k-nR_0} = 2^{-n(R_0-\frac{k}{n})} \tag{1.17}$$

This result may be interpreted as providing the number of information bits of the code is less than the length of the code times the cut-off rate, then the probability of decoder error will approach zero as the length of the code approaches infinity. Alternatively, provided the rate of the code, $\frac{k}{n}$, is less than the cut-off rate, R_0, then the probability of decoder error will approach zero as the length of the code approaches infinity. The cut-off rate R_0, particularly in the period from the late 1950s to the 1970s was used as a practical measure of the code rate of an achievable error-correction system [11, 20–22]. However, plotting the exact expression for probability of decoder error, Eq. (1.10), in comparison to the cut-off rate approximation Eq. (1.17), shows a significant difference in performance, as shown in Fig. 1.1. The codes shown are the $(128, 2^{64})$, $(256, 2^{128})$ and $(512, 2^{256})$ code ensembles of nonlinear, random binary codes. It is recommended that the exact expression, Eq. (1.10) be evaluated unless the code in question is a long code. As a consequence, in the following sections we shall only use the exact Gallager bound.

Shown in Fig. 1.2 is the sphere packing lower bound, offset by the loss attributable to binary transmission and the Gallager upper bound for the $(128, 2^{64})$, $(256, 2^{128})$ and $(512, 2^{256})$ nonlinear binary codes. For each code, the exact Gallager upper bound given by (1.10), is shown. One reason why Gallager's bound is some way

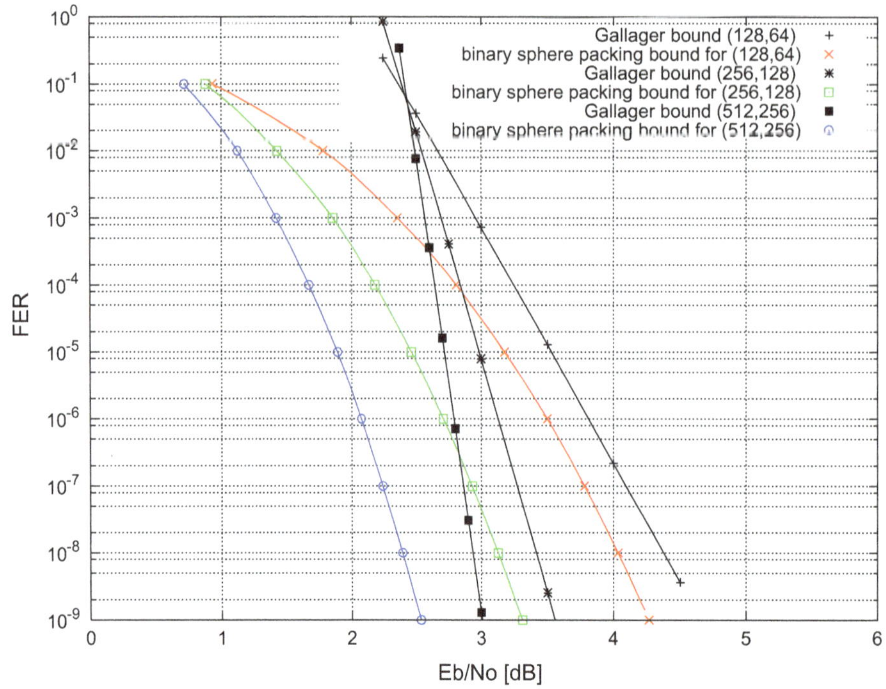

Fig. 1.2 Sphere packing and Gallager bounds for $(128, 2^{64})$, $(256, 2^{128})$ and $(512, 2^{256})$ nonlinear binary codes

from the sphere packing lower bound as shown in Fig. 1.2 is that the bound is based on the union bound and counts all error events as if these are independent. Except for orthogonal codes, this produces increasing inaccuracy as the $\frac{E_b}{N_0}$ is reduced. Equivalently expressed, double counting is taking place since some codewords include the support of other codewords. It is shown in the next section that for linear codes the Gallager bound may be improved by considering the erasure correcting capability of codes, viz. no (n, k) code can correct more than $n - k$ erasures.

1.1.1 Linear Codes with a Binomial Weight Distribution

The weight enumerator polynomial of a code is defined as $A(z)$ which is given by

$$A(z) = \sum_{i=0}^{n} A_i \, z^i \tag{1.18}$$

For many good and exceptional, linear, binary codes including algebraic and quasi-cyclic codes, the weight distributions of the codes closely approximates to a binomial distribution where,

$$A(z) = \frac{1}{2^{n-k}} \sum_{i=0}^{n} \frac{n!}{(n-i)!i!} z^i \tag{1.19}$$

with coefficients A_i given by

$$A_i = \frac{1}{2^{n-k}} \frac{n!}{(n-i)!i!} = \frac{1}{2^{n-k}} \binom{n}{i}. \tag{1.20}$$

Tables of the best-known linear codes have been published from time to time [3, 10, 13, 16, 19] and a regularly updated database is maintained by Markus Grassl [5]. Remembering that for a linear code, the difference between any two codewords is also a codeword, and hence the distribution of the Hamming distances between a codeword and all other codewords is the same as the weight distribution of the code. Accordingly, the overall probability of decoder error, for the same system as before using a bank of 2^k matched filters with each filter matched to a codeword is upper bounded by

$$p_{\text{overall}} < \frac{1}{2} \sum_{d=0}^{n} A_d \text{erfc}\left(\sqrt{d \frac{k}{n} \frac{E_b}{N_0}}\right) \tag{1.21}$$

For codes having a binomial weight distribution

$$p_{\text{overall}} < \frac{1}{2} \sum_{d=0}^{n} \frac{1}{2^{n-k}} \binom{n}{d} \text{erfc}\left(\sqrt{d \frac{k}{n} \frac{E_b}{N_0}}\right) \tag{1.22}$$

which becomes

$$p_{\text{overall}} < \frac{2^k}{2^{n+1}} \sum_{d=0}^{n} \binom{n}{d} \text{erfc}\left(\sqrt{d \frac{k}{n} \frac{E_b}{N_0}}\right). \tag{1.23}$$

It will be noticed that this equation is identical to Eq. (1.10). This leads to the somewhat surprising conclusion that the decoder error probability performance of some of the best-known, linear, binary codes is the same as the average performance of the ensemble of all randomly chosen, binary nonlinear codes having the same values for n and k. Moreover, some of the nonlinear codes must have better performance than their average, and hence some nonlinear codes must be better than the best-known linear codes.

A tighter upper bound than the Gallager bound may be obtained by considering the erasure correcting capability of the code. It is shown in Chap. 14 that for the erasure channel, given a probability of erasure, p, the probability of decoder error, $P_{\text{code}}(p)$, is bounded by

$$P_{\text{code}}(p) < \sum_{s=d_{min}}^{n-k} \sum_{j=d_{min}}^{s} A_j \frac{(n-j)!\,(n-s)!}{(s-j)!} p^s (1-p)^{(n-s)} + \sum_{s=n-k+1}^{n} p^s (1-p)^{(n-s)}. \tag{1.24}$$

In Eq. (1.24), the first term depends upon the weight distribution of the code while the second term is independent of the code. The basic principle in the above equation is that an erasure decoder error is caused if an erasure pattern includes the support of a codeword. Since no erasure pattern can be corrected if it contains more than $n-k$ errors, only codewords with weight less than or equal to $n-k$ are involved. Consequently, a much tighter bound is obtained than a bound based on the union bound as there is less likelihood of double counting error events.

Considering the maximum likelihood decoder consisting of a bank of correlators, a decoder error occurs if one correlator has a higher output than the correlator corresponding to the correct codeword where the two codewords differ in s bit positions. To the decoder, it makes no difference if the decoder error event is due to erasures, from the erasure channel, or Gaussian noise from the AWGN channel; the outcome is the same. For the erasure channel, the probability of this error event due to erasures, $P_{\text{erasure}}(p)$ is

$$P_{\text{erasure}}(p) = p^s \tag{1.25}$$

The probability of this error event due to noise, $P_{\text{noise}}\left(\frac{E_b}{N_0}\right)$ is

$$P_{\text{noise}}\left(\frac{E_b}{N_0}\right) = \frac{1}{2}\text{erfc}\left(\sqrt{s\frac{k}{n}\frac{E_b}{N_0}}\right) \tag{1.26}$$

Equating Eqs. (1.25) to (1.26), for these probabilities gives a relationship between the erasure probability, p and $\frac{E_b}{N_0}$ and the Hamming distance, s.

$$p^s = \frac{1}{2}\text{erfc}\left(\sqrt{s\frac{k}{n}\frac{E_b}{N_0}}\right) \tag{1.27}$$

For many codes, the erasure decoding performance is determined by a narrow range of Hamming distances and the variation in $\frac{E_b}{N_0}$ as a function of s is insignificant. This is illustrated in Fig. 1.3 which shows the variation in $\frac{E_s}{N_0}$ as a function of s and p.

It is well known that the distance distribution for many linear, binary codes including BCH codes, Goppa codes, self-dual codes [7, 8, 10, 14] approximates to a binomial distribution. Accordingly,

$$A_j \approx \frac{n!}{(n-j)!\,j!\,2^{n-k}}. \tag{1.28}$$

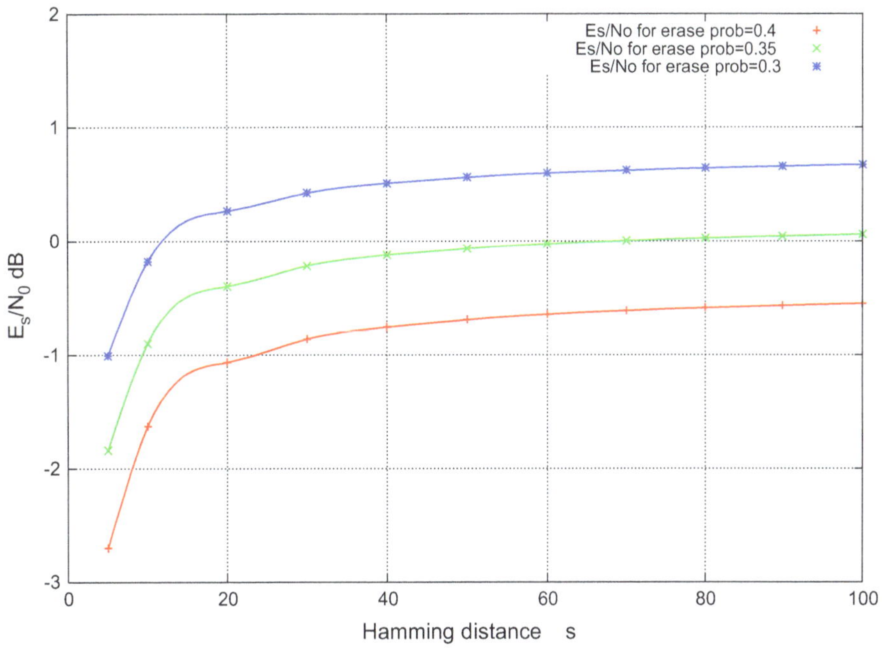

Fig. 1.3 $\frac{E_s}{N_0}$ as a function of Hamming distance s and erasure probability p

Substituting this into Eq. (1.24) produces

$$P_{\text{code}}(p) < \sum_{s=1}^{n-k} \frac{2^s - 1}{2^{n-k}} \binom{n}{s} p^s (1-p)^{(n-s)} + \sum_{s=n-k+1}^{n} p^s (1-p)^{(n-s)} \tag{1.29}$$

With the assumption of a binomial weight distribution, an upper bound may be determined for the erasure performance of any (n, k) code, and in turn, equating Eq. (1.25) with Eq. (1.26) produces an upper bound for the AWGN channel. For example, Fig. 1.4 shows an upper bound of the erasure decoding performance of a $(128, 64)$ code with a binomial weight distribution.

Using Eq. (1.27), the decoding performance may be expressed in terms of $\frac{E_b}{N_0}$ and Fig. 1.5 shows the upper bound of the decoding performance of the same code against Gaussian noise, as a function of $\frac{E_b}{N_0}$.

The comparison of the sphere packing bound and the Gallager bounds is shown in Fig. 1.6. Also shown in Fig. 1.6 is the performance of the BCH (128, 64, 22) code evaluated using the modified Dorsch decoder. It can be seen from Fig. 1.6 that the erasure-based upper bound is very close to the sphere packing lower bound and tighter than the Gallager bound.

Figure 1.7 gives the bounds for the (512, 256) and (256, 128) codes. It will be noticed that the gap between the sphere packing bound and the erasure-based upper bound increases with code length, but is tighter than the Gallager bound.

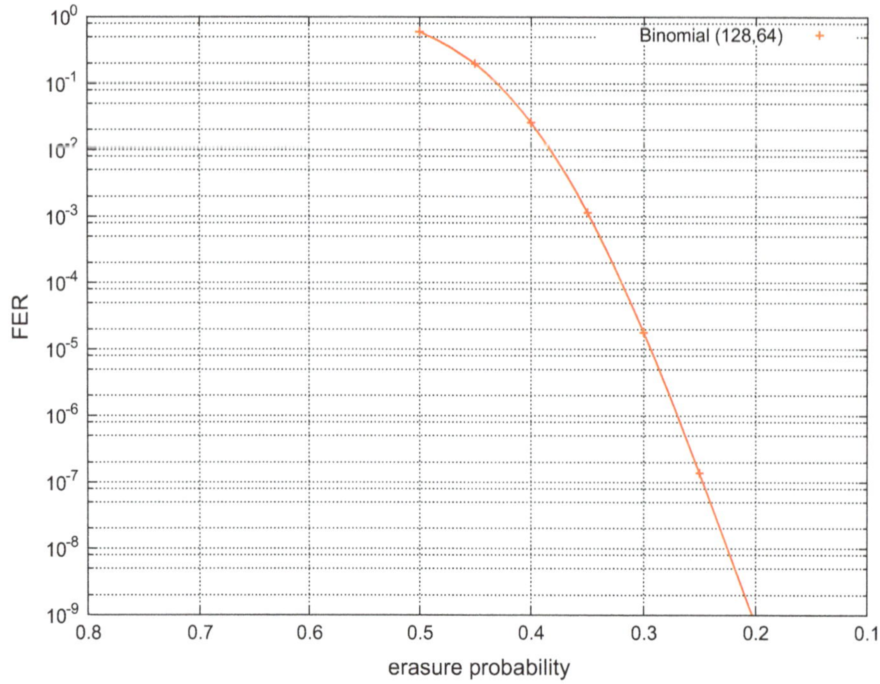

Fig. 1.4 Erasure decoding performance of a (128, 64) code with a binomial weight distribution

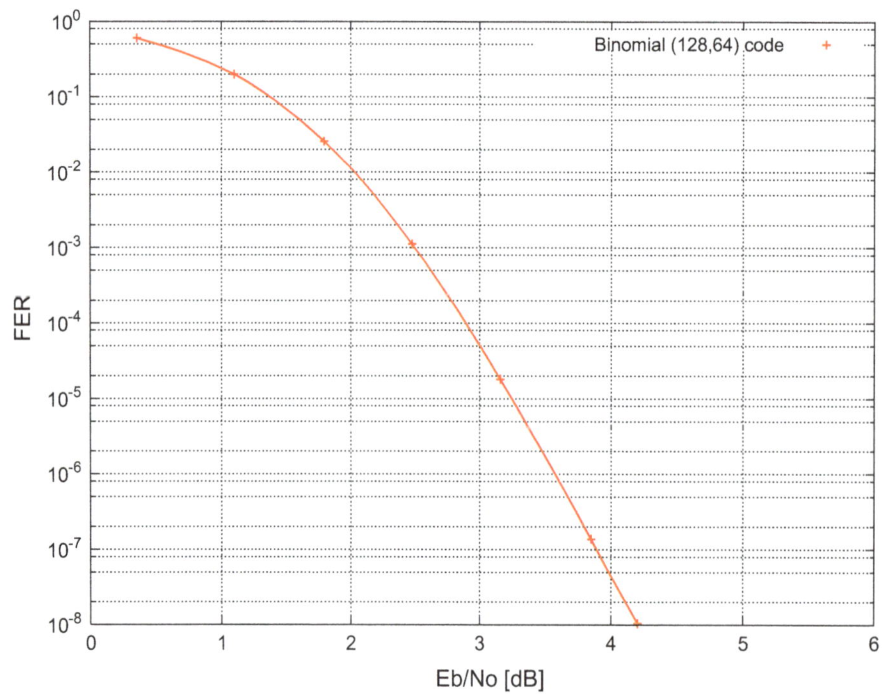

Fig. 1.5 Decoding performance of a (128, 64) code with a binomial weight distribution for Gaussian noise

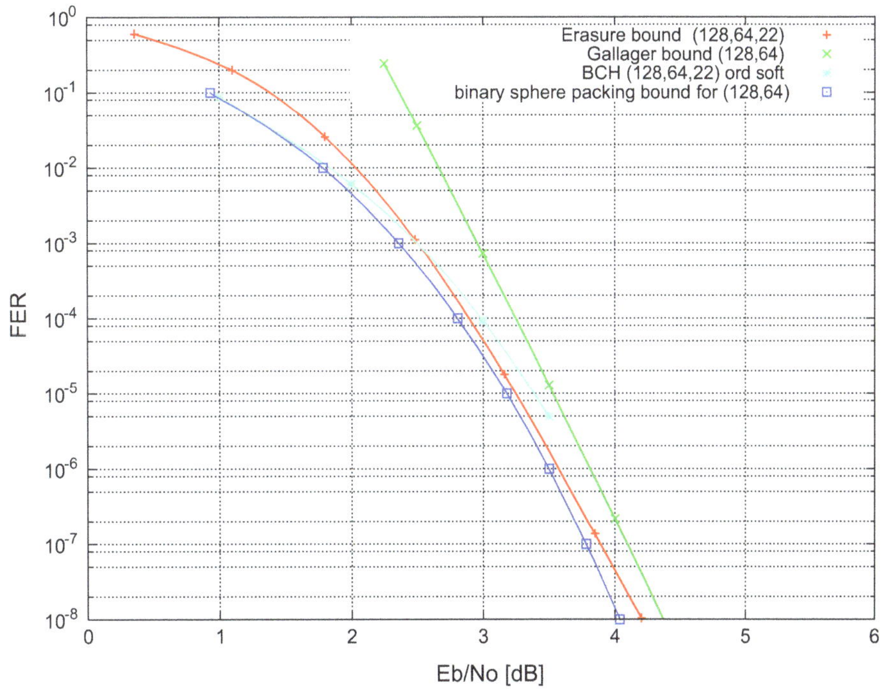

Fig. 1.6 Comparison of sphere packing and Gallager bounds to the upper bound based on erasure performance for the (128, 64) code with a binomial weight distribution

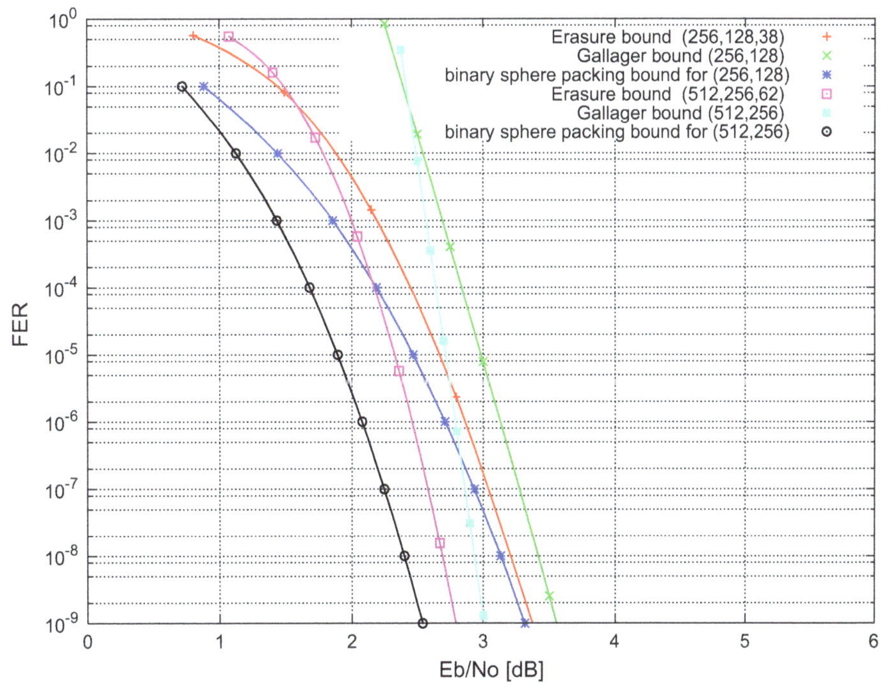

Fig. 1.7 Comparison of sphere packing and Gallager bounds to the upper bound based on erasure performance for (256, 128) and (512, 256) codes with a binomial weight distribution

1.1.2 Covering Radius of Codes

The covering radius of a code, c_r if it is known, together with the weight spectrum of the low-weight codewords may be used to tighten the Union bound upper bound on decoder performance given by Eq. (1.23). The covering radius of a code is defined as the minimum radius which when placed around each codeword includes all possible q^n vectors. Equivalently, the covering radius is the maximum number of hard decision errors that are correctable by the code. For a perfect code, such as the Hamming codes, the covering radius is equal to $\frac{d_{min}-1}{2}$. For the $[2^m-1, 2^m-m-1, 3]$ Hamming codes, the covering radius is equal to 1 and for the (23, 12, 7) Golay code the covering radius is equal to 3. As a corollary, for any received vector in Euclidean space, there is always a codeword within a Euclidean distance of $c_r + 0.5$. It follows that the summation in Eq. (1.23) may be limited to codewords of weight $2c_r + 1$ to produce

$$p_{\text{overall}} < \frac{2^k}{2^{n+1}} \sum_{d=0}^{2c_r+1} \binom{n}{d} \text{erfc}\left(\sqrt{d\frac{k}{n}\frac{E_b}{N_0}}\right). \tag{1.30}$$

1.1.3 Usefulness of Bounds

The usefulness of bounds may be realised from Fig. 1.8 which shows the performance of optimised codes and decoders all (512, 256) codes for a turbo code, LDPC code and a concatenated code.

1.2 Bounds on the Construction of Error-Correcting Codes

A code (linear or nonlinear), $\mathscr{C}$, defined in a finite field of size q can be described with its length n, number of codewords[1] M and minimum distance d. We use $(n, M, d)_q$ to denote these four important parameters of a code. Given any number of codes defined in a field of size q with the same length n and distance d, the code with the maximum number of codewords M is the most desirable. Equivalently, one may choose to fix n, M and q and maximise d or fix M, d and q and maximise n. As a result, it is of interest in coding theory to determine the maximum number of codewords possible of any code defined in a field of size q, with minimum distance d and length n. This number is denoted by $A_q(n, d)$. Bounds on $A_q(n, d)$ are indicators to the maximum performance achievable from any code with parameters $(n, M, d)_q$. As a result, these bounds are especially useful when one constructs good error-correcting codes. The tables in [5] contain the best-known upper and lower bounds on $A_q(n, d)$ for linear codes. The tables in [9] contain bounds on $A_2(n, d)$ for nonlinear binary codes.

[1]Where the code dimension $k = \log_q M$.

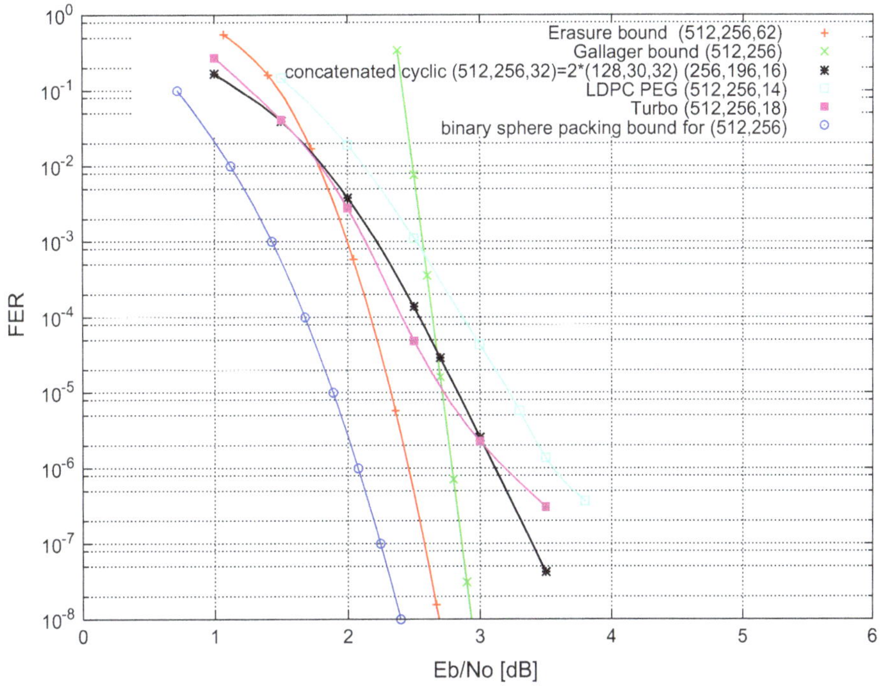

Fig. 1.8 Comparison of sphere packing, Gallager and erasure-based bounds to the performance realised for a (512, 256, 18) turbo code, (512, 256, 14) LDPC code and (512, 256, 32) concatenated code

Lower bounds on $A_q(n, d)$ tend to be code specific; however, there are several generic upper bounds. As an example, consider the best-known upper and lower bounds on $A_2(128, d)$ obtained from the tables in [5]. These are shown in Fig. 1.9 for the range $1 \leq d \leq 128$. Optimal codes of length $n = 128$ are codes whose lower and upper bounds on $A_2(128, d)$ coincide. The two curves coincide when k is small and d is large or vice versa. The gap between the upper and lower bounds that exists for other values of k and d suggests that one can construct good codes with a larger number of codewords and improve the lower bounds. An additional observation is that extended BCH codes count as some of the known codes with the most number of codewords.

It is often useful to see the performance of codes as their code lengths become arbitrarily large. We define the information rate

$$\alpha_q(\delta) = \lim_{n \to \infty} \frac{\log_q(A_q(n, \delta n))}{n}, \tag{1.31}$$

where $\delta = \frac{d}{n}$ is called the relative distance. Since the dimension of the code is defined as $k = \log_q(A_q(n, \delta n))$, then a bound on the information rate $\alpha_q(\delta)$ is a bound on $\frac{k}{n}$, as $n \to \infty$.

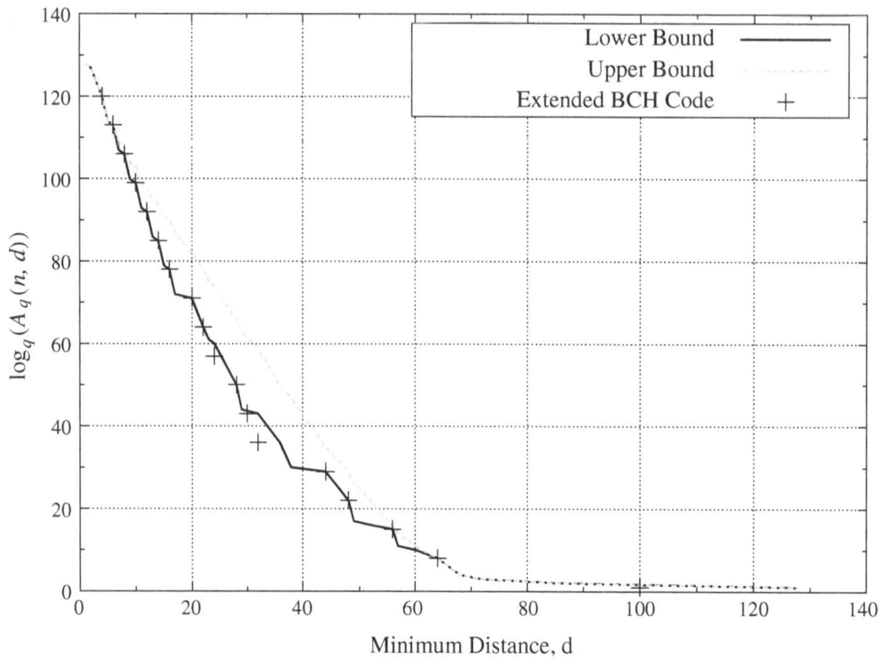

Fig. 1.9 Upper and lower bounds on $A_2(128, d)$

1.2.1 *Upper Bounds*

1.2.1.1 Sphere Packing (Hamming) Bound

Let $V_q(n, t)$ represent the number of vectors in each sphere then,

$$V_q(n,t) = \sum_{i=0}^{t} \binom{n}{i} (q-1)^i. \tag{1.32}$$

Theorem 1.1 (Sphere Packing Bound) *The maximum number of codewords $A_q(n, d)$ is upper bounded by,*

$$A_q(n,d) \leq \frac{q^n}{\sum\limits_{i=0}^{t} \binom{n}{i} (q-1)^i}$$

Proof A code $\mathscr{C}$ is a subset of a vector space $\mathrm{GF}(q)^n$. Each codeword of $\mathscr{C}$ has only those vectors $\mathrm{GF}(q)^n$ but not in $\mathscr{C}$ lying at a hamming distance $t = \left\lfloor \frac{d-1}{2} \right\rfloor$ from it since codewords are spaced at least d places apart. In other words, no codewords lie in a sphere of radius t around any codeword of $\mathscr{C}$. As such, for counting purposes, these spheres can represent individual codewords. The Hamming bound counts the number of such non-overlapping spheres in the vector space $\mathrm{GF}(q)^n$.

Codes that meet this bound are called *perfect* codes. In order to state the asymptotic sphere packing bound, we first define the $q$ary entropy function, $H_q(x)$, for the values $0 \le x \le r$,

$$H_q(x) = \begin{cases} 0 & \text{if } \ x = 0 \\ x \log_q(q-1) - x \log_q x - (1-x)\log_q(1-x) & \text{if } \ 0 < x \le r \end{cases} \tag{1.33}$$

Theorem 1.2 (Asymptotic Sphere Packing Bound) *The information rate of a code* $\alpha_q(\delta)$ *is upper bounded by,*

$$\alpha_q(\delta) \le 1 - H_q\left(\frac{\delta}{2}\right)$$

for the range $0 < \delta \le 1 - q^{-1}$.

1.2.1.2 Plotkin Bound

Theorem 1.3 (Plotkin Bound) *Provided* $d > \theta n$, *where* $\theta = 1 - q^{-1}$, *then,*

$$A_q(n, d) \le \left\lfloor \frac{d}{d - \theta n} \right\rfloor$$

Proof Let $S = \sum d(\mathbf{x}, \mathbf{y})$ for all codewords $\mathbf{x}, \mathbf{y} \in \mathscr{C}$, and $\mathbf{x} \ne \mathbf{y}$, and $d(\mathbf{x}, \mathbf{y})$ denotes the hamming distance between codewords $\mathbf{x}$ and $\mathbf{y}$. Assume that all the codewords of $\mathscr{C}$ are arranged in an $M \times n$ matrix D. Since $d(\mathbf{x}, \mathbf{y}) \ge d$,

$$S \ge \frac{M!}{(M-2)!} d = M(M-1)d. \tag{1.34}$$

Let $n_{i,\alpha}$ be the number of times an element α in the defining field of the code GF(q) occurs in the ith column of the matrix D. Then, $\sum_{\alpha \in \mathrm{GF}(q)} n_{i,\alpha} = M$. For each $n_{i,\alpha}$ there are $M - n_{i,\alpha}$ entries of the matrix D in column i that have elements other than α. These entries are a hamming distance 1 from the $n_{i,\alpha}$ entries and there are n possible columns. Thus,

$$\begin{aligned} S &= n \sum_{i=1}^{n} \sum_{\alpha \in \mathrm{GF}(q)} n_{i,\alpha}(M - n_{i,\alpha}) \\ &= nM^2 - \sum_{i=1}^{n} \sum_{\alpha \in \mathrm{GF}(q)} n_{i,\alpha}^2. \end{aligned} \tag{1.35}$$

From the Cauchy–Schwartz inequality,

$$\left(\sum_{\alpha\in\mathrm{GF}(q)} n_{i,\alpha}\right)^2 \leq q \sum_{\alpha\in\mathrm{GF}(q)} n_{i,\alpha}^2. \tag{1.36}$$

Equation (1.35) becomes,

$$S \leq nM^2 - \sum_{i=1}^{n} q^{-1}\left(\sum_{\alpha\in\mathrm{GF}(q)} n_{i,\alpha}\right)^2 \tag{1.37}$$

Let $\theta = 1 - q^{-1}$,

$$\begin{aligned} S &\leq nM^2 - \sum_{i=1}^{n} q^{-1}\left(\sum_{\alpha\in\mathrm{GF}(q)} n_{i,\alpha}\right)^2 \\ &\leq nM^2 - q^{-1}nM^2 \\ &\leq n\theta M^2. \end{aligned} \tag{1.38}$$

Thus from (1.34) and (1.38) we have,

$$M(M-1)d \leq S \leq n\theta M^2 \tag{1.39}$$

$$M \leq \left\lfloor \frac{d}{d-\theta n} \right\rfloor \tag{1.40}$$

and clearly $d > \theta n$.

Corollary 1.1 (Asymptotic Plotkin Bound) *The asymptotic Plotkin bound is given by,*

$$\begin{aligned} \alpha_q(\delta) &= 0 && \text{if} \quad \theta \leq \delta \leq 1 \\ \alpha_q(\delta) &\leq 1 - \frac{\delta}{\theta} && \text{if} \quad 0 \leq \delta \leq \theta. \end{aligned}$$

1.2.1.3 Singleton Bound

Theorem 1.4 (Singleton Bound) *The maximum number of codewords $A_q(n,d)$ is upper bounded by,*

$$A_q(n,d) \leq q^{n-d+1}.$$

Codes that meet this bound with equality, i.e. $d = n - k + 1$, are called maximum distance separable codes (MDS). The asymptotic Singleton bound is given Theorem 1.5.

Theorem 1.5 (Asymptotic Singleton Bound) *The information rate $\alpha_q(\delta)$ is upper bounded by,*

$$\alpha_q(n, \delta) \leq 1 - \delta.$$

The asymptotic Singleton bound does not depend on the field size q and is a straight line with a negative slope in a plot of $\alpha_q(\delta)$ against δ for every field.

1.2.1.4 Elias Bound

Another upper bound is the Elias bound [17]. This bound was discovered by P. Elias but was never published by the author. We only state the bound here as the proof is beyond the scope of this text. For a complete treatment see [6, 10].

Theorem 1.6 (Elias Bound) *A code $\mathscr{C}$ of length n with codewords having weight at most w, $w < \theta n$ with $\theta = 1 - q^{-1}$ has,*

$$d \leq \frac{Mw}{M-1}\left(2 - \frac{w}{\theta n}\right)$$

Theorem 1.7 (Asymptotic Elias Bound) *The information rate $\alpha_q(\delta)$ is upper bounded by,*

$$\alpha_q(\delta) \leq 1 - H_q(\theta - \sqrt{\theta(\theta - \delta)})$$

provided $0 < \delta < \theta$ *where* $\theta = 1 - q^{-1}$.

1.2.1.5 MRRW Bounds

The McEliece–Rodemich–Rumsey–Welch (MRRW) bounds are asymptotic bounds obtained using linear programming.

Theorem 1.8 (Asymptotic MRRW Bound I) *Provided* $0 < r < \theta$, $\theta = 1 - q^{-1}$ *then,*

$$\alpha_q(\delta) \leq H_q\left(\frac{1}{q}(q - 1 - (q-2)\delta - 2\sqrt{\delta(1-\delta)(q-1)})\right)$$

The second MRRW bound applies to the case when $q = 2$.

Theorem 1.9 (MRRW Bound II) *Provided* $0 < \delta < \frac{1}{2}$ *and* $q = 2$ *then,*

$$\alpha_2(\delta) \leq \min_{0 \leq u \leq 1-2\delta}\{1 + g(u^2) - g(u^2 + 2\delta u + 2\delta)\}$$

where

$$g(x) = H_2\left(\frac{1 - \sqrt{1-x}}{2}\right).$$

The MRRW bounds are the best-known upper bound on the information rate for the binary case. The MRRW-II bound is better than the MRRW-I bound when δ is small and $q = 2$. An in depth treatment and proofs of the bounds can be found in [12].

1.2.2 Lower Bounds

1.2.2.1 Gilbert–Varshamov Bound

Theorem 1.10 (Gilbert–Varshamov Bound) *The maximum number of codewords $A_q(n, d)$ is lower bounded by,*

$$A_q(n, d) \geq \frac{q^n}{V_q(n, d-1)} = \frac{q^n}{\sum_{i=0}^{d-1} \binom{n}{i} (q-1)^i}.$$

Proof We know that $V_q(n, d-1)$ represents the volume of a sphere centred on a codeword of $\mathscr{C}$ of radius $d-1$. Suppose $\mathscr{C}$ has $A_q(n, d)$ codewords. Every vector $\mathbf{v} \in \mathbb{F}_q^n$ lies within a sphere of volume $V_q(n, d-1)$ centred at a codeword of $\mathscr{C}$ as such,

$$\left| \bigcup_{i=1}^{A_q(n,d)} S_i \right| = |\mathbb{F}_q^n|,$$

where S_i is a set containing all vectors in a sphere of radius $d-1$ centred on a codeword of $\mathscr{C}$. The spheres S_i are not mutually disjoint. If we assume S_i are mutually disjoint then,

$$A_q(n, d)V_q(n, d-1) \geq |\mathbb{F}_q^n|.$$

Theorem 1.11 *The information rate of a code is lower bounded by,*

$$\alpha_q(\delta) \geq 1 - H_q(\delta)$$

for $0 \leq \delta \leq \theta$, $\theta = 1 - q^{-1}$.

Figures 1.10 and 1.11 show the asymptotic upper and lower bounds for the cases where $q = 2$ and $q = 32$, respectively. Figure 1.11 shows that the MRRW bounds are the best-known upper bounds when $q = 2$. Observe that the Plotkin bound is the best upper bound for the case when $q = 32$.

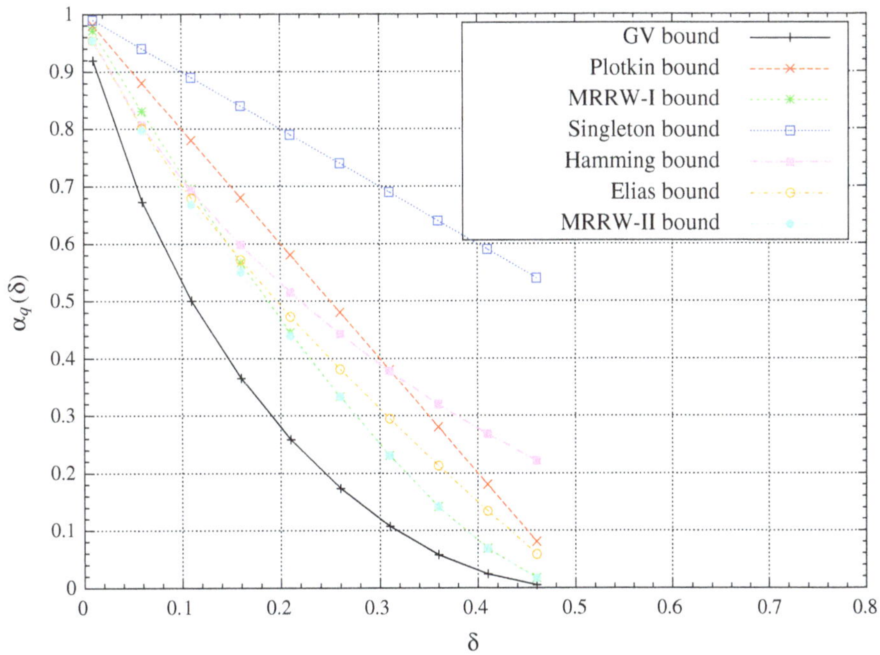

Fig. 1.10 $\alpha_q(\delta)$ against δ for $q = 2$

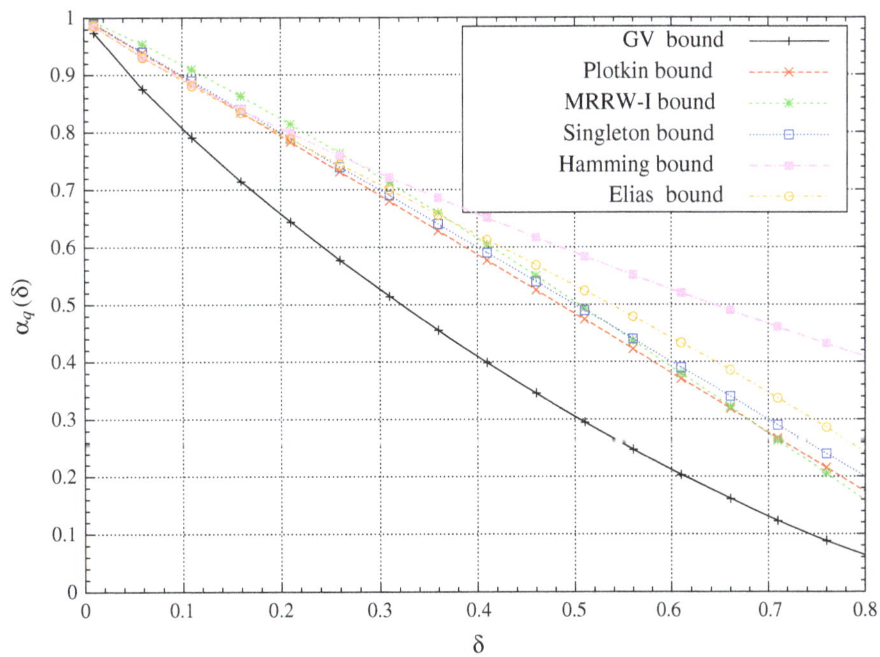

Fig. 1.11 $\alpha_q(\delta)$ against δ for $q = 32$

Table 1.1 Ranges for codes

Finite field	Range
$\mathbb{F}_2$	$1 \leq k \leq n \leq 256$
$\mathbb{F}_3$	$1 \leq k \leq n \leq 243$
$\mathbb{F}_4$	$1 \leq k \leq n \leq 256$
$\mathbb{F}_5$	$1 \leq k \leq n \leq 130$
$\mathbb{F}_7$	$1 \leq k \leq n \leq 100$
$\mathbb{F}_8$	$1 \leq k \leq n \leq 130$
$\mathbb{F}_9$	$1 \leq k \leq n \leq 130$

1.2.3 Lower Bounds from Code Tables

Tables of best-known codes are maintained such that if a code defined in a field q is constructed with an evaluated and verifiable minimum Hamming distance d that exceeds a previously best-known code with the same length n and dimension, the dimension of the new code is a lower bound on $A_q(n, d)$. The first catalogue of best-known codes was presented by Calabi and Myrvaagnes [2] containing binary codes of length n and dimension k in the range $1 \leq k \leq n \leq 24$. Brouwer and Verhoeff [1] subsequently presented a comprehensive update to the tables which included codes with finite fields up to size 9 with the ranges for k and n.

At present, Grassl [5] maintains a significantly updated version of the tables in [1]. The tables now contain codes with k and n in ranges from Table 1.1. Finally, Schimd and Shurer [15] provide an online database for optimal parameters of (t, m, s)-nets, (t, s)-sequences, orthogonal arrays, linear codes and ordered orthogonal arrays. These are relatively new tables and give the best-known codes up to finite fields of size 256. The search for codes whose dimension exceeds the best-known lower bounds on $A_q(n, d)$ is an active area of research with the research community constantly finding improvements.

1.3 Summary

In this chapter we discussed the theoretical performance of binary codes for the additive white Gaussian noise (AWGN) channel. In particular the usefulness of Gallager's coding theorem for binary codes was explored. By assuming a binomial weight distribution for linear codes, it was shown that the decoder error probability performance of some of the best, known linear, binary codes is the same as the average performance of the ensemble of all randomly chosen, binary nonlinear codes having the same length and dimension. Assuming a binomial weight distribution, an upper bound was determined for the erasure performance of any code, and it was shown that this can be translated into an upper bound for code performance in the AWGN channel. Different theoretical bounds on the construction of error-correction codes were discussed. For the purpose of constructing good error-correcting codes, theoretical upper bounds provide fundamental limits beyond which no improvement is possible.

References

1. Brouwer, A., Verhoeff, T.: An updated table of minimum-distance bounds for binary linear codes. IEEE Trans. Inf. Theory **39**(2), 662–677 (1993)
2. Calabi, L., Myrvaagnes, E.: On the minimal weight of binary group codes. IEEE Trans. Inf. Theory **10**(4), 385–387 (1964)
3. Chen, C.L.: Computer results on the minimum distance of some binary cyclic codes. IEEE Trans. Inf. Theory **16**(3), 359–360 (1970)
4. Gallager, R.G.: A simple derivation of the coding theorem and some applications. IEEE Trans. Inf. Theory **11**(1), 459–470 (1960)
5. Grassl, M.: Code Tables: Bounds on the parameters of various types of codes, http://www.codetables.de (2007)
6. Huffman, W.C., Pless, V.S.: Fundamentals of Error-Correcting Codes. Cambridge University Press, Cambridge (2003). ISBN 0 521 78280 5
7. Krasikov, I., Litsyn, S.: On spectra of BCH codes. IEEE Trans. Inf. Theory **41**(3), 786–788 (1995)
8. Krasikov, I., Litsyn, S.: On the accuracy of the binomial approximation to the distance distribution of codes. IEEE Trans. Inf. Theory **41**(5), 1472–1474 (1995)
9. Litsyn, S.: Table of nonlinear binary codes, http://www2.research.att.com/~njas/codes/And/ (1999)
10. MacWilliams, F.J., Sloane, N.J.A.: The Theory of Error-Correcting Codes. North-Holland, Amsterdam (1977)
11. Massey, J.: Coding and modulation in digital communication. In: Proceedings of International Zurich Seminar on Digital Communication, pp. E2(1)–E2(24) (1974)
12. McEliece, R., Rodemich, E., Rumsey, H., Welch, L.: New upper bounds on the rate of a code via the delsarte-macwilliams inequalities. IEEE Trans. Inf. Theory **23**(2), 157–166 (1977)
13. Promhouse, G., Tavares, S.E.: The minimum distance of all binary cyclic codes of odd lengths from 69 to 99. IEEE Trans. Inf. Theory **24**(4), 438–442 (1978)
14. Roychowdhury, V.P., Vatan, F.: Bounds for the weight distribution of weakly self-dual codes. IEEE Trans. Inf. Theory **47**(1), 393–396 (2001)
15. Schimd, W., Shurer, R.: Mint: a database for optimal net parameters, http://mint.sbg.ac.at (2004)
16. Schomaker, D., Wirtz, M.: On binary cyclic codes of odd lengths from 101 to 127. IEEE Trans. Inf. Theory **38**(2), 516–518 (1992)
17. Shannon, C., Gallager, R., Berlekamp, E.: Lower bounds to error probability for coding on discrete memoryless channels, i. Inf. Control **10**(1), 65–103 (1967)
18. Shannon, C.E.: Probability of error for optimal codes in a Gaussian channel. Bell Syst. Tech. J. **38**(3), 611–656 (1959)
19. Tjhai, C., Tomlinson, M.: Results on binary cyclic codes. Electron. Lett. **43**(4), 234–235 (2007)
20. Wozencraft, J.: Sequential decoding for reliable communications. Technical Report No. 325 Research Laboratory of Electronics, MIT (1957)

21. Wozencraft, J., Jacobs, I.: Principles of Communication Engineering. Wiley, New York (1965)
22. Wozencraft, J., Kennedy, R.: Modulation and demodulation for probabilistic coding. IEEE Trans. Inf. Theory IT **12**, 291–297 (1966)

Chapter 2
Soft and Hard Decision Decoding Performance

2.1 Introduction

This chapter is concerned with the performance of binary codes under maximum likelihood soft decision decoding and maximum likelihood hard decision decoding. Maximum likelihood decoding gives the best performance possible for a code and is therefore used to assess the quality of the code. In practice, maximum likelihood decoding of codes is computationally difficult, and as such, theoretical bounds on the performance of codes are used instead. These bounds are in lower and upper form and the expected performance of the code is within the region bounded by the two. For hard decision decoding, lower and upper bounds on maximum likelihood decoding are computed using information on the coset weight leader distribution. For maximum likelihood soft decision decoding, the bounds are computed using the weight distribution of the codes. The union bound is a simple and well-known bound for the performance of codes under maximum likelihood soft decision decoding. The union bound can be expressed as both an upper and lower bound. Using these bounds, we see that as the SNR per bit becomes large the performance of the codes can be completely determined by the lower bound. However, this is not the case with the bounds on maximum likelihood hard decision decoding of codes. In general, soft decision decoding has better performance than hard decision decoding and being able to estimate the performance of codes under soft decision decoding is attractive. Computation of the union bound requires the knowledge of the weight distribution of the code. In Sect. 2.3.1, we use a binomial approximation for the weight distribution of codes for which the actual computation of the weight distribution is prohibitive. As a result, it possible to calculate within an acceptable degree of error the region in which the performance of codes can be completely predicted.

M. Tomlinson et al., *Error-Correction Coding and Decoding*,
Signals and Communication Technology,
DOI 10.1007/978-3-319-51103-0_2

2.2 Hard Decision Performance

2.2.1 *Complete and Bounded Distance Decoding*

Hard decision decoding is concerned with decoding of the received sequence in hamming space. Typically, the real-valued received sequence is quantised using a threshold to a binary sequence. A bounded distance decoder is guaranteed to correct all t errors or less, where t is called the packing radius and is given by:

$$t = \left\lfloor \frac{d-1}{2} \right\rfloor$$

and d is the minimum hamming distance of the code. Within a sphere centred around a codeword in the hamming space of radius t there is no other codeword, and the received sequence in this sphere is closest to the codeword. Beyond the packing radius, some error patterns may be corrected. A complete decoder exhaustively matches all codewords to the received sequence and selects the codeword with minimum hamming distance. A complete decoder is also called a minimum distance decoder or maximum likelihood decoder. Thus, a complete decoder corrects some patterns of error beyond the packing radius. The complexity of implementing a complete decoder is known to be NP-complete [3]. Complete decoding can be accomplished using a standard array. In order to discuss standard array decoding, we first need to define cosets and coset leaders.

Definition 2.1 A coset of a code $\mathscr{C}$ is a set containing all the codewords of $\mathscr{C}$ corrupted by a single sequence $\mathbf{a} \in \mathbb{F}_q^n \setminus \mathscr{C} \cup \{\mathbf{0}\}$.

A coset of a binary code contains 2^k sequences and there are 2^{n-k} possible cosets. Any sequence of minimum hamming weight in a coset can be chosen as a coset leader. In order to use a standard array, the coset leaders of all the cosets of a code must be known. We illustrate complete decoding with an example. Using a (7, 3) dual Hamming code with the following generator matrix

$$G = \begin{bmatrix} 1000111 \\ 0101011 \\ 0011101 \end{bmatrix}$$

This code has codewords

$$C = \begin{Bmatrix} 0000000 \\ 1000111 \\ 0101011 \\ 0011101 \\ 1101100 \\ 0110110 \\ 1011010 \\ 1110001 \end{Bmatrix}$$

Coset Leaders							
0000000	**1000111**	**0101011**	**0011101**	**1101100**	**0110110**	**1011010**	**1110001**
0000001	1000110	0101010	0011100	1101101	0110111	1011011	1110000
0000010	1000101	0101001	0011111	1101110	0110100	1011000	1110011
0000100	1000011	0101111	0011001	1101000	0110010	1011110	1110101
0001000	1001111	0100011	0010101	1100100	0111110	1010010	1111001
0010000	1010111	0111011	0001101	1111100	0111110	1001010	1100001
0100000	1100111	0001011	0111101	1001100	0010110	1111010	1010001
1000000	0000111	1101011	1011101	0101100	1110110	0011010	0110001
0000011	1000100	0101000	0011110	1101111	0110101	1011001	1110010
0000110	1000001	0101101	0011011	1101010	0110000	1011100	1110111
0001100	1001011	0100111	0010001	1100000	0111010	1010110	1111101
0011000	1011111	0110011	0000101	1110100	0101110	1000010	1101001
0001010	1001101	0100001	0010111	1100110	0111100	1010000	1111011
0010100	1010011	0111111	0001001	1111000	0100010	1001110	1100101
0010010	1010101	0111001	0001111	1111110	0100100	1001000	1100011
0001110	1001001	0100101	0010011	1100010	0111000	1010100	1111111

Fig. 2.1 Standard array for the (7, 3, 4) binary code

Complete decoding can be accomplished using standard array decoding. The example code is decoded using standard array decoding as follows, The top row of the array in Fig. 2.1 in bold contains the codewords of the (7, 3, 4) code.[1] Subsequent rows contain all the other cosets of the code with the array arranged so that the coset leaders are in the first column. The decoder finds the received sequence on a row in the array and then subtracts the coset leader corresponding to that row from it to obtain a decoded sequence. The standard array is partitioned based on the weight of the coset leaders. Received sequences on rows with coset leaders of weight less than or equal to $t = \frac{3-1}{2} = 1$ are all corrected. Some received sequences on rows with coset leaders with weight greater than t are also corrected. Examining the standard array, it can be seen that the code can correct all single error sequences, some two error sequences and one three error sequence. The coset weight $\mathbb{C}_i$ distribution is

$$\mathbb{C}_0 = 1$$
$$\mathbb{C}_1 = 7$$
$$\mathbb{C}_2 = 7$$
$$\mathbb{C}_3 = 1$$

The covering radius of the code is the weight of the largest coset leader (in this example it is 3).

[1]It is worth noting that a code itself can be considered as a coset with the sequence **a** an all zero sequence.

2.2.2 The Performance of Codes on the Binary Symmetric Channel

Consider a real-valued sequence received from a transmission through an AWGN channel. If a demodulator makes hard decisions at the receiver, the channel may be modelled as a binary symmetric channel. Assuming the probability of bit error for the BSC is p, the probability of decoding error with a bounded distance decoder is given by,

$$P_{\text{BDD}}(e) = 1 - \sum_{i=0}^{t} \mathbb{C}_i p^i (1-p)^{n-i} \tag{2.1}$$

where $\mathbb{C}_i$ is the number of coset leaders with weight i. $\mathbb{C}_i$ known for $0 \leq i \leq t$ and is given by,

$$\mathbb{C}_i = \binom{n}{i} \quad 0 \leq i \leq t.$$

However, $\mathbb{C}_i$, $i > t$ need to be computed for individual codes. The probability of error after full decoding is

$$P_{\text{Full}}(e) = 1 - \sum_{i=0}^{n} \mathbb{C}_i p^i (1-p)^{n-i}. \tag{2.2}$$

Figure 2.2 shows the performance of the bounded distance decoder and the full decoder for different codes. The bounds are computed using (2.1) and (2.2). As expected, there is significant coding gain between unencoded and coded transmission (bounded distance and full decoding) for all the cases. There is a small coding gain between bounded distance and full decoders. This coding gain depends on the coset leader weight distribution $\mathbb{C}_i$ for $i > t$ of the individual codes. The balance between complexity and performance for full and bounded distance decoders[2] ensures that the latter are preferred in practice. Observe that in Fig. 2.2 that the complete decoder consistently outperforms the bounded distance decoder as the probability of error decreases and $\frac{E_b}{N_0}$ increases. We will see in Sect. 2.3 that a similar setup using soft decision decoding in Euclidean space produces different results.

2.2.2.1 Bounds on Decoding on the BSC Channel

Suppose s is such that $\mathbb{C}_s$ is the maximum non-zero value for a code then s is the covering radius of the code. If the covering radius s of a code is known and $\mathbb{C}_i$, $i > t$ are not known, then the probability of error after decoding can be bounded by

[2]Bounded distance decoders usually have polynomial complexity, e.g. the Berlekamp Massey decoder for BCH codes has complexity $O(t^2)$ [1].

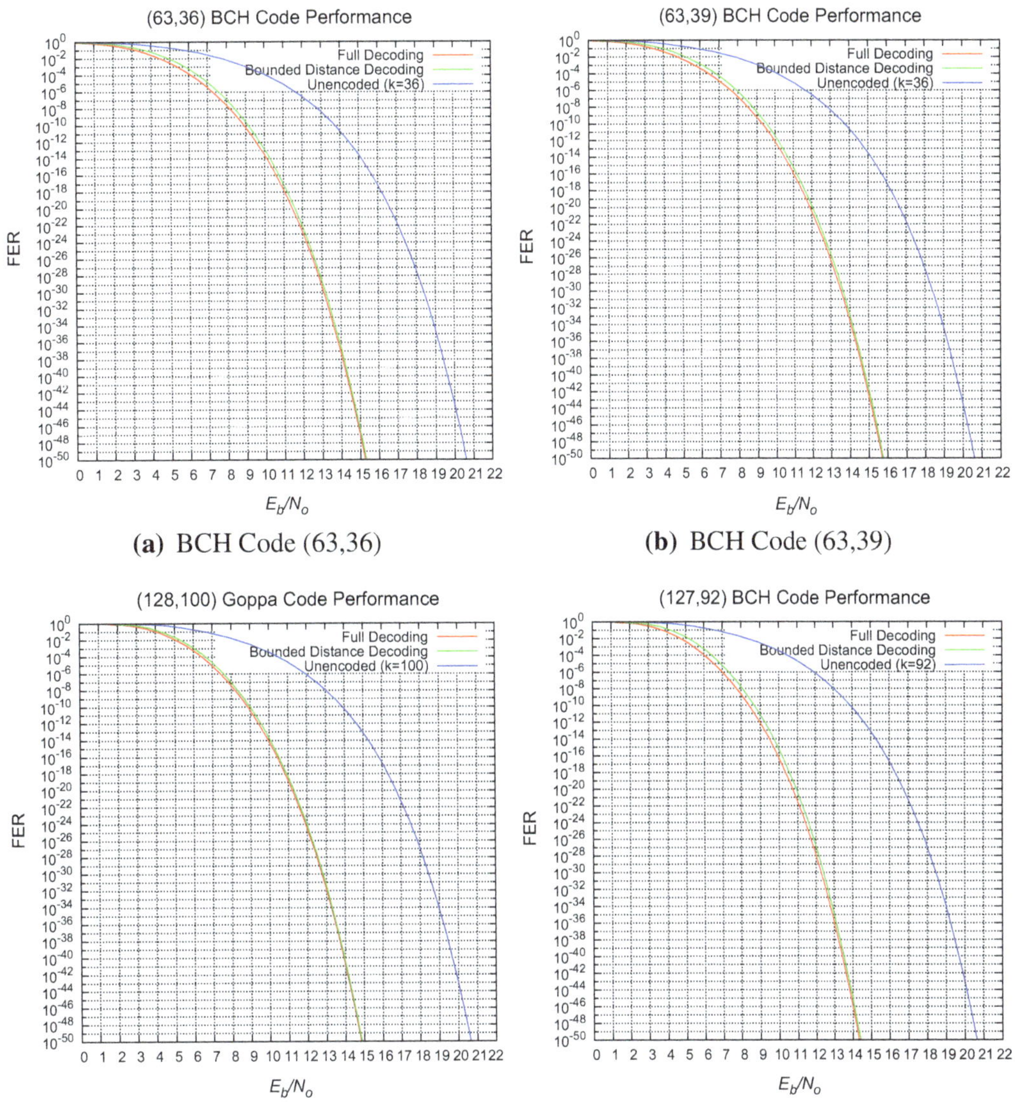

(a) BCH Code (63,36) **(b)** BCH Code (63,39)

(c) Goppa Code (128,100) **(d)** BCH Code (127,92)

Fig. 2.2 BCH code BDD and full decoder performance, frame error rate (FER) against $\frac{E_b}{N_0}$

$$P_e \geq 1 - \left[\sum_{i=0}^{t} \binom{n}{i} p^i (1-p)^{n-i} + p^s (1-p)^{n-s} \right] \tag{2.3}$$

$$\leq 1 - \left[\sum_{i=0}^{t} \binom{n}{i} p^i (1-p)^{n-i} + \mathbb{W}_s p^s (1-p)^{n-s} \right] \tag{2.4}$$

assuming the code can correct t errors and

$$\mathbb{W}_s = 2^{n-k} - \sum_{i=0}^{t} \binom{n}{i}.$$

The lower bound assumes that there is a single coset leader of weight s, and hence the term $p^s(1-p)^{n-s}$ while the upper bound assumes that all the coset leaders of weight greater than t have weight equal to the covering radius s. For the lower bound to hold, $\mathbb{W}_s > 1$. The lower bound can be further tightened by assuming that the $\mathbb{W}_s - 1$ cosets have weight of $t+1, t+2, \ldots$ until they can all be accounted for.[3]

2.3 Soft Decision Performance

The union bound for the probability of sequence error using maximum likelihood soft decoding performance on binary codes with BPSK modulation in the AWGN channel is given by [2],

$$P_s \leq \frac{1}{2}\sum_{j=1}^{n} A_j \operatorname{erfc}\left(\sqrt{\frac{E_b}{N_0}Rj}\right) \tag{2.5}$$

where R is the code rate, A_j is the number of codewords of weight j and $\frac{E_b}{N_0}$ is the SNR per bit. The union bound is obtained by assuming that events in which the received sequence is closer in euclidean distance to a codeword of weight j are independent as such the probability of error is the sum of all these events. A drawback to the exact computation of the union bound is the fact that the weight distribution A_j, $0 \leq j \leq n$ of the code is required. Except for a small number of cases, the complete weight distribution of many codes is not known due to complexity limitations. Since $A_j = 0$ for $1 \leq j < d$ where d is the minimum distance of the code we can express (2.5) as,

$$P_s \leq \frac{1}{2}\sum_{j=d}^{n} A_j \operatorname{erfc}\left(\sqrt{\frac{E_b}{N_0}Rj}\right) \tag{2.6}$$

$$\leq \frac{1}{2} A_d \operatorname{erfc}\left(\sqrt{\frac{E_b}{N_0}Rd}\right) + \frac{1}{2}\sum_{j=d+1}^{n} A_j \operatorname{erfc}\left(\sqrt{\frac{E_b}{N_0}Rj}\right) \tag{2.7}$$

A lower bound on the probability of error can be obtained if it is assumed that error events occur only when the received sequence is closer in euclidean distance to codewords at a distance d from the correct codeword.

$$P_s \geq \frac{1}{2} A_d \operatorname{erfc}\left(\sqrt{\frac{E_b}{N_0}Rd}\right) \tag{2.8}$$

[3]This can be viewed as the code only has one term at the covering radius, and all other terms are at $t+1$.

where

$$\frac{1}{2}\sum_{j=d+1}^{n} A_j \,\mathrm{erfc}\left(\sqrt{\frac{\mathrm{E_b}}{\mathrm{N_0}}\mathrm{Rj}}\right) = 0. \tag{2.9}$$

As such,

$$\frac{1}{2} A_d \,\mathrm{erfc}\left(\sqrt{\frac{\mathrm{E_b}}{\mathrm{N_0}}\mathrm{Rd}}\right) \le \mathrm{P_s} \le \frac{1}{2}\sum_{\mathrm{j=d}}^{\mathrm{n}} \mathrm{A_j}\,\mathrm{erfc}\left(\sqrt{\frac{\mathrm{E_b}}{\mathrm{N_0}}\mathrm{Rj}}\right) \tag{2.10}$$

Therefore, the practical soft decision performance of a binary code lies between the upper and lower Union bound. It will be instructive to observe the union bound performance for actual codes using their computed weight distributions as the SNR per bit $\frac{E_b}{N_0}$ increases. By allowing $\frac{E_b}{N_0}$ to become large (and P_s to decrease) simulations for several codes suggest that at a certain *intersection* value of $\frac{E_b}{N_0}$ the upper bound equals the lower bound. Consider Figs. 2.3, 2.4 and 2.5 which show the frame error rate against the SNR per bit for three types of codes. The upper bounds in the figures are obtained using the complete weight distribution of the codes with Eq. (2.5). The lower bounds are obtained using only the number of codewords of minimum weight of the codes with Eq. (2.8). It can be observed that as $\frac{E_b}{N_0}$ becomes large, the upper bound meets and equals the lower bound. The significance of this observation is that for $\frac{E_b}{N_0}$ values above the point where the two bounds intersect the performance of the codes under soft decision can be completely determined by the lower bound (or the upper bound). In this region where the bounds agree, when errors occur they do so because the received sequence is closer to codewords a distance d away from the correct codeword. The actual performance of the codes before this region is somewhere between the upper and lower bounds. As we have seen earlier, the two bounds agree when the sum in (2.9) approaches 0. It may be useful to consider an approximation of the complementary error function (erfc),

$$\mathrm{erfc(x)} < \mathrm{e}^{-x^2}$$

in which case the condition becomes

$$\frac{1}{2}\sum_{j=d+1}^{n} A_j \,\mathrm{e}^{-\frac{\mathrm{E_b}}{\mathrm{N_0}}\mathrm{Rj}} \approx 0. \tag{2.11}$$

Clearly, the sum approximates to zero if each term in the sum also approximates to zero. It is safe to assume that the term $A_j\,\mathrm{erfc}\left(\sqrt{\frac{\mathrm{E_b}}{\mathrm{N_0}}\mathrm{Rj}}\right)$ decreases as j increases since $\mathrm{erfc}\left(\sqrt{\frac{\mathrm{E_b}}{\mathrm{N_0}}\mathrm{Rj}}\right)$ reduces exponentially with j and A_j increases in a binomial (in most cases). The size of the gap between the lower and upper bounds is also

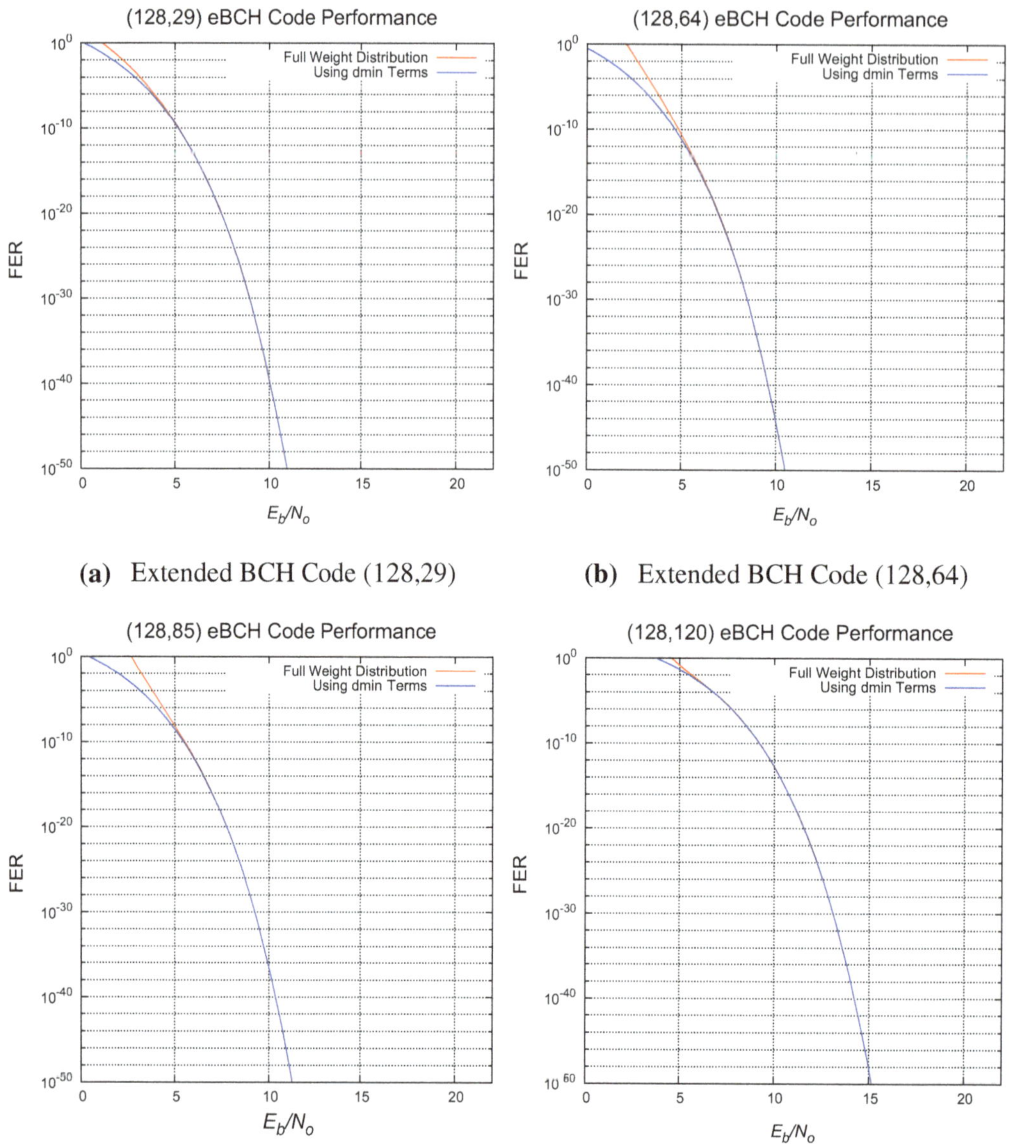

(a) Extended BCH Code (128,29) **(b)** Extended BCH Code (128,64)

(c) Extended BCH Code (128,85) **(d)** Extended BCH Code (128,120)

Fig. 2.3 Extended BCH code lower and upper union bound performance, frame error rate (FER) against $\frac{E_b}{N_0}$

determined by these terms. Each term $A_j\, e^{-\frac{E_b}{N_0}Rj}$ becomes small if one or both of the following conditions are met,

(a) Some of the A_j, $j > d$ are zero. This is common in low rate binary codes with a small number of codewords.
(b) The product $\frac{E_b}{N_0}Rj$ for $j > d$ becomes very large.

Observing Fig. 2.3, 2.4 and 2.5, it can be seen that at small values of $\frac{E_b}{N_0}$ and for low rate codes for which $R = \frac{k}{n}$ is small have some $A_j = 0$, $j > d$ and as such the gaps

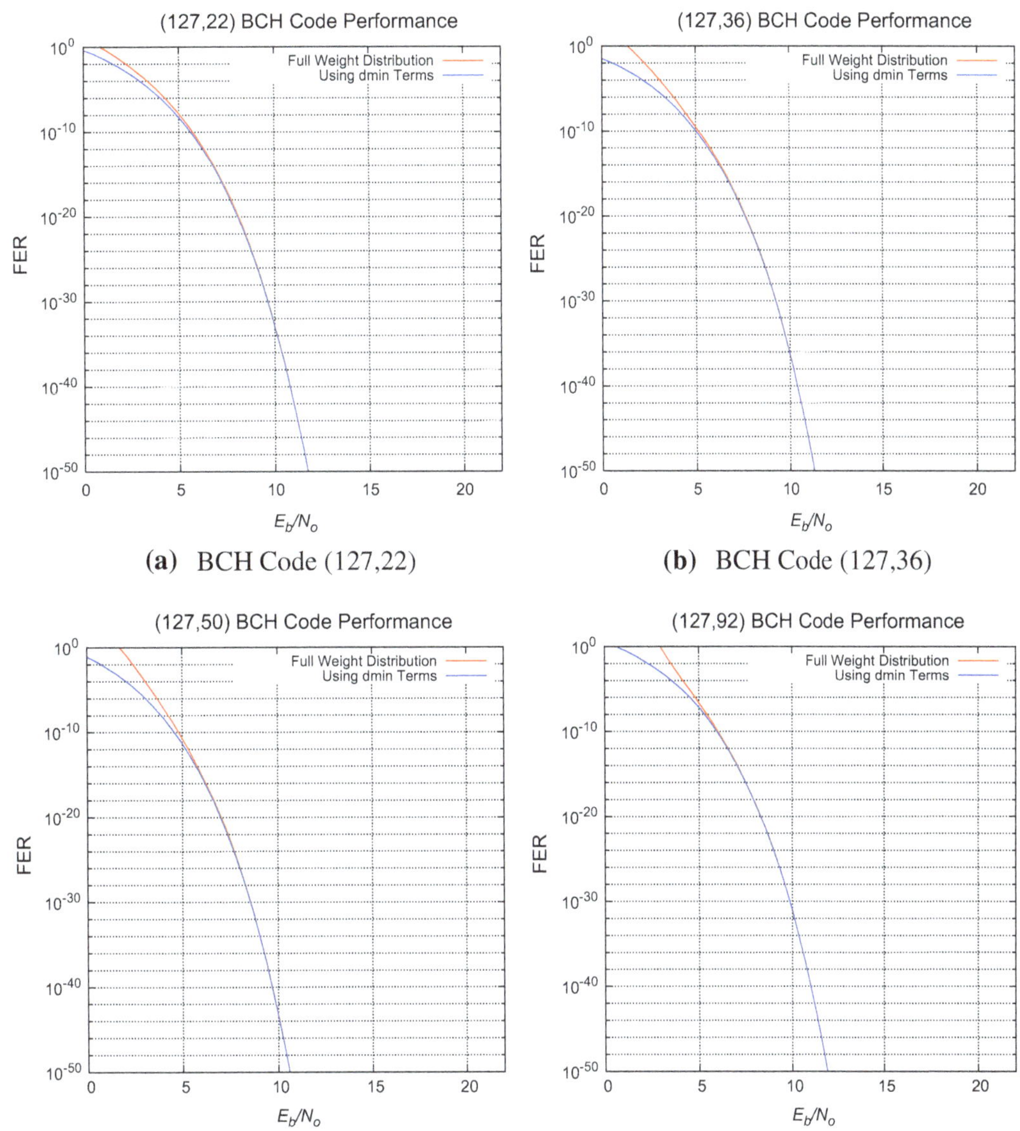

(a) BCH Code (127,22) **(b)** BCH Code (127,36)

(c) BCH Code (127,50) **(d)** BCH Code(127,92)

Fig. 2.4 BCH code lower and upper union bound performance, frame error rate (FER) against $\frac{E_b}{N_0}$

between the upper and lower bounds are small. As an example consider the low rate (127, 22, 47) BCH code in Fig. 2.4a which has,

$$A_j = 0 \quad j \in \{49 \ldots 54\} \cup \{57 \ldots 62\} \cup \{65 \ldots 70\} \cup \{73 \ldots 78\} \cup \{81 \ldots 126\}.$$

For the high rate codes, R is large so that the product $\frac{E_b}{N_0} Rj$ becomes very large therefore the gaps between the upper and lower bounds are small.

Figure 2.6 compares bounded distance decoding and full decoding with maximum likelihood soft decision decoding of the (63, 39) and (63, 36) BCH codes. It can be seen from the figure that whilst the probability of error for maximum likelihood

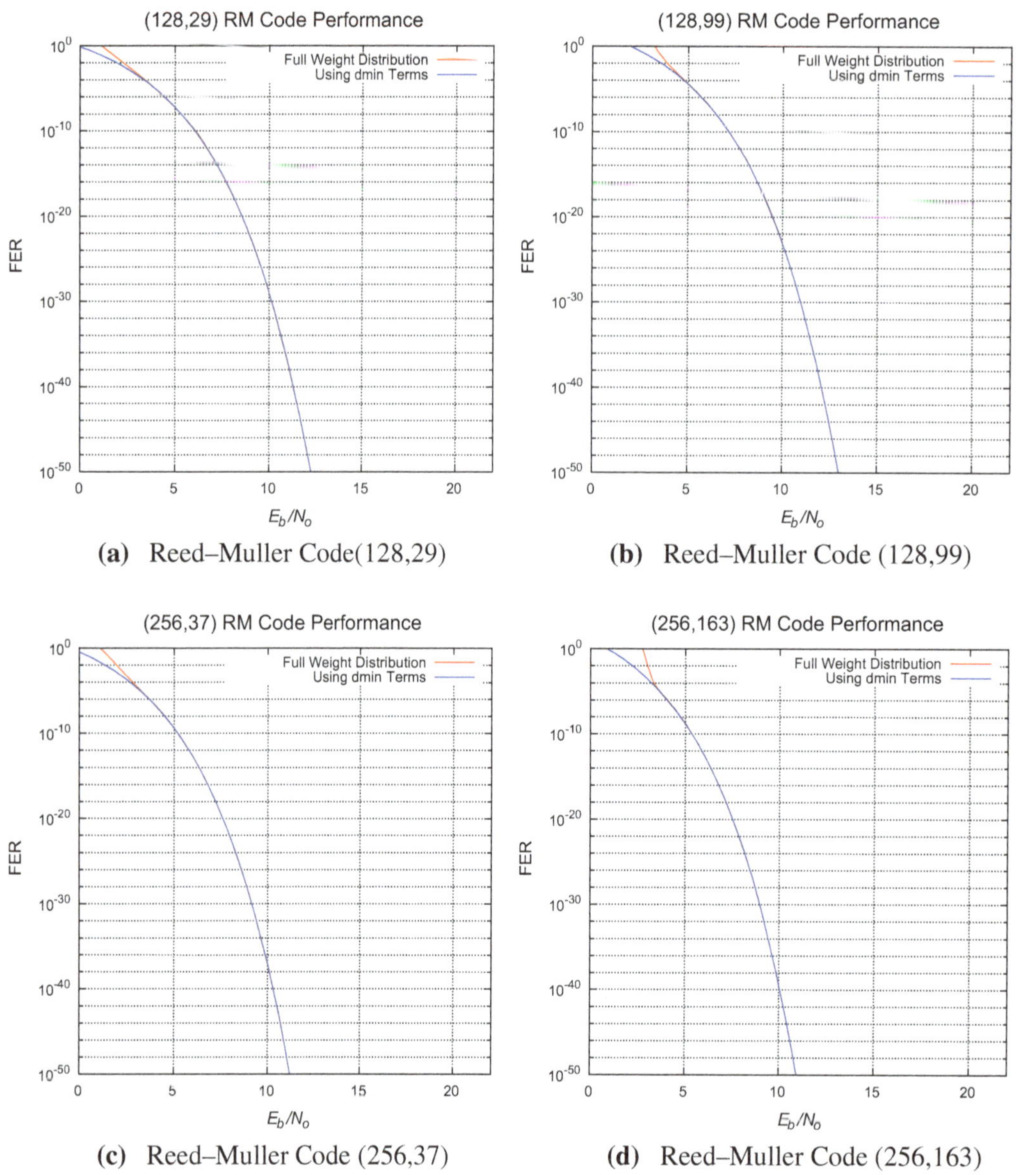

(a) Reed–Muller Code(128,29) **(b)** Reed–Muller Code (128,99)

(c) Reed–Muller Code (256,37) **(d)** Reed–Muller Code (256,163)

Fig. 2.5 Reed–Muller code lower and upper union bound performance, frame error rate (FER) against $\frac{E_b}{N_0}$

hard decision decoding is smaller than that of bounded distance decoding for all the values of $\frac{E_b}{N_0}$, the upper bound on the probability of error for maximum likelihood soft decision decoding agrees with the lower bound from certain values of $\frac{E_b}{N_0}$. This suggests that for soft decision decoding, the probability of error can be accurately determined by the lower union bound from a certain value of $\frac{E_b}{N_0}$. Computing the lower union bound from (2.10) requires only the knowledge of the minimum distance of the code d and the multiplicity of the minimum weight terms A_d. In practice, A_d is much easier to obtain than the complete weight distribution of the code.

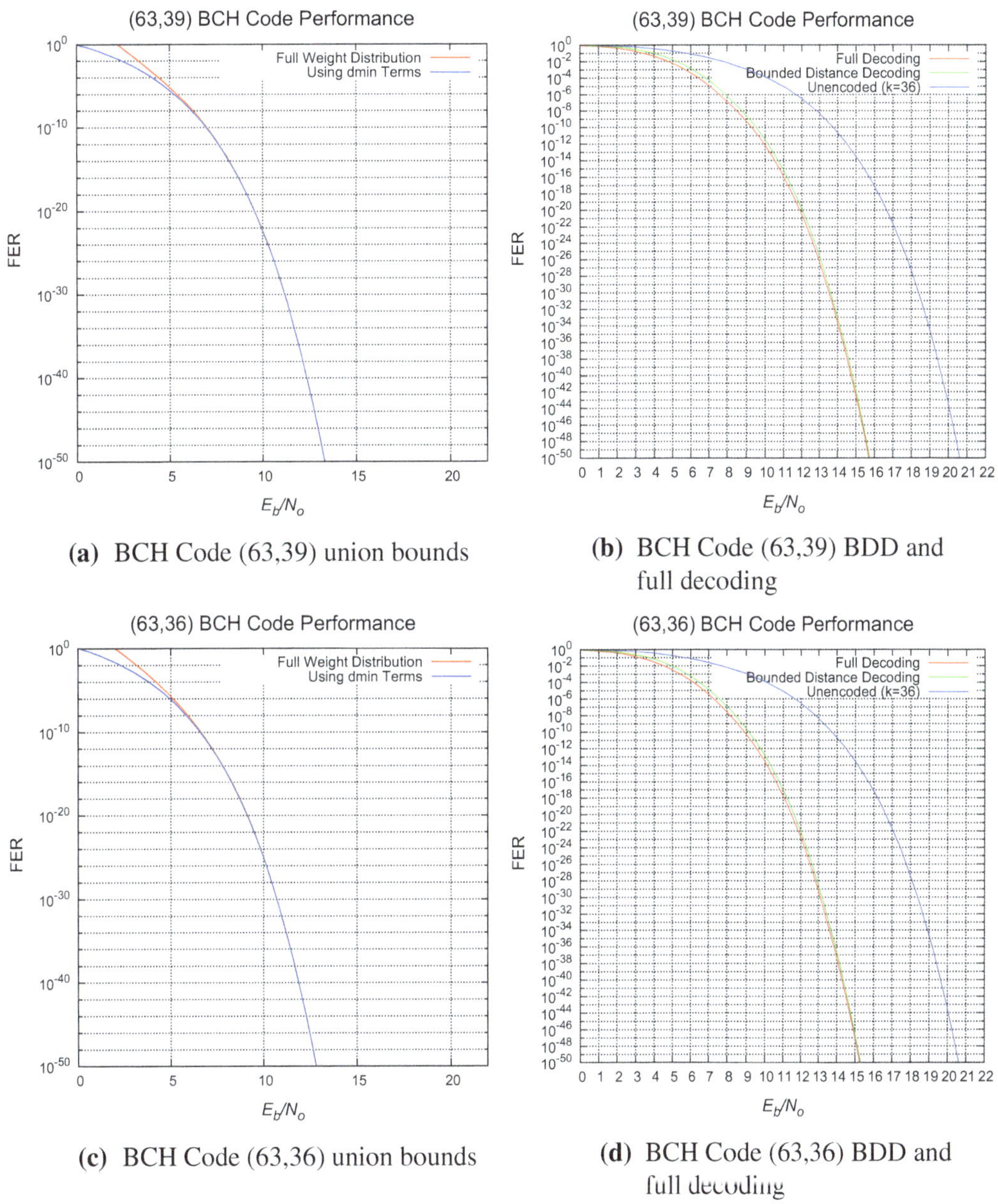

(a) BCH Code (63,39) union bounds

(b) BCH Code (63,39) BDD and full decoding

(c) BCH Code (63,36) union bounds

(d) BCH Code (63,36) BDD and full decoding

Fig. 2.6 BCH code: Bounded distance, full and maximum likelihood soft decoding

2.3.1 Performance Assuming a Binomial Weight Distribution

Evaluating the performance of long codes with many codewords using the union upper bound is difficult since one needs to compute the complete weight distribution of the codes. For many good linear binary codes, the weight distributions of the codes closely approximates to a binomial distribution. Computing the weight distribution of a binary code is known to be NP-complete [3]. Let $\left(\frac{E_b}{N_0}\right)_\delta$ be defined as,

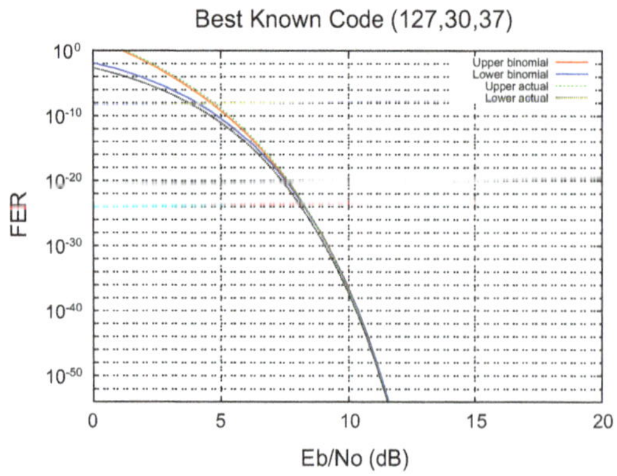

(a) (127,30,37) binomial and actual weight distributions

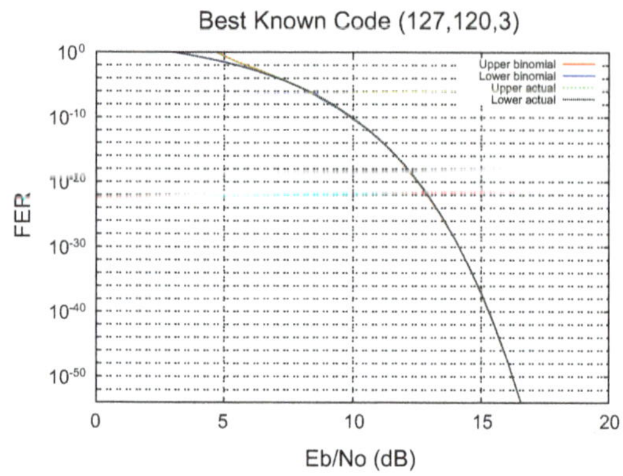

(b) (127,120,3) binomial and actual weight distributions

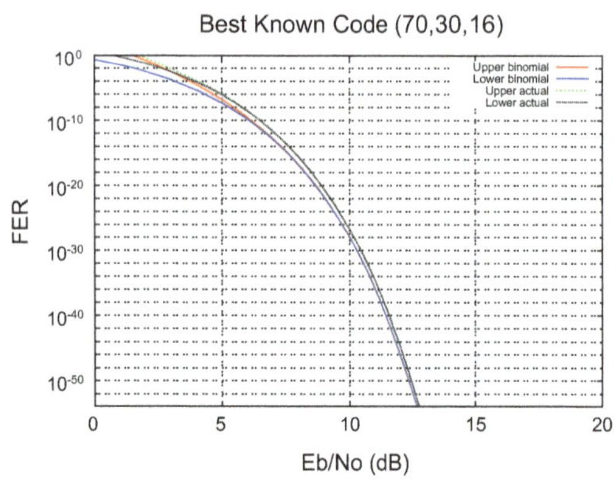

(c) (70,30,16) binomial and actual weight distributions

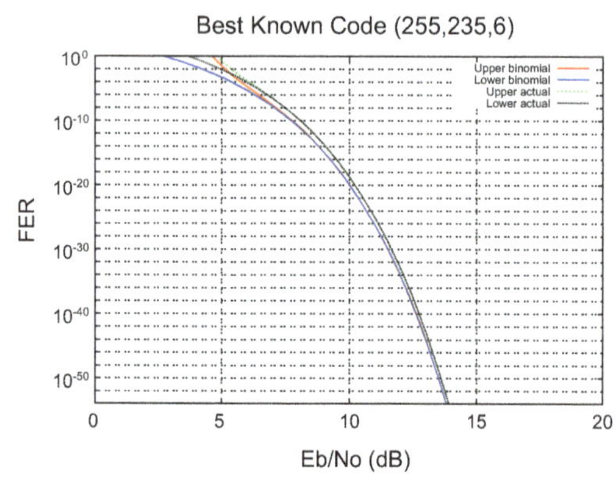

(d) (255,235,6) binomial and actual weight distributions

Fig. 2.7 Union bounds using binomial and actual weight distributions (WD) for best known codes

$$\frac{1}{2} A_d \,\mathrm{erfc}\left(\sqrt{\frac{\mathrm{E_b}}{\mathrm{N_0}}\mathrm{Rd}}\right)\Bigg|_{\frac{\mathrm{E_b}}{\mathrm{N_0}}=\left(\frac{\mathrm{E_b}}{\mathrm{N_0}}\right)_\delta} \approx \frac{1}{2}\sum_{\mathrm{j=d}}^{\mathrm{n}} \mathrm{A_j}\,\mathrm{erfc}\left(\sqrt{\frac{\mathrm{E_b}}{\mathrm{N_0}}\mathrm{Rj}}\right)\Bigg|_{\frac{\mathrm{E_b}}{\mathrm{N_0}}=\left(\frac{\mathrm{E_b}}{\mathrm{N_0}}\right)_\delta}. \tag{2.12}$$

Hence, $\left(\frac{E_b}{N_0}\right)_\delta$ is the SNR per bit at which the difference between upper and lower union bound for the code is very small. It is worth noting that equality is only possible when $\frac{E_b}{N_0}$ approaches infinity in (2.12) since $\lim_{x\to\infty} \mathrm{erfc(x)} = 0$. To find $\left(\frac{E_b}{N_0}\right)_\delta$ for a binary code (n, k, d) we simply assume a binomial weight distribution for the code so that,

$$A_i = \frac{2^k}{2^n}\binom{n}{i} \tag{2.13}$$

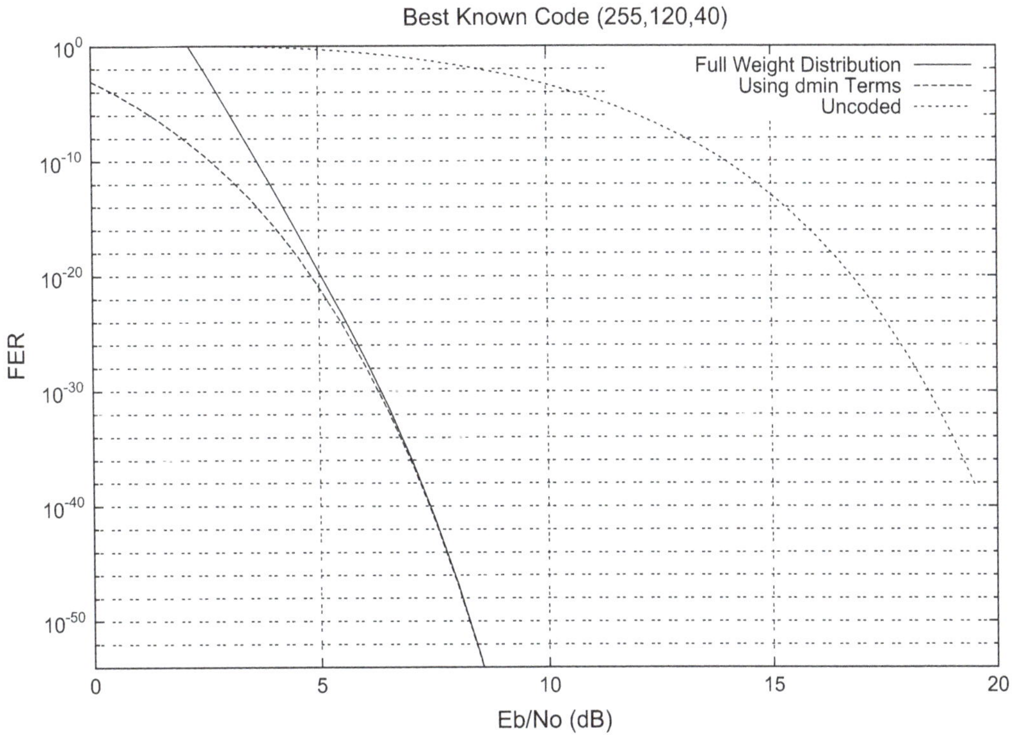

Fig. 2.8 Union bounds using binomial and actual weight distributions (WD) for the (255, 120, 40) best known code

and compute an $\frac{E_b}{N_0}$ value that satisfies (2.12). It must be noted that $\left(\frac{E_b}{N_0}\right)_\delta$ obtained using this approach is only an estimate. The accuracy of $\left(\frac{E_b}{N_0}\right)_\delta$ depends on how closely the weight distribution of the code approximates to a binomial and how small the difference between the upper and lower union bounds $P_{\text{upper}} - P_{\text{lower}}$ is. Consider Fig. 2.7 that show the upper and lower union bounds using binomial weight distributions and the actual weight distributions of the codes. From Fig. 2.7a, it can be seen that for the low rate code (127, 30, 37) the performance of the code using the binomial approximation of the weight distribution does not agree with the performance using the actual weight distribution at low values of $\frac{E_b}{N_0}$. Interestingly Fig. 2.7b–d show that as the rate of the codes increases the actual weight distribution of the codes approximates to a binomial. The difference in the performance of the codes using the binomial approximation and actual weight distribution decreases as $\frac{E_b}{N_0}$ increases. Figure 2.8 shows the performance of the (255, 120, 40) using a binomial weight distribution. An estimate for $\left(\frac{E_b}{N_0}\right)_\delta$ from the figure is 5.2 dB. Thus for $\frac{E_b}{N_0} \geq 5.2$ dB, we can estimate the performance of the (255, 120, 40) code under maximum likelihood soft decision decoding in the AWGN channel using the lower union bound.

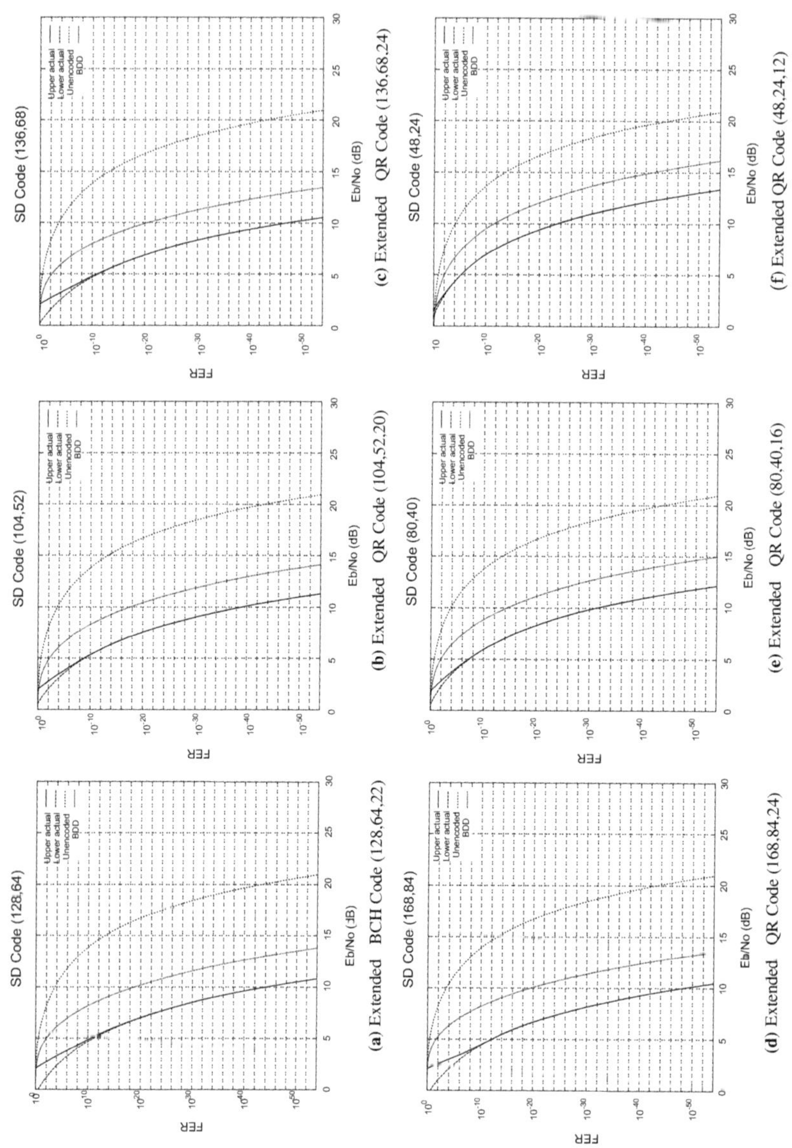

(a) Extended BCH Code (128,64,22)

(b) Extended QR Code (104,52,20)

(c) Extended QR Code (136,68,24)

(d) Extended QR Code (168,84,24)

(e) Extended QR Code (80,40,16)

(f) Extended QR Code (48,24,12)

Fig. 2.9 Performance of self-dual codes

2.3.2 Performance of Self-dual Codes

A self-dual code $\mathscr{C}$ has the property that it is its own dual such that,

$$\mathscr{C} = \mathscr{C}^{\perp}.$$

Self-dual codes are always half rate with parameters $(n, \frac{1}{2}n, d)$. These codes are known to meet the Gilbert–Varshamov bound and some of the best known codes are self-dual codes. Self-dual codes form a subclass of formally self-dual codes which have the property that,

$$W(\mathscr{C}) = W(\mathscr{C}^{\perp}).$$

where $W(\mathscr{C})$ means the weight distribution of $\mathscr{C}$. The weight distribution of certain types of formally self-dual codes can be computed without enumerating all the codewords of the code. For this reason, these codes can readily be used for analytical purposes. The fact that self-dual codes have the same code rate and good properties makes them ideal for performance evaluation of codes of varying length. Consider Fig. 2.9 which shows the performance of binary self-dual (and formally self-dual) codes of different lengths using the upper and lower union bounds with actual weight distributions, bounded distance decoding and unencoded transmission. Figure 2.10

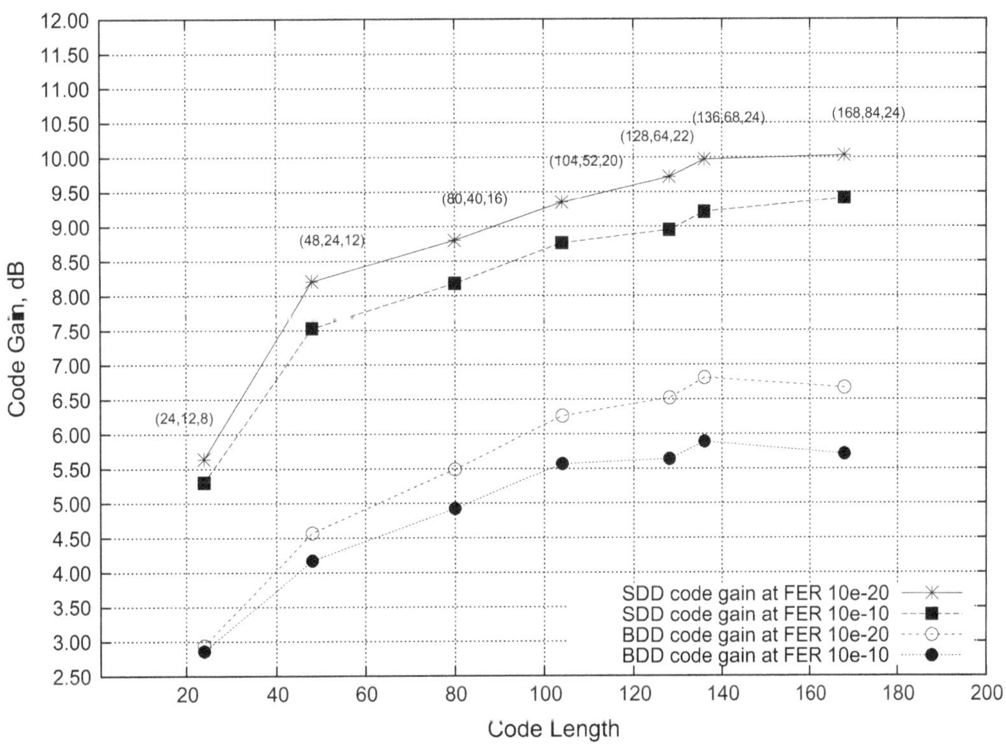

Fig. 2.10 Coding gain against code length for self-dual codes at FER 10^{-10} and 10^{-20}

shows the coding gain of the self-dual codes at frame error rates (FER) 10^{-10} and 10^{-20} for soft decision decoding (SDD) and bounded distance decoding (BDD). The coding gain represents the difference in dB between the SDD/BDD performance and unencoded transmission. The coding gain is a measure of the power saving obtainable from a coded system relative to an unencoded system in dB at a certain probability of error. The SDD performance of codes with length 168, 136 and 128 at FER 10^{-10} are obtained from the union upper bound because the upper and lower bound do not agree at this FER. Thus, the coding gain for these cases is a lower bound. It is instructive to note that the difference between the coding gain for SDD and BDD at the two values of FER increases as the length of the code increases. At FER of 10^{-20} SDD gives 3.36 dB coding gain over BDD for the code of length 168 and 2.70 dB for the code of length 24. At a FER of 10^{-10}, SDD gives 3.70 dB coding gain over BDD for the code of length 168 and 2.44 dB for the code of length 24.

2.4 Summary

In this chapter, we discussed the performance of codes under hard and soft decision decoding. For hard decision decoding, the performance of codes in the binary symmetric channel was discussed and numerically evaluated results for the bounded distance decoder compared to the full decoder were presented for a range of codes whose coset leader weight distribution is known. It was shown that as the SNR per information bit increases there is still an observable difference between bounded distance and full decoders. A lower and upper bound for decoding in the BSC was also given for cases where the covering radius of the code is known. For soft decision decoding, the performance of a wide range of specific codes was evaluated numerically using the union bounds. The upper and lower union bounds were shown to converge for all codes as the SNR per information bit increases. It was apparent that for surprisingly low values of $\frac{E_b}{N_0}$ the performance of a linear code can be predicted by only using knowledge of the multiplicity of codewords of minimum weight. It was also shown for those codes whose weight distribution is difficult to compute, a binomial weight distribution can be used instead.

References

1. Moon, T.K.: Error Correction Coding: Mathematical Methods and Algorithms. Wiley, New Jersey (2005)
2. Proakis, J.: Digital Communications, 4th edn. McGraw-Hill, New York (2001)
3. Vardy, A.: The intractability of computing the minimum distance of a code. IEEE Trans. Inf. Theory **43**, 1759–1766 (1997)

Chapter 3
Soft Decision and Quantised Soft Decision Decoding

3.1 Introduction

The use of hard decision decoding results in a decoding loss compared to soft decision decoding. There are several references that have quantified the loss which is a function of the operating $\frac{E_b}{N_0}$ ratio, the error-correcting code and the quantisation of the soft decisions. Wozencraft and Jacobs [6] give a detailed analysis of the effects of soft decision quantisation on the probability of decoding error, P_{ec}, for the ensemble of all binary codes of length n without restriction of the choice of code. Their analysis follows from the Coding Theorem, presented by Gallager for the ensemble of random binary codes [3].

3.2 Soft Decision Bounds

There are 2^n possible binary combinations for each codeword, which in terms of the n-dimensional signal space hypercube corresponds to one vertex taken from 2^n possible vertices. There are 2^k codewords and therefore 2^{nk} different possible codes. The receiver is considered to be composed of 2^k matched filters, one for each codeword, and a decoder error occurs if any of the matched filter receivers has a larger output than the matched filter receiver corresponding to the transmitted codeword. Consider this matched filter receiver and another different matched filter receiver, and consider that the two codewords differ in d bit positions. The Hamming distance between the two codewords is d. The energy per transmitted bit is $E_s = \frac{k}{n}E_b$, where E_b is the energy per information bit. The noise variance per matched filtered received bit, $\sigma^2 = \frac{N_0}{2}$, where N_0 is the single sided noise spectral density. In the absence of noise, the output of the matched filter receiver for the transmitted codeword is $n\sqrt{E_s}$, and the output of the other codeword matched filter receiver is $(n-2d)\sqrt{E_s}$. The noise voltage at the output of the matched filter receiver for the transmitted codeword

M. Tomlinson et al., *Error-Correction Coding and Decoding*,
Signals and Communication Technology,
DOI 10.1007/978-3-319-51103-0_3

is denoted as $n_c - n_1$, and the noise voltage at the output of the other matched filter receiver will be $n_c + n_1$. The common noise voltage n_c arises from correlation of the bits common to both codewords with the received noise, and the noise voltages $-n_1$ and n_1 arise respectively from correlation of the other d bits with the received noise.

A decoder error occurs if

$$(n - 2d)\sqrt{E_s} + n_c + n_1 > n\sqrt{E_s} + n_c - n_1, \tag{3.1}$$

that is, a decoder error occurs when $2n_1 > 2d\sqrt{E_s}$.

The average noise power associated with n_1 is $d\sigma^2 = d\frac{N_0}{2}$, and as the noise is Gaussian distributed, the probability of decoder error, p_d, is given by

$$p_d = \frac{1}{\sqrt{\pi d N_0}} \int_{d\sqrt{E_s}}^{\infty} e^{\frac{-x^2}{dN_0}} dx. \tag{3.2}$$

This may be expressed in terms of the complementary error function

$$\text{erfc}(y) = 2\frac{1}{\sqrt{2\pi}} \int_{y}^{\infty} e^{\frac{-x^2}{2}} dx \tag{3.3}$$

and leads to

$$p_d = \frac{1}{2}\text{erfc}\left(\sqrt{d\frac{k}{n}\frac{E_b}{N_0}}\right) \tag{3.4}$$

Each of the other $2^k - 2$ codewords may also cause a decoder error but the weight distribution of the code C_i is unknown. However, by averaging over all possible codes, knowledge of the weight distribution of a particular code is not required. The probability of two codewords of a code C_i, differing in d bit positions, $p(d|C_i)$ is given by the Binomial distribution

$$p(d|C_i) = \frac{\frac{n!}{(n-d)!d!}}{2^n} \tag{3.5}$$

A given linear code C_i cannot have codewords of arbitrary weight, because the sum of a sub-set of codewords is also a codeword. However, for non linear codes, p_d may be averaged over all of the codes without this constraint.

$$\overline{p_C} = \sum_{i=1}^{2^{n2^k}} p(d|C_i)p(C_i) < \frac{1}{2^{n2^k}} \sum_{d=0}^{n} \sum_{i=1}^{2^{n2^k}} \frac{\frac{n!}{(n-d)!d!}}{2^{n+1}} \text{erfc}\left(\sqrt{d\frac{k}{n}\frac{E_b}{N_0}}\right) \tag{3.6}$$

rearranging the order of summation

$$\overline{p_C} < \frac{1}{2^{n2^k}} \sum_{i=1}^{2^{n2^k}} \sum_{d=0}^{n} \frac{\frac{n!}{(n-d)!d!}}{2^{n+1}} \operatorname{erfc}\left(\sqrt{d \frac{k}{n} \frac{E_b}{N_0}}\right) \tag{3.7}$$

and

$$\overline{p_C} < \frac{1}{2^{n+1}} \sum_{d=0}^{n} \frac{n!}{(n-d)!d!} \operatorname{erfc}\left(\sqrt{d \frac{k}{n} \frac{E_b}{N_0}}\right) \tag{3.8}$$

Remembering that any of the $2^k - 1$ matched filters may cause a decoder error, the overall probability of decoder error averaged over all possible binary codes $\overline{p_{\text{overall}}}$, is

$$\overline{p_{\text{overall}}} = 1 - (1 - \overline{p_C})^{2^k - 1} < 2^k \overline{p_C} \tag{3.9}$$

and

$$\overline{p_{\text{overall}}} < \frac{2^k}{2^{n+1}} \sum_{d=0}^{n} \frac{n!}{(n-d)!d!} \operatorname{erfc}\left(\sqrt{d \frac{k}{n} \frac{E_b}{N_0}}\right) \tag{3.10}$$

An analytic solution may be obtained by observing that $\frac{1}{2}\operatorname{erfc}(y)$ is upper bounded by e^{-y^2},

$$\overline{p_{\text{overall}}} < \frac{2^k}{2^n} \sum_{d=0}^{n} \frac{n!}{(n-d)!d!} e^{-d \frac{k}{n} \frac{E_b}{N_0}} \tag{3.11}$$

and as observed by Wozencraft and Jacobs [6],

$$(1 + e^{-\frac{k}{n} \frac{E_b}{N_0}})^n = \sum_{d=0}^{n} \frac{n!}{(n-d)!d!} e^{-d \frac{k}{n} \frac{E_b}{N_0}} \tag{3.12}$$

and

$$\overline{p_C} < \frac{1}{2^n} (1 + e^{-\frac{k}{n} \frac{E_b}{N_0}})^n \tag{3.13}$$

$$\overline{p_{\text{overall}}} < \frac{2^k}{2^n} (1 + e^{-\frac{k}{n} \frac{E_b}{N_0}})^n \tag{3.14}$$

Traditionally, a cut-off rate R_0 is defined after observing that

$$\frac{2^k}{2^n}(1+e^{-\frac{k}{n}\frac{E_b}{N_0}})^n = 2^k\left(\frac{1+e^{-\frac{k}{n}\frac{E_b}{N_0}}}{2}\right)^n \tag{3.15}$$

with

$$2^{R_0} = \left(\frac{2}{1+e^{-\frac{k}{n}\frac{E_b}{N_0}}}\right), \tag{3.16}$$

then

$$\overline{p_{\text{overall}}} < 2^k 2^{-nR_0} = 2^{k-nR_0} = 2^{-n(R_0-\frac{k}{n})} \tag{3.17}$$

This result may be interpreted as, providing the number of information bits of the code is less than the length of the code times the cut-off rate, then the probability of decoder error will approach zero as the length of the code approaches infinity. Alternatively, provided the rate of the code, $\frac{k}{n}$, is less than the cut-off rate, R_0, then the probability of decoder error will approach zero as the length of the code approaches infinity.

When s quantised soft decisions are used with integer levels 0 to $2s-1$, for s even and integer levels 0 to $s-1$ for s odd, the transmitted binary signal has levels 0 and $2(s-1)$, for s even and levels 0 and $s-1$, for s odd and the probability distribution of the quantised signal (bit) plus noise, after matched filtering, has probability p_i, $i=0$ to $s-1$, represented as

$$p(z) = \sum_{i=0}^{s-1} p_i z^{-2i}, \text{ for } s \text{ even} \tag{3.18}$$

and

$$p(z) = \sum_{i=0}^{s-1} p_i z^{-i}, \text{ for } s \text{ odd} \tag{3.19}$$

A decoder error occurs if

$$s(n-2d) + n_c + n_1 > sn + n_c - n_1 \tag{3.20}$$

and occurs when

$$n_1 > sd \tag{3.21}$$

and has probability 0.5 when

$$n_1 = sd \tag{3.22}$$

The probability of decoder error may be determined from a summation of terms from the overall probability distribution for the sum of d independent, quantised noise samples, and is given by a polynomial $q_d(z)$ at $z = 0$, where $q_d(z)$ is given by

$$q_d(z) = p(z)^d \left(\frac{1 - z^{(s-1)d+1}}{1 - z} - 0.5z^{(s-1)d} \right), \text{ for } s \text{ even} \tag{3.23}$$

The $0.5z^{(s-1)d}$ term corresponds to $n_1 = sd$ when the probability of decoder error is 0.5.

$$q_d(z) = p(z)^d \left(\frac{1 - z^{\frac{s-1}{2}d+1}}{1 - z} - 0.5z^{\frac{s-1}{2}d} \right), \text{ when } s \text{ is odd} \tag{3.24}$$

and the $0.5z^{\frac{s-1}{2}d}$ term corresponds to $n_1 = sd$ when the probability of decoder error is 0.5.

The probability of decoder error is given by $q_d(z)$ when $z = 0$,

$$p_d = q_d(0) \tag{3.25}$$

The evaluation of the average probability of decoder error for quantised soft decisions, $\overline{p_{C_Q}}$ is given, as before by averaging over all codes and rearranging the order of summation

$$\overline{p_{C_Q}} < \frac{1}{2^{n2^k}} \sum_{i=1}^{2^{n2^k}} \sum_{d=0}^{n} \frac{\frac{n!}{(n-d)!d!}}{2^n} q_d(0) \tag{3.26}$$

Simplifying

$$\overline{p_{C_Q}} < \sum_{d=0}^{n} \frac{\frac{n!}{(n-d)!d!}}{2^n} q_d(0) \tag{3.27}$$

When hard decisions are used, the probability of each transmitted bit being received in error is given by

$$p_b = 0.5\text{erfc}\left(\sqrt{\frac{k}{n} \frac{E_b}{N_0}} \right) \tag{3.28}$$

Accordingly,

$$p(z) = 1 - p_b + p_b z^{-2} \tag{3.29}$$

and $q_d(z)$ for hard decisions becomes

$$q_d(z) = (1 - p_b + p_b z^{-2})^d \left(\frac{1 - z^{d+1}}{1 - z} - 0.5z^d \right) \tag{3.30}$$

giving

$$\overline{p_{C_Q}} < \sum_{d=0}^{n} \frac{\frac{n!}{(n-d)!d!}}{2^n} (1 - p_b + p_b z^{-2})^d \left(\frac{1 - z^{d+1}}{1 - z} - 0.5z^d \right) \quad \text{for } z = 0 \tag{3.31}$$

As before, any of the $2^k - 1$ matched filters may cause a decoder error, the overall probability of decoder error averaged over all possible binary codes $\overline{p_{overall_Q}}$, is

$$\overline{p_{overall_Q}} < 1 - (1 - \overline{p_{C_Q}})^{2^k - 1} < 2^k \overline{p_{C_Q}} \tag{3.32}$$

and

$$\overline{p_{overall_Q}} < \frac{2^k}{2^n} \sum_{d=0}^{n} \frac{n!}{(n-d)!d!} (1 - p_b + p_b z^{-2})^d \left(\frac{1 - z^{d+1}}{1 - z} - 0.5z^d \right), \text{ for } z = 0 \tag{3.33}$$

When three-level quantisation is used for the received signal plus noise, a threshold, v_{thresh} is defined, whereby, if the magnitude of the received signal plus noise is less than v_{thresh}, an erasure is declared otherwise a hard decision is made. The probability of an erasure, p_{erase} is given by

$$p_{\text{erase}} = \frac{2}{\sqrt{\pi N_0}} \int_0^{\sqrt{\frac{k}{n} E_b} - v_{thresh}} e^{\frac{-x^2}{N_0}} dx \tag{3.34}$$

The probability of a bit error for the hard decision, p_b, is now given by

$$p_b = \frac{1}{\sqrt{\pi N_0}} \int_{\sqrt{\frac{k}{n} E_b} + v_{thresh}}^{\infty} e^{\frac{-x^2}{N_0}} dx \tag{3.35}$$

Accordingly, $p(z)$ becomes

$$p(z) = 1 - p_b - p_{\text{erase}} + p_{\text{erase}} z^{-1} + p_b z^{-2} \tag{3.36}$$

and $q_d(z)$ for three-level soft decisions is

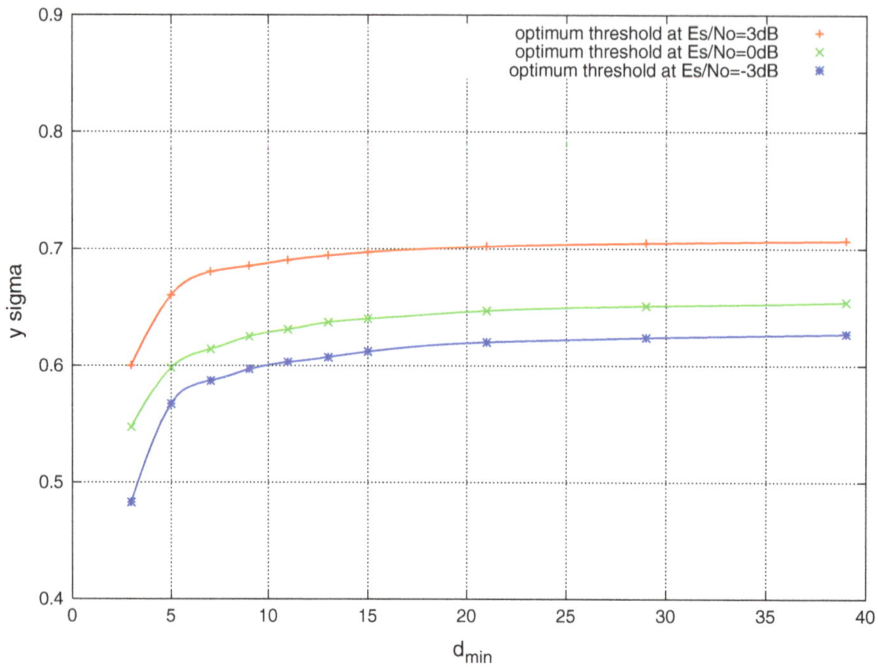

Fig. 3.1 Optimum threshold $\sqrt{E_s} - y \times \sigma$ with $y \times \sigma$ plotted as a function of $\frac{E_s}{N_0} = \frac{k}{n}\frac{E_b}{N_0}$ and d_{min}

$$q_d(z) = (1 - p_b - p_{\text{erase}} + p_{\text{erase}}z^{-1} + p_b z^{-2})^d \left(\frac{1 - z^{d+1}}{1 - z} - 0.5z^d\right) \quad (3.37)$$

giving

$$\overline{p_{overall_Q}} < \frac{2^k}{2^n}\sum_{d=0}^{n}\left(\frac{n!}{(n-d)!d!}(1 - p_b - p_{\text{erase}} + p_{\text{erase}}z^{-1} + p_b z^{-2})^d \left(\frac{1 - z^{d+1}}{1 - z} - 0.5z^d\right)\right) \quad \text{for } z = 0 \quad (3.38)$$

There is a best choice of v_{thresh} which minimises $\overline{p_{overall_Q}}$ and this is dependent on the code parameters, (n, k), and $\frac{E_b}{N_0}$. However, v_{thresh} is not an unduly sensitive parameter and best values typically range from 0.6 to 0.7σ. The value of 0.65σ is mentioned in Wozencraft and Jacobs [6]. Optimum values of v_{thresh} are given in Fig. 3.1.

3.3 Examples

The overall probability of decoder error averaged over all possible binary codes has been evaluated for $\frac{k}{n} = \frac{1}{2}$ for soft decisions, using Eq. (3.10), the approximation given by Eq. (3.14) and for hard decisions, using Eq. (3.38), for various code lengths. Results are shown in Fig. 3.2 for the ensemble of (100, 50) binary codes. The difference between the exact random coding bound, Eq. (3.10), and the original, approximate, random coding bound, Eq. (3.14) is about 0.5 dB for (100, 50) codes. The loss due to hard decisions is around 2.1 dB (at 1×10^{-5} it is 2.18 dB), and for three-level quantisation is around 1 dB (at 1×10^{-5} it is 1.03 dB). Also shown in Fig. 3.2 is the sphere packing bound offset by the loss associated with binary transmission.

Results are shown in Fig. 3.3 for the ensemble of (200, 100) binary codes. The difference between the exact random coding bound, Eq. (3.10), and the original, approximate, random coding bound, Eq. (3.14) is about 0.25 dB for (200, 100) codes. The loss due to hard decisions is around 2.1 dB, (at 1×10^{-5} it is 2.15 dB) and for three-level quantisation is around 1 dB, (at 1×10^{-5} it is 0.999 dB). Also shown in Fig. 3.3 is the sphere packing bound offset by the loss associated with binary transmission. The exact random coding bound is now much closer to the sphere packing bound, offset by the loss associated with binary transmission, with a gap of about 0.2 dB at 10^{-8}. It should be noted that the sphere packing bound is a lower bound whilst the random binary code bound is an upper bound.

Instead of considering random codes, the effect of soft decision quantisation is analysed for codes with a given weight spectrum. The analysis is restricted to two-level and three-level quantisation because these are the most common. In other cases, the quantisation is chosen such that near ideal soft decision decoding is realised. The

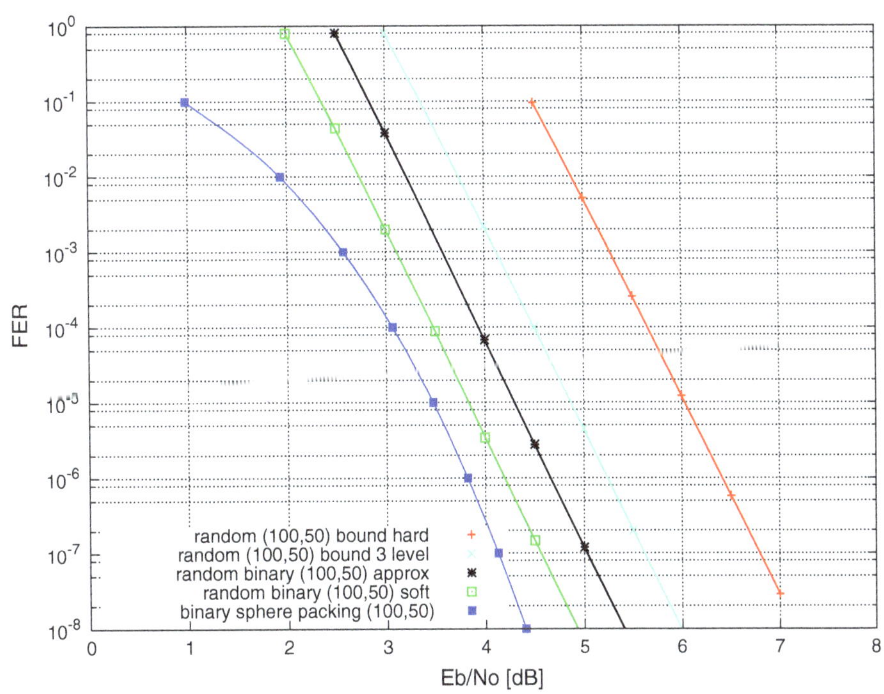

Fig. 3.2 Exact and approximate random coding bounds for [100, 50] binary codes and quantised decisions

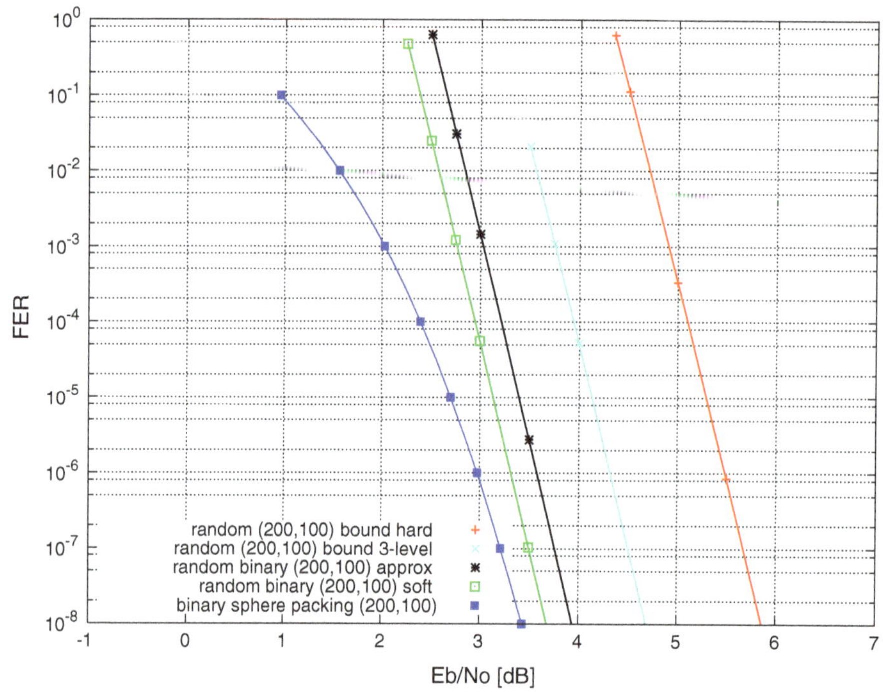

Fig. 3.3 Exact and approximate random coding bounds for [200, 100] binary codes and quantised decisions

analysis starts with a hypothetical code in which the Hamming distance between all codewords is the same, d_{min}. The probability of decoder error due to a single matched filter having a greater output than the correct matched filter follows immediately from Eq. (3.4) and the code parameters may be eliminated by considering $\frac{E_s}{N_0}$ instead of $\frac{E_b}{N_0}$.

$$p_d = \frac{1}{2}\text{erfc}\left(\sqrt{d_{min}\frac{E_s}{N_0}}\right) \tag{3.39}$$

For hard decisions and three-level quantisation, p_d is given by

$$p_d = (1 - p_b - p_{\text{erase}} + p_{\text{erase}}z^{-1} + p_b z^{-2})^d_{min}\left(\frac{1 - z^{d_{min}+1}}{1 - z} - 0.5z^{d_{min}}\right), \text{for } z = 0 \tag{3.40}$$

For hard decisions, p_{erase} is set equal to zero and p_b is given by Eq. (3.28). For three-level quantisation, p_{erase} is expressed in terms of $\frac{E_{Qs}}{N_0}$, the $\frac{E_{Qs}}{N_0}$ ratio required when quantised soft decision decoding is used.

$$p_{\text{erase}} = \frac{2}{\sqrt{\pi N_0}} \int_0^{\sqrt{E_{Qs}}-v_{thresh}} e^{\frac{-x^2}{N_0}} dx \tag{3.41}$$

Similarly, the probability of a bit error for the hard decision, p_b is given by

$$p_b = \frac{1}{\sqrt{\pi N_0}} \int_{\sqrt{E_{Qs}}+v_{thresh}}^{\infty} e^{\frac{-x^2}{N_0}} dx \tag{3.42}$$

By equating Eq. (3.39) with Eq. (3.40), the $\frac{E_{Qs}}{N_0}$ required for the same decoder error probability may be determined as a function of $\frac{E_{Qs}}{N_0}$ and d_{min}. The loss, in dB, due to soft decision quantisation may be defined as

$$Loss_Q = 10 \times log_{10}\frac{E_{Qs}}{N_0} - 10 \times log_{10}\frac{E_s}{N_0} \tag{3.43}$$

Figure 3.4 shows the soft decision quantisation loss, $Loss_Q$, as a function of d_{min} and $\frac{E_s}{N_0}$ for hard decisions. For low d_{min}, the loss is around 1.5 dB but rises rapidly with d_{min} to around 2 dB. For $\frac{E_s}{N_0}$ = 3 dB, practical systems operate with d_{min} less than 15 or so because the decoder error rate is so very low (at d_{min} = 15, the decoder error rate is less than $1 \times 10_{-20}$). Most practical systems will operate where the loss is around 2 dB. Low code rate systems ($\frac{1}{3}$ or less) operate with negative $\frac{E_s}{N_0}$ ratios with d_{min} in the range 25 to 40 whereas $\frac{1}{2}$ code rate systems with d_{min} in the range 20 to

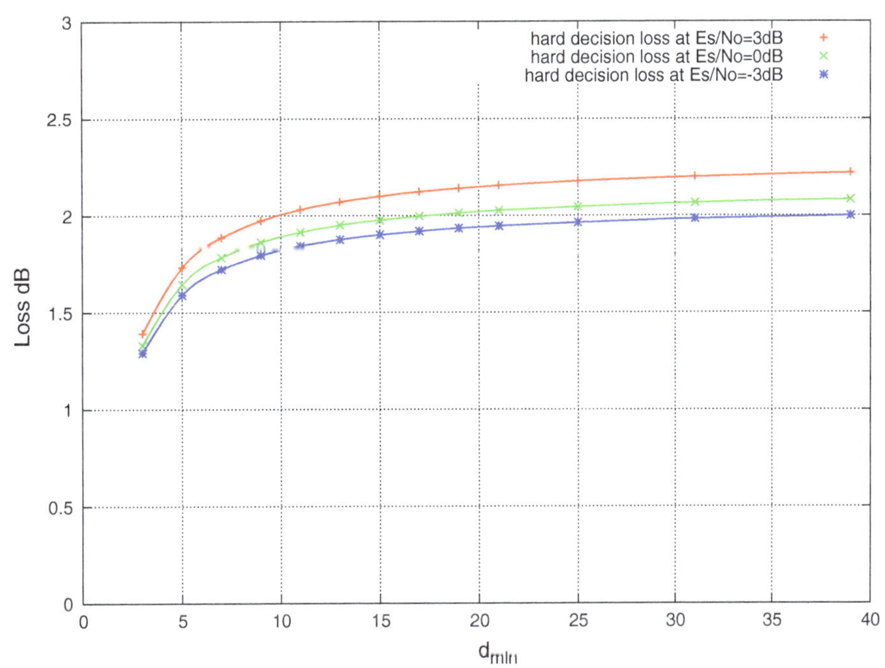

Fig. 3.4 Loss due to hard decisions as a function of $\frac{E_s}{N_0}$ and d_{min}

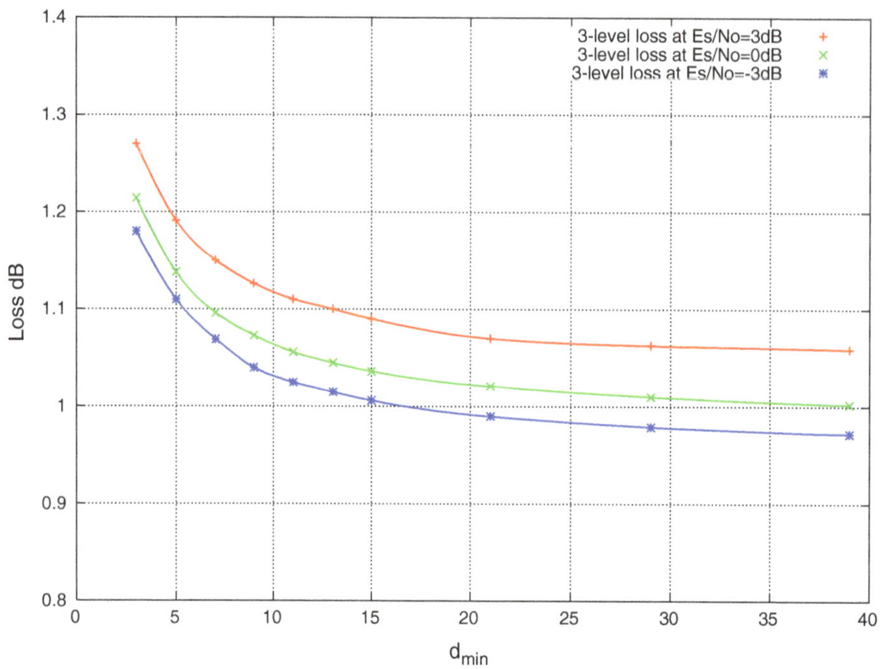

Fig. 3.5 Loss due to three-level soft decisions (erasures) as a function of $\frac{E_s}{N_0}$ and d_{min}

30 will typically operate at $\frac{E_s}{N_0}$ around 0 dB. Of course not all decoder error events are d_{min} events, but the asymptotic nature of the loss produces an average loss of around 2 dB.

Figure 3.5 shows the soft decision quantisation loss, $Loss_Q$, as a function of d_{min} and $\frac{E_s}{N_0}$ for three-level soft decisions. An optimum threshold has been determined for each value of d_{min} and $\frac{E_s}{N_0}$, and these threshold values are in terms of $\sqrt{E_s} - y \times \sigma$ with $y \times \sigma$ plotted against d_{min} in Fig. 3.1. Unlike the hard decision case, for three-level quantisation the lowest loss occurs at high d_{min} values. In common with hard decisions, the lowest loss is for the smallest $\frac{E_s}{N_0}$ values, which are negative when expressed in dB. In absolute terms, the lowest loss is less than 1 dB for $\frac{E_s}{N_0} = -3$ dB and high d_{min}. This corresponds to low-rate codes with code rates of $\frac{1}{3}$ or $\frac{1}{4}$. The loss for three-level quantisation is so much better than hard decisions that it is somewhat surprising that three-level quantisation is not found more often in practical systems. The erasure channel is much underrated.

3.4 A Hard Decision Dorsch Decoder and BCH Codes

The effects of soft decision quantisation on the decoding performance of BCH codes may be explored using the extended Dorsch decoder (see Chap. 15) and by a bounded distance, hard decision decoder, first devised by Peterson [5], refined by Chien [2], Berlekamp [1] and Massey [4]. The extended Dorsch decoder may be used directly on the received three-level quantised soft decisions and of course, on the received unquantised soft decisions. It may also be used on the received hard decisions, to form a near maximum likelihood decoder which is a non bounded distance, hard decision decoder, but requires some modification.

The first stage of the extended Dorsch decoder is to rank the received signal samples in order of likelihood. For hard decisions, all signal samples have equal likelihood and no ranking is possible. However, a random ranking of k, independent bits may be substituted for the ranked k most reliable, independent bits. Provided the number of bit errors contained in these k bits is within the search space of the decoder, the most likely, or the correct codeword, will be found by the decoder. Given the received hard decisions contain t errors, and assuming the search space of the decoder can accommodate m errors, the probability of finding the correct codeword, or a more likely codeword, p_f is given by

$$p_f = \sum_{i=0}^{m} \frac{n!}{(n-i)!\, i!} \left(\frac{t}{n}\right)^{i} \left(1 - \frac{t}{n}\right)^{n-i} \tag{3.44}$$

This probability may be improved by repeatedly carrying out a random ordering of the received samples and running the decoder. With N such orderings, the probability of finding the correct codeword, or a more likely codeword, p_{Nf} becomes more likely and is given by

$$p_{Nf} = 1 - \left(1 - \sum_{i=0}^{m} \frac{n!}{(n-i)!\, i!} \left(\frac{t}{n}\right)^{i} \left(1 - \frac{t}{n}\right)^{n-i}\right)^{N} \tag{3.45}$$

Increasing N gives

$$\left(1 - \sum_{i=0}^{m} \frac{n!}{(n-i)!\, i!} \left(\frac{t}{n}\right)^{i} \left(1 - \frac{t}{n}\right)^{n-i}\right)^{N} \simeq 0 \tag{3.46}$$

and

$$p_{Nf} \simeq 1 \tag{3.47}$$

Of course there is a price to be paid because the complexity of the decoder increases with N. The parity check matrix needs to be solved N times. On the other hand, the size of the search space may be reduced because the repeated decoding allows several chances for the correct codeword to be found.

The modified Dorsch decoder and a bounded distance hard decision BCH decoder have been applied to the [63, 36, 11] BCH code and the simulation results are shown in Fig. 3.6. The decoder search space was set to search 1×10^6 codewords for each received vector which ensures that quasi maximum likelihood decoding is obtained. Also shown in Fig. 3.6 is the sphere packing bound for a (63, 36) code offset by

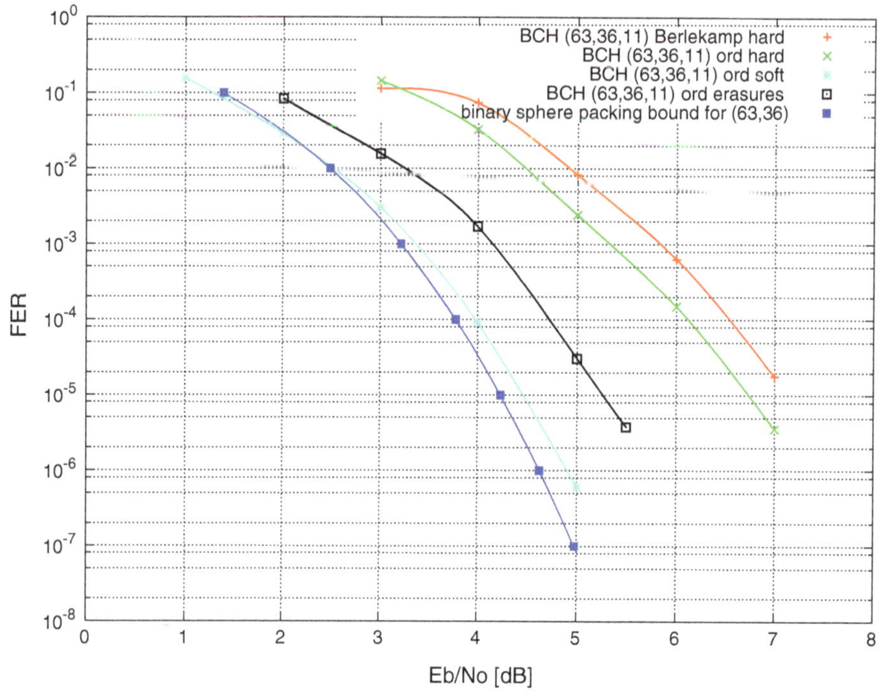

Fig. 3.6 Soft decision decoding of the (63, 36, 11) BCH code compared to hard decision decoding

the binary transmission loss. As can be seen, the unquantised soft decision decoder produces a performance close to the offset sphere packing bound. The three-level quantisation decoder results are offset approximately 0.9 dB at 1×10^{-5} from the unquantised soft decision performance. For hard decisions, the modified Dorsch decoder has a performance approximately 2 dB at 1×10^{-3} from the unquantised soft decision performance and approximately 2.2 dB at 1×10^{-5}. Interestingly, this hard decision performance is approximately 0.4 dB better than the bounded distance BCH decoder correcting up to and including 5 errors.

The results for the BCH (127, 92, 11) code are shown in Fig. 3.7. These results are similar to those of the (63, 36, 11) BCH code. At 1×10^{-5} Frame Error Rate (FER), the unquantised soft decision decoder produces a performance nearly 0.2 dB from the offset sphere packing bound. The three-level quantisation decoder results are offset approximately 1.1 dB at 1×10^{-5} from the unquantised soft decision performance. This is a higher rate code than the (63, 36, 11) code, and at 1×10^{-5} the $\frac{E_s}{N_0}$ ratio is 4.1 dB. Figure 3.5 for a d_{min} of 11 and an $\frac{E_s}{N_0}$ ratio of 3 dB indicates a loss of 1.1 dB, giving good agreement to the simulation results. For hard decisions, the modified Dorsch decoder has a performance approximately 2 dB at 1×10^{-3} from the unquantised soft decision performance, and approximately 2.1 dB at 1×10^{-5}. This is consistent with the theoretical hard decision losses shown in Fig. 3.4. As before, the hard decision performance obtained with the modified Dorsch decoder is better than the bounded distance BCH decoder correcting up to and including five errors, and shows almost 0.5 dB improvement.

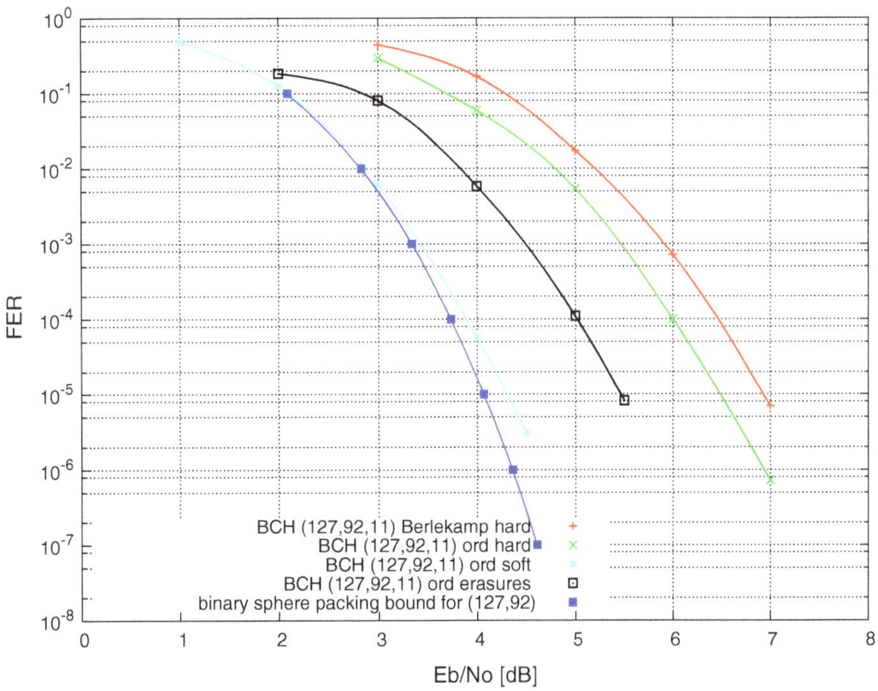

Fig. 3.7 Soft decision decoding of the (127, 92, 11) BCH code compared to hard decision decoding

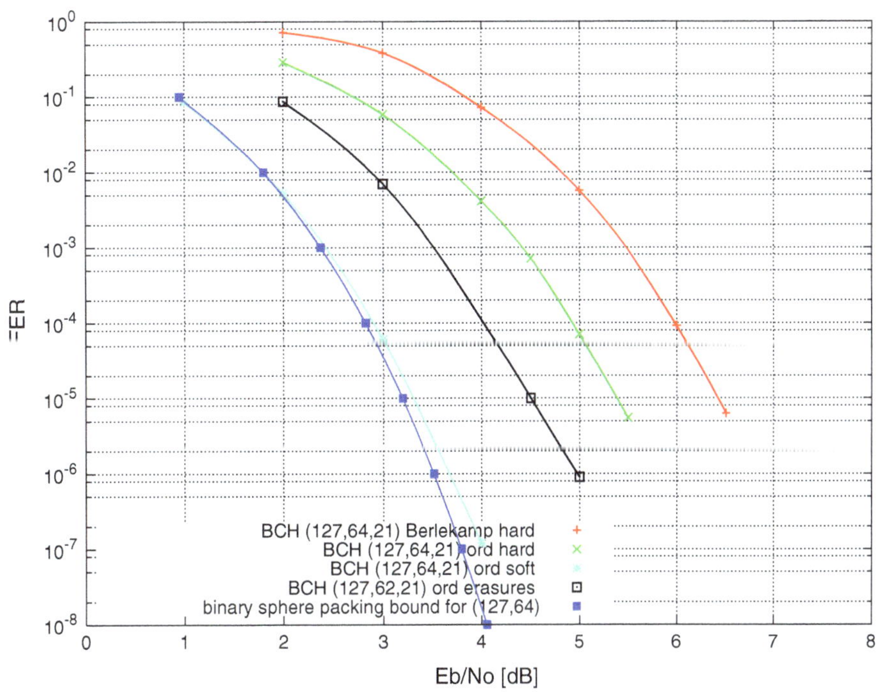

Fig. 3.8 Soft decision decoding of the (127, 64, 21) BCH code compared to hard decision decoding

The results for the BCH (127, 64, 21) code are shown in Fig. 3.8. This is an outstanding code, and consequently the unquantised soft decision decoding performance is very close to the offset sphere packing bound, being almost 0.1 dB away from the bound at 1×10^{-5}. However, a list size of 10^7 codewords was used in order to ensure that near maximum likelihood performance was obtained by the modified Dorsch decoder. Similar to before the three-level quantisation decoder results are offset approximately 1.1 dB at 1×10^{-5} from the unquantised soft decision performance. However, 3×10^7 codewords were necessary in order to obtain near maximum likelihood performance was obtained by the modified Dorsch decoder operating on the three-level quantised decisions. The BCH bounded distance decoder is approximately 3 dB offset from the unquantised soft decision decoding performance and 1 dB from the modified Dorsch decoder operating on the quantised hard decisions.

These simulation results for the losses due to quantisation of the soft decisions show a very close agreement to the losses anticipated from the theoretical analysis.

3.5 Summary

In this chapter, we derived both approximate and exact bounds on the performance of soft decision decoding compared to hard decision decoding as a function of code parameters. The effects of soft decision quantisation were explored showing the decoding performance loss as a function of number of quantisation levels. Results were presented for the ensembles of all (100, 50) and (200, 100) codes. It was shown that the loss due to quantisation is a function of both d_{min} and SNR. Performance graphs showing the relationship were presented.

It was shown that the near maximum likelihood decoder, the Dorsch decoder described in Chap. 15, may be adapted for hard decision decoding in order to produce better performance than bounded distance decoding. Performance graphs were presented for some BCH codes showing the performance achieved compared to bounded distance decoding.

References

1. Berlekamp, E.: On decoding binary Bose-Chadhuri-Hocquenghem codes. IEEE Trans. Inf. Theory **11**(4), 577–579 (1965)
2. Chien, R.: Cyclic decoding procedures for Bose-Chaudhuri-Hocquenghem codes. IEEE Trans. Inf. Theory **10**(4), 357–363 (1964)
3. Gallager, R.G.: A simple derivation of the coding theorem and some applications. IEEE Trans. Inf. Theory **11**(1), 459–470 (1960)
4. Massey, J.: Shift register synthesis and BCH decoding. IEEE Trans. Inf. Theory **15**(1), 122–127 (1969)
5. Peterson, W.: Encoding and error-correction procedures for the Bose-Chaudhuri codes. IRE Trans. Inf. Theory **6**(4), 459–470 (1960)
6. Wozencraft, J., Jacobs, I.: Principles of Communication Engineering. Wiley, New York (1965)

Part II
Code Construction

This part of the book deals with the construction of error-correcting codes having good code properties. With an emphasis on binary codes, a wide range of different code constructions are described including cyclic codes, double circulant codes, quadratic residue codes, Goppa codes, Lagrange codes, BCH codes and Reed–Solomon codes. Code combining constructions such as Construction X are also included. For shorter codes, typically less than 512 symbols long, the emphasis is on the highest minimum Hamming distance for a given length and code rate. The construction of some outstanding codes is described in detail together with the derivation of the weight distributions of the codes. For longer codes, the emphasis is on the best code design for a given type of decoder, such as the iterative decoder. Binary convolutional codes are discussed from the point of view of their historical performance in comparison to the performance realised with modern best decoding techniques. Convolutional codes, designed for space communications in the 1960s, are implemented as tail-biting block codes. The performance realised with near maximum likelihood decoding, featuring the modifed Dorsch decoder described in Chap. 15, is somewhat surprising.

Chapter 4
Cyclotomic Cosets, the Mattson–Solomon Polynomial, Idempotents and Cyclic Codes

4.1 Introduction

Much of the pioneering research on cyclic codes was carried out by Prange [5] in the 1950s and considerably developed by Peterson [4] in terms of generator and parity-check polynomials. MacWilliams and Sloane [2] showed that cyclic codes could be generated from idempotents and the Mattson–Solomon polynomial, first introduced by Mattson and Solomon in 1961 [3]. The binary idempotent polynomials follow directly from cyclotomic cosets.

4.2 Cyclotomic Cosets

Consider the expansion of polynomial $a(x) = \prod_{i=0}^{m-1}(x - \alpha^{2^i})$. The coefficients of $a(x)$ are a cyclotomic coset of powers of α or a sum of cyclotomic cosets of powers of α. For example, if $m = 4$

$$a(x) = (x - \alpha)(x - \alpha^2)(x - \alpha^4)(x - \alpha^8) \tag{4.1}$$

and expanding $a(x)$ produces

$$\begin{aligned} a(x) = x^4 - (\alpha + \alpha^2 + \alpha^4 + \alpha^8)x^3 + (\alpha^3 + \alpha^6 + \alpha^{12} + \alpha^9 + \alpha^5 + \alpha^{10})x^2 \\ + (\alpha^7 + \alpha^{14} + \alpha^{13} + \alpha^{11})x + \alpha^{15}. \end{aligned} \tag{4.2}$$

Definition 4.1 (*Cyclotomic Coset*) Let s be a positive integer, and the 2−cyclotomic coset of $s \pmod n$ is given by

M. Tomlinson et al., *Error-Correction Coding and Decoding*,
Signals and Communication Technology,
DOI 10.1007/978-3-319-51103-0_4

$$C_s = \{2^i s \pmod{n} \mid 0 \le i \le t\},$$

where s is the smallest element in the set C_s and t is the smallest positive integer such that $2^{t+1}s \equiv s \pmod{n}$.

For convenience, we will use the term cyclotomic coset to refer to 2−cyclotomic coset. If $\mathcal{N}$ is the set consisting of the smallest elements of all possible cyclotomic cosets, then it follows that

$$C = \bigcup_{s \in \mathcal{N}} C_s = \{0, 1, 2, \ldots, n-1\}.$$

Example 4.1 The entire cyclotomic cosets of 15 are as follows:

$$\begin{aligned} C_0 &= \{0\} \\ C_1 &= \{1, 2, 4, 8\} \\ C_3 &= \{3, 6, 12, 9\} \\ C_5 &= \{5, 10\} \\ C_7 &= \{7, 14, 13, 11\} \end{aligned}$$

and $\mathcal{N} = \{0, 1, 3, 5, 7\}$.

It can be seen that for $GF(2^4)$ above, Eq. (4.2), the coefficients of $a(x)$ are a cyclotomic coset of powers of α or a sum of cyclotomic cosets of powers of α. For example, the coefficient of x^3 is the sum of powers of α from cyclotomic coset C_1.

In the next step of the argument we note that there is an important property of Galois fields.

Theorem 4.1 *For a Galois field $GF(p^m)$, then*

$$\big(b(x) + c(x)\big)^p = b(x)^p + c(x)^p.$$

Proof Expanding $\big(b(x) + c(x)\big)^p$ produces

$$\begin{aligned} \big(b(x) + c(x)\big)^p = {} & b(x)^p + \binom{p}{1} b(x)^{p-1}c(x) + \binom{p}{2} b(x)^{p-2}c(x)^2 + \\ & \ldots + \binom{p}{p-1} b(x)c(x)^{p-1} + c(x)^p. \end{aligned} \tag{4.3}$$

As p modulo $p = 0$, then all of the binomial coefficients $\binom{p}{r} = 0$ and

$$\big(b(x) + c(x)\big)^p = b(x)^p + c(x)^p.$$

Another theorem follows.

Theorem 4.2 *The sum of powers of α that are from a cyclotomic coset C_i is equal to either 1 or 0.*

Proof The sum of powers of α that are from a cyclotomic coset C_i must equal to a field element, some power, j of α, α^j or 0. Also, from Theorem 1.1,

$$\left(\sum \alpha^{C_i}\right)^2 = \sum \alpha^{C_i}.$$

If the sum of powers of α is non-zero then

$$\left(\sum \alpha^{C_i}\right)^2 = \alpha^{2j} = \sum \alpha^{C_i} = \alpha^j.$$

The only non-zero field element that satisfies $\alpha^{2j} = \alpha^j$ is $\alpha^0 = 1$. Hence, the sum of powers of α that are from a cyclotomic coset C_i is equal to either 1 or 0.

In the example of C_1 from $GF(2^4)$ we have

$$(\alpha + \alpha^2 + \alpha^4 + \alpha^8)^2 = \alpha^2 + \alpha^4 + \alpha^8 + \alpha^{16} = \alpha^2 + \alpha^4 + \alpha^8 + \alpha$$

and so

$$\alpha + \alpha^2 + \alpha^4 + \alpha^8 = 0 \text{ or } 1.$$

Returning to the expansion of polynomial $a(x) = \prod_{i=0}^{m-1}(x - \alpha^{2^i})$. Since the coefficients of $a(x)$ are a cyclotomic coset of powers of α or a sum of cyclotomic cosets of powers of α, the coefficients of $a(x)$ must be 0 or 1 and $a(x)$ must have binary coefficients after noting that the coefficient of x^0 is $\prod_{i=0}^{m-1} \alpha^{2^i} = \alpha^{2^m-1} = 1$, the maximum order of α. Considering the previous example of $m = 4$ ($GF(2^4)$), since $a(x)$ is constrained to have binary coefficients, we have the following possible identities:

$$\begin{aligned} \alpha^{15} &= 1 \\ \alpha + \alpha^2 + \alpha^4 + \alpha^8 &= 0 \text{ or } 1 \\ \alpha^7 + \alpha^{14} + \alpha^{13} + \alpha^{11} &= 0 \text{ or } 1 \\ \alpha^3 + \alpha^6 + \alpha^{12} + \alpha^9 + \alpha^5 + \alpha^{10} &= 0 \text{ or } 1. \end{aligned} \tag{4.4}$$

These identities are determined by the choice of primitive polynomial used to generate the extension field. This can be seen from the Trace function, $T_m(x)$, defined as

$$T_m(x) = \sum_{i=0}^{m-1} x^{2^i} \tag{4.5}$$

and expanding the product of $T_m(x)\big(1 + T_m(x)\big)$ produces the identity

$$T_m(x)\big(1 + T_m(x)\big) = x(1 - x^n). \tag{4.6}$$

α is a root of $(1 - x^n)$ and so α is a root of either $T_m(x)$ or $\big(1 + T_m(x)\big)$, and so either $T_m(\alpha) = 0$ or $\big(1 + T_m(\alpha)\big) = 0$. For $GF(2^4)$

$$T_m(x) = \sum_{i=0}^{3} x^{2^i} = x + x^2 + x^4 + x^8. \tag{4.7}$$

Factorising produces

$$x + x^2 + x^4 + x^8 = x(1 + x)(1 + x + x^2)(1 + x + x^4), \tag{4.8}$$

and

$$1 + T_m(x) = 1 + \sum_{i=0}^{3} x^{2^i} = 1 + x + x^2 + x^4 + x^8. \tag{4.9}$$

Factorising produces

$$1 + x + x^2 + x^4 + x^8 = (1 + x^3 + x^4)(1 + x + x^2 + x^3 + x^4). \tag{4.10}$$

It may be verified that

$$\begin{aligned} T_m(x)\big(1 + T_m(x)\big) &= (x + x^2 + x^4 + x^8)(1 + x + x^2 + x^4 + x^8) \\ &= x(1 + x)(1 + x + x^2)(1 + x + x^4)(1 + x^3 + x^4) \\ &\quad (1 + x + x^2 + x^3 + x^4) \\ &= x(1 - x^{15}). \end{aligned}$$

Consequently, if $1 + x + x^4$ is used to generate the extension field $GF(16)$ then $\alpha + \alpha^2 + \alpha^4 + \alpha^8 = 0$ and if $1 + x^3 + x^4$ is used to generate the extension field $GF(16)$, then $1 + \alpha + \alpha^2 + \alpha^4 + \alpha^8 = 0$.

Taking the case that $a(x) = 1 + x + x^4$ is used to generate the extension field $GF(16)$ by comparing the coefficients given by Eq. (4.2), we can solve the identities of (4.4) after noting that $\alpha^5 + \alpha^{10}$ must equal 1 otherwise the order of α is equal to 5, contradicting α being a primitive root. All of the identities of the sum for each cyclotomic coset of powers of α are denoted by $S_{i\,m}$ and these are

$$
\begin{aligned}
S_{0\,4} &= \alpha^0 = 1 \\
S_{1\,4} &= \alpha + \alpha^2 + \alpha^4 + \alpha^8 = 0 \\
S_{3\,4} &= \alpha^3 + \alpha^6 + \alpha^{12} + \alpha^9 = 1 \\
S_{5\,4} &= \alpha^5 + \alpha^{10} = 1 \\
S_{7\,4} &= \alpha^7 + \alpha^{14} + \alpha^{13} + \alpha^{11} = 1 \\
S_{15\,4} &= \alpha^{15} = 1.
\end{aligned} \tag{4.11}
$$

The lowest degree polynomial that has β as a root is traditionally known as a minimal polynomial [2], and is denoted as $M_{i\,m}$ where $\beta = \alpha^i$. With $M_{i\,m}$ having binary coefficients

$$M_{i\,m} = \prod_{j=0}^{m-1} (x - \alpha^{i2^j}). \tag{4.12}$$

For $GF(2^4)$ and considering $M_{3\,4}$ for example,

$$M_{3\,4} = (x - \alpha^3)(x - \alpha^6)(x - \alpha^{12})(x - \alpha^9), \tag{4.13}$$

and expanding leads to

$$
\begin{aligned}
M_{3\,4} = x^4 &- (\alpha^3 + \alpha^6 + \alpha^{12} + \alpha^9)x^3 + (\alpha^9 + \alpha^3 + \alpha^6 + \alpha^{12})x^2 \\
&+ (\alpha^6 + \alpha^{12} + \alpha^9 + \alpha^3)x + 1.
\end{aligned} \tag{4.14}
$$

It will be noticed that this is the same as Eq. (4.2) with α replaced with α^3. Using the identities of Eq. (4.11), it is found that

$$M_{3\,4} = x^4 + x^3 + x^2 + x + 1. \tag{4.15}$$

Similarly, it is found that for $M_{5\,4}$ substitution produces $x^4 + x^2 + 1$ which is $(x^2 + x + 1)^2$, and so

$$M_{5\,4} = x^2 + x + 1; \tag{4.16}$$

similarly, it is found that

$$M_{7\,4} = x^4 + x^3 + 1 \tag{4.17}$$

for $M_{0\,4}$ with $\beta = 15$, and substitution produces $x^4 + 1 = (1 + x)^4$ and

$$M_{0\,4} = x + 1. \tag{4.18}$$

It will be noticed that all of the minimal polynomials correspond to the factors of $1 + x^{15}$ given above. Also, it was not necessary to generate a table of $GF(2^4)$ field elements in order to determine all of the minimal polynomials once $M_{1\,4}$ was chosen.

A recurrence relation exists for the cyclotomic cosets with increasing m for

$$M_{i\,m+1} = \Big(\prod_{j=0}^{m-1}(x - \alpha^{i2^j})\Big)x - \alpha^{i2^m}. \tag{4.19}$$

For $m = 4$,

$$M_{1\,4} = x^4 + S_{1\,4}x^3 + (S_{3\,4} + S_{5\,4})x^2 + S_{7\,4}x + \alpha^{15} \tag{4.20}$$

and so

$$M_{1\,5} = \Big(x^4 + S_{1\,4}x^3 + (S_{3\,4} + S_{5\,4})x^2 + S_{7\,4}x + \alpha^{15}\Big)(x + \alpha^{16}) \tag{4.21}$$

and

$$\begin{aligned} M_{1\,5} =& x^5 + (\alpha^{16} + S_{1\,4})x^4 + (\alpha^{16}S_{1\,4} + (S_{3\,4} + S_{5\,4}))x^3 \\ &+ \big(\alpha^{16}(S_{3\,4} + S_{5\,4}) + S_{7\,4}\big)x^2 + (\alpha^{16}S_{7\,4} + \alpha^{15})x + \alpha^{31} \end{aligned} \tag{4.22}$$

and we find that

$$\begin{aligned} M_{1\,5} =& x^5 + S_{1\,5}x^4 + (S_{3\,5} + S_{5\,5})x^3 \\ &+ (S_{7\,5} + S_{11\,5})x^2 + S_{15\,5}x + \alpha^{31}. \end{aligned} \tag{4.23}$$

We have the following identities, linking the cyclotomic cosets of $GF(2^4)$ to $GF(2^5)$

$$\begin{aligned} S_{3\,5} + S_{5\,5} &= \alpha^{16}S_{1\,4} + S_{3\,4} + S_{5\,4} \\ S_{7\,5} + S_{11\,5} &= \alpha^{16}(S_{3\,4} + S_{5\,4}) + S_{7\,4} \\ S_{15\,5} &= \alpha^{16}S_{7\,4} + \alpha^{15}. \end{aligned}$$

With $1 + x^2 + x^5$ used to generate the extension field $GF(32)$, then $\alpha + \alpha^2 + \alpha^4 + \alpha^8 + \alpha^{16} = 0$. Evaluating the cyclotomic cosets of powers of α produces

$$\begin{aligned} S_{0\,5} &= \alpha^0 = 1 \\ S_{1\,5} &= \alpha + \alpha^2 + \alpha^4 + \alpha^8 + \alpha^{16} = 0 \\ S_{3\,5} &= \alpha^3 + \alpha^6 + \alpha^{12} + \alpha^{24} + \alpha^{17} = 1 \\ S_{5\,5} &= \alpha^5 + \alpha^{10} + \alpha^{20} + \alpha^9 + \alpha^{18} = 1 \\ S_{7\,5} &= \alpha^7 + \alpha^{14} + \alpha^{28} + \alpha^{25} + \alpha^{19} = 0 \end{aligned}$$

$$
\begin{aligned}
S_{11\,5} &= \alpha^{11} + \alpha^{22} + \alpha^{13} + \alpha^{26} + \alpha^{21} = 1 \\
S_{15\,5} &= \alpha^{15} + \alpha^{30} + \alpha^{29} + \alpha^{27} + \alpha^{23} = 0.
\end{aligned}
\tag{4.24}
$$

Substituting for the minimal polynomials, $M_{i,5}$ produces

$$
\begin{aligned}
M_{0\,5} &= x + 1 \\
M_{1\,5} &= x^5 + x^2 + 1 \\
M_{3\,5} &= x^5 + x^4 + x^3 + x^2 + 1 \\
M_{5\,5} &= x^5 + x^4 + x^2 + x + 1 \\
M_{7\,5} &= x^5 + x^3 + x^2 + x + 1 \\
M_{11\,5} &= x^5 + x^4 + x^3 + x + 1 \\
M_{15\,5} &= x^5 + x^3 + 1.
\end{aligned}
\tag{4.25}
$$

For $GF(2^5)$, the order of a root of a primitive polynomial is 31, a prime number. Moreover, 31 is a Mersenne prime $(2^p - 1)$ and the first 12 Mersenne primes correspond to $p = 2, 3, 5, 7, 13, 17, 19, 31, 61, 89, 107$ and 127. Interestingly, only 49 Mersenne primes are known. The last known Mersenne prime being $2^{74207281} - 1$, discovered in January 2016. As $(2^5 - 1)$ is prime, each of the minimal polynomials in Eq. (4.25) is primitive.

If α is a root of $T_m(x)$ and m is even, then $1 + T_{2m}(x) = 1 + T_m(x) + \left(1 + T_m(x)\right)^{2^m}$ and $\alpha^{\frac{2^{2m}-1}{2^m-1}}$ is a root of $x^{2^{2m}}$. For example, if α is a root of $1 + x + x^2$, α is of order 3 and α^5 is a root of $x + x^2 + x^4 + x^8$. Correspondingly, $1 + x + x^2$ is a factor of $1 + x^3$ and also a factor of$1 + x^{15}$ and necessarily $2^{2m} - 1$ cannot be prime. Similarly, if m is not a prime and $m = ab$, then

$$
\frac{2^m - 1}{2^a - 1} = 2^{b(a-1)} + 2^{b(a-2)} + 2^{b(a-3)} \ldots + 1 \tag{4.26}
$$

and so

$$
2^m - 1 = (2^{b(a-1)} + 2^{b(a-2)} + 2^{b(a-3)} \ldots + 1)2^a - 1. \tag{4.27}
$$

Similarly

$$
2^m - 1 = (2^{a(b-1)} + 2^{a(b-2)} + 2^{a(b-3)} \ldots + 1)2^b - 1. \tag{4.28}
$$

As a consequence

$$
M_{(2^{b(a-1)}+2^{b(a-2)}+2^{b(a-3)}\ldots+1)\times j\;m} = M_{j\;a} \tag{4.29}
$$

for all minimal polynomials of $x^{2^a-1}-1$, and

$$M_{(2^{a(b-1)}+2^{a(b-2)}+2^{a(b-3)}\dots+1)\times j\, m} = M_{j\, b} \tag{4.30}$$

for all minimal polynomials of $x^{2^b-1}-1$.

For $M_{1\,6}$, following the same procedure,

$$\begin{aligned} M_{1\,6} = x^6 + S_{1\,6}x^5 + (S_{3\,6} + S_{5\,6} + S_{9\,6})x^4 + (S_{7\,6} + S_{11\,6} + S_{13\,6} + S_{21\,6})x^3 \\ + (S_{15\,6} + S_{23\,6} + S_{27\,6})x^2 + S_{15\,6}x^2 + S_{31\,6}x + \alpha^{63}. \end{aligned} \tag{4.31}$$

Substituting for the minimal polynomials, $M_{i,6}$ produces

$$\begin{aligned} M_{0\,6} &= x + 1 \\ M_{1\,6} &= x^6 + x + 1 \\ M_{3\,6} &= x^6 + x^4 + x^2 + x + 1 \\ M_{5\,6} &= x^6 + x^5 + x^2 + x + 1 \\ M_{7\,6} &= x^6 + x^3 + 1 \\ M_{9\,6} &= x^3 + x^2 + 1 \\ M_{11\,6} &= x^6 + x^5 + x^3 + x^2 + 1 \\ M_{13\,6} &= x^6 + x^4 + x^3 + x + 1 \\ M_{15\,6} &= x^6 + x^5 + x^4 + x^2 + 1 \\ M_{21\,6} &= x^2 + x + 1 \\ M_{23\,6} &= x^6 + x^5 + x^4 + x + 1 \\ M_{27\,6} &= x^3 + x + 1 \\ M_{31\,6} &= x^6 + x^5 + 1. \end{aligned} \tag{4.32}$$

Notice that $M_{9\,6} = M_{3\,4}$ because $\alpha^9 + \alpha^{18} + \alpha^{36} = 1$ and $M_{27\,6} = M_{1\,4}$ because $\alpha^9 + \alpha^{18} + \alpha^{36} = 0$. $M_{21\,6} = M_{1\,3}$ because $\alpha^{21} + \alpha^{42} = 1$. The order of α is 63 which factorises to $7 \times 3 \times 3$ and so $x^{63} - 1$ will have roots of order 7 (α^9) and roots of order 3 (α^21). Another way of looking at this is the factorisation of $x^{63} - 1$. $x^7 - 1$ is a factor and $x^3 - 1$ is a factor

$$\begin{aligned} x^{63} - 1 = (x^7 - 1)(1 + x^7 + x^{14} + x^{21} \\ + x^{28} + x^{35} + x^{42} + x^{49} + x^{56}) \end{aligned} \tag{4.33}$$

also

$$\begin{aligned} x^{63} - 1 = (x^3 - 1)(1 + x^3 + x^6 + x^9 + x^{12} + x^{15} + x^{18} + x^{21} \\ + x^{24} + x^{27} + x^{30} + x^{33} + x^{36} + x^{39} + x^{42} + x^{45} \\ + x^{48} + x^{51} + x^{54} + x^{57} + x^{60}) \end{aligned} \tag{4.34}$$

and

$$\begin{aligned}
x^3 - 1 &= (x+1)(x^2+x+1) \\
x^7 - 1 &= (x+1)(x^3+x+1)(x^3+x^2+1) \\
x^{63} - 1 &= (x+1)(x^2+x+1)(x^3+x+1)(x^3+x^2+1)(x^6+x+1) \\
&\quad (x^6+x^4+x^2+x+1)\ldots(x^6+x^5+1).
\end{aligned} \tag{4.35}$$

For $M_{1\,7}$

$$\begin{aligned}
M_{1\,7} = x^7 &+ S_{1\,7}x^6 + (S_{3\,7} + S_{5\,7} + S_{9\,7})x^4 + (S_{7\,7} + S_{11\,7} + S_{13\,7} + S_{19\,7} + S_{21\,7})x^3 \\
&+ (S_{15\,7} + S_{23\,7} + S_{27\,7} + S_{29\,7})x^3 + (S_{15\,7} + S_{31\,7} + S_{43\,7} + S_{47\,7} + S_{55\,7})x^2 \\
&+ S_{63\,7}x + \alpha^{127}.
\end{aligned} \tag{4.36}$$

Although the above procedure using the sums of powers of α from the cyclotomic cosets may be used to generate the minimal polynomials $M_{i\,m}$ for any m, the procedure becomes tedious with increasing m, and it is easier to use the Mattson Polynomial or combinations of the idempotents as described in Sect. 4.4.

4.3 The Mattson–Solomon Polynomial

The Mattson–Solomon polynomial is very useful for it can be conveniently used to generate minimal polynomials and idempotents. It also may be used to design cyclic codes, RS codes and Goppa codes as well as determining the weight distribution of codes. The Mattson–Solomon polynomial [2] of a polynomial $a(x)$ is a linear transformation of $a(x)$ to $A(z)$. The Mattson–Solomon polynomial is the same as the inverse Discrete Fourier Transform over a finite field. The polynomial variables x and z are used to distinguish the polynomials in either domain.

Let the splitting field of $x^n - 1$ over $\mathbb{F}_2$ be $\mathbb{F}_{2^m}$, where n is an odd integer and $m > 1$, and let a generator of $\mathbb{F}_{2^m}$ be α and an integer $r = (2^m - 1)/n$. Let $a(x)$ be a polynomial of degree at most $n - 1$ with coefficients over $\mathbb{F}_{2^m}$.

Definition 4.2 (*Mattson–Solomon polynomial*) The Mattson–Solomon polynomial of $a(x)$ is the linear transformation of $a(x)$ to $A(z)$ and is defined by [2]

$$A(z) = \mathrm{MS}(a(x)) = \sum_{j=0}^{n-1} a(\alpha^{-rj})z^j. \tag{4.37}$$

The inverse Mattson–Solomon transformation or Fourier transform is

Table 4.1 $GF(16)$ extension field defined by $1+\alpha+\alpha^4=0$

$\alpha^0 = 1$
$\alpha^1 = \alpha$
$\alpha^2 = \alpha^2$
$\alpha^3 = \alpha^3$
$\alpha^4 = 1+\alpha$
$\alpha^5 = \alpha+\alpha^2$
$\alpha^6 = \alpha^2+\alpha^3$
$\alpha^7 = 1+\alpha+\alpha^3$
$\alpha^8 = 1+\alpha^2$
$\alpha^9 = \alpha+\alpha^3$
$\alpha^{10} = 1+\alpha+\alpha^2$
$\alpha^{11} = \alpha+\alpha^2+\alpha^3$
$\alpha^{12} = 1+\alpha+\alpha^2+\alpha^3$
$\alpha^{13} = 1+\alpha^2+\alpha^3$
$\alpha^{14} = 1+\alpha^3$

$$a(x) = \text{MS}^{-1}(A(z)) = \frac{1}{n}\sum_{i=0}^{n-1} A(\alpha^{ri})x^i. \tag{4.38}$$

The integer r comes into play when $2^m - 1$ is not a prime, that is, $2^m - 1$ is not a Mersenne prime, otherwise $r = 1$. As an example, we will consider $\mathbb{F}_{2^4}$ and the extension field table of non-zero elements is given in Table 4.1 with $1+\alpha+\alpha^4 = 0$, modulo $1+x^{15}$.

Consider the polynomial $a(x)$ denoted as

$$a(x) = \sum_{i=0}^{n-1} a_i x^i = 1 + x^3 + x^4. \tag{4.39}$$

We will evaluate the Mattson–Solomon polynomial coefficient by coefficient:

$$A(0) = a_0 + a_3 + a_4 = 1+1+1 = 1$$
$$A(1) = a_0 + a_3\alpha^{-3} + a_4\alpha^{-4} = 1+\alpha^{12}+\alpha^{11} = 1+1+\alpha+\alpha^2+\alpha^3+\alpha+\alpha^2+\alpha^3 = 0$$
$$A(2) = a_0 + a_3\alpha^{-6} + a_4\alpha^{-8} = 1+\alpha^9+\alpha^7 = 1+\alpha+\alpha^3+1+\alpha+\alpha^3 = 0$$
$$A(3) = a_0 + a_3\alpha^{-9} + a_4\alpha^{-12} = 1+\alpha^6+\alpha^3 = 1+\alpha^2+\alpha^3+\alpha^3 = \alpha^8$$
$$A(4) = a_0 + a_3\alpha^{-12} + a_4\alpha^{-16} = 1+\alpha^3+\alpha^{14} = 1+\alpha^3+1+\alpha^3 = 0$$
$$A(5) = a_0 + a_3\alpha^{-15} + a_4\alpha^{-20} = 1+1+\alpha^{10} = \alpha^{10}$$
$$A(6) = a_0 + a_3\alpha^{-18} + a_4\alpha^{-24} = 1+\alpha^{12}+\alpha^6 = \alpha$$
$$A(7) = a_0 + a_3\alpha^{-21} + a_4\alpha^{-28} = 1+\alpha^9+\alpha^2 = 1+\alpha+\alpha^3+\alpha^2 = \alpha^{12}$$
$$A(8) = a_0 + a_3\alpha^{-24} + a_4\alpha^{-32} = 1+\alpha^6+\alpha^{13} = 0$$

$$\begin{aligned} A(9) &= a_0 + a_3\alpha^{-27} + a_4\alpha^{-36} = 1 + \alpha^3 + \alpha^9 = 1 + \alpha = \alpha^4 \\ A(10) &= a_0 + a_3\alpha^{-30} + a_4\alpha^{-40} = 1 + 1 + \alpha^5 = \alpha^5 \\ A(11) &= a_0 + a_3\alpha^{-33} + a_4\alpha^{-44} = 1 + \alpha^{12} + \alpha = \alpha^6 \\ A(12) &= a_0 + a_3\alpha^{-36} + a_4\alpha^{-48} = 1 + \alpha^9 + \alpha^{12} = \alpha^2 \\ A(13) &= a_0 + a_3\alpha^{-39} + a_4\alpha^{-52} = 1 + \alpha^6 + \alpha^8 = \alpha^3 \\ A(14) &= a_0 + a_3\alpha^{-42} + a_4\alpha^{-56} = 1 + \alpha^3 + \alpha^4 = \alpha^9. \end{aligned} \tag{4.40}$$

It can be seen that $A(z)$ is

$$\begin{aligned} A(z) = 1 &+ \alpha^8 z^3 + \alpha^{10} z^5 + \alpha z^6 + \alpha^{12} z^7 + \alpha^4 z^9 + \alpha^5 z(10) + \alpha^6 z^{11} + \alpha^2 z^{12} \\ &+ \alpha^3 z^{13} + \alpha^9 z^{14}. \end{aligned}$$

A(z) has four zeros corresponding to the roots α^{-1}, α^{-2}, α^{-4} and α^{-8}, and these are the roots of $1 + x^3 + x^4$. These are also 4 of the 15 roots of $1 + x^{15}$. Factorising $1 + x^{15}$ produces the identity

$$1 + x^{15} = (1 + x)(1 + x + x^2)(1 + x + x^4)(1 + x^3 + x^4)(1 + x + x^2 + x^3 + x^4). \tag{4.41}$$

It can be seen that $1 + x^3 + x^4$ is one of the factors of $1 + x^{15}$.

Another point to notice is that $A(z) = A(z)^2$ and $A(z)$ is an idempotent. The reason for this is that the inverse Mattson–Solomon polynomial of $A(z)$ will produce $a(x)$ a polynomial that has binary coefficients. Let $\cdot$ denote the dot product of polynomials, i.e.

$$\left(\sum A_i z^i\right) \cdot \left(\sum B_i z^i\right) = \sum A_i B_i z^i.$$

It follows from the Mattson–Solomon polynomial that with $a(x)b(x) = c(x)$, $\sum C_i z^i = \sum A_i B_i z^i$.

This concept is analogous to multiplication and convolution in the time and frequency domains, where the Fourier and inverse Fourier transforms correspond to the inverse Mattson–Solomon and Mattson–Solomon polynomials, respectively. In the above example, $A(z)$ is an idempotent which leads to the following lemma.

Lemma 4.1 *The Mattson–Solomon polynomial of a polynomial having binary coefficients is an idempotent.*

Proof Let $c(x) = a(x) \cdot b(x)$. The Mattson–Solomon polynomial of $c(x)$ is $C(z) = A(z)B(z)$. Setting $b(x) = a(x)$ then $C(z) = A(z)A(z) = A(z)^2$. If $a(x)$ has binary coefficients, then $c(x) = a(x) \cdot a(x) = a(x)$ and $A(z)^2 = A(z)$. Therefore $A(z)$ is an idempotent.

Of course the reverse is true.

Lemma 4.2 *The Mattson–Solomon polynomial of an idempotent is a polynomial having binary coefficients.*

Proof Let $c(x) = a(x)b(x)$. The Mattson–Solomon polynomial of $c(x)$ is $C(z) = A(z)B(z)$. Setting $b(x) = a(x)$ then $C(z) = A(z) \cdot A(z)$. If $a(x)$ is an idempotent then $c(x) = a(x)^2 = a(x)$ and $A(z) = A(z) \cdot A(z)$. The only values for the coefficients of $A(z)$ that satisfy this constraint are the values 0 and 1. Hence, the Mattson Solomon polynomial, $A(z)$, has binary coefficients.

A polynomial that has binary coefficients and is an idempotent is a binary idempotent, and combining Lemmas 4.1 and 4.2 produces the following lemma.

Lemma 4.3 *The Mattson–Solomon polynomial of a binary idempotent is also a binary idempotent.*

Proof The proof follows immediately from the proofs of Lemmas 4.1 and 4.2. As $a(x)$ is an idempotent, then from Lemma 4.1, $A(z)$ has binary coefficients. As $a(x)$ also has binary coefficients, then from Lemma 4.2, $A(z)$ is an idempotent. Hence, $A(z)$ is a binary idempotent.

As an example consider the binary idempotent $a(x)$ from $GF(16)$ listed in Table 4.1:

$$a(x) = x + x^2 + x^3 + x^4 + x^6 + x^8 + x^9 + x^{12}.$$

The Mattson–Solomon polynomial $A(z)$ is

$$A(z) = z^7 + z^{11} + z^{13} + z^{14},$$

which is also a binary idempotent.

Since the Mattson polynomial of $a(x^{-1})$ is the same as the inverse Mattson polynomial of $a(x)$ consider the following example:

$$a(x) = x^{-7} + x^{-11} + x^{-13} + x^{-14} = x + x^2 + x^4 + x^4.$$

The Mattson–Solomon polynomial $A(z)$ is the binary idempotent

$$A(z) = z + z^2 + z^3 + z^4 + z^6 + z^8 + z^9 + z^{12}.$$

This is the reverse of the first example above.

The polynomial $1 + x + x^3$ has no roots of $1 + x^{15}$ and so defining $b(x)$

$$b(x) = (1 + x + x^3)(1 + x^3 + x^4) = 1 + x + x^5 + x^6 + x^7. \tag{4.42}$$

When the Mattson–Solomon polynomial is evaluated, $B(z)$ is given by

$$B(z) = 1 + z + z^5 + z^6 + z^7. \tag{4.43}$$

4.4 Binary Cyclic Codes Derived from Idempotents

In their book, MacWilliams and Sloane [2] describe the Mattson–Solomon polynomial and show that cyclic codes may be constructed straightforwardly from idempotents. An idempotent is a polynomial $\theta(x)$ with coefficients from a base field $GF(p)$ that has the property that $\theta^p(x) = \theta(x)$. The family of Bose–Chaudhuri–Hocquenghem (BCH) cyclic codes may be constructed directly from the Mattson–Solomon polynomial. From the idempotents, other cyclic codes may be constructed which have low-weight dual-code codewords or equivalently sparseness of the parity-check matrix (see Chap. 12).

Definition 4.3 (*Binary Idempotent*) Consider $e(x) \in T(x)$, $e(x)$ is an idempotent if the property of $e(x) = e^2(x) = e(x^2) \bmod (x^n - 1)$ is satisfied.

An (n, k) binary cyclic code may be described by the generator polynomial $g(x) \in T(x)$ of degree $n-k$ and the parity-check polynomial $h(x) \in T(x)$ of degree k, such that $g(x)h(x) = x^n - 1$. According to [2], as an alternative to $g(x)$, an idempotent may also be used to generate cyclic codes. Any binary cyclic code can be described by a unique idempotent $e_g(x) \in T(x)$ which consists of a sum of primitive idempotents. The unique idempotent $e_g(x)$ is known as the *generating idempotent* and as the name implies, $g(x)$ is a divisor of $e_g(x)$, and to be more specific $e_g(x) = m(x)g(x)$, where $m(x) \in T(x)$ contains repeated factors or non-factors of $x^n - 1$.

Lemma 4.4 *If $e(x) \in T(x)$ is an idempotent, $E(z) = MS(e(x)) \in T(z)$.*

Proof Since $e(x) = e(x)^2 \pmod{x^n - 1}$, from (4.37) it follows that $e(\alpha^{-rj}) = e(\alpha^{-rj})^2$ for $j = \{0, 1, \ldots, n-1\}$ and some integer r. Clearly $e(\alpha^{-rj}) \in \{0, 1\}$ implying that $E(z)$ is a binary polynomial.

Definition 4.4 (*Cyclotomic Coset*) Let s be a positive integer, and the 2−cyclotomic coset of $s \pmod n$ is given by

$$C_s = \left\{2^i s \pmod n \mid 0 \le i \le t\right\},$$

where we shall always assume that the subscript s is the smallest element in the set C_s and t is the smallest positive integer such that $2^{t+1}s \equiv s \pmod n$.

For convenience, we will use the term cyclotomic coset to refer to 2−cyclotomic coset throughout this book. If $\mathscr{N}$ is the set consisting of the smallest elements of all possible cyclotomic cosets, then it follows that

$$C = \bigcup_{s \in \mathscr{N}} C_s = \{0, 1, 2, \ldots, n-1\}.$$

Definition 4.5 (*Binary Cyclotomic Idempotent*) Let the polynomial $e_s(x) \in T(x)$ be given by

$$e_s(x) = \sum_{0 \le i \le |C_s|-1} x^{C_{s,i}}, \tag{4.44}$$

where $|C_s|$ is the number of elements in C_s and $C_{s,i} = 2^i s \pmod{n}$, the $(i+1)$th element of C_s. The polynomial $e_s(x)$ is called a binary cyclotomic idempotent.

Example 4.2 The entire cyclotomic cosets of 63 and their corresponding binary cyclotomic idempotents are as follows:

$$\begin{aligned}
C_0 &= \{0\} & e_0(x) &= 1\\
C_1 &= \{1, 2, 4, 8, 16, 32\} & e_1(x) &= x + x^2 + x^4 + x^8 + x^{16} + x^{32}\\
C_3 &= \{3, 6, 12, 24, 48, 33\} & e_3(x) &= x^3 + x^6 + x^{12} + x^{24} + x^{33} + x^{48}\\
C_5 &= \{5, 10, 20, 40, 17, 34\} & e_5(x) &= x^5 + x^{10} + x^{17} + x^{20} + x^{34} + x^{40}\\
C_7 &= \{7, 14, 28, 56, 49, 35\} & e_7(x) &= x^7 + x^{14} + x^{28} + x^{35} + x^{49} + x^{56}\\
C_9 &= \{9, 18, 36\} & e_9(x) &= x^9 + x^{18} + x^{36}\\
C_{11} &= \{11, 22, 44, 25, 50, 37\} & e_{11}(x) &= x^{11} + x^{22} + x^{25} + x^{37} + x^{44} + x^{50}\\
C_{13} &= \{13, 26, 52, 41, 19, 38\} & e_{13}(x) &= x^{13} + x^{19} + x^{26} + x^{38} + x^{41} + x^{52}\\
C_{15} &= \{15, 30, 60, 57, 51, 39\} & e_{15}(x) &= x^{15} + x^{30} + x^{39} + x^{51} + x^{57} + x^{60}\\
C_{21} &= \{21, 42\} & e_{21}(x) &= x^{21} + x^{42}\\
C_{23} &= \{23, 46, 29, 58, 53, 43\} & e_{23}(x) &= x^{23} + x^{29} + x^{43} + x^{46} + x^{53} + x^{58}\\
C_{27} &= \{27, 54, 45\} & e_{27}(x) &= x^{27} + x^{45} + x^{54}\\
C_{31} &= \{31, 62, 61, 59, 55, 47\} & e_{31}(x) &= x^{31} + x^{47} + x^{55} + x^{59} + x^{61} + x^{62}
\end{aligned}$$

and $\mathscr{N} = \{0, 1, 3, 5, 7, 9, 11, 13, 15, 21, 23, 27, 31\}$.

Definition 4.6 (*Binary Parity-Check Idempotent*) Let $\mathscr{M} \subseteq \mathscr{N}$ and let the polynomial $u(x) \in T(x)$ be defined by

$$u(x) = \sum_{s \in \mathscr{M}} e_s(x), \tag{4.45}$$

where $e_s(x)$ is an idempotent. The polynomial $u(x)$ is called a binary parity-check idempotent.

The binary parity-check idempotent $u(x)$ can be used to describe an $[n, k]$ cyclic code. Since $\mathrm{GCD}(u(x), x^n - 1) = h(x)$, the polynomial $\bar{u}(x) = x^{\deg(u(x))} u(x^{-1})$ and its n cyclic shifts $\pmod{x^n - 1}$ can be used to define the parity-check matrix of a binary cyclic code. In general, $\mathrm{wt}_H(\bar{u}(x))$ is much lower than $\mathrm{wt}_H(h(x))$, and therefore a sparse parity-check matrix can be derived from $\bar{u}(x)$. This is important for cyclic codes designed to be used as low-density parity-check (LDPC) codes, see Chap. 12.

4.4.1 Non-Primitive Cyclic Codes Derived from Idempotents

The factors of $2^m - 1$ dictate the degrees of the minimal polynomials through the order of the cyclotomic cosets. Some relatively short non-primitive cyclic codes have minimal polynomials of high degree which makes it tedious to derive the generator polynomial or parity-check polynomial using the Mattson–Solomon polynomial. The prime factors of $2^m - 1$ for $m \leq 43$ are tabulated below in Table 4.2.

The Mersenne primes shown in Table 4.2 are $2^3 - 1$, $2^5 - 1$, $2^7 - 1$, $2^{13} - 1$, $2^{17} - 1$, $2^{19} - 1$, $2^{23} - 1$ and $2^{31} - 1$, and cyclic codes of these lengths are primitive cyclic codes. Non-primitive cyclic codes have lengths corresponding to factors of $2^m - 1$ which are not Mersenne primes. Also it may be seen in Table 4.2 that for m even, 3 is a common factor. Where m is congruent to 5, with $m = 5 \times s$, 31 is a common factor and all $M_{j\,5}$ minimal polynomials will be contained in the set, $M_{j\,5\times s}$ of minimal polynomials.

As an example of how useful Table 4.2 can be, consider a code of length 113. Table 4.2 shows that $2^{28} - 1$ contains 113 as a factor. This means that there is a polynomial of degree 28 that has a root β of order 113. In fact, $\beta = \alpha^{2375535}$, where α is a primitive root, because $2^{28} - 1 = 2375535 \times 113$.

The cyclotomic cosets of 113 are as follows:

$$
\begin{aligned}
C_0 &= \{0\} \\
C_1 &= \{1, 2, 4, 8, 16, 32, 64, 15, 30, 60, 7, 14, 28, 56, \\
&\quad 112, 111, 109, 105, 97, 81, 49, 98, 83, 53, 106, 99, 85, 57\} \\
C_3 &= \{3, 6, 12, 24, 48, 96, 79, 45, 90, 67, 21, 42, 84, \\
&\quad 55, 110, 107, 101, 89, 65, 17, 34, 68, 23, 46, 92, 71, 29, 58\} \\
C_5 &= \{5, 10, 20, 40, 80, 47, 94, 75, 37, 74, 35, 70, 27, \\
&\quad 54, 108, 103, 93, 73, 33, 66, 19, 38, 76, 39, 78, 43, 86, 59\} \\
C_7 &= \{9, 18, 36, 72, 31, 62, 11, 22, 44, 88, 63, 13, 26, \\
&\quad 52, 104, 95, 77, 41, 82, 51, 102, 91, 69, 25, 50, 100, 87, 61\}.
\end{aligned}
$$

Each coset apart from C_0 may be used to define 28 roots from a polynomial having binary coefficients and of degree 28. Alternatively, each cyclotomic coset may be used to define the non-zero coefficients of a polynomial, a minimum weight idempotent (see Sect. 4.4). Adding together any combination of the 5 minimum weight idempotents generates a cyclic code of length 113. Consequently, there are only $2^5 - 2 = 30$ non-trivial, different cyclic codes of length 113 and some of these will be equivalent codes. Using Euclid's algorithm, it is easy to find the common factors of each idempotent combination and $x^{113} - 1$. The resulting polynomial may be used as the generator polynomial, or the parity-check polynomial of the cyclic code.

Table 4.2 Prime factors of $2^m - 1$

m	$2^m - 1$	Factors	m	$2^m - 1$	Factors
2	3	3	23	8388607	47 × 178481
3	7	7	24	16777215	3 × 3 × 5 × 7 × 13 × 17 × 241
4	15	5 × 3	25	33554431	31 × 601 × 1801
5	31	31	26	67108863	3 × 2731 × 8191
6	63	3 × 3 × 7	27	134217727	7 × 73 × 262657
7	127	127	28	268435455	3 × 5 × 29 × 43 × 113 × 127
8	255	3 × 5 × 17	29	536870911	233 × 1103 × 2089
9	511	7 × 73	30	1073741823	3 × 3 × 7 × 11 × 31 × 151 × 331
10	1023	3 × 11 × 31	31	2147483647	2147483647
11	2047	23 × 89	32	4294967295	3 × 5 × 17 × 257 × 65537
12	4095	3 × 3 × 5 × 7 × 13	33	8589934591	7 × 23 × 89 × 599479
13	8191	8191	34	17179869183	3 × 43691 × 131071
14	16383	3 × 43 × 127	35	34359738367	31 × 71×127 × 122921
15	32767	7 × 31 × 151	36	68719476735	3 × 3 × 3 × 5 × 7 × 13 × 19 × 37 × 73 × 109
16	65535	3 × 5 × 17 × 257	37	137438953471	223 × 616318177
17	131071	131071	38	274877906943	3 × 174763 × 524287
18	262143	3 × 3 × 3 × 7 × 19 × 73	39	549755813887	7 × 79 × 8191 × 121369
19	524287	524287	40	1099511627775	3 × 5 × 5 × 11 × 17 × 31 × 41 × 61681
20	1048575	3 × 5 × 5 × 11 × 31 × 41	41	2199023255551	13367 × 164511353
21	2097151	7 × 7 × 127 × 337	42	4398046511103	3 × 3 × 7 × 7 × 43 × 127 × 337 × 5419
22	4194303	3 × 23 × 89 × 683	43	8796093022207	431 × 9719 × 2099863

For example, consider the GCD of $C_1 + C_3 = x + x^2 + x^3 + x^4 + x^6 + x^8 + \ldots + x^{109} + x^{110} + x^{111} + x^{112}$ and $x^{113} - 1$. This is the polynomial, $u(x)$, which turns out to have degree 57

$$u(x) = 1 + x + x^2 + x^3 + x^5 + x^6 + x^7 + x^{10} + x^{13} \\ \ldots + x^{51} + x^{52} + x^{54} + x^{55} + x^{56} + x^{57}.$$

Using $u(x)$ as the parity-check polynomial of the cyclic code produces a (113, 57, 18) code. This is quite a good code as the very best (113, 57) code has a minimum Hamming distance of 19.

As another example of using this method for non-primitive cyclic code construction, consider the factors of $2^{39} - 1$ in Table 4.2. It will be seen that 79 is a factor and so a cyclic code of length 79 may be constructed from polynomials of degree 39. The cyclotomic cosets of 79 are as follows:

$$C_0 = \{0\}$$
$$C_1 = \{1, 2, 4, 8, 16, 32, 64, 49, 19, 38, 76, 73, \ldots 20, 40\}$$
$$C_3 = \{3, 6, 12, 24, 48, 17, 34, 68, 57, 35, 70, \ldots 60, 41\}.$$

The GCD of the idempotent sum given by the cyclotomic cosets $C_0 + C_1$ and $x^{79} - 1$ is the polynomial, $u(x)$, of degree 40:

$$u(x) = 1 + x + x^3 + x^5 + x^8 + x^{11} + x^{12} + x^{16} \\ \ldots + x^{28} + x^{29} + x^{34} + x^{36} + x^{37} + x^{40}.$$

Using $u(x)$ as the parity-check polynomial of the cyclic code produces a (79, 40, 15) code. This is the quadratic residue cyclic code for the prime number 79 and is a best-known code.

In a further example Table 4.2 shows that $2^{37} - 1$ has 223 as a factor. The GCD of the idempotent given by the cyclotomic coset C_3 $x^3 + x^6 + x^{12} + x^{24} + x^{48} + \ldots + x^{198} + x^{204}$ and $x^{223} - 1$ is the polynomial, $u(x)$, of degree 111

$$u(x) = 1 + x^2 + x^3 + x^5 + x^8 + x^9 + x^{10} + x^{12} \\ \ldots + x^{92} + x^{93} + x^{95} + x^{103} + x^{107} + x^{111}.$$

Using $u(x)$ as the parity-check polynomial of the cyclic code produces a (223, 111, 32) cyclic code.

4.5 Binary Cyclic Codes of Odd Lengths from 129 to 189

Since many of the best-known codes are cyclic codes, it is useful to have a table of the best cyclic codes. The literature already contains tables of the best cyclic codes up to length 127 and so the following table starts at 129. All possible binary cyclic codes up to length 189 have been constructed and their minimum Hamming distance has been evaluated.

The highest minimum distance attainable by all binary cyclic codes of odd lengths $129 \leq n \leq 189$ is tabulated in Table 4.3. The column "Roots of $g(x)$" in Table 4.3 denotes the exponents of roots of the generator polynomial $g(x)$, excluding the conjugate roots. All cyclic codes with generator polynomials $1+x$ and $(x^n-1)/(1+x)$, since they are trivial codes, are excluded in Table 4.3 and since primes $n = 8m \pm 3$ contain these trivial cyclic codes only, there is no entry in the table for these primes. The number of permutation inequivalent and non-degenerate cyclic codes, excluding the two trivial codes mentioned earlier, for each odd integer n is given by $N_{\mathscr{C}}$. The primitive polynomial $m(x)$ defining the field is given in octal. Full details describing the derivation of Table 4.3 are provided in Sect. 5.3.

In Table 4.3, there is no cyclic code that improves the lower bound given by Brouwer [1], but there are 134 cyclic codes that meet this lower bound and these codes are printed in bold.

4.6 Summary

The important large family of binary cyclic codes has been explored in this chapter. Starting with cyclotomic cosets, the minimal polynomials were introduced. The Mattson–Solomon polynomial was described and it was shown to be an inverse discrete Fourier transform based on a primitive root of unity. The usefulness of the Mattson–Solomon polynomial in the design of cyclic codes was demonstrated. The relationship between idempotents and the Mattson–Solomon polynomial of a polynomial that has binary coefficients was described with examples given. It was shown how binary cyclic codes may be easily derived from idempotents and the cyclotomic cosets. In particular, a method was described based on cyclotomic cosets for the design of high-degree non-primitive binary cyclic codes. Code examples using the method were presented.

A table listing the complete set of the best binary cyclic codes, having the highest minimum Hamming distance, has been included for all code lengths from 129 to 189 bits.

Table 4.3 The highest attainable minimum distance of binary cyclic codes of odd lengths from 129 to 189

k	d	Roots of $g(x)$	k	d	Roots of $g(x)$	k	d	Roots of $g(x)$
$n = 129$, $m(x) = 77277$, $N_{\mathscr{C}} = 388$								
127	**2**	**43**	**84**	**14**	**0, 1, 19, 21, 43**	42	30	0, 1, 3, 7, 9, 11, 19, 43
115	3	1	73	15	1, 3, 7, 19	31	32	1, 7, 9, 11, 13, 19, 21
114	**6**	**0, 1**	**72**	**18**	**0, 1, 7, 9, 19**	**30**	**38**	**0, 1, 3, 7, 9, 11, 13, 19**
113	4	3, 43	71	17	1, 3, 7, 19, 43	29	37	1, 3, 7, 11, 13, 19, 21, 43
112	**6**	**0, 1, 43**	70	18	0, 1, 3, 7, 19, 43	28	40	0, 1, 3, 7, 11, 13, 19, 21, 43
101	**8**	**1, 9**	59	22	1, 3, 7, 9, 19	17	43	1, 3, 7, 9, 11, 13, 19, 21
100	**10**	**0, 1, 3**	58	22	0, 1, 3, 7, 9, 19	**16**	**52**	**0, 1, 3, 7, 9, 11, 13, 19, 21**
99	8	1, 9, 43	57	22	1, 3, 7, 9, 19, 43	15	54	1, 3, 7, 9, 11, 13, 19, 21, 43
98	**10**	**0, 1, 3, 43**	**56**	**24**	**0, 1, 5, 9, 19, 21, 43**	14	54	0, 1, 3, 7, 9, 11, 13, 19, 21, 43
87	**13**	**1, 13, 21**	**45**	**29**	**1, 3, 7, 9, 11, 19**	**2**	**86**	**0, 1, 3, 5, 7, 9, 11, 13, 19, 21**
86	**14**	**0, 1, 19, 21**	**44**	**30**	**0, 1, 3, 7, 9, 11, 19**			
85	13	1, 19, 21, 43	43	29	1, 3, 7, 9, 11, 19, 43			
$n = 133$, $m(x) = 1334325$, $N_{\mathscr{C}} = 198$								
130	**2**	**57**	91	8	1, 7, 19, 57	43	19	1, 7, 9, 15, 31
129	**2**	**0, 57**	90	10	0, 1, 19, 31, 57	42	28	0, 1, 5, 7, 9, 31
127	**2**	**19, 57**	79	14	1, 7, 31	**40**	**32**	**1, 5, 7, 9, 31, 57**
126	**2**	**0, 19, 57**	78	14	0, 1, 5, 9	**39**	**32**	**0, 1, 5, 7, 9, 31, 57**
115	3	1	76	16	1, 7, 31, 57	37	32	1, 5, 7, 9, 19, 31, 57
114	4	0, 1	75	16	0, 1, 7, 31, 57	36	32	0, 1, 5, 7, 9, 19, 31, 57
112	**6**	**31, 57**	73	16	1, 7, 19, 31, 57	25	19	1, 3, 5, 7, 9, 31
111	6	0, 31, 57	72	16	0, 1, 7, 19, 31, 57	24	38	0, 1, 3, 5, 7, 9, 31
109	6	1, 19, 57	61	19	1, 7, 9, 31	22	44	1, 5, 7, 9, 15, 31, 57
108	6	0, 1, 19, 57	**60**	**24**	**0, 1, 3, 7, 9**	21	44	0, 1, 5, 7, 9, 15, 31, 57

(continued)

Table 4.3 (continued)

k	d	Roots of $g(x)$	k	d	Roots of $g(x)$	k	d	Roots of $g(x)$
97	7	1, 31	**58**	**24**	**1, 7, 9, 31, 57**	19	48	1, 3, 5, 7, 9, 19, 31, 57
96	10	0, 1, 31	**57**	**24**	**0, 1, 7, 9, 31, 57**	18	48	0, 1, 3, 5, 7, 9, 19, 31, 57
94	8	7, 31, 57	55	24	1, 7, 9, 19, 31, 57	4	57	1, 3, 5, 7, 9, 15, 31, 57
93	10	0, 1, 31, 57	54	24	0, 1, 7, 9, 19, 31, 57	**3**	**76**	**0, 1, 3, 5, 7, 9, 15, 31, 57**
$n = 135$, $m(x) = 1000000001001$, $N_{\mathscr{C}} = 982$								
133	**2**	**45**	89	6	1, 15, 63	45	10	1, 7, 21, 45, 63
132	**2**	**0, 45**	88	6	0, 1, 15, 63	44	10	0, 1, 7, 21, 45, 63
131	**2**	**63**	87	6	1, 15, 45, 63	43	10	1, 7, 15, 21, 45
130	**2**	**0, 63**	86	6	0, 1, 15, 45, 63	42	10	0, 1, 7, 15, 21, 45
129	**2**	**45, 63**	85	6	1, 21, 45	41	10	1, 7, 15, 21, 63
128	**2**	**0, 45, 63**	84	6	0, 1, 21, 45	40	10	0, 1, 7, 15, 21, 63
127	2	15, 45	83	6	1, 15, 27, 45, 63	39	10	1, 7, 15, 21, 45, 63
126	2	0, 15, 45	82	6	0, 1, 21, 63	38	10	0, 1, 7, 15, 21, 45, 63
125	2	15, 63	81	6	1, 21, 45, 63	37	10	1, 3, 7, 21, 45
124	2	0, 15, 63	80	6	0, 1, 21, 45, 63	36	10	0, 1, 3, 7, 21, 45
123	2	15, 45, 63	79	6	1, 15, 21, 45	35	12	1, 5, 7, 15, 63
122	2	0, 15, 45, 63	78	6	0, 1, 15, 21, 45	34	12	0, 1, 5, 7, 15, 63
121	2	21, 45	77	6	1, 5, 63	33	12	1, 5, 7, 15, 45, 63
120	2	0, 21, 45	76	6	0, 1, 5, 63	32	12	0, 1, 5, 7, 15, 45, 63
119	2	21, 63	75	6	1, 5, 45, 63	31	12	1, 5, 7, 21, 45
118	2	0, 21, 63	74	6	0, 1, 5, 45, 63	30	12	0, 1, 5, 7, 21, 45
117	2	21, 45, 63	73	6	1, 3, 21, 45	29	15	1, 5, 7, 21, 63
116	2	0, 21, 45, 63	72	6	0, 1, 3, 21, 45	28	18	0, 1, 5, 7, 21, 63
115	2	5, 45	71	8	1, 5, 15, 63	27	18	1, 5, 7, 21, 45, 63
114	2	0, 5, 45	70	8	0, 1, 5, 15, 63	26	18	0, 1, 5, 7, 21, 45, 63

(continued)

Table 4.3 (continued)

k	d	Roots of $g(x)$	k	d	Roots of $g(x)$	k	d	Roots of $g(x)$
113	4	5, 63	69	8	1, 5, 15, 45, 63	25	15	1, 5, 7, 21, 27, 63
112	4	0, 5, 63	68	8	0, 1, 5, 15, 45, 63	24	18	0, 1, 5, 7, 21, 27, 63
111	4	5, 45, 63	67	8	1, 5, 21, 45	23	21	1, 5, 7, 15, 21, 63
110	4	0, 5, 45, 63	66	8	0, 1, 5, 21, 45	22	24	0, 1, 5, 7, 15, 21, 63
109	4	5, 27, 63	65	8	1, 5, 15, 27, 45, 63	21	24	1, 5, 7, 15, 21, 45, 63
108	4	0, 5, 27, 63	64	8	0, 1, 5, 21, 63	20	24	0, 1, 5, 7, 15, 21, 45, 63
107	4	5, 15, 63	63	8	1, 5, 21, 45, 63	19	21	1, 5, 7, 15, 21, 27, 63
106	4	0, 5, 15, 63	62	8	0, 1, 5, 21, 45, 63	18	24	0, 1, 5, 7, 15, 21, 27, 63
105	4	5, 15, 45, 63	61	8	1, 5, 15, 21, 45	17	24	1, 5, 7, 15, 21, 27, 45, 63
104	4	0, 5, 15, 45, 63	60	8	0, 1, 5, 15, 21, 45	16	30	0, 1, 3, 5, 7, 21, 63
103	4	5, 21, 45	59	8	1, 5, 15, 21, 63	15	30	1, 3, 5, 7, 21, 27, 45
102	4	0, 5, 21, 45	58	8	0, 1, 5, 15, 21, 63	14	30	0, 1, 3, 5, 7, 21, 45, 63
101	4	5, 21, 63	57	8	1, 5, 15, 21, 45, 63	13	24	1, 5, 7, 9, 15, 21, 27, 45, 63
100	4	0, 5, 21, 63	56	8	0, 1, 5, 15, 21, 45, 63	12	30	0, 1, 3, 5, 7, 21, 27, 63
99	4	5, 21, 45, 63	55	8	1, 3, 5, 21, 45	11	30	1, 3, 5, 7, 21, 27, 45, 63
98	4	0, 5, 21, 45, 63	54	8	0, 1, 3, 5, 21, 45	10	36	0, 1, 3, 5, 7, 15, 21, 63
97	4	1, 45	53	10	1, 7, 15, 63	9	36	1, 3, 5, 7, 15, 21, 27, 45
96	4	0, 1, 45	52	10	0, 1, 7, 15, 63	8	36	0, 1, 3, 5, 7, 15, 21, 45, 63
95	5	1, 63	51	10	1, 7, 15, 45, 63	7	45	1, 3, 5, 7, 15, 21, 27, 63
94	6	0, 1, 63	50	10	0, 1, 7, 15, 45, 63	6	54	0, 1, 3, 5, 7, 15, 21, 27, 63
93	6	1, 45, 63	49	10	1, 7, 21, 45	5	63	1, 3, 5, 7, 15, 21, 27, 45, 63
92	6	0, 1, 45, 63	48	10	0, 1, 7, 21, 45	**4**	**72**	**0, 1, 3, 5, 7, 15, 21, 27, 45, 63**
91	5	1, 27, 63	47	10	1, 7, 15, 27, 45, 63			
90	6	0, 1, 27, 63	46	10	0, 1, 7, 21, 63			

(continued)

Table 4.3 (continued)

k	d	Roots of $g(x)$	k	d	Roots of $g(x)$	k	d	Roots of $g(x)$
$n = 137$, $m(x) = 67357330373267606675673$, $N_{\mathscr{C}} = 2$								
69	21	1	68	22	0, 1			
$n = 141$, $m(x) = 2146417666311013$, $N_{\mathscr{C}} = 30$								
139	**2**	**47**	93	4	3, 15, 47	47	24	1, 3, 15, 47
138	**2**	**0, 47**	92	6	0, 1, 47	46	24	0, 1, 3, 15, 47
118	2	3	72	21	3, 5	26	33	1, 3, 5
117	2	0, 3	**71**	**22**	**0, 3, 5**	25	36	0, 1, 3, 5
116	4	3, 47	70	21	3, 5, 47	24	33	1, 3, 5, 47
115	4	0, 3, 47	**69**	**24**	**0, 3, 5, 47**	23	36	0, 1, 3, 5, 47
95	3	1	49	22	1, 3, 15			
94	6	0, 1	48	22	0, 1, 3, 15			
$n = 143$, $m(x) = 145236760547324505061$, $N_{\mathscr{C}} = 16$								
133	2	13	83	11	1	61	24	1, 11, 13
132	2	0, 13	82	12	0, 1	60	24	0, 1, 11, 13
131	2	11	73	11	1, 13	23	11	1, 5
130	2	0, 11	72	16	0, 1, 13	22	22	0, 1, 5
121	4	11, 13	71	13	1, 11			
120	4	0, 11, 13	70	18	0, 1, 11			
$n = 145$, $m(x) = 3572445367$, $N_{\mathscr{C}} = 40$								
141	**2**	**29**	89	14	1, 5	57	26	1, 5, 11, 29
140	**2**	**0, 29**	88	14	0, 1, 5	56	26	0, 1, 5, 11, 29
117	5	1	85	14	1, 5, 29	33	29	1, 3, 5, 11
116	**8**	**0, 1**	84	14	0, 1, 5, 29	**32**	**44**	**0, 1, 3, 5, 11**
113	5	1, 29	61	24	1, 5, 11	29	46	1, 3, 5, 11, 29
112	**10**	**0, 1, 29**	60	24	0, 1, 5, 11	28	46	0, 1, 3, 5, 11, 29

(continued)

Table 4.3 (continued)

k	d	Roots of $g(x)$	k	d	Roots of $g(x)$	k	d	Roots of $g(x)$
$n = 147, m(x) = 100002000040201, N_{\mathscr{C}} = 488$								
145	**2**	**49**	96	4	0, 1, 35, 49	48	8	1, 3, 7, 9, 21, 35
144	**2**	**0, 49**	95	4	0, 1, 21, 35	47	8	0, 1, 3, 7, 9, 21, 35
143	**2**	**0, 21**	94	4	1, 21, 35, 49	46	8	1, 3, 7, 9, 21, 35, 49
142	**2**	**21, 49**	93	4	0, 1, 21, 35, 49	45	8	0, 1, 3, 7, 9, 21, 35, 49
141	**2**	**35**	92	4	0, 1, 7, 35	44	8	0, 1, 3, 7, 9, 21, 35, 63
140	**2**	**0, 35**	91	4	1, 21, 35, 49, 63	43	8	1, 3, 7, 9, 21, 35, 49, 63
139	2	35, 49	90	4	1, 7, 21, 35	42	8	0, 1, 3, 7, 9, 21, 35, 49, 63
138	2	0, 35, 49	89	4	0, 1, 7, 21, 35	40	9	1, 5, 9, 49
137	2	0, 7, 21	88	4	1, 7, 21, 35, 49	39	12	0, 1, 5, 9, 49
136	2	21, 35, 49	87	4	0, 1, 7, 21, 35, 49	38	10	0, 1, 5, 9, 21
135	2	7, 35	86	4	0, 1, 7, 21, 35, 63	37	12	1, 5, 9, 21, 49
134	2	0, 7, 35	85	4	1, 7, 21, 35, 49, 63	36	12	0, 1, 5, 9, 21, 49
133	2	21, 35, 49, 63	84	4	0, 1, 7, 21, 35, 49, 63	35	12	0, 1, 5, 9, 35
132	2	7, 21, 35	82	5	5, 9, 49	34	12	1, 5, 9, 21, 49, 63
131	2	0, 7, 21, 35	81	8	0, 5, 9, 49	33	12	0, 1, 5, 9, 35, 49
130	2	7, 21, 35, 49	80	6	0, 5, 9, 21	32	12	0, 1, 5, 9, 21, 35
129	2	0, 7, 21, 35, 49	79	8	5, 9, 21, 49	31	12	1, 5, 9, 21, 35, 49
127	2	7, 21, 35, 49, 63	78	8	0, 5, 9, 21, 49	30	12	0, 1, 5, 9, 21, 35, 49
126	2	0, 7, 21, 35, 49, 63	77	8	0, 5, 9, 35	29	12	0, 1, 5, 7, 9, 35
124	3	9, 49	76	8	5, 9, 21, 49, 63	28	12	1, 5, 9, 21, 35, 49, 63
123	4	0, 9, 49	75	8	0, 5, 9, 35, 49	27	12	1, 5, 7, 9, 21, 35
122	2	0, 9, 21	74	8	0, 5, 9, 21, 35	26	12	0, 1, 5, 7, 9, 21, 35
121	4	9, 21, 49	73	8	5, 9, 21, 35, 49	25	12	1, 5, 7, 9, 21, 35, 49
120	4	0, 9, 21, 49	72	8	0, 5, 9, 21, 35, 49	24	12	0, 1, 5, 7, 9, 21, 35, 49

(continued)

Table 4.3 (continued)

k	d	Roots of $g(x)$	k	d	Roots of $g(x)$	k	d	Roots of $g(x)$
119	4	0, 9, 35	71	8	0, 5, 7, 9, 35	23	12	0, 1, 5, 7, 9, 21, 35, 63
118	4	9, 21, 49, 63	70	8	5, 9, 21, 35, 49, 63	22	12	1, 5, 7, 9, 21, 35, 49, 63
117	4	0, 9, 35, 49	69	8	5, 7, 9, 21, 35	21	12	0, 1, 5, 7, 9, 21, 35, 49, 63
116	4	0, 9, 21, 35	68	8	0, 5, 7, 9, 21, 35	19	14	1, 3, 5, 9, 49
115	4	9, 21, 35, 49	67	8	5, 7, 9, 21, 35, 49	18	14	0, 1, 3, 5, 9, 49
114	4	0, 9, 21, 35, 49	66	8	0, 5, 7, 9, 21, 35, 49	17	14	0, 1, 3, 5, 9, 21
113	4	0, 7, 9, 35	65	8	0, 5, 7, 9, 21, 35, 63	16	21	1, 3, 5, 9, 21, 49
112	4	9, 21, 35, 49, 63	64	8	5, 7, 9, 21, 35, 49, 63	15	28	0, 1, 3, 5, 9, 21, 49
111	4	7, 9, 21, 35	63	8	0, 5, 7, 9, 21, 35, 49, 63	14	28	0, 1, 3, 5, 9, 35
110	4	0, 7, 9, 21, 35	61	8	1, 3, 9, 49	13	28	1, 3, 5, 9, 21, 49, 63
109	4	7, 9, 21, 35, 49	60	8	0, 1, 3, 9, 49	12	35	1, 3, 5, 7, 9, 21
108	4	0, 7, 9, 21, 35, 49	59	6	0, 1, 5, 21	11	42	0, 1, 3, 5, 7, 9, 21
107	4	0, 7, 9, 21, 35, 63	58	8	1, 3, 9, 21, 49	10	35	1, 3, 5, 7, 9, 21, 49
106	4	7, 9, 21, 35, 49, 63	57	8	1, 3, 9, 35	9	56	0, 1, 3, 5, 7, 9, 21, 49
105	4	0, 7, 9, 21, 35, 49, 63	56	8	0, 1, 3, 9, 35	8	42	0, 1, 3, 5, 7, 9, 35
103	4	3, 9, 49	55	8	1, 3, 9, 35, 49	7	56	1, 3, 5, 9, 21, 35, 49, 63
102	4	0, 1, 49	54	8	0, 1, 3, 9, 35, 49	6	56	0, 1, 3, 5, 9, 21, 35, 49, 63
101	4	0, 1, 21	53	8	0, 1, 3, 9, 21, 35	5	70	0, 1, 3, 5, 7, 9, 21, 35
100	4	1, 21, 49	52	8	1, 3, 9, 21, 35, 49	4	63	1, 3, 5, 7, 9, 21, 35, 49
99	4	0, 1, 21, 49	51	8	1, 3, 7, 9, 35	**3**	**84**	**0, 1, 3, 5, 7, 9, 21, 35, 49**
98	4	0, 1, 35	50	8	0, 1, 3, 7, 9, 35			
97	4	1, 21, 49, 63	49	8	1, 3, 9, 21, 35, 49, 63			
					$n = 151$, $m(x) = 166761$, $N_{\mathscr{C}} = 212$			
136	**5**	**1**	**91**	**17**	**1, 5, 15, 37**	46	31	1, 5, 7, 11, 15, 23, 37

(continued)

Table 4.3 (continued)

k	d	Roots of $g(x)$	k	d	Roots of $g(x)$	k	d	Roots of $g(x)$
135	**6**	**0, 1**	**90**	**18**	**0, 1, 5, 15, 37**	**45**	**36**	**0, 1, 5, 7, 11, 15, 23, 37**
121	**8**	**1, 5**	**76**	**23**	**1, 5, 15, 35, 37**	**31**	**47**	**1, 5, 7, 11, 15, 17, 23, 37**
120	8	0, 1, 5	**75**	**24**	**0, 1, 5, 15, 35, 37**	**30**	**48**	**0, 1, 5, 7, 11, 15, 17, 23, 37**
106	**13**	**1, 3, 5**	**61**	**31**	**1, 3, 5, 11, 15, 37**	16	60	1, 5, 7, 11, 15, 17, 23, 35, 37
105	**14**	**0, 1, 3, 5**	**60**	**32**	**0, 1, 3, 5, 11, 15, 37**	15	60	0, 1, 5, 7, 11, 15, 17, 23, 35, 37
$n = 153$, $m(x) = 110110001$, $N_{\mathscr{C}} = 2114$								
151	**2**	**51**	99	8	1, 9, 15, 17, 27	51	19	1, 5, 9, 11, 15, 17, 27
150	**2**	**0, 51**	98	8	0, 1, 9, 15, 17, 27	50	24	0, 1, 5, 9, 11, 15, 17, 27
145	2	9	97	9	1, 5, 15	49	24	1, 5, 9, 11, 15, 17, 27, 51
144	2	0, 9	96	10	0, 1, 5, 15	48	24	0, 1, 5, 9, 11, 15, 17, 27, 51
143	2	9, 51	95	10	1, 5, 9, 51	47	18	1, 5, 9, 11, 15, 27, 33, 51
142	2	0, 9, 51	94	10	0, 1, 5, 9, 51	46	18	0, 1, 5, 9, 11, 15, 27, 33, 51
139	**4**	**9, 17**	91	9	1, 5, 15, 17	43	19	1, 5, 9, 11, 15, 17, 27, 33
138	**4**	**0, 9, 17**	90	10	0, 1, 5, 15, 17	42	24	0, 1, 5, 9, 11, 15, 17, 27, 33
137	4	9, 17, 51	89	13	1, 5, 9, 57	41	24	1, 5, 9, 11, 15, 17, 27, 33, 51
136	4	0, 9, 17, 51	88	14	0, 1, 5, 9, 57	40	24	0, 1, 5, 9, 11, 15, 17, 27, 33, 51
135	2	9, 27, 51	87	14	1, 5, 9, 51, 57	39	18	1, 5, 9, 11, 15, 19, 51
134	2	0, 9, 27, 51	86	14	0, 1, 5, 9, 51, 57	38	18	0, 1, 5, 9, 11, 15, 19, 51
131	4	9, 17, 27	83	15	1, 5, 9, 17, 57	35	19	1, 5, 9, 11, 15, 17, 27, 33, 57
130	4	0, 9, 17, 27	82	16	0, 1, 5, 9, 17, 57	34	24	0, 1, 5, 9, 11, 15, 17, 27, 33, 57
129	4	9, 17, 27, 51	81	16	1, 5, 9, 17, 51, 57	33	24	1, 5, 9, 11, 15, 17, 27, 33, 51, 57
128	4	0, 9, 17, 27, 51	80	16	0, 1, 5, 9, 17, 51, 57	32	30	0, 1, 5, 9, 11, 15, 19, 57
127	4	1, 51	79	14	1, 5, 9, 15, 27, 51	31	30	1, 5, 9, 11, 15, 19, 51, 57
126	4	0, 1, 51	78	14	0, 1, 5, 9, 15, 27, 51	30	30	0, 1, 5, 9, 11, 15, 19, 51, 57

(continued)

Table 4.3 (continued)

k	d	Roots of $g(x)$	k	d	Roots of $g(x)$	k	d	Roots of $g(x)$
123	4	9, 15, 17, 27	75	16	1, 5, 9, 15, 17, 27	27	27	1, 5, 9, 11, 15, 17, 19, 57
122	4	0, 9, 15, 17, 27	74	16	0, 1, 5, 9, 15, 17, 27	26	30	0, 1, 5, 9, 11, 15, 17, 19, 57
121	5	1, 9	73	16	1, 5, 9, 15, 17, 27, 51	25	30	1, 5, 9, 11, 15, 17, 19, 51, 57
120	6	0, 1, 9	72	16	0, 1, 5, 9, 15, 17, 27, 51	24	34	0, 1, 5, 9, 11, 15, 19, 27, 57
119	6	1, 9, 51	71	14	1, 5, 9, 15, 27, 33, 51	23	34	1, 5, 9, 11, 15, 19, 27, 33, 51
118	6	0, 1, 9, 51	70	14	0, 1, 5, 9, 15, 27, 33, 51	22	34	0, 1, 5, 9, 11, 15, 19, 27, 33, 51
115	6	1, 9, 17	67	16	1, 5, 9, 15, 17, 27, 33	19	42	1, 5, 9, 11, 15, 17, 19, 27, 57
114	6	0, 1, 9, 17	66	16	0, 1, 5, 9, 15, 17, 27, 33	18	42	0, 1, 5, 9, 11, 15, 17, 19, 27, 57
113	8	1, 9, 57	65	16	1, 5, 9, 15, 17, 27, 33, 51	17	48	1, 5, 9, 11, 15, 17, 19, 27, 51, 57
112	8	0, 1, 9, 57	64	18	0, 1, 5, 9, 11, 57	16	48	0, 1, 5, 9, 11, 15, 17, 19, 27, 51, 57
111	8	1, 9, 27, 51	63	18	1, 5, 9, 19, 51, 57	15	34	1, 5, 9, 11, 15, 19, 27, 33, 51, 57
110	8	0, 1, 9, 27, 51	62	18	0, 1, 5, 9, 11, 51, 57	14	34	0, 1, 5, 9, 11, 15, 19, 27, 33, 51, 57
107	8	1, 9, 17, 57	59	16	1, 5, 9, 15, 17, 27, 33, 57	11	51	1, 5, 9, 11, 15, 17, 19, 27, 33, 57
106	8	0, 1, 9, 17, 57	58	18	0, 1, 5, 9, 11, 17, 57	10	54	0, 1, 5, 9, 11, 15, 17, 19, 27, 33, 57
105	8	1, 9, 15, 27	57	18	1, 5, 9, 11, 17, 51, 57	9	57	1, 5, 9, 11, 15, 17, 19, 27, 33, 51, 57
104	8	0, 1, 9, 15, 27	56	18	0, 1, 5, 9, 11, 15, 27	8	72	0, 1, 5, 9, 11, 15, 17, 19, 27, 33, 51, 57
103	8	1, 9, 15, 27, 51	55	18	1, 5, 9, 11, 15, 27, 51	7	34	1, 3, 5, 9, 11, 15, 19, 27, 33, 51, 57
102	8	0, 1, 9, 15, 27, 51	54	18	0, 1, 5, 9, 11, 15, 27, 51	6	34	0, 1, 3, 5, 9, 11, 15, 19, 27, 33, 51, 57
$n = 155$, $m(x) = 7154113$, $N_{\mathscr{C}} = 2768$								
151	**2**	**31**	101	12	1, 3, 25, 31, 75	51	24	1, 3, 9, 23, 25, 31, 35, 55, 75
150	**2**	**0, 31**	100	12	0, 1, 9, 25, 31, 75	50	24	0, 1, 3, 9, 23, 25, 31, 35, 55, 75
149	**2**	**0, 25**	99	10	0, 1, 9, 25, 35, 75	49	22	0, 1, 3, 5, 11, 23, 25, 35, 55, 75
146	**4**	**25, 31**	96	12	1, 9, 25, 31, 35, 75	46	24	1, 3, 5, 11, 23, 25, 31, 35, 55, 75
145	**4**	**0, 25, 31**	95	12	0, 1, 9, 25, 31, 35, 75	45	25	1, 3, 9, 11, 23, 25, 75
144	2	0, 25, 75	94	10	0, 1, 11, 25, 35, 55, 75	44	28	0, 1, 3, 9, 11, 23, 25, 75

(continued)

Table 4.3 (continued)

k	d	Roots of $g(x)$	k	d	Roots of $g(x)$	k	d	Roots of $g(x)$
141	**4**	**25, 31, 75**	91	12	1, 11, 25, 31, 35, 55, 75	41	25	1, 3, 9, 11, 23, 25, 31, 75
140	**4**	**0, 25, 31, 75**	90	12	0, 1, 11, 25, 31, 35, 55, 75	40	30	0, 1, 3, 9, 11, 23, 25, 31, 75
139	2	0, 25, 35, 75	89	12	0, 1, 3, 11, 25	39	30	0, 1, 3, 9, 11, 23, 25, 35, 75
136	4	25, 31, 35, 75	86	12	9, 11, 23, 25, 31	36	31	1, 3, 9, 11, 23, 25, 31, 35, 75
135	4	0, 25, 31, 35, 75	85	14	1, 3, 9, 25, 75	35	32	0, 1, 3, 9, 11, 23, 25, 31, 35, 75
134	4	0, 1	84	14	0, 1, 3, 9, 25, 75	34	30	0, 1, 3, 9, 11, 23, 25, 35, 55, 75
131	4	1, 31	81	16	1, 3, 9, 25, 31, 75	31	32	1, 3, 9, 11, 23, 25, 31, 35, 55, 75
130	5	1, 25	80	16	0, 1, 3, 9, 25, 31, 75	30	32	0, 1, 3, 9, 11, 23, 25, 31, 35, 55, 75
129	6	0, 1, 25	79	14	0, 1, 3, 9, 25, 35, 75	29	30	0, 1, 3, 5, 9, 11, 23, 25, 35, 55, 75
126	6	1, 25, 31	76	16	1, 3, 9, 25, 31, 35, 75	26	32	1, 3, 5, 9, 11, 23, 25, 31, 35, 55, 75
125	6	0, 1, 25, 31	75	16	0, 1, 3, 9, 25, 31, 35, 75	25	32	0, 1, 3, 5, 9, 11, 23, 25, 31, 35, 55, 75
124	6	0, 1, 25, 75	74	14	0, 1, 3, 9, 25, 35, 55, 75	24	30	0, 1, 3, 7, 9, 11, 23, 25, 75
121	8	1, 25, 31, 75	71	16	1, 9, 11, 25, 31, 35, 55, 75	21	32	1, 3, 5, 9, 11, 15, 23, 25, 31, 35, 55, 75
120	8	0, 1, 25, 31, 75	70	16	0, 1, 9, 11, 25, 31, 35, 55, 75	20	32	0, 1, 3, 5, 9, 11, 15, 23, 25, 31, 35, 55, 75
119	6	0, 1, 25, 35, 75	69	16	0, 1, 9, 11, 23, 25	19	40	0, 1, 3, 7, 9, 11, 23, 25, 35, 75
116	8	1, 25, 31, 35, 75	66	16	1, 5, 9, 11, 25, 31, 35, 55, 75	16	35	1, 3, 7, 9, 11, 23, 25, 31, 35, 75
115	8	0, 1, 25, 31, 35, 75	65	16	0, 1, 3, 9, 11, 25, 31	15	40	0, 1, 3, 7, 9, 11, 23, 25, 31, 35, 75
114	6	0, 1, 11	64	20	0, 1, 9, 11, 23, 25, 55	14	60	0, 1, 3, 7, 9, 11, 23, 25, 35, 55, 75
111	8	1, 25, 31, 35, 55, 75	61	20	1, 3, 9, 23, 25, 31, 75	11	55	1, 3, 7, 9, 11, 23, 25, 31, 35, 55, 75
110	8	0, 1, 25, 31, 35, 55, 75	60	22	1, 3, 9, 23, 25, 35, 75	10	60	0, 1, 3, 7, 9, 11, 23, 25, 31, 35, 55, 75
109	8	0, 1, 11, 25	59	22	0, 1, 3, 9, 23, 25, 35, 75	9	62	0, 1, 3, 5, 7, 9, 11, 23, 25, 35, 55, 75
106	8	1, 11, 25, 31	56	24	1, 3, 9, 23, 25, 31, 35, 75	6	75	1, 3, 5, 7, 9, 11, 23, 25, 31, 35, 55, 75
105	10	1, 3, 25, 75	55	24	0, 1, 3, 9, 23, 25, 31, 35, 75	**5**	**80**	**0, 1, 3, 5, 7, 9, 11, 23, 25, 31, 35, 55, 75**
104	10	0, 1, 9, 25, 75	54	22	0, 1, 3, 9, 23, 25, 35, 55, 75			

(continued)

Table 4.3 (continued)

k	d	Roots of $g(x)$	k	d	Roots of $g(x)$	k	d	Roots of $g(x)$
$n = 157$, $m(x) = 352125723713652127$, $N_{\mathscr{C}} = 4$								
105	13	1	53	26	1, 3			
104	**14**	**0, 1**	52	26	0, 1, 3			
$n = 159$, $m(x) = 303667410520550411$, $N_{\mathscr{C}} = 16$								
157	**2**	**53**	105	4	3, 53	53	32	1, 3, 53
156	**2**	**0, 53**	104	6	0, 1, 53	52	32	0, 1, 3, 53
107	3	1	55	30	1, 3			
106	6	0, 1	54	30	0, 1, 3			
$n = 161$, $m(x) = 150536353761$, $N_{\mathscr{C}} = 156$								
158	**2**	**23**	106	4	1, 7, 35	56	7	1, 3, 5, 23, 69
157	**2**	**0, 23**	105	4	0, 1, 7, 35	55	14	0, 1, 3, 5, 23, 69
155	**2**	**23, 69**	103	8	5, 7, 23, 35	51	23	1, 5, 11, 35
154	**2**	**0, 23, 69**	102	8	0, 5, 7, 23, 35	50	28	0, 1, 5, 11, 35
150	2	35	100	8	1, 7, 23, 35, 69	48	32	3, 5, 11, 23, 35
149	2	0, 35	99	8	0, 1, 7, 23, 35, 69	47	32	0, 3, 5, 11, 23, 35
147	**4**	**23, 35**	95	7	1, 5	45	32	1, 5, 11, 23, 35, 69
146	**4**	**0, 23, 35**	94	14	0, 1, 5	44	32	0, 1, 5, 11, 23, 35, 69
144	4	23, 35, 69	92	7	1, 3, 23	40	23	1, 3, 5, 7, 35
143	4	0, 23, 35, 69	91	14	0, 1, 5, 23	39	28	0, 1, 3, 5, 7, 35
139	2	7, 35	89	7	1, 5, 23, 69	37	32	1, 3, 5, 7, 23, 35
138	2	0, 7, 35	88	14	0, 1, 5, 23, 69	36	32	0, 1, 3, 5, 7, 23, 35
136	4	7, 23, 35	84	14	1, 5, 35	34	32	1, 3, 5, 7, 23, 35, 69
135	4	0, 7, 23, 35	83	14	0, 1, 5, 35	33	32	0, 1, 3, 5, 7, 23, 35, 69
133	4	7. 23, 35, 69	81	16	5, 11, 23, 35	29	7	1, 3, 5, 11
132	4	0. 7, 23, 35, 69	80	18	0, 3, 11, 23, 35	28	14	0, 1, 3, 5, 11

(continued)

Table 4.3 (continued)

k	d	Roots of $g(x)$	k	d	Roots of $g(x)$	k	d	Roots of $g(x)$
128	3	1	78	18	1, 5, 23, 35, 69	26	7	1, 3, 5, 11, 23
127	4	0, 1	77	18	0, 1, 5, 23, 35, 69	25	14	0, 1, 3, 5, 11, 23
125	6	5, 23	73	23	1, 5, 7, 35	18	23	1, 3, 5, 11, 35
124	6	0, 5, 23	72	24	0, 1, 5, 7, 35	17	46	0, 1, 3, 5, 11, 35
122	6	1, 23, 69	70	24	1, 5, 7, 23, 35	15	49	1, 3, 5, 11, 23, 35
121	6	0, 1, 23, 69	69	24	0, 1, 5, 7, 23, 35	14	56	0, 1, 3, 5, 11, 23, 35
117	4	1, 35	**67**	**28**	**1, 5, 7, 23, 35, 69**	12	49	1, 3, 5, 11, 23, 35, 69
116	4	0, 1, 35	66	28	0, 1, 5, 7, 23, 35, 69	11	56	0, 1, 3, 5, 11, 23, 35, 69
114	8	5, 23, 35	62	7	1, 3, 5	4	69	1, 3, 5, 7, 11, 23, 35
113	8	0, 5, 23, 35	61	14	0, 1, 3, 5	**3**	**92**	**0, 1, 3, 5, 7, 11, 23, 35**
111	8	1, 23, 35, 69	59	7	1, 3, 5, 23			
110	8	0, 1, 23, 35, 69	58	14	0, 1, 3, 5, 23			
$n = 165, m(x) = 6223427, N_{\mathscr{C}} = 4800$								
163	**2**	**55**	109	12	5, 9, 29, 55, 77	55	32	1, 5, 7, 9, 15, 29, 33, 55, 77
162	**2**	**0, 55**	108	12	0, 5, 9, 29, 55, 77	54	32	0, 1, 5, 7, 9, 15, 29, 33, 55, 77
161	**2**	**77**	107	12	5, 9, 29, 33, 77	53	32	1, 5, 7, 9, 11, 15, 29, 33, 77
160	**2**	**0, 77**	106	12	0, 5, 9, 29, 33, 77	52	32	0, 1, 5, 7, 9, 11, 15, 29, 33, 77
159	**2**	**55, 77**	105	12	5, 9, 29, 33, 55, 77	51	32	1, 5, 7, 9, 11, 15, 29, 33, 55, 77
158	**2**	**0, 55, 77**	104	12	0, 5, 9, 29, 33, 55, 77	50	32	0, 1, 5, 7, 9, 11, 15, 29, 33, 55, 77
157	2	33, 77	103	12	1, 5, 9, 11, 33, 77	49	28	1, 5, 7, 9, 15, 25, 29, 55, 77
156	2	0, 33, 77	102	12	0, 1, 9, 29, 55	48	30	0, 1, 3, 5, 7, 9, 29, 55, 77
155	2	5	101	12	5, 9, 15, 29, 77	47	32	1, 5, 7, 9, 15, 25, 29, 33, 77
154	2	0, 5	100	12	0, 1, 9, 29, 77	46	32	0, 1, 5, 7, 9, 15, 25, 29, 33, 77
153	2	5, 55	99	12	1, 9, 29, 33, 55	45	32	1, 5, 7, 9, 15, 25, 29, 33, 55, 77
152	2	0, 5, 55	98	12	0, 5, 9, 15, 29, 55, 77	44	32	0, 1, 5, 7, 9, 15, 25, 29, 33, 55, 77

(continued)

Table 4.3 (continued)

k	d	Roots of $g(x)$	k	d	Roots of $g(x)$	k	d	Roots of $g(x)$
151	**4**	**15, 77**	97	16	5, 9, 15, 29, 33, 77	43	32	1, 5, 7, 9, 11, 15, 25, 29, 33, 77
150	**4**	**0, 5, 77**	96	16	0, 5, 9, 15, 29, 33, 77	42	32	0, 1, 5, 7, 9, 11, 15, 25, 29, 33, 77
149	4	15, 55, 77	95	16	5, 9, 15, 29, 33, 55, 77	41	33	1, 3, 5, 7, 9, 15, 29, 77
148	4	0, 5, 55, 77	94	16	0, 5, 9, 15, 29, 33, 55, 77	40	38	0, 1, 3, 5, 7, 9, 15, 29, 77
147	4	5, 33, 77	93	16	1, 3, 5, 7, 55	39	39	1, 5, 7, 9, 15, 19, 29, 33, 55
146	4	0, 5, 33, 77	92	16	0, 1, 5, 9, 29, 55	**38**	**44**	**0, 1, 3, 5, 7, 9, 15, 29, 55, 77**
145	4	5, 33, 55, 77	91	16	5, 9, 19, 29, 77	37	40	1, 3, 5, 7, 9, 15, 29, 33, 77
144	4	0, 1	90	18	0, 1, 5, 9, 29, 33	36	44	0, 1, 5, 7, 9, 15, 19, 29, 33, 77
143	4	9, 55	89	19	1, 3, 7, 15, 55, 77	35	44	1, 3, 5, 7, 9, 15, 29, 33, 55, 77
142	4	0, 1, 55	88	20	0, 1, 5, 9, 29, 55, 77	34	44	0, 1, 3, 5, 7, 9, 15, 29, 33, 55, 77
141	5	1, 33	87	16	5, 9, 15, 25, 29, 33, 77	33	44	1, 3, 5, 7, 9, 11, 15, 29, 33, 77
140	6	0, 29, 77	86	20	0, 1, 5, 9, 29, 33, 77	32	44	0, 1, 3, 5, 7, 9, 11, 15, 29, 33, 77
139	6	1, 33, 55	85	20	1, 3, 7, 15, 33, 55, 77	31	44	1, 3, 5, 7, 9, 11, 15, 29, 33, 55, 77
138	6	0, 29, 55, 77	84	20	0, 1, 5, 9, 29, 33, 55, 77	30	44	0, 1, 3, 5, 7, 9, 15, 25, 29, 77
137	5	29, 33, 77	83	17	1, 5, 9, 15, 29, 55	29	44	1, 5, 7, 9, 15, 19, 25, 29, 33, 55
136	6	0, 1, 33, 77	82	20	0, 1, 5, 9, 15, 29, 55	28	44	0, 1, 3, 5, 7, 9, 15, 25, 29, 55, 77
135	6	1, 33, 55, 77	81	21	1, 5, 9, 15, 29, 77	27	48	1, 3, 5, 7, 9, 15, 25, 29, 33, 77
134	6	0, 1, 33, 55, 77	**80**	**24**	**0, 1, 3, 5, 7, 15, 77**	26	48	0, 1, 3, 5, 7, 9, 15, 25, 29, 33, 77
133	5	1, 11, 33, 77	79	23	1, 3, 5, 7, 15, 55, 77	25	48	1, 3, 5, 7, 9, 15, 25, 29, 33, 55, 77
132	6	0, 1, 11, 33, 77	78	24	0, 1, 3, 5, 7, 15, 55, 77	24	48	0, 1, 3, 5, 7, 9, 15, 25, 29, 33, 55, 77
131	7	3, 5, 77	77	24	1, 5, 9, 15, 29, 33, 77	23	48	1, 3, 5, 7, 9, 11, 15, 25, 29, 33, 77
130	8	0, 5, 9, 77	76	24	0, 1, 5, 9, 15, 29, 33, 77	22	48	0, 1, 3, 5, 7, 9, 11, 15, 25, 29, 33, 77
129	8	1, 15, 33, 55	75	24	1, 5, 9, 15, 29, 33, 55, 77	21	48	1, 3, 5, 7, 9, 11, 15, 25, 29, 33, 55, 77
128	8	0, 5, 9, 55, 77	74	24	0, 1, 5, 9, 15, 29, 33, 55, 77	20	48	0, 1, 3, 5, 7, 9, 11, 15, 25, 29, 33, 55, 77

(continued)

Table 4.3 (continued)

k	d	Roots of $g(x)$	k	d	Roots of $g(x)$	k	d	Roots of $g(x)$
127	8	5, 29, 33, 77	73	24	1, 5, 9, 11, 15, 29, 33, 77	19	44	1, 3, 5, 7, 9, 15, 19, 29, 33, 55
126	8	0, 1, 5, 33, 77	72	24	0, 1, 5, 9, 11, 15, 29, 33, 77	18	44	0, 1, 3, 5, 7, 9, 15, 19, 29, 55, 77
125	8	5, 29, 33, 55, 77	71	24	1, 3, 5, 7, 15, 25, 77	17	50	1, 3, 5, 7, 9, 15, 19, 29, 33, 77
124	8	0, 1, 5, 33, 55, 77	70	24	0, 3, 5, 7, 19, 29, 77	16	50	0, 1, 3, 5, 7, 9, 15, 19, 29, 33, 77
123	8	1, 9, 55	69	24	1, 7, 9, 15, 29, 55, 77	15	55	1, 3, 5, 7, 9, 15, 19, 29, 33, 55, 77
122	8	0, 1, 9, 55	68	24	0, 1, 5, 7, 9, 29, 55, 77	14	60	0, 1, 3, 5, 7, 9, 15, 19, 29, 33, 55, 77
121	8	5, 9, 15, 77	67	24	1, 5, 9, 15, 25, 29, 33, 77	13	50	1, 3, 5, 7, 9, 11, 15, 19, 29, 33, 77
120	10	0, 7, 9, 77	66	26	0, 1, 5, 9, 19, 29, 33, 77	12	50	0, 1, 3, 5, 7, 9, 11, 15, 19, 29, 33, 77
119	10	7, 9, 55, 77	65	24	1, 5, 9, 15, 25, 29, 33, 55, 77	11	55	1, 3, 5, 7, 9, 11, 15, 19, 29, 33, 55, 77
118	10	0, 7, 9, 55, 77	64	28	0, 1, 5, 9, 19, 29, 33, 55, 77	10	60	0, 1, 3, 5, 7, 9, 11, 15, 19, 29, 33, 55, 77
117	8	1, 5, 15, 33, 77	63	24	1, 5, 7, 9, 15, 29, 55	9	44	1, 3, 5, 7, 9, 15, 19, 25, 29, 33, 55
116	10	0, 9, 29, 33, 77	62	28	0, 1, 5, 9, 11, 19, 29, 33, 77	8	44	0, 1, 3, 5, 7, 9, 15, 19, 25, 29, 55, 77
115	10	9, 29, 33, 55, 77	61	27	1, 5, 7, 9, 15, 29, 77	7	55	1, 3, 5, 7, 9, 15, 19, 25, 29, 33, 77
114	10	0, 9, 29, 33, 55, 77	60	28	0, 1, 5, 7, 9, 15, 29, 77	6	66	0, 1, 3, 5, 7, 9, 15, 19, 25, 29, 33, 77
113	8	1, 9, 15, 55	59	28	1, 5, 7, 9, 15, 29, 55, 77	5	77	1, 3, 5, 7, 9, 15, 19, 25, 29, 33, 55, 77
112	10	0, 1, 9, 11, 33, 77	58	28	0, 1, 5, 7, 9, 15, 29, 55, 77	**4**	**88**	**0, 1, 3, 5, 7, 9, 15, 19, 25, 29, 33, 55, 77**
111	11	5, 7, 9, 77	57	32	1, 5, 7, 9, 15, 29, 33, 77			
110	12	0, 5, 7, 9, 77	56	32	0, 1, 5, 7, 9, 15, 29, 33, 77			
$n = 167$, $m(x) = 5122622544667121565742432523$, $N_{\mathscr{C}} = 2$								
84	23	1	**83**	**24**	**0, 1**			
$n = 169$, $m(x) = 10000400020001000040002000100004000200010000400020001$, $N_{\mathscr{C}} = 2$								
157	2	13	12	26	0, 1			
$n = 171$, $m(x) = 1167671$, $N_{\mathscr{C}} = 802$								
169	**2**	**57**	111	12	1, 3, 9, 19	57	35	1, 3, 5, 7, 9, 13, 19

(continued)

Table 4.3 (continued)

k	d	Roots of $g(x)$	k	d	Roots of $g(x)$	k	d	Roots of $g(x)$
168	**2**	**0, 57**	**110**	**16**	**0, 1, 3, 5, 19**	**56**	**36**	**0, 1, 3, 5, 7, 9, 13, 19**
163	2	19, 57	109	12	1, 3, 9, 19, 57	55	36	1, 3, 5, 7, 9, 13, 19, 57
162	2	0, 19, 57	108	16	0, 1, 3, 5, 19, 57	54	36	0, 1, 3, 5, 7, 9, 13, 19, 57
153	3	1	99	18	1, 3, 9, 13	45	19	1, 3, 5, 7, 9, 15, 17
152	**6**	**0, 1**	98	18	0, 1, 3, 9, 13	44	38	0, 1, 3, 5, 7, 9, 15, 17
151	5	1, 57	97	18	1, 3, 9, 13, 57	43	38	1, 3, 5, 7, 9, 15, 17, 57
150	**6**	**0, 1, 57**	96	18	0, 1, 3, 5, 7, 57	42	38	0, 1, 3, 5, 7, 9, 13, 17, 57
147	4	9, 19	93	20	1, 3, 5, 9, 19	39	45	1, 3, 5, 7, 9, 15, 17, 19
146	6	0, 1, 19	92	20	0, 1, 3, 5, 9, 19	**38**	**48**	**0, 1, 3, 5, 7, 9, 15, 17, 19**
145	6	1, 19, 57	91	21	1, 3, 5, 9, 19, 57	**37**	**48**	**1, 3, 5, 7, 9, 15, 17, 19, 57**
144	6	0, 1, 19, 57	**90**	**22**	**0, 1, 3, 5, 9, 19, 57**	36	48	0, 1, 3, 5, 7, 9, 15, 17, 19, 57
135	9	1, 3	81	19	1, 3, 5, 7, 9	27	19	1, 3, 5, 7, 9, 13, 15, 17
134	**10**	**0, 1, 3**	**80**	**26**	**0, 1, 3, 5, 7, 9**	26	38	0, 1, 3, 5, 7, 9, 13, 15, 17
133	9	1, 3, 57	79	23	1, 3, 5, 9, 17, 57	25	38	1, 3, 5, 7, 9, 13, 15, 17, 57
132	10	0, 1, 3, 57	78	26	0, 1, 3, 5, 7, 9, 57	24	38	0, 1, 3, 5, 7, 9, 13, 15, 17, 57
129	9	1, 3, 19	75	27	1, 3, 5, 9, 17, 19	21	55	1, 3, 5, 7, 9, 13, 15, 17, 19
128	10	0, 1, 3, 19	**74**	**28**	**0, 1, 3, 5, 9, 17, 19**	20	64	0, 1, 3, 5, 7, 9, 13, 15, 17, 19
127	10	1, 9, 19, 57	73	28	1, 3, 5, 9, 17, 19, 57	**19**	**68**	**1, 3, 5, 7, 9, 13, 15, 17, 19, 57**
126	**12**	**0, 1, 15, 19, 57**	72	28	0, 1, 3, 5, 9, 17, 19, 57	**18**	**68**	**0, 1, 3, 5, 7, 9, 13, 15, 17, 19, 57**
117	10	1, 3, 9	63	19	1, 3, 5, 7, 9, 25	7	38	1, 3, 5, 7, 9, 13, 15, 17, 25, 57
116	14	0, 1, 3, 5	62	32	0, 1, 3, 5, 9, 17, 25	6	38	0, 1, 3, 5, 7, 9, 13, 15, 17, 25, 57
115	12	1, 7, 9, 57	61	32	1, 3, 5, 9, 17, 25, 57			
114	14	0, 1, 7, 9, 57	60	32	0, 1, 3, 5, 9, 17, 25, 57			
$n = 175$, $m(x) = 100041020400004002041$, $N_{\mathscr{C}} = 242$								

(continued)

Table 4.3 (continued)

k	d	Roots of $g(x)$	k	d	Roots of $g(x)$	k	d	Roots of $g(x)$
172	**2**	**25**	112	6	3, 25	60	8	0, 1, 5, 7, 15, 25, 35, 75
171	**2**	**0, 25**	111	6	0, 3, 25	52	7	1, 3, 25
170	**2**	**0, 35**	110	4	0, 1, 35	51	10	0, 1, 3, 25
169	**2**	**25, 75**	109	6	1, 25, 75	50	10	0, 1, 3, 35
168	**2**	**0, 25, 75**	108	6	0, 1, 25, 75	49	7	1, 3, 25, 75
167	2	0, 25, 35	107	6	0, 3, 25, 35	48	14	1, 3, 25, 35
165	2	25, 35, 75	105	6	1, 25, 35, 75	47	14	0, 1, 3, 25, 35
164	2	0, 25, 35, 75	104	6	0, 1, 25, 35, 75	45	14	1, 3, 25, 35, 75
163	2	5	103	6	1, 5	44	14	0, 1, 3, 25, 35, 75
162	2	0, 5	102	6	0, 1, 5	43	7	1, 3, 5
160	2	5, 25	100	6	1, 5, 25	42	14	0, 1, 3, 5
159	2	0, 5, 25	99	6	0, 1, 5, 25	40	7	1, 3, 5, 25
158	2	0, 5, 35	98	6	0, 1, 5, 35	39	14	0, 1, 3, 5, 25
157	2	5, 25, 75	97	6	1, 5, 25, 75	38	14	0, 1, 3, 5, 35
156	2	0, 5, 25, 75	96	6	0, 1, 5, 25, 75	37	7	1, 3, 5, 25, 75
155	2	0, 5, 25, 35	95	6	0, 1, 5, 25, 35	36	14	0, 1, 3, 5, 25, 75
153	2	5, 25, 35, 75	93	6	1, 5, 25, 35, 75	35	14	0, 1, 3, 5, 25, 35
152	4	7, 25	92	7	3, 7, 25	33	14	1, 3, 5, 25, 35, 75
151	4	0, 7, 25	91	8	0, 3, 7, 25	32	14	0, 1, 3, 5, 25, 35, 75
150	2	0, 7, 35	90	6	0, 1, 5, 15	31	10	0, 1, 3, 7, 25
149	4	7, 25, 75	89	8	1, 7, 25, 75	30	14	0, 1, 3, 5, 15
148	4	0, 7, 25, 75	88	8	0, 1, 7, 25, 75	29	10	1, 3, 7, 25, 75
147	4	0, 7, 25, 35	87	8	0, 3, 7, 25, 35	28	20	1, 3, 7, 25, 35
145	4	7, 25, 35, 75	86	6	0, 1, 5, 15, 35	27	20	0, 1, 3, 7, 25, 35
144	4	0, 7, 25, 35, 75	85	8	1, 7, 25, 35, 75	26	14	0, 1, 3, 5, 15, 35

(continued)

Table 4.3 (continued)

k	d	Roots of $g(x)$	k	d	Roots of $g(x)$	k	d	Roots of $g(x)$
143	4	5, 7	84	8	0, 1, 7, 25, 35, 75	25	20	1, 3, 7, 25, 35, 75
142	4	0, 5, 7	83	7	1, 5, 7	24	20	0, 1, 3, 7, 25, 35, 75
141	2	5, 15, 25, 35, 75	82	8	0, 1, 5, 7	23	15	1, 3, 5, 7
140	4	5, 7, 25	81	6	1, 5, 15, 25, 35, 75	22	20	0, 1, 3, 5, 7
139	4	0, 5, 7, 25	80	8	1, 5, 7, 25	21	14	1, 3, 5, 15, 25, 35, 75
138	4	0, 5, 7, 35	79	8	0, 1, 5, 7, 25	20	30	1, 3, 5, 7, 25
137	4	5, 7, 25, 75	78	8	0, 1, 5, 7, 35	19	30	0, 1, 3, 5, 7, 25
136	4	0, 5, 7, 25, 75	77	8	1, 5, 7, 25, 75	18	20	0, 1, 3, 5, 7, 35
135	4	0, 5, 7, 25, 35	76	8	0, 1, 5, 7, 25, 75	17	30	1, 3, 5, 7, 25, 75
133	4	5, 7, 25, 35, 75	75	8	0, 1, 5, 7, 25, 35	16	35	1, 3, 5, 7, 25, 35
132	4	0, 5, 7, 25, 35, 75	73	8	1, 5, 7, 25, 35, 75	15	40	0, 1, 3, 5, 7, 25, 35
131	4	5, 7, 15	72	8	0, 1, 5, 7, 25, 35, 75	13	40	1, 3, 5, 7, 25, 35, 75
130	4	0, 5, 7, 15	71	8	1, 5, 7, 15	12	40	0, 1, 3, 5, 7, 25, 35, 75
128	4	5, 7, 15, 25	70	8	0, 1, 5, 7, 15	11	25	1, 3, 5, 7, 15
127	4	0, 5, 7, 15, 25	68	8	1, 5, 7, 15, 25	10	50	0, 1, 3, 5, 7, 15
126	4	0, 5, 7, 15, 35	67	8	0, 1, 5, 7, 15, 25	8	35	1, 3, 5, 7, 15, 25
124	4	5, 7, 15, 25, 35	66	8	0, 1, 5, 7, 15, 35	7	70	0, 1, 3, 5, 7, 15, 25
123	4	0, 5, 7, 15, 25, 35	64	8	1, 5, 7, 15, 25, 35	4	75	1, 3, 5, 7, 15, 25, 35
121	4	5, 7, 15, 25, 35, 75	63	8	0, 1, 5, 7, 15, 25, 35	**3**	**100**	**0, 1, 3, 5, 7, 15, 25, 35**
120	4	0, 5, 7, 15, 25, 35, 75	61	8	1, 5, 7, 15, 25, 35, 75			
$n = 177, m(x) = 23563311065422331671, N_{\mathscr{C}} = 16$								
175	**2**	**59**	117	4	3, 59	59	30	1, 3, 59
174	**2**	**0, 59**	116	6	0, 1, 59	58	30	0, 1, 3, 59
119	3	1	61	28	1, 3			
118	6	0, 1	60	28	0, 1, 3			

(continued)

Table 4.3 (continued)

k	d	Roots of $g(x)$	k	d	Roots of $g(x)$	k	d	Roots of $g(x)$
$n = 183$, $m(x) = 131010354441637571637$, $N_{\mathscr{C}} = 16$								
181	**2**	**61**	121	4	3, 61	61	36	1, 3, 61
180	**2**	**0, 61**	120	6	0, 1, 61	60	36	0, 1, 3, 61
123	3	1	63	34	1, 3			
122	6	0, 1	62	34	0, 1, 3			
$n = 185$, $m(x) = 1761557733077$, $N_{\mathscr{C}} = 40$								
181	**2**	**37**	113	14	1, 5	73	32	1, 3, 5, 37
180	**2**	**0, 37**	112	14	0, 1, 5	72	32	0, 1, 3, 5, 37
149	5	1	109	16	1, 5, 37	41	37	1, 3, 5, 19
148	8	0, 1	108	16	0, 1, 5, 37	40	48	0, 1, 3, 5, 19
145	5	1, 37	77	28	1, 3, 5	37	37	1, 3, 5, 19, 37
144	8	0, 1, 37	76	28	0, 1, 3, 5	36	54	0, 1, 3, 5, 19, 37
$n = 187$, $m(x) = 36000132706473$, $N_{\mathscr{C}} = 78$								
179	2	33	129	12	3, 17, 33	59	17	1, 3, 9, 33
178	2	0, 33	128	12	0, 3, 17, 33	58	30	0, 1, 3, 23, 33
177	2	17	121	12	1, 11, 17, 33	57	11	1, 3, 9, 17
176	2	0, 17	120	12	0, 1, 11, 17, 33	56	22	0, 1, 3, 9, 17
171	2	11, 33	107	11	1, 3	51	17	1, 3, 9, 11, 33
170	2	0, 11, 33	106	14	0, 1, 3	50	34	0, 1, 3, 9, 11, 33
169	4	17, 33	99	17	1, 3, 33	49	38	1, 3, 9, 17, 33
168	4	0, 17, 33	98	22	0, 1, 3, 33	48	38	0, 1, 3, 9, 17, 33
161	4	11, 17, 33	97	11	1, 3, 17	41	48	1, 3, 9, 11, 17, 33
160	4	0, 11, 17, 33	96	16	0, 1, 3, 17	40	48	0, 1, 3, 9, 11, 17, 33
147	5	1	91	17	1, 3, 11, 33	27	11	1, 3, 9, 23

(continued)

Table 4.3 (continued)

k	d	Roots of $g(x)$	k	d	Roots of $g(x)$	k	d	Roots of $g(x)$
146	6	0, 1	90	22	0, 1, 3, 11, 33	26	22	0, 1, 3, 9, 23
139	9	3, 33	89	24	1, 3, 17, 33	19	17	1, 3, 9, 23, 33
138	10	0, 3, 33	88	24	0, 1, 3, 17, 33	18	34	0, 1, 3, 9, 23, 33
137	6	1, 17	81	24	1, 3, 11, 17, 33	9	55	1, 3, 9, 17, 23, 33
136	6	0, 1, 17	80	24	0, 1, 3, 11, 17, 33	8	66	0, 1, 3, 9, 17, 23, 33
131	10	1, 11, 33	67	11	1, 3, 9			
130	10	0, 1, 11, 33	66	22	0, 1, 3, 9			
$n = 189$, $m(x) = 1100111$, $N_{\mathscr{C}} = 175286$								
187	**2**	**63**	125	12	0, 1, 3, 5, 31, 81	63	24	0, 1, 3, 5, 7, 11, 13, 31, 39, 63, 81
186	**2**	**0, 63**	124	12	1, 3, 5, 7, 63, 81	62	24	0, 1, 3, 5, 7, 9, 11, 13, 15, 31
185	**2**	**0, 81**	123	12	0, 1, 3, 5, 7, 63, 81	61	24	1, 3, 5, 7, 11, 13, 21, 23, 27, 63, 81
184	**2**	**63, 81**	122	12	0, 1, 3, 5, 7, 9	60	24	0, 1, 3, 5, 7, 9, 11, 13, 31, 39, 63
183	**2**	**3**	121	12	1, 3, 5, 7, 9, 63	59	24	0, 1, 3, 5, 7, 9, 11, 13, 15, 31, 81
182	**2**	**0, 3**	120	12	0, 1, 3, 5, 7, 9, 63	58	24	1, 3, 5, 7, 11, 13, 21, 31, 63, 69, 81
181	2	3, 63	119	12	0, 1, 3, 5, 7, 69, 81	57	27	1, 3, 5, 7, 9, 11, 13, 15, 21, 23
180	2	0, 3, 63	118	14	1, 3, 5, 31, 39, 63, 81	56	28	0, 1, 3, 5, 7, 9, 11, 13, 15, 21, 23
179	2	0, 3, 81	117	14	0, 1, 3, 5, 31, 39, 63, 81	55	27	1, 3, 5, 7, 9, 11, 13, 21, 23, 39, 63
178	2	3, 63, 81	116	14	0, 1, 3, 5, 9, 31, 39	54	31	1, 3, 5, 7, 11, 13, 21, 23, 39, 45, 81
177	2	3, 69	115	14	1, 3, 5, 9, 31, 39, 63	53	32	0, 1, 3, 5, 7, 11, 13, 21, 23, 39, 45, 81
176	2	0, 3, 69	114	14	0, 1, 3, 5, 9, 31, 39, 63	52	32	1, 3, 5, 7, 11, 13, 21, 23, 39, 45, 63, 81
175	2	3, 63, 69	113	14	0, 1, 3, 5, 31, 39, 69, 81	51	32	0, 1, 3, 5, 7, 11, 13, 21, 23, 39, 45, 63, 81
174	2	0, 3, 63, 69	112	14	1, 3, 5, 31, 39, 63, 69, 81	50	32	0, 1, 3, 5, 7, 9, 11, 13, 15, 21, 23, 27, 81
173	2	0, 3, 69, 81	111	14	0, 1, 3, 5, 31, 39, 63, 69, 81	49	32	1, 3, 5, 7, 11, 13, 21, 23, 27, 39, 63, 69, 81

(continued)

Table 4.3 (continued)

k	d	Roots of $g(x)$	k	d	Roots of $g(x)$	k	d	Roots of $g(x)$
172	2	3, 63, 69, 81	110	14	0, 1, 3, 5, 9, 31, 39, 69	48	32	0, 1, 3, 5, 7, 11, 13, 21, 23, 27, 39, 63, 69, 81
171	2	3, 21, 69	109	14	1, 3, 5, 9, 31, 39, 63, 69	47	32	0, 1, 3, 5, 7, 11, 13, 21, 23, 39, 45, 69, 81
170	2	0, 3, 21, 69	108	14	0, 1, 3, 5, 9, 31, 39, 63, 69	46	32	1, 3, 5, 7, 11, 13, 21, 23, 39, 45, 63, 69, 81
169	4	1, 63	107	14	0, 1, 3, 11, 13, 21, 39, 69, 81	45	32	0, 1, 3, 5, 7, 11, 13, 21, 23, 39, 45, 63, 69, 81
168	4	0, 1, 63	106	14	1, 3, 11, 13, 21, 39, 63, 69, 81	44	32	0, 1, 3, 5, 7, 9, 11, 13, 21, 23, 39, 45, 69
167	4	0, 1, 81	105	14	1, 3, 9, 11, 13, 21, 39, 45	43	32	1, 3, 5, 7, 9, 11, 13, 21, 23, 27, 39, 63, 69, 81
166	4	1, 63, 81	104	14	0, 1, 3, 5, 7, 9, 11	42	32	0, 1, 3, 5, 7, 9, 11, 13, 21, 23, 39, 45, 63, 69
165	5	1, 3	103	14	1, 3, 5, 7, 9, 11, 63	41	32	0, 1, 3, 5, 7, 9, 11, 13, 21, 23, 39, 45, 69, 81
164	6	0, 1, 3	102	15	1, 3, 5, 7, 11, 21, 81	40	32	1, 3, 5, 7, 9, 11, 13, 21, 23, 39, 45, 63, 69, 81
163	6	1, 3, 63	101	16	0, 1, 3, 5, 7, 11, 21, 81	39	32	0, 1, 3, 5, 7, 9, 11, 13, 21, 23, 39, 45, 63, 69, 81
162	6	0, 1, 3, 63	100	16	1, 3, 5, 7, 11, 21, 63, 81	38	32	0, 1, 3, 5, 7, 9, 11, 13, 15, 21, 23, 39, 45, 69
161	6	0, 1, 3, 81	99	17	1, 3, 7, 11, 27, 31, 39, 81	37	32	1, 3, 5, 7, 9, 11, 13, 15, 21, 23, 27, 39, 63, 69, 81
160	6	1, 3, 63, 81	98	18	0, 1, 3, 7, 11, 27, 31, 39, 81	36	33	1, 3, 5, 7, 11, 13, 21, 23, 31, 39, 45, 81
159	6	3, 9, 13	97	18	1, 3, 7, 11, 27, 31, 39, 63, 81	35	36	0, 1, 3, 5, 7, 11, 13, 23, 31, 45, 69, 81, 93
158	6	0, 1, 3, 69	96	18	0, 1, 3, 7, 11, 27, 31, 39, 63, 81	34	33	1, 3, 5, 7, 11, 13, 23, 31, 39, 45, 63, 69, 81
157	6	1, 3, 63, 69	95	18	0, 1, 3, 7, 11, 21, 31, 39, 81	33	36	0, 1, 3, 5, 7, 11, 13, 23, 31, 39, 45, 63, 69, 81
156	6	0, 1, 3, 63, 69	94	18	1, 3, 7, 11, 21, 31, 39, 63, 81	32	36	0, 1, 3, 5, 7, 11, 13, 21, 23, 31, 39, 45, 69
155	6	0, 1, 3, 69, 81	93	18	3, 5, 7, 9, 11, 13, 39, 45	31	36	1, 3, 5, 7, 11, 13, 21, 23, 27, 31, 39, 63, 69, 81
154	6	1, 3, 63, 69, 81	92	18	0, 1, 3, 7, 9, 11, 21, 31, 39	30	39	1, 3, 5, 7, 11, 13, 21, 23, 31, 39, 45, 69, 81
153	6	3, 9, 13, 69	91	18	1, 3, 7, 9, 11, 21, 31, 39, 63	29	42	0, 1, 3, 5, 7, 11, 13, 21, 23, 31, 39, 45, 69, 81
152	6	0, 1, 3, 21, 69	90	18	0, 1, 3, 7, 9, 11, 21, 31, 39, 63	28	45	1, 3, 5, 7, 11, 13, 21, 23, 31, 39, 45, 63, 69, 81
151	6	1, 31, 63	89	18	0, 1, 3, 7, 11, 21, 31, 39, 69, 81	27	48	0, 1, 3, 5, 7, 11, 13, 21, 23, 31, 39, 45, 63, 69, 81
150	6	0, 1, 31, 63	88	18	1, 3, 7, 11, 21, 31, 39, 63, 69, 81	26	42	0, 1, 3, 5, 7, 9, 11, 13, 21, 23, 31, 39, 45, 69
149	6	0, 1, 5, 81	87	18	1, 3, 7, 9, 11, 21, 31, 39, 45	25	45	1, 3, 5, 7, 9, 11, 13, 21, 23, 27, 31, 39, 63, 69, 81
148	6	1, 5, 63, 81	86	18	0, 1, 3, 7, 11, 13, 31, 93	24	48	0, 1, 3, 5, 7, 9, 11, 13, 21, 23, 27, 31. 39, 63, 69, 81

(continued)

Table 4.3 (continued)

k	d	Roots of $g(x)$	k	d	Roots of $g(x)$	k	d	Roots of $g(x)$
147	7	1, 3, 5	85	18	1, 3, 7, 9, 11, 21, 31, 39, 63, 69	23	48	0, 1, 3, 5, 7, 9, 11, 13, 21, 23, 31, 39, 45, 69, 81
146	8	0, 1, 3, 5	84	18	0, 1, 3, 7, 11, 13, 31, 39, 63	22	48	1, 3, 5, 7, 9, 11, 13, 15, 21, 23, 31, 45, 63, 69, 81
145	8	1, 3, 5, 63	83	18	0, 1, 3, 5, 11, 13, 21, 31, 81	21	54	0, 1, 3, 5, 7, 9, 11, 13, 21, 23, 31, 39, 45, 63, 81, 93
144	8	0, 1, 3, 5, 63	82	20	1, 3, 5, 7, 11, 13, 39, 63, 81	20	54	0, 1, 3, 5, 7, 9, 11, 13, 15, 21, 23, 31, 45, 69, 93
143	8	0, 1, 3, 5, 81	81	21	1, 3, 5, 7, 9, 11, 13, 15	19	57	1, 3, 5, 7, 9, 11, 13, 15, 21, 23, 27, 31, 63, 69, 81, 93
142	8	1, 3, 5, 63, 81	80	22	0, 1, 3, 5, 7, 9, 11, 13, 15	18	63	1, 3, 5, 7, 9, 11, 13, 15, 21, 23, 31, 39, 45, 69, 81
141	8	1, 3, 7, 39	79	21	1, 3, 5, 7, 9, 11, 13, 39, 63	17	66	0, 1, 3, 5, 7, 9, 11, 13, 15, 21, 23, 31, 39, 45, 69, 81
140	10	0, 1, 3, 7, 39	78	22	0, 1, 3, 5, 7, 9, 11, 13, 39, 63	16	69	1, 3, 5, 7, 9, 11, 13, 15, 21, 23, 31, 39, 45, 63, 69, 81
139	10	1, 3, 7, 39, 63	77	22	0, 1, 3, 5, 7, 9, 11, 13, 15, 81	15	72	0, 1, 3, 5, 7, 9, 11, 13, 15, 21, 23, 31, 39, 45, 63, 69, 81
138	10	0, 1, 3, 7, 39, 63	76	24	1, 3, 5, 7, 11, 13, 21, 39, 63, 81	14	66	0, 1, 3, 5, 7, 9, 11, 13, 15, 21, 23, 27, 31, 39, 45, 69, 81
137	10	0, 1, 3, 7, 39, 81	75	24	0, 1, 3, 5, 7, 11, 13, 21, 39, 63, 81	13	72	1, 3, 5, 7, 9, 11, 13, 15, 21, 23, 27, 31, 39, 63, 69, 81, 93
136	10	1, 3, 7, 39, 63, 81	74	24	0, 1, 3, 5, 7, 9, 11, 13, 15, 21	12	72	0, 1, 3, 5, 7, 9, 11, 13, 15, 21, 23, 27, 31, 33, 39, 63, 69, 81
135	10	1, 3, 9, 31, 39	73	24	1, 3, 5, 7, 9, 11, 13, 21, 39, 63	11	78	0, 1, 3, 5, 7, 9, 11, 13, 15, 21, 23, 31, 33, 39, 45, 69, 81
134	10	0, 1, 3, 31, 39, 69	72	24	0, 1, 3, 5, 7, 9, 11, 13, 21, 39, 63	10	81	1, 3, 5, 7, 9, 11, 13, 15, 21, 23, 31, 39, 45, 63, 69, 81, 93
133	10	1, 3, 31, 39, 63, 69	71	24	0, 1, 3, 5, 7, 9, 11, 13, 15, 21, 81	9	84	0, 1, 3, 5, 7, 9, 11, 13, 15, 21, 23, 31, 33, 39, 45, 63, 69, 81
132	10	0, 1, 3, 31, 39, 63, 69	70	24	1, 3, 5, 7, 11, 13, 21, 39, 63, 69, 81	8	78	0, 1, 3, 5, 7, 9, 11, 13, 15, 21, 23, 27, 31, 33, 39, 45, 69, 81
131	10	0, 1, 3, 31, 39, 69, 81	69	24	1, 3, 7, 9, 11, 13, 21, 31, 39, 45	7	93	1, 3, 5, 7, 9, 11, 13, 15, 21, 23, 27, 31, 33, 39, 45, 63, 69, 81
130	10	1, 3, 31, 39, 63, 69, 81	68	24	0, 1, 3, 5, 7, 9, 11, 13, 21, 39, 69	**6**	**96**	**0, 1, 3, 5, 7, 9, 11, 13, 15, 21, 23, 27, 31, 33, 39, 45, 63, 69, 81**
129	10	1, 3, 9, 21, 31, 45	67	24	1, 3, 5, 7, 9, 11, 13, 21, 39, 63, 69	5	90	0, 1, 3, 5, 7, 9, 11, 13, 15, 21, 23, 31, 33, 39, 45, 69, 81, 93
128	10	0, 1, 3, 5, 7	66	24	0, 1, 3, 5, 7, 9, 11, 13, 21, 39, 63, 69	4	81	1, 3, 5, 7, 9, 11, 13, 15, 21, 23, 31, 33, 39, 45, 63, 69, 81, 93
127	10	1, 3, 5, 7, 63	65	24	0, 1, 3, 5, 7, 9, 11, 13, 21, 39, 69, 81	**3**	**108**	**0, 1, 3, 5, 7, 9, 11, 13, 15, 21, 23, 31, 33, 39, 45, 63, 69, 81, 93**
126	10	0, 1, 3, 5, 7, 63	64	24	1, 3, 5, 7, 9, 11, 13, 21, 39, 63, 69, 81			

References

1. Brouwer, A.E.: Bounds on the size of linear codes. In: Pless, V.S., Huffman, W.C. (eds.) Handbook of Coding Theory, pp. 295–461. Elsevier, North Holland (1998)
2. MacWilliams, F.J., Sloane, N.J.A.: The Theory of Error-Correcting Codes. North-Holland, Amsterdam (1977)
3. Mattson, H.F., Solomon, G.: A new treatment of Bose-Chaudhuri codes. J. Soc. Ind. Appl. Math. **9**(4), 654–669 (1961). doi:10.1137/0109055
4. Peterson, W.W.: Error-Correcting Codes. MIT Press, Cambridge (1961)
5. Prange, E.: Cyclic error-correcting codes in two symbols. Technical Report TN-58–103, Air Force Cambridge Research Labs, Bedford, Massachusetts, USA (1957)

Chapter 5
Good Binary Linear Codes

5.1 Introduction

Two of the important performance indicators for a linear code are the minimum Hamming distance and the weight distribution. Efficient algorithms for computing the minimum distance and weight distribution of linear codes are explored below. Using these methods, the minimum distances of all binary cyclic codes of length 129–189 have been enumerated. The results are presented in Chap. 4. Many improvements to the database of best-known codes are described below. In addition, methods of combining known codes to produce good codes are explored in detail. These methods are applied to cyclic codes, and many new binary codes have been found and are given below.

The quest of achieving Shannon's limit for the AWGN channel has been approached in a number of different ways. Here we consider the problem formulated by Shannon of the construction of good codes which maximise the difference between the error rate performance for uncoded transmission and coded transmission. For uncoded, bipolar transmission with matched filtered reception, it is well known (see for example Proakis [20]) that the bit error rate, p_b, is given by

$$p_b = \frac{1}{2}\text{erfc}\left(\sqrt{\frac{E_b}{N_0}}\right). \tag{5.1}$$

Comparing this equation with the equation for the probability of error when using coding, viz. the probability of deciding on one codeword rather than another, Eq. (1.4) given in Chap. 1, it can be seen that the improvement due to coding, the coding gain is indicated by the term $d_{min}.\frac{k}{n}$, the product of the minimum distance between codewords and the code rate. This is not the end of the story in calculating the overall probability of decoder error because this error probability needs to be multiplied by the number of codewords distance d_{min} apart.

For a linear binary code, the Hamming distance between two codewords is equal to the Hamming weight of the codeword formed by adding the two codewords together. Moreover, as the probability of decoder error at high $\frac{E_b}{N_0}$ values depends on the minimum Hamming distance between codewords, for a linear binary code,

M. Tomlinson et al., *Error-Correction Coding and Decoding*,
Signals and Communication Technology,
DOI 10.1007/978-3-319-51103-0_5

the performance of the code depends on the minimum Hamming weight codewords of the code, the d_{min} of the code and the number of codewords with this weight (the multiplicity). For a given code rate ($\frac{k}{n}$) and length n, the higher the weight of the minimum Hamming weight codewords of the code, the better the performance, assuming the multiplicity is not too high. It is for this reason that a great deal of research effort has been extended, around the world in determining codes with the highest minimum Hamming weight for a given code rate ($\frac{k}{n}$) and length n. These codes are called the best-known codes with parameters (n, k, d), where d is understood to be the d_{min} of the code, and the codes are tabulated in a database available online [12] with sometimes a brief description or reference to their method of construction.[1]

In this approach, it is assumed that a decoding algorithm either exists or will be invented which realises the full performance of a best-known code. For binary codes of length less than 200 bits the Dorsch decoder described in Chap. 15 does realise the full performance of the code.

Computing the minimum Hamming weight codewords of a linear code is, in general, a Nondeterministic Polynomial-time (NP) hard problem, as conjectured by [2] and later proved by [24]. Nowadays, it is a common practice to use a multi-threaded algorithm which runs on multiple parallel computers (grid computing) for minimum Hamming distance evaluation. Even then, it is not always possible to evaluate the exact minimum Hamming distance for large codes. For some algebraic codes, however, there are some shortcuts that make it possible to obtain the lower and upper bounds on this distance. But knowing these bounds are not sufficient as the whole idea is to know explicitly the exact minimum Hamming distance of a specific constructed code. As a consequence, algorithms for evaluating the minimum Hamming distance of a code are very important in this subject area and these are described in the following section.

It is worth mentioning that a more accurate benchmark of how good a code is, in fact its Hamming weight distribution. Whilst computing the minimum Hamming distance of a code is in general NP-hard, computing the Hamming weight distribution of a code is even more complex. In general, for two codes of the same length and dimension but of different minimum Hamming distance, we can be reasonably certain that the code with the higher distance is the superior code. Unless we are required to decide between two codes with the same parameters, including minimum Hamming distance, it is not necessary to go down the route of evaluating the Hamming weight distribution of both codes.

[1]Multiplicities are ignored in the compiling of the best, known code Tables with the result that sometimes the best, known code from the Tables is not the code that has the best performance.

5.2 Algorithms to Compute the Minimum Hamming Distance of Binary Linear Codes

5.2.1 The First Approach to Minimum Distance Evaluation

For a $[n, k, d]$ linear code over $\mathbb{F}_2$ with a reduced-echelon generator matrix $\boldsymbol{G}_{sys} = [\boldsymbol{I}_k|\boldsymbol{P}]$, where $\boldsymbol{I}_k$ and $\boldsymbol{P}$ are $k \times k$ identity and $k \times (n-k)$ matrices respectively, a codeword of this linear code can be generated by taking a linear combination of some rows of $\boldsymbol{G}_{sys}$. Since the minimum Hamming distance of a linear code is the minimum non-zero weight among all of the 2^k codewords, a brute-force method to compute the minimum distance is to generate codewords by taking

$$\binom{k}{1}, \binom{k}{2}, \binom{k}{3}, \ldots, \binom{k}{k-1}, \text{ and } \binom{k}{k}$$

linear combinations of the rows in $\boldsymbol{G}_{sys}$, noting the weight of each codeword generated and returning the minimum weight codeword of all $2^k - 1$ non-zero codewords. This method gives not only the minimum distance, but also the weight distribution of a code. It is obvious that as k grows larger this method becomes infeasible. However, if $n - k$ is not too large, the minimum distance can still be obtained by evaluating the weight distribution of the $[n, n-k, d']$ dual code and using the MacWilliams Identities to compute the weight distribution of the code. It should be noted that the whole weight distribution of the $[n, n-k, d']$ dual code has to be obtained, not just the minimum distance of the dual code.

In direct codeword evaluation, it is clear that there are too many unnecessary codeword enumerations involved. A better approach which avoids enumerating large numbers of unnecessary codewords can be devised. Let

$$\boldsymbol{c} = (\boldsymbol{i}|\boldsymbol{p}) = (c_0, c_1, \ldots, c_{k-1}|c_k, \ldots, c_{n-2}, c_{n-1})$$

be a codeword of a binary linear code of minimum distance d. Let $\boldsymbol{c}' = (\boldsymbol{i}'|\boldsymbol{p}')$ be a codeword of weight d, then if $\text{wt}_H(\boldsymbol{i}') = w$ for some integer $w < d$, $\text{wt}_H(\boldsymbol{p}') = d - w$. This means that at most

$$\sum_{w=1}^{\min\{d-1,k\}} \binom{k}{w} \tag{5.2}$$

codewords are required to be enumerated.

In practice, d is unknown and an upper bound d_{ub} on the minimum distance is required during the evaluation and the minimum Hamming weight found thus far can be used for this purpose. It is clear that once all $\sum_{w'=1}^{w} \binom{k}{w'}$ codewords of information weight w' are enumerated,

- we know that we have considered all possibilities of $d \leq w$; and
- if $w < d_{ub}$, we also know that the minimum distance of the code is at least $w + 1$.

Therefore, having an upper bound, a lower bound $d_{lb} = w + 1$ on the minimum distance can also be obtained. The evaluation continues until the condition $d_{lb} \geq d_{ub}$ is met and in this event, d_{ub} is the minimum Hamming distance.

5.2.2 Brouwer's Algorithm for Linear Codes

There is an apparent drawback of the above approach. In general, the minimum distance of a low-rate linear code is greater than its dimension. This implies that $\sum_{w=1}^{k} \binom{k}{w}$ codewords would need to be enumerated. A more efficient algorithm was attributed to Brouwer[2] and the idea behind this approach is to use a collection of generator matrices of mutually disjoint information sets [11].

Definition 5.1 (*Information Set*) Let the set $S = \{0, 1, 2, \ldots, n-\}$ be the coordinates of an $[n, k, d]$ binary linear code with generator matrix $\boldsymbol{G}$. The set $\mathscr{I} \subseteq S$ of k elements is an information set if the corresponding coordinates in the generator matrix is linearly independent and the submatrix corresponding to the coordinates in $\mathscr{I}$ has rank k, hence, it can be transformed into a $k \times k$ identity matrix.

In other words, we can say, in relation to a codeword, the k symbols user message is contained at the coordinates specified by $\mathscr{I}$ and the redundant symbols are stored in the remaining $n - k$ positions.

An information set corresponds to a reduced-echelon generator matrix and it may be obtained as follows. Starting with a reduced-echelon generator matrix $\boldsymbol{G}_{sys}^{(1)} = \boldsymbol{G}_{sys} = [\boldsymbol{I}_k|\boldsymbol{P}]$, Gaussian elimination is applied to submatrix $\boldsymbol{P}$ so that it is transformed to reduced-echelon form.

The resulting generator matrix now becomes $\boldsymbol{G}_{sys}^{(2)} = [\boldsymbol{A}|\boldsymbol{I}_k|\boldsymbol{P}']$, where $\boldsymbol{P}'$ is a $k \times (n - 2k)$ matrix. Next, submatrix $\boldsymbol{P}'$ is put into reduced-echelon form and the process continues until there exists a $k \times (n - lk)$ submatrix of rank less than k, for some integer l. Note that column permutations may be necessary during the transformation to maximise the number of disjoint information sets.

Let $\mathscr{G}$ be a collection of m reduced-echelon generator matrices of disjoint information sets, $\mathscr{G} = \{\boldsymbol{G}_{sys}^{(1)}, \boldsymbol{G}_{sys}^{(2)}, \ldots, \boldsymbol{G}_{sys}^{(m)}\}$.

Using these m matrices means that after $\sum_{w'=1}^{w} \binom{k}{w'}$ enumerations

- all possibilities of $d \leq mw$ have been considered; and
- if $mw < d_{ub}$, the minimum distance of the code is at least $m(w + 1)$, i.e. $d_{lb} = m(w + 1)$.

We can see that the lower bound has been increased by a factor of m, instead of 1 compared to the previous approach. For $w \leq k/2$, we know that $\binom{k}{w} \gg \binom{k}{w-1}$ and this lower bound increment reduces the bulk of computations significantly.

If d is the minimum distance of the code, the total number of enumerations required is given by

[2] Zimmermann attributed this algorithm to Brouwer in [25].

$$\sum_{w=1}^{\min\{\lceil d/m\rceil-1,k\}} m\binom{k}{w}. \tag{5.3}$$

Example 5.1 (*Disjoint Information Sets*) Consider the $[55, 15, 20]_2$ optimal binary linear, a shortened code of the Goppa code discovered by [15]. The reduced-echelon generator matrices of disjoint information sets are given by

$$\boldsymbol{G}_{sys}^{(1)} = \left[\begin{array}{c|c}
100000000000000 & 0101101111110101001110101010101010000000\\
010000000000000 & 1000111001101011110100100111001000110000\\
001000000000000 & 1011001111011010111010111111001001000001\\
000100000000000 & 1010101011101101101111110100010010101010\\
000010000000000 & 0011110100111110110110100011000101011101\\
000001000000000 & 0101000010100101111110001110001010001110\\
000000100000000 & 1001001110100010100110011001010010100111\\
000000010000000 & 1011000100110100000001110010110011110101\\
000000001000000 & 1010110101111110101001001011101110100001\\
000000000100000 & 1010000101001011000001110101111101100010\\
000000000010000 & 1101101011110001001011111011100101010100\\
000000000001000 & 1101101110101111110011101111111000001011\\
000000000000100 & 0000010001010011101110010011001011101101\\
000000000000010 & 0101100000011101011110010001100111000011\\
000000000000001 & 0011111001011000111001010001000000111011
\end{array}\right],$$

$$\boldsymbol{G}_{sys}^{(2)} = \left[\begin{array}{c|c|c}
101101001011001 & 100000000000000 & 0011110011110111110110000\\
000000011000110 & 010000000000000 & 1011000111011110001111010\\
001111100100011 & 001000000000000 & 0011101001010011100000101\\
010110100111101 & 000100000000000 & 1100101001010101110011011\\
111111001000100 & 000010000000000 & 0110011001100101101000100\\
111110010101001 & 000001000000000 & 0000000100011111100110001\\
111100100011110 & 000000100000000 & 1010010110110101100011101\\
000001100111111 & 000000010000000 & 1101000100101011100010001\\
000000101000001 & 000000001000000 & 1110110000011111100111101\\
111001100100100 & 000000000100000 & 1100100111100111011010111\\
100011111001111 & 000000000010000 & 0100001100100001000101110\\
010110000110111 & 000000000001000 & 1101110101101101011100100\\
001011011111111 & 000000000000100 & 0101010011001011111111110\\
100101011001011 & 000000000000010 & 0110000111001000110010011\\
110100101110101 & 000000000000001 & 0010100011000100001111100
\end{array}\right],$$

and

$$\boldsymbol{G}_{sys}^{(3)} = \left[\begin{array}{c|c|c}
010010100110100010111110001110 & 100000000000000 & 1100010001\\
111101011110000111110111110011 & 010000000000000 & 1000110111\\
011101101111001011001110110100 & 001000000000000 & 1101000101\\
001100010010101010010001111101 & 000100000000000 & 0110011100\\
000111101101010110110110101011 & 000010000000000 & 1011000101\\
101110101001000100110101001000 & 000001000000000 & 0110101011\\
010101000110100000010111100110 & 000000100000000 & 0011001111\\
100101101110111100001101101010 & 000000010000000 & 0100000001\\
100100110101011110011110011000 & 000000001000000 & 1101111001\\
011001100011100111110111011111 & 000000000100000 & 0111111010\\
101110001111100011101101101111 & 000000000010000 & 0111010111\\
100001101101010000101110110110 & 000000000001000 & 1000110010\\
011101010000010001011101101000 & 000000000000100 & 1101100011\\
010011111010001100010001011001 & 000000000000010 & 1101111111\\
110100111100111001100111101101 & 000000000000001 & 0000011110
\end{array}\right].$$

Brouwer's algorithm requires 9948 codewords to be evaluated to prove the minimum distance of this code is 20. In contrast, for the same proof, 32767 codewords would need to be evaluated if only one generator matrix is employed.

5.2.3 Zimmermann's Algorithm for Linear Codes and Some Improvements

A further refinement to the minimum distance algorithm is due to Zimmermann [25]. Similar to Brouwer's approach, a set of reduced-echelon generator matrices are required. While in Brouwer's approach the procedure is stopped once a non-full-rank submatrix is reached; Zimmermann's approach proceeds further to obtain submatrices with overlapping information sets. Let $\boldsymbol{G}_{sys}^{(m)} = [\boldsymbol{A}_m|\boldsymbol{I}_k|\boldsymbol{B}_{m+1}]$ be the last generator matrix which contains a disjoint information set. To obtain matrices with overlapping information sets, Gaussian elimination is performed on the submatrix $\boldsymbol{B}_{m+1}$ and this yields

$$\boldsymbol{G}_{sys}^{(m+1)} = \left[\hat{\boldsymbol{A}}_m \left| \begin{array}{c|c} \boldsymbol{0} & \boldsymbol{I}_{r_{m+1}} \\ \hline \boldsymbol{I}_{k-r_{m+1}} & \boldsymbol{0} \end{array} \right| \boldsymbol{B}_{m+2} \right],$$

where $r_{m+1} = \text{Rank}\,(\boldsymbol{B}_{m+1})$. Next, $\boldsymbol{G}_{sys}^{(m+2)}$ is produced by carrying out Gaussian elimination on the submatrix $\boldsymbol{B}_{m+2}$ and so on.

From $\boldsymbol{G}_{sys}^{(3)}$ of Example 5.1, we can see that the last 10 coordinates do not form an information set since the rank of this submatrix is clearly less than k. Nonetheless, a "partial" reduced-echelon generator matrix can be obtained from $\boldsymbol{G}_{sys}^{(3)}$,

$$\boldsymbol{G}_{sys}^{(4)} = \left[\begin{array}{c|c|c}
0111111001011010010100110001100101010000 & 00000 & 1000000000 \\
1100110000010110000010111110000110101110 & 00000 & 0100000000 \\
1010010000110001011100111001110100001110 & 00000 & 0010000000 \\
0010010010001100001001001001111011101111 & 00000 & 0001000000 \\
1011110010101000101110010011100001100010 & 00000 & 0000100000 \\
1010001011011101111000001000101101010100 & 00000 & 0000010000 \\
0100000100011011101111011111110010001111 & 00000 & 0000001000 \\
1011101011001011101101001111111100111011 & 00000 & 0000000100 \\
0111010111111100010011111101000111110111 & 00000 & 0000000010 \\
0101101011111001000001100100100110101010 & 00000 & 0000000001 \\
1100001101000000010101100101100100111101 & 10000 & 0000000000 \\
1001001100000111011010111010001110010001 & 01000 & 0000000000 \\
0111000001101001110100010110011000101110 & 00100 & 0000000000 \\
0001001111000011011101000010101011001110 & 00010 & 0000000000 \\
1111111100111111110000011110110100010111 & 00001 & 0000000000
\end{array} \right].$$

From $\boldsymbol{G}_{sys}^{(4)}$, we can see that the last k columns is also an information set, but $k - \text{Rank}\left(\boldsymbol{G}_{sys}^{(4)}\right)$ coordinates of which overlap with those in $\boldsymbol{G}_{sys}^{(3)}$. The generator matrix $\boldsymbol{G}_{sys}^{(4)}$ then may be used to enumerate codewords with condition that the effect of overlapping information set has to be taken into account.

Assuming that all codewords with information weight $\leq w$ have been enumerated, we know that

- for all $\boldsymbol{G}_{sys}^{(i)}$ of full-rank, say there are m of these matrices, all cases of $d \leq mw$ have been considered and each contributes to the lower bound. As a result, the lower bound becomes $d_{lb} = m(w+1)$.
- for each $\boldsymbol{G}_{sys}^{(i)}$ that do not have full-rank, we can join $\boldsymbol{G}_{sys}^{(i)}$ with column subsets of $\boldsymbol{G}_{sys}^{(j)}$, for $j < i$, so that we have an information set $\mathscr{I}_i$, which of course overlaps with information set $\mathscr{I}_j$. Therefore, for all of these matrices, say M, all cases of $d \leq Mw$ have been considered, but some of which are attributed to other information sets, and considering these would result in double counting.

According to Zimmermann [25], for each matrix $G_{sys}^{(m+j)}$ with an overlapping information set unless $w \geq k - \text{Rank}\left(B_{m+j}\right)$ for which the lower bound becomes $d_{lb} = d_{lb} + \left\{w - \left(k - \text{Rank}\left(B_{m+j}\right)\right) + 1\right\}$, there is no contribution to the lower bound.

Let the collection of full rank-reduced echelon matrices be denoted by, as before, $\mathcal{G} = \left\{G_{sys}^{(1)}, G_{sys}^{(2)}, \ldots, G_{sys}^{(m)}\right\}$, and let $\mathcal{G}'$ denote the collection of M rank matrices with overlapping information sets $\mathcal{G}' = \left\{G_{sys}^{(m+1)}, G_{sys}^{(m+2)}, \ldots, G_{sys}^{(m+M)}\right\}$. All $m + M$ generator matrices are needed by the [25] algorithm. Clearly, if the condition $w \geq k - \text{Rank}\left(B_{m+j}\right)$ is never satisfied throughout the enumeration, the corresponding generator matrix contributes nothing to the lower bound and, hence, can be excluded [11]. In order to accommodate this improvement, we need to know w_{max} the maximum information weight that would need to be enumerated before the minimum distance is found. This can be accomplished as follows: Say at information weight w, a lower weight codeword is found, i.e. new d_{ub}, starting from $w' = w$, we let $\mathcal{X} = \mathcal{G}'$, set $d_{lb} = m(w'+1)$ and then increment it by $(w' - (k - \text{Rank}(B_{m+j})) + 1)$ for each matrix in $\mathcal{G}'$ that satisfies $w' \geq k - \text{Rank}\left(B_{m+j}\right)$. Each matrix that satisfies this condition is also excluded from $\mathcal{X}$. The weight w' is incremented, d_{lb} is recomputed and at the point when $d_{lb} \geq d_{ub}$, we have w_{max} and all matrices in $\mathcal{X}$ are those to be excluded from codeword enumeration.

In some cases, it has been observed that while enumerating codewords of information weight w, a codeword, whose weight coincides with the lower bound obtained at enumeration step $w - 1$, appears. Clearly, this implies that the newly found codeword is indeed a minimum weight codeword; any other codeword of lower weight, if they exist, would have been found in the earlier enumeration steps. This suggests that the enumeration at step w may be terminated immediately. Since the bulk of computation time increases exponentially as the information weight is increased, this termination may result in a considerable saving of time.

Without loss of generality, we can assume that $\text{Rank}(B_{m+j}) > \text{Rank}(B_{m+j+1})$. With this consideration, we can implement the Zimmermann approach to minimum distance evaluation of linear code over $\mathbb{F}_2$–with the improvements, in Algorithm 5.1. The procedure to update w_{max} and $\mathcal{X}$ is given in Algorithm 5.2.

If there is additional code structure, the computation time required by Algorithm 5.1 can be reduced. For example, in some cases it is known that the binary code considered has even weight codewords only, then at the end of codeword enumeration at each step, the lower bound d_{lb} that we obtained may be rounded down to the next multiple of 2. Similarly, for codes where the weight of every codeword is divisible by 4, the lower bound may be rounded down to the next multiple of 4.

5.2.4 Chen's Algorithm for Cyclic Codes

Binary cyclic codes, which were introduced by Prange [19], form an important class of block codes over $\mathbb{F}_2$. Cyclic codes constitute many well-known error-

Algorithm 5.1 Minimum distance algorithm: improved Zimmermann's approach

Input: $\mathscr{G} = \left\{\boldsymbol{G}_{sys}^{(1)}, \boldsymbol{G}_{sys}^{(2)}, \ldots, \boldsymbol{G}_{sys}^{(m)}\right\}$ where $|\mathscr{G}| = m$
Input: $\mathscr{G}' = \left\{\boldsymbol{G}_{sys}^{(m+1)}, \boldsymbol{G}_{sys}^{(m+2)}, \ldots, \boldsymbol{G}_{sys}^{(m+M)}\right\}$ where $|\mathscr{G}'| = M$
Output: d (minimum distance)
1: $d' \leftarrow d_{ub} \leftarrow w_{max} \leftarrow k$
2: $d_{lb} \leftarrow w \leftarrow 1$
3: $\mathscr{X} = \emptyset$
4: **repeat**
5:     $M \leftarrow M - |\mathscr{X}|$
6:     **for all** $\boldsymbol{i} \in \mathbb{F}_2^k$ where $\text{wt}_H(\boldsymbol{i}) = w$ **do**
7:         **for** $1 \leq j \leq m$ **do**
8:             $d' \leftarrow \text{wt}_H(\boldsymbol{i} \cdot \boldsymbol{G}_{sys}^{(j)})$
9:             **if** $d' < d_{ub}$ **then**
10:                 $d_{ub} \leftarrow d'$
11:                 **if** $d_{ub} \leq d_{lb}$ **then**
12:                     Goto Step 36
13:                 **end if**
14:                 $w_{max}, \mathscr{X} \leftarrow$ Update w_{max} and $\mathscr{X}$ $\left(d_{ub}, k, m, \mathscr{G}'\right)$
15:             **end if**
16:         **end for**
17:         **for** $1 \leq j \leq M$ **do**
18:             $d' \leftarrow \text{wt}_H(\boldsymbol{i} \cdot \boldsymbol{G}_{sys}^{(m+j)})$
19:             **if** $d' < d_{ub}$ **then**
20:                 $d_{ub} \leftarrow d'$
21:                 **if** $d_{ub} \leq d_{lb}$ **then**
22:                     Goto Step 36
23:                 **end if**
24:                 $w_{max}, \mathscr{X} \leftarrow$ Update w_{max} and $\mathscr{X}$ $\left(d_{ub}, k, m, \mathscr{G}'\right)$
25:             **end if**
26:         **end for**
27:     **end for**
28:     $d_{lb} \leftarrow m(w + 1)$
29:     **for** $1 \leq j \leq M$ **do**
30:         **if** $w \geq \left\{k - \text{Rank}\left(\boldsymbol{B}_{m+j}\right)\right\}$ **then**
31:             $d_{lb} = d_{lb} + \left\{w - \left(k - \text{Rank}\left(\boldsymbol{B}_{m+j}\right)\right) + 1\right\}$
32:         **end if**
33:     **end for**
34:     $w \leftarrow w + 1$
35: **until** $d_{lb} \geq d_{ub}$ OR $w > k$
36: $d \Leftarrow d_{ub}$

correcting codes, such as the quadratic-residue codes and the commonly used in practice Bose–Chaudhuri–Hocquenghem (BCH) and Reed–Solomon (RS) codes. A binary cyclic code of length n, where n is necessarily odd, has the property that if $c(x) = \sum_{i=0}^{n-1} c_i x^i$, where $c_i \in \mathbb{F}_2$ is a codeword of the cyclic code, then $x^j c(x)$ $\pmod{x^n - 1}$, for some integer j, is also a codeword of that cyclic code. That is to say that the automorphism group of a cyclic code contains the coordinate permutation $i \rightarrow i + 1 \pmod{n}$.

Algorithm 5.2 $w_{max}, \mathscr{X} = \text{Update } w_{max} \text{ and } \mathscr{X}\left(d_{ub}, k, m, \mathscr{G}'\right)$

Input: $d_{ub},\ k,\ m$
Input: $\mathscr{G}'\left\{\boldsymbol{G}_{sys}^{(m+1)}, \boldsymbol{G}_{sys}^{(m+2)}, \ldots, \boldsymbol{G}_{sys}^{(m+M)}\right\}$
Output: w_{max} and $\mathscr{X}$

$\mathscr{X} \leftarrow \mathscr{G}'$
$w_{max} \leftarrow 1$
repeat
  $d_{lb} \leftarrow m(w_{max}+1)$
  for $1 \le j \le |\mathscr{G}'|$ **do**
    if $w_{max} \ge \{k - \text{Rank}(\boldsymbol{B}_{m+j})\}$ **then**
      Remove $\boldsymbol{G}_{sys}^{(m+j)}$ from $\mathscr{X}$ if $\boldsymbol{G}_{sys}^{(m+j)} \in \mathscr{X}$
      $d_{lb} = d_{lb} + \{w_{max} - (k - \text{Rank}(\boldsymbol{B}_{m+j})) + 1\}$
    end if
  end for
  $w_{max} \leftarrow w_{max} + 1$
until $d_{lb} \ge d_{ub}$ OR $w_{max} > k$
return w_{max} and $\mathscr{X}$

An $[n, k, d]$ binary cyclic code is defined by a generator polynomial $g(x)$ of degree $n-k$, and a parity-check polynomial $h(x)$ of degree k, such that $g(x)h(x) = 0 \pmod{x^n - 1}$. Any codeword of this cyclic code is a multiple of $g(x)$, that is $c(x) = u(x)g(x)$, where $u(x)$ is any polynomial of degree less than k. The generator matrix $\boldsymbol{G}$ can be simply formed from the cyclic shifts of $g(x)$, i.e.

$$\boldsymbol{G} = \begin{bmatrix} g(x) \pmod{x^n - 1} \\ xg(x) \pmod{x^n - 1} \\ \vdots \\ x^{k-1}g(x) \pmod{x^n - 1} \end{bmatrix}. \tag{5.4}$$

Since for some integer i, $x^i = q_i(x)g(x) + r_i(x)$ where $r_i(x) = x^i \pmod{g(x)}$, we can write

$$x^k\left(x^{n-k+i} - r_{n-k+i}(x)\right) = x^k q_i(x) g(x)$$

and based on this, a reduced-echelon generator matrix $\boldsymbol{G}_{sys}$ of a cyclic code is obtained:

$$\boldsymbol{G}_{sys} = \left[\begin{array}{c|c} & -x^{n-k} \pmod{g(x)} \\ & -x^{n-k+1} \pmod{g(x)} \\ \boldsymbol{I}_k & -x^{n-k+2} \pmod{g(x)} \\ & \vdots \\ & -x^{n-1} \pmod{g(x)} \end{array}\right]. \tag{5.5}$$

The matrix $\boldsymbol{G}_{sys}$ in (5.5) may contain several mutually disjoint information sets. But because each codeword is invariant under a cyclic shift, a codeword generated by information set $\mathscr{I}_i$ can be obtained from information set $\mathscr{I}_j$ by means of a simple cyclic shift. For an $[n, k, d]$ cyclic code, there always exists $\lfloor n/k \rfloor$ mutually disjoint information sets. As a consequence of this, using a single information set is sufficient to improve the lower bound to $\lfloor n/k \rfloor (w + 1)$ at the end of enumeration step w. However, Chen [7] showed that this lower bound could be further improved by noting that the average number of non-zeros of a weight w_0 codeword in an information set is $w_0 k/n$. After enumerating $\sum_{i=1}^{w} \binom{k}{i}$ codewords, we know that the weight of a codeword restricted to the coordinates specified by an information set is at least $w + 1$. Relating this to the average weight of codeword in an information set, we have an improved lower bound of $d_{lb} = \lceil (w+1)n/k \rceil$. Algorithm 5.3 summarises Chen's [7] approach to minimum distance evaluation of a binary cyclic code. Note that Algorithm 5.3 takes into account the early termination condition suggested in Sect. 5.2.3.

Algorithm 5.3 Minimum distance algorithm for cyclic codes: Chen's approach

Input: $\boldsymbol{G}_{sys} = [\boldsymbol{I}_k | \boldsymbol{P}]$ {see (5.5)}
Output: d (minimum distance)
1: $d_{ub} \leftarrow k$
2: $d_{lb} \leftarrow 1$
3: $w \leftarrow 1$
4: **repeat**
5:     $d' \leftarrow k$
6:     **for all** $\boldsymbol{i} \in \mathbb{F}_2^k$ where $\mathrm{wt}_H(\boldsymbol{i}) = w$ **do**
7:         $d' \leftarrow \mathrm{wt}_H(\boldsymbol{i} \cdot \boldsymbol{G}_{sys})$
8:         **if** $d' < d_{ub}$ **then**
9:             $d_{ub} \leftarrow d'$
10:             **if** $d_{ub} \leq d_{lb}$ **then**
11:                 Goto Step 18
12:             **end if**
13:         **end if**
14:     **end for**
15:     $d_{lb} \leftarrow \left\lceil \frac{n}{k}(w + 1) \right\rceil$
16:     $w \leftarrow w + 1$
17: **until** $d_{lb} \geq d_{ub}$ OR $w > k$
18: $d \Leftarrow d_{ub}$

It is worth noting that both minimum distance evaluation algorithms of Zimmermann [25] for linear codes and that of Chen [7] for cyclic codes may be used to compute the number of codewords of a given weight. In evaluating the minimum distance d, we stop the algorithm after enumerating all codewords having information weight i to w, where w is the smallest integer at which the condition $d_{lb} \geq d$ is reached. To compute the number of codewords of weight d, in addition to enumerating all codewords of weight i to w in their information set, all codewords having weight $w+1$ in their information set, also need to be enumerated. For Zimmermann's method, we use all of the available information sets, including those that overlap, and store all codewords whose weight matches d. In contrast to Chen's algorithm, we use only a single information set but for each codeword of weight d found, we accumulate this codeword and all of the $n-1$ cyclic shifts. In both approaches, it is necessary to remove the doubly-counted codewords at the end of the enumeration stage.

5.2.5 *Codeword Enumeration Algorithm*

The core of all minimum distance evaluation and codeword counting algorithms lies in the codeword enumeration. Given a reduced-echelon generator matrix, codewords can be enumerated by taking linear combinations of the rows in the generator matrix. This suggests the need for an efficient algorithm to generate combinations.

One of the most efficient algorithm for this purpose is the revolving-door algorithm, see [4, 13, 17]. The efficiency of the revolving-door algorithm arises from the property that in going from one combination pattern to the next, there is only one element that is exchanged. An efficient implementation of the revolving-door algorithm is given in [13], called *Algorithm R*, which is attributed to [18].[3]

In many cases, using a single-threaded program to either compute the minimum distance, or count the number of codewords of a given weight, of a linear code may take a considerable amount of computer time and can take several weeks.

For these long codes, we may resort to a multi-threaded approach by splitting the codeword enumeration task between multiple computers. The revolving-door algorithm has a nice property that allows such splitting to be neatly realised. Let $a_t a_{t-1} \dots a_2 a_1$, where $a_t > a_{t-1} > \dots > a_2 > a_1$ be a pattern of an t out of s combinations–C_t^s. A pattern is said to have rank i if this pattern appears as the $(i+1)$th element in the list of all C_t^s combinations.[4] Let $\text{Rank}(a_t a_{t-1} \dots a_2 a_1)$ be the rank of pattern $a_t a_{t-1} \dots a_2 a_1$, the revolving-door algorithm has the property that [13]

$$\text{Rank}(a_t a_{t-1} \dots a_2 a_1) = \left[\binom{a_t + 1}{t} - 1\right] - \text{Rank}(a_{t-1} \dots a_2 a_1) \tag{5.6}$$

and, for each integer N, where $0 \le N \le \binom{s}{t} - 1$, we can represent it uniquely with an ordered pattern $a_t a_{t-1} \dots a_2 a_1$. As an implication of this and (5.6), if all $\binom{k}{t}$ codewords need to be enumerated, we can split the enumeration into $\left\lceil \binom{k}{t} / M \right\rceil$ blocks, where in each block only at most M codewords need to be generated. In

[3] This is the version that the authors implemented to compute the minimum distance and to count the number of codewords of a given weight of a binary linear code.

[4] Here it is assume that the first element in the list of all C_t^s combinations has rank 0.

C_4^6

Rank	$a_4\,a_3\,a_2\,a_1$	
0	3 2 1 0	→ C_4^4
1	4 3 1 0	
2	4 3 2 1	→ C_4^5
3	4 3 2 0	
4	4 2 1 0	
5	5 4 1 0	
6	5 4 2 1	
7	5 4 2 0	→ C_4^6
8	5 4 3 2	
9	5 4 3 1	
10	5 4 3 0	
11	5 3 1 0	
12	5 3 2 1	
13	5 3 2 0	
14	5 2 1 0	

C_5^7

Rank	$a_5\,a_4\,a_3\,a_2\,a_1$	
0	4 3 2 1 0	→ C_5^5
1	5 4 2 1 0	
2	5 4 3 2 0	→ C_5^6
3	5 4 3 2 1	
4	5 4 3 1 0	
5	5 3 2 1 0	
6	6 5 2 1 0	
7	6 5 3 2 0	
8	6 5 3 2 1	
9	6 5 3 1 0	
10	6 5 4 3 0	→ C_5^7
11	6 5 4 3 1	
12	6 5 4 3 2	
13	6 5 4 2 0	
14	6 5 4 2 1	
15	6 5 4 1 0	
16	6 4 2 1 0	
17	6 4 3 2 0	
18	6 4 3 2 1	
19	6 4 3 1 0	
20	6 3 2 1 0	

Fig. 5.1 C_4^6 and C_5^7 revolving-door combination patterns

this way, we can do the enumeration of each block on a separate computer and this allows a parallelisation of the minimum distance evaluation, as well as the counting of the number of codewords of a given weight. We know that at the ith block, the enumeration would start from rank $(i-1)M$, and the corresponding pattern can be easily obtained following (5.6) and Lemma 5.1 below.

All $a_t a_{t-1} \ldots a_2 a_1$ revolving-door patterns of C_t^s satisfy the property that if the values in position a_t grow in an increasing order, then for fixed a_t, the values in position a_{t-1} grow in a decreasing order, moreover for fixed $a_t a_{t-1}$ the values in position a_{t-2} grow in an increasing order, and so on in an alternating order. This behaviour is evident by observing all revolving-door patterns of C_4^6 (left) and C_5^7 (right) shown in Fig. 5.1.

From this figure, we can also observe that

$$C_t^s \supset C_t^{s-1} \supset \ldots \supset C_t^{t+1} \supset C_t^t , \tag{5.7}$$

and this suggests the following lemma.

Lemma 5.1 (Maximum and Minimum Ranks) *Consider the $a_t a_{t-1} \ldots a_2 a_1$ revolving-door combination pattern, if we consider patterns with fixed a_t, the maximum and minimum ranks of such pattern are respectively given by*

$$\binom{a_t + 1}{t} - 1 \quad \textit{and} \quad \binom{a_t}{t} .$$

Example 5.2 (*Maximum and Minimum Ranks*) Say, if we consider all C_4^6 revolving-door combination patterns (left portion of Fig. 5.1) where $a_4 = 4$. From Lemma 5.1, we have a maximum rank of $\binom{5}{4} - 1 = 4$, and a minimum rank of $\binom{4}{4} = 1$. We can see that these rank values are correct from Fig. 5.1.

Example 5.3 (*The Revolving-Door Algorithm*) Consider combinations C_5^7 generated by the revolving-door algorithm, we would like to determine the rank of combination pattern 17. We know that the combination pattern takes the ordered form of $a_5a_4a_3a_2a_1$, where $a_i > a_{i-1}$. Starting from a_5, which can takes values from 0 to 6, we need to find a_5 such that the inequality $\binom{a_5}{5} \le 17 \le \binom{a_5+1}{5} - 1$ is satisfied (Lemma 5.1). It follows that $a_5 = 6$ and using (5.6), we have

$$\begin{aligned} 17 &= \text{Rank}(6a_4a_3a_2a_1) \\ &= \left[\binom{6+1}{5} - 1\right] - \text{Rank}(a_4a_3a_2a_1) \\ \text{Rank}(a_4a_3a_2a_1) &= 20 - 17 = 3 . \end{aligned}$$

Next, we consider a_4 and as before, we need to find $a_4 \in \{5, 4, 3, 2, 1, 0\}$ such that the inequality $\binom{a_4}{4} \le \text{Rank}(a_4a_3a_2a_1) \le \binom{a_4+1}{4} - 1$ is satisfied. It follows that $a_4 = 4$ and from (5.6), we have

$$\begin{aligned} 3 &= \text{Rank}(4a_3a_2a_1) \\ &= \left[\binom{4+1}{4} - 1\right] - \text{Rank}(a_3a_2a_1) \\ \text{Rank}(a_3a_2a_1) &= 4 - 3 = 1 . \end{aligned}$$

Next, we need to find a_3, which can only take a value less than 4, such that the inequality $\binom{a_3}{3} \le \text{Rank}(a_3a_2a_1) \le \binom{a_3+1}{3} - 1$ is satisfied. It follows that $a_3 = 3$ and from (5.6), $\text{Rank}(a_2a_1) = \left[\binom{3+1}{3} - 1\right] - 1 = 2$.

So far we have $643a_2a_1$, only a_2 and a_1 are unknown. Since $a_3 = 3$, a_2 can only take a value less than 3. The inequality $\binom{a_2}{2} \le \text{Rank}(a_2a_1) \le \binom{a_2+1}{2} - 1$ is satisfied if $a_2 = 2$ and correspondingly, $\text{Rank}(a_1) = \left[\binom{2+1}{2} - 1\right] - 2 = 0$.

For the last case, the inequality $\binom{a_1}{1} \le \text{Rank}(a_1) \le \binom{a_1+1}{1} - 1$ is true if and only if $a_1 = 0$. Thus, we have 64320 as the rank 17 C_5^7 revolving-door pattern. Cross-checking this with Fig. 5.1, we can see that 64320 is indeed of rank 17.

From (5.6) and Example 5.3, we can see that given a rank N, where $0 \le N \le \binom{s}{t} - 1$, we can construct an ordered pattern of C_t^s revolving-door combinations $a_ta_{t-1}\ldots a_2a_1$, recursively. A software realisation of this recursive approach is given in Algorithm 5.4.

Algorithm 5.4 Recursively Compute a_i $(\text{Rank}(a_ia_{i-1}\ldots a_2a_1), i)$

Input: i and $\text{Rank}(a_ia_{i-1}\ldots a_2a_1)$
Output: a_i

Find a_i, where $0 \le a_i < a_{i+1}$, such that $\binom{a_i}{i} \le \text{Rank}(a_ia_{i-1}\ldots a_2a_1) \le \left[\binom{a_i+1}{i} - 1\right]$
if $i > i$ **then**
  Compute $\text{Rank}(a_{i-1}\ldots a_2a_1) = \left[\binom{a_i+1}{i} - 1\right] - \text{Rank}(a_ia_{i-1}\ldots a_2a_1)$
  RecursiveCompute a_i $(\text{Rank}(a_{i-1}\ldots a_2a_1), i - 1)$
end if
return a_i

5.3 Binary Cyclic Codes of Lengths $129 \le n \le 189$

The minimum distance of all binary cyclic codes of lengths less than or equal to 99 has been determined by Chen [7, 8] and Promhouse et al. [21].

This was later extended to longer codes with the evaluation of the minimum distance of binary cyclic codes of lengths from 101 to 127 by Schomaker et al. [22]. We extend this work to include all cyclic codes of odd lengths from 129 to 189 in this book. The aim was to produce a Table of codes as a reference source for the highest minimum distance, with the corresponding roots of the generator polynomial, attainable by all cyclic codes over $\mathbb{F}_2$ of odd lengths from 129 to 189. It is well known that the coordinate permutation $\sigma : i \to \mu i$, where μ is an integer relatively prime to n, produces equivalent cyclic codes [3, p. 141f]. With respect to this property, we construct a list of generator polynomials $g(x)$ of all inequivalent and non-degenerate [16, p. 223f] cyclic codes of $129 \le n \le 189$ by taking products of the irreducible factors of $x^n - 1$. Two trivial cases are excluded, namely $g(x) = x + 1$ and $g(x) = (x^n - 1)/(x + 1)$, since these codes have trivial minimum distance and exist for any odd integer n. The idea is for each $g(x)$ of cyclic codes of odd length n; the systematic generator matrix is formed and the minimum distance of the code is determined using Chen's algorithm (Algorithm 5.3). However, due to the large number of cyclic codes and the fact that we are only interested in those of largest minimum distance for given n and k, we include a threshold distance d_{th} in Algorithm 5.3. Say, for given n and k, we have a list of generator polynomials $g(x)$ of all inequivalent cyclic codes. Starting from the top of the list, the minimum distance of the corresponding cyclic code is evaluated. If a codeword of weight less than or equal to d_{th} is found during the enumeration, the computation is terminated immediately and the next $g(x)$ is then processed. The threshold d_{th}, which is initialised with 0, is updated with the largest minimum distance found so far for given n and k.

Table 4.3 in Sect. 4.5 shows the highest attainable minimum distance of all binary cyclic codes of odd lengths from 129 to 189. The number of inequivalent and non-degenerate cyclic codes for a given odd integer n, excluding the two trivial cases mentioned above, is denoted by $N_{\mathscr{C}}$.

Note that Table 4.3 does not contain entries for primes $n = 8m \pm 3$. This is because for these primes, 2 is not a quadratic residue modulo n and hence, $\text{ord}_2(n) = n - 1$. As a consequence, $x^n - 1$ factors into two irreducible polynomials only, namely $x + 1$ and $(x^n - 1)/(x + 1)$ which generate trivial codes. Let β be a primitive nth root of unity, the roots of $g(x)$ of a cyclic code (excluding the conjugate roots) are given in terms of the exponents of β. The polynomial $m(x)$ is the minimal polynomial of β and it is represented in octal format with most significant bit on the left. That is, $m(x) = 166761$, as in the case for $n = 151$, represents $x^{15} + x^{14} + x^{13} + x^{11} + x^{10} + x^8 + x^7 + x^6 + x^5 + x^4 + 1$.

5.4 Some New Binary Cyclic Codes Having Large Minimum Distance

Constructing an $[n, k]$ linear code possessing the largest minimum distance is one of the main problems in coding theory. There exists a database containing the lower and upper bounds of minimum distance of binary linear codes of lengths $1 \le n \le 256$. This database appears in [6] and the updated version is accessible online.[5]

The lower bound corresponds to the largest minimum distance for a given $[n, k]_q$ code that has been found to date. Constructing codes which improves Brouwer's lower bounds is an on-going research activity in coding theory. Recently, Tables of lower- and upper-bounds of not only codes over finite-fields, but also quantum error-correcting codes, have been published by Grassl [12]. These bounds for codes over finite-fields, which are derived from MAGMA [5], appear to be more up-to-date than those of Brouwer.

We have presented in Sect. 5.3, the highest minimum distance attainable by all binary cyclic codes of odd lengths from 129 to 189 and found none of these cyclic codes has larger minimum distance than the corresponding Brouwer's lower bound for the same n and k. The next step is to consider longer length cyclic codes, $191 \le$

[5]The database is available at http://www.win.tue.nl/~aeb/voorlincod.html.

Note that, since 12th March 2007, A. Brouwer has stopped maintaining his database and hence it is no longer accessible. This database is now superseded by the one maintained by Grassl [12].

$n \leq 255$. For these lengths, unfortunately, we have not been able to repeat the exhaustive approach of Sect. 5.3 in a reasonable amount of time. This is due to the computation time to determine the minimum distance of these cyclic codes and also, for some lengths (e.g. 195 and 255), there are a tremendous number of inequivalent cyclic codes. Having said that, we can still search for improvements from lower rate cyclic codes of these lengths for which the minimum distance computation can be completed in a reasonable time. We have found many new cyclic codes that improve Brouwer's lower bound and before we present these codes, we should first consider the evaluation procedure.

As before, let β be a primitive nth root of unity and let Λ be a set containing all distinct (excluding the conjugates) exponents of β. The polynomial $x^n - 1$ can be factorised into irreducible polynomials $f_i(x)$ over $\mathbb{F}_2$, $x^n - 1 = \prod_{i \in \Lambda} f_i(x)$. For notational purposes, we denote the irreducible polynomial $f_i(x)$ as the minimal polynomial of β^i. The generator and parity-check polynomials, denoted by $g(x)$ and $h(x)$ respectively, are products of $f_i(x)$. Given a set $\Gamma \subseteq \Lambda$, a cyclic code $\mathscr{C}$ which has β^i, $i \in \Gamma$, as the non-zeros can be constructed. This means the parity-check polynomial $h(x)$ is given by

$$h(x) = \prod_{i \in \Gamma} f_i(x)$$

and the dimension k of this cyclic code is $\sum_{i \in \Gamma} \deg(f_i(x))$, where $\deg(f(x))$ denotes the degree of $f(x)$. Let $\Gamma' \subseteq \Lambda \setminus \{0\}$, $h'(x) = \prod_{i \in \Gamma'} f_i(x)$ and $h(x) = (1+x)h'(x)$. Given $\mathscr{C}$ with parity-check polynomial $h(x)$, there exists an $[n, k-1, d']$ expurgated cyclic code, $\mathscr{C}'$, which has parity-check polynomial $h'(x)$. For this cyclic code, $\mathrm{wt}_H(\boldsymbol{c}) \equiv 0 \pmod 2$ for all $\boldsymbol{c} \in \mathscr{C}'$. For convenience, we call $\mathscr{C}$ the augmented code of $\mathscr{C}'$.

Consider an $[n, k-1, d']$ expurgated cyclic code $\mathscr{C}'$, let the set $\boldsymbol{\Gamma} = \{\Gamma_1, \Gamma_2, \ldots, \Gamma_r\}$ where, for $1 \leq j \leq r$, $\Gamma_j \subseteq \Lambda \setminus \{0\}$ and $\sum_{i \in \Gamma_j} \deg(f_i(x)) = k - 1$. For each $\Gamma_j \in \boldsymbol{\Gamma}$, we compute $h'(x)$ and construct $\mathscr{C}'$. Having constructed the expurgated code, the augmented code can be easily obtained as shown below. Let $\boldsymbol{G}$ be a generator matrix of the augmented code $\mathscr{C}$, and without loss of generality, it can be written as

$$\boldsymbol{G} = \left[\begin{array}{c} \boldsymbol{G}' \\ \hline \mathbf{v} \end{array}\right] \tag{5.8}$$

where $\boldsymbol{G}'$ is a generator matrix of $\mathscr{C}'$ and the vector $\mathbf{v}$ is a coset of $\mathscr{C}'$ in $\mathscr{C}$. Using the arrangement in (5.8), we evaluate d' by enumerating codewords $c \in \mathscr{C}'$ from $\boldsymbol{G}'$. The minimum distance of $\mathscr{C}$, denoted by d, is simply $\min_{c \in \mathscr{C}'}\{d', \mathrm{wt}_H(\boldsymbol{c} + \mathbf{v})\}$ for all codewords $\boldsymbol{c}$ enumerated. We follow Algorithm 5.3 to evaluate d'. Let $d_{Brouwer}$

and $d'_{Brouwer}$ denote the lower bounds of [6] for linear codes of the same length and dimension as those of $\mathscr{C}$ and $\mathscr{C}'$ respectively. During the enumerations, as soon as $d \leq d_{Brouwer}$ and $d' \leq d'_{Brouwer}$, the evaluation is terminated and the next Γ_j in $\boldsymbol{\Gamma}$ is then processed. However, if $d < d_{Brouwer}$ and $d' > d'_{Brouwer}$, only the evaluation for $\mathscr{C}$ is discarded. Nothing is discarded if both $d' > d'_{Brouwer}$ and $d > d_{Brouwer}$. This procedure continues until an improvement is obtained; or the set in $\boldsymbol{\Gamma}$ has been exhausted, which means that there does not exist $[n, k-1]$ and $[n, k]$ cyclic codes which are improvements to Brouwer's lower bounds. In cases where the minimum distance computation is not feasible using a single computer, we switch to a parallel version using grid computers.

Table 5.1 presents the results of the search for new binary cyclic codes having lengths $195 \leq n \leq 255$. The cyclic codes in this table are expressed in terms of the parity-check polynomial $h(x)$, which is given in the last column by the exponents of β (excluding the conjugates). Note that the polynomial $m(x)$, which is given in octal with the most significant bit on the left, is the minimal polynomial of β. In many cases, the entries of $\mathscr{C}$ and $\mathscr{C}'$ are combined in a single row and this is indicated by "a/b" where the parameters a and b are for $\mathscr{C}'$ and $\mathscr{C}$, respectively. The notation "[0]" indicates that the polynomial $(1 + x)$ is to be excluded from the parity-check polynomial of $\mathscr{C}'$.
Some of these tabulated cyclic codes have a minimum Hamming distance which coincides with the lower bounds given in [12]. These are presented in Table 5.1 with the indicative mark "†".

In the late 1970s, computing the minimum distance of extended Quadratic Residue (QR) codes was posed as an open research problem by MacWilliams and Sloane [16]. Since then, the minimum distance of the extended QR code for the prime 199 has remained an open question. For this code, the bounds of the minimum distance were given as 16 – 32 in [16] and the lower bound was improved to 24 in [9]. Since $199 \equiv -1 \pmod 8$, the extended code is a doubly even self-dual code and its automorphism group contains a projective special linear group, which is known to be doubly transitive [16]. As a result, the minimum distance of the binary [199, 100] QR code is odd, i.e. $d \equiv 3 \pmod 4$, and hence, $d = 23$, 27 or 31. Due to the cyclic property and the rate of this QR code [7], we can safely assume that a codeword of weight d has maximum information weight of $\lfloor d/2 \rfloor$. If a weight d codeword does not satisfy this property, there must exist one of its cyclic shifts that does. After enumerating all codewords up to (and including) information weight 13 using grid computers, no codeword of weight less than 31 was found, implying that d of this binary [199, 100] QR code is indeed 31.

Without exploiting the property that $d \equiv 3 \pmod 4$, an additional $\binom{100}{14} + \binom{100}{15}$ codewords (88,373,885,354,647,200 codewords) would need to be enumerated in order to establish the same result and beyond available computer resources. Accordingly, we now know that there exists the [199, 99, 32] expurgated QR code and the [200, 100, 32] extended QR code.

It is interesting to note that many of the code improvements are contributed by low-rate cyclic codes of length 255 and there are 16 cases of this. Furthermore, it is also interesting that Table 5.1 includes a [255, 55, 70] cyclic code and a [255, 63, 65]

Table 5.1 New binary cyclic codes

$[m(x)]_8$	n	k	d	$d_{Brouwer}$	$h(x)$
17277	195	† 66/67	42/41	40/40	[0], 3, 5, 9, 19, 39, 65, 67
		† 68/69	40/39	39/38	[0], 1, 3, 13, 19, 35, 67, 91
		† 73	38	37	0, 3, 7, 19, 33, 35, 47
		† 74/75	38/37	36/36	[0], 3, 7, 19, 33, 35, 47, 65
		78	36	35	3, 7, 9, 11, 19, 35, 39, 65
13237042705-30057231362-555070452551	199	99/100	32/31	28/28	[0], 1
6727273	205	† 60	48	46	5, 11, 31
		† 61	46	44	0, 3, 11, 31
3346667657	215	70/71	46/46	44/44	[0], 3, 13, 35
3705317547055	223	74/75	48/47	46/45	[0], 5, 9
3460425444467-7544446504147	229	76	48	46	1
6704436621	233	† 58/59	60/60	56/56	[0], 3, 29
150153013	241	† 49	68	65	0, 1, 21
		73	54	53	0, 1, 3, 25
435	255	48/49	76/75	75/72	[0], 47, 55, 91, 95, 111, 127
		50/51	74/74	72/72	[0], 9, 13, 23, 47, 61, 85, 127
		52/53	72/72	71/68	[0], 7, 9, 17, 47, 55, 111, 127
		54/55	70/70	68/68	[0], 3, 7, 23, 47, 55, 85, 119, 127
		56/57	68/68	67/65	[0], 7, 27, 31, 45, 47, 55, 127
		58	66	64	7, 39, 43, 45, 47, 55, 85, 127
		60	66	64	7, 17, 23, 39, 45, 47, 55, 127
		62/63	66/65	64/63	[0], 11, 21, 47, 55, 61, 85, 87, 119, 127
		64/65	64/63	62/62	[0], 19, 31, 39, 47, 55, 63, 91, 127

cyclic code, which are superior to the BCH codes of the same length and dimension. Both of these BCH codes have minimum distance 63 only.

5.5 Constructing New Codes from Existing Ones

It is difficult to explicitly construct a new code with large minimum distance. However, the alternative approach, which starts from a known code which already has large minimum distance, seems to be more fruitful. Some of these methods are described below and in the following subsections, we present some new binary codes constructed using these methods, which improve on Brouwer's lower bound.

Theorem 5.1 (Construction X) *Let $\mathscr{B}_1$ and $\mathscr{B}_2$ be $[n, k_1, d_1]$ and $[n, k_2, d_2]$ linear codes over $\mathbb{F}_q$ respectively, where $\mathscr{B}_1 \supset \mathscr{B}_2$ ($\mathscr{B}_2$ is a subcode of $\mathscr{B}_1$). Let $\mathscr{A}$*

be an $[n', k_3 = k_1 - k_2, d']$ *auxiliary code over the same field. There exists an* $[n + n', k_1, \min\{d_2, d_1 + d'\}]$ *code* $\mathscr{C}_X$ *over* $\mathbb{F}_q$.

Construction X is due to Sloane et al. [23] and it basically adds a tail, which is a codeword of the auxiliary code $\mathscr{A}$, to $\mathscr{B}_1$ so that the minimum distance is increased. The effect of Construction X can be visualised as follows. Let $\boldsymbol{G}_{\mathscr{C}}$ be the generator matrix of code $\mathscr{C}$. Since $\mathscr{B}_1 \supset \mathscr{B}_2$, we may express $\boldsymbol{G}_{\mathscr{B}_1}$ as

$$\boldsymbol{G}_{\mathscr{B}_1} = \left[\begin{array}{c} \boldsymbol{G}_{\mathscr{B}_2} \\ \hline \boldsymbol{V} \end{array} \right],$$

where $\boldsymbol{V}$ is a $(k_1 - k_2) \times n$ matrix which contains the cosets of $\mathscr{B}_2$ in $\mathscr{B}_1$. We can see that the code generated by $\boldsymbol{G}_{\mathscr{B}_2}$ has minimum distance d_2, and the set of codewords $\{\mathbf{v} + \boldsymbol{c}_2\}$, for all $\mathbf{v} \in \boldsymbol{V}$ and all codewords $\boldsymbol{c}_2$ generated by $\boldsymbol{G}_{\mathscr{B}_2}$, have minimum weight of d_1. By appending non-zero weight codewords of $\mathscr{A}$ to the set $\{\mathbf{v} + \boldsymbol{c}_2\}$, and all zeros codeword to each codeword of $\mathscr{B}_2$, we have a lengthened code of larger minimum distance, $\mathscr{C}_X$, whose generator matrix is given by

$$\boldsymbol{G}_{\mathscr{C}_X} = \left[\begin{array}{c|c} \boldsymbol{G}_{\mathscr{B}_2} & \boldsymbol{0} \\ \hline \boldsymbol{V} & \boldsymbol{G}_{\mathscr{A}} \end{array} \right]. \tag{5.9}$$

We can see that, for binary cyclic linear codes of odd minimum distance, code extension by annexing an overall parity-check bit is an instance of Construction X. In this case, $\mathscr{B}_2$ is the even-weight subcode of $\mathscr{B}_1$ and the auxiliary code $\mathscr{A}$ is the trivial $[1, 1, 1]_2$ code.

Construction X given in Theorem 5.1 considers a chain of two codes only. There also exists a variant of Construction X, called Construction XX, which makes use of Construction X twice and it was introduced by Alltop [1].

Theorem 5.2 (Construction XX) *Consider three linear codes of the same length,* $\mathscr{B}_1 = [n, k_1, d_1]$, $\mathscr{B}_2 = [n, k_2, d_2]$ *and* $\mathscr{B}_3 = [n, k_3, d_3]$ *where* $\mathscr{B}_2 \subset \mathscr{B}_1$ *and* $\mathscr{B}_3 \subset \mathscr{B}_1$. *Let* $\mathscr{B}_4$ *be an* $[n, k_4, d_4]$ *linear code which is the intersection code of* $\mathscr{B}_2$ *and* $\mathscr{B}_3$, *i.e.* $\mathscr{B}_4 = \mathscr{B}_2 \cap \mathscr{B}_3$. *Using auxiliary codes* $\mathscr{A}_1 = [n_1, k_1 - k_2, d_1']$ *and* $\mathscr{A}_2 = [n_2, k_1 - k_3, d_2']$, *there exists an* $[n + n_1 + n_2, k_1, d]$ *linear code* $\mathscr{C}_{XX}$, *where* $d = \min\{d_4, d_3 + d_1', d_2 + d_2', d_1 + d_1' + d_2'\}$.

The relationship among $\mathscr{B}_1$, $\mathscr{B}_2$, $\mathscr{B}_3$ and $\mathscr{B}_4$ can be illustrated as a lattice shown below [11].

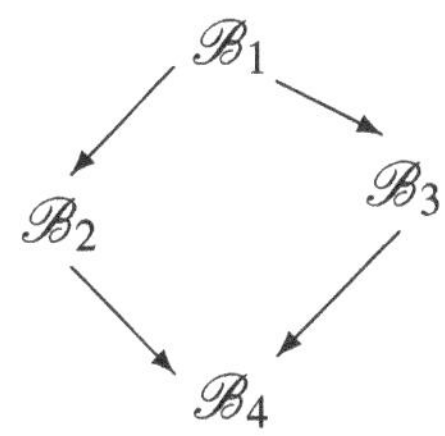

Since $\mathscr{B}_1 \supset \mathscr{B}_2$, $\mathscr{B}_1 \supset \mathscr{B}_3$, $\mathscr{B}_4 \subset \mathscr{B}_2$ and $\mathscr{B}_4 \subset \mathscr{B}_3$, the generator matrices of $\mathscr{B}_2$, $\mathscr{B}_3$ and $\mathscr{B}_1$ can be written as

$$\boldsymbol{G}_{\mathscr{B}_2} = \left[\begin{array}{c} \boldsymbol{G}_{\mathscr{B}_4} \\ \hline \boldsymbol{V}_2 \end{array}\right], \quad \boldsymbol{G}_{\mathscr{B}_3} = \left[\begin{array}{c} \boldsymbol{G}_{\mathscr{B}_4} \\ \hline \boldsymbol{V}_3 \end{array}\right]$$

$$\text{and} \quad \boldsymbol{G}_{\mathscr{B}_1} = \left[\begin{array}{c} \boldsymbol{G}_{\mathscr{B}_4} \\ \hline \boldsymbol{V}_2 \\ \hline \boldsymbol{V}_3 \\ \hline \boldsymbol{V} \end{array}\right]$$

respectively, where $\boldsymbol{V}_i$, $i = 2, 3$, is the coset of $\mathscr{B}_4$ in $\mathscr{B}_i$, and $\boldsymbol{V}$ contains the cosets of $\mathscr{B}_2$ and $\mathscr{B}_3$ in $\mathscr{B}_1$. Construction XX starts by applying Construction X to the pair of codes $\mathscr{B}_1 \supset \mathscr{B}_2$ using $\mathscr{A}_1$ as the auxiliary code. The resulting code is $\mathscr{C}_X = [n + n_1, k_1, \min\{d_2, d_1 + d_1'\}]$, whose generator matrix is given by

$$\boldsymbol{G}_{\mathscr{C}_X} = \left[\begin{array}{c|c} \boldsymbol{G}_{\mathscr{B}_4} & \boldsymbol{0} \\ \cline{1-1} \boldsymbol{V}_2 & \\ \hline \boldsymbol{V}_3 & \\ \cline{1-1} \boldsymbol{V} & \boldsymbol{G}_{\mathscr{A}_1} \end{array}\right].$$

This generator matrix can be rearranged such that the codewords formed from the first n coordinates are cosets of $\mathscr{B}_3$ in $\mathscr{B}_1$. This rearrangement results in the following generator matrix of $\mathscr{C}_X$,

$$\boldsymbol{G}_{\mathscr{C}_X} = \left[\begin{array}{c|c} \boldsymbol{G}_{\mathscr{B}_4} & \boldsymbol{0} \\ \hline \boldsymbol{V}_3 & \boldsymbol{G}^{(1)}_{\mathscr{A}_1} \\ \hline \boldsymbol{V}_2 & \boldsymbol{0} \\ \hline \boldsymbol{V} & \boldsymbol{G}^{(2)}_{\mathscr{A}_1} \end{array}\right],$$

where $\boldsymbol{G}_{\mathscr{A}_1} = \begin{bmatrix} \boldsymbol{G}^{(1)}_{\mathscr{A}_1} \\ \hline \boldsymbol{G}^{(2)}_{\mathscr{A}_1} \end{bmatrix}$. Next, using $\mathscr{A}_2$ as the auxiliary code, applying Construction X to the pair $\mathscr{B}_1 \supset \mathscr{B}_3$ with the rearrangement above, we obtain $\mathscr{C}_{XX}$ whose generator matrix is

$$\boldsymbol{G}_{\mathscr{C}_{XX}} = \left[\begin{array}{c|c|c} \boldsymbol{G}_{\mathscr{B}_4} & \mathbf{0} & \mathbf{0} \\ \hline \boldsymbol{V}_3 & \boldsymbol{G}^{(1)}_{\mathscr{A}_1} & \\ \hline \boldsymbol{V}_2 & \mathbf{0} & \\ \cline{1-1} \boldsymbol{V} & \boldsymbol{G}^{(2)}_{\mathscr{A}_1} & \smash{\raisebox{1.5ex}{$\boldsymbol{G}_{\mathscr{A}_2}$}} \end{array}\right].$$

While Constructions X and XX result in a code with increased length, there also exists a technique to obtain a shorter code with known minimum distance lower bounded from a longer code whose minimum distance and also that of its dual code are known explicitly. This technique is due to Sloane et al. [23] and it is called Construction Y1.

Theorem 5.3 (Construction Y1) *Given an* $[n, k, d]$ *linear code* $\mathscr{C}$*, which has an* $[n, n-k, d^{\perp}]$ $\mathscr{C}^{\perp}$ *as its dual, an* $[n-d^{\perp}, k-d^{\perp}+1, \geq d]$ *code* $\mathscr{C}'$ *can be constructed.*

Given an $[n, k, d]$ code, with standard code shortening, we obtain an $[n-i, k-i, \geq d]$ code where i indicates the number of coordinates to shorten. With Construction Y1, however, we can gain an additional dimension in the resulting shortened code. This can be explained as follows. Without loss of generality, we can assume the parity-check matrix of $\mathscr{C}$, which is also the generator matrix of $\mathscr{C}^{\perp}$, $\boldsymbol{H}$ contains a codeword $\boldsymbol{c}^{\perp}$ of weight $d^{\perp}$. If we delete the coordinates which form the support of $\boldsymbol{c}^{\perp}$ from $\boldsymbol{H}$, now $\boldsymbol{H}$ becomes an $(n-k) \times n - d^{\perp}$ matrix and there is a row which contains all zeros among these $n-k$ rows. Removing this all zeros row, we have an $(n-k-1) \times (n-d^{\perp})$ matrix $\boldsymbol{H}'$, which is the parity-check matrix of an $[n-d^{\perp}, n-d^{\perp}-(n-k-1), \geq d] = [n-d^{\perp}, k-d^{\perp}+1, \geq d]$ code $\mathscr{C}'$.

5.5.1 New Binary Codes from Cyclic Codes of Length 151

Amongst all of the cyclic codes in Table 4.3, those of length 151 have minimum distances that were found to have the highest number of matches against Brouwer's [6] lower bounds. This shows that binary cyclic codes of length 151 are indeed good codes. Since 151 is a prime, cyclic codes of this length are special as all of the irreducible factors of $x^{151}-1$, apart from $1+x$, have a fixed degree of 15. Having a fixed degree implies that duadic codes [14], which includes the quadratic residue codes, also exist for this length. Due to their large minimum distance, they are good candidate component codes for Constructions X and XX.

Table 5.2 Order of β in an optimum chain of $[151, k_i, d_i]$ cyclic codes

i	k_i	d_i	Roots of $g(x)$, excluding conjugate roots									
1	150	2	β^0									
2	135	6	β^0	β^1								
3	120	8	β^0	β^1	β^3							
4	105	14	β^0	β^1	β^3	β^5						
5	90	18	β^0	β^1	β^3	β^5		β^{11}				
6	75	24	β^0	β^1	β^3	β^5		β^{11}	β^{15}			
7	60	32	β^0	β^1	β^3	β^5		β^{11}	β^{15}			β^{37}
8	45	36	β^0	β^1	β^3	β^5		β^{11}	β^{15}	β^{23}		β^{37}
9	30	48	β^0	β^1	β^3	β^5		β^{11}	β^{15}	β^{23}	β^{35}	β^{37}
10	15	60	β^0	β^1	β^3	β^5	β^7	β^{11}	β^{15}	β^{23}	β^{35}	β^{37}

Definition 5.2 (*Chain of Cyclic Codes*) A pair of cyclic codes, $\mathscr{C}_1 = [n, k_1, d_1]$ and $\mathscr{C}_2 = [n, k_2, d_2]$ where $k_1 > k_2$, is nested, denoted $\mathscr{C}_1 \supset \mathscr{C}_2$, if all roots of $\mathscr{C}_1$ are contained in $\mathscr{C}_2$. Here, the roots refer to those of the generator polynomial. By appropriate arrangement of their roots, cyclic codes of the same length may be partitioned into a sequence of cyclic codes $\mathscr{C}_1 \supset \mathscr{C}_2 \supset \ldots \supset \mathscr{C}_t$. This sequence of codes is termed a chain of cyclic codes.

Given all cyclic codes of the same length, it is important to order the roots of these cyclic codes so that an optimum chain can be obtained. For all cyclic codes of length 151 given in Table 4.3, whose generator polynomial contains $1+x$ as a factor, an ordering of roots (excluding the conjugate roots) shown in Table 5.2 results in an optimum chain arrangement. Here β is a primitive 151st root of unity. Similarly, a chain which contains cyclic codes, whose generator polynomial does not divide $1 + x$, can also be obtained.

All the constituent codes in the chain $\mathscr{C}_1 \supset \mathscr{C}_2 \supset \ldots \supset \mathscr{C}_{10}$ of Table 5.2 are cyclic. Following Grassl [10], a chain of non-cyclic subcodes may also be constructed from a chain of cyclic codes. This is because for a given generator matrix of an $[n, k, d]$ cyclic code (not necessarily in row-echelon form), removing the last i rows of this matrix will produce an $[n, k - i, \geq d]$ code which will no longer be cyclic. As a consequence, with respect to Table 5.2, there exists $[151, k, d]$ linear codes, for $15 \leq k \leq 150$.

Each combination of pairs of codes in the $[151, k, d]$ chain is a nested pair which can be used as component codes for Construction X to produce another linear code with increased distance. There is a chance that the minimum distance of the resulting linear code is larger than that of the best-known codes for the same length and dimension. In order to find the existence of such cases, the following exhaustive approach has been taken. There are $\binom{150-15+1}{2} = \binom{136}{2}$ distinct pair of codes in the above chain of linear codes, and each pair say $\mathscr{C}_1 = [n, k_1, d_1] \supset \mathscr{C}_2 = [n, k_2, d_1]$, is combined using Construction X with an auxiliary code $\mathscr{A}$, which is an $[n', k_1 - k_2, d']$ best-known linear code. The minimum distance of the resulting code $\mathscr{C}_X$ is then compared to that of the best-known linear code for the same length and dimension to check for a possible improvement. Two improvements were obtained and they are tabulated in in the top half of Table 5.3.

In the case where $k_1 - k_2$ is small, the minimum distance of $\mathscr{C}_1$, i.e. d_1, obtained from a chain of linear codes, can be unsatisfactory. We can improve d_1 by augmenting $\mathscr{C}_1$ with a vector $\mathbf{v}$ of length n, i.e. add $\mathbf{v}$ as an additional row in $\boldsymbol{G}_{\mathscr{C}_2}$. In finding a vector $\mathbf{v}$ that can maximise the minimum distance of the enlarged code, we have adopted the following procedure. Choose a code $\mathscr{C}_2 = [n, k_2, d_2]$ that has sufficiently high minimum distance.

Assuming that $\boldsymbol{G}_{\mathscr{C}_2}$ is in reduced-echelon format, generate a vector $\mathbf{v}$ which satisfies the following conditions:

1. $v_i = 0$ for $0 \le i \le k - 1$ where v_i is the ith element of $\mathbf{v}$,
2. $\text{wt}_H(\mathbf{v}) > d_1$, and
3. $\text{wt}_H(\mathbf{v} + \boldsymbol{G}_r) > d_1$ for all $r \in \{0, 1, \ldots, k_2 - 1\}$ where $\boldsymbol{G}_{\mathscr{C}_2,r}$ denotes the rth row of $\boldsymbol{G}_{\mathscr{C}_2}$.

The vector $\mathbf{v}$ is then appended to $\boldsymbol{G}_{\mathscr{C}_2}$ as an additional row. The minimum distance of the resulting code is computed using Algorithm 5.1. A threshold is applied during the minimum distance evaluation and a termination is called whenever: $d_{ub} \le d_1$, in which case a different $\mathbf{v}$ is chosen and Algorithm 5.1 is restarted; or $d_1 < d_{ub} \le d_{lb}$ which means that an improvement has been found.

Using this approach, we found two new linear codes, $[151, 77, 20]$ and $[151, 62, 27]$, which have higher minimum distance than the corresponding codes obtained from a chain of nested cyclic codes. These two codes are obtained starting from the cyclic code $[151, 76, 23]$–which has roots $\{\beta, \beta^5, \beta^{15}, \beta^{35}, \beta^{37}\}$ and the cyclic code $[151, 61, 31]$–which has roots $\{\beta, \beta^3, \beta^5, \beta^{11}, \beta^{15}, \beta^{37}\}$, respectively and therefore

$$[151, 77, 20] \supset [151, 76, 23]$$

and

$$[151, 62, 27] \supset [151, 61, 31].$$

The second half of Table 5.3 shows the foundation codes for these new codes.

Note that when searching for the $[151, 62, 27]$ code, we exploited the property that the $[152, 61, 32]$ code obtained by extending the $[151, 61, 31]$ cyclic code is doubly even. We chose the additional vector $\mathbf{v}$ such that extending the enlarged code $[151, 62, d_1]$ yields again a doubly even code. This implies the congruence $d_1 = 0, 3 \bmod 4$ for the minimum distance of the enlarged code. Hence, it is sufficient to establish a lower bound $d_{lb} = 25$ using Algorithm 5.1 to show that $d_1 \ge 27$.

Furthermore, we also derived two different codes, $\mathscr{C}_2 = [151, 62, 27] \subset \mathscr{C}_1$ and $\mathscr{C}_3 = [151, 62, 27] \subset \mathscr{C}_1$, where $\mathscr{C}_1 = [151, 63, 23]$ and $\mathscr{C}_4 = \mathscr{C}_2 \cap \mathscr{C}_3 = [151, 61, 31]$. Using Construction XX, a $[159, 63, 31]$ code is obtained, see Table 5.4.

Table 5.3 New binary codes from Construction X and cyclic codes of length 151

$\mathscr{C}_1$	$\mathscr{C}_2$	$\mathscr{A}$	$\mathscr{C}_X$
Using chain of linear codes			
[151,72,24]	[151,60,32]	[23,12,7]	[174,72,31]
[151,60,32]	[151,45,36]	[20,15,3]	[171,60,35]
Using an improved subcode			
[151,77,20]	[151,76,23]	[3,1,3]	[154,77,23]
[151,62,27]	[151,61,31]	[4,1,4]	[155,62,31]

Table 5.4 New binary code from Construction XX and cyclic codes of length 151

$\mathscr{C}_1$	$\mathscr{C}_2$	$\mathscr{C}_3$	$\mathscr{C}_4 = \mathscr{C}_2 \cap \mathscr{C}_3$	$\mathscr{A}_1$	$\mathscr{A}_2$	$\mathscr{C}_{XX}$
[151, 63, 23]	[151, 62, 27]	[151, 62, 27]	[151, 61, 31]	[4, 1, 4]	[4, 1, 4]	[159, 63, 31]

5.5.2 *New Binary Codes from Cyclic Codes of Length* ≥ 199

We know from Table 5.1 that there exists an outstanding $[199, 100, 31]$ cyclic code. The extended code, obtained by annexing an overall parity-check bit, is a $[200, 100, 32]$ doubly even self-dual code. As the name implies, being self-dual we know that the dual code has minimum distance 32. By using Construction Y1 (Theorem 5.3), a $[168, 69, 32]$ new, improved binary code is obtained. The minimum distance of the $[168, 69]$ previously considered best-known binary linear code is 30.

Considering cyclic codes of length 205, in addition to a $[205, 61, 46]$ cyclic code (see Table 5.1), there also exists a $[205, 61, 45]$ cyclic code which contains a $[205, 60, 48]$ cyclic code as its even-weight subcode. Applying Construction X (Theorem 5.1) to the $[205, 61, 45] \supset [205, 60, 48]$ pair of cyclic codes with a repetition code of length 3 as the auxiliary code, a $[208, 61, 48]$ new binary linear code is constructed, which improves Brouwer's lower bound distance by 2.

Furthermore, by analysing the dual codes of the $[255, 65, 63]$ cyclic code in Table 5.1 and its $[255, 64, 64]$ even weight subcode it was found that both have minimum distance of 8. Applying Construction Y1 (Theorem 5.3), we obtain the $[247, 57, 64]$ and the $[247, 58, 63]$ new binary linear codes, which improves on Brouwer's lower bound distances by 2 and 1, respectively.

5.6 Concluding Observations on Producing New Binary Codes

In the search for error-correcting codes with large minimum distance, having a fast, efficient algorithm to compute the exact minimum distance of a linear code is important. The evolution of various algorithms to evaluate the minimum distance of a binary linear code, from the naive approach to Zimmermann's efficient approach, have been explored in detail. In addition to these algorithms, Chen's approach in computing the minimum distance of binary cyclic codes is a significant breakthrough.

The core basis of a minimum distance evaluation algorithm is codeword enumeration. As we increase the weight of the information vector, the number of codewords grows exponentially. Zimmermann's very useful algorithm may be improved by omitting generator matrices with overlapping information sets that never contribute to the lower bound throughout the enumeration. Early termination is important in the event that a new minimum distance is found that meets the lower bound value of the previous enumeration step. In addition, if the code under consideration has the property that every codeword weight is divisible by 2 or 4, the number of codewords that need to be enumerated can be considerably reduced.

With some simple modifications, these algorithms can also be used to collect and hence, count all codewords of a given weight to determine all or part of the weight spectrum of a code.

Given a generator matrix, codewords may be efficiently generated by taking linear combinations of rows of this matrix. This implies the faster we can generate the combinations, the less time the minimum distance evaluation algorithm will take. One such efficient algorithm to generate these combinations is called the revolving-door algorithm. The revolving-door algorithm has a nice property that allows the problem of generating combinations to be readily implemented in parallel. Having an efficient minimum distance computation algorithm, which can be computed in parallel on multiple computers has allowed us to extend earlier research results [8, 21, 22] in the evaluation of the minimum distance of cyclic codes. In this way, we obtained the highest minimum distance attainable by all binary cyclic codes of odd lengths from 129 to 189. We found that none of these cyclic codes has a minimum distance that exceeds the minimum distance of the best-known linear codes of the same length and dimension, which are given as lower bounds in [6]. However there are 134 cyclic codes that meet the lower bounds, see Sect. 5.3 and encoders and decoders may be easier to implement for the cyclic codes.

Having an efficient, multiple computer based, minimum distance computation algorithm also allowed us to search for the existence of binary cyclic codes of length longer than 189 which are improvements to Brouwer's lower bounds. We found 35 of these cyclic codes, namely

$$\begin{array}{llllll}
[195,66,42], & [195,67,41], & [195,68,40], & [195,69,39], & [195,73,38], & [195,74,38],\\
[195,75,37], & [195,78,36], & [199,99,32], & [199,100,32], & [205,60,48], & [205,61,46],\\
[215,70,46], & [215,71,46], & [223,74,48], & [223,75,47], & [229,76,48], & [233,58,60],\\
[233,59,60], & [255,48,76], & [255,49,75], & [255,50,74], & [255,51,74], & [255,52,72],\\
[255,53,72], & [255,54,70], & [255,55,70], & [255,56,68], & [255,57,68], & [255,58,66],\\
[255,60,66], & [255,62,66], & [255,63,65], & [255,64,64], & [255,65,63]. &
\end{array}$$

From the cyclic codes above, using Construction X to lengthen the code or Construction Y1 to shorten the code, four additional improvements to [6] lower bound are found, namely

Table 5.5 Updated minimum distance lower bounds of linear codes $\mathscr{C} = [n, k]$ for $153 \le n \le 174$ and $58 \le k \le 77$

$n\backslash k$	58	59	60	61	62	63	64	65	66	67	68	69	70	71	72	73	74	75	76	77	k/n
153	32	32	32	32	29^{E}	28	28	27	26	26	26	26	25	24	24	24	24	24	24	22	153
154	32	32	32	32	30^{E}	28	28	28	27	26	26	26	26	24	24	24	24	24	24	23^{X}	154
155	32	32	32	32	31^{E}	28	28	28	28	27	26	26	26	25	24	24	24	24	24	24^{E}	155
156	32	32	32	32	32^{E}	28	28	28	28	28	27	26	26	26	25	24	24	24	24	24^{E}	156
157	32	32	32	32	32^{E}	29	28	28	28	28	28	26	26	26	26	24	24	24	24	24^{E}	157
158	32	32	32	32	32^{E}	30	29	28	28	28	28	26	26	26	26	25	24	24	24	24	158
159	32	32	32	32	32^{E}	31^{XX}	30	29	28	28	28	27	26	26	26	26	25	24	24	24	159
160	32	32	32	32	32^{E}	32^{E}	30	30	28	28	28	28	26	26	26	26	26	25	24	24	160
161	32	32	32	32	32^{E}	32^{E}	30	30	29	28	28	28	27	26	26	26	26	26	25	24	161
162	33	32	32	32	32	32^{E}	31^{S}	30	30	29	28	28	28	27	26	26	26	26	26	24	162
163	34	33	32	32	32	32	32^{S}	31^{S}	30	30	29	28	28	28	27	26	26	26	26	25	163
164	34	34	33	32	32	32	32	32^{S}	31^{S}	30	30	29	28	28	28	27	26	26	26	26	164
165	34	34	34	33	32	32	32	32	32^{S}	31^{S}	30	30	28	28	28	28	27	26	26	26	165
166	34	34	34	34	32	32	32	32	32^{S}	32^{S}	31^{S}	30	28	28	28	28	28	27	26	26	166
167	34	34	34	34	32	32	32	32	32^{S}	32^{S}	32^{S}	31^{P}	29	28	28	28	28	28	27	26	167
168	34	34	34	34	32	32	32	32	32^{S}	32^{S}	32^{S}	32^{Y1}	30	29^{S}	28	28	28	28	28	26	168
169	35^{S}	34	34	34	32	32	32	32	32^{E}	32^{S}	32^{S}	32^{E}	31^{S}	30^{S}	29^{S}	28	28	28	28	27	169
170	36^{E}	35^{S}	34	34	33	32	32	32	32	32^{E}	32^{S}	32^{E}	32^{S}	31^{S}	30^{S}	29^{S}	28	28	28	28	170
171	36	36^{E}	35^{X}	34	34	33	32	32	32	32	32^{E}	32^{E}	32^{S}	32^{S}	31^{S}	30^{S}	29^{S}	28	28	28	171
172	36	36	36^{E}	34	34	34	33	32	32	32	32	32^{E}	32^{S}	32^{S}	32^{S}	31^{S}	30^{S}	29^{S}	28	28	172
173	36	36	36	35	34	34	34	33	32	32	32	32	32^{E}	32^{S}	32^{S}	32^{S}	31^{S}	30^{S}	29^{S}	28	173
174	36	36	36	36	34	34	34	34	32	32	32	32	32	32^{E}	32^{S}	32^{S}	32^{S}	31^{S}	30^{S}	29^{S}	174

Table 5.6 Updated minimum distance lower bounds of linear codes $\mathscr{C} = [n, k]$ for $175 \le n \le 224$ and $56 \le k \le 78$

$n \backslash k$	56	57	58	59	60	61	62	63	64	65	66	67	68	69	70	71	72	73	74	75	76	77	78	k/n
175	38	36	36	36	36	36	34	34	34	34	33	32	32	32	32	32^E	32^E	32^S	32^S	32^S	31^S	30^S	29^S	175
176	38	37	36	36	36	36	35	34	34	34	34	33	32	32	32	32	32^E	32^S	32^S	32^S	32^S	31^S	30^S	176
177	38	38	37	36	36	36	36	35	34	34	34	34	33	32	32	32	32	32^S	32^S	32^S	32^S	32^S	31^S	177
178	38	38	38	37	36	36	36	36	35	34	34	34	34	33	32	32	32	32	32^S	32^S	32^S	32^S	32^S	178
179	39	38	38	38	37	36	36	36	36	35	34	34	34	34	33	32	32	32	32	32^S	32^S	32^S	32^S	179
180	40	38	38	38	38	36	36	36	36	36	34	34	34	34	34	32	32	32	32	32	32^S	32^S	32^S	180
181	40	39	38	38	38	37	36	36	36	36	35	34	34	34	34	33	32	32	32	32	32^S	32^S	32^S	181
182	40	40	39	38	38	38	37	36	36	36	36	35	34	34	34	34	33	32	32	32	32	32^S	32^S	182
183	40	40	40	39	38	38	38	37	36	36	36	36	35	34	34	34	34	33	32	32	32	32	32^S	183
184	41	40	40	40	39	38	38	38	37	36	36	36	36	35	34	34	34	34	33	32	32	32	32	184
185	42	41^S	40	40	40	38	38	38	38	37	36	36	36	36	35	34	34	34	34	33	32	32	32	185
186	42	42^S	41^S	40	40	39	38	38	38	38	37	36	36	36	36	34	34	34	34	34	32	32	32	186
187	42	42	42^S	41^S	40	40	39	38	38	38	38	37	36	36	36	35	34	34	34	34	33	32	32	187
188	42	42	42	42^S	41^S	40	40	39	38	38	38	38	37	36	36	36	35	34	34	34	34	33	32	188
189	43	42	42	42	42^S	41^S	40	40	39	38	38	38	38	37	36	36	36	35	34	34	34	34	33	189
190	44	42	42	42	42	42^S	41^S	40	40	38	38	38	38	38	37^S	36	36	36	35	34	34	34	34	190
191	44	43	42	42	42	42	42^S	41^S	40	39	38	38	38	38	38^S	37^S	36	36	36	35	34	34	34	191
192	44	44	43	42	42	42	42	42^S	41^S	40	39^P	38	38	38	38	38^S	37^S	36	36	36	35^S	34	34	192
193	44	44	44	43	42	42	42	42	42^S	41^S	40^P	39^P	38	38	38	38^S	38^S	37^S	36	36	36^S	35^S	34	193
194	44	44	44	44	43	42	42	42	42	42^S	41^P	40^P	39^P	38	38	38	38^S	38^S	37^P	36	36	36^S	35^P	194
195	44	44	44	44	44	43	42	42	42	42	42^C	41^C	40^C	39^C	38	38	38	38^C	38^C	37^C	36	36	36^C	195
196	44	44	44	44	44	44	42	42	42	42	42^E	42^E	40	40^E	38	38	38	38	38^E	38^E	36	36	36	196
197	45	44	44	44	44	44	42	42	42	42	42^E	42^E	40	40	39	38	38	38	38	38^E	36	36	36	197
198	46	44	44	44	44	44	42	42	42	42	42^E	42^E	40	40	40	38	38	38	38	38	36	36	36	198
199	46	45^S	44	44	44	44	43	42	42	42	42^E	42^E	40	40	40	38	38	38	38	38	36	36	36	199

(continued)

Table 5.6 (continued)

$n\backslash k$	56	57	58	59	60	61	62	63	64	65	66	67	68	69	70	71	72	73	74	75	76	77	78	k/n
200	47^S	46^S	45^S	44	44	44	44	42	42	42	42^E	42^E	40	40	40	38	38	38	38	38	37	36	36	200
201	48^S	47^S	46^S	45^S	44	44	44	42	42	42	42	42^E	40	40	40	38	38	38	38	38	38	37	36	201
202	48^S	48^S	47^S	46^S	45^P	44	44	43	42	42	42	42^E	40	40	40	39	38	38	38	38	38	38	37	202
203	48^S	48^S	48^S	47^S	46^P	44	44	44	43	42	42	42^E	40	40	40	40	39	38	38	38	38	38	38	203
204	48^S	48^S	48^S	48^S	47^P	45^P	44	44	44	43	42	42	41	40	40	40	40	39	38	38	38	38	38	204
205	48	48^S	48^S	48^S	48^C	46^C	45^S	44	44	44	42	42	42	41	40	40	40	40	39	38	38	38	38	205
206	48	48^S	48^S	48^S	48^E	46^E	46^S	45^S	44	44	43	42	42	42	41	40	40	40	40	39	38	38	38	206
207	48	48	48^S	48^S	48^E	47^P	46^S	46^S	45^S	44	44	43	42	42	42	41	40	40	40	40	38	38	38	207
208	48	48	48	48^S	48^E	48^X	46	46^S	46^S	45^S	44	44	43	42	42	42	41	40	40	40	39	38	38	208
209	49	48	48	48	48^E	48^E	46	46	46^S	46^S	45^S	44	44	43	42	42	42	41	40	40	40	39	38	209
210	50	48	48	48	48	48^E	47^S	46	46	46^S	46^S	45^S	44	44	43	42	42	42	40	40	40	40	39	210
211	50	49	48	48	48	48^E	48^S	47^S	46	46^S	46^S	46^S	45^S	44	44	43	42	42	41	40	40	40	40	211
212	50	50	49	48	48	48	48^S	48^S	47^S	46	46^S	46^S	46^S	45^S	44	44	43	42	42	41	40	40	40	212
213	50	50	50	49	48	48	48	48^S	48^S	47^S	46	46^S	46^S	46^S	45^S	44	44	43	42	42	41	40	40	213
214	51	50	50	50	49	48	48	48	48^S	48^S	47^S	46	46^S	46^S	46^S	45^P	44	44	43	42	42	41	40	214
215	52	50	50	50	50	48	48	48	48	48^S	48^S	47^S	46	46^S	46^C	46^C	44	44	44	43	42	42	40	215
216	52	51	50	50	50	49	48	48	48	48^S	48^S	48^S	47^S	46^S	46^E	46^E	44	44	44	44	43	42	41	216
217	52	52	51	50	50	50	49	48	48	48	48^S	48^S	48^S	47^S	46^E	46^E	44	44	44	44	44	43	42	217
218	52	52	52	51	50	50	50	49	48	48	48	48^S	48^S	48^S	47^S	46^E	45^S	44	44	44	44	44	43	218
219	53	52	52	52	51	50	50	50	49	48	48	48	48^S	48^S	48^S	47^S	46^S	45^S	44	44	44	44	44	219
220	54	52	52	52	52	50	50	50	50	48	48	48	48	48^S	48^S	48^S	47^S	46^S	45^P	44	44	44	44	220
221	54	53	52	52	52	51	50	50	50	49	48	48	48	48^S	48^S	48^S	48^S	47^S	46^P	45^P	44	44	44	221
222	54	54	53	52	52	52	51	50	50	50	49	48	48	48	48^S	48^S	48^S	48^S	47^P	46^P	44	44	44	222
223	54	54	54	53	52	52	52	51	50	50	50	49	48	48	48	48^S	48^S	48^S	48^C	47^C	44	44	44	223
224	55	54	54	54	53	52	52	52	51	50	50	50	49	48	48	48	48^S	48^S	48^E	48^E	45	44	44	224

Table 5.7 Updated minimum distance lower bounds of linear codes $\mathscr{C} = [n, k]$ for $175 \le n \le 224$ and $79 \le k \le 100$

$n\backslash k$	79	80	81	82	83	84	85	86	87	88	89	90	91	92	93	94	95	96	97	98	99	100	k/n
175	28	28	28	27	26	26	26	26	25	24	24	24	24	23	22	22	22	22	22	22	22	21	175
176	29^S	28	28	28	27	26	26	26	26	25	24	24	24	24	23	22	22	22	22	22	22	22	176
177	30^S	29^S	28	28	28	27	26	26	26	26	25	24	24	24	24	23	22	22	22	22	22	22	177
178	31^S	30^S	29^S	28	28	28	27	26	26	26	26	25	24	24	24	24	23	22	22	22	22	22	178
179	32^S	31^S	30^S	29^S	28	28	28	27	26	26	26	26	25	24	24	24	24	23	22	22	22	22	179
180	32^S	32^S	31^S	30^S	29^{a}	28	28	28	26	26	26	26	26	24	24	24	24	24	23	22	22	22	180
181	32^S	32^S	32^S	31^S	30^{a}	29^S	28	28	27	26	26	26	26	25	24	24	24	24	24	23	22	22	181
182	32^S	32^S	32^S	32^S	31^{a}	30^S	29^S	28	28	27	26	26	26	26	25	24	24	24	24	24	23	22	182
183	32^S	32^S	32^S	32^S	32^{a}	31^S	30^S	29^S	28	28	27	26	26	26	26	25	24	24	24	24	24	23	183
184	32^S	32^S	32^S	32^S	32^{a}	32^S	31^S	30^S	29^S	28	28	27	26	26	26	26	25	24	24	24	24	24	184
185	32	32^S	32^S	32^S	32^{a}	32^S	32^S	31^S	30^S	29^S	28	28	27	26	26	26	26	25	24	24	24	24	185
186	32	32	32^S	32^S	32^{a}	32^S	32^S	32^S	31^S	30^S	29^S	28	28	26	26	26	26	26	24	24	24	24	186
187	32	32	32	32^S	32^{a}	32^S	32^S	32^S	32^S	31^S	30^S	29^S	28	27	26	26	26	26	25	24	24	24	187
188	32	32	32	32	32^{a}	32^S	32^S	32^S	32^S	32^S	31^S	30^S	29^S	28	27	26	26	26	26	25	24	24	188
189	32	32	32	32	32	32^S	32^S	32^S	32^S	32^S	32^S	31^S	30^S	29^S	28	27	26	26	26	26	25	24	189
190	33	32	32	32	32	32	32^S	32^S	32^S	32^S	32^S	32^S	31^S	30^S	29^S	28	27	26	26	26	26	25	190
191	34	33	32	32	32	32	32	32^S	32^S	32^S	32^S	32^S	32^S	31^S	30^S	29^S	28	27	26	26	26	26	191
192	34	34	32	32	32	32	32	32	32^S	32^S	32^S	32^S	32^S	32^S	31^S	30^S	29^S	28	27^S	26	26	26	192
193	34	34	33	32	32	32	32	32	32^S	32^S	32^S	32^S	32^S	32^S	32^S	31^S	30^S	29^S	28^S	27^S	26	26	193
194	34	34	34	33	32	32	32	32	32	32^S	32^S	32^S	32^S	32^S	32^S	32^S	31^S	30^S	29^S	28^S	27^P	26	194
195	34	34	34	34	33	32	32	32	32	32	32^S	32^S	32^S	32^S	32^S	32^S	32^S	31^S	30^S	29^S	28^P	27^P	195
196	35	34	34	34	34	33	32	32	32	32	32	32^S	32^S	32^S	32^S	32^S	32^S	32^S	31^S	30^S	29^P	28^P	196
197	36	35	34	34	34	34	33	32	32	32	32	32	32^S	32^S	32^S	32^S	32^S	32^S	32^S	31^S	30^P	29^P	197
198	36	36	34	34	34	34	34	33	32	32	32	32	32	32^S	32^S	32^S	32^S	32^S	32^S	32^S	31^P	30^P	198
199	36	36	34	34	34	34	34	34	32	32	32	32	32	32^S	32^S	32^S	32^S	32^S	32^S	32^S	32^C	31^C	199

(continued)

Table 5.7 (continued)

$n\backslash k$	79	80	81	82	83	84	85	86	87	88	89	90	91	92	93	94	95	96	97	98	99	100	k/n
200	36	36	35	34	34	34	34	34	32	32	32	32	32	32^S	32^S	32^S	32^S	32^S	32^S	32^S	32^E	32^E	200
201	36	36	36	34	34	34	34	34	32	32	32	32	32	32^S	32^S	32^S	32^S	32^S	32^S	32^S	32^E	32^E	201
202	36	36	36	34	34	34	34	34	32	32	32	32	32	32^S	32^S	32^S	32^S	32^S	32^S	32^S	32^E	32^E	202
203	37	36	36	35	34	34	34	34	33	32	32	32	32	32	32^S	32^S	32^S	32^S	32^S	32^S	32^E	32^E	203
204	38	37	36	36	35	34	34	34	34	33	32	32	32	32	32	32^S	32^S	32^S	32^S	32^S	32^E	32^E	204
205	38	38	37	36	36	35	34	34	34	34	33	32	32	32	32	32	32^S	32^S	32^S	32^S	32^E	32^E	205
206	38	38	38	37	36	36	35	34	34	34	34	33	32	32	32	32	32	32^S	32^S	32^S	32^E	32^E	206
207	38	38	38	38	37	36	36	35	34	34	34	34	33	32	32	32	32	32	32^S	32^S	32^E	32^E	207
208	38	38	38	38	38	37	36	36	34	34	34	34	34	32	32	32	32	32	32^S	32^S	32^E	32^E	208
209	38	38	38	38	38	38	37	36	35	34	34	34	34	32	32	32	32	32	32	32^S	32^E	32^E	209
210	38	38	38	38	38	38	38	37	36	35	34	34	34	32	32	32	32	32	32	32^S	32^E	32^E	210
211	39	38	38	38	38	38	38	38	37	36	35	34	34	33	32	32	32	32	32	32^S	32^E	32^E	211
212	40	39	38	38	38	38	38	38	38	37	36	35	34	34	33	32	32	32	32	32^S	32^E	32^E	212
213	40	40	39	38	38	38	38	38	38	38	37	36	35	34	34	33	32	32	32	32	32^E	32^E	213
214	40	40	40	39	38	38	38	38	38	38	38	37	36	35	34	34	33	32	32	32	32	32^E	214
215	40	40	40	40	39	38	38	38	38	38	38	38	37	36	35	34	34	33	32	32	32	32	215
216	40	40	40	40	40	39	38	38	38	38	38	38	38	37	36	35	34	34	32	32	32	32	216
217	41	40	40	40	40	40	39	38	38	38	38	38	38	38	37	36	35	34	33	32	32	32	217
218	42	41	40	40	40	40	40	39	38	38	38	38	38	38	38	37	36	35	34	33	32	32	218
219	43	42	41	40	40	40	40	40	39	38	38	38	38	38	38	38	37	36	35	34	33	32	219
220	44	43	42	41	40	40	40	40	40	39	38	38	38	38	38	38	38	37	36	35	34	33	220
221	44	44	43	42	41	40	40	40	40	40	39	38	38	38	38	38	38	38	37	36	35	34	221
222	44	44	44	43	42	41	40	40	40	40	40	39	38	38	38	38	38	38	38	37	36	35	222
223	44	44	44	44	43	42	41	40	40	40	40	40	39	38	38	38	38	38	38	38	37	36	223
224	44	44	44	44	44	43	42	41	40	40	40	40	40	39	38	38	38	38	38	38	38	37	224

Table 5.8 Updated minimum distance lower bounds of linear codes $\mathscr{C} = [n, k]$ for $225 \le n \le 256$ and $48 \le k \le 62$

$n\backslash k$	48	49	50	51	52	53	54	55	56	57	58	59	60	61	62	k/n
225	60	60^S	60^S	60^S	59^S	58^S	57^S	56	56	54	54	54	54	52	52	225
226	60	60	60^S	60^S	60^S	59^S	58^S	57^S	56	55^S	54	54	54	52	52	226
227	60	60	60^S	60^S	60^S	60^S	59^S	58^S	57^S	56^S	55^S	54	54	52	52	227
228	61	60	60^S	60^S	60^S	60^S	60^S	59^S	58^S	57^S	56^S	55^P	54	53^S	52	228
229	62	60	60^S	60^S	60^S	60^S	60^S	60^S	59^S	58^S	57^S	56^P	54	54^S	53^S	229
230	62	60	60	60^S	60^S	60^S	60^S	60^S	60^S	59^S	58^S	57^P	54	54	54^S	230
231	63	61	60	60	60^S	60^S	60^S	60^S	60^S	60^S	59^S	58^P	54	54	54	231
232	64	62	60	60	60	60^S	60^S	60^S	60^S	60^S	60^S	59^P	54	54	54	232
233	64	62	60	60	60	60^S	60^S	60^S	60^S	60^S	60^C	60^C	54	54	54	233
234	64	62	61	60	60	60	60^S	60^S	60^S	60^S	60^E	60^E	55	54	54	234
235	64	63	62	61	60	60	60	60^S	60^S	60^S	60^E	60^E	56	55	54	235
236	65	64	62	62	61	60	60	60	60^S	60^S	60^E	60^E	56	56	54	236
237	66	64	63	62	62	61	60	60	60	60^S	60^E	60^E	56	56	55	237
238	66	65^P	64	63	62	62	61	60	60	60	60^E	60^E	57	56	56	238
239	67	66^P	64	64	63	62	62	61	60	60	60	60^E	58	57	56	239
240	68	67^P	64	64	64	62	62	62	61	60	60	60	58	58	56	240
241	68	68^C	64	64	64	62	62	62	62	61	60	60	58	58	57	241
242	68	68^E	65	64	64	63^S	62	62	62	62	61	60	59	58	58	242
243	68	68^E	66	65	64	64^S	63^S	62	62	62	62	61	60	59	58	243
244	69	68	66	66	65	64	64^S	63^S	62	62	62	62	61	60	59	244
245	70	68	67	66	66	65^S	64^S	64^S	63^S	62	62	62	62	61	60	245
246	70	68	68	67	66	66^S	65^S	64^S	64^S	63^P	62	62	62	62	61	246
247	71	68	68	68	67	66^S	66^S	65^S	64^S	64^{Y1}	63^{Y1}	62	62	62	62	247

(continued)

Table 5.8 (continued)

$n\backslash k$	48	49	50	51	52	53	54	55	56	57	58	59	60	61	62	k/n
248	72	69^S	68	68	68	66^S	66^S	66^S	65^S	64^E	64^E	62	62	62	62	248
249	72	70^S	69^S	68	68	66^P	66^S	66^S	66^S	65^S	64^E	63^S	62	62	62	249
250	72	71^S	70^S	69^P	68	67^P	66^S	66^S	66^S	66^S	65^S	64^S	63^S	62	62	250
251	73^S	72^S	71^S	70^P	69^S	68^P	67^S	66^P	66^S	66^S	66^S	65^S	64^S	63^S	62	251
252	74^S	73^S	72^S	71^P	70^S	69^P	68^S	67^P	66^S	66^S	66^S	66^S	65^S	64^S	63^P	252
253	74	74^S	73^S	72^P	71^S	70^P	69^S	68^P	67^S	66^P	66^S	66^S	66^S	65^S	64^P	253
254	75^P	74^P	74^S	73^P	72^S	71^P	70^S	69^P	68^S	67^P	66^S	66^S	66^S	66^S	65^P	254
255	76^C	75^C	74^C	74^C	72^C	72^C	70^C	70^C	68^C	68^C	66^C	66^S	66^C	66^S	66^C	255
256	76	76^E	74^E	74^E	72	72^E	70^E	70^E	68	68^E	66^E	66^S	66^E	66^S	66^E	256

Table 5.9 Updated minimum distance lower bounds of linear codes $\mathscr{C} = [n, k]$ for $225 \le n \le 256$ and $63 \le k \le 76$

$n \backslash k$	63	64	65	66	67	68	69	70	71	72	73	74	75	76	k/n
225	52	52	50	50	50	50	48	48	48	48	48^S	48^E	48^E	46	225
226	52	52	50	50	50	50	48	48	48	48	48^S	48^E	48^E	46	226
227	52	52	50	50	50	50	48	48	48	48	48^S	48^E	48^E	46	227
228	52	52	50	50	50	50	48	48	48	48	48	48^E	48^E	47^P	228
229	52	52	51	50	50	50	49	48	48	48	48	48^E	48^E	48^C	229
230	53^S	52	52	51	50	50	50	48	48	48	48	48^E	48^E	48^E	230
231	54^S	53^S	52	52	51	50	50	48	48	48	48	48	48^E	48^E	231
232	54	54^S	53^S	52	52	51	50	49	48	48	48	48	48	48^E	232
233	54	54	54^S	53^S	52	52	51	50	49	48	48	48	48	48	233
234	54	54	54	54^S	53^S	52	52	51	50	49	48	48	48	48	234
235	54	54	54	54	54^S	53^S	52	52	51	50	49	48	48	48	235
236	54	54	54	54	54	54^S	53^S	52	52	51	50	49	48	48	236
237	54	54	54	54	54	54	54^S	53^S	52	52	51	50	49	48	237
238	55	54	54	54	54	54	54	54^S	53^S	52	52	51	50	49	238
239	56	55	54	54	54	54	54	54	54^S	53^S	52	52	51	50	239
240	56	56	54	54	54	54	54	54	54	54^S	53^P	52	52	51	240
241	56	56	55	54	54	54	54	54	54	54	54^C	52	52	52	241
242	57	56	56	55	54	54	54	54	54	54	54	53	52	52	242
243	58	57	56	56	55	54	54	54	54	54	54	54	53	52	243
244	58	58	56	56	56	55	54	54	54	54	54	54	54	53	244
245	59	58	57	56	56	56	55	54	54	54	54	54	54	54	245
246	60	59	58	57	56	56	56	55	54	54	54	54	54	54	246
247	61	60	59	58	57	56	56	56	55	54	54	54	54	54	247
248	62	61	60	59	58	57	56	56	56	55	54	54	54	54	248
249	62	62	61	60	59	58	57	56	56	56	55	54	54	54	249
250	62	62	62	61	60	59	58	57	56	56	56	55	54	54	250
251	62	62	62	62	61	60	59	58	57	56	56	56	55	54	251
252	62	62	62	62	62	61	60	59	58	56	56	56	56	55	252
253	63^P	62	62	62	62	62	61	60	59	56	56	56	56	56	253
254	64^P	63^P	62	62	62	62	62	61	60	57	56	56	56	56	254
255	65^C	64^C	63^C	62	62	62	62	62	61	58	57	56	56	56	255
256	66^E	64^E	64^E	62	62	62	62	62	62	58	58	56	56	56	256

$$[168, 69, 32], [208, 61, 48], [247, 57, 64], [247, 58, 63].$$

Five new linear codes, which are derived from cyclic codes of length 151, have also been constructed. These new codes, which are produced by Constructions X and XX, are

$$[154, 77, 23], [155, 62, 31], [159, 63, 31], [171, 60, 35], [174, 72, 31].$$

Given an $[n, k, d]$ code $\mathscr{C}$, where d is larger than the minimum distance of the best-known linear code of the same n and k, it is possible to obtain more codes, whose minimum distance is still larger than that of the corresponding best-known linear code, by recursively extending (annexing parity-checks), puncturing and/or shortening $\mathscr{C}$. For example, consider the new code [168, 69, 32] as a starting point. New codes can be obtained by annexing parity-check bits $[168 + i, 69, 32]$, for $1 \leq i \leq 3$. With puncturing by one bit a [167, 69, 31] new code is obtained by shortening $[168 - i, 69 - i, 32]$, for $1 \leq i \leq 5$, 5 new codes are obtained with a minimum distance of 32. More improvements are also obtained by shortening these extended and punctured codes. Overall, with all of the new codes described and presented in this chapter, there are some 901 new binary linear codes which improve on Brouwer's lower bounds. The updated lower bounds are tabulated in Tables 5.5, 5.6, 5.7, 5.8 and 5.9 in Appendix "Improved Lower Bounds of the Minimum Distance of Binary Linear Codes".

5.7 Summary

Methods have been described and presented which may be used to determine the minimum Hamming distance and weight distribution of a linear code. These are the main tools for testing new codes which are candidates for improvements to currently known, best codes. Several efficient algorithms for computing the minimum distance and weight distribution of linear codes have been explored in detail. The many different methods of constructing codes have been described, particularly those based on using known good or outstanding codes as a construction basis. Using such methods, several hundred new codes have been presented or described which are improvements to the public database of best, known codes.

For cyclic codes, which have implementation advantages over other codes, many new outstanding codes have been presented including the determination of a table giving the code designs and highest attainable minimum distance of all binary cyclic codes of odd lengths from 129 to 189. It has been shown that outstanding cyclic codes may be used as code components to produce new codes that are better than the previously thought best codes, for the same code length and code rate.

Appendix

Improved Lower Bounds of the Minimum Distance of Binary Linear Codes

The following tables list the updated lower bounds of minimum distance of linear codes over $\mathbb{F}_2$. These improvements—there are 901 of them in total—are due to the new binary linear codes described above. In the tables, entries marked with C refer to cyclic codes, those marked with X, XX and $Y1$ refer to codes obtained from Constructions X, XX and Y1, respectively. Similarly, entries marked with E, P and S denote $[n, k, d]$ codes obtained by extending (annexing an overall parity-check bit) to $(n-1, k, d')$ codes, puncturing $(n+1, k, d+1)$ codes and shortening $(n+1, k+1, d)$ codes, respectively. Unmarked entries are the original lower bounds of Brouwer [6].

References

1. Alltop, W.O.: A method of extending binary linear codes. IEEE Trans. Inf. Theory **30**(6), 871–872 (1984)
2. Berlekamp, E., McEliece, R., van Tilborg, H.: On the inherent intractability of certain coding problems. IEEE Trans. Inf. Theory **24**, 384–386 (1978)
3. Berlekamp, E.R.: Algebraic Coding Theory. Aegean Park Press, Laguna Hills (1984). ISBN 0 894 12063 8
4. Bitner, J.R., Ehrlich, G., Reingold, E.M.: Efficient generation of the binary reflected gray code and its applications. Commun. ACM **19**(9), 517–521 (1976)
5. Bosma, W., Cannon, J.J., Playoust, C.P.: Magma Algebra Syst. User Lang. **24**, 235–266 (1997)
6. Brouwer, A.E.: Bounds on the size of linear codes. In: Pless, V.S., Huffman, W.C. (eds.) Handbook of Coding Theory, pp. 295–461. Elsevier, North Holland (1998)
7. Chen C.L. (1969) Some results on algebraically structured error-correcting codes. Ph.D Dissertation, University of Hawaii, USA
8. Chen, C.L.: Computer results on the minimum distance of some binary cyclic codes. IEEE Trans. Inf. Theory **16**(3), 359–360 (1970)
9. Grassl, M.: On the minimum distance of some quadratic residue codes. In: Proceedings of the IEEE International Symposium on Information and Theory, Sorento, Italy, p. 253 (2000)
10. Grassl, M.: New binary codes from a chain of cyclic codes. IEEE Trans. Inf. Theory **47**(3), 1178–1181 (2001)
11. Grassl, M.: Searching for linear codes with large minimum distance. In: Bosma, W., Cannon, J. (eds.) Discovering Mathematics with MAGMA - Reducing the Abstract to the Concrete, pp. 287–313. Springer, Heidelberg (2006)
12. Grassl M.: Code tables: bounds on the parameters of various types of codes. http://www.codetables.de
13. Knuth D.E. (2005) The Art of Computer Programming, Vol. 4: Fascicle 3: Generating All Combinations and Partitions, 3rd edn. Addison-Wesley, ISBN 0 201 85394 9
14. Leon, J.S., Masley, J.M., Pless, V.: Duadic codes. IEEE Trans. Inf. Theory **30**(5), 709–713 (1984)
15. Loeloeian, M., Conan, J.: A [55,16,19] binary Goppa code. IEEE Trans. Inf. Theory **30**, 773 (1984)

16. MacWilliams, F.J., Sloane, N.J.A.: The Theory of Error-Correcting Codes. North-Holland, Amsterdam (1977)
17. Nijenhuis, A., Wilf, H.S.: Combinatorial Algorithms for Computers and Calculators, 2nd edn. Academic Press, London (1978)
18. Payne, W.H., Ives, F.M.: Combination generators. ACM Trans. Math. Softw. **5**(2), 163–172 (1979)
19. Prange E.: Cyclic error-correcting codes in two symbols. Technical report TN-58-103, Air Force Cambridge Research Labs, Bedford, Massachusetts, USA (1957)
20. Proakis, J.G.: Digital Communications, 3rd edn. McGraw-Hill, New York (1995)
21. Promhouse, G., Tavares, S.E.: The minimum distance of all binary cyclic codes of odd lengths from 69 to 99. IEEE Trans. Inf. Theory **24**(4), 438–442 (1978)
22. Schomaker, D., Wirtz, M.: On binary cyclic codes of odd lengths from 101 to 127. IEEE Trans. Inf. Theory **38**(2), 516–518 (1992)
23. Sloane, N.J., Reddy, S.M., Chen, C.L.: New binary codes. IEEE Trans. Inf. Theory **IT–18**, 503–510 (1972)
24. Vardy, A.: The intractability of computing the minimum distance of a code. IEEE Trans. Inf. Theory **43**, 1759–1766 (1997)
25. Zimmermann, K.H.: Integral hecke modules, integral generalized reed-muller codes, and linear codes. Technical report 3–96, Technische Universität Hamburg-Harburg, Hamburg, Germany (1996)

Chapter 6
Lagrange Codes

6.1 Introduction

Joseph Louis Lagrange was a famous eighteenth century Italian mathematician [1] credited with minimum degree polynomial interpolation amongst his many other achievements. Polynomial interpolation may be applied straightforwardly using Galois Fields and provides the basis for an extensive family of error-correcting codes. For a Galois Field $GF(2^m)$, the maximum code length is 2^{m+1}, consisting of 2^m data symbols and 2^m parity symbols. Many of the different types of codes originated by Goppa [3, 4] may be linked to Lagrange codes.

6.2 Lagrange Interpolation

The interpolation polynomial, $p(z)$, is constructed such that the value of the polynomial for each element of $GF(2^m)$ is equal to a data symbol x_i also from $GF(2^m)$. Thus,

$$\begin{bmatrix} p(0) & = & x_0 \\ p(1) & = & x_1 \\ p(\alpha^1) & = & x_2 \\ p(\alpha^2) & = & x_3 \\ \dots & \dots & \dots \\ p(\alpha^{2^m-3}) & = & x_{2^m-2} \\ p(\alpha^{2^m-2}) & = & x_{2^m-1} \end{bmatrix}$$

Using the method of Lagrange, the interpolation polynomial is constructed as a summation of 2^m polynomials, each of degree $2^m - 1$. Thus,

M. Tomlinson et al., *Error-Correction Coding and Decoding*,
Signals and Communication Technology,
DOI 10.1007/978-3-319-51103-0_6

Table 6.1 $GF(8)$ extension field defined by $1+\alpha^1+\alpha^3=0$

$\alpha^0 = 1$
$\alpha^1 = \alpha$
$\alpha^2 = \alpha^2$
$\alpha^3 = 1+\alpha$
$\alpha^4 = \alpha+\alpha^2$
$\alpha^5 = 1+\alpha+\alpha^2$
$\alpha^6 = 1+\alpha^2$

$$p(z) = \sum_{i=0}^{2^m-1} p_i(z) \tag{6.1}$$

where

$$p_i(z) = x_i \frac{z}{\alpha^i} \prod_{j=0, j\neq i}^{j=2^m-2} \frac{z-\alpha^j}{\alpha^i-\alpha^j} \quad \text{for} \quad i \neq 0 \tag{6.2}$$

and

$$p_0(z) = x_0 \prod_{j=0}^{j=2^m-2} \frac{z-\alpha^j}{(-\alpha^j)} \tag{6.3}$$

The idea is that each of the $p_i(z)$ polynomials has a value of zero for z equal to each element of $GF(2^m)$, except for the one element corresponding to i (namely α^{i-1} except for $i=0$).

A simpler form for the polynomials $p_i(z)$ is given by

$$p_i(z) = x_i \frac{(\alpha^i-\alpha^j)}{\alpha^i(\alpha^i-1)} \frac{z(z^{2^m-1}-1)}{z-\alpha^j} \quad \text{for} \quad i \neq 0 \tag{6.4}$$

and

$$p_0(z) = -x_0(z^{2^m-1}-1) \tag{6.5}$$

In an example using $GF(2^3)$, where all the nonzero field elements may express as a power of a primitive root α of the primitive polynomial $1+x+x^3$, modulo $1+x^7$. The nonzero field elements are tabulated in Table 6.1.

All of the 8 polynomials $p_i(z)$ are given below

$$\begin{aligned}
p_0(z) &= x_0(z^7 + 1)\\
p_1(z) &= x_1(z^7 + z^6 + z^5 + z^4 + z^3 + z^2 + z)\\
p_2(z) &= x_2(z^7 + \alpha z^6 + \alpha^2 z^5 + \alpha^3 z^4 + \alpha^4 z^3 + \alpha^5 z^2 + \alpha^6 z)\\
p_3(z) &= x_3(z^7 + \alpha^2 z^6 + \alpha^4 z^5 + \alpha^6 z^4 + \alpha z^3 + \alpha^3 z^2 + \alpha^5 z)\\
p_4(z) &= x_4(z^7 + \alpha^3 z^6 + \alpha^6 z^5 + \alpha^2 z^4 + \alpha^5 z^3 + \alpha z^2 + \alpha^4 z)\\
p_5(z) &= x_5(z^7 + \alpha^4 z^6 + \alpha z^5 + \alpha^5 z^4 + \alpha^2 z^3 + \alpha^6 z^2 + \alpha^3 z)\\
p_6(z) &= x_6(z^7 + \alpha^5 z^6 + \alpha^3 z^5 + \alpha z^4 + \alpha^6 z^3 + \alpha^4 z^2 + \alpha^2 z)\\
p_7(z) &= x_7(z^7 + \alpha^6 z^6 + \alpha^5 z^5 + \alpha^4 z^4 + \alpha^3 z^3 + \alpha^2 z^2 + \alpha z)
\end{aligned}$$

These polynomials are simply summed to produce the Lagrange interpolation polynomial $p(z)$

$$\begin{aligned}
p(z) = {} & z^7(x_0 + x_1 + x_2 + x_3 + x_4 + x_5 + x_6 + x_7)\\
& + z^6(\alpha x_1 + \alpha^2 x_2 + \alpha^3 x_3 + \alpha^4 x_4 + \alpha^5 x_5 + \alpha^6 x_6 + x_7)\\
& + z^5(\alpha^2 x_1 + \alpha^4 x_2 + \alpha^6 x_3 + \alpha x_4 + \alpha^3 x_5 + \alpha^5 x_6 + x_7)\\
& + z^4(\alpha^3 x_1 + \alpha^6 x_2 + \alpha^2 x_3 + \alpha^5 x_4 + \alpha x_5 + \alpha^4 x_6 + x_7)\\
& + z^3(\alpha^4 x_1 + \alpha x_2 + \alpha^5 x_3 + \alpha^2 x_4 + \alpha^6 x_5 + \alpha^3 x_6 + x_7)\\
& + z^2(\alpha^5 x_1 + \alpha^3 x_2 + \alpha x_3 + \alpha^6 x_4 + \alpha^4 x_5 + \alpha^2 x_6 + x_7)\\
& + z(\alpha^6 x_1 + \alpha^5 x_2 + \alpha^4 x_3 + \alpha^3 x_4 + \alpha^2 x_5 + \alpha x_6 + x_7)\\
& + x_0
\end{aligned} \tag{6.6}$$

This can be easily verified by evaluating $p(z)$ for each element of $GR(2^3)$ to produce

$$\begin{aligned}
p(0) &= x_0\\
p(1) &= x_1\\
p(\alpha) &= x_2\\
p(\alpha^2) &= x_3\\
p(\alpha^3) &= x_4\\
p(\alpha^4) &= x_5\\
p(\alpha^5) &= x_6\\
p(\alpha^6) &= x_7
\end{aligned}$$

6.3 Lagrange Error-Correcting Codes

The interpolation polynomial $p(z)$ may be expressed in terms of its coefficients and used as a basis for defining error-correcting codes.

$$p(z) = \sum_{i=0}^{2^m-1} \mu_i z^i \tag{6.7}$$

It is clear that an interpolation equation and a parity check equation are equivalent, and for the 8 identities given by the interpolation polynomial we may define 8 parity check equations:

$$\begin{aligned} x_0 + p(0) &= 0 \\ x_1 + p(1) &= 0 \\ x_2 + p(\alpha) &= 0 \\ x_3 + p(\alpha^2) &= 0 \\ x_4 + p(\alpha^3) &= 0 \\ x_5 + p(\alpha^4) &= 0 \\ x_6 + p(\alpha^5) &= 0 \\ x_7 + p(\alpha^6) &= 0 \end{aligned} \tag{6.8}$$

The 8 parity check equations become

$$\begin{aligned} &x_0 + \mu_0 &&= 0 \\ &x_1 + \mu_1 + \mu_2 + \mu_3 + \mu_4 + \mu_5 + \mu_6 + \mu_7 &&= 0 \\ &x_2 + \alpha\mu_1 + \alpha^2\mu_2 + \alpha^3\mu_3 + \alpha^4\mu_4 + \alpha^5\mu_5 + \alpha^6\mu_6 + \mu_7 &&= 0 \\ &x_3 + \alpha^2\mu_1 + \alpha^4\mu_2 + \alpha^6\mu_3 + \alpha\mu_4 + \alpha^3\mu_5 + \alpha^5\mu_6 + \mu_7 &&= 0 \\ &x_4 + \alpha^3\mu_1 + \alpha^6\mu_2 + \alpha^2\mu_3 + \alpha^5\mu_4 + \alpha\mu_5 + \alpha^4\mu_6 + \mu_7 &&= 0 \\ &x_5 + \alpha^4\mu_1 + \alpha\mu_2 + \alpha^5\mu_3 + \alpha^2\mu_4 + \alpha^6\mu_5 + \alpha^3\mu_6 + \mu_7 &&= 0 \\ &x_6 + \alpha^5\mu_1 + \alpha^3\mu_2 + \alpha\mu_3 + \alpha^6\mu_4 + \alpha^4\mu_5 + \alpha^2\mu_6 + \mu_7 &&= 0 \\ &x_7 + \alpha^6\mu_1 + \alpha^5\mu_2 + \alpha^4\mu_3 + \alpha^3\mu_4 + \alpha^2\mu_5 + \alpha\mu_6 + \mu_7 &&= 0 \end{aligned} \tag{6.9}$$

A number of different codes may be derived from these equations. Using the first 4 equations, apart from the first, and setting x_2 and x_3 equal to 0, the following parity check matrix is obtained, producing a (9, 5) code:

$$\mathbf{H}_{9,5} = \begin{bmatrix} 1 & 0 & 1 & 1 & 1 & 1 & 1 & 1 & 1 \\ 0 & 0 & \alpha & \alpha^2 & \alpha^3 & \alpha^4 & \alpha^5 & \alpha^6 & 1 \\ 0 & 0 & \alpha^2 & \alpha^4 & \alpha^6 & \alpha & \alpha^3 & \alpha^5 & 1 \\ 0 & 1 & \alpha^3 & \alpha^6 & \alpha^2 & \alpha^5 & \alpha & \alpha^4 & 1 \end{bmatrix}$$

Rearranging the order of the columns produces a parity check matrix, $\hat{\mathbf{H}}$ identical to the MDS (9, 5, 5) code based on the doubly extended Reed–Solomon code [7].

$$\hat{\mathbf{H}}_{(9,5,5)} = \begin{bmatrix} 1 & 1 & 1 & 1 & 1 & 1 & 1 & 1 & 0 \\ 1 & \alpha & \alpha^2 & \alpha^3 & \alpha^4 & \alpha^5 & \alpha^6 & 0 & 0 \\ 1 & \alpha^2 & \alpha^4 & \alpha^6 & \alpha & \alpha^3 & \alpha^5 & 0 & 0 \\ 1 & \alpha^3 & \alpha^6 & \alpha^2 & \alpha^5 & \alpha & \alpha^4 & 0 & 1 \end{bmatrix}$$

Correspondingly, we know that the code with parity check matrix, $\mathbf{H}_{9,5}$ derived from the Lagrange interpolating polynomial is MDS and has a minimum Hamming distance of 5. Useful, longer codes can also be obtained. Adding the first row of (6.9) to the second equation of the above example and setting x_0 equal to x_1, a parity check matrix for a (10, 6) code is obtained:

$$\mathbf{H_{10,6}} = \begin{bmatrix} 0 & 1 & 0 & 1 & 1 & 1 & 1 & 1 & 1 & 1 \\ 1 & 1 & 0 & \alpha & \alpha^2 & \alpha^3 & \alpha^4 & \alpha^5 & \alpha^6 & 1 \\ 0 & 0 & 0 & \alpha^2 & \alpha^4 & \alpha^6 & \alpha & \alpha^3 & \alpha^5 & 1 \\ 0 & 0 & 1 & \alpha^3 & \alpha^6 & \alpha^2 & \alpha^5 & \alpha & \alpha^4 & 1 \end{bmatrix}$$

It is straightforward to map any code with $GF(2^m)$ symbols into a binary code by simply mapping each $GF(2^m)$ symbol into a $m \times m$ binary matrix using the $GF(2^m)$ table of field elements. If the codeword coordinate is α^i, the coordinate is replaced with the matrix, where each column is the binary representation of the $GF(2^m)$ symbol:

$$\begin{bmatrix} \alpha^i & \alpha^{i+1} & \alpha^{i+2} & \ldots & \alpha^{i+m-1} \end{bmatrix}$$

As an example for $GF(2^3)$, if the codeword coordinate is α^3, the symbol is replaced with the binary matrix whose columns are the binary values of α^3, α^4, and α^5 using Table 6.1.

$$\begin{bmatrix} 1 & 0 & 1 \\ 1 & 1 & 1 \\ 0 & 1 & 1 \end{bmatrix}$$

In another example the symbol α^0 produces the identity matrix

$$\begin{bmatrix} 1 & 0 & 0 \\ 0 & 1 & 0 \\ 0 & 0 & 1 \end{bmatrix}$$

The (10, 6) GF(8) code above forms a (30, 18) binary code with parity check matrix

$$\mathbf{H_{30,18}} = \left[\begin{array}{cccccccccccccccccccccccccccccc}
0&0&0&1&0&0&0&0&0&1&0&0&1&0&0&1&0&0&1&0&0&1&0&0&1&0&0&1&0&0 \\
0&0&0&0&1&0&0&0&0&0&1&0&0&1&0&0&1&0&0&1&0&0&1&0&0&1&0&0&1&0 \\
0&0&0&0&0&1&0&0&0&0&0&1&0&0&1&0&0&1&0&0&1&0&0&1&0&0&1&0&0&1 \\[1ex]
1&0&0&1&0&0&0&0&0&0&0&1&0&1&0&1&0&1&0&1&1&1&1&1&1&1&0&1&0&0 \\
0&1&0&0&1&0&0&0&0&1&0&1&0&1&1&1&1&1&1&1&0&1&0&0&0&0&1&0&1&0 \\
0&0&1&0&0&1&0&0&0&0&1&0&1&0&1&0&1&1&1&1&1&1&1&0&1&0&0&0&0&1 \\[1ex]
0&0&0&0&0&0&0&0&0&0&1&0&0&1&1&1&1&0&0&0&1&1&0&1&1&1&1&1&0&0 \\
0&0&0&0&0&0&0&0&0&0&1&1&1&1&0&0&0&1&1&0&1&1&1&1&1&0&0&0&1&0 \\
0&0&0&0&0&0&0&0&0&1&0&1&1&1&1&1&0&0&0&1&0&0&1&1&1&1&0&0&0&1 \\[1ex]
0&0&0&0&0&0&1&0&0&1&0&1&1&1&0&0&1&0&1&1&1&0&0&1&0&1&1&1&0&0 \\
0&0&0&0&0&0&0&1&0&1&1&1&0&0&1&0&1&1&1&0&0&1&0&1&1&1&0&0&1&0 \\
0&0&0&0&0&0&0&0&1&0&1&1&1&0&0&1&0&1&1&1&0&0&1&0&1&1&0&0&0&1
\end{array}\right]$$

The minimum Hamming distance of this code has been evaluated and it turns out to be 4. Methods for evaluating the minimum Hamming distance are described in Chap. 5. Consequently, extending the length of the code by one symbol has reduced the d_{min} by 1. The d_{min} may be increased by 2 by adding an overall parity bit to the first two symbols plus an overall parity bit to all bits to produce a (32, 18, 6) code with parity check matrix

$$
\mathbf{H}_{32,18} = \begin{bmatrix}
0\,0\,0\,1\,0\,0\,0\,0\,0\,1\,0\,0\,1\,0\,0\,1\,0\,0\,1\,0\,0\,1\,0\,0\,1\,0\,0\,1\,0\,0\,0\,0 \\
0\,0\,0\,0\,1\,0\,0\,0\,0\,0\,1\,0\,0\,1\,0\,0\,1\,0\,0\,1\,0\,0\,1\,0\,0\,1\,0\,0\,1\,0\,0\,0 \\
0\,0\,0\,0\,0\,1\,0\,0\,0\,0\,0\,1\,0\,0\,1\,0\,0\,1\,0\,0\,1\,0\,0\,1\,0\,0\,1\,0\,0\,1\,0\,0 \\[6pt]
1\,0\,0\,1\,0\,0\,0\,0\,0\,0\,0\,1\,0\,1\,0\,1\,0\,1\,0\,1\,1\,1\,1\,1\,1\,1\,0\,1\,0\,0\,0\,0 \\
0\,1\,0\,0\,1\,0\,0\,0\,0\,1\,0\,1\,0\,1\,1\,1\,1\,1\,1\,1\,0\,1\,0\,0\,0\,0\,1\,0\,1\,0\,0\,0 \\
0\,0\,1\,0\,0\,1\,0\,0\,0\,0\,1\,0\,1\,0\,1\,0\,1\,1\,1\,1\,1\,1\,1\,0\,1\,0\,0\,0\,0\,1\,0\,0 \\
1\,1\,1\,1\,1\,1\,0\,0\,0\,0\,0\,0\,0\,0\,0\,0\,0\,0\,0\,0\,0\,0\,0\,0\,0\,0\,0\,0\,0\,0\,1\,0 \\[6pt]
0\,0\,0\,0\,0\,0\,0\,0\,0\,0\,1\,0\,0\,1\,1\,1\,1\,0\,0\,0\,1\,1\,0\,1\,1\,1\,1\,1\,0\,0\,0\,0 \\
0\,0\,0\,0\,0\,0\,0\,0\,0\,0\,1\,1\,1\,1\,0\,0\,0\,1\,1\,0\,1\,1\,1\,1\,1\,0\,0\,0\,1\,0\,0\,0 \\
0\,0\,0\,0\,0\,0\,0\,0\,0\,1\,0\,1\,1\,1\,1\,1\,0\,0\,0\,1\,0\,0\,1\,1\,1\,1\,0\,0\,0\,1\,0\,0 \\[6pt]
0\,0\,0\,0\,0\,0\,1\,0\,0\,1\,0\,1\,1\,1\,0\,0\,1\,0\,1\,1\,1\,0\,0\,1\,0\,1\,1\,1\,0\,0\,0\,0 \\
0\,0\,0\,0\,0\,0\,0\,1\,0\,1\,1\,1\,0\,0\,1\,0\,1\,1\,1\,0\,0\,1\,0\,1\,1\,1\,0\,0\,1\,0\,0\,0 \\
0\,0\,0\,0\,0\,0\,0\,0\,1\,0\,1\,1\,1\,0\,0\,1\,0\,1\,1\,1\,0\,0\,1\,0\,1\,1\,0\,0\,0\,1\,0\,0 \\
1\,1\,1\,1\,1\,1\,1\,1\,1\,1\,1\,1\,1\,1\,1\,1\,1\,1\,1\,1\,1\,1\,1\,1\,1\,1\,1\,1\,1\,1\,1\,1
\end{bmatrix}
$$

This is a good code as weight spectrum analysis shows that it has the same minimum Hamming distance as the best known (32, 18, 6) code [5]. It is interesting to note that in extending the length of the code beyond the MDS length of 9 symbols for $GF(2^3)$, two *weak* symbols are produced but these are counterbalanced by adding an overall parity bit to these two symbols.

6.4 Error-Correcting Codes Derived from the Lagrange Coefficients

In another approach, we may set some of the equations defining the Lagrange polynomial coefficients to zero, and then use these equations to define parity checks for the code. As an example, using $GF(2^3)$, from Eq. (6.6) we may set coefficients μ_7, μ_6, μ_5, μ_4 and μ_3 equal to zero. The parity check equations become

$$
\begin{array}{llllllll}
x_0 & +x_1 & +x_2 & +x_3 & +x_4 & +x_5 & +x_6 & +x_7 = 0 \\
 & \alpha x_1 & +\alpha^2 x_2 & +\alpha^3 x_3 & +\alpha^4 x_4 & +\alpha^5 x_5 & +\alpha^6 x_6 & +x_7 = 0 \\
 & \alpha^2 x_1 & +\alpha^4 x_2 & +\alpha^6 x_3 & +\alpha x_4 & +\alpha^3 x_5 & +\alpha^5 x_6 & +x_7 = 0 \\
 & \alpha^3 x_1 & +\alpha^6 x_2 & +\alpha^2 x_3 & +\alpha^5 x_4 & +\alpha x_5 & +\alpha^4 x_6 & +x_7 = 0 \\
 & \alpha^4 x_1 & +\alpha x_2 & +\alpha^5 x_3 & +\alpha^2 x_4 & +\alpha^6 x_5 & +\alpha^3 x_6 & +x_7 = 0
\end{array}
\tag{6.10}
$$

and the corresponding parity check matrix is

$$\mathbf{H_{8,3}} = \begin{bmatrix} 1 & 1 & 1 & 1 & 1 & 1 & 1 & 1 \\ 0 & \alpha & \alpha^2 & \alpha^3 & \alpha^4 & \alpha^5 & \alpha^6 & 1 \\ 0 & \alpha^2 & \alpha^4 & \alpha^6 & \alpha & \alpha^3 & \alpha^5 & 1 \\ 0 & \alpha^3 & \alpha^6 & \alpha^2 & \alpha^5 & \alpha & \alpha^4 & 1 \\ 0 & \alpha^4 & \alpha & \alpha^5 & \alpha^2 & \alpha^6 & \alpha^3 & 1 \end{bmatrix} \tag{6.11}$$

As a $GF(2^3)$ code, this code is MDS with a d_{min} of 6 and equivalent to the extended Reed–Solomon code. As a binary code with the following parity check matrix a (24, 9, 8) code is obtained. This is a good code as it has the same minimum Hamming distance as the best known (24, 9, 8) code [5].

$$\mathbf{H_{24,9}} = \begin{bmatrix} 1\,0\,0\,1\,0\,0\,1\,0\,0\,1\,0\,0\,1\,0\,0\,1\,0\,0\,1\,0\,0\,1\,0\,0 \\ 0\,1\,0\,0\,1\,0\,0\,1\,0\,0\,1\,0\,0\,1\,0\,0\,1\,0\,0\,1\,0\,0\,1\,0 \\ 0\,0\,1\,0\,0\,1\,0\,0\,1\,0\,0\,1\,0\,0\,1\,0\,0\,1\,0\,0\,1\,0\,0\,1 \\ \\ 0\,0\,0\,0\,0\,1\,0\,1\,0\,1\,0\,1\,0\,1\,1\,1\,1\,1\,1\,1\,0\,1\,0\,0 \\ 0\,0\,0\,1\,0\,1\,0\,1\,1\,1\,1\,1\,1\,1\,0\,1\,0\,0\,0\,0\,1\,0\,1\,0 \\ 0\,0\,0\,0\,1\,0\,1\,0\,1\,0\,1\,1\,1\,1\,1\,1\,1\,0\,1\,0\,0\,0\,0\,1 \\ \\ 0\,0\,0\,0\,1\,0\,0\,1\,1\,1\,1\,0\,0\,0\,1\,1\,0\,1\,1\,1\,1\,1\,0\,0 \\ 0\,0\,0\,0\,1\,1\,1\,1\,0\,0\,0\,1\,1\,0\,1\,1\,1\,1\,1\,0\,0\,0\,1\,0 \\ 0\,0\,0\,1\,0\,1\,1\,1\,1\,1\,0\,0\,0\,1\,0\,0\,1\,1\,1\,1\,0\,0\,0\,1 \\ \\ 0\,0\,0\,1\,0\,1\,1\,1\,0\,0\,1\,0\,1\,1\,1\,0\,0\,1\,0\,1\,1\,1\,0\,0 \\ 0\,0\,0\,1\,1\,1\,0\,0\,1\,0\,1\,1\,1\,0\,0\,1\,0\,1\,1\,1\,0\,0\,1\,0 \\ 0\,0\,0\,0\,1\,1\,1\,0\,0\,1\,0\,1\,1\,1\,0\,0\,1\,0\,1\,1\,0\,0\,0\,1 \\ \\ 0\,0\,0\,0\,1\,1\,0\,0\,1\,1\,1\,1\,0\,1\,0\,1\,1\,0\,1\,0\,1\,1\,0\,0 \\ 0\,0\,0\,1\,1\,0\,1\,0\,1\,1\,0\,0\,0\,1\,1\,0\,0\,1\,1\,1\,1\,0\,1\,0 \\ 0\,0\,0\,1\,1\,1\,0\,1\,0\,1\,1\,0\,1\,0\,1\,1\,0\,0\,0\,1\,1\,0\,0\,1 \end{bmatrix}$$

6.5 Goppa Codes

So far codes have been constructed using the Lagrange interpolating polynomial in a rather ad hoc manner. Goppa defined a family of codes [3] in terms of the Lagrange interpolating polynomial, where the coordinates of each codeword $\{c_0, c_1, c_2, \ldots c_{2^m-1}\}$ with $\{c_0 = x_0, c_1 = x_1, c_2 = x_2, \ldots c_{2^m-1} = x_{2^m-1}\}$ satisfy the congruence $p(z)$ *modulo* $g(z) = 0$ where $g(z)$ is known as the Goppa polynomial.

Goppa codes have coefficients from $GF(2^m)$ and provided $g(z)$ has no roots which are elements of $GF(2^m)$ (which is straightforward to achieve) the Goppa codes have parameters $(2^m, k, 2^m - k + 1)$. These codes are MDS codes and satisfy the Singleton bound [8]. Goppa codes as binary codes, provided that $g(z)$ has no roots which are elements of $GF(2^m)$ and has no repeated roots, have parameters $(2^m, 2^m - mt, d_{min})$ where $d_{min} \geq 2t + 1$, the Goppa code bound on minimum Hamming distance. Most binary Goppa codes have equality for the bound and t is the number of correctable errors for hard decision, bounded distance decoding. Primitive binary BCH codes have parameters $(2^m - 1, 2^m - mt - 1, d_{min})$, where $d_{min} \geq 2t + 1$ and so binary Goppa codes usually have the advantage over binary BCH codes of an additional information bit for the same minimum Hamming distance. However, depending on the cyclotomic cosets, many cases of BCH codes can be found having either $k > 2^m - mt - 1$ for a given t, or $d_{min} > 2t + 1$, giving BCH codes the advantage for these cases.

For a Goppa polynomial of degree r, there are r parity check equations derived from the congruence $p(z)$ *modulo* $g(z) = 0$. Denoting $g(z)$ by

$$g(z) = g_r z^r + g_{r-1} z^{r-1} + g_{r-2} z^{r-2} + \cdots + g_1 z + g_0 \tag{6.12}$$

$$\sum_{i=0}^{2^m-1} \frac{c_i}{z - \alpha_i} = 0 \quad \text{modulo}\, g(z) \tag{6.13}$$

Since (6.13) is modulo $g(z)$ then $g(z)$ is equivalent to 0, and we can add $g(z)$ to the numerator. Noting that

$$g(z) = (z - \alpha_i) q_i(z) + r_m \tag{6.14}$$

where r_m is the remainder, an element of $GF(2^m)$ after dividing $g(z)$ by $z - \alpha_i$. Dividing each term $z - \alpha_i$ into $1 + g(z)$ produces the following:

$$\frac{g(z) + 1}{z - \alpha_i} = q_i(z) + \frac{r_m + 1}{z - \alpha_i} \tag{6.15}$$

As r_m is a scalar, we may simply pre-weight $g(z)$ by $\frac{1}{r_m}$ so that the remainder cancels with the other numerator term which is 1.

$$\frac{\frac{g(z)}{r_m} + 1}{z - \alpha_i} = \frac{q_i(z)}{r_m} + \frac{\frac{r_m}{r_m} + 1}{z - \alpha_i} = \frac{q_i(z)}{r_m} \tag{6.16}$$

As a result of

$$g(z) = (z - \alpha_i) q_i(z) + r_m$$

when $z = \alpha_i$, $r_m = g(\alpha_i)$

Substituting for r_m in (6.16) produces

$$\frac{\frac{g(z)}{g(\alpha_i)}+1}{z-\alpha_i}=\frac{q_i(z)}{g(\alpha_i)} \tag{6.17}$$

Since $\frac{g(z)}{g(\alpha_i)}$ modulo $g(z)=0$

$$\frac{1}{z-\alpha_i}=\frac{q_i(z)}{g(\alpha_i)} \tag{6.18}$$

The quotient polynomial $q_i(z)$ is a polynomial of degree $r-1$, with coefficients which are a function of α_i and the Goppa polynomial coefficients. Denoting $q_i(z)$ as

$$q_i(z)=q_{i,0}+q_{i,1}z+q_{i,2}z^2+q_{i,3}z^3+\cdots+q_{i,(r-1)}z^{r-1} \tag{6.19}$$

Since the coefficients of each power of z sum to zero, the r parity check equations are given by

$$\sum_{i=0}^{2^m-1}\frac{c_i q_{i,j}}{g(\alpha_i)}=0 \quad \text{for} \quad j=0 \quad \text{to} \quad r-1 \tag{6.20}$$

If the Goppa polynomial has any roots which are elements of $GF(2^m)$, say α_j, then the codeword coordinate c_j has to be permanently set to zero in order to satisfy the parity check equations. Effectively, the code length is shortened by the number of roots of $g(z)$ which are elements of $GF(2^m)$. Usually, the Goppa polynomial is chosen to have distinct roots which are not in $GF(2^m)$.

Consider an example of a Goppa (32, 28, 5) code. There are 4 parity check symbols defined by the 4 parity check equations and the Goppa polynomial has degree 4. Choosing somewhat arbitrarily the polynomial $1+z+z^4$ which has roots in $GF(16)$ but not in $GF(32)$, we determine $q_i(z)$ by dividing by $z-\alpha_i$.

$$q_i(z)=z^3+\alpha_i z^2+\alpha_i^2 z+(1+\alpha_i^3) \tag{6.21}$$

The 4 parity check equations are

$$\sum_{i=0}^{31}\frac{c_i}{g(\alpha_i)}=0 \tag{6.22}$$

$$\sum_{i=0}^{31}\frac{c_i\alpha_i}{g(\alpha_i)}=0 \tag{6.23}$$

$$\sum_{i=0}^{31}\frac{c_i\alpha_i^2}{g(\alpha_i)}=0 \tag{6.24}$$

$$\sum_{i=0}^{31} \frac{c_i(1+\alpha_i^3)}{g(\alpha_i)} = 0 \tag{6.25}$$

Using Table 6.2 to evaluate the different terms for $GF(2^5)$, the parity check matrix is

$$\mathbf{H}_{(32,28,5)} = \begin{bmatrix} 1 & 1 & \alpha^{14} & \alpha^{20} & \alpha^{25} & \dots & \alpha^{10} \\ 0 & 1 & \alpha^{15} & \alpha^{22} & \alpha^{28} & \dots & \alpha^{9} \\ 0 & 1 & \alpha^{16} & \alpha^{24} & 1 & \dots & \alpha^{8} \\ 0 & 1 & \alpha^{17} & \alpha^{26} & \alpha^{3} & \dots & \alpha^{7} \end{bmatrix} \tag{6.26}$$

To implement the Goppa code as a binary code, the symbols in the parity check matrix are replaced with their m-bit binary column representations of each respective $GF(2^m)$ symbol. For the (32, 28, 5) Goppa code above, each of the 4 parity symbols will be represented as a 5 bit symbol from Table 6.2. The parity check matrix will now have 20 rows for the binary code. The minimum Hamming distance of the binary Goppa code is improved from $r+1$ to $2r+1$, namely from 5 to 9. Correspondingly, the binary Goppa code becomes a (32, 12, 9) code with parity check matrix

Table 6.2 $GF(32)$ nonzero extension field elements defined by $1+\alpha^2+\alpha^5=0$

$\alpha^0 = 1$	$\alpha^{16} = 1+\alpha+\alpha^3+\alpha^4$
$\alpha^1 = \alpha$	$\alpha^{17} = 1+\alpha+\alpha^4$
$\alpha^2 = \alpha^2$	$\alpha^{18} = 1+\alpha$
$\alpha^3 = \alpha^3$	$\alpha^{19} = \alpha+\alpha^2$
$\alpha^4 = \alpha^4$	$\alpha^{20} = \alpha^2+\alpha^3$
$\alpha^5 = 1+\alpha^2$	$\alpha^{21} = \alpha^3+\alpha^4$
$\alpha^6 = \alpha+\alpha^3$	$\alpha^{22} = 1+\alpha^2+\alpha^4$
$\alpha^7 = \alpha^2+\alpha^4$	$\alpha^{23} = 1+\alpha+\alpha^2+\alpha^3$
$\alpha^8 = 1+\alpha^2+\alpha^3$	$\alpha^{24} = \alpha+\alpha^2+\alpha^3+\alpha^4$
$\alpha^9 = \alpha+\alpha^3+\alpha^4$	$\alpha^{25} = 1+\alpha^3+\alpha^4$
$\alpha^{10} = 1+\alpha^4$	$\alpha^{26} = 1+\alpha+\alpha^2+\alpha^4$
$\alpha^{11} = 1+\alpha+\alpha^2$	$\alpha^{27} = 1+\alpha+\alpha^3$
$\alpha^{12} = \alpha+\alpha^2+\alpha^3$	$\alpha^{28} = \alpha+\alpha^2+\alpha^4$
$\alpha^{13} = \alpha^2+\alpha^3+\alpha^4$	$\alpha^{29} = 1+\alpha^3$
$\alpha^{14} = 1+\alpha^2+\alpha^3+\alpha^4$	$\alpha^{30} = \alpha+\alpha^4$
$\alpha^{15} = 1+\alpha+\alpha^2+\alpha^3+\alpha^4$	

$$\mathbf{H}_{(32,12,9)} = \begin{bmatrix} 1\,1\,1\,0\,1 \ldots 1 \\ 0\,0\,0\,0\,0 \ldots 0 \\ 0\,0\,1\,1\,0 \ldots 0 \\ 0\,0\,1\,1\,1 \ldots 0 \\ 0\,0\,1\,0\,1 \ldots 1 \\ 0\,1\,1\,1\,0 \ldots 0 \\ 0\,0\,1\,0\,1 \ldots 1 \\ 0\,0\,1\,1\,1 \ldots 0 \\ 0\,0\,1\,0\,0 \ldots 1 \\ 0\,0\,1\,1\,1 \ldots 1 \\ 0\,1\,1\,0\,1 \ldots 1 \\ 0\,0\,1\,1\,0 \ldots 0 \\ 0\,0\,0\,1\,0 \ldots 1 \\ 0\,0\,1\,1\,0 \ldots 1 \\ 0\,0\,1\,1\,0 \ldots 0 \\ 0\,1\,1\,1\,0 \ldots 0 \\ 0\,0\,1\,1\,0 \ldots 0 \\ 0\,0\,0\,1\,0 \ldots 1 \\ 0\,0\,0\,0\,1 \ldots 0 \\ 0\,0\,1\,1\,0 \ldots 1 \end{bmatrix} \tag{6.27}$$

6.6 BCH Codes as Goppa Codes

Surprisingly, the family of Goppa codes includes as a subset the family of BCH codes with codeword coefficients from $GF(2^m)$ and parameters $(2^m - 1, 2^m - 1 - t, t + 1)$. As binary codes, using codeword coefficients $\{0, 1\}$, the BCH codes have parameters $(2^m - 1, 2^m - 1 - mt, 2t + 1)$.

For a nonbinary BCH code to correspond to a Goppa code, the Goppa polynomial, $g(z)$, is given by

$$g(z) = z^t \tag{6.28}$$

There are t parity check equations relating to the codeword coordinates $\{c_0, c_1, c_2, \ldots, c_{2^m-2}\}$ and these are given by

$$\sum_{i=0}^{2^m-2} \frac{c_i}{z - \alpha^i} = 0 \quad \text{modulo}\, z^t \tag{6.29}$$

Dividing 1 by $z - \alpha^i$ starting with α^i produces

$$\frac{1}{z - \alpha^i} = \alpha^{-i} + \alpha^{-2i}z + \alpha^{-3i}z^2 + \alpha^{-3i}z^3 + \cdots + \alpha^{-ti}z^{t-1} + \frac{\alpha^{-(t+1)i}z^t}{z - \alpha^i} \tag{6.30}$$

As $\alpha^{-(t+1)i}z^t$ modulo $z^t = 0$, the t parity check equations are given by

$$\sum_{i=0}^{2^m-2} c_i(\alpha^{-i} + \alpha^{-2i}z + \alpha^{-3i}z^2 + \alpha^{-4i}z^3 + \cdots + \alpha^{-ti}z^{t-1}) = 0 \tag{6.31}$$

Every coefficient of z^0 through to z^{t-1} is equated to zero, producing t parity check equations. The corresponding parity check matrix is

$$\mathbf{H}_{(2^m-1,\,2^m-t,\,t+1)} = \begin{bmatrix} 1 & \alpha^{-1} & \alpha^{-2} & \alpha^{-3} & \alpha^{-4} & \ldots & \alpha^{-(2^m-2)} \\ 1 & \alpha^{-2} & \alpha^{-4} & \alpha^{-6} & \alpha^{-8} & \ldots & \alpha^{-2(2^m-2)} \\ 1 & \alpha^{-3} & \alpha^{-6} & \alpha^{-9} & \alpha^{-12} & \ldots & \alpha^{-3(2^m-2)} \\ \ldots & \ldots & \ldots & \ldots & \ldots & \ldots & \\ 1 & \alpha^{-t} & \alpha^{-2t} & \alpha^{-3t} & \alpha^{-4t} & \ldots & \alpha^{-t(2^m-2)} \end{bmatrix} \tag{6.32}$$

To obtain the binary BCH code, as before, the $GF(2^m)$ symbols are replaced with their m-bit binary column representations for each corresponding $GF(2^m)$ value for each symbol. As a result, only half of the parity check equations are independent and the dependent equations may be deleted. To keep the same number of independent parity check equations as before, the degree of the Goppa polynomial is doubled. The Goppa polynomial is now given by

$$g(z) = z^{2t} \tag{6.33}$$

The parity check matrix for the binary Goppa BCH code is

$$\mathbf{H}_{(2^m-1,\,2^m-mt,\,2t+1)} = \begin{bmatrix} 1 & \alpha^{-1} & \alpha^{-2} & \alpha^{-3} & \alpha^{-4} & \ldots & \alpha^{-(2^m-2)} \\ 1 & \alpha^{-3} & \alpha^{-6} & \alpha^{-9} & \alpha^{-12} & \ldots & \alpha^{-3(2^m-2)} \\ 1 & \alpha^{-5} & \alpha^{-10} & \alpha^{-15} & \alpha^{-20} & \ldots & \alpha^{-5(2^m-2)} \\ & \ldots & \ldots & \ldots & \ldots & \ldots & \\ 1 & \alpha^{-2t-1} & \alpha^{-2(2t-1)} & \alpha^{-3(2t-1)} & \alpha^{-4(2t-1)} & \ldots & \alpha^{-(2t-1)(2^m-2)} \end{bmatrix}$$

For binary codes, any parity check equation may be squared and the resulting parity check equation will still be satisfied. As a consequence, only one parity check equation is needed for each representative from each respective cyclotomic coset. This is clearer with an example.

The cyclotomic cosets of 31, expressed as negative integers for convenience, are as follows

$$\begin{aligned} C_0 &= \{0\} \\ C_{-1} &= \{-1, -2, -4, -8, -16\} \\ C_{-3} &= \{-3, -6, -12, -24, -17\} \\ C_{-5} &= \{-5, -10, -20, -9, -18\} \\ C_{-7} &= \{-7, -14, -28, -25, -19\} \\ C_{-11} &= \{-11, -22, -13, -26, -21\} \\ C_{-15} &= \{-15, -30, -29, -27, -23\} \end{aligned}$$

To construct the $GF(32)$ nonbinary (31, 27) BCH code, the Goppa polynomial is $g(z) = z^4$ and there are 4 parity check equations with parity check matrix:

$$\mathbf{H}_{(31,27,5)} = \begin{bmatrix} 1 & \alpha^{-1} & \alpha^{-2} & \alpha^{-3} & \alpha^{-4} & \dots & \alpha^{-30} \\ 1 & \alpha^{-2} & \alpha^{-4} & \alpha^{-6} & \alpha^{-8} & \dots & \alpha^{-29} \\ 1 & \alpha^{-3} & \alpha^{-6} & \alpha^{-9} & \alpha^{-12} & \dots & \alpha^{-28} \\ 1 & \alpha^{-4} & \alpha^{-8} & \alpha^{-12} & \alpha^{-16} & \dots & \alpha^{-27} \end{bmatrix} \quad (6.34)$$

As a binary code with binary codeword coefficients, the parity check matrix has only two independent rows. To construct the binary parity check matrix, each $GF(32)$ symbol is replaced with its 5-bit column vector so that each parity symbol will require 5 rows of the binary parity check matrix. The code becomes a $(31, 21, 5)$ binary code. The parity check matrix for the binary code after removing the dependent rows is given by

$$\mathbf{H}_{(31,21,5)} = \begin{bmatrix} 1 & \alpha^{-1} & \alpha^{-2} & \alpha^{-3} & \alpha^{-4} & \dots & \alpha^{-30} \\ 1 & \alpha^{-3} & \alpha^{-6} & \alpha^{-9} & \alpha^{-12} & \dots & \alpha^{-28} \end{bmatrix} \quad (6.35)$$

To maintain 4 independent parity check equations for the binary code, the Goppa polynomial is doubled in degree to become $g(z) = z^8$. Replacing each $GF(32)$ symbol with its 5-bit column vector will produce a (31, 11) binary code. The parity check matrix for the binary code is given by:

$$\mathbf{H}_{(31,11,,9)} = \begin{bmatrix} 1 & \alpha^{-1} & \alpha^{-2} & \alpha^{-3} & \alpha^{-4} & \dots & \alpha^{-30} \\ 1 & \alpha^{-3} & \alpha^{-6} & \alpha^{-9} & \alpha^{-12} & \dots & \alpha^{-28} \\ 1 & \alpha^{-5} & \alpha^{-10} & \alpha^{-15} & \alpha^{-20} & \dots & \alpha^{-26} \\ 1 & \alpha^{-7} & \alpha^{-14} & \alpha^{-21} & \alpha^{-28} & \dots & \alpha^{-24} \end{bmatrix} \quad (6.36)$$

Looking at the cyclotomic cosets for 31, it will be noticed that α^{-9} is in the same coset as α^{-5}, and for codewords with binary coefficients, we may use the Goppa polynomial $g(z) = z^{10}$ with the corresponding parity check matrix

$$\mathbf{H}_{(31,11,11)} = \begin{bmatrix} 1 & \alpha^{-1} & \alpha^{-2} & \alpha^{-3} & \alpha^{-4} & \alpha^{-5} & \alpha^{-6} & \dots & \alpha^{-30} \\ 1 & \alpha^{-3} & \alpha^{-6} & \alpha^{-9} & \alpha^{-12} & \alpha^{-15} & \alpha^{-18} & \dots & \alpha^{-28} \\ 1 & \alpha^{-7} & \alpha^{-14} & \alpha^{-21} & \alpha^{-28} & \alpha^{-4} & \alpha^{-11} & \dots & \alpha^{-24} \\ 1 & \alpha^{-9} & \alpha^{-18} & \alpha^{-27} & \alpha^{-5} & \alpha^{-14} & \alpha^{-23} & \dots & \alpha^{-22} \end{bmatrix} \quad (6.37)$$

Alternatively, we may use Goppa polynomial $g(z) = z^8$ with parity check matrix given by (6.36). The result is the same code. From this analysis we can see why the d_{min} of this BCH code is greater by 2 than the BCH code bound because the degree of the Goppa polynomial is 10.

To find other exceptional BCH codes we need to look at the cyclotomic cosets to find similar cases where a row of the parity check matrix is equivalent to a higher degree row. Consider the construction of the $(31, 6, 2t + 1)$ BCH code which will have 5 parity check equations. From the cyclotomic cosets for 31, it will be noticed that α^{-13} is in the same coset as α^{-11}, and so we may use the Goppa polynomial $g(z) = z^{14}$ and obtain a (31, 6, 15) binary BCH code. The BCH bound indicates a minimum Hamming distance of 11. Another example is evident from the cyclotomic cosets of 127 where α^{-17} is in the same coset as α^{-9}. Setting the Goppa polynomial $g(z) = z^{30}$ produces the (127, 71, 19) BCH code, whilst the BCH bound indicates a minimum Hamming distance of 17.

To see the details in the construction of the parity check matrix for the binary BCH code, we will consider the (31, 11, 11) code with parity check matrix given by matrix (6.37). Each $GF(32)$ symbol is replaced with the binary representation given by Table 6.2, as a 5-bit column vector, where α is a primitive root of the polynomial $1 + x^2 + x^5$.

The binary parity check matrix that is obtained is given by matrix (6.38).

$$
\mathbf{H}_{(31,11,11)} = \begin{bmatrix}
1\,0\,1\,0\,1\,1\,1 \ldots 0 \\
0\,1\,0\,1\,1\,1\,0 \ldots 1 \\
0\,0\,0\,1\,0\,1\,0 \ldots 0 \\
0\,0\,1\,0\,1\,0\,1 \ldots 0 \\
0\,1\,0\,1\,0\,1\,1 \ldots 0 \\
\\
1\,0\,1\,1\,0\,1\,0 \ldots 0 \\
0\,1\,0\,0\,1\,1\,0 \ldots 0 \\
0\,1\,0\,1\,1\,0\,1 \ldots 0 \\
0\,0\,1\,0\,0\,1\,1 \ldots 1 \\
0\,1\,1\,1\,0\,1\,1 \ldots 0 \\
\\
1\,1\,0\,1\,1\,0\,0 \ldots 1 \\
0\,1\,0\,1\,1\,1\,1 \ldots 0 \\
0\,1\,0\,0\,1\,0\,0 \ldots 1 \\
0\,0\,1\,1\,0\,1\,0 \ldots 0 \\
0\,1\,1\,1\,0\,0\,0 \ldots 0 \\
\\
1\,1\,0\,0\,1\,1\,1 \ldots 0 \\
0\,0\,0\,0\,1\,1\,0 \ldots 1 \\
0\,1\,1\,0\,1\,0\,1 \ldots 0 \\
0\,0\,1\,0\,0\,0\,1 \ldots 1 \\
0\,1\,1\,1\,1\,1\,0 \ldots 1
\end{bmatrix} \tag{6.38}
$$

Evaluating the minimum Hamming distance of this code confirms that it is 11, an increase of 2 over the BCH bound for the minimum Hamming distance.

6.7 Extended BCH Codes as Goppa Codes

In a short paper in 1971 [4], Goppa showed how a binary Goppa code could be constructed with parameters $(2^m + (m-1)t, 2^m - t, 2t+1)$. Each parity check symbol, m bits long has a Forney concatenation [2], i.e. an overall parity bit on each symbol. In a completely novel approach by Goppa, each parity symbol, apart from 1 bit in each symbol, is external to the code as if these are additional parity symbols. These symbols are also independent of each other extending the length of the code and, importantly, increasing the d_{min} of the code. Sugiyama et al. [9, 10] described a construction technique mixing the standard Goppa code construction with the Goppa external parity check construction. We give below a simpler construction method applicable to BCH codes and to more general Goppa codes.

Consider a binary BCH code constructed as a Goppa code with Goppa polynomial $g(z) = z^{2t}$ but extended by including an additional root α_0, an element of $GF(2^m)$. The Goppa polynomial is now $g(z) = (z^{2t+1} + \alpha_0 z^{2t})$. The parity check equations are given by

$$\sum_{i=0}^{2^m-2} \frac{c_i}{z-\alpha^i} = 0 \quad \text{modulo } g(z) \quad \alpha^i \neq \alpha_0 \tag{6.39}$$

Substituting for r_m and $q(z)$, as in Sect. 6.5

$$\frac{1}{z-\alpha^i} \quad \text{modulo } g(z) = \frac{g(z)+g(\alpha^i)}{g(\alpha^i)(z-\alpha^i)} \tag{6.40}$$

For the extended binary BCH code with Goppa polynomial $g(z) = (z^{2t+1} + \alpha z^{2t})$ the parity check equations are given by

$$\begin{aligned} \sum_{i=1}^{2^m-2} \frac{c_i}{z-\alpha^i} &= \sum_{i=1}^{2^m-2} c_i \left(\frac{z^{2t}}{\alpha^{i2t}(\alpha^i+\alpha_0)} + \frac{z^{2t-1}}{\alpha^{i2t}} + \frac{z^{2t-2}}{\alpha^{i(2t-1)}} + \frac{z^{2t-3}}{\alpha^{i(2t-2)}} + \cdots + \frac{1}{\alpha^i} \right) \\ &= 0 \end{aligned} \tag{6.41}$$

Equating each coefficient of powers of z to zero and using only the independent parity check equations (as it is a binary code) produces $t+1$ independent parity check equations with parity check matrix

$$\mathbf{H}_{(2^m-2,\, 2^m-2-mt-m)} = \begin{bmatrix} \alpha^{-1} & \alpha^{-2} & \alpha^{-3} & \ldots & \alpha^{-(2^m-2)} \\ \alpha^{-3} & \alpha^{-6} & \alpha^{-9} & \ldots & \alpha^{-3(2^m-2)} \\ \alpha^{-5} & \alpha^{-10} & \alpha^{-15} & \ldots & \alpha^{-5(2^m-2)} \\ \ldots & \ldots & \ldots & \ldots & \ldots \\ \alpha^{-2t+1} & \alpha^{-2(2t-1)} & \alpha^{-3(2t-1)} & \ldots & \alpha^{-(2t-1)(2^m-2)} \\ \frac{\alpha^{-2t}}{\alpha+\alpha_0} & \frac{\alpha^{-4t}}{\alpha^2+\alpha_0} & \frac{\alpha^{-6t}}{\alpha^3+\alpha_0} & \ldots & \frac{\alpha^{-2t(2^m-2)}}{\alpha^{2^m-2}+\alpha_0} \end{bmatrix} \tag{6.42}$$

The last row may be simplified by noting that

$$\frac{1+\alpha_0^{-2t}\alpha^{2t}}{(\alpha_0+\alpha)\alpha^{2t}}=\frac{\alpha_0^{-1}}{\alpha^{2t-1}}+\frac{\alpha_0^{-2}}{\alpha^{2t-2}}+\frac{\alpha_0^{-3}}{\alpha^{2t-3}}+\cdots+\frac{\alpha_0^{-2t+1}}{\alpha} \tag{6.43}$$

Rearranging produces

$$\frac{1}{(\alpha_0+\alpha)\alpha^{2t}}=\frac{\alpha_0^{-2t}\alpha^{2t}}{(\alpha_0+\alpha)\alpha^{2t}}+\frac{\alpha_0^{-1}}{\alpha^{2t-1}}+\frac{\alpha_0^{-2}}{\alpha^{2t-2}}+\frac{\alpha_0^{-3}}{\alpha^{2t-3}}+\cdots+\frac{\alpha_0^{-2t+1}}{\alpha} \tag{6.44}$$

and

$$\frac{\alpha^{-2t}}{(\alpha_0+\alpha)}=\frac{\alpha_0^{-2t}}{(\alpha_0+\alpha)}+\frac{\alpha_0^{-1}}{\alpha^{2t-1}}+\frac{\alpha_0^{-2}}{\alpha^{2t-2}}+\frac{\alpha_0^{-3}}{\alpha^{2t-3}}+\cdots+\frac{\alpha_0^{-2t+1}}{\alpha} \tag{6.45}$$

The point here is because of the above equality, the last parity check equation in (6.42) may be replaced with a simpler equation to produce the same Cauchy style parity check given by Goppa in his 1971 paper [4]. The parity check matrix becomes

$$\mathbf{H}_{(2^m-2,\,2^m-2-mt-m)}=\begin{bmatrix} \alpha^{-1} & \alpha^{-2} & \alpha^{-3} & \ldots & \alpha^{-(2^m-2)} \\ \alpha^{-3} & \alpha^{-6} & \alpha^{-9} & \ldots & \alpha^{-3(2^m-2)} \\ \alpha^{-5} & \alpha^{-10} & \alpha^{-15} & \ldots & \alpha^{-5(2^m-2)} \\ \ldots & \ldots & \ldots & \ldots & \ldots \\ \alpha^{-2t+1} & \alpha^{-2(2t-1)} & \alpha^{-3(2t-1)} & \ldots & \alpha^{-(2t-1)(2^m-2)} \\ \frac{1}{\alpha+\alpha_0} & \frac{1}{\alpha^2+\alpha_0} & \frac{1}{\alpha^3+\alpha_0} & \ldots & \frac{1}{\alpha^{2^m-2}+\alpha_0} \end{bmatrix} \tag{6.46}$$

The justification for this is that from (6.45), the last row of (6.42) is equal to a scalar weighted linear combination of the rows of the parity check matrix (6.46), so that these rows will produce the same code as the parity check matrix (6.42). By induction, other roots of $GF(2^m)$ may be used to produce similar parity check equations to increase the distance of the code producing parity check matrices of the form:

$$\mathbf{H}=\begin{bmatrix} \alpha^{-1} & \alpha^{-2} & \alpha^{-3} & \alpha^{-4} & \ldots & \alpha^{-(2^m-2)} \\ \alpha^{-3} & \alpha^{-6} & \alpha^{-9} & \alpha^{-12} & \ldots & \alpha^{-3(2^m-2)} \\ \alpha^{-5} & \alpha^{-10} & \alpha^{-15} & \alpha^{-20} & \ldots & \alpha^{-5(2^m-2)} \\ \ldots & \ldots & \ldots & \ldots & \ldots & \\ \alpha^{-2t+1} & \alpha^{-2(2t-1)} & \alpha^{-3(2t-1)} & \alpha^{-4(2t-1)} & \ldots & \alpha^{-(2t-1)(2^m-2)} \\ \frac{1}{\alpha+\alpha_0} & \frac{1}{\alpha^2+\alpha_0} & \frac{1}{\alpha^3+\alpha_0} & \frac{1}{\alpha^4+\alpha_0} & \ldots & \frac{1}{\alpha^{2^m-2}+\alpha_0} \\ \frac{1}{\alpha+\alpha_1} & \frac{1}{\alpha^2+\alpha_1} & \frac{1}{\alpha^3+\alpha_1} & \frac{1}{\alpha^4+\alpha_1} & \ldots & \frac{1}{\alpha^{2^m-2}+\alpha_1} \\ \ldots & \ldots & \ldots & \ldots & \ldots & \\ \frac{1}{\alpha+\alpha_{s_0-1}} & \frac{1}{\alpha^2+\alpha_{s_0-1}} & \frac{1}{\alpha^3+\alpha_{s_0-1}} & \frac{1}{\alpha^4+\alpha_{s_0-1}} & \ldots & \frac{1}{\alpha^{2^m-2}+\alpha_{s_0-1}} \end{bmatrix} \tag{6.47}$$

The parity symbols given by the last s_0 rows of this matrix are in the Cauchy matrix style [7] and will necessarily reduce the length of the code for each root of the

Goppa polynomial which is an element of $GF(2^m)$. However, Goppa was the first to show [4] that the parity symbols may be optionally placed external to the code, without decreasing the length of the code. For binary codes the length of the code increases as will be shown below. Accordingly, with external parity symbols, the parity check matrix becomes

$$\mathbf{H} = \begin{bmatrix} \alpha^{-1} & \alpha^{-2} & \alpha^{-3} & \alpha^{-4} & \dots & \alpha^{-(2^m-2)} & 0 & 0 & 0 & 0 \\ \alpha^{-3} & \alpha^{-6} & \alpha^{-9} & \alpha^{-12} & \dots & \alpha^{-3(2^m-2)} & 0 & 0 & 0 & 0 \\ \alpha^{-5} & \alpha^{-10} & \alpha^{-15} & \alpha^{-20} & \dots & \alpha^{-5(2^m-2)} & 0 & 0 & 0 & 0 \\ \dots & \dots & \dots & \dots & \dots & \dots & \dots & \dots & \dots & \dots \\ \alpha^{-2t+1} & \alpha^{-2(2t-1)} & \alpha^{-3(2t-1)} & \alpha^{-4(2t-1)} & \dots & \alpha^{-(2t-1)(2^m-2)} & 0 & 0 & 0 & 0 \\ \frac{1}{\alpha+\alpha_0} & \frac{1}{\alpha^2+\alpha_0} & \frac{1}{\alpha^3+\alpha_0} & \frac{1}{\alpha^4+\alpha_0} & \dots & \frac{1}{\alpha^{2^m-2}+\alpha_0} & 1 & 0 & 0 & 0 \\ \frac{1}{\alpha+\alpha_1} & \frac{1}{\alpha^2+\alpha_1} & \frac{1}{\alpha^3+\alpha_1} & \frac{1}{\alpha^4+\alpha_1} & \dots & \frac{1}{\alpha^{2^m-2}+\alpha_1} & 0 & 1 & 0 & 0 \\ \dots & \dots & \dots & \dots & \dots & \dots & \dots & \dots & \dots & \dots \\ \frac{1}{\alpha+\alpha_{s_0-1}} & \frac{1}{\alpha^2+\alpha_{s_0-1}} & \frac{1}{\alpha^3+\alpha_{s_0-1}} & \frac{1}{\alpha^4+\alpha_{s_0-1}} & \dots & \frac{1}{\alpha^{2^m-2}+\alpha_{s_0-1}} & 0 & 0 & 0 & 1 \end{bmatrix} \tag{6.48}$$

As an example of the procedure, consider the (31, 11, 11) binary BCH code described in Sect. 6.6. We shall add one external parity symbol to this code according to the parity check matrix in (6.48) and eventually produce a (36, 10, 13) binary BCH code. Arbitrarily, we shall choose $\alpha_0 = 1$. This means that the first column of the parity check matrix for the (31, 11, 11) code given in (6.38) is deleted and there is one additional parity check row. The parity check matrix for this (35, 10, 12) extended BCH code is given below. Note we will add later an additional parity bit in a Forney concatenation of the external parity symbol to produce the (36, 10, 13) code as a last step.

$$\mathbf{H}_{(35,10,12)} = \begin{bmatrix} \alpha^{-1} & \alpha^{-2} & \alpha^{-3} & \alpha^{-4} & \alpha^{-5} & \alpha^{-6} & \dots & \alpha^{-30} & 0 \\ \alpha^{-3} & \alpha^{-6} & \alpha^{-9} & \alpha^{-12} & \alpha^{-15} & \alpha^{-18} & \dots & \alpha^{-28} & 0 \\ \alpha^{-5} & \alpha^{-10} & \alpha^{-15} & \alpha^{-20} & \alpha^{-25} & \alpha^{-30} & \dots & \alpha^{-26} & 0 \\ \alpha^{-9} & \alpha^{-18} & \alpha^{-27} & \alpha^{-5} & \alpha^{-14} & \alpha^{-23} & \dots & \alpha^{-22} & 0 \\ \frac{1}{\alpha+1} & \frac{1}{\alpha^2+1} & \frac{1}{\alpha^3+1} & \frac{1}{\alpha^4+1} & \frac{1}{\alpha^5+1} & \frac{1}{\alpha^6+1} & \dots & \frac{1}{\alpha^{29}+1} & 1 \end{bmatrix} \tag{6.49}$$

Evaluating the last row by carrying out the additions, and inversions, referring to the table of $GF(32)$ symbols in Table 6.2 produces the resulting parity check matrix

$$\mathbf{H}_{(35,10,12)} = \begin{bmatrix} \alpha^{-1} & \alpha^{-2} & \alpha^{-3} & \alpha^{-4} & \alpha^{-5} & \alpha^{-6} & \dots & \alpha^{-30} & 0 \\ \alpha^{-3} & \alpha^{-6} & \alpha^{-9} & \alpha^{-12} & \alpha^{-15} & \alpha^{-18} & \dots & \alpha^{-28} & 0 \\ \alpha^{-5} & \alpha^{-10} & \alpha^{-15} & \alpha^{-20} & \alpha^{-25} & \alpha^{-30} & \dots & \alpha^{-26} & 0 \\ \alpha^{-9} & \alpha^{-18} & \alpha^{-27} & \alpha^{-5} & \alpha^{-14} & \alpha^{-23} & \dots & \alpha^{-22} & 0 \\ \alpha^{-13} & \alpha^{-26} & \alpha^{-2} & \alpha^{-21} & \alpha^{-29} & \alpha^{-4} & \dots & \alpha^{-14} & 1 \end{bmatrix} \tag{6.50}$$

The next step is to determine the binary parity check matrix for the code by replacing each $GF(32)$ symbol by its corresponding 5-bit representation using Table 6.2, but as a 5-bit column vector. Also we will add an additional parity check row to implement the Forney concatenation of the external parity symbol. The resulting binary parity check matrix in (6.51) is obtained. Evaluating the minimum Hamming distance of this code using one of the methods described in Chap. 5 verifies that it is indeed 13.

Adding the external parity symbol has increased the minimum Hamming distance by 2, but at the cost of one data symbol. Instead of choosing $\alpha_0 = 1$, a good idea is to choose $\alpha_0 = 0$, since 0 is a multiple root of the Goppa polynomial $g(z) = z^{10}$ which caused the BCH code to be shortened from length 2^m to $2^m - 1$ in the first place. (The length of a Goppa code with Goppa polynomial $g(z)$ having no roots in $GF(2^m)$ is 2^m). The resulting parity check matrix is given in (6.52).

$$
\mathbf{H}_{(\mathbf{36},\,\mathbf{10},\,\mathbf{13})} = \begin{bmatrix}
0\,1\,0\,1\,1\,1 \ldots 0\,0\,0\,0\,0\,0\,0 \\
1\,0\,1\,1\,1\,0 \ldots 1\,0\,0\,0\,0\,0\,0 \\
0\,0\,1\,0\,1\,0 \ldots 0\,0\,0\,0\,0\,0\,0 \\
0\,1\,0\,1\,0\,1 \ldots 0\,0\,0\,0\,0\,0\,0 \\
1\,0\,1\,0\,1\,1 \ldots 0\,0\,0\,0\,0\,0\,0 \\
\\
0\,1\,1\,0\,1\,0 \ldots 0\,0\,0\,0\,0\,0\,0 \\
1\,0\,0\,1\,1\,0 \ldots 0\,0\,0\,0\,0\,0\,0 \\
1\,0\,1\,1\,0\,1 \ldots 0\,0\,0\,0\,0\,0\,0 \\
0\,1\,0\,0\,1\,1 \ldots 1\,0\,0\,0\,0\,0\,0 \\
1\,1\,1\,0\,1\,1 \ldots 0\,0\,0\,0\,0\,0\,0 \\
\\
1\,0\,1\,1\,0\,0 \ldots 1\,0\,0\,0\,0\,0\,0 \\
1\,0\,1\,1\,1\,1 \ldots 0\,0\,0\,0\,0\,0\,0 \\
1\,0\,0\,1\,0\,0 \ldots 1\,0\,0\,0\,0\,0\,0 \\
0\,1\,1\,0\,1\,0 \ldots 0\,0\,0\,0\,0\,0\,0 \\
1\,1\,1\,0\,0\,0 \ldots 0\,0\,0\,0\,0\,0\,0 \\
\\
1\,0\,0\,1\,1\,1 \ldots 0\,0\,0\,0\,0\,0\,0 \\
0\,0\,0\,1\,1\,0 \ldots 1\,0\,0\,0\,0\,0\,0 \\
1\,1\,0\,1\,0\,1 \ldots 0\,0\,0\,0\,0\,0\,0 \\
0\,1\,0\,0\,0\,1 \ldots 1\,0\,0\,0\,0\,0\,0 \\
1\,1\,1\,1\,1\,0 \ldots 1\,0\,0\,0\,0\,0\,0 \\
\\
1\,1\,1\,1\,0\,1 \ldots 1\,1\,0\,0\,0\,0\,0 \\
1\,0\,0\,0\,0\,1 \ldots 1\,0\,1\,0\,0\,0\,0 \\
0\,1\,0\,0\,1\,0 \ldots 0\,0\,0\,1\,0\,0\,0 \\
0\,0\,1\,0\,0\,1 \ldots 0\,0\,0\,0\,1\,0\,0 \\
0\,0\,0\,1\,0\,0 \ldots 1\,0\,0\,0\,0\,1\,0 \\
0\,0\,0\,0\,0\,0 \ldots 0\,1\,1\,1\,1\,1\,1
\end{bmatrix} \tag{6.51}
$$

$$\mathbf{H}_{(36,11)} = \begin{bmatrix} 1 & \alpha^{-1} & \alpha^{-2} & \alpha^{-3} & \alpha^{-4} & \alpha^{-5} & \alpha^{-6} & \ldots & \alpha^{-30} & 0 \\ 1 & \alpha^{-3} & \alpha^{-6} & \alpha^{-9} & \alpha^{-12} & \alpha^{-15} & \alpha^{-18} & \ldots & \alpha^{-28} & 0 \\ 1 & \alpha^{-5} & \alpha^{-10} & \alpha^{-15} & \alpha^{-20} & \alpha^{-25} & \alpha^{-30} & \ldots & \alpha^{-26} & 0 \\ 1 & \alpha^{-9} & \alpha^{-18} & \alpha^{-27} & \alpha^{-5} & \alpha^{-14} & \alpha^{-23} & \ldots & \alpha^{-22} & 0 \\ 1 & \alpha^{-1} & \alpha^{-2} & \alpha^{-3} & \alpha^{-4} & \alpha^{-5} & \alpha^{-6} & \ldots & \alpha^{-30} & 1 \end{bmatrix} \tag{6.52}$$

The problem with this is that the minimum Hamming distance is still 11 because the last row of the parity check matrix is the same as the first row, apart from the external parity symbol because 0 is a root of the Goppa polynomial. The solution is to increase the degree of the Goppa polynomial but still retain the external parity symbol. Referring to the cyclotomic cosets of 31, see (6.35), we should use $g(z) = z^{12}$ to produce the parity check matrix

$$\mathbf{H}_{(36,11)} = \begin{bmatrix} 1 & \alpha^{-1} & \alpha^{-2} & \alpha^{-3} & \alpha^{-4} & \alpha^{-5} & \alpha^{-6} & \ldots & \alpha^{-30} & 0 \\ 1 & \alpha^{-3} & \alpha^{-6} & \alpha^{-9} & \alpha^{-12} & \alpha^{-15} & \alpha^{-18} & \ldots & \alpha^{-28} & 0 \\ 1 & \alpha^{-5} & \alpha^{-10} & \alpha^{-15} & \alpha^{-20} & \alpha^{-25} & \alpha^{-30} & \ldots & \alpha^{-26} & 0 \\ 1 & \alpha^{-9} & \alpha^{-18} & \alpha^{-27} & \alpha^{-5} & \alpha^{-14} & \alpha^{-23} & \ldots & \alpha^{-22} & 0 \\ 1 & \alpha^{-11} & \alpha^{-22} & \alpha^{-2} & \alpha^{-13} & \alpha^{-24} & \alpha^{-4} & \ldots & \alpha^{-20} & 1 \end{bmatrix} \tag{6.53}$$

As before, the next step is to determine the binary parity check matrix for the code from this matrix by replacing each $GF(32)$ symbol by its corresponding 5 bit representation using Table 6.2 as a 5 bit column vector. Also we will add an additional parity check row to implement the Forney concatenation of the external parity symbol. The resulting binary parity check matrix is obtained

$$\begin{bmatrix} 1\,0\,1\,0\,1\,1\,1 \ldots 0\,0\,0\,0\,0\,0\,0 \\ 0\,1\,0\,1\,1\,1\,0 \ldots 1\,0\,0\,0\,0\,0\,0 \\ 0\,0\,0\,1\,0\,1\,0 \ldots 0\,0\,0\,0\,0\,0\,0 \\ 0\,0\,1\,0\,1\,0\,1 \ldots 0\,0\,0\,0\,0\,0\,0 \\ 0\,1\,0\,1\,0\,1\,1 \ldots 0\,0\,0\,0\,0\,0\,0 \\ \\ 1\,0\,1\,1\,0\,1\,0 \ldots 0\,0\,0\,0\,0\,0\,0 \\ 0\,1\,0\,0\,1\,1\,0 \ldots 0\,0\,0\,0\,0\,0\,0 \\ 0\,1\,0\,1\,1\,0\,1 \ldots 0\,0\,0\,0\,0\,0\,0 \\ 0\,0\,1\,0\,0\,1\,1 \ldots 1\,0\,0\,0\,0\,0\,0 \\ 0\,1\,1\,1\,0\,1\,1 \ldots 0\,0\,0\,0\,0\,0\,0 \\ \\ 1\,1\,0\,1\,1\,0\,0 \ldots 1\,0\,0\,0\,0\,0\,0 \\ 0\,1\,0\,1\,1\,1\,1 \ldots 0\,0\,0\,0\,0\,0\,0 \\ 0\,1\,0\,0\,1\,0\,0 \ldots 1\,0\,0\,0\,0\,0\,0 \\ 0\,0\,1\,1\,0\,1\,0 \ldots 0\,0\,0\,0\,0\,0\,0 \\ 0\,1\,1\,1\,0\,0\,0 \ldots 0\,0\,0\,0\,0\,0\,0 \end{bmatrix}$$

$$\mathbf{H}_{(37,11,13)} = \begin{bmatrix} 1\,1\,0\,0\,1\,1\,1 \ldots 0\,0\,0\,0\,0\,0\,0 \\ 0\,0\,0\,0\,1\,1\,0 \ldots 1\,0\,0\,0\,0\,0\,0 \\ 0\,1\,1\,0\,1\,0\,1 \ldots 0\,0\,0\,0\,0\,0\,0 \\ 0\,0\,1\,0\,0\,0\,1 \ldots 1\,0\,0\,0\,0\,0\,0 \\ 0\,1\,1\,1\,1\,1\,0 \ldots 1\,0\,0\,0\,0\,0\,0 \\ \\ 1\,0\,0\,1\,1\,0\,1 \ldots 1\,1\,0\,0\,0\,0\,0 \\ 0\,0\,1\,0\,1\,0\,1 \ldots 1\,0\,1\,0\,0\,0\,0 \\ 0\,1\,0\,0\,0\,1\,0 \ldots 1\,0\,0\,1\,0\,0\,0 \\ 0\,1\,1\,1\,0\,0\,1 \ldots 0\,0\,0\,0\,1\,0\,0 \\ 0\,0\,1\,0\,0\,1\,0 \ldots 0\,0\,0\,0\,0\,1\,0 \\ 0\,0\,0\,0\,0\,0\,0 \ldots 0\,1\,1\,1\,1\,1\,1 \end{bmatrix} \tag{6.54}$$

Weight spectrum analysis of this code confirms that the d_{min} is indeed 13. One or more Cauchy style parity check equations may be added to this code to increase the d_{min} of the code. For example, with one more parity check equation again with the choice of $\alpha_0 = 1$, the parity check matrix for the (42,10) code is

$$\mathbf{H}_{(42,10)} = \begin{bmatrix} \alpha^{-1} & \alpha^{-2} & \alpha^{-3} & \alpha^{-4} & \alpha^{-5} & \alpha^{-6} & \ldots \alpha^{-30} & 0\,0 \\ \alpha^{-3} & \alpha^{-6} & \alpha^{-9} & \alpha^{-12} & \alpha^{-15} & \alpha^{-18} & \ldots \alpha^{-28} & 0\,0 \\ \alpha^{-5} & \alpha^{-10} & \alpha^{-15} & \alpha^{-20} & \alpha^{-25} & \alpha^{-30} & \ldots \alpha^{-26} & 0\,0 \\ \alpha^{-9} & \alpha^{-18} & \alpha^{-27} & \alpha^{-5} & \alpha^{-14} & \alpha^{-23} & \ldots \alpha^{-22} & 0\,0 \\ \alpha^{-11} & \alpha^{-22} & \alpha^{-2} & \alpha^{-13} & \alpha^{-24} & \alpha^{-4} & \ldots \alpha^{-20} & 1\,0 \\ \alpha^{-18} & \alpha^{-5} & \alpha^{-29} & \alpha^{-10} & \alpha^{-2} & \alpha^{-27} & \ldots \alpha^{-17} & 0\,1 \end{bmatrix} \tag{6.55}$$

Replacing each $GF(32)$ symbol by its corresponding 5 bit representation using Table 6.2 as a 5-bit column vector and adding an additional parity check row to each external parity symbol produces the binary parity check matrix for the (42, 10, 15) code.

$$\begin{bmatrix} 0\,1\,0\,1\,1\,1 \ldots 0\,0\,0\,0\,0\,0\,0\,0\,0\,0\,0\,0\,0\,0 \\ 1\,0\,1\,1\,1\,0 \ldots 1\,0\,0\,0\,0\,0\,0\,0\,0\,0\,0\,0\,0\,0 \\ 0\,0\,1\,0\,1\,0 \ldots 0\,0\,0\,0\,0\,0\,0\,0\,0\,0\,0\,0\,0\,0 \\ 0\,1\,0\,1\,0\,1 \ldots 0\,0\,0\,0\,0\,0\,0\,0\,0\,0\,0\,0\,0\,0 \\ 1\,0\,1\,0\,1\,1 \ldots 0\,0\,0\,0\,0\,0\,0\,0\,0\,0\,0\,0\,0\,0 \\ \\ 0\,1\,1\,0\,1\,0 \ldots 0\,0\,0\,0\,0\,0\,0\,0\,0\,0\,0\,0\,0\,0 \\ 1\,0\,0\,1\,1\,0 \ldots 0\,0\,0\,0\,0\,0\,0\,0\,0\,0\,0\,0\,0\,0 \\ 1\,0\,1\,1\,0\,1 \ldots 0\,0\,0\,0\,0\,0\,0\,0\,0\,0\,0\,0\,0\,0 \\ 0\,1\,0\,0\,1\,1 \ldots 1\,0\,0\,0\,0\,0\,0\,0\,0\,0\,0\,0\,0\,0 \\ 1\,1\,1\,0\,1\,1 \ldots 0\,0\,0\,0\,0\,0\,0\,0\,0\,0\,0\,0\,0\,0 \\ \\ 1\,0\,1\,1\,0\,0 \ldots 1\,0\,0\,0\,0\,0\,0\,0\,0\,0\,0\,0\,0\,0 \\ 1\,0\,1\,1\,1\,1 \ldots 0\,0\,0\,0\,0\,0\,0\,0\,0\,0\,0\,0\,0\,0 \\ 1\,0\,0\,1\,0\,0 \ldots 1\,0\,0\,0\,0\,0\,0\,0\,0\,0\,0\,0\,0\,0 \\ 0\,1\,1\,0\,1\,0 \ldots 0\,0\,0\,0\,0\,0\,0\,0\,0\,0\,0\,0\,0\,0 \\ 1\,1\,1\,0\,0\,0 \ldots 0\,0\,0\,0\,0\,0\,0\,0\,0\,0\,0\,0\,0\,0 \end{bmatrix}$$

$$\mathbf{H}_{(42,10,15)} = \begin{bmatrix} 1\,0\,0\,1\,1\,1 \ldots 0\,0\,0\,0\,0\,0\,0\,0\,0\,0\,0\,0\,0\,0 \\ 0\,0\,0\,1\,1\,0 \ldots 1\,0\,0\,0\,0\,0\,0\,0\,0\,0\,0\,0\,0 \\ 1\,1\,0\,1\,0\,1 \ldots 0\,0\,0\,0\,0\,0\,0\,0\,0\,0\,0\,0\,0\,0 \\ 0\,1\,0\,0\,0\,1 \ldots 1\,0\,0\,0\,0\,0\,0\,0\,0\,0\,0\,0\,0 \\ 1\,1\,1\,1\,1\,0 \ldots 1\,0\,0\,0\,0\,0\,0\,0\,0\,0\,0\,0\,0 \\ \\ 0\,0\,1\,1\,0\,1 \ldots 1\,1\,0\,0\,0\,0\,0\,0\,0\,0\,0\,0\,0 \\ 0\,1\,0\,1\,0\,1 \ldots 1\,0\,1\,0\,0\,0\,0\,0\,0\,0\,0\,0\,0 \\ 1\,0\,0\,0\,1\,0 \ldots 1\,0\,0\,1\,0\,0\,0\,0\,0\,0\,0\,0\,0 \\ 1\,1\,1\,0\,0\,1 \ldots 0\,0\,0\,0\,1\,0\,0\,0\,0\,0\,0\,0\,0 \\ 0\,1\,0\,0\,1\,0 \ldots 0\,0\,0\,0\,0\,1\,0\,0\,0\,0\,0\,0\,0 \\ 0\,0\,0\,0\,0\,0 \ldots 0\,1\,1\,1\,1\,1\,1\,0\,0\,0\,0\,0\,0 \\ \\ 0\,1\,0\,0\,1\,0 \ldots 1\,0\,0\,0\,0\,0\,0\,1\,0\,0\,0\,0\,0 \\ 0\,1\,0\,0\,0\,0 \ldots 0\,0\,0\,0\,0\,0\,0\,0\,1\,0\,0\,0\,0 \\ 1\,1\,1\,0\,0\,0 \ldots 1\,0\,0\,0\,0\,0\,0\,0\,0\,1\,0\,0\,0 \\ 1\,0\,0\,1\,1\,0 \ldots 1\,0\,0\,0\,0\,0\,0\,0\,0\,0\,1\,0\,0 \\ 1\,1\,0\,1\,0\,1 \ldots 1\,0\,0\,0\,0\,0\,0\,0\,0\,0\,0\,1\,0 \\ 0\,0\,0\,0\,0\,0 \ldots 0\,0\,0\,0\,0\,0\,0\,1\,1\,1\,1\,1\,1 \end{bmatrix} \tag{6.56}$$

Weight spectrum analysis of this code confirms that the d_{min} is indeed 15. In this construction the information bit coordinate corresponding to $\alpha_0 = 1$ is deleted, reducing the dimension of the code by 1. This is conventional practice when the Goppa polynomial $g(z)$ contains a root that is in $GF(2^m)$. However, on reflection, this is not essential. Certainly, in the parity check symbol equations of the constructed code, there will be one parity check equation where the coordinate is missing, but additional parity check equations may be used to compensate for the missing coordinate(s).

Consider the (42, 10) code above, given by parity check matrix (6.55) without the deletion of the first coordinate. The parity check matrix for the (42, 11) code becomes

$$\mathbf{H}_{(42,11)} = \begin{bmatrix} 1 & \alpha^{-1} & \alpha^{-2} & \alpha^{-3} & \alpha^{-4} & \alpha^{-5} & \alpha^{-6} & \ldots & \alpha^{-30} & 0 & 0 \\ 1 & \alpha^{-3} & \alpha^{-6} & \alpha^{-9} & \alpha^{-12} & \alpha^{-15} & \alpha^{-18} & \ldots & \alpha^{-28} & 0 & 0 \\ 1 & \alpha^{-5} & \alpha^{-10} & \alpha^{-15} & \alpha^{-20} & \alpha^{-25} & \alpha^{-30} & \ldots & \alpha^{-26} & 0 & 0 \\ 1 & \alpha^{-9} & \alpha^{-18} & \alpha^{-27} & \alpha^{-5} & \alpha^{-14} & \alpha^{-23} & \ldots & \alpha^{-22} & 0 & 0 \\ 1 & \alpha^{-11} & \alpha^{-22} & \alpha^{-2} & \alpha^{-13} & \alpha^{-24} & \alpha^{-4} & \ldots & \alpha^{-20} & 1 & 0 \\ 0 & \alpha^{-18} & \alpha^{-5} & \alpha^{-29} & \alpha^{-10} & \alpha^{-2} & \alpha^{-27} & \ldots & \alpha^{-17} & 0 & 1 \end{bmatrix} \tag{6.57}$$

It will be noticed that the first coordinate is not in the last parity check equation. Constructing the binary code as before by replacing each $GF(32)$ symbol by its corresponding 5-bit representation using Table 6.2 as a 5-bit column vector and adding an additional parity check row to each external parity symbol produces a (42, 11, 13) binary code. There is no improvement in the d_{min} of the (42, 11, 13) binary code compared to the (37, 11, 13) binary code despite the 5 additional parity bits. However, weight spectrum analysis of the (42, 11, 13) binary code shows that there is only 1 codeword of weight 13 and only 3 codewords of weight 14. All of these low weight codewords contain the first coordinate which is not surprising. Two more parity check equations containing the first coordinate need to be added to the parity check matrix to compensate for the coordinate not being in the last equation of the parity check symbol matrix (6.57).

It turns out that the coordinate in question can always be inserted into the overall parity check equation to each external parity symbol without any loss, so that only one additional parity check equation is required for each root of $g(z)$ that is in $GF(2^m)$.

This produces the following binary parity check matrix for the (43, 11, 15) code.

$$\mathbf{H}_{(43,11,15)} = \begin{bmatrix} 1\,0\,1\,0\,1\,1\,1 \ldots 0\,0\,0\,0\,0\,0\,0\,0\,0\,0\,0\,0\,0\,0 \\ 0\,1\,0\,1\,1\,1\,0 \ldots 1\,0\,0\,0\,0\,0\,0\,0\,0\,0\,0\,0\,0\,0 \\ 0\,0\,0\,1\,0\,1\,0 \ldots 0\,0\,0\,0\,0\,0\,0\,0\,0\,0\,0\,0\,0\,0 \\ 0\,0\,1\,0\,1\,0\,1 \ldots 0\,0\,0\,0\,0\,0\,0\,0\,0\,0\,0\,0\,0\,0 \\ 0\,1\,0\,1\,0\,1\,1 \ldots 0\,0\,0\,0\,0\,0\,0\,0\,0\,0\,0\,0\,0\,0 \\ \\ 1\,0\,1\,1\,0\,1\,0 \ldots 0\,0\,0\,0\,0\,0\,0\,0\,0\,0\,0\,0\,0\,0 \\ 0\,1\,0\,0\,1\,1\,0 \ldots 0\,0\,0\,0\,0\,0\,0\,0\,0\,0\,0\,0\,0\,0 \\ 0\,1\,0\,1\,1\,0\,1 \ldots 0\,0\,0\,0\,0\,0\,0\,0\,0\,0\,0\,0\,0\,0 \\ 0\,0\,1\,0\,0\,1\,1 \ldots 1\,0\,0\,0\,0\,0\,0\,0\,0\,0\,0\,0\,0\,0 \\ 0\,1\,1\,1\,0\,1\,1 \ldots 0\,0\,0\,0\,0\,0\,0\,0\,0\,0\,0\,0\,0\,0 \\ \\ 1\,1\,0\,1\,1\,0\,0 \ldots 1\,0\,0\,0\,0\,0\,0\,0\,0\,0\,0\,0\,0\,0 \\ 0\,1\,0\,1\,1\,1\,1 \ldots 0\,0\,0\,0\,0\,0\,0\,0\,0\,0\,0\,0\,0\,0 \\ 0\,1\,0\,0\,1\,0\,0 \ldots 1\,0\,0\,0\,0\,0\,0\,0\,0\,0\,0\,0\,0\,0 \\ 0\,0\,1\,1\,0\,1\,0 \ldots 0\,0\,0\,0\,0\,0\,0\,0\,0\,0\,0\,0\,0\,0 \\ 0\,1\,1\,1\,0\,0\,0 \ldots 0\,0\,0\,0\,0\,0\,0\,0\,0\,0\,0\,0\,0\,0 \\ \\ 1\,1\,0\,0\,1\,1\,1 \ldots 0\,0\,0\,0\,0\,0\,0\,0\,0\,0\,0\,0\,0\,0 \\ 0\,0\,0\,0\,1\,1\,0 \ldots 1\,0\,0\,0\,0\,0\,0\,0\,0\,0\,0\,0\,0\,0 \\ 0\,1\,1\,0\,1\,0\,1 \ldots 0\,0\,0\,0\,0\,0\,0\,0\,0\,0\,0\,0\,0\,0 \\ 0\,0\,1\,0\,0\,0\,1 \ldots 1\,0\,0\,0\,0\,0\,0\,0\,0\,0\,0\,0\,0\,0 \\ 0\,1\,1\,1\,1\,1\,0 \ldots 1\,0\,0\,0\,0\,0\,0\,0\,0\,0\,0\,0\,0\,0 \\ \\ 1\,0\,0\,1\,1\,0\,1 \ldots 1\,1\,0\,0\,0\,0\,0\,0\,0\,0\,0\,0\,0\,0 \\ 0\,0\,1\,0\,1\,0\,1 \ldots 1\,0\,1\,0\,0\,0\,0\,0\,0\,0\,0\,0\,0\,0 \\ 0\,1\,0\,0\,0\,1\,0 \ldots 1\,0\,0\,1\,0\,0\,0\,0\,0\,0\,0\,0\,0\,0 \\ 0\,1\,1\,1\,0\,0\,1 \ldots 0\,0\,0\,0\,1\,0\,0\,0\,0\,0\,0\,0\,0\,0 \\ 0\,0\,1\,0\,0\,1\,0 \ldots 0\,0\,0\,0\,0\,1\,0\,0\,0\,0\,0\,0\,0\,0 \\ 0\,0\,0\,0\,0\,0\,0 \ldots 0\,1\,1\,1\,1\,1\,1\,0\,0\,0\,0\,0\,0\,0 \\ \\ 0\,0\,1\,0\,0\,1\,0 \ldots 1\,0\,0\,0\,0\,0\,0\,1\,0\,0\,0\,0\,0\,0 \\ 0\,0\,1\,0\,0\,0\,0 \ldots 0\,0\,0\,0\,0\,0\,0\,0\,1\,0\,0\,0\,0\,0 \end{bmatrix} \tag{6.58}$$

$$\begin{bmatrix} 0\,1\,1\,1\,0\,0\,0 \ldots 1\,0\,0\,0\,0\,0\,0\,0\,0\,1\,0\,0\,0\,0 \\ 0\,1\,0\,0\,1\,1\,0 \ldots 1\,0\,0\,0\,0\,0\,0\,0\,0\,0\,1\,0\,0\,0 \\ 0\,1\,1\,0\,1\,0\,1 \ldots 1\,0\,0\,0\,0\,0\,0\,0\,0\,0\,0\,1\,0\,0 \\ 1\,0\,0\,0\,0\,0\,0 \ldots 0\,0\,0\,0\,0\,0\,0\,1\,1\,1\,1\,1\,1\,0 \\ 1\,0\,0\,0\,0\,0\,0 \ldots 0\,0\,0\,0\,0\,0\,0\,0\,0\,0\,0\,0\,0\,1 \end{bmatrix}$$

It will be noticed that the last but one row is the Forney concatenation on the last $GF(32)$ symbol of parity check matrix (6.57), the overall parity check on parity bits 36–41. Bit 0 has been added to this equation. Also, the last row of the binary parity check matrix is simply a repeat of bit 0. In this way, bit 0 has been fully compensated for not being in the last row of parity check symbol matrix (6.57).

BCH codes extended in length in this way can be very competitive compared to the best known codes [5]. The most efficient extensions of BCH codes are for $g(z)$ having only multiple roots of $z = 0$ because no additional deletions of information bits are necessary nor are compensating parity check equations necessary. However, n does need to be a Mersenne prime, and the maximum extension is 2 symbols with $2m + 2$ additional, overall parity bits, increasing the d_{min} by 4. Where n is not a Mersenne prime the maximum extension is 1 symbol with $m + 1$ additional, overall parity bits, increasing the d_{min} by 2.

However regardless of n being a Mersenne prime or not, multiple symbol extensions may be carried out if $g(z)$ has additional roots from $GF(2^m)$, increasing the d_{min} by 2 for each additional root. The additional root can also be $z = 0$.

As further examples, a (37, 11, 13) code and a (43, 11, 15) code can be constructed in this way by extending the (31, 11, 11) BCH code. Also a (135, 92, 13) code and a (143, 92, 15) code can be constructed by extending the (127, 92, 11) BCH code. A (135, 71, 21) code and a (143, 71, 23) code can be constructed by extending the (127, 71, 19) BCH code.

For more than 2 extended symbols for Mersenne primes, or more than 1 extended symbol for non-Mersenne primes, it is necessary to use mixed roots of $g(z)$ from $GF(2^m)$ and have either deletions of information bits or compensating parity check equations or both. As examples of these code constructions there are:

- An example of a non Mersenne prime, the (76, 50, 9) code constructed from the BCH (63, 51, 5) code with additional roots of $g(z)$ at $z = 0$ and α^0 deleting the first information bit.
- The (153, 71, 25) code extended from the (127, 71, 19) code with additional roots of $g(z)$ at $z = 0$, α^0 and α^1 with 2 additional, compensating parity check bits.
- The (151, 70, 25) code extended from the (127, 71, 19) code with additional roots of $g(z)$ at $z = 0$, α^0 and α^1 with the first coordinate deleted reducing the dimension by 1 and one additional, compensating parity check bit.
- The (160, 70, 27) code extended from the (127, 71, 19) code with additional roots of $g(z)$ at $z = 0$, α^0, α^1 and α^2 with the first coordinate deleted reducing the dimension by 1 and with 2 additional, compensating parity check bits.
- The (158, 69, 27) code extended from the (127, 71, 19) code with additional roots of $g(z)$ at $z = 0$, α^0, α^1, α^2 and α^3 with the first 2 coordinates deleted reducing

the dimension by 2 and one additional, compensating parity check bit. All of these codes are best known codes [5].

6.8 Binary Codes from MDS Codes

The Goppa codes and BCH codes, which are a subset of Goppa codes, when constructed as codes with symbols from $GF(q)$ are all MDS codes and are examples of generalised Reed–Solomon codes [7]. MDS codes are exceptional codes and there are not many construction methods for these codes. For (n, k) MDS codes the repetition code, having $k = 1$, can have any length of n independently of the field size q. For values $k = 3$ and $k = q - 1$ and with q even the maximum value of n is $n = q + 2$ [7]. For all other cases, the maximum value of n is $n = q + 1$ with a construction known as the doubly extended Reed–Solomon codes. The parity check matrix for a $(q + 1, k)$ doubly extended Reed–Solomon code is

$$\mathbf{H}_{\mathbf{RS}+} = \begin{bmatrix} 1 & 1 & 1 & 1 & 1 & 1 & 1 & \dots & 1 & 1 & 0 \\ 1 & \alpha_1 & \alpha_2 & \alpha_3 & \alpha_4 & \alpha_5 & \alpha_6 & \dots & \alpha_{q-2} & 0 & 0 \\ 1 & \alpha_1^2 & \alpha_2^2 & \alpha_3^2 & \alpha_4^2 & \alpha_5^2 & \alpha_6^2 & \dots & \alpha_{q-2}^2 & 0 & 0 \\ 1 & \alpha_1^3 & \alpha_2^3 & \alpha_3^3 & \alpha_4^3 & \alpha_5^3 & \alpha_6^3 & \dots & \alpha_{q-2}^3 & 0 & 0 \\ 1 & \alpha_1^4 & \alpha_2^4 & \alpha_3^4 & \alpha_4^4 & \alpha_5^4 & \alpha_6^4 & \dots & \alpha_{q-2}^4 & 0 & 0 \\ \dots & \dots & \dots & \dots & \dots & \dots & \dots & \dots & \dots & \dots & \dots \\ 1 & \alpha_1^{q-k} & \alpha_2^{q-k} & \alpha_3^{q-k} & \alpha_4^{q-k} & \alpha_5^{q-k} & \alpha_6^{q-k} & \dots & \alpha_{q-2}^{q-k} & 0 & 1 \end{bmatrix} \quad (6.59)$$

where the q elements of $GF(q)$ are $\{0, 1, \alpha_1, \alpha_2, \alpha_3, \dots, \alpha_{q-1}\}$.

As the codes are MDS, the minimum Hamming distance is $q + 2 - k$, forming a family of $(q + 1, k, q + 2 - k)$ codes meeting the Singleton bound [8].

The MDS codes may be used as binary codes simply by restricting the data symbols to values of {0 and 1} to produce a subfield subcode. Alternatively for $GF(2^m)$ each symbol may be replaced with a $m \times m$ binary matrix to produce the family of $((2^m + 1)m, mk, 2^m + 2 - k)$ of binary codes. As an example, with $m = 4$ and $k = 12$, the result is a $(68, 48, 5)$ binary code. This is not a very competitive code because the equivalent best known code [5], the $(68, 48, 8)$ code, has much better minimum Hamming distance.

However, using the Forney concatenation [2] on each symbol almost doubles the minimum Hamming distance with little increase in redundancy and produces the family of $((2^m + 1)(m + 1), mk, 2(2^m + 1 - k) + 1)$ of binary codes. With the same example values for m and k the $(85, 48, 11)$ binary code is produced. Kasahara [6] noticed that it is sometimes possible with this code construction to add an additional information bit by adding the all 1's codeword to the generator matrix of the code. Equivalently expressed, all of the codewords may be complemented without degrading the minimum Hamming distance. It is possible to go further depending on the length of the code and the minimum Hamming distance. Since the binary parity of each symbol is always even, then if $m+1$ is an odd number, then adding the all 1's pattern to each symbol will produce weight of at least 1 per symbol. For the $(85, 48, 11)$ constructed binary code $m+1=5$, an odd number and the number of symbols is 17. Hence, adding the all 1's pattern (i.e. 85 1's) to each codeword will produce a minimum weight of at least 17. Accordingly, a $(85, 49, 11)$ code is produced. Adding an overall parity bit to each codeword increases the minimum Hamming distance to 12 producing a $(86, 49, 12)$ code and shortening the code by deleting one information bit produces a $(85, 48, 12)$ code. This is a good code because the corresponding best known code is also a $(85, 48, 12)$ code. However, the construction method is different because the best known code is derived from the $(89, 56, 11)$ cyclic code.

Looking at constructing binary codes from MDS codes by simply restricting the data symbols to values of {0 and 1}, consider the example of the extended Reed–Solomon code of length 16 using $GF(2^4)$ with 2 parity symbols. The code is the MDS $(16, 14, 3)$ code. The parity check matrix is

$$\mathbf{H}_{(16,14)} = \begin{bmatrix} 1 & \alpha^1 & \alpha^2 & \alpha^3 & \alpha^4 & \alpha^5 & \alpha^6 & \alpha^7 & \alpha^8 & \alpha^9 & \alpha^{10} & \alpha^{11} & \alpha^{12} & \alpha^{13} & \alpha^{14} & 0 \\ 1 & \alpha^3 & \alpha^6 & \alpha^9 & \alpha^{12} & 1 & \alpha^3 & \alpha^6 & \alpha^9 & \alpha^{12} & 1 & \alpha^3 & \alpha^6 & \alpha^9 & \alpha^{12} & 1 \end{bmatrix} \tag{6.60}$$

With binary codeword coordinates, denoted as c_i the first parity check equation from the first row of the parity check matrix is

$$\sum_{i=0}^{14} c_i \alpha^i = 0 \tag{6.61}$$

Squaring both sides of this equation produces

$$\sum_{i=0}^{14} c_i^2 \alpha^{2i} = 0 \tag{6.62}$$

As the codeword coordinates are binary, $c_i^2 = c_i$ and so any codeword satisfying the equations of (6.58) satisfies all of the following equations by induction from (6.60)

$$\mathbf{H}_{(16,14)} = \begin{bmatrix} 1 & \alpha^1 & \alpha^2 & \alpha^3 & \alpha^4 & \alpha^5 & \alpha^6 & \alpha^7 & \alpha^8 & \alpha^9 & \alpha^{10} & \alpha^{11} & \alpha^{12} & \alpha^{13} & \alpha^{14} & 0 \\ 1 & \alpha^2 & \alpha^4 & \alpha^6 & \alpha^8 & \alpha^{10} & \alpha^{12} & \alpha^{14} & \alpha^1 & \alpha^3 & \alpha^5 & \alpha^7 & \alpha^9 & \alpha^{11} & \alpha^{13} & 0 \\ 1 & \alpha^3 & \alpha^6 & \alpha^9 & \alpha^{12} & 1 & \alpha^3 & \alpha^6 & \alpha^9 & \alpha^{12} & 1 & \alpha^3 & \alpha^6 & \alpha^9 & \alpha^{12} & 1 \\ 1 & \alpha^4 & \alpha^8 & \alpha^{12} & \alpha^1 & \alpha^5 & \alpha^9 & \alpha^{13} & \alpha^2 & \alpha^4 & \alpha^{10} & \alpha^{14} & \alpha^3 & \alpha^7 & \alpha^{11} & 0 \\ 1 & \alpha^6 & \alpha^{12} & \alpha^3 & \alpha^9 & 1 & \alpha^6 & \alpha^{12} & \alpha^3 & \alpha^9 & 1 & \alpha^6 & \alpha^{12} & \alpha^3 & \alpha^9 & 1 \\ & \dots & \dots & \dots & \dots & \dots & \dots & \dots & \dots & \dots & \dots & \dots & \dots & \dots & \dots & \end{bmatrix} \tag{6.63}$$

There are 4 consecutive zeros of the parent Reed–Solomon code from the first 4 rows of the parity check matrix indicating that the minimum Hamming distance may be 5

Table 6.3 $GF(16)$ extension field defined by $1 + \alpha^1 + \alpha^4 = 0$

$\alpha^0 = 1$
$\alpha^1 = \alpha$
$\alpha^2 = \alpha^2$
$\alpha^3 = \alpha^3$
$\alpha^4 = 1 + \alpha$
$\alpha^5 = \alpha + \alpha^2$
$\alpha^6 = \alpha^2 + \alpha^3$
$\alpha^7 = 1 + \alpha + \alpha^3$
$\alpha^8 = 1 + \alpha^2$
$\alpha^9 = \alpha + \alpha^3$
$\alpha^{10} = 1 + \alpha + \alpha^2$
$\alpha^{11} = \alpha + \alpha^2 + \alpha^3$
$\alpha^{12} = 1 + \alpha + \alpha^2 + \alpha^3$
$\alpha^{13} = 1 + \alpha^2 + \alpha^3$
$\alpha^{14} = 1 + \alpha^3$

for the binary code. However, comparing the last column of this matrix with (6.57) indicates that this column is not correct.

Constructing the binary check matrix from the parity check equations, (6.58) using Table 6.3 substituting the respective 4 bit vector for each column vector of each nonzero $GF(16)$ symbol, (0 in $GF(16)$ is 0000) produces the following binary check matrix

$$\mathbf{H}_{(16,8)} = \begin{bmatrix} 1\,0\,0\,0\,1\,0\,0\,1\,1\,0\,1\,0\,1\,1\,1\,0 \\ 0\,1\,0\,0\,1\,1\,0\,1\,0\,0\,1\,0\,1\,0\,0\,0 \\ 0\,0\,1\,0\,0\,1\,1\,0\,1\,0\,1\,1\,1\,1\,0\,0 \\ 0\,0\,0\,1\,0\,0\,1\,1\,0\,1\,0\,1\,1\,1\,1\,0 \\ \\ 1\,0\,0\,0\,1\,1\,0\,0\,0\,1\,1\,0\,0\,0\,1\,1 \\ 0\,0\,0\,1\,1\,0\,0\,0\,1\,1\,0\,0\,0\,1\,1\,0 \\ 0\,0\,1\,0\,1\,0\,0\,1\,0\,1\,0\,0\,1\,0\,1\,0 \\ 0\,1\,1\,1\,1\,0\,1\,1\,1\,1\,0\,1\,1\,1\,1\,0 \end{bmatrix} \tag{6.64}$$

Weight spectrum analysis indicates the minimum Hamming distance of this code is 4 due to a single codeword of weight 4,{0, 5, 10, 15}. Deleting the last column of the parity check matrix produces a (15, 8, 5) code. Another approach is needed to go from the MDS code to a binary code without incurring a loss in the minimum Hamming distance.

It is necessary to use the generalised Reed–Solomon MDS code. Here, each column of the parity check matrix is multiplied by a nonzero element of the $GF(2^m)$ field defined as $\{\mu_0, \mu_1, \mu_2, \mu_3, \ldots, \mu_{2^m}\}$. It is not necessary for these to be distinct, just to have a multiplicative inverse. The parity check matrix for the $(q + 1, k)$ generalised Reed–Solomon MDS code is

$$\mathbf{H}_{\mathbf{GRS1}} = \begin{bmatrix} v_0 & v_1 & v_2 & v_3 & v_4 & v_5 & \dots & v_{q-2} & v_{q-1} & 0 \\ v_0 & v_1\alpha_1 & v_2\alpha_2 & v_3\alpha_3 & v_4\alpha_4 & v_5\alpha_5 & \dots & v_{q-2}\alpha_{q-2} & 0 & 0 \\ v_0 & v_1\alpha_1^2 & v_2\alpha_2^2 & v_3\alpha_3^2 & v_4\alpha_4^2 & v_5\alpha_5^2 & \dots & v_{q-2}\alpha_{q-2}^2 & 0 & 0 \\ v_0 & v_1\alpha_1^3 & v_2\alpha_2^3 & v_3\alpha_3^3 & v_4\alpha_4^3 & v_5\alpha_5^3 & \dots & v_{q-2}\alpha_{q-2}^3 & 0 & 0 \\ v_0 & v_1\alpha_1^4 & v_2\alpha_2^4 & v_3\alpha_3^4 & v_4\alpha_4^4 & v_5\alpha_5^4 & \dots & v_{q-2}\alpha_{q-2}^4 & 0 & 0 \\ & \dots & \dots & \dots & \dots & \dots & \dots & \dots & \dots & \dots \\ v_0 & v_1\alpha_1^{q-k} & v_2\alpha_2^{q-k} & v_3\alpha_3^{q-k} & v_4\alpha_4^{q-k} & v_5\alpha_5^{q-k} & \dots & v_{q-2}\alpha_{q-2}^{q-k} & 0 & v_q \end{bmatrix}$$

It is clear that as a nonbinary code with codeword coefficients from $GF(2^m)$, the distance properties will remain unchanged as the generalised Reed–Solomon is still an MDS code. Depending on the coordinate position each nonzero element value has a unique mapping to another nonzero element value. It is as subfield subcodes that the generalised Reed–Solomon codes have an advantage. It should be noted that Goppa codes are examples of a generalised Reed–Solomon code.

Returning to the relatively poor (16, 8, 4) binary code derived from the (16, 14, 3) MDS code, consider the generalised (16, 14, 3) Reed–Solomon code with parity check matrix.

$$\mathbf{H}_{(16,14)} = \begin{bmatrix} v_0 & v_1 & v_2 & v_3 & v_4 & v_5 & v_6 & \dots & v_{13} & v_{14} & v_{15} \\ v_0 & v_1\alpha^1 & v_2\alpha^2 & v_3\alpha^3 & v_4\alpha^4 & v_5\alpha^5 & v_6\alpha^6 & \dots & v_{13}\alpha^{13} & v_{14}\alpha^{14} & 0 \end{bmatrix} \tag{6.65}$$

Setting the vector v to

$$\{\alpha^{12}, \alpha^4, \alpha^3, \alpha^9, \alpha^4, \alpha^1, \alpha^8, \alpha^6, \alpha^3, \alpha^6, \alpha^1, \alpha^2, \alpha^2, \alpha^8, \alpha^9, \alpha^{12}\}$$

Constructing the binary check matrix from these parity check equations using Table 6.3 by substituting the respective 4 bit vector for each column vector of each nonzero $GF(16)$ symbol, (0 in $GF(16)$ is 0000) produces the following binary check matrix

$$\mathbf{H}_{(16,8,5)} = \begin{bmatrix} 1&1&0&0&1&0&1&0&0&0&0&0&0&1&0&1 \\ 1&1&0&1&1&1&0&0&0&0&1&0&0&0&1&1 \\ 1&0&0&0&0&0&1&1&0&1&0&1&1&1&0&1 \\ 1&0&1&1&0&0&0&1&1&1&0&0&0&0&1&1 \\ \\ 1&0&0&1&1&0&1&1&0&1&0&1&1&0&1&0 \\ 1&1&1&1&0&0&0&0&1&0&1&0&0&0&0&0 \\ 1&1&1&1&1&1&0&1&1&0&1&1&0&1&1&0 \\ 1&0&0&1&0&1&1&1&1&0&1&1&1&1&0&0 \end{bmatrix} \tag{6.66}$$

Weight spectrum analysis indicates that the minimum Hamming distance of this code is 5 and achieves the aim of deriving a binary code from an MDS code without loss of minimum Hamming distance. Moreover, the additional symbol of 1, the last column in (6.59), may be appended to produce the following check matrix for the (17, 9, 5) binary code:

$$\mathbf{H}_{(17,9,5)} = \begin{bmatrix} 1&1&0&0&1&0&1&0&0&0&0&0&0&1&0&1&0 \\ 1&1&0&1&1&1&0&0&0&0&1&0&0&0&1&1&0 \\ 1&0&0&0&0&0&1&1&0&1&0&1&1&1&0&1&0 \\ 1&0&1&1&0&0&0&1&1&1&0&0&0&0&1&1&0 \\ \\ 1&0&0&1&1&0&1&1&0&1&0&1&1&0&1&0&1 \\ 1&1&1&1&0&0&0&0&1&0&1&0&0&0&0&0&0 \\ 1&1&1&1&1&1&0&1&1&0&1&1&0&1&1&0&0 \\ 1&0&0&1&0&1&1&1&1&0&1&1&1&1&0&0&0 \end{bmatrix} \tag{6.67}$$

Not surprisingly, this code has the same parameters as the best known code [5]. The reader will be asking, how is the vector ν chosen?

Using trial and error methods, it is extremely difficult, and somewhat tiresome to find a suitable vector ν, even for such a short code. Also weight spectrum analysis has to be carried out for each trial code.

The answer is that the vector ν is constructed from an irreducible Goppa polynomial of degree 2 with $g(z) = \alpha^3 + z + z^2$. Referring to Table 6.3, the reader may verify using all elements of $GF(16)$, that ν is given by $g(\alpha_i)^{-1}$ for $i = 0$ to 15.

Unfortunately the technique is only valid for binary codes with minimum Hamming distance of 5 and also m has to be even. Weight spectrum analysis has confirmed that the (65, 53, 5), (257, 241, 5), (1025, 1005, 5) and (4097, 4073, 5) codes can be constructed in this way from doubly extended, generalised Reed–Solomon, MDS codes.

6.9 Summary

It has been shown that interpolation plays an important, mostly hidden role in algebraic coding theory. The Reed–Solomon codes, BCH codes, and Goppa codes are all codes that may be constructed via interpolation. It has also been demonstrated that all of these codes form part of a large family of generalised MDS codes. The encoding of BCH and Goppa codes has been explored from the viewpoint of classical Lagrange interpolation. It was shown in detail how Goppa codes are designed and constructed starting from first principles. The parity check matrix of a BCH code was derived as a Goppa code proving that BCH codes are a subset of Goppa codes. Following from this result and using properties of the cyclotomic cosets it was explained how the minimum Hamming distance of some BCH codes is able to exceed the BCH bound producing outstanding codes. It was shown how these exceptional BCH codes can be identified and constructed. A little known paper by Goppa was discussed and as a result it was shown how Goppa codes and BCH codes may be extended in length with additional parity check bits resulting in increased minimum Hamming distance of the code. Several examples were given of the technique which results in some outstanding codes. Reed–Solomon codes were explored as a means of constructing binary codes resulting in improvements to the database of best known codes.

References

1. Bell, E.T.: Men of Mathematics: The Lives and Achievements of the Great Mathematicians from Zeno to Poincar. Simon and Schuster, New York (1986)
2. Forney Jr., G.D.: Concatenated Codes. MIT Press, Cambridge (1966)
3. Goppa, V.D.: A new class of linear error-correcting codes. Probl Inform Transm **6**, 24–30 (1970)
4. Goppa, V.D.: Rational representation of codes and (l; g)-codes. Probl Inform Transm **7**, 41–49 (1970)
5. Grassl, M.: Code Tables: Bounds on the parameters of various types of codes (2007). http://www.codetables.de
6. Kasahara, M., Sugiyama, Y., Hirasawa, S., Namekawa, T.: New classes of binary codes constructed on the basis of concatenated codes and product codes. IEEE Trans. Inf. Theory **IT–22**(4), 462–468 (1976)
7. MacWilliams, F.J., Sloane, N.J.A.: The Theory of Error-Correcting Codes. North-Holland, New York (1977)
8. Singleton, R.C.: Maximum distance q-ary codes. IEEE Trans. Inf. Theory **IT–10**, 116–118 (1964)
9. Sugiyama, Y., Kasahara, M., Namekawa, T.: Some efficient binary codes constructed using srivastava codes. IEEE Trans. Inf. Theory **IT–21**(5), 581–582 (1975)
10. Sugiyama, Y., Kasahara, M., Namekawa, T.: Further results on goppa codes and their applications to constructing efficient binary codes. IEEE Trans. Inf. Theory **IT–22**(5), 518–526 (1976)

Chapter 7
Reed–Solomon Codes and Binary Transmission

7.1 Introduction

Reed–Solomon codes named after Reed and Solomon [9] following their publication in 1960 have been used together with hard decision decoding in a wide range of applications. Reed–Solomon codes are maximum distance separable (MDS) codes and have the highest possible minimum Hamming distance. The codes have symbols from $\mathbb{F}_q$ with parameters $(q-1,\ k,\ q-k)$. They are not binary codes but frequently are used with $q = 2^m$, and so there is a mapping of residue classes of a primitive polynomial with binary coefficients [6] and each element of $\mathbb{F}_{2^m}$ is represented as a binary m-tuple. Thus, binary codes with code parameters $(m[2^m-1],\ km,\ 2^m-k)$ can be constructed from Reed–Solomon codes. Reed–Solomon codes can be extended in length by up to two symbols and in special cases extended in length by up to three symbols. In terms of applications, they are probably the most popular family of codes.

Researchers over the years have tried to come up with an efficient soft decision decoding algorithm and a breakthrough in hard decision decoding in 1997 by Madhu Sudan [10], enabled more than $\frac{2^m-k}{2}$ errors to be corrected with polynomial time complexity. The algorithm was limited to low rate Reed–Solomon codes. An improved algorithm for all code rates was discovered by Gursuswami and Sudan [3] and led to the Guruswami and Sudan algorithm being applied in a soft decision decoder by Kötter and Vardy [5]. A very readable, tutorial style explanation of the Guruswami and Sudan algorithm is presented by McEliece [7]. Many papers followed, discussing soft decision decoding of Reed–Solomon codes [1] mostly featuring simulation results of short codes such as the (15, 11, 5) and the (31, 25, 7) code. Binary transmission using baseband bipolar signalling or binary phase shift keying (BPSK) [8] and the additive white gaussian noise (AWGN) channel is most common. Some authors have used quadrature amplitude modulation (QAM) [8] with 2^m levels to map to each $\mathbb{F}_{2^m}$ symbol [5]. In either case, there is a poor match between

M. Tomlinson et al., *Error-Correction Coding and Decoding*,
Signals and Communication Technology,
DOI 10.1007/978-3-319-51103-0_7

the modulation method and the error-correcting code. The performance achieved is not competitive compared to other error-correcting code arrangements. For binary transmission, a binary error-correcting code should be used and not a symbol-based error-correcting code. For QAM and other multilevel signalling, better performance is obtained by applying low-rate codes to the least significant bits of received symbols and high-rate codes to the most significant bits of received symbols. Applying a fixed-rate error-correcting code to all symbol bits is the reason for the inefficiency in using Reed–Solomon codes on binary channels.

Still, these modulation methods do provide a means of comparing different decoder arrangements for RS codes. This theme is explored later in Sect. 7.3 where soft decision decoding of RS codes is explored.

7.2 Reed–Solomon Codes Used with Binary Transmission-Hard Decisions

Whilst RS codes are very efficient codes, being MDS codes, they are not particularly well suited to the binary channel as it will become apparent from the results presented below. Defining the RS code over $\mathbb{F}_{2^m}$, RS codes extended with a single symbol are considered with length $n = 2^m$, with k information symbols, and with $d_{min} = n-k+1$. The length in bits, $n_b = mn$ and there are k_b information bits with $k_b = km$.

The probability of a symbol error with binary transmission and the AWGN channel is

$$p_s = 1 - \left(1 - \frac{1}{2} erfc\left(\sqrt{\frac{k}{n}\frac{E_b}{N_0}}\right)\right)^m \tag{7.1}$$

The RS code can correct t errors where $t = \left\lfloor \frac{n-k+1}{2} \right\rfloor$. Accordingly, a decoder error occurs if there are more than t symbol errors and the probability of decoder error, p_C is given by

$$p_C = \sum_{i=t+1}^{n} \frac{n!}{(n-i)!i!} p_s^i (1-p_s)^{n-i} \tag{7.2}$$

As a practical example, we will consider the (256, 234, 23) extended RS code. Representing each $\mathbb{F}_{2^8}$ symbol as a binary 8 tuple the RS code becomes a (2048, 1872, 23) binary code. The performance with hard decisions is shown in Fig. 7.1 as a function of $\frac{E_b}{N_0}$. This code may be directly compared to the binary (2048, 1872, 33) Goppa code since their lengths and code rates are identical. The decoder error probability for the binary Goppa code is given by

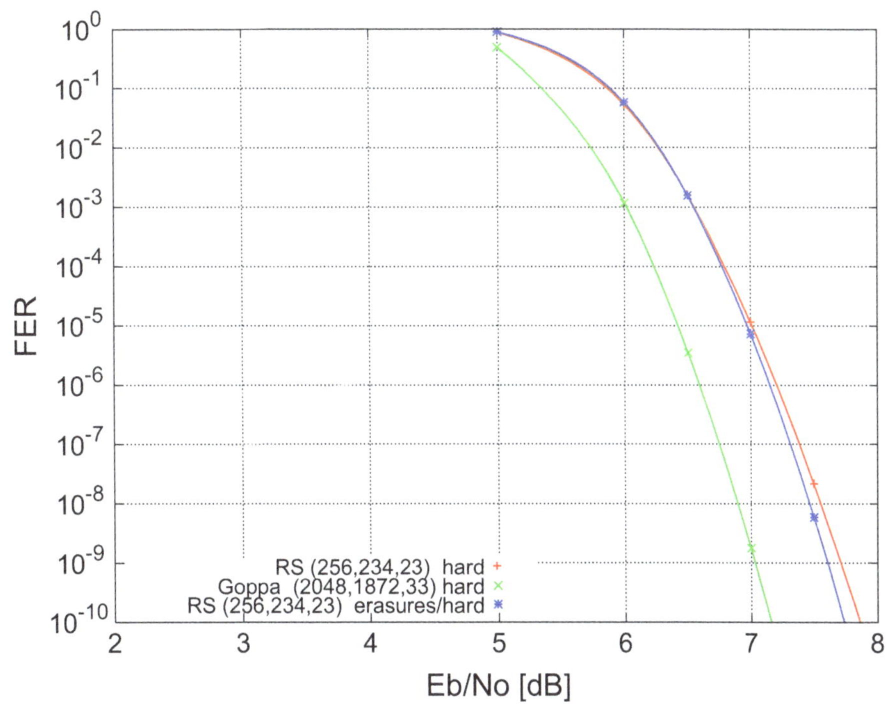

Fig. 7.1 Comparison of hard decision decoding of the (256, 234, 23) RS code compared to the (2048, 1872, 33) Goppa code (same code length in bits and code rate)

$$p_C = \sum_{i=t_G+1}^{nm} \frac{(nm)!}{(nm-i)!i!}\left(\frac{1}{2}erfc\sqrt{\frac{k}{n}\frac{E_b}{N_0}}\right)^{i}\left(1-\frac{1}{2}erfc\sqrt{\frac{k}{n}\frac{E_b}{N_0}}\right)^{nm-i} \tag{7.3}$$

where $t_G = \left\lfloor \frac{d_{min}+1}{2} \right\rfloor$ for the binary Goppa code.

The comparison in performance is shown in Fig. 7.1 and it can be seen that the Goppa code is approximately 0.75dB better than the RS code at 1×10^{-10} frame error rate.

It is interesting to speculate whether the performance of the RS code could be improved by using 3-level quantisation of the channel bits and erasing symbols if any of the bits within a symbol are erased. The probabilities of a bit erasure p_{erase} and bit error p_b for 3-level quantisation are given in Chap. 3, Eqs. (3.41) and (3.42) respectively, but note that a lower threshold needs to be used for best performance with these code parameters, $\sqrt{E_s} - 0.2 \times \sigma$ instead of $\sqrt{E_s} - 0.65 \times \sigma$. The probability of a symbol erasure, $p_{S\,erase}$ is given by

$$p_{S\,erase} = 1 - (1 - p_{erase})^m \tag{7.4}$$

and the probability of a symbol error, $p_{S\,error}$ is given by

$$p_{S\,error} = 1 - \left(1 - (1 - p_{erase})^m\right) - (1 - p_b)^m \tag{7.5}$$

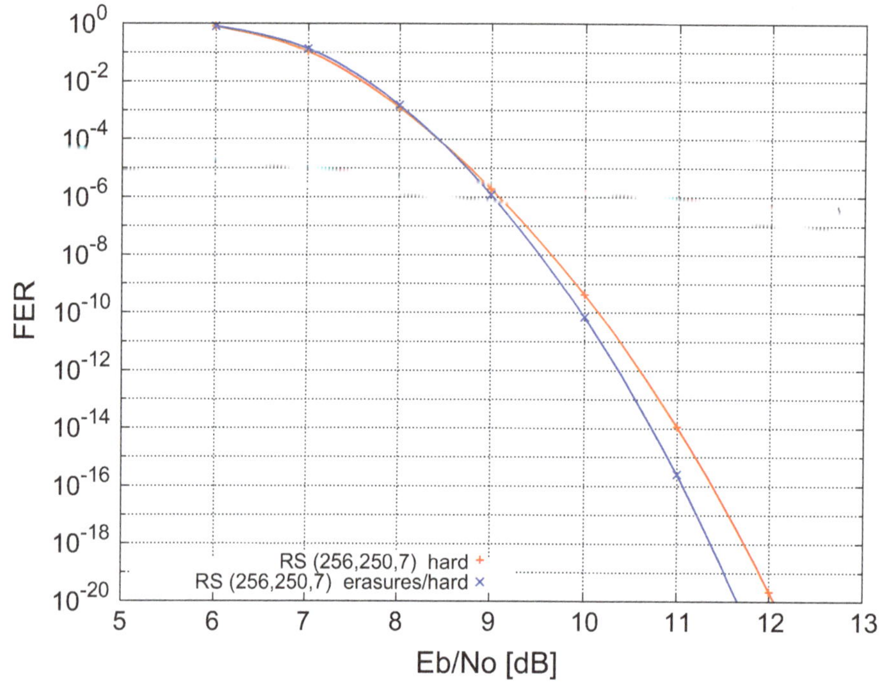

Fig. 7.2 Comparison of hard decision and erasure decoding of the (256, 250, 7) RS code for the binary channel

and

$$p_{S\ error} = (1 - p_{erase})^m - (1 - p_b)^m \tag{7.6}$$

For each received vector, provided the number of errors t and the number of erasures s such that $2t + s \leq n - k$, then the received vector will be decoded correctly. A decoder error occurs if $2t + s > n - k$.

The probability distribution of errors and erasures in the received vector, $e(z)$ may be easily found by defining a polynomial $p(z)$ and raising it to the power of n, the number of symbols in a codeword.

$$e(z) = \left(1 - p_{S\ error} - p_{S\ erase} + p_{S\ erase}z^{-1} + p_{S\ error}z^{-2}\right)^n \tag{7.7}$$

The probability of decoder error p_C is simply found from $e(z)$ by summing all coefficients of z^{-i} where i is greater than $n - k$. This is very straightforward with a symbolic mathematics program such as Mathematica. The results for the RS (256, 234, 23) code are shown in Fig. 7.1. It can be seen that there is an improvement over the hard decision case but it is rather marginal.

A rather more convincing case is shown in Fig. 7.2 for the RS (256, 250, 7) code where the performance is shown down to frame error rates of 1×10^{-20}. In this case, there is an improvement of approximately 0.4 dB.

It has already been established that for the binary transmission channel, the RS codes based on $GF(2^m)$, do not perform as well as a binary designed code with the same code parameters. The problem is that bit errors occur independently and it only takes a single bit error to cause a symbol error. Thus, the code structure, being symbol based, is not well matched to the transmission channel. Another way of looking at this is to consider the Hamming distance. For the binary (2048, 1872) codes considered previously, the RS-based code turns out to have a binary Hamming distance of 23 whilst the binary Goppa code has a Hamming distance of 33. However, there is a simple method of modifying RS codes to produce good binary codes as discussed in Chap. 6. It is a code concatenation method best suited for producing symbol-based binary codes whereby a single overall binary parity check is added to each binary m-tuple representing each symbol. Starting with a RS $(n, k, n-k-1)$ code, adding the overall binary parity checks produces a $(n[m+1], km, 2[n-k-1])$ binary code. Now the minimum weight of each symbol is 2, producing a binary code with twice the minimum Hamming distance of the original RS code. Kasahara [4] realised that in some cases an additional information bit may be added by adding the all $1's$ codeword to the generator matrix. Some best known codes are constructed in this way as discussed in Chap. 6. One example is the $(161, 81, 23)$ binary code [6].

7.3 Reed–Solomon Codes and Binary Transmission Using Soft Decisions

RS codes applied to the binary transmission channel will now be considered using unquantised soft decision decoding. The best decoder to use is the modified Dorsch decoder, discussed in Chap. 15, because it provides near maximum likelihood decoding. However when used with codes having a significant coding gain, the code length needs to be typically less than 200 bits.

We will consider augmented, extended RS codes constructed from $GF(2^m)$. The length is $2^m + 1$ and these are Maximum Distance Separable (MDS) codes with parameters $(2^m + 1, k, 2^{m+1} - k)$. Moreover, the general case is that augmented, extended RS codes may be constructed using any Galois Field $GF(q)$ with parameters $(q+1, k, q+2-k)$ [6]. Denoting the q field elements as $0,\ \alpha_0,\ \alpha_1,\ \alpha_2, \ldots\ \alpha_{q-2}$, the parity-check matrix is given by

$$\mathbf{H} = \begin{bmatrix} \alpha_0^j & \alpha_1^j & \alpha_2^j & \ldots & \alpha_{q-2}^j & 1 & 0 \\ \alpha_0^{j+1} & \alpha_1^{j+1} & \alpha_2^{j+1} & \ldots & \alpha_{q-2}^{j+1} & 0 & 0 \\ \alpha_0^{j+2} & \alpha_1^{j+2} & \alpha_2^{j+2} & \ldots & \alpha_{q-2}^{j+2} & 0 & 0 \\ \ldots & \ldots & \ldots & \ldots & \ldots & \ldots & \ldots \\ \alpha_0^{j+q-k-1} & \alpha_1^{j+q-k-1} & \alpha_2^{j+q-k-1} & \ldots & \alpha_{q-2}^{j+q-k-1} & 0 & 0 \\ \alpha_0^{j+q-k} & \alpha_1^{j+q-k} & \alpha_2^{j+q-k} & \ldots & \alpha_{q-2}^{j+q-k} & 0 & 1 \end{bmatrix}$$

Table 7.1 $GF(32)$ non-zero extension field elements defined by $1+\alpha^2+\alpha^5=0$

$\alpha^0 = 1$	$\alpha^{16} = 1+\alpha+\alpha^3+\alpha^4$
$\alpha^1 = \alpha$	$\alpha^{17} = 1+\alpha+\alpha^4$
$\alpha^2 = \alpha^2$	$\alpha^{18} = 1+\alpha$
$\alpha^3 = \alpha^3$	$\alpha^{19} = \alpha+\alpha^2$
$\alpha^4 = \alpha^4$	$\alpha^{20} = \alpha^2+\alpha^3$
$\alpha^5 = 1+\alpha^2$	$\alpha^{21} = \alpha^3+\alpha^4$
$\alpha^6 = \alpha+\alpha^3$	$\alpha^{22} = 1+\alpha^2+\alpha^4$
$\alpha^7 = \alpha^2+\alpha^4$	$\alpha^{23} = 1+\alpha+\alpha^2+\alpha^3$
$\alpha^8 = 1+\alpha^2+\alpha^3$	$\alpha^{24} = \alpha+\alpha^2+\alpha^3+\alpha^4$
$\alpha^9 = \alpha+\alpha^3+\alpha^4$	$\alpha^{25} = 1+\alpha^3+\alpha^4$
$\alpha^{10} = 1+\alpha^4$	$\alpha^{26} = 1+\alpha+\alpha^2+\alpha^4$
$\alpha^{11} = 1+\alpha+\alpha^2$	$\alpha^{27} = 1+\alpha+\alpha^3$
$\alpha^{12} = \alpha+\alpha^2+\alpha^3$	$\alpha^{28} = \alpha+\alpha^2+\alpha^4$
$\alpha^{13} = \alpha^2+\alpha^3+\alpha^4$	$\alpha^{29} = 1+\alpha^3$
$\alpha^{14} = 1+\alpha^2+\alpha^3+\alpha^4$	$\alpha^{30} = \alpha+\alpha^4$
$\alpha^{15} = 1+\alpha+\alpha^2+\alpha^3+\alpha^4$	

There are $q-k+1$ rows of the matrix corresponding to the $q-k+1$ parity symbols of the code. Any of the $q-k+1$ columns form a Vandermonde matrix and the matrix is non-singular which means that any set of $q-k+1$ symbols of a codeword may be erased and solved using the parity-check equations. Thus, the code is MDS. The columns of the parity-check matrix may be permuted into any order and any set of s symbols of a codeword may be defined as parity symbols and permanently erased. Thus, their respective columns of $\mathbf{H}$ may be deleted to form a shortened $(2^m+1-s, k, 2^{m+1}-s-k)$ MDS code. This is an important property of MDS codes, particularly for their practical realisation in the form of augmented, extended RS codes because it enables efficient implementation in applications such as incremental redundancy systems, discussed in Chap. 17, and network coding. Using the first $q-1$ columns of $\mathbf{H}$, and setting $\alpha_0,\ \alpha_1,\ \alpha_2,\dots\ \alpha_{q-2}$ equal to $\alpha^0,\ \alpha^1,\ \alpha^2,\dots\ \alpha^{q-2}$, where α is a primitive element of $GF(q)$ a cyclic code may be constructed, which has advantages for encoding and decoding implementation.

We will consider the shortened RS code (30, 15, 16) constructed from the $GF(2^5)$ extension field with $\mathbf{H}$ constructed using $j=0$ and α being the primitive root of $1+x^2+x^5$. The $GF(32)$ extension field table is given in Table 7.1 based on the primitive polynomial $1+x^2+x^5$ so that $1+\alpha^2+\alpha^5=0$, modulo $1+x^{31}$.

The first step in the construction of the binary code is to construct the parity-check matrix for the shortened RS code (30, 15, 16) which is

$$\mathbf{H}_{(30,15)} = \begin{bmatrix} 1 & 1 & 1 & \dots & 1 \\ 1 & \alpha & \alpha^2 & \dots & \alpha^{29} \\ 1 & \alpha^2 & \alpha^4 & \dots & \alpha^{27} \\ 1 & \alpha^3 & \alpha^6 & \dots & \alpha^{25} \\ \dots & \dots & \dots & \dots & \dots \\ 1 & \alpha^{13} & \alpha^{26} & \dots & \alpha^5 \\ 1 & \alpha^{14} & \alpha^{28} & \dots & \alpha^3 \end{bmatrix}$$

Each element of this parity-check matrix is to be replaced with a 5×5 matrix in terms of the base field, which in this case is binary. First, the number of rows are expanded to form $\mathbf{H}_{(30,75)}$ given by matrix (7.8). The next step is to expand the columns in terms of the base field by substituting for powers of α using Table 7.1. For example, if an element of the parity-check matrix $\mathbf{H}_{(30,75)}$ is, say α^{26}, then this is replaced by $1+\alpha+\alpha^2+\alpha^4$ which in binary is 11101. Proceeding in this way the binary matrix $\mathbf{H}_{(150,75)}$ is produced (some entries have been left as they were to show the procedure partly completed) as in matrix (7.9).

$$\mathbf{H}_{(30,75)} = \begin{bmatrix} 1 & 1 & 1 & \dots & 1 \\ \alpha & \alpha & \alpha & \dots & \alpha \\ \alpha^2 & \alpha^2 & \alpha^2 & \dots & \alpha^2 \\ \alpha^3 & \alpha^3 & \alpha^3 & \dots & \alpha^3 \\ \alpha^4 & \alpha^4 & \alpha^4 & \dots & \alpha^4 \\ 1 & \alpha & \alpha^2 & \dots & \alpha^{29} \\ \alpha & \alpha^2 & \alpha^3 & \dots & \alpha^{30} \\ \alpha^2 & \alpha^3 & \alpha^4 & \dots & 1 \\ \alpha^3 & \alpha^4 & \alpha^5 & \dots & \alpha \\ \alpha^4 & \alpha^5 & \alpha^6 & \dots & \alpha^2 \\ 1 & \alpha^2 & \alpha^4 & \dots & \alpha^{27} \\ \alpha & \alpha^3 & \alpha^5 & \dots & \alpha^{28} \\ \alpha^2 & \alpha^4 & \alpha^6 & \dots & \alpha^{29} \\ \alpha^3 & \alpha^5 & \alpha^7 & \dots & \alpha^{30} \\ \alpha^4 & \alpha^6 & \alpha^8 & \dots & 1 \\ 1 & \alpha^3 & \alpha^6 & \dots & \alpha^{25} \\ \alpha & \alpha^4 & \alpha^7 & \dots & \alpha^{26} \\ \alpha^2 & \alpha^5 & \alpha^8 & \dots & \alpha^{27} \\ \alpha^3 & \alpha^6 & \alpha^9 & \dots & \alpha^{28} \\ \alpha^4 & \alpha^7 & \alpha^{10} & \dots & \alpha^{27} \\ \dots & \dots & \dots & \dots & \dots \\ 1 & \alpha^{14} & \alpha^{28} & \dots & \alpha^3 \\ \alpha & \alpha^{15} & \alpha^{29} & \dots & \alpha^4 \\ \alpha^2 & \alpha^{16} & \alpha^{30} & \dots & \alpha^5 \\ \alpha^3 & \alpha^{17} & 1 & \dots & \alpha^6 \\ \alpha^4 & \alpha^{18} & \alpha & \dots & \alpha^7 \end{bmatrix} \tag{7.8}$$

$$
\mathbf{H}_{(150,75)} = \left[\begin{array}{ccccc}
10000 & 10000 & 10000 & \ldots & 10000 \\
01000 & 01000 & 01000 & \ldots & 01000 \\
00100 & 00100 & 00100 & \ldots & 00100 \\
00010 & 00010 & 00010 & \ldots & 00010 \\
00001 & 00001 & 00001 & \ldots & 00001 \\
10000 & 01000 & 00100 & \ldots & 10010 \\
01000 & 00100 & 00010 & \ldots & \alpha^{30} \\
00100 & 00010 & 00001 & \ldots & 10000 \\
00010 & 00001 & 10100 & \ldots & 01000 \\
00001 & 10100 & 01010 & \ldots & 00100 \\
10000 & 00100 & 00001 & \ldots & 11010 \\
01000 & 00010 & 10100 & \ldots & 01101 \\
00100 & 00001 & 01010 & \ldots & 10010 \\
00010 & 10100 & 00101 & \ldots & \alpha^{30} \\
00001 & 01010 & 10110 & \ldots & 10000 \\
10000 & 00010 & 01010 & \ldots & 10011 \\
01000 & 00001 & 00101 & \ldots & 11101 \\
00100 & 10100 & 10110 & \ldots & 11010 \\
00010 & 01010 & 01011 & \ldots & 01101 \\
00001 & 00101 & \alpha^{10} & \ldots & 11010 \\
\vdots & \vdots & \vdots & \vdots & \vdots \\
10000 & \alpha^{14} & 01101 & \ldots & 00010 \\
01000 & \alpha^{15} & 10010 & \ldots & 00001 \\
00100 & \alpha^{16} & \alpha^{30} & \ldots & 10100 \\
00010 & \alpha^{17} & 1 & \ldots & 01010 \\
00001 & \alpha^{18} & \alpha & \ldots & 00101
\end{array}\right] \tag{7.9}
$$

The resulting binary code is a (150, 75, 16) code with the d_{min} the same as the symbol-based RS (30, 15, 16) code. As observed by MacWilliams [6], changing the basis can increase the d_{min} of the resulting binary code, and making $j = 3$ in the RS parity-check matrix above produces a (150, 75, 19) binary code.

A (150, 75, 22) binary code with increased d_{min} can be constructed using the overall binary parity-check concatenation as discussed above. Starting with the (25, 15, 11) RS code, an overall parity check is added to each symbol, producing a parity-check matrix, $\mathbf{H}_{(150,75,22)}$ given by matrix (7.10). We have constructed two binary (150, 75) codes from RS codes. It is interesting to compare these codes to the known best code of length 150 and rate $\frac{1}{2}$. The known, best codes are to be found in a database [2] and the best (150, 75) code has a d_{min} of 23 and is derived by shortening by one bit (by deleting the x^{150} coordinate from the $\mathbf{G}$ matrix) of the (151, 76, 23) cyclic code whose generator polynomial is

$$
\left[\begin{array}{ccccc}
100001 & 100001 & 100001 & \ldots & 100001 \\
010001 & 010001 & 010001 & \ldots & 010001 \\
001001 & 001001 & 001001 & \ldots & 001001 \\
000101 & 000101 & 000101 & \ldots & 000101 \\
000011 & 000011 & 000011 & \ldots & 000011
\end{array}\right]
$$

$$\mathbf{H}_{(150,75,22)} = \begin{bmatrix} 100001\ 010001\ 001001 \ldots 100100 \\ 010001\ 001001\ 000101 \ldots 010010 \\ 001001\ 000101\ 000011 \ldots 100001 \\ 000101\ 000011\ 101000 \ldots 010001 \\ 000011\ 101000\ 010100 \ldots 001001 \\ 100001\ 001001\ 000011 \ldots 110101 \\ 010001\ 000101\ 101000 \ldots 011011 \\ 001001\ 000011\ 010100 \ldots 100100 \\ 000101\ 101000\ 001010 \ldots 010010 \\ 000011\ 010100\ 101101 \ldots 100001 \\ 100001\ 000101\ 010100 \ldots 100111 \\ 010001\ 000011\ 001010 \ldots 111010 \\ 001001\ 101000\ 101101 \ldots 110101 \\ 000101\ 010100\ 010111 \ldots 011011 \\ 000011\ 001010\ 100010 \ldots 110101 \\ \ldots \quad \ldots \quad \ldots \quad \ldots \quad \ldots \\ 100001\ 101110\ 011011 \ldots 000101 \\ 010001\ 111111\ 100100 \ldots 000011 \\ 001001\ 110110\ 010010 \ldots 101000 \\ 000101\ 110011 \quad 1 \quad \ldots 010100 \\ 000011\ 110000\ 010001 \ldots 001010 \end{bmatrix} \tag{7.10}$$

$$\begin{aligned} g(x) = {} & 1 + x^3 + x^5 + x^8 + x^{10} + x^{11} + x^{14} + x^{15} + x^{17} + x^{19} + x^{20} + x^{22} \\ & + x^{25} + x^{27} + x^{28} + x^{30} + x^{31} + x^{34} + x^{36} + x^{37} + x^{39} + x^{40} + x^{45} + x^{46} \\ & + x^{48} + x^{50} + x^{52} + x^{59} + x^{60} + x^{63} + x^{67} + x^{70} + x^{73} + x^{74} + x^{75} \end{aligned} \tag{7.11}$$

These three binary codes, the RS-based (150, 75, 19) and (150, 75, 22) codes together with the (150, 75, 23) shortened cyclic code have been simulated using binary transmission for the AWGN channel. The decoder used is a modified Dorsch decoder set to evaluate 2×10^7 codewords per received vector. This is a large number of codewords and is sufficient to ensure that quasi-maximum likelihood performance is obtained. In this way, the true performance of each code is revealed rather than any shortcomings of the decoder.

The results are shown in Fig. 7.3. Also shown in Fig. 7.3, for comparison purposes, is the sphere packing bound and the erasure-based binomial bound discussed in Chap. 1. Interestingly, all three codes have very good performance and are very close to the erasure-based binomial bound. Although not close to the sphere packing bound, this bound is for non-binary codes and there is an asymptotic loss of 0.187 dB for rate $\frac{1}{2}$ binary codes in comparison to the sphere packing bound as the code length extends towards ∞.

Comparing the three codes, no code has the best overall performance over the entire range of $\frac{E_b}{N_0}$, and, surprisingly the d_{min} of the code is no guide. The reason for this can be seen from the Hamming distances of the codewords decoded in error for

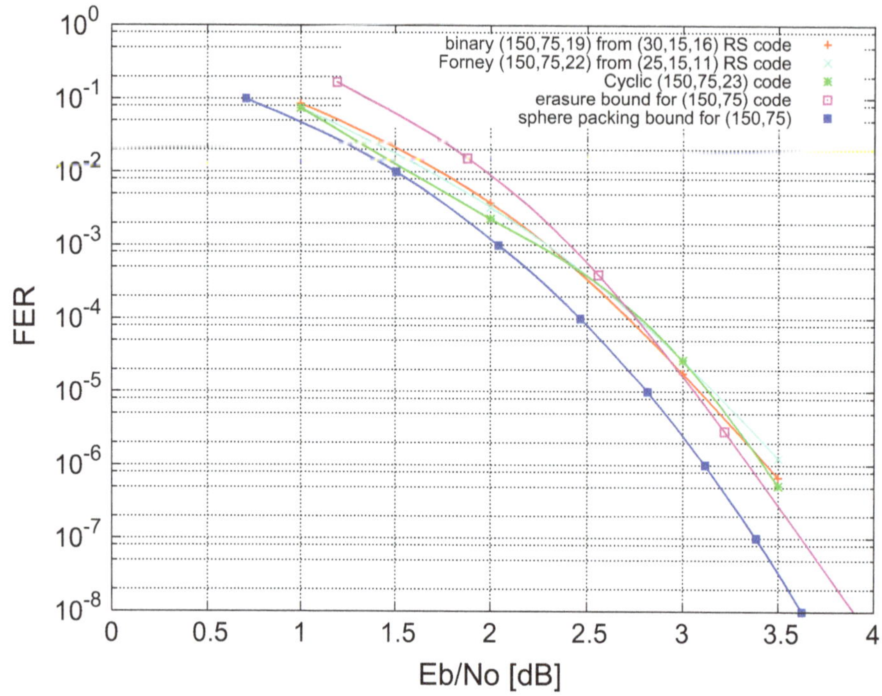

Fig. 7.3 Comparison of the (150, 75, 19) code derived from the RS(30, 15, 16) code, the concatenated (150, 75, 22) code and the known, best (150, 75, 23) code derived by shortening the (151, 76, 23) cyclic code

the three codes after 100 decoder error events. The results are shown in Table 7.2 at $\frac{E_b}{N_0} = 3$ dB. From Table 7.2 it can be seen that the concatenated code (150, 75, 22) has more error events with Hamming distances in the range 22–32, but the (150, 75, 23) known, best code has more error events for Hamming distances up to 36 compared to the (150, 75, 19) RS derived code, and this is the best code at $\frac{E_b}{N_0} = 3$ dB.

The distribution of error events is illustrated by the cumulative distribution of error events plotted in Fig. 7.4 as a function of Hamming distance. The weakness of the (150, 75, 22) code at $\frac{E_b}{N_0} = 3$ dB is apparent.

At higher values of $\frac{E_b}{N_0}$, the higher d_{min} of the (150, 75, 23) known, best code causes it to have the best performance as can be seen from Fig. 7.3.

7.4 Summary

This chapter studied further the Reed–Solomon codes which are ideal symbol-based codes because they are Maximum Distance Separable (MDS) codes. These codes are not binary codes but were considered for use as binary codes in this chapter. The performance of Reed–Solomon codes when used on a binary channel was explored and compared to codes which are designed for binary transmission. The construction of the parity-check matrices of RS codes for use as binary codes was described

Table 7.2 Hamming distances and multiplicities of 100 error events for each of the (150, 75) codes at $\frac{E_b}{N_0} = 3$ dB

Hamming distance	(150, 75, 19) Code number	(150, 75, 22) Code number	(150, 75, 23) Code number
22	0	4	0
24	0	4	0
25	1	0	0
26	1	9	0
27	3	0	7
28	5	7	6
29	4	0	0
30	8	22	0
31	7	0	15
32	7	19	20
33	14	0	0
34	8	14	0
35	5	0	19
36	8	13	18
37	3	0	0
38	8	5	0
39	7	0	9
40	6	1	2
41	2	0	0
42	1	2	0
43	1	0	1
44	1	0	1
47	0	0	1
48	0	0	1

in detail for specific code examples. The performance results of three differently constructed (150, 75) codes simulated for the binary AWGN channel, using a near maximum likelihood decoder, were presented. Surprisingly the best performing code at 10^{-4} error rate is not the best, known (150, 75, 23) code. Error event analysis was presented which showed that this was due to the higher multiplicities of weight 32–36 codeword errors. However, beyond 10^{-6} error rates the best, known (150, 75, 23) code was shown to be the best performing code.

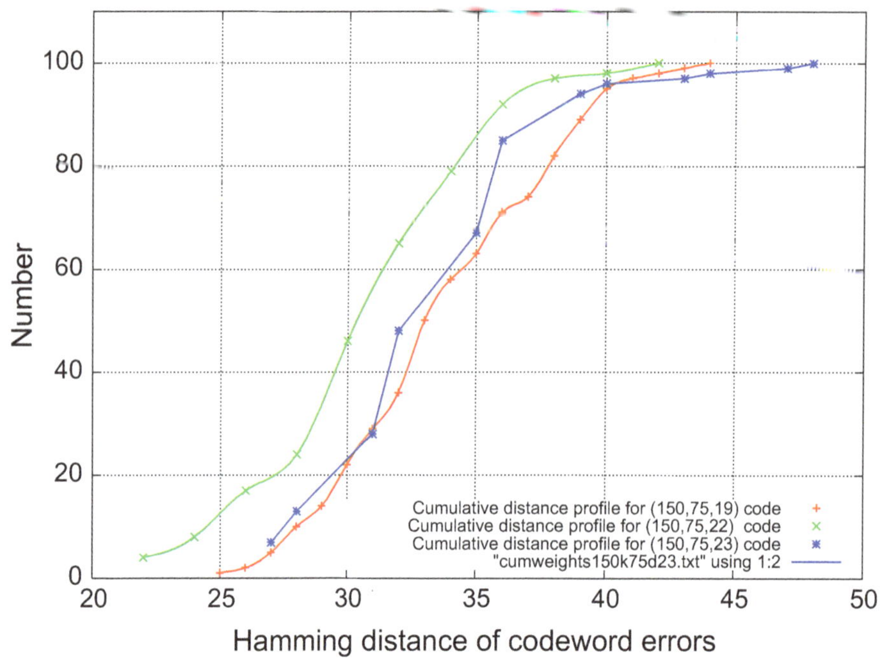

Fig. 7.4 Cumulative distribution of Hamming distance error events for the (150, 75, 19) code derived from the RS(30, 15, 16) code, the RS binary parity-check concatenated (150, 75, 22) code and the known, best (150, 75, 23) code derived by shortening the (151, 76, 23) cyclic code

References

1. El-Khamy, M., McEliece, R.J.: Iterative algebraic soft decision decoding of Reed–Solomon codes. Int. Symp. Inf. Theory Appl. **2004**, 1456–1461 (2004)
2. Grassl, M.: Code Tables: bounds on the parameters of various types of codes. http://www.codetables.de (2007)
3. Guruswami, V., Sudan, M.: Improved decoding of Reed–Solomon and algebraic-geometry codes. IEEE Trans. Inf. Theory **45**(6), 1757–1767 (1999)
4. Kasahara, M., Sugiyama, Y., Hirasawa, S., Namekawa, T.: New classes of binary codes constructed on the basis of concatenated codes and product codes. IEEE Trans. Inf. Theory IT-22 **4**, 462–468 (1976)
5. Koetter, R., Vardy, A.: Algebraic soft-decision decoding of Reed–Solomon codes. IEEE Trans. Inf. Theory **49**(11), 2809–2825 (2003)
6. MacWilliams, F.J., Sloane N.J.A.: The Theory of Error-Correcting Codes. North-Holland (1977)
7. McEliece, R.J.: The Guruswami–Sudan decoding algorithm for Reed–Solomon codes. JPL TDA Prog. Rep. **42**(153), 1–60 (2003)
8. Proakis, J.: Digital Communications, 4th edn. McGraw-Hill, New York (2001)
9. Reed, I., Solomon, G.: Polynomial codes over certain finite fields. J. Soc. Ind. Appl. Math. **8**, 300–304 (1960)
10. Sudan, M.: Decoding of Reed–Solomon codes beyond the error-correction bound. J. Complex. **13**, 180–193 (1997)

Chapter 8
Algebraic Geometry Codes

8.1 Introduction

In order to meet channel capacity, as Shannon postulated, long error-correction codes with large minimum distances need to be found. A large effort in research has been dedicated to finding algebraic codes with good properties and efficient decoding algorithms. Reed–Solomon (RS) codes are a product of this research and have over the years found numerous applications, the most noteworthy being their implementation in satellite systems, compact discs, hard drives and modern, digitally based communications. These codes are defined with non-binary alphabets and have the maximum achievable minimum distance for codes of their lengths. A generalisation of RS codes was introduced by Goppa using a unique construction of codes from algebraic curves. This development led to active research in that area so that currently the complexity of encoding and decoding these codes has been reduced greatly from when they were first presented. These codes are algebraic geometry (AG) codes and have much greater lengths than RS codes with the same alphabets. Furthermore these codes can be improved if curves with desirable properties can be found. AG codes have good properties and some families of these codes have been shown to be asymptotically superior as they exceed the well-known Gilbert–Varshamov bound [16] when the defining finite field $\mathbb{F}_q$ has size $q \geq 49$ with q always a square.

8.2 Motivation for Studying AG Codes

Aside from their proven superior asymptotic performance when the field size $q^2 > 49$, AG codes defined in much smaller fields have very good parameters. A closer look at tables of best-known codes in [8, 15] shows that algebraic geometry codes feature as the best-known linear codes for an appreciable range of code lengths for

M. Tomlinson et al., *Error-Correction Coding and Decoding*,
Signals and Communication Technology,
DOI 10.1007/978-3-319-51103-0_8

different field sizes q. To demonstrate a comparison the parameters of AG codes with shortened BCH codes in fields with small sizes and characteristic 2 is given. AG codes of length n, dimension k have minimum distance $d = n - k - g + 1$ where g is called the genus . Notice that $n - k + 1$ is the distance of a maximum distance (MDS) separable code. The genus g is then the Singleton defect s of an AG code. The Singleton defect is simply the difference between the distance of a code and the distance some hypothetical MDS code of the same length and dimension. Similarly a BCH code is a code with length n, dimension k, and distance $d = n - k - s + 1$ where s is the Singleton defect and number of non-consecutive roots of the BCH code.

Consider Table 8.1, which compares the parameters of AG codes from three curves with genera 3, 7, and 14 with shortened BCH codes with similar code rates. At high rates, BCH codes tend to have better minimum distances or smaller Singleton defects. This is because the roots of the BCH code with high rates are usually cyclically consecutive and thus contribute to the minimum distance. For rates close to half, AG codes are better than BCH codes since the number of non-consecutive roots of the BCH code is increased as a result of conjugacy classes. The AG codes benefit from the fact that their Singleton defect or genus remains fixed for all rates. As a consequence AG codes significantly outperform BCH codes at lower rates. However, the genera of curves with many points in small finite fields are usually large and as the length of the AG codes increases in $\mathbb{F}_8$, the BCH codes beat AG codes in performance. Tables 8.2 and 8.3 show the comparison between AG and BCH codes in fields $\mathbb{F}_{16}$ and $\mathbb{F}_{32}$, respectively. With larger field sizes, curves with many points and small genera can be used, and AG codes do much better than BCH codes. It is worth noting that Tables 8.1, 8.2 and 8.3 show codes in fields with size less than 49.

8.2.1 Bounds Relevant to Algebraic Geometry Codes

Bounds on the performance of codes that are relevant to AG codes are presented in order to show the performance of these codes. Let $A_q(n, d)$ represent the number of codewords in the code space of a code $\mathscr{C}$ with length n, minimum distance d and defined over a field of size q. Let the information rate be $R = k/n$ and the relative minimum distance be $\delta = d/n$ for $0 \leq \delta \leq 1$, then

$$\alpha_q(\delta) = \lim_{n\to\infty} \frac{1}{n} A_q(n, \delta n)$$

which represents the k/n such that there exists a code over a field of size q that has d/n converging to δ [18]. The q-ary entropy function is given by

$$H_q(x) = \begin{cases} 0, & x = 0 \\ x\log_q(q-1) - x\log_q x - (1-x)\log_q(1-x), & 0 < x \leq \theta \end{cases}$$

Table 8.1 Comparison between BCH and AG codes in $\mathbb{F}_8$

Rate	AG code in $\mathbb{F}_{2^3}$	Number of points	Genus	Shortened BCH code in $\mathbb{F}_{2^3}$	BCH code in $\mathbb{F}_{2^3}$
0.2500	[23, 5, 16]	24	3	[23, 5, 12]	[63, 45, 12]
0.3333	[23, 7, 14]	24	3	[23, 7, 11]	[63, 47, 11]
0.5000	[23, 11, 10]	24	3	[23, 10, 8]	[63, 50, 8]
0.6667	[23, 15, 6]	24	3	[23, 14, 6]	[63, 54, 6]
0.7500	[23, 17, 4]	24	3	[23, 16, 5]	[63, 56, 5]
0.8500	[23, 19, 2]	24	3	[23, 18, 4]	[63, 58, 4]
0.2500	[33, 8, 19]	34	7	[33, 7, 16]	[63, 37, 16]
0.3333	[33, 11, 16]	34	7	[33, 11, 14]	[63, 41, 14]
0.5000	[33, 16, 11]	34	7	[33, 15, 12]	[63, 45, 12]
0.6667	[33, 22, 5]	34	7	[33, 22, 7]	[63, 52, 7]
0.7500	[33, 24, 3]	34	7	[33, 24, 6]	[63, 54, 6]
0.2500	[64, 16, 35]	65	14	[64, 16, 37]	[63, 15, 37]
0.3333	[64, 21, 30]	65	14	[64, 20, 31]	[63, 19, 31]
0.5000	[64, 32, 19]	65	14	[64, 31, 22]	[63, 30, 22]
0.6667	[64, 42, 9]	65	14	[64, 42, 14]	[63, 41, 14]
0.7500	[64, 48, 3]	65	14	[64, 48, 11]	[63, 47, 11]

Table 8.2 Comparison between BCH and AG codes in $\mathbb{F}_{16}$

Rate	AG code in $\mathbb{F}_{2^4}$	Number of points	Genus	Shortened BCH code in $\mathbb{F}_{2^4}$	BCH code in $\mathbb{F}_{2^4}$
0.2500	[23, 5, 18]	24	1	[23, 4, 11]	[255, 236, 11]
0.3333	[23, 7, 16]	24	1	[23, 6, 10]	[255, 238, 10]
0.5000	[23, 11, 12]	24	1	[23, 10, 8]	[255, 242, 8]
0.6667	[23, 15, 8]	24	1	[23, 14, 6]	[255, 246, 6]
0.7500	[23, 17, 6]	24	1	[23, 16, 5]	[255, 248, 5]
0.8500	[23, 19, 4]	24	1	[23, 18, 4]	[255, 250, 4]
0.2500	[64, 16, 43]	65	6	[64, 16, 27]	[255, 207, 27]
0.3333	[64, 21, 38]	65	6	[64, 20, 25]	[255, 211, 25]
0.5000	[64, 32, 27]	65	6	[64, 32, 19]	[255, 223, 19]
0.6667	[64, 42, 17]	65	6	[64, 41, 13]	[255, 232, 13]
0.7500	[64, 48, 11]	65	6	[64, 47, 10]	[255, 238, 10]
0.8500	[64, 54, 5]	65	6	[64, 53, 7]	[255, 244, 7]
0.2500	[126, 31, 76]	127	20	[126, 30, 57]	[255, 159, 57]
0.3333	[126, 42, 65]	127	20	[126, 41, 48]	[255, 170, 48]
0.5000	[126, 63, 44]	127	20	[126, 63, 37]	[255, 192, 37]
0.6667	[126, 84, 23]	127	20	[126, 84, 24]	[255, 213, 24]
0.7500	[126, 94, 13]	127	20	[126, 94, 19]	[255, 223, 19]

Table 8.3 Comparison between BCH and AG codes in $\mathbb{F}_{32}$

Rate	AG code in $\mathbb{F}_{2^4}$	Number of points	Genus	Shortened BCH code in $\mathbb{F}_{2^4}$	BCH code in $\mathbb{F}_{2^4}$
0.2500	[43, 10, 33]	44	1	[43, 10, 18]	[1023, 990, 18]
0.3333	[43, 14, 29]	44	1	[43, 14, 16]	[1023, 994, 16]
0.5000	[43, 21, 22]	44	1	[43, 20, 13]	[1023, 1000, 13]
0.6667	[43, 28, 15]	44	1	[43, 28, 9]	[1023, 1008, 9]
0.7500	[43, 32, 11]	44	1	[43, 32, 7]	[1023, 1012, 7]
0.8500	[43, 36, 7]	44	1	[43, 36, 5]	[1023, 1016, 5]
0.2500	[75, 18, 53]	76	5	[75, 18, 30]	[1023, 966, 30]
0.3333	[75, 25, 46]	76	5	[75, 24, 27]	[1023, 972, 27]
0.5000	[75, 37, 34]	76	5	[75, 36, 21]	[1023, 984, 21]
0.6667	[75, 50, 21]	76	5	[75, 50, 14]	[1023, 998, 14]
0.7500	[75, 56, 15]	76	5	[75, 56, 11]	[1023, 1004, 11]
0.8500	[75, 63, 8]	76	5	[75, 62, 8]	[1023, 1010, 8]
0.2500	[103, 25, 70]	104	9	[103, 25, 42]	[1023, 945, 42]
0.3333	[103, 34, 61]	104	9	[103, 33, 38]	[1023, 953, 38]
0.5000	[103, 51, 44]	104	9	[103, 50, 28]	[1023, 970, 28]
0.6667	[103, 68, 27]	104	9	[103, 68, 19]	[1023, 988, 19]
0.7500	[103, 77, 18]	104	9	[103, 76, 15]	[1023, 996, 15]
0.8500	[103, 87, 8]	104	9	[103, 86, 10]	[1023, 1006, 10]

The asymptotic Gilbert–Varshamov lower bound on $\alpha_q(\delta)$ is given by,

$$\alpha_q(\delta) \geq 1 - H_q(\delta) \quad \text{for } 0 \leq \delta \leq \theta$$

The Tsfasman–Vladut–Zink bound is a lower bound on $\alpha_q(\delta)$ and holds true for certain families of AG codes, it is given by

$$\alpha_q(\delta) \geq 1 - \delta - \frac{1}{\sqrt{q} - 1} \quad \text{where } \sqrt{q} \in \mathbb{N}/0$$

The supremacy of AG codes lies in the fact that the TVZ bound ensures that these codes have better performance when q is a perfect square and $q \geq 49$.

The Figs. 8.1, 8.2 and 8.3 show the R vs δ plot of these bounds for some range of q.

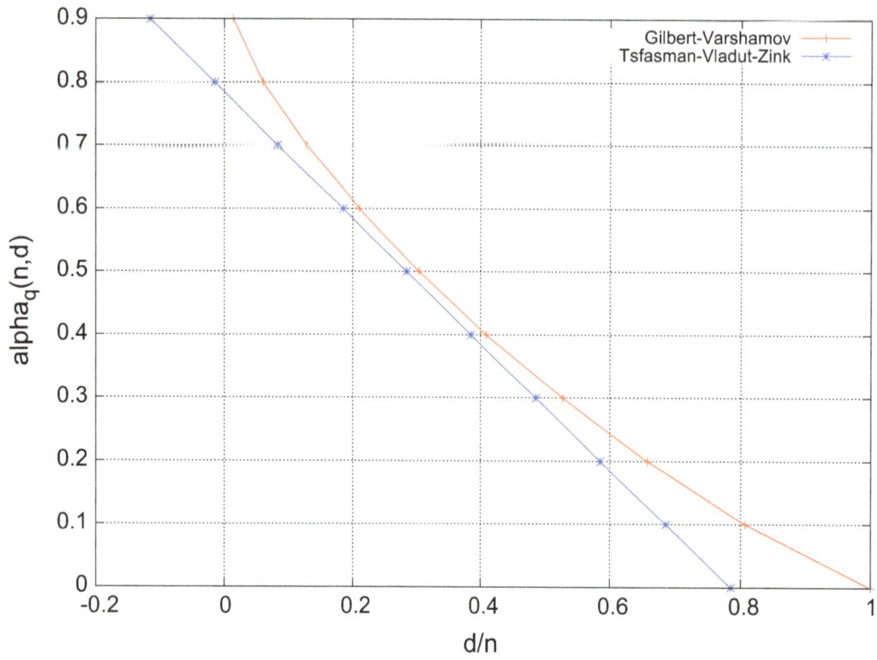

Fig. 8.1 Tsfasman–Vladut–Zink and Gilbert–Varshamov bound for $q = 32$

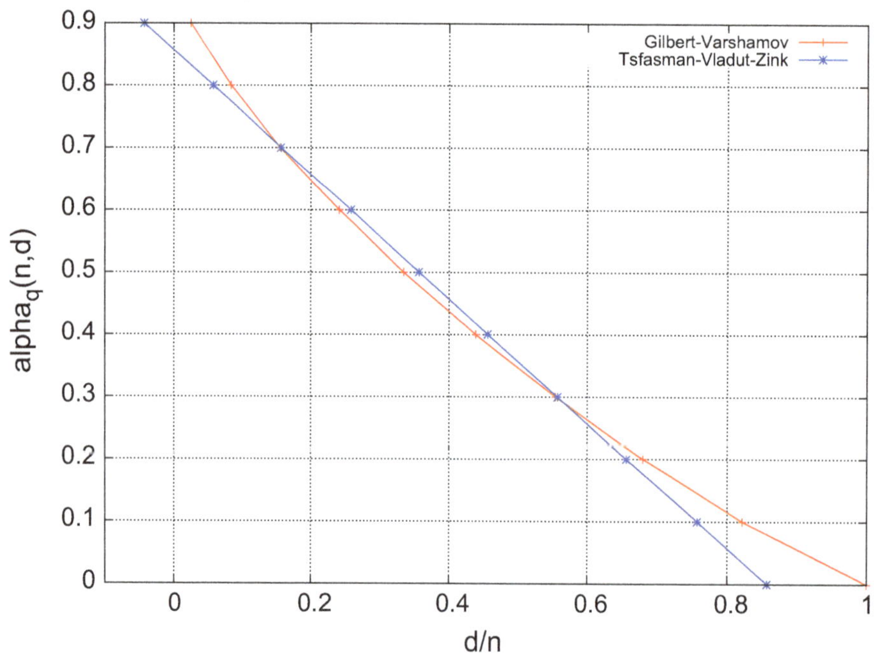

Fig. 8.2 Tsfasman–Vladut–Zink and Gilbert–Varshamov bound for $q = 64$

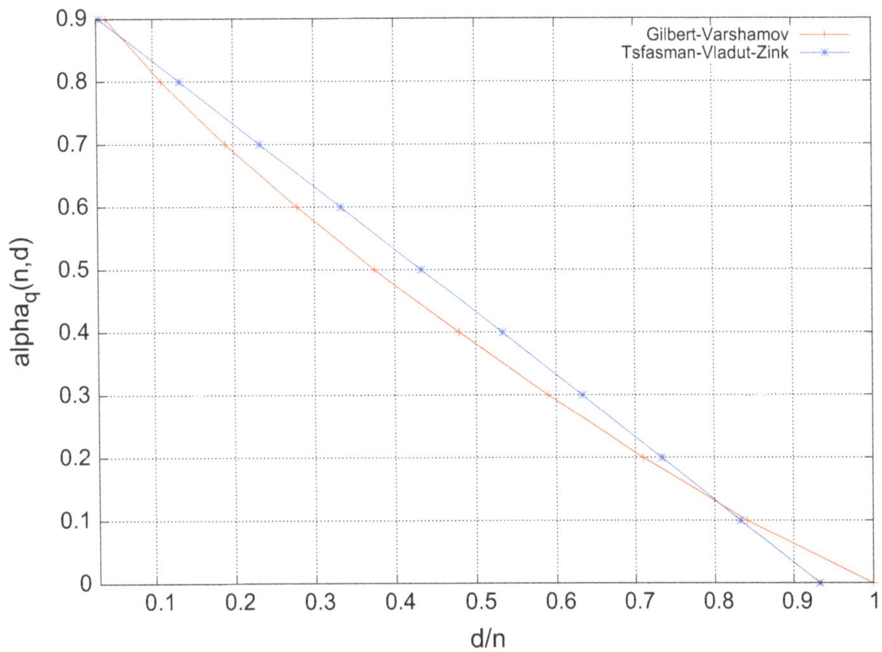

Fig. 8.3 Tsfasman–Vladut–Zink and Gilbert–Varshamov bound for $q = 256$

8.3 Curves and Planes

In this section, the notion of curves and planes are introduced. Definitions and discussions are restricted to two-dimensional planes and all polynomials are assumed to be defined with coefficients in the finite field $\mathbb{F}_q$. The section draws from the following sources [2, 12, 17, 18]. Let $f(x, y)$ be a polynomial in the bivariate ring $\mathbb{F}_q[x, y]$.

Definition 8.1 (*Curve*) A curve is the set of points for which the polynomial $f(x, y)$ vanishes to zero. Mathematically, a curve $\mathscr{X}$ is associated with a polynomial $f(x, y)$ so that $f(P) = \{0 | P \in \mathscr{X}\}$.

A curve is a subset of a plane. There are two main types of planes; the affine plane and the projective plane. These planes are multidimensional, however, we restrict our discussion to two-dimensional planes only.

Definition 8.2 (*Affine Plane*) A two-dimensional affine plane denoted by $\mathbb{A}^2(\mathbb{F}_q)$ is a set of points,

$$\mathbb{A}^2(\mathbb{F}_q) = \{(\alpha, \beta) : \alpha, \beta \in \mathbb{F}_q\} \tag{8.1}$$

which has cardinality q^2.

A curve $\mathscr{X}$ is called an affine curve if $\mathscr{X} \subset \mathbb{A}^2(\mathbb{F}_q)$.

Definition 8.3 (*Projective Plane*) A two-dimensional projective plane $\mathbb{P}^2(\mathbb{F}_q)$ is the algebraic closure of $\mathbb{A}^2$ and is defined as the set of equivalence points,

$$\mathbb{P}^2(\mathbb{F}_q) = \{(\alpha : \beta : 1) : \alpha, \beta \in \mathbb{F}_q\} \bigcup \{(\alpha : 1 : 0) : \alpha \in \mathbb{F}_q\} \bigcup \{(1 : 0 : 0)\}.$$

A curve $\mathscr{X}$ is said to lie in the projective plane if $\mathscr{X} \subset \mathbb{P}^2(\mathbb{F}_q)$. The points in the projective plane are called equivalence points since for any point $P \in \mathbb{P}^2(\mathbb{F}_q)$,

$$\text{if } f(x_0, y_0, z_0) = 0, \quad \text{then } f(\alpha x_0, \alpha y_0, \alpha z_0) = 0 \qquad \alpha \in \mathbb{F}_q^*, \; P = (x_0 : y_0 : z_0)$$

because $f(x, y, z)$ is homogeneous. The colons in the notation of a projective point $(x : y : z)$ represents this equivalence property.

The affine polynomial $f(x, y)$ is in two variables, in order to define a projective polynomial in three variables, *homogenisation* is used,

$$f(x, y, z) = z^d f\left(\frac{x}{z}, \frac{y}{z}\right) \qquad d = \text{Degree of } f(x, y)$$

which turns $f(x, y)$ into a homogeneous[1] polynomial in three variables. An n-dimensional projective polynomial has $n + 1$ variables. The affine space $\mathbb{A}^2(\mathbb{F}_q)$ is a subset of $\mathbb{P}^2(\mathbb{F}_q)$ and is given by,

$$\mathbb{A}^2(\mathbb{F}_q) = \{(\alpha : \beta : 1) : \alpha, \beta \in \mathbb{F}_q\} \subset \mathbb{P}^2(\mathbb{F}_q).$$

A projective curve can then be defined as a set of points,

$$\mathscr{X} = \{P : f(P) = 0, \; P \in \mathbb{P}^2(\mathbb{F}_q)\}.$$

Definition 8.4 (*Point at Infinity*) A point on a projective curve $\mathscr{X}$ that coincides with any of the points of $\mathbb{P}^2(\mathbb{F}_q)$ of the form,

$$\{(\alpha : 1 : 0) : \alpha \in \mathbb{F}_q\} \cup \{(1 : 0 : 0)\}$$

i.e. points $(x_0 : y_0 : z_0)$ for which $z_0 = 0$ is called a point at infinity.

A third plane, called the bicyclic plane [1], is a subset of the $\mathbb{A}^2(\mathbb{F}_q)$ and consists of points,

$$\{(\alpha, \beta) : \alpha, \beta \in \mathbb{F}_q \setminus \{0\}\}.$$

This plane was defined so as to adapt the Fourier transform to AG codes since the inverse Fourier transform is undefined for zero coordinates.

Example 8.1 Consider the two-dimensional affine plane $\mathbb{A}^2(\mathbb{F}_4)$. Following the definition of $\mathbb{A}^2(\mathbb{F}_4)$ we have,

---

[1]Each term in the polynomial has degree equal to d.

$$\begin{array}{cccc} (0,0) & (0,1) & (1,0) & (1,1) \\ (1,\alpha) & (\alpha,1) & (1,\alpha^2) & (\alpha^2,1) \\ (\alpha^2,\alpha) & (\alpha,\alpha^2) & (0,\alpha^2) & (0,\alpha) \\ (\alpha^2,0) & (\alpha,0) & (\alpha^2,\alpha^2) & (\alpha,\alpha) \end{array}$$

where α is the primitive element of the finite field $\mathbb{F}_4$. The two-dimensional projective plane $\mathbb{P}^2(\mathbb{F}_4)$ is given by,

Affine Points				Points at Infinity	
$(0:0:1)$	$(0:1:1)$	$(1:0:1)$	$(1:1:1)$	$(0:1:0)$	$(1:0:0)$
$(1:\alpha:1)$	$(\alpha:1:1)$	$(1:\alpha^2:1)$	$(\alpha^2:1:1)$	$(\alpha:1:0)$	
$(\alpha^2:\alpha:1)$	$(\alpha:\alpha^2:1)$	$(0:\alpha^2:1)$	$(0:\alpha:1)$	$(\alpha^2:1:0)$	
$(\alpha^2:0:1)$	$(\alpha:0:1)$	$(\alpha^2:\alpha^2:1)$	$(\alpha:\alpha:1)$	$(1:1:0)$	

Definition 8.5 (*Irreducible Curve*) A curve associated with a polynomial $f(x, y, z)$ that cannot be reduced or factorised is called *irreducible*.

Definition 8.6 (*Singular Point*) A point on a curve is singular if its evaluation on all partial derivatives of the defining polynomial with respect to each indeterminate is zero.

Suppose f_x, f_y, and f_z denote partial derivatives of $f(x, y, z)$ with respect to x, y, and z respectively. A point $P \in \mathscr{X}$ is singular if,

$$\frac{\partial f(x,y,z)}{\partial x} = f_x, \; \frac{\partial f(x,y,z)}{\partial y} = f_y, \; \frac{\partial f(x,y,z)}{\partial z} = f_z$$
$$f_x(P) = f_y(P) = f_z(P) = 0.$$

Definition 8.7 (*Smooth Curve*) A curve $\mathscr{X}$ is nonsingular or smooth does not contain any singular points.

To obtain AG codes, it is required that the defining curve is both irreducible and smooth.

Definition 8.8 (*Genus*) The genus of a curve can be seen as a measure of how many bends a curve has on its plane. The genus of a smooth curve defined by $f(x, y, z)$ is given by the Plücker formula,

$$g = \frac{(d-1)(d-2)}{2}, \quad d = \text{Degree of } f(x, y, z)$$

The genus plays an important role in determining the quality of AG codes. It is desirable for curves that define AG codes to have small genera.

Example 8.2 Consider the Hermitian curve in $\mathbb{F}_4$ defined as,

$$f(x, y) = x^3 + y^2 + y \quad \text{affine}$$
$$f(x, y, z) = x^3 + y^2z + yz^2 \quad \text{projective via homogenisation}$$

It is straightforward to verify that the curve is irreducible. The curve has the following projective points,

$$(0:0:1) \quad (0:1:1) \quad (\alpha:\alpha:1) \; (\alpha:\alpha^2:1)$$
$$(\alpha^2:\alpha:1) \; (\alpha^2:\alpha^2:1) \; (1:\alpha:1) \; (1:\alpha^2:1) \; (0:1:0)$$

Notice the curve has a single point at infinity $P_\infty = (0:1:0)$. One can easily check that the curve has no singular points and is thus smooth.

8.3.1 Important Theorems and Concepts

The length of an AG code is utmost the number of points on the defining curve. Since it is desirable to obtain codes that are as long as possible, it is desirable to know what the maximum number of points attainable from a curve, given a genus is.

Theorem 8.1 (Hasse–Weil with Serre's Improvement [2]) *The Hasse–Weil theorem with Serre's improvement says that the number of rational points*[2] *of an irreducible curve, n, with genus g in* $\mathbb{F}_q$ *is upper bounded by,*

$$n \leq q + 1 + g\lfloor 2\sqrt{q} \rfloor.$$

Curves that meet this bound are called *maximal* curves. The Hermitian curves are examples of maximal curves. Bezout's theorem is an important theorem, and is used to determine the minimum distance of algebraic geometry codes. It describes the size of the set which is the intersection of two curves in the projective plane.

Theorem 8.2 (Bezout's Theorem [2]) *Any two curves* $\mathcal{X}_a$ *and* $\mathcal{X}_b$ *with degrees of their associated polynomials as m and n respectively, have utmost* $m \times n$ *common roots in the projective plane counted with multiplicity.*

Definition 8.9 (*Divisor*) A divisor on a curve $\mathcal{X}$ is a formal sum associated with the points of the curve.

$$D = \sum_{P \in \mathcal{X}} n_p P$$

where n_p are integers.

---

[2] A rational point is a point of degree one. See Sect. 8.4 for the definition of the degree of point on a curve.

A zero divisor is one that has $n_p = 0$ for all $P \in \mathscr{X}$. A divisor is called effective if it is not a zero divisor. The support of a divisor is a subset of $\mathscr{X}$ for which $n_p \neq 0$. The degree of a divisor is given as,

$$deg(D) = \sum_{P \in \mathscr{X}} n_p \, deg(P)$$

For simplicity, it is assumed that the degree of points $P \in \mathscr{X}$, i.e. $deg(P)$ is 1 (points of higher degree are discussed in Sect. 8.4). Addition of two divisors $D_1 = \sum_{P \in \mathscr{X}} n_p P$ and $D_2 = \sum_{P \in \mathscr{X}} \acute{n}_p P$ is so defined,

$$D_1 + D_2 = \sum_{P \in \mathscr{X}} (n_p + \acute{n}_p) P.$$

Divisors are simply book-keeping structures that store information on points of a curve. Below is an example the intersection divisor of two curves.

Example 8.3 Consider the Hermitian curve in $\mathbb{F}_4$ defined as,

$$f_1(x, y, z) = x^3 + y^2 z + y z^2$$

with points given in Example 8.2 and the curve defined by

$$f_2(x, y, z) = x$$

with points

$$(0:0:1)\ (0:1:1)\ (0:\alpha:1)\ (0:\alpha^2:1)\ (0:1:0)$$

These two curves intersect at 3 points below all with multiplicity 1,

$$(0:0:1)\ (0:1:0)\ (0:1:1).$$

Alternatively, this may be represented using a divisor D,

$$D = (0:0:1) + (0:1:0) + (0:1:1)$$

with n_p the multiplicity, equal to 1 for all the points. Notice that the two curves meet at exactly $deg(f_1)deg(f_2) = 3$ points in agreement with Bezout's theorem.

For rational functions with denominators, points in divisor with $n_p < 0$ are poles. For example, $D = P_1 - 2P_2$ will denote an intersection divisor of two curves that have one zero P_1 and pole P_2 with multiplicity two in common. Below is the formal definition of the field of fractions of a curve $\mathscr{X}$.

Definition 8.10 (*Field of fractions*) The field of fractions $\mathbb{F}_q(\mathscr{X})$ of a curve $\mathscr{X}$ defined by a polynomial $f(x, y, z)$ contains all rational functions of the form

$$\frac{g(x, y, z)}{h(x, y, z)}$$

with the restriction that $g(x, y, z)$ and $h(x, y, z)$ are homogeneous polynomials that have the same degree and are not divisible by $f(x, y, z)$.

A subset (Riemann–Roch space) of the field of fractions of $\mathscr{X}$ meeting certain conditions are evaluated at points of the curve $\mathscr{X}$ to form codewords of an AG code. Thus, there is a one-to-one mapping between rational functions in this subset and codewords of an AG code. The Riemann–Roch theorem defines this subset and gives a lower bound on the dimension of AG codes. The definition of a Riemann–Roch space is given.

Definition 8.11 (*Riemann–Roch Space*) The Riemann–Roch space associated with a divisor D is given by,

$$L(D) = \{t \in \mathbb{F}_q(\mathscr{X}) | (t) + D \geq 0\} \cup 0$$

where $\mathbb{F}_q(\mathscr{X})$ is the field of fractions and (t) is the intersection divisor[3] of the rational function t and the curve $\mathscr{X}$.

Essentially, the Riemann–Roch space associated with a divisor D is a set of functions of the form t from the field of fractions $\mathbb{F}_q(\mathscr{X})$ such that the divisor sum $(t) + D$ has no poles, i.e. $(t) + D \geq 0$.

The rational functions in $L(D)$ are functions from the field of fractions $\mathbb{F}_q(\mathscr{X})$ that must have poles only in the zeros (positive terms) contained in the divisor D, each pole occurring with utmost the multiplicity defined in the divisor D and most have zeros only in the poles (negative terms) contained in the divisor D, each zero occurring with at least the multiplicity defined in the divisor D.

Example 8.4 Suppose a hypothetical curve $\mathscr{X}$ has points of degree one,

$$\mathscr{X} = \{P_1, P_2, P_3, P_4\}$$

We choose a divisor $D = 2P_1 - 5P_2$ with degree -3, and define a Riemann–Roch space $L(D)$. If we randomly select three functions t_1, t_2, and t_3 from the field of fractions $\mathbb{F}_q(\mathscr{X})$ such that they have divisors,

$$(t_1) = -3P_1 + 5P_2 + 4P_4 \; (t_2) = 2P_1 + 4P_2 \; (t_3) = -P_1 + 8P_2 + P_3.$$

$t_1 \notin L(D)$ since $(t_1) + D = -P_1 + 4P_4$ contains negative terms or poles. Also, $t_2 \notin L(D)$ since $(t_2) + D = 4P_1 - P_2$ contains negative terms. However, $t_3 \in L(D)$ since $(t_3) + D = P_1 + 3P_2 + P_3$ contains no negative terms. Any function $t \in \mathbb{F}_q(\mathscr{X})$ is also in $L(D)$ if it has a pole at P_1 with multiplicity at most 2 (with no other poles in common with $\mathscr{X}$) and a zero at P_2 with multiplicity at least 5.

[3] An intersection divisor is a divisor that contains information on the points of intersection of two curves.

The Riemann–Roch space is a vector space (with rational functions as elements) thus, a set of basis functions. The size of this set is the dimension of the space.

Theorem 8.3 (Riemann Roch Theorem [2]) *Let $\mathscr{X}$ be a curve with genus g and D any divisor with degree $(D) > 2g - 2$, then the dimension of the Riemann–Roch space associated with D, denoted by $l(D)$ is,*

$$l(D) = degree(D) - g + 1$$

Algebraic geometry codes are the image of an evaluation map of a Riemann–Roch space associated with a divisor D so that

$$L(D) \rightarrow \mathbb{F}_q^n$$

$$t \rightarrow (t(P_1), t(P_2), \ldots, t(P_n))$$

where $\mathscr{X} = \{P_1, P_2, \ldots, P_n, P_x\}$ is a smooth irreducible projective curve of genus g defined over $\mathbb{F}_q$. The divisor D must have no points in common with a divisor T associated with $\mathscr{X}$, i.e. it has support disjoint from T. For example, if the divisor T is of the form

$$T = P_1 + P_2 + \cdots + P_n$$

then, $D = mP_x$.

Codes defined by the divisors T and $D = mP_x$ are called one-point AG codes (since the divisor D has a support containing only one point), and AG codes are predominantly defined as so since the parameters of such codes are easily determined [10].

8.3.2 Construction of AG Codes

The following steps are necessary in order to construct a generator matrix of an AG code,

1. Find the points of a smooth irreducible curve and its genus.
2. Choose divisors D and $T = P_1 + \cdots + P_n$. From the Riemann–Roch theorem determine the dimension of the Riemann–Roch space $L(D)$ associated with divisor D. This dimension $l(D)$ is the dimension of the AG code.
3. Find $k = l(D)$ linearly independent rational functions from $L(D)$. These form the basis functions of $L(D)$.
4. Evaluate all k basis functions on the points in the support of T to form the k rows of a generator matrix of the AG code.

Example 8.5 Consider again the Hermitian curve defined in $\mathbb{F}_4$ as,

$$f(x, y, z) = x^3 + y^2z + yz^2$$

1. In Example 8.2 this curve was shown to have 8 affine points and one point at infinity. The genus of this curve is given by the Plücker formula,

$$g = \frac{(r-1)(r-2)}{2} = 1$$

   where $r = 3$ is the degree of $f(x, y, z)$.
2. Let $D = 5P_\infty$ where $P_\infty = (0\colon 1\colon 0)$ and T be the sum of all 8 affine points. The dimension of the Riemann–Roch space is then given by,

$$l(5P_\infty) = 5 - 1 + 1 = 5$$

   thus, the AG code has dimension $k = 5$.
3. The basis functions for the space $L(5P_\infty)$ are

$$\{t_1, \ldots, t_k\} = \left\{1, \frac{x}{z}, \frac{x^2}{z^2}, \frac{y}{z}, \frac{xy}{z^2}\right\}$$

   By examining the basis, it is clear that $t_1 = 1$ has no poles, thus, $(t_1) + D$ has no poles also. Basis functions with denominator z have $(t_i) = S - P_\infty$, where S is a divisor of the numerator. Thus, $(t_i) + D$ has no poles. Basis functions with denominator z^2 have $(t_j) = S - 2P_\infty$, where S is a divisor of the numerator. Thus, $(t_j) + D$ also has no poles.
4. The generator matrix of the Hermitian code defined with divisor $D = 5P_\infty$ is thus,

$$G = \begin{bmatrix} t_1(P_1) & \cdots & t_1(P_n) \\ \vdots & \ddots & \vdots \\ t_k(P_1) & \cdots & t_k(P_n) \end{bmatrix} = \begin{bmatrix} 1\,0\,0\,0\,0 & \alpha^2 & \alpha^2 & 1 \\ 0\,1\,0\,0\,0 & \alpha^2 & \alpha & 0 \\ 0\,0\,1\,0\,0 & \alpha & 1 & \alpha \\ 0\,0\,0\,1\,0 & \alpha & 0 & \alpha^2 \\ 0\,0\,0\,0\,1 & 1 & 1 & 1 \end{bmatrix}$$

Example 8.6 Consider the curve defined in $\mathbb{F}_8$ as,

$$f(x, y, z) = x$$

1. This curve is a straight line and has 8 affine points of the form $(0 : \beta : 1)$ and one point at infinity $(0 : 1 : 0)$. The curve is both irreducible and smooth. The genus of this curve is given by the Plücker formula,

$$g = \frac{(r-1)(r-2)}{2} = 0$$

where $r = 1$ is the degree of $f(x, y, z)$. Clearly, the genus is zero since the curve is straight line and has no bends.
2. Let $D = 5P_\infty$, where $P_\infty = (0{:}\ 1{:}\ 0)$ and T be the sum of all 8 affine points. The dimension of the Riemann–Roch space is then given by,

$$l(5P_\infty) = 5 - 0 + 1 = 6$$

thus, the AG code has dimension $k = 6$.
3. The basis functions for the space $L(5P_\infty)$ are

$$\{t_1, \ldots, t_k\} = \left\{1, \frac{y}{z}, \frac{y^2}{z^2}, \frac{y^3}{z^3}, \frac{y^4}{z^4}, \frac{y^5}{z^5}\right\}$$

By examining the basis, it is clear that $t_1 = 1$ has no poles, thus, $(t_1) + D$ has no poles also. Basis functions with denominator z have $(t_1) = S - P_\infty$ where $S = (0 : 0 : 1)$ is a divisor of the numerator. The denominator polynomial z evaluates to zero at the point at infinity P_∞ of the divisor D, thus, $(t_1) + D$ has no poles. Basis functions with denominator z^2 have $(t_2) = S - 2P_\infty$ where $S = 2 \times (0 : 0 : 1)$ is a divisor of the numerator. The denominator polynomial z^2 evaluates to zero at the point at infinity P_∞ of the divisor D with multiplicity 2, thus, $(t_2) + D$ has no poles. Basis functions with denominator z^3 have $(t_3) = S - 3P_\infty$ where $S = 3 \times (0 : 0 : 1)$ is a divisor of the numerator. Thus, $(t_3) + D$ also has no poles. And so on.
4. The generator matrix of the code defined with divisor $D = 5P_\infty$ is thus,

$$G = \begin{bmatrix} t_1(P_1) & \cdots & t_1(P_n) \\ \vdots & \ddots & \vdots \\ t_k(P_1) & \cdots & t_k(P_n) \end{bmatrix}$$
$$= \begin{bmatrix} 1 & 1 & 1 & 1 & 1 & 1 & 1 & 1 \\ 0 & \alpha & \alpha^2 & \alpha^3 & \alpha^4 & \alpha^5 & \alpha^6 & 1 \\ 0 & \alpha^2 & \alpha^4 & \alpha^6 & \alpha & \alpha^3 & \alpha^5 & 1 \\ 0 & \alpha^3 & \alpha^6 & \alpha^2 & \alpha^5 & \alpha & \alpha^4 & 1 \\ 0 & \alpha^4 & \alpha & \alpha^5 & \alpha^2 & \alpha^6 & \alpha^3 & 1 \\ 0 & \alpha^5 & \alpha^3 & \alpha & \alpha^6 & \alpha^4 & \alpha^2 & 1 \end{bmatrix}$$

Clearly, this is a generator matrix of an extended Reed–Solomon code with parameters $[3, 6, 8]_8$.

Theorem 8.4 (From [2]) *The minimum distance of an AG code is given by,*

$$d \geq n - degree(D)$$

Thus, the Hermitian code defined by $D = 5P_\infty$ is a $[8, 5, 3]_4$ code. The dual of an AG code has parameters [17],

$$\text{Dimension, } k^\perp = n - degree(D) + g - 1$$
$$\text{Distance, } d^\perp \geq degree(D) - 2g + 2$$

8.4 Generalised AG Codes

Algebraic geometry codes and codes obtained from them feature prominently in the databases of best-known codes [8, 15] for an appreciable range of code lengths for different field sizes q. Generalised algebraic geometry codes were first presented by Niederreiter et al. [21], Xing et al. [13]. A subsequent paper by Ozbudak and Stichtenoth [14] shed more light on the construction. AG codes as defined by Goppa utilised places of degree one or rational places. Generalised AG codes however were constructed by Xing et al. using places of higher degree (including places of degree one). In [20], the authors presented a method of constructing generalised AG codes which uses a concatenation concept. The paper showed that best-known codes were obtainable via this construction. In [4] it was shown that the method can be effective in constructing new codes and the authors presented 59 codes in finite fields $\mathbb{F}_4$, $\mathbb{F}_8$ and $\mathbb{F}_9$ better than the codes in [8]. In [11], the authors presented a construction method based on [20] that uses a subfield image concept and obtained new binary codes as a result. In [19] the authors presented some new curves as well as 129 new codes in $\mathbb{F}_8$ and $\mathbb{F}_9$.

8.4.1 *Concept of Places of Higher Degree*

Recall from Chap. 8 that a two-dimensional affine space $\mathbb{A}^2(\mathbb{F}_q)$ is given by the set of points

$$\{(\alpha, \beta) : \alpha, \beta \in \mathbb{F}_q\}$$

while its projective closure $\mathbb{P}^2(\mathbb{F}_q)$ is given by the set of equivalence points

$$\{\{(\alpha : \beta : 1)\} \cup \{(\alpha : 1 : 0)\} \cup \{(1 : 0 : 0)\} : \alpha, \beta \in \mathbb{F}_q\}.$$

Given a homogeneous polynomial $F(x, y, z)$, a curve $\mathscr{X}/\mathbb{F}_q$ defined in $\mathbb{P}^2(\mathbb{F}_q)$ is a set of distinct points

$$\mathscr{X}/\mathbb{F}_q = \{T \in \mathbb{P}^2(\mathbb{F}_q) : F(T) = 0\}$$

Let $\mathbb{F}_{q^\ell}$ be an extension of the field $\mathbb{F}_q$, the Frobenius automorphism is given as

$$\begin{aligned} \phi_{q,\ell} &: \mathbb{F}_{q^\ell} \to \mathbb{F}_{q^\ell} \\ \phi_{q,\ell}(\beta) &= \beta^q \qquad \beta \in \mathbb{F}_{q^\ell} \end{aligned}$$

and its action on a projective point $(x : y : z)$ in $\mathbb{F}_{q^\ell}$ is

$$\phi_{q,\ell}((x : y : z)) = (x^q : y^q : z^q).$$

Definition 8.12 (*Place of Degree from* [18]) A place of degree ℓ is a set of ℓ points of a curve defined in the extension field $\mathbb{F}_{q^\ell}$ denoted by $\{T_0, T_1, \ldots, T_{\ell-1}\}$ where each $T_i = \phi^i_{q,l}(T_0)$. Places of degree one are called rational places.

Example 8.7 Consider the curve in $\mathbb{F}_4$ defined as,

$$F(x, y, z) = x$$

The curve has the following projective rational points (points of degree 1),

$$\begin{array}{l} (0:0:1)\ (0:1:1)\ (0:\alpha:1)\ (0:\alpha^2:1) \\ (0:1:0) \end{array}$$

where α is the primitive polynomial of $\mathbb{F}_4$. The curve has the following places of degree 2,

$$\begin{array}{cc} \{(0:\beta:1), (0:\beta^4:1)\} & \{(0:\beta^2:1), (0:\beta^8:1)\} \\ \{(0:\beta^3:1), (0:\beta^{12}:1)\} & \{(0:\beta^6:1), (0:\beta^9:1)\} \\ \{(0:\beta^7:1), (0:\beta^{13}:1)\} & \{(0:\beta^{11}:1), (0:\beta^{14}:1)\} \end{array}$$

where β is the primitive element of $\mathbb{F}_{16}$.

8.4.2 Generalised Construction

This section gives details of the construction of generalised AG codes as described in [21]. Two maps that are useful in the construction of generalised AG codes are now described. Observe that $\mathbb{F}_q$ is a subfield of $\mathbb{F}_{q^\ell}$ for all $\ell \geq 2$. It is then possible to map $\mathbb{F}_{q^\ell}$ to an ℓ-dimensional vector space with elements from $\mathbb{F}_q$ using a suitable basis. The map π_ℓ is defined as such,

$$\pi_\ell : \mathbb{F}_{q^\ell} \rightarrow \mathbb{F}_q^\ell$$
$$\pi_\ell(\beta) = (c_1 c_2 \dots c_\ell) \qquad \beta \in \mathbb{F}_{q^\ell},\ c_i \in \mathbb{F}_q.$$

Suppose $(\gamma_1, \gamma_2, \dots, \gamma_\ell)$ forms a suitable basis of the vector space $\mathbb{F}_q^\ell$, then $\beta = c_1\gamma_1 + c_2\gamma_2 + \dots + c_\ell\gamma_\ell$. Finally, the map $\sigma_{\ell,n}$ is used to represent an encoding map from an ℓ-dimensional message space in $\mathbb{F}_q$ to an n-dimensional code space,

$$\sigma_{\ell,n} : \mathbb{F}_q^\ell \rightarrow \mathbb{F}_q^n$$

with $\ell \leq n$.

A description of generalised AG codes as presented in [4, 13, 21] is now presented. Let $F = F(x, y, z)$ be a homogeneous polynomial defined in $\mathbb{F}_q$. Let g be the genus of a smooth irreducible curve $\mathscr{X}/\mathbb{F}_q$ corresponding to the polynomial F. Also, let $P_1, P_2, \dots, P_r$ be r distinct places of $\mathscr{X}/\mathbb{F}_q$ and $k_i = deg(P_i)$ (deg is degree of). W is a divisor of the curve $\mathscr{X}/\mathbb{F}_q$ such that

$$W = P_1 + P_2 + \dots + P_r$$

and another divisor G such that the two do not intersect.[4] Specifically, the divisor $G = m(Q - R)$ where $deg(Q) = deg(R) + 1$ for arbitrary[5] divisors Q and R. As mentioned earlier, associated with the divisor G is a Riemann–Roch space $\mathscr{L}(G)$ with $m = deg(G))$ an integer, $m \geq 0$. From the Riemann–Roch theorem (Theorem 8.3) it is known that the dimension of $\mathscr{L}(G)$ is given by $l(G)$ and

$$l(G) \geq m - g + 1.$$

Also, associated with each P_i is a q-ary code C_i with parameters $[n_i, k_i = deg(P_i), d_i]_q$ with the restriction that $d_i \leq k_i$. Let

$$\{f_1, f_2, ..., f_k : f_l \in \mathscr{L}(G)\}$$

denote a set of k linearly independent elements of $\mathscr{L}(G)$ that form a basis. A generator matrix for a generalised AG code is given as such,

$$M = \begin{bmatrix} \sigma_{k_1,n_1}(\pi_{k_1}(f_1(P_1))) & \cdots\cdots & \sigma_{k_r,n_r}(\pi_{k_r}(f_1(P_r))) \\ \sigma_{k_1,n_1}(\pi_{k_1}(f_2(P_1))) & \cdots\cdots & \sigma_{k_r,n_r}(\pi_{k_r}(f_2(P_r))) \\ \vdots & \ddots & \vdots \\ \sigma_{k_1,n_1}(\pi_{k_1}(f_k(P_1))) & \cdots\cdots & \sigma_{k_r,n_r}(\pi_{k_r}(f_k(P_r))) \end{bmatrix}$$

---

[4]This is consistent with the definition of AG codes. The two divisors should have no points in common.

[5]These are randomly chosen places such that the difference between their degrees is 1 and G does not intersect W.

where $f_l(P_i)$ is an evaluation of a polynomial and basis element f_l at a place P_i, π_{k_i} is a mapping from $\mathbb{F}_{q^{k_i}}$ to $\mathbb{F}_q$ and σ_{k_i,n_i} is the encoding of a message vector in $\mathbb{F}_q^{k_i}$ to a code vector in $\mathbb{F}_q^{n_i}$. This is a 3 step process. The place P_i is first evaluated at f_l resulting in an element of $\mathbb{F}_q^{k_i}$. The result is then mapped to a vector of length k_i in the subfield $\mathbb{F}_q$. Finally, this vector is encoded with code with parameters $[n_i, k_i, d_i]_q$.

It is desirable to choose the maximum possible minimum distance for all codes C_i so that $d_i = k_i$ [21]. The same code is used in the map σ_{k_i,n_i} for all points of the same degree k_i, i.e. the code C_j has parameters $[n_j, j, d_j]_q$ for a place of degree j. Let A_j be an integer denoting the number of places of degree j and B_j be an integer such that $0 \leq B_j \leq A_j$.

If t is the maximum degree of any place P_i that is chosen in the construction, then the generalised AG code is represented as a

$$C_1(k; t; B_1, B_2, \ldots, B_t; d_1, d_2, \ldots, d_t).$$

Let $[n, k, d]_q$ represent a linear code in $\mathbb{F}_q$ with length n, dimension k, and minimum distance d, then a generalised AG code is given by the parameters [21],

$$\begin{aligned} k &= l(G) \geq m - g + 1 \\ n &= \sum_{i=1}^{r} n_i = \sum_{j=1}^{t} B_j n_j \\ d &\geq \sum_{i=1}^{r} d_i - g - k + 1 = \sum_{j=1}^{t} B_j d_j - g - k + 1. \end{aligned}$$

Below are two examples showing the construction of generalised AG codes.

Example 8.8 Let $F(x, y, z) = x^3 + xyz + xz^2 + y^2 z$ [21] be a polynomial in $\mathbb{F}_2$. The curve $\mathscr{X}/\mathbb{F}_2$ has genus $g = 1$ and $A_1 = 4$ places of degree 1 and $A_2 = 2$ places of degree 2.

Table 8.4 gives the places of $\mathscr{X}/\mathbb{F}_2$ up degree 2. The field $\mathbb{F}_{2^2}$ is defined by a primitive polynomial $s^2 + s + 1$ with α as its primitive element. Points

$$R = (1 : a^3 + a^2 : 1)$$

as a place of degree 4 and

$$Q = (1 : b^4 + b^3 + b^2 : 1)$$

as a place of degree 5 are also chosen arbitrarily while a and b are primitive elements of $\mathbb{F}_{2^4}$ (defined by the polynomial $s^4 + s^3 + s^2 + s + 1$) and $\mathbb{F}_{2^5}$ (defined by the polynomial $s^5 + s^2 + 1$),g respectively. The divisor W is

$$W = P_1 + \cdots + P_6.$$

Table 8.4 Places of $\mathcal{X}/\mathbb{F}_2$

#	P_i	$deg(P_i)$
P_1	$(0:1:0)$	1
P_2	$(0:0:1)$	1
P_3	$(1:0:1)$	1
P_4	$(1:1:1)$	1
P_5	$\{(\alpha:1:1),(\alpha^2:1:1)\}$	2
P_6	$\{(\alpha:\alpha+1:1),(\alpha^2:\alpha:1)\}$	2

The basis of the Riemann–Roch space $\mathcal{L}(2D)$ with $D = Q - R$ and $m = 2$ is obtained with computer algebra software MAGMA [3] as,

$$\begin{aligned} f_1 &= (x^7 + x^3 + x)/(x^{10} + x^4 + 1)y \\ &\quad + (x^{10} + x^9 + x^7 + x^6 + x^5 + x + 1)/(x^{10} + x^4 + 1) \\ f_2 &= (x^8 + x^7 + x^4 + x^3 + x + 1)/(x^{10} + x^4 + 1)y \\ &\quad + (x^8 + x^4 + x^2)/(x^{10} + x^4 + 1) \end{aligned}$$

For the map σ_{k_i,n_i} the codes; c_1 a $[1, 1, 1]_2$ cyclic code for places of degree 1 and c_2 a $[3, 2, 2]_2$ cyclic code places of degree 2 are used. For the map π_2 which applies to places of degree 2, a polynomial basis $[\gamma_1, \gamma_2] = [1, \alpha]$ is used. Only the first point in the place P_i for $deg(P_i) = 2$ in the evaluation of f_1 and f_2 at P_i is utilised. The generator matrix M of the resulting $[10, 2, 6]_2$ generalised AG code over $\mathbb{F}_2$ is,

$$M = \begin{bmatrix} 1\,1\,0\,1\,0\,1\,1\,0\,1\,1 \\ 0\,0\,1\,1\,1\,1\,0\,1\,0\,1 \end{bmatrix}$$

Example 8.9 Consider again the polynomial

$$F(x, y, z) = x^3 + xyz + xz^2 + y^2 z$$

with coefficients from $\mathbb{F}_2$ whose curve (with genus equal to 1) has places up to degree 2 as in Table 8.4. An element f of the Riemann–Roch space defined by the divisor $G = (R - Q)$ with

$$Q = (a : a^3 + a^2 : 1)$$

and

$$R = (b : b^4 + b^3 + b^2 + b + 1 : 1)$$

where a and b primitive elements of $\mathbb{F}_{2^4}$ and $\mathbb{F}_{2^5}$ (since the curve has no place of degree 3) respectively, is given by,

$$f = (x^3x + x^2z^2 + z^4)y/(x^5 + x^3z^2 + z^5) \\ + (x^5 + x^4z + x^3z^2 + z^3x^2 + xz^4 + z^5)/(x^5 + x^3z^2 + z^5)$$

Evaluating f at all the 5 places P_i from the Table 8.4 and using the map $\pi_{\deg(P_i)}$ that maps all evaluations to $\mathbb{F}_2$ results in,

$$[\ \overbrace{1 \mid 1 \mid 0 \mid 1 \mid}^{f(P_i)\ |_{\deg(P_i)=1}} \qquad \underbrace{1 \mid \alpha^2}_{f(P_i)\ |_{\deg(P_i)=2}} \quad]$$

This forms the code $[6, 1, 5]_4$.[6] In $\mathbb{F}_2$ this becomes,

$$[\ 1 \mid 1 \mid 0 \mid 1 \mid \underbrace{1\ 0}_{1} \mid \underbrace{1\ 1}_{\alpha^2}]$$

which forms the code $[8, 1, 5]_2$. Short auxiliary codes $[1, 1, 1]_2$ to encode $f(P_i)\ |_{\deg(P_i)=1}$ and $[3, 2, 2]_2$ to encode $f(P_i)\ |_{\deg(P_i)=2}$ are used. The resulting codeword of a generalised AG code is,

$$[\ 1 \mid 1 \mid 0 \mid 1 \mid 1\ \ 0\ \ 1 \mid 1\ \ 1\ \ 0\ ].$$

This forms the code $[10, 1, 7]_2$.

Three polynomials and their associated curves are used to obtain codes in $\mathbb{F}_{16}$ better than the best-known codes in [15]. The three polynomials are given in Table 8.5, while Table 8.6 gives a summary of the properties of their associated curves (with $t = 4$). w is the primitive element of $\mathbb{F}_{16}$. The number of places of degree j, A_j, is determined by computer algebra system MAGMA [3]. The best-known linear codes from [15] over $\mathbb{F}_{16}$ with $j = d_j$ for $1 \le j \le 4$ are

$$[1, 1, 1]_{16} \quad [3, 2, 2]_{16} \quad [5, 3, 3]_{16} \quad [7, 4, 4]_{16}$$

which correspond to C_1, C_2, C_3 and C_4, respectively. Since $t = 4$ for all the codes in this paper and

$$[d_1, d_2, d_3, d_4] = [1, 2, 3, 4]$$

The representation $C_1(k; t; B_1, B_2, \ldots, B_t; d_1, d_2, \ldots, d_t)$ is shortened as such,

$$C_1(k; t; B_1, B_2, \ldots, B_t; d_1, d_2, \ldots, d_t) \equiv C_1(k; B_1, B_2, \ldots, B_t).$$

Tables 8.7 to 8.9 show improved codes from generalised AG codes with better minimum distance than codes in [15]. It is also worth noting that codes of the form

[6]From Bezout's $d_{min} = n - m = n - k - g + 1$.

Table 8.5 Polynomials in $\mathbb{F}_{16}$

$F_1 = x^5 + y^4 z + yz^4$
$F_2 = x^{16} + x^4 y^{15} + x^4 + xy^{15} + w^4 y^{15} + w^4$
$F_3 = x^{28} + wx^{20} + x^{18} + w^{10}x^{17} + w^{10}x^{15} + w^4x^{14} + w^3x^{13} + w^3x^{12} + wx^{11} + x^{10} + w^{11}x^9 + w^{12}x^8 + w^{14}x^7 + w^{13}x^6y^2 + w^9x^6y + w^6x^6 + w^2x^5y^2 + w^{13}x^5y + w^{14}x^5 + w^{14}x^4y^4 + w^7x^4y^2 + w^6x^4y + w^9x^4 + w^8x^3y^4 + w^{11}x^3y + w^4x^3 + w^{11}x^2y^4 + w^{11}x^2y^2 + wx^2y + w^5x^2 + w^8xy^4 + w^6xy^2 + w^9xy + w^{11}y^8 + y^4 + w^2y^2 + w^3y$

Table 8.6 Properties of $\mathscr{X}_i/\mathbb{F}_{16}$

Curve	Genus	A_1	A_2	A_3	A_4	Reference
$\mathscr{X}_1$	6	65	0	1600	15600	
$\mathscr{X}_2$	40	225	0	904	16920	[5]
$\mathscr{X}_3$	13	97	16	1376	15840	[6] via [9]

Table 8.7 New codes from $\mathscr{X}_1/\mathbb{F}_{16}$

Codes	k Range	Description	#
$[70, k, d \geq 63 - k]_{16}$	$10 \leq k \leq 50$	$C_1(k; [65, 0, 1, 0])$	41

Table 8.8 New codes from $\mathscr{X}_2/\mathbb{F}_{16}$

Code	k Range	Description	#
$[232, k, 190 - k]$	$102 \geq k \geq 129$	$C_1(k; [225, 0, 0, 1])$	28
$[230, k, 189 - k]$	$100 \geq k \geq 129$	$C_1(k; [225, 0, 1, 0])$	30
$[235, k, 192 - k]$	$105 \geq k \geq 121$	$C_1(k; [225, 0, 2, 0])$	17

$C_1(k; N, 0, 0, 0)$ are simply Goppa codes (defined with only rational points). The symbol # in the Tables 8.7 to 8.9 denotes the number of new codes from each generalised AG code $C_1(k; B_1, B_2, \ldots, B_t)$. The tables in [7] contain curves known to have the most number of rational points for a given genus. The curve $\mathscr{X}_2/\mathbb{F}_{16}$ is defined by the well-known Hermitian polynomial [5].

Table 8.9 New codes from $\mathscr{X}_3/\mathbb{F}_{16}$

Codes	k Range	Description	#
$[102, k, 88 - k]$	$8 \leq k \leq 66$	$C(k; [97, 0, 1, 0])$	59
$[103, k, 89 - k]$	$8 \leq k \leq 68$	$C(k; [97, 2, 0, 0])$	61
$[106, k, 91 - k]$	$k = 8$	$C(k; [97, 3, 0, 0])$	1

8.5 Summary

Algebraic geometry codes are codes obtained from curves. First, the motivation for studying these codes was given. From an asymptotic point of view, some families of AG codes have superior performance than the previous best known bound on the performance of linear codes, the Gilbert–Varshamov bound. For codes of moderate length, AG codes have better minimum distances than their main competitors, non-binary BCH codes with the same rate defined in the same finite fields. Theorems and definitions as a precursor to AG codes was given. Key theorems are Bezout's and Riemann–Roch. Examples using the well-known Hermitian code in a finite field of cardinality 4 were then discussed. The concept of place of higher degrees of curves was presented. This notion was used in the construction of generalised AG codes.

References

1. Blahut, R.E.: Algebraic Codes on Lines, Planes and Curves. Cambridge (2008)
2. Blake, I., Heegard, C., Hoholdt, T., Wei, V.: Algebraic-geometry codes. IEEE Trans. Inf. Theory **44**(6), 2596–2618 (1998)
3. Bosma, W., Cannon, J.J., Playoust, C.P.: The Magma algebra system I: The user language **24**, 235–266 (1997)
4. Ding, C., Niederreiter, H., Xing, C.: Some new codes from algebraic curves. IEEE Trans. Inf. Theory **46**(7), 2638–2642 (2000)
5. Garcia, A., Quoos, L.: A construction of curves over finite fields. ACTA Arithmetica **98**(2), (2001)
6. van der Geer, G., van der Vlugt, M.: Kummer covers with many points. Finite Fields Appl. **6**(4), 327–341 (2000)
7. van der Geer, G., et al.: Manypoints: A Table of Curves with Many Points (2009). http://www.manypoints.org
8. Grassl, M.: Code Tables: Bounds on the Parameters of Various Types of Codes (2007). http://www.codetables.de
9. Grassl, M.: Private Communication (2010)
10. Lachaud, G., Tsfasman, M., Justesen, J., Wei, V.W.: Introduction to the special issue on algebraic geometry codes. IEEE Trans. Inf. Theory **41**(6), 1545 (1995)
11. Leung, K.H., Ling, S., Xing, C.: New binary linear codes from algebraic curves. IEEE Trans. Inf. Theory **48**(1), 285–287 (2002)
12. Massimo, G.: Notes on Algebraic-geometric Codes. Lecture Notes (2003). http://www.math.kth.se/math/forskningsrapporter/Giulietti.pdf
13. Niederreiter, H., Xing, C., Lam, K.Y.: A new construction of algebraic-geometry codes. Appl. Algebra Eng. Commun. Comput. **9**(5), (1999)
14. Ozbudak, F., Stichtenoth, H.: Constructing codes from algebraic curves. IEEE Trans. Inf. Theory **45**(7) (1999)
15. Schimd, W., Shurer, R.: Mint: A Database for Optimal Net Parameters (2004). http://mint.sbg.ac.at
16. Tsfasman, M., Vladut, S., Zink, T.: On Goppa codes which are better than the Varshamov-Gilbert bound. Math. Nacr. **109**, 21–28 (1982)
17. Van-Lint, J.: Algebraic geometry codes. In: Ray-Chaudhari, D. (ed.) Coding theory and design theory: Part I: Coding Theory, p. 137. Springer, New York (1990)
18. Walker, J.L.: Codes and Curves. American Mathematical Society, Rhode Island (2000)

19. Xing, C., Ling, S.: A class of linear codes with good parameters from algebraic curves. IEEE Trans. Inf. Theory **46**(4), 1527–1532 (2000)
20. Xing, C., Niederreiter, H., Lam, K.: A generalization of algebraic-geometry codes. IEEE Trans. Inf. Theory **45**(7), 2498–2501 (1999a)
21. Xing, C., Niederreiter, H., Lam, K.Y.: Constructions of algebraic-geometry codes. IEEE Trans. Inf. Theory **45**(4), 1186–1193 (1999b)

Chapter 9
Algebraic Quasi Cyclic Codes

9.1 Introduction

Binary self-dual codes have an interesting structure and some are known to have the best possible minimum Hamming distance of any known codes. Closely related to the self-dual codes are the double-circulant codes. Many good binary self-dual codes can be constructed in double-circulant form. Double-circulant codes as a class of codes have been the subject of a great deal of attention, probably because they include codes, or the equivalent codes, of some of the most powerful and efficient codes known to date. An interesting family of binary, double-circulant codes, which includes self-dual and formally self-dual codes, is the family of codes based on primes. A classic paper for this family was published by Karlin [9] in which double-circulant codes based on primes congruent to ± 1 and ± 3 modulo 8 were considered. Self-dual codes are an important category of codes because there are bounds on their minimal distance [**?**]. The possibilities for their weight spectrum are constrained, and known ahead of the discovery, and analysis of the codes themselves. This has created a great deal of excitement among researchers in the rush to be the first in finding some of these codes. A paper summarising the state of knowledge of these codes was given by Dougherty et al. [1]. Advances in high-speed digital processors now make it feasible to implement near maximum likelihood, soft decision decoders for these codes and thus, make it possible to approach the predictions for frame error rate (FER) performance for the additive white Gaussian noise channel made by Claude Shannon back in 1959 [16].

This chapter considers the binary double-circulant codes based on primes, especially in analysis of their Hamming weight distributions. Section 9.2 introduces the notation used to describe double-circulant codes and gives a review of double-circulant codes based on primes congruent to ± 1 and ± 3 modulo 8. Section 9.4 describes the construction of double-circulant codes for these primes and Sect. 9.5 presents an improved algorithm to compute the minimum Hamming distance and also

M. Tomlinson et al., *Error-Correction Coding and Decoding*,
Signals and Communication Technology,
DOI 10.1007/978-3-319-51103-0_9

the number of codewords of a given Hamming weight for certain double-circulant codes. The algorithm presented in this section requires the enumeration of less codewords than that of the commonly used technique [4, 18] e.g. Sect. 9.6 considers the Hamming weight distribution of the double-circulant codes based on primes. A method to provide an independent verification to the number of codewords of a given Hamming weight in these double-circulant codes is also discussed in this section. In the last section of this chapter, Sect. 9.7, a probabilistic method—based on its automorphism group, to determine the minimum Hamming distance of these double-circulant codes is described.

Note that, as we consider Hamming space only in this chapter, we shall omit the word "Hamming" when we refer to Hamming weight and distance.

9.2 Background and Notation

A code $\mathscr{C}$ is called *self-dual* if,

$$\mathscr{C} = \mathscr{C}^{\perp}$$

where $\mathscr{C}^{\perp}$ is the dual of $\mathscr{C}$. There are two types of self-dual code: *doubly even* or Type-II for which the weight of every codeword is divisible by 4; *singly even* or Type-I for which the weight of every codeword is divisible by 2. Furthermore, the code length of a Type-II code is divisible by 8. On the other hand, formally self-dual (FSD) codes are codes that have

$$\mathscr{C} \neq \mathscr{C}^{\perp},$$

but satisfy $A_{\mathscr{C}}(z) = A_{\mathscr{C}^{\perp}}(z)$, where $A(\mathscr{C})$ denotes the weight distribution of the code $\mathscr{C}$. A self-dual, or FSD, code is called *extremal* if its minimum distance is the highest possible given its parameters. The bound of the minimum distance of the extremal codes is [15]

$$d \leq 4\left\lfloor \frac{n}{24} \right\rfloor + 4 + \varepsilon, \tag{9.1}$$

where

$$\varepsilon = \begin{cases} -2 & \text{if } \mathscr{C} \text{ is Type-I with } n = 2, 4, \text{ or } 6, \\ 2, & \text{if } \mathscr{C} \text{ is Type-I with } n \equiv 22 \pmod{24}, \text{ or} \\ 0, & \text{if } \mathscr{C} \text{ is Type-I or Type-II with } n \not\equiv 22 \pmod{24}. \end{cases} \tag{9.2}$$

for an extremal FSD code with length n and minimum distance d. For an FSD code, the minimum distance of the extremal case is upper bounded by [4]

$$d \leq 2\left\lfloor \frac{n}{8} \right\rfloor + 2. \tag{9.3}$$

As a consequence of this upper bound, extremal FSD codes are known to only exist for lengths $n \leq 30$ and $n \neq 16$ and $n \neq 26$ [7]. Databases of best-known, not necessary extremal, self-dual codes are given in [3, 15]. A table of binary self-dual double-circulant codes is also provided in [15].

As a class, double-circulant codes are (n, k) linear codes, where $k = n/2$, whose generator matrix $\boldsymbol{G}$ consists of two circulant matrices.

Definition 9.1 (*Circulant Matrix*) A circulant matrix is a square matrix in which each row is a cyclic shift of the adjacent row. In addition, each column is also a cyclic shift of the adjacent column and the number of non-zeros per column is equal to those per row.

A circulant matrix is completely characterised by a polynomial formed by its first row

$$r(x) = \sum_{i=0}^{m-1} r_i x^i,$$

which is called the *defining polynomial.*

Note that the algebra of polynomials modulo $x^m - 1$ is isomorphic to that of circulants [13]. Let the polynomial $r(x)$ have a maximum degree of m, and the corresponding circulant matrix $\boldsymbol{R}$ is an $m \times m$ square matrix of the form

$$\boldsymbol{R} = \begin{bmatrix} r(x) \pmod{x^m - 1} \\ xr(x) \pmod{x^m - 1} \\ \vdots \\ x^i r(x) \pmod{x^m - 1} \\ \vdots \\ x^{m-1} r(x) \pmod{x^m - 1} \end{bmatrix} \tag{9.4}$$

where the polynomial in each row can be represented by an m-dimensional vector, which contains the coefficients of the corresponding polynomial.

9.2.1 *Description of Double-Circulant Codes*

A double-circulant binary code is an $(n, \frac{n}{2})$ code in which the generator matrix is defined by two circulant matrices, each matrix being $\frac{n}{2}$ by $\frac{n}{2}$ bits. Circulant consists of cyclically shifted rows, modulo $\frac{n}{2}$, of a generator polynomial. These generator polynomials are defined as $r_1(x)$ and $r_2(x)$. Each codeword consists of two parts: the information data, defined as $u(x)$, convolved with $r_1(x)$ modulo $(1 + x^{\frac{n}{2}})$ adjoined with $u(x)$ and convolved with $r_2(x)$ modulo $(1 + x^{\frac{n}{2}})$. The code is the same as a non-systematic, tail-biting convolutional code of rate one half. Each codeword is $[u(x)r_1(x), u(x)r_2(x)]$. If $r_1(x)$ [or $r_2(x)$] has no common factors of $(1 + x^{\frac{n}{2}})$, then the respective circulant matrix is non-singular and may be inverted. The inverted circulant matrix becomes an identity matrix, and each codeword is defined by $u(x), u(x)r(x)$, where $r(x) = \frac{r_1(x)}{r_2(x)}$ modulo $(1 + x^{\frac{n}{2}})$, [or $r(x) = \frac{r_2(x)}{r_1(x)}$ modulo $(1 + x^{\frac{n}{2}})$, respectively]. The code is now the same as a systematic, tail-biting convolutional code of rate one half.

For double-circulant codes where one circulant matrix is non-singular and may be inverted, the codes can be put into two classes, namely *pure*, and *bordered* double-circulant codes, whose generator matrices $\boldsymbol{G}_p$ and $\boldsymbol{G}_b$ are shown in (9.5a)

$$\boldsymbol{G}_p = \left[\begin{array}{c|c} \boldsymbol{I}_k & \boldsymbol{R} \end{array}\right] \tag{9.5a}$$

and (9.5b),

$$\boldsymbol{G}_b = \left[\begin{array}{c|c|c} & 1 \ldots 1 & \alpha \\ \cline{2-3} \boldsymbol{I}_k & \boldsymbol{R} & \begin{matrix} 1 \\ \vdots \\ 1 \end{matrix} \end{array}\right] \tag{9.5b}$$

respectively. Here, $\boldsymbol{I}_k$ is a k-dimensional identity matrix, and $\alpha \in \{0, 1\}$.

Definition 9.2 (*Quadratic Residues*) Let α be a generator of the finite field $\mathbb{F}_p$, where p be an odd prime, $r \equiv \alpha^2 \pmod{p}$ is called a quadratic residue modulo p and so is $r^i \in \mathbb{F}_p$ for some integer i. Because the element α has (multiplicative) order $p - 1$ over $\mathbb{F}_p$, $r = \alpha^2$ has order $\frac{1}{2}(p - 1)$. A set of quadratic residues modulo p, Q and non-quadratic residues modulo p, N, are defined as

$$\begin{aligned} Q &= \{r, r^2, \ldots, r^i, \ldots, r^{\frac{p-3}{2}}, r^{\frac{p-1}{2}} = 1\} \\ &= \{\alpha^2, \alpha^4, \ldots, \alpha^{2i} \ldots, \alpha^{p-3}, \alpha^{p-1} = 1\} \end{aligned} \tag{9.6a}$$

and

$$\begin{aligned} N &= \{n : \forall n \in \mathbb{F}_p,\ n \neq Q \text{ and } n \neq 0\} \\ &= \{nr, nr^2, \ldots, nr^i, \ldots, nr^{\frac{p-3}{2}}, n\} \\ &= \{\alpha^{2i+1}\ :\ 0 \leq i \leq \tfrac{p-3}{2}\} \end{aligned} \tag{9.6b}$$

respectively.

As such $R \cup Q \cup \{0\} = \mathbb{F}_p$. It can be seen from the definition of Q and N that, if $r \in Q$, $r = \alpha^e$ for even e; and if $n \in N$, $n = \alpha^e$ for odd e. Hence, if $n \in N$ and

$r \in Q$, $rn = \alpha^{2i}\alpha^{2j+1} = \alpha^{2(i+j)+1} \in N$. Similarly, $rr = \alpha^{2i}\alpha^{2j} = \alpha^{2(i+j)} \in Q$ and $nn = \alpha^{2i+1}\alpha^{2j+1} = \alpha^{2(i+j+1)} \in Q$.

Furthermore,

- $2 \in Q$ if $p \equiv \pm 1 \pmod 8$, and $2 \in N$ if $p \equiv \pm 3 \pmod 8$
- $-1 \in Q$ if $p \equiv 1 \pmod 8$ or $p \equiv -3 \pmod 8$, and $-1 \in N$ if $p \equiv -1 \pmod 8$ and $p \equiv 3 \pmod 8$

9.3 Good Double-Circulant Codes

9.3.1 Circulants Based Upon Prime Numbers Congruent to ±3 Modulo 8

An important category is circulants whose length is equal to a prime number, p, which is congruent to ± 3 modulo 8. For many of these prime numbers, there is only a single cyclotomic coset, apart from zero. In these cases, $1 + x^p$ factorises into the product of two irreducible polynomials, $(1 + x)(1 + x + x^2 + x^3 + \cdots + x^{p-1})$. Apart from the polynomial, $(1 + x + x^2 + x^3 + \cdots + x^{p-1})$, all of the other $2^p - 2$ non-zero polynomials of degree less than p are in one of two sets: The set of 2^{p-1} even weight, polynomials with $1 + x$ as a factor, denoted as $\mathbf{S_f}$, and the set of 2^{p-1} odd weight polynomials which are relatively prime to $1 + x^p$, denoted as $\mathbf{S_r}$. The multiplicative order of each set is $2^{p-1} - 1$, and each forms a ring of polynomials modulo $1 + x^p$. Any non-zero polynomial apart from $(1 + x + x^2 + x^3 + \cdots + x^{p-1})$ is equal to $\alpha(x)^i$ for some integer i if the polynomial is in $\mathbf{S_f}$ or is equal to $a(x)^i$ for some integer i if in $\mathbf{S_r}$. An example for $p = 11$ is given in Appendix "Circulant Analysis $p = 11$". In this table, $\alpha(x) = 1 + x + x^2 + x^4$ and $a(x) = 1 + x + x^3$. For these primes, as the circulant length is equal to p, the generator polynomial $r(x)$ can either contain $1 + x$ as a factor, or not contain $1 + x$ as a factor, or be equal to $(1 + x + x^2 + x^3 + \cdots + x^{p-1})$. For the last case, this is not a good choice for r(x) as the minimum codeword weight is 2, which occurs when $u(x) = 1 + x$. In this case, $r(x)u(x) = 1 + x^p = 0$ modulo $1 + x^p$ and the codeword is $[1 + x,\ 0]$, a weight of 2.

When $r(x)$ is in the ring $\mathbf{S_f}$, $u(x)r(x)$ must also be in $\mathbf{S_f}$ and therefore, be of even weight, except when $u(x) = (1 + x + x^2 + x^3 + \cdots + x^{p-1})$.

In this case $u(x)r(x) = 0$ modulo $1 + x^p$ and the codeword is $[1 + x + x^2 + x^3 + \cdots + x^{p-1}, 0]$ of weight p. When $u(x)$ has even weight, the resulting codewords are doubly even. When $u(x)$ has odd weight, the resulting codewords consist of two parts, one with odd weight and the other with even weight. The net result is the codewords that always have odd weight. Thus, there are both even and odd weight codewords when $u(x)$ is from $\mathbf{S_f}$.

When $r(x)$ is in the ring $\mathbf{S_r}$, $u(x)r(x)$ is always non-zero and is in $\mathbf{S_f}$ (even weight) only when $u(x)$ has even weight, and the resulting codewords are doubly even. When $u(x)$ has odd weight, $u(x) = a(x)^j$ and $u(x)r(x) = a(x)^j a(x)^i =$

$a(x)^{i+j}$ and hence is in the ring $\mathbf{S_f}$ and has odd weight. The resulting codewords have even weight since they consist of two parts, each with odd weight. Thus, when $r(x)$ is from $\mathbf{S_r}$ all of the codewords have even weight. Furthermore, since $r(x) = a(x)^i$, $r(x)a(x)^{2^{(p-1)}-1-i} = a(x)^{2^{(p-1)}-1} = 1$ and hence, the inverse of $r(x)$, $\frac{1}{g(x)} = a(x)^{2^{(p-1)}-1-i}$.

By constructing a table (or sampled table) of $\mathbf{S_r}$, it is very straightforward to design non-singular double-circulant codes. The minimum codeword weight of the code d_{min} cannot exceed the weight of $r(x) + 1$. Hence, the weight of $a(x)^i$ needs to be at least $d_{min} - 1$ to be considered as a candidate for $r(x)$. The weight of the inverse of $r(x)$, $a(x)^{2^{(p-1)}-1-i}$ also needs to be at least $d_{min} - 1$. For odd weight $u(x) = a(x)^j$ and $u(x)r(x) = a(x)^j a(x)^i = a(x)^{(j+i)}$. Hence, the weight of $u(x)r(x)$ can be found simply by looking up the weight of $a(x)^{i+j}$ from the table. Self-dual codes are those with $\frac{1}{r(x)} = r(x^{-1})$. With a single cyclotomic coset $2^{\frac{(p-1)}{2}} = -1$, and it follows that $a(x)^{2^{\frac{(p-1)}{2}}} = a(x^{-1})$. With $r(x) = a(x)^i$, $r(x^{-1}) = a(x)^{2^{\frac{(p-1)}{2}} i}$.

In order that $\frac{1}{r(x)} = r(x^{-1})$, it is necessary that

$$a(x)^{2^{(p-1)}-1-i} = a(x)^{2^{\frac{(p-1)}{2}} i}. \tag{9.7}$$

Equating the exponents, modulo $2^{(p-1)} - 1$, gives

$$2^{\frac{(p-1)}{2}} i = m(2^{(p-1)} - 1) - i, \tag{9.8}$$

where m is an integer. Solving for i:

$$i = \frac{m(2^{(p-1)} - 1)}{(2^{\frac{(p-1)}{2}} + 1)}. \tag{9.9}$$

Hence, the number of distinct self-dual codes is equal to $(2^{\frac{(p-1)}{2}} + 1)$.

For the example, $p = 13$ as above,

$$i = m\frac{2^{(p-1)} - 1}{2^{\frac{(p-1)}{2}} + 1} = m\frac{4095}{65} = 63m$$

and there are $2^{\frac{(p-1)}{2}} + 1 = 65$ self-dual codes for $1 \leq j \leq 65$ and these are $a(x)^{63}$, $a(x)^{126}$, $a(x)^{189}, \ldots, a(x)^{4095}$.

As p is congruent to ± 3, the set $(u(x)r(x))^{2^t}$ maps to $(u(x)r(x))$ for $t = 1 \rightarrow r$, where r is the size of the cyclotomic cosets of $2^{\frac{(p-1)}{2}} + 1$. In the case of $p = 13$ above, there are 4 cyclotomic cosets of 65, three of length 10 and one of length 2. This implies that there on 4 non-equivalent self-dual codes.

For p congruent to -3 modulo 8, $(2^{\frac{(p-1)}{2}} + 1)$ is not divisible by 3. This means that the pure double-circulant quadratic residue code is not self-dual. Since the quadratic residue code has multiplicative order 3, this means that for p congruent to -3 modulo 8, the quadratic residue, pure double-circulant code is self-orthogonal, and $r(x) = r(x^{-1})$.

For p congruent to 3, $(2^{\frac{(p-1)}{2}} + 1)$ is divisible by 3 and the pure double-circulant quadratic residue code is self-dual. In this case, $a(x)$ has multiplicative order of $2^{(p-1)} - 1$, and $a(x)^{\frac{(2^{(p-1)}-1)}{3}}$ must have exponents equal to the quadratic residues (or non-residues). The inverse polynomial is $a(x)^{\frac{2(2^{(p-1)}-1)}{3}}$ with exponents equal to the non-residues (or residues, respectively), and defines a self-dual circulant code. As an example, for $p = 11$ as listed in Appendix "Circulant Analysis $p = 11$", $2^{(p-1)} - 1 = 1023$ and $a(x)^{341} = x + x^3 + x^4 + x^5 + x^9$, the quadratic non-residues of 11 are $1, 4, 5, 9$ and 3. $a(x)^{682} = x^2 + x^6 + x^7 + x^8 + x^{10}$ corresponding to the quadratic residues: 2, 8, 10, 7 and 6 as can be seen from Appendix "Circulant Analysis $p = 11$". Section 9.4.3 discusses in more detail pure double-circulant codes for these primes.

9.3.2 *Circulants Based Upon Prime Numbers Congruent to ±1 Modulo 8: Cyclic Codes*

MacWilliams and Sloane [13] discuss the Automorphism group of the extended cyclic quadratic residue (eQR) codes and show that this includes the projective special linear group $PSL_2(p)$. They describe a procedure in which a double-circulant code may be constructed from a codeword of the eQR code. It is fairly straightforward. The projective special linear group $PSL_2(p)$ for a prime p is defined by the permutation $y \to \frac{ay+b}{cy+d}$ mod p, where the integers a, b, c, d are such that two cyclic groups of order $\frac{p+1}{2}$ are obtained. A codeword of the $(p + 1, \frac{p+1}{2})$ eQR code is obtained and the non-zero coordinates of the codeword placed in each cyclic group. This splits the codeword into two cyclic parts each of which defines a circulant polynomial.

The procedure is best illustrated with an example. Let $\alpha \in \mathbb{F}_{p^2}$ be a primitive $(p^2 - 1)^{\text{ti}}$ root of unity; then, $\beta = \alpha^{2p-2}$ is a primitive $\frac{1}{2}(p + 1)^{\text{TA}}$ root of unity since $p^2 - 1 = \frac{1}{2}(2p - 2)(p - 1)$. Let $\lambda = 1/(1 + \beta)$ and $a = \lambda^2 - \lambda$; then, the permutation π_1 on a coordinate y is defined as

$$\pi_1 : y \mapsto \frac{y+1}{ay} \qquad \text{mod } p$$

where $\pi_1 \in \text{PSL}_2(p)$ (see Sect. 9.4.3 for the definition of $\text{PSL}_2(p)$). As an example, consider the prime $p = 23$. The permutation $\pi_1 : y \to \frac{y+1}{5y}$ mod p produces the two cyclic groups

$$(1, 5, 3, 11, 9, 13, 8, 10, 20, 17, 4, 6)$$

and

$$(2, 21, 7, 16, 12, 19, 22, 0, 23, 14, 15, 18).$$

There are 3 cyclotomic cosets for $p = 23$ as follows:

$$C_0 = \{0\}$$
$$C_1 = \{1, 2, 4, 8, 16, 9, 18, 13, 3, 6, 12\}$$
$$C_5 = \{5, 10, 20, 17, 11, 22, 21, 19, 15, 7, 14\}.$$

The idempotent given by C_1 may be used to define a generator polynomial, $r(x)$, which defines the (23, 12, 7) cyclic quadratic residue code:

$$r(x) = x + x^2 + x^3 + x^4 + x^6 + x^8 + x^9 + x^{12} + x^{13} + x^{16} + x^{18}. \quad (9.10)$$

Codewords of the (23, 12, 7) cyclic code are given by $u(x)r(x)$ modulo $1 + x^{23}$ and with $u(x) = 1$ the non-zero coordinates of the codeword obtained are

$$(1, 2, 4, 8, 16, 9, 18, 13, 3, 6, 12)$$

the cyclotomic coset C_1.

The extended code has an additional parity check using coordinate 23 to produce the corresponding codeword of the extended (24, 12, 8) code with the non-zero coordinates:

$$(1, 2, 4, 8, 16, 9, 18, 13, 3, 6, 12, 23).$$

Mapping these coordinates to the cyclic groups with 1 in the position, where each coordinate is in the respective cyclic group and 0 otherwise, produces

$$(1, 0, 1, 0, 1, 1, 1, 0, 0, 0, 1, 1)$$

and

$$(1, 0, 0, 1, 1, 0, 0, 0, 1, 0, 0, 1)$$

which define the two circulant polynomials, $r_1(x)$ and $r_2(x)$, for the (24, 12, 8) pure double-circulant code

$$r_1(x) = 1 + x^2 + x^4 + x^5 + x^6 + x^{10} + x^{11}$$
$$r_2(x) = 1 + x^3 + x^4 + x^8 + x^{11}. \quad (9.11)$$

The inverse of $r_1(x)$ modulo $(1 + x^{12})$ is $\psi(x)$, where

$$\psi(x) = 1 + x + x^2 + x^6 + x^7 + x^8 + x^{10},$$

and this may be used to produce an equivalent (24, 12, 8) pure double-circulant code which has the identity matrix as the first circulant

Table 9.1 Double-circulant codes mostly based upon quadratic residues of prime numbers

Prime (p)	p mod 8	Circulant codes $(2p, p)$	Circulant codes $(2p+2, p+1)$	Circulant codes $(p+1, \frac{p+1}{2})$	d_{min}
7	−1			(8, 4, 4)	4
17	1			(18, 9, 6)	6
11	3	[a](22, 11, 7) $\beta(x)$	(24, 12, 8)		8
23	−1			[a](24, 12, 8)	8
13	−3	(26, 13, 7) $b(x)$			7
31	−1			(32, 16, 8)	8
19	3	(38, 19, 8) $b(x)$			8
41	1	(82, 41, 14)		(42, 21, 10)	10
47	−1			[a](48, 24, 12)	12
29	−3	(58, 29, 11) $\beta(x)$	(60, 30, 12)		12
71	−1			(72, 36, 12)	12
				[b](72, 36, 14)	14
73	1			(74, 37, 14)	14
37	−3	(74, 37, 12) $b(x)$			12
79	−1			[a](80, 40, 16)	16
43	3	(86, 43, 16) $\beta(x)$	(88, 44, 16)		16
97	1			(98, 49, 16)	16
103	−1			[a](104, 52, 20)	20
53	−3	(106, 53, 19) $\beta(x)$	(108, 54, 20)		20
113	1			(114, 57, 16)	16
59	3	(118, 59, 19) $\beta(x)$	(120, 60, 20)		20
61	−3	(122, 61, 19) $\beta(x)$	(124, 62, 20)		20
127	−1			(128, 64, 20)	20
67	3	[a](134, 67, 23) $\beta(x)$	(136, 68, 24)		24
137	1			(138, 69, 22)	22
151	−1			(152, 76, 20)	20
83	3	(166, 83, 23) $\beta(x)$	(168, 84, 24)		24
191	−1			(192, 96, 28)	28
193	1			(194, 97, 28)	28
199	−1			[a](200, 100, 32)	32
101	−3	(202, 101, 23) $\beta(x)$	(204, 102, 24)		24
107	3	(214, 107, 23) $\beta(x)$	(216, 108, 24)		24
109	−3	(218, 109, 30) $b(x)$			30
223	−1			(224, 112, 32)	32
233	1			(234, 117, 26)	26
239	−1			(240, 120, 32)	32
241	1			(242, 121, 32?)	32?
131	3	[a](262, 131, 38?) $b(x)$			38?

[a]Codes with outstanding d_{min}

[b]Codes not based on quadratic residues

The best $(2p, p)$ circulant polynomial either contains the factor $1+x$: $\beta(x)$ or is relatively prime to $1+x^n$: $b(x)$

$\beta(x)$ circulants can be bordered to produce $(2p+2, p+1)$ circulants

$$\hat{r}_1(x) = (1 + x^2 + x^4 + x^5 + x^6 + x^{10} + x^{11})\psi(x) \text{ modulo } (1 + x^{12})$$
$$\hat{r}_2(x) = (1 + x^3 + x^4 + x^8 + x^{11})\psi(x) \textit{ modulo } (1 + x^{12}).$$

After evaluating terms, the two circulant polynomials are found to be

$$\begin{aligned}\hat{r}_1(x) &= 1\\ \hat{r}_2(x) &= 1 + x + x^2 + x^4 + x^5 + x^9 + x^{11},\end{aligned} \tag{9.12}$$

and it can be seen that the first circulant will produce the identity matrix of dimension 12. Jenson [8] lists the circulant codes for primes $p < 200$ that can be constructed in this way. There are two cases, $p = 89$ and $p = 167$, where a systematic double-circulant construction is not possible. A non-systematic double-circulant code is possible for all cases but the existence of a systematic code depends upon one of the circulant matrices being non-singular. Apart from $p = 89$ and $p = 167$ (for $p < 200$) a systematic circulant code can always be constructed in each case.

Table 9.1 lists the best circulant codes as a function of length. Most of these codes are well known and have been previously published but not necessarily as circulant codes. Moreover, the d_{min} of some of the longer codes have only been bounded and have not been explicitly stated in the literature. Nearly, all of the best codes are codes

Table 9.2 Generator polynomials for pure double-circulant codes

Code	Circulant generator polynomial exponents
(8, 4, 4)	0, 1, 2
(24, 12, 8)	0, 1, 3, 4, 5, 6, 8
(48, 24, 12)	0, 1, 2, 3, 4, 5, 6, 8, 10, 11, 13, 14, 16, 17, 18
(80, 40, 16)	0, 1, 5, 7, 9, 10, 11, 14, 15, 19, 23, 25, 27, 30, 38
(104, 52, 20)	0, 2, 5, 7, 10, 13, 14, 17, 18, 22, 23, 25, 26, 27, 28, 37, 38, 39, 40, 41, 42, 44, 45, 46, 47, 48, 49
(122, 61, 20)	0, 1, 3, 4, 5, 9, 12, 13, 14, 15, 16, 19, 20, 22, 25, 27, 34, 36, 39, 41, 42, 45, 46, 47, 48, 49, 52, 56, 57, 58, 60
(134, 67, 23)	0, 1, 4, 6, 9, 10, 14, 15, 16, 17, 19, 21, 22, 23, 24, 25, 26, 29, 33, 35, 36, 37, 39, 40, 47, 49, 54, 55, 56, 59, 60, 62, 64, 65
(156, 78, 22)	0, 2, 3, 4, 8, 9, 11, 12, 14, 16, 17, 18, 20, 22, 24, 26, 27, 29, 33, 38, 39, 41, 42, 43, 44, 45, 46, 48, 49, 50, 52, 55, 56, 60, 64, 66, 68, 71, 72, 73, 74, 75, 77
(166, 83, 24)	1, 3, 4, 7, 9, 10, 11, 12, 16, 17, 21, 23, 25, 26, 27, 28, 29, 30, 31, 33, 36, 37, 38, 40, 41, 44, 48, 49, 51, 59, 61, 63, 64, 65, 68, 69, 70, 75, 77, 78, 81
(180, 90, 26)	0, 3, 5, 6, 7, 8, 9, 11, 12, 13, 14, 17, 18, 19, 21, 22, 23, 28, 36, 37, 41, 45, 50, 51, 53, 55, 58, 59, 60, 61, 62, 63, 67, 68, 69, 72, 75, 76, 78, 81, 82, 83, 84, 85, 88
(200, 100, 32)	0, 1, 2, 5, 6, 8, 9, 10, 11, 15, 16, 17, 18, 19, 20, 26, 27, 28, 31, 34, 35, 37, 38, 39, 42, 44, 45, 50, 51, 52, 53, 57, 58, 59, 64, 66, 67, 70, 73, 75, 76, 77, 80, 82, 85, 86, 89, 92, 93, 97, 98

based upon the two types of quadratic residue circulant codes. For codes based upon $p = \pm 3 \quad mod\, 8$, it is an open question whether a better circulant code exists than that given by the quadratic residues. For $p = \pm 1 \quad mod\, 8$, there are counter examples. For example, the (72, 36, 14) code in Table 9.1 is better than the (72, 36, 12) circulant code which is based upon the extended cyclic quadratic residue code of length 71. The circulant generator polynomial g(x) for all of the codes of Table 9.1 is given in Table 9.2.

In Table 9.1, where the best $(2p, p)$ code is given as $b(x)$, the $(2p + 2, p + 1)$ circulant code can still be constructed from $\beta(x)$ but this code has the same d_{min} as the pure, double-circulant, shorter code. For example, for the prime 109, $b(x)$ produces a double-circulant (218, 109, 30) code. The polynomial $\beta(x)$ produces a double-circulant (218, 109, 29) code, which bordered becomes a (220, 110, 30) code. It should be noted that $\beta(x)$ need not have the overall parity bit border added. In this case, a $(2p + 1, p + 1)$ code is produced but with the same d_{min} as the $\beta(x)$ code. In the latter example, a (219, 110, 29) code is produced.

9.4 Code Construction

Two binary linear codes, $\mathscr{A}$ and $\mathscr{B}$, are *equivalent* if there exists a permutation π on the coordinates of the codewords which maps the codewords of $\mathscr{A}$ onto codewords of $\mathscr{B}$. We shall write this as $\mathscr{B} = \pi(\mathscr{A})$. If π transforms $\mathscr{C}$ into itself, then we say that π fixes the code, and the set of all permutations of this kind forms the automorphism group of $\mathscr{C}$, denoted as Aut($\mathscr{C}$). MacWilliams and Sloane [13] gives some necessary but not sufficient conditions on the equivalence of double-circulant codes, which are restated for convenience in the lemma below.

Lemma 9.1 (cf. [13, Problem 7, Chap. 16]) *Let $\mathscr{A}$ and $\mathscr{B}$ be double-circulant codes with generator matrices $[\boldsymbol{I}_k|\boldsymbol{A}]$ and $[\boldsymbol{I}_k|\boldsymbol{B}]$, respectively. Let the polynomials $a(x)$ and $b(x)$ be the defining polynomials of **A** and **B**. The codes $\mathscr{A}$ and $\mathscr{B}$ are equivalent if any of the following conditions holds:*

(i) $\boldsymbol{B} = \boldsymbol{A}^T$, *or*
(ii) $b(x)$ *is the reciprocal of* $a(x)$, *or*
(iii) $a(x)b(x) = 1 \pmod{x^m - 1}$, *or*
(iv) $b(x) = a(x^u)$, *where m and u are relatively prime.*

Proof

(i) We can clearly see that $b(x) = \sum_{i=0}^{m-1} a_i x^{m-i}$. It follows that $b(x) = \pi(a(x))$, where $\pi : i \rightarrow m - i \pmod m$ and hence, $\mathscr{A}$ and $\mathscr{B}$ are equivalent.
(ii) Given a polynomial $a(x)$, its reciprocal polynomial can be written as $\bar{a}(x) = \sum_{i=0}^{m-1} a_i x^{m-i-1}$. It follows that $\bar{a}(x) = \pi(a(x))$, where $\pi : i \rightarrow m - i - 1 \pmod m$.

(iii) Consider the code $\mathscr{A}$, since $b(x)$ has degree less than m, it can be one of the possible data patterns and in this case, the codeword of $\mathscr{A}$ has the form $|b(x)|1|$. Clearly, this is a permuted codeword of $\mathscr{B}$.
(iv) If $(u, m) = 1$, then $\pi : i \to iu \pmod m$ is a permutation on $\{0, 1, \ldots, m-1\}$. So $b(x) = a(x^u)$ is in the code $\mu(\mathscr{A})$.

Consider an (n, k, d) pure double-circulant code, we can see that for a given user message, represented by a polynomial $u(x)$ of degree at most $k-1$, a codeword of the double-circulant code has the form $(u(x)|u(x)r(x) \pmod{x^m - 1})$. The defining polynomial $r(x)$ characterises the resulting double-circulant code. Before the choice of $r(x)$ is discussed, consider the following lemmas and corollary.

Lemma 9.2 *Let $a(x)$ be a polynomial over $\mathbb{F}_2$ of degree at most $m-1$, i.e. $a(x) = \sum_{i=0}^{m-1} a_i x^i$ where $a_i \in \{0, 1\}$. The weight of the polynomial $(1+x)a(x) \pmod{x^m - 1}$, denoted by $\mathrm{wt}_H((1+x)a(x))$ is even.*

Proof Let $w = \mathrm{wt}_H(a(x)) = \mathrm{wt}_H(xa(x))$ and $S = \{i : a_{i+1 \bmod m} = a_i \neq 0,\ 0 \leq i \leq m-1\}$:

$$\begin{aligned}\mathrm{wt}_H((1+x)a(x)) &= \mathrm{wt}_H(a(x)) + \mathrm{wt}_H(xa(x)) - 2|S| \\ &= 2(w - |S|),\end{aligned}$$

which is even.

Lemma 9.3 *An $m \times m$ circulant matrix $\boldsymbol{R}$ with defining polynomial $r(x)$ is non-singular if and only if $r(x)$ is relatively prime to $x^m - 1$.*

Proof If $r(x)$ is not relatively prime to $x^m - 1$, i.e. GCD $(r(x), x^m - 1) = t(x)$ for some polynomial $t(x) \neq 1$, then from the extended Euclidean algorithm, it follows that, for some unique polynomials $a(x)$ and $b(x)$, $r(x)a(x) + (x^m - 1)b(x) = 0$, and therefore $\boldsymbol{R}$ is singular.

If $r(x)$ is relatively prime to $x^m - 1$, i.e. GCD $(r(x), x^m - 1) = 1$, then from the extended Euclidean algorithm, it follows that, for some unique polynomials $a(x)$ and $b(x)$, $r(x)a(x) + (x^m - 1)b(x) = 1$, which is equivalent to $r(x)a(x) = 1 \pmod{x^m - 1}$. Hence $\boldsymbol{R}$ is non-singular, being invertible with a matrix inverse whose defining polynomial is $a(x)$.

Corollary 9.1 *From Lemma 9.3,*

(i) if $\boldsymbol{R}$ is non-singular, $\boldsymbol{R}^{-1}$ is an $m \times m$ circulant matrix with defining polynomial $r(x)^{-1}$, and
(ii) the weight of $r(x)$ or $r(x)^{-1}$ is odd.

Proof The proof for the first case is obvious from the proof of Lemma 9.3. For the second case, if the weight of $r(x)$ is even then $r(x)$ is divisible by $1+x$. Since $1+x$ is a factor of $x^m - 1$ then $r(x)$ is not relatively prime to $x^m - 1$ and the weight of $r(x)$ is necessarily odd. The inverse of $r(x)^{-1}$ is $r(x)$ and for this to exist $r(x)^{-1}$ must be relatively prime to $x^m - 1$ and the weight of $r(x)^{-1}$ is necessarily odd.

Lemma 9.4 *Let p be an odd prime, and then*

(i) $p \mid 2^{p-1} - 1$, *and*
(ii) the integer q for $pq = 2^{p-1} - 1$ is odd.

Proof From Fermat's little theorem, we know that for any integer a and a prime p, $a^{p-1} \equiv 1 \pmod{p}$. This is equivalent to $a^{p-1} - 1 = pq$ for some integer q. Let $a = 2$, we have

$$q = \frac{2^{p-1} - 1}{p}$$

which is clearly odd since neither denominator nor numerator contains 2 as a factor.

Lemma 9.5 *Let p be a prime and $j(x) = \sum_{i=0}^{p-1} x^i$; then*

$$(1 + x)^{2^{p-1}-1} = 1 + j(x) \; mod \; (x^p - 1).$$

Proof We can write $(1 + x)^{2^{p-1}-1}$ as

$$\begin{aligned} (1 + x)^{2^{p-1}-1} &= \frac{(1 + x)^{2^{p-1}}}{1 + x} = \frac{1 + x^{2^{p-1}}}{1 + x} \\ &= \sum_{i=0}^{2^{p-1}-1} x^i. \end{aligned}$$

From Lemma 9.4, we know that the integer $q = (2^{p-1} - 1)/p$ and is odd. We can then write $\sum_{i=0}^{2^{p-1}-1} x^i$ in terms of $j(x)$ as follows:

$$\begin{aligned} \sum_{i=0}^{2^{p-1}-1} x^i = 1 + x \underbrace{\left(1 + x + \cdots + x^{p-1}\right)}_{j(x)} + x^{p+1} \underbrace{\left(1 + x + \cdots + x^{p-1}\right)}_{j(x)} + \ldots + \\ x^{(q-3)p+1} \underbrace{\left(1 + x + \cdots + x^{p-1}\right)}_{j(x)} + x^{(q-2)p+1} \underbrace{\left(1 + x + \cdots + x^{p-1}\right)}_{j(x)} + \\ x^{(q-1)p+1} \underbrace{\left(1 + x + \cdots + x^{p-1}\right)}_{j(x)} \\ = 1 + \underbrace{x j(x)(1 + x^p) + x^{2p+1} j(x)(1 + x^p) + \ldots + x^{(q-3)p+1} j(x)(1 + x^p)}_{J(x)} + \\ x^{(q-1)p+1} j(x) \end{aligned}$$

Since $(1 + x^p) \pmod{x^p - 1} = 0$ for a binary polynomial, $J(x) = 0$ and we have

$$\sum_{i=0}^{2^{p-1}-1} x^i = 1 + x x^{(q-1)p} j(x) \pmod{x^p - 1}.$$

Because $x^{ip} \pmod{x^p - 1} = 1$,

$$\begin{aligned}\sum_{i=0}^{2^{p-1}-1} x^i &= 1 + x j(x) \pmod{x^p - 1}\\ &= 1 + j(x) \pmod{x^p - 1}.\end{aligned}$$

For the rest of this chapter, we consider the bordered case only and for convenience, unless otherwise stated, we shall assume that the term double-circulant code refers to (9.5b). Furthermore, we call the double-circulant codes based on primes congruent to ± 1 modulo 8, the $[p + 1, \frac{1}{2}(p + 1), d]$ extended quadratic residue (QR) codes since these exist only for $p \equiv \pm 1 \pmod 8$.

Following Gaborone [2], we call those double-circulant codes based on primes congruent to ± 3 modulo 8 the $[2(p + 1), p + 1, d]$ quadratic double-circulant (QDC) codes, i.e. $p \equiv \pm 3 \pmod 8$.

9.4.1 Double-Circulant Codes from Extended Quadratic Residue Codes

The following is a summary of the extended QR codes as double-circulant codes [8, 9, 13].

Binary QR codes are cyclic codes of length p over $\mathbb{F}_2$. For a given p, there exist four QR codes:

1. $\bar{\mathscr{L}}_p, \bar{\mathscr{N}}_p$ which are equivalent and have dimension $\frac{1}{2}(p - 1)$, and
2. $\mathscr{L}_p, \mathscr{N}_p$ which are equivalent and have dimension $\frac{1}{2}(p + 1)$.

The $(p + 1, \frac{1}{2}(p + 1), d)$ extended quadratic residue code, denoted by $\hat{\mathscr{L}}_p$ (resp. $\hat{\mathscr{N}}_p$), is obtained by annexing an overall parity check to $\mathscr{L}_p$ (resp. $\mathscr{N}_p$). If $p \equiv -1 \pmod 8$, $\hat{\mathscr{L}}_p$ (resp. $\hat{\mathscr{N}}_p$) is Type-II; otherwise it is FSD.

It is well known that[1] $\mathrm{Aut}(\hat{\mathscr{L}}_p)$ contains the projective special linear group denoted by $\mathrm{PSL}_2(p)$ [13]. If r is a generator of the cyclic group Q, then $\sigma : i \rightarrow \pmod p$ is a member of $\mathrm{PSL}_2(p)$. Given $n \in N$, the cycles of σ can be written as

[1]Since $\hat{\mathscr{L}}_p$ and $\hat{\mathscr{N}}_p$ are equivalent, considering either one is sufficient.

$$(\infty)(n, nr, nr^2, \ldots, nr^t)(1, r, r^2, \ldots, r^t)(0), \tag{9.13}$$

where $t = \frac{1}{2}(p-3)$. Due to this property, $\boldsymbol{G}$, the generator matrix of $\hat{\mathscr{L}}_p$ can be arranged into circulants as shown in (9.14),

$$\boldsymbol{G} = \begin{array}{c|c|cccccc|cccccc|c|} & \infty & n & nr & \ldots & nr^{t-1} & nr^t & & 1 & r & \ldots & r^{t-1} & r^t & & 0 \\ \hline \infty & 1 & 1 & 1 & \ldots & 1 & 1 & & 1 & 1 & \ldots & 1 & 1 & & 1 \\ \hline \beta & 0 & & & & & & & & & & & & & 1 \\ \beta r & 0 & & & & & & & & & & & & & 1 \\ \vdots & \vdots & & & \boldsymbol{L} & & & & & & \boldsymbol{R} & & & & \vdots \\ \beta r^{t-1} & 0 & & & & & & & & & & & & & 1 \\ \beta r^t & 0 & & & & & & & & & & & & & 1 \\ \hline \end{array}, \tag{9.14}$$

where $\boldsymbol{L}$ and $\boldsymbol{R}$ are $\frac{1}{2}(p-1) \times \frac{1}{2}(p-1)$ circulant matrices. The rows $\beta, \beta r, \ldots, \beta r^t$ in the above generator matrix contain $\bar{e}_\beta(x), \bar{e}_{\beta r}(x), \ldots, \bar{e}_{\beta r^t}(x)$, where $\bar{e}_i(x) = x^i e(x)$ whose coordinates are arranged in the order of (9.13). Note that (9.14) can be transformed to (9.5b) as follows:

$$\begin{bmatrix} 1 & \boldsymbol{J} \\ \boldsymbol{0}^T & \boldsymbol{L}^{-1} \end{bmatrix} \times \left[\begin{array}{cc|cc} 1 & \boldsymbol{J} & \boldsymbol{J} & 1 \\ \boldsymbol{0}^T & \boldsymbol{L} & \boldsymbol{R} & \boldsymbol{J}^T \end{array}\right] = \left[\begin{array}{cc|cc} 1 & \boldsymbol{J} + \mathbf{w}(\boldsymbol{L}^T) & \boldsymbol{J} + \mathbf{w}(\boldsymbol{R}^T) & \frac{1}{2}(p+1) \\ \boldsymbol{0}^T & \boldsymbol{I}_{\frac{1}{2}(p-1)} & \boldsymbol{L}^{-1}\boldsymbol{R} & \mathbf{w}(\boldsymbol{L}^{-1})^T \end{array}\right] \tag{9.15}$$

where $\boldsymbol{J}$ is an all-ones vector and $\mathbf{w}(\boldsymbol{A}) = [\mathrm{wt}_H(\boldsymbol{A}_0) \pmod 2, \mathrm{wt}_H(\boldsymbol{A}_1) \pmod 2, \ldots]$, $\boldsymbol{A}_i$ being the ith row vector of matrix $\boldsymbol{A}$. The multiplication in (9.15) assumes that $\boldsymbol{L}^{-1}$ exists and following Corollary 9.1, $\mathrm{wt}_H(l^{-1}(x)) = \mathrm{wt}_H(l(x))$ is odd. Therefore, (9.15) becomes

$$\boldsymbol{G} = \left[\begin{array}{c|c|c} & \boldsymbol{J} + \mathbf{w}(\boldsymbol{R}^T) & \frac{1}{2}(p+1) \\ \cline{2-3} \boldsymbol{I}_{\frac{1}{2}(p+1)} & & 1 \\ & \boldsymbol{L}^{-1}\boldsymbol{R} & \vdots \\ & & 1 \end{array}\right]. \tag{9.16}$$

In relation to (9.14), consider extended QR codes for the classes of primes:

1. $p = 8m + 1$, the idempotent $e(x) = \sum_{n \in N} x^n$ and $\beta \in N$. Following [13, Theorem 24, Chap. 16], we know that $\bar{e}_{\beta r^i}(x)$ where $\beta r^i \in N$, for $0 \le i \le t$, contains $2m + 1$ quadratic residues modulo p (including 0) and $2m - 1$ non-quadratic residues modulo p. As a consequence, $\mathrm{wt}_H(r(x))$ is even, implying $\mathbf{w}(\boldsymbol{R}^T) = \boldsymbol{0}$ and $r(x)$ is not invertible (cf. Corollary 9.1); and $\mathrm{wt}_H(l(x))$ is odd and $l(x)$ may be invertible over polynomial modulo $x^{\frac{1}{2}(p-1)} - 1$ (cf. Corollary 9.1). Furthermore, referring to (9.5b), we have $\alpha = \frac{1}{2}(p+1) = 4m + 1 = 1 \bmod 2$.

2. $p = 8m - 1$, the idempotent $e(x) = 1 + \sum_{n \in N} x^n$ and $\beta \in Q$. Following [13, Theorem 24, Chap. 16], if we have a set S containing 0 and $4m - 1$ non-quadratic residues modulo p, the set $\beta + S$ contains $2m + 1$ quadratic residues modulo p (including 0) and $2m - 1$ non-quadratic residues modulo p. It follows that $\bar{e}_{\beta r^i}(x)$, where $\beta r^i \in Q$, for $0 \le i \le t$, contains $2m$ quadratic residues modulo p (excluding 0), implying that $\boldsymbol{R}$ is singular (cf. Corollary 9.1); and $2m - 1$ non-quadratic residues modulo p, implying $\boldsymbol{L}^{-1}$ may exist (cf. Corollary 9.1). Furthermore, $\mathbf{w}(\boldsymbol{R}^T) = \mathbf{0}$ and referring to (9.5b), we have $\alpha = \frac{1}{2}(p+1) = 4m = 0 \bmod 2$.

For many $\hat{\mathscr{L}}_p$, $\boldsymbol{L}$ is invertible and Karlin [9] has shown that $p = 73, 97, 127, 137, 241$ are the known cases where the canonical form (9.5b) cannot be obtained.

Consider the case for $p = 73$, with $\beta = 5 \in N$, we have $l(x)$, the defining polynomial of the left circulant, given by

$$l(x) = x^2 + x^3 + x^4 + x^5 + x^6 + x^{11} + x^{15} + x^{16} + x^{18} + \\ x^{20} + x^{21} + x^{25} + x^{30} + x^{31} + x^{32} + x^{33} + x^{34}.$$

The polynomial $l(x)$ contains some irreducible factors of $x^{\frac{1}{2}(p-1)} - 1 = x^{36} - 1$, i.e. GCD $(l(x), x^{36} - 1) = 1 + x^2 + x^4$, and hence, it is not invertible. In addition to form (9.5b), $\boldsymbol{G}$ can also be transformed to (9.5a), and Jenson [8] has shown that, for $7 \le p \le 199$, except for $p = 89, 167$, the canonical form (9.5a) exists.

9.4.2 *Pure Double-Circulant Codes for Primes ±3 Modulo 8*

Recall that $\mathbf{S_r}$ is a multiplicative group of order $2^{p-1} - 1$ containing all polynomials of odd weight (excluding the all-ones polynomial) of degree at most $p - 1$, where p is a prime. We assume that $a(x)$ is a generator of $\mathbf{S_r}$. For $p \equiv \pm 3 \pmod 8$, we have the following lemma.

Lemma 9.6 *For* $p \equiv \pm 3 \pmod 8$, *let the polynomials* $q(x) = \sum_{i \in Q} x^i$ *and* $n(x) = \sum_{i \in N} x^i$. *Self-dual pure double-circulant codes with* $r(x) = q(x)$ *or* $r(x) = n(x)$ *exist if and only if* $p \equiv 3 \pmod 8$.

Proof For self-dual codes the condition $q(x)^T = n(x)$ must be satisfied where $q(x)^T = q(x^{-1}) = \sum_{i \in Q} x^{-i}$. Let $r(x) = q(x)$, for the case when $p \equiv \pm 3 \pmod 8$, $2 \in N$ we have $q(x)^2 = \sum_{i \in Q} x^{2i} = n(x)$. We know that $1 + q(x) + n(x) = j(x)$, therefore, $q(x)^3 = q(x)^2 q(x) = n(x)q(x) = (1 + q(x) + j(x))q(x) = q(x) + n(x) + j(x) = 1$. Then, $\frac{q(x)^2}{q(x)^3} = q(x)^2$ and $q(x)^2 = n(x) = q(x)^{-1} = q(x^{-1})$. On the other hand, $-1 \in N$ if $p \equiv 3 \pmod 8$ and thus $q(x)^T = n(x)$. If $p \equiv -3 \pmod 8$, $-1 \in Q$, we have $q(x)^T = q(x)$. For $r(x) = n(x)$, the same arguments follow.

Let $\mathscr{P}_p$ denote a $(2p, p, d)$ pure double-circulant code for $p \equiv \pm 3 \pmod 8$. The properties of $\mathscr{P}_p$ can be summarised as follows:

1. For $p \equiv 3 \pmod 8$, since $q(x)^3 = 1$ and $a^{2^{p-1}-1} = 1$, we have $q(x) = a(x)^{(2^{p-1}-1)/3}$ and $q(x)^T = a(x)^{(2^p-2)/3}$. There are two full-rank generator matrices with mutually disjoint information sets associated with $\mathscr{P}_p$ for these primes. Let $\boldsymbol{G}_1$ be a reduced echelon generator matrix of $\mathscr{P}_p$, which has the form of (9.5a) with $\boldsymbol{R} = \boldsymbol{B}$, where $\boldsymbol{B}$ is a circulant matrix with defining polynomial $b(x) = q(x)$. The other full-rank generator matrix $\boldsymbol{G}_2$ can be obtained as follows:

$$\boldsymbol{G}_2 = \begin{array}{|c|}\hline \boldsymbol{B}^T \\ \hline\end{array} \times \boldsymbol{G}_1 = \begin{array}{|c|c|}\hline \boldsymbol{B}^T & \boldsymbol{I}_p \\ \hline\end{array}\,. \tag{9.17}$$

The self-duality of this pure double-circulant code is obvious from $\boldsymbol{G}_2$.
2. For $p \equiv -3 \pmod 8$, $(p-1)/2$ is even and hence, neither $q(x)$ nor $n(x)$ is invertible, which means that if this polynomial was chosen as the defining polynomial for $\mathscr{P}_p$, there exists only one full-rank generator matrix. However, either $1 + q(x)$ (resp. $1 + n(x)$) is invertible and the inverse is $1 + n(x)$ (resp. $1 + q(x)$), i.e.

$$\begin{aligned}(1 + q(x))(1 + n(x)) &= 1 + q(x) + n(x) + q(x)n(x) \\ &= 1 + q(x) + n(x) + q(x)(1 + j(x) + q(x)) \\ &= 1 + q(x) + n(x) + q(x) + q(x)j(x) + q(x)^2,\end{aligned}$$

and since $q(x)j(x) = 0$ and $q(x)^2 = n(x)$ under polynomial modulo $x^p - 1$, it follows that

$$(1 + q(x))(1 + n(x)) = 1 \pmod{x^p - 1}.$$

Let $\boldsymbol{G}_1$ be the first reduced echelon generator matrix, which has the form of (9.5a) where $\boldsymbol{R} = \boldsymbol{I}_p + \boldsymbol{Q}$. The other full-rank generator matrix with disjoint information sets $\boldsymbol{G}_2$ can be obtained as follows:

$$\boldsymbol{G}_2 = \begin{array}{|c|}\hline \boldsymbol{I}_p + \boldsymbol{N} \\ \hline\end{array} \times \boldsymbol{G}_1 = \begin{array}{|c|c|}\hline \boldsymbol{I}_p + \boldsymbol{N} & \boldsymbol{I}_p \\ \hline\end{array}\,. \tag{9.18}$$

Since $-1 \in Q$ for this prime, $(\boldsymbol{I}_p + \boldsymbol{Q})^T = \boldsymbol{I}_p + \boldsymbol{Q}$ implying that the $(2p, p, d)$ pure double-circulant code is FSD, i.e. the generator matrix of $\mathscr{P}_p^{\perp}$ is given by

$$\boldsymbol{G}^{\perp} = \begin{array}{|c|c|}\hline \boldsymbol{I}_p + \boldsymbol{Q} & \boldsymbol{I}_p \\ \hline\end{array}\,.$$

A bordered double-circulant construction based on these primes—commonly known as the *quadratic double-circulant* construction—also exists, see Sect. 9.4.3 below.

9.4.3 Quadratic Double-Circulant Codes

Let p be a prime that is congruent to ± 3 modulo 8. A $(2(p+1), p+1, d)$ binary quadratic double-circulant code, denoted by $\mathscr{B}_p$, can be constructed using the defining polynomial

$$b(x) = \begin{cases} 1+q(x) & \text{if } p \equiv 3 \pmod 8, \text{ and} \\ q(x) & \text{if } p \equiv -3 \pmod 8 \end{cases} \tag{9.19}$$

where $q(x) = \sum_{i \in Q} x^i$. Following [13], the generator matrix $\boldsymbol{G}$ of $\mathscr{B}_p$ is

$$\boldsymbol{G} = \begin{array}{c} l_\infty \; l_0 \; \ldots \; l_{p-1} \; r_\infty \; r_0 \; \ldots \; r_{p-1} \\ \left[\begin{array}{c|ccc|c|ccc} 1 & & & & 0 & & & \\ \vdots & & \boldsymbol{I}_p & & \vdots & & \boldsymbol{B} & \\ 1 & & & & 0 & & & \\ \hline 0 & 0 & \ldots & 0 & 1 & 1 & \ldots & 1 \end{array}\right] \end{array} \tag{9.20}$$

which is, if the last row of $\boldsymbol{G}$ is rearranged as the first row, the columns indexed by l_∞ and r_∞ are rearranged as the last and the first columns, respectively, equivalent to (9.5b) with $\alpha = 0$ and $k = p+1$. Let $j(x) = 1 + x + x^2 + \cdots + x^{p-1}$, and the following are some properties of $\mathscr{B}_p$ [9]:

1. for $p \equiv 3 \pmod 8$, $b(x)^3 = (1+q(x))^2(1+q(x)) = (1+n(x))(1+q(x)) = 1 + j(x)$, since $q(x)^2 = n(x)$ ($2 \in N$ for this prime) and $q(x)j(x) = n(x)j(x) = j(x)$ ($\text{wt}_H(q(x)) = \text{wt}_H(n(x))$ is odd). Also, $(b(x)+j(x))^3 = (1+q(x)+j(x))^2(1+q(x)+j(x)) = n(x)^2(1+q(x)+j(x)) = q(x)+n(x)+j(x) = 1$ because $n(x)^2 = q(x)$. Since $-1 \in N$ and we have $b(x)^T = 1 + \sum_{i \in Q} x^{-i} = 1 + n(x)$ and thus, $b(x)b(x)^T = (1+q(x))(1+n(x)) = 1 + j(x)$.
   There are two generator full-rank matrices with disjoint information sets for $\mathscr{B}_p$. This is because, although $b(x)$ has no inverse, $b(x)+j(x)$ does, and the inverse is $(b(x)+j(x))^2$.
   Let $\boldsymbol{G}_1$ has the form of (9.5b) where $\boldsymbol{R} = \boldsymbol{B}$, and the other full-rank generator matrix $\boldsymbol{G}_2$ can be obtained as follows:

$$\boldsymbol{G}_2 = \left[\begin{array}{c|c} 1 & 1 \ \ldots \ 1 \\ \hline 0 & \\ \vdots & \boldsymbol{B}^T \\ 0 & \end{array}\right] \times \boldsymbol{G}_1 = \left[\begin{array}{c|c|c|c} 0 & 1 \ \ldots \ 1 & 1 & 0 \ \ldots \ 0 \\ \hline 1 & & 0 & \\ \vdots & \boldsymbol{B}^T & \vdots & \boldsymbol{I}_p \\ 1 & & 0 & \end{array}\right]. \tag{9.21}$$

It is obvious that $\boldsymbol{G}_2$ is identical to the generator matrix of $\mathscr{B}_p^\perp$ and hence, it is self-dual.

2. for $p \equiv -3 \pmod 8$, $b(x)^3 = n(x)q(x) = (1 + j(x) + q(x))q(x) = 1 + j(x)$ since $q(x)^2 = n(x)$ ($2 \in N$ for this prime) and $q(x)j(x) = n(x)j(x) = 0$ ($\mathrm{wt}_H(q(x)) = \mathrm{wt}_H(n(x))$ is even). Also, $(b(x) + j(x))^3 = (q(x) + j(x))^2(1 + n(x)) = q(x)^2 + q(x)^2n(x) + j(x)^2 + j(x)^2n(x) = n(x) + q(x) + j(x) = 1$ because $n(x)^2 = q(x)$. Since $-1 \in Q$ and we have $b(x)^T = \sum_{i \in Q} x^{-i} = b(x)$ and it follows that $\mathscr{B}_p$ is FSD, i.e. the generator matrix of $\mathscr{B}_p^\perp$ is given by

$$\boldsymbol{G}^\perp = \left[\begin{array}{c|c|c|c} 0 & 1 \ \ldots \ 1 & 1 & 0 \ \ldots \ 0 \\ \hline 1 & & 0 & \\ \vdots & \boldsymbol{B} & \vdots & \boldsymbol{I}_p \\ 1 & & 0 & \end{array}\right]$$

Since $(b(x) + j(x))^{-1} = (b(x) + j(x))^2$, there exist full-rank two generator matrices of disjoint information sets for $\mathscr{B}_p$. Let $\boldsymbol{G}_1$ has the form of (9.5b) where $\boldsymbol{R} = \boldsymbol{B}$, and the other full-rank generator matrix $\boldsymbol{G}_2$ can be obtained as follows:

$$\boldsymbol{G}_2 = \left[\begin{array}{c|c} 1 & 1 \ \ldots \ 1 \\ \hline 0 & \\ \vdots & \boldsymbol{B}^2 \\ 0 & \end{array}\right] \times \boldsymbol{G}_1 = \left[\begin{array}{c|c|c|c} 0 & 1 \ \ldots \ 1 & 1 & 0 \ \ldots \ 0 \\ \hline 1 & & 0 & \\ \vdots & \boldsymbol{B}^2 & \vdots & \boldsymbol{I}_p \\ 1 & & 0 & \end{array}\right] \tag{9.22}$$

Codes of the form $\mathscr{B}_p$ form an interesting family of double-circulant codes. In terms of self-dual codes, the family contains the longest extremal Type-II code known, $n = 136$. Probably, it is the longest extremal code that exists, see Sect. 9.7. Moreover, $\mathscr{B}_p$ is the binary image of the extended QR code over $\mathbb{F}_4$ [10].

The $(p + 1, \frac{1}{2}(p + 1), d)$ double-circulant codes for $p \equiv \pm 1 \pmod 8$ are fixed by $\mathrm{PSL}_2(p)$, see Sect. 9.4.1. This linear group $\mathrm{PSL}_2(p)$ is generated by the set of all permutations to the coordinates $(\infty, 0, 1, \ldots, p - 1)$ of the form

$$y \to \frac{ay + b}{cy + d}, \tag{9.23}$$

where $a, b, c, d \in \mathbb{F}_p$, $ad - bc = 1$, $y \in \mathbb{F}_p \cup \{\infty\}$, and it is assumed that $\pm\frac{1}{0} = \infty$ and $\pm\frac{1}{\infty} = 0$ in the arithmetic operations.

We know from [13] that this form of permutation is generated by the following transformations:

$$\begin{aligned} S &: y \rightarrow y+1 \\ V &: y \rightarrow \alpha^2 y \\ T &: y \rightarrow -\frac{1}{y}, \end{aligned} \tag{9.24}$$

where α is a primitive element of $\mathbb{F}_p$. In fact, V is redundant since it can be obtained from S and T, i.e.

$$V = TS^{\alpha}TS^{\mu}TS^{\alpha} \tag{9.25}$$

for[2] $\mu = \alpha^{-1} \in \mathbb{F}_p$.

The linear group $\mathrm{PSL}_2(p)$ fixes not only the $(p+1, \frac{1}{2}(p+1), d)$ binary double-circulant codes, for $p \equiv \pm 1 \pmod 8$, but also the $(2(p+1), p+1, d)$ binary quadratic double-circulant codes, as shown as follows. Consider the coordinates $(\infty, 0, 1, \ldots, p-1)$ of a circulant, the transformation S leaves the coordinate ∞ invariant and introduces a cyclic shift to the rest of the coordinates and hence S fixes a circulant. Let $\boldsymbol{R}_i$ and $\boldsymbol{L}_i$ denote the ith row of the right and left circulants of (9.20), respectively (we assume that the index starts with 0), and let $\boldsymbol{J}$ and $\boldsymbol{J}'$ denote the last row of the right and left circulant of (9.20), respectively.

Consider the primes $p = 8m+3$, $\boldsymbol{R}_0 = \left(0 \mid 1 + \sum_{i \in Q} x^i\right)$. Let e_i and f_j, for some integers i and j, be even and odd integers, respectively. If $i \in Q$, $-1/i = -1 \times \alpha^{p-1}/\alpha^{e_1} = \alpha^{f_1} \times \alpha^{e_2 - e_1} \in N$ since $-1 \in N$ for these primes. Therefore, the transformation T interchanges residues to non-residues and vice versa. In addition, we also know that T interchanges coordinates ∞ and 0. Applying transformation T to $\boldsymbol{R}_0$, $T(\boldsymbol{R}_0)$, results in

$$T(\boldsymbol{R}_0) = \left(1 \mid \sum_{j \in N} x^j\right) = \boldsymbol{R}_0 + \boldsymbol{J}.$$

Similarly, for the first row of $\boldsymbol{L}$, which has 1 at coordinates ∞ and 0 only, i.e. $\boldsymbol{L}_0 = (1 \mid 1)$

$$T(\boldsymbol{L}_0) = \boldsymbol{L}_0 + \boldsymbol{J}.$$

---

[2] $TS^{\alpha}TS^{\mu}TS^{\alpha}(y) = TS^{\alpha}TS^{\mu}T(y+\alpha) = TS^{\alpha}TS^{\mu}(-y^{-1}+\alpha) = TS^{\alpha}T\left(-\frac{1}{y+\mu}+\alpha\right) = TS^{\alpha}T$
$\left(\frac{\alpha y+\alpha\mu-1}{y+\mu}\right) = TS^{\alpha}\left(\frac{-\alpha y^{-1}+\alpha\mu-1}{-y^{-1}+\mu}\right) = T\left(\frac{-\alpha(y+\alpha)^{-1}+\alpha\mu-1}{-(y+\alpha)^{-1}+\mu}\right) = T\left(\frac{(\alpha\mu-1)y+\alpha(\alpha\mu-1)-\alpha}{\mu y+(\alpha\mu-1)}\right) =$
$\left(\frac{(-\alpha\mu-1)y^{-1}+\alpha(\alpha\mu-1)-\alpha}{-\mu y^{-1}+(\alpha\mu-1)}\right) = \left(\frac{-\alpha}{-\mu y^{-1}}\right) = \alpha^2 y = V(y).$

Let $s \in Q$ and let the set $\hat{Q} = Q \cup \{0\}$, $\boldsymbol{R}_s = \left(0 \mid \sum_{i \in \hat{Q}} x^{s+i}\right)$ and $T\left(\sum_{i \in \hat{Q}} x^{s+i}\right) = \sum_{i \in \hat{Q}} x^{-1/(s+i)}$. Following MacWilliams and Sloane [13, Theorem 24, Chap. 16], we know that the exponents of $\sum_{i \in \hat{Q}} x^{s+i}$ contain $2m+1$ residues and $2m+1$ non-residues. Note that $s+i$ produces no 0.[3] It follows that $-1/(s+i)$ contains $2m+1$ non-residues and $2m+1$ residues. Now consider $\boldsymbol{R}_{-1/s} = \left(0 \mid \sum_{i \in \hat{Q}} x^{i-1/s}\right)$, $i - 1/s$ contains[4] 0 $i, s \in Q$, $2m$ residues and $2m+1$ non-residues. We can write $-1/(s+i)$ as

$$-\frac{1}{s+i} = \frac{i/s}{s+i} - \frac{1}{s} = z - \frac{1}{s}.$$

Let $I \subset \hat{Q}$ be a set of all residues such that for all $i \in I$, $i - 1/s \in N$. If $-1/(s+i) \in N$, $z \in \hat{Q}$ and we can see that z must belong to I such that $z - 1/s \in N$. This means these non-residues cancel each other in $T(\boldsymbol{R}_s) + \boldsymbol{R}_{-1/s}$. On the other hand, if $-1/(s+i) \in Q$, $z \in N$ and it is obvious that $z - 1/s \neq i - 1/s$ for all $i \in \hat{Q}$, implying that all $2m+1$ residues in $T(\boldsymbol{R}_s)$ are disjoint from all $2m+1$ residues (including 0) in $\boldsymbol{R}_{-1/s}$. Therefore, $T(\boldsymbol{R}_s) + \boldsymbol{R}_{-1/s} = \left(0 \mid \sum_{i \in \hat{Q}} x^i\right)$, i.e.

$$T(\boldsymbol{R}_s) = \boldsymbol{R}_{-1/s} + \boldsymbol{R}_0.$$

Similarly, $T(\boldsymbol{L}_s) = \left(0 \mid 1 + x^{-1/s}\right)$ and $\boldsymbol{L}_{-1/s} = \left(1 \mid x^{-1/s}\right)$, which means

$$T(\boldsymbol{L}_s) = \boldsymbol{L}_{-1/s} + \boldsymbol{L}_0.$$

Let $t \in N$, $\boldsymbol{R}_t = \left(0 \mid \sum_{i \in \hat{Q}} x^{t+i}\right)$ and $T\left(\sum_{i \in \hat{Q}} x^{t+i}\right) = \sum_{i \in \hat{Q}} x^{-1/(t+i)}$. We know that $t+i$ contains 0, $2m$ residues and $2m+1$ non-residues [13, Theorem 24, Chap. 16], and correspondingly $-1/(t+i)$ contains ∞, $2m$ non-residues and $2m+1$ residues. As before, now consider $\boldsymbol{R}_{-1/t} = \left(0 \mid \sum_{i \in \hat{Q}} x^{i-1/t}\right)$. There are $2m+1$ residues (excluding 0) and $2m+1$ non-residues in $i - 1/t$, and let $I' \subset \hat{Q}$ be a set of all residues such that, for all $i \in I'$, $i - 1/t \in Q$. As before, we can write $-1/(t+i)$ as $z - 1/t$, where $z = (i/t)/(t+i)$. If $-1/(t+i) \in Q$, $z \in I'$ and hence, the $2m+1$ residues from $-1/(t+i)$ are identical to those from $i - 1/t$. If $-1/(t+i) \in N$, $z \in N$ and hence, all of the $2m$ non-residues of $-1/(t+i)$ are disjoint from all $2m+1$ non-residues of $i - 1/t$. Therefore, $T(\boldsymbol{R}_t) + R_{-1/t} = \left(1 \mid \sum_{i \in N} x^i\right)$, i.e.

$$T(\boldsymbol{R}_t) = \boldsymbol{R}_{-1/t} + \boldsymbol{R}_0 + \boldsymbol{J}.$$

[3] Consider a prime $p = \pm 3 \pmod 8$, $q \in Q$ and an integer a where $(a, p) = 1$. In order for $q + a = 0$ to happen, $a = -q$. The integer a is a residue if $p = 8m - 3$ and a non-residue if $p = 8m + 3$.

[4] This is because all $i \in Q$ are considered and $1/s \in Q$.

Similarly, $T(\boldsymbol{L}_t) = \left(0 \mid 1 + x^{-1/t}\right)$ and $\boldsymbol{L}_{-1/t} = \left(1 \mid x^{-1/t}\right)$, which means

$$T(\boldsymbol{L}_t) = \boldsymbol{L}_{-1/t} + \boldsymbol{L}_0 + \boldsymbol{J}'.$$

For primes $p = 8m - 3$, $\boldsymbol{R}_0 = \left(0 \mid \sum_{i \in Q} x^i\right)$ and since $-1 \in Q$, $-1/i \in Q$ for $i \in Q$. Thus,

$$T(\boldsymbol{R}_0) = \left(0 \mid \sum_{i \in Q} x^{-1/i}\right) = \boldsymbol{R}_0.$$

Similarly, for $\boldsymbol{L}_0$, which contains 1 at coordinates 0 and ∞,

$$T\left(\boldsymbol{L}_0\right) = \boldsymbol{L}_0.$$

Consider $\boldsymbol{R}_s = \left(0 \mid \sum_{i \in Q} x^{s+i}\right)$, for $s \in Q$, $T\left(\sum_{i \in Q} x^{s+i}\right) = \sum_{i \in Q} x^{-1/(s+i)}$. There are 0 (when $i = -s \in Q$), $2m - 2$ residues and $2m - 1$ non-residues in the set $s + i$ [13, Theorem 24, Chap. 16]. Correspondingly, $-1/(s+i) = z - 1/s$, where $z = (i/s)/(s+i)$, contains ∞, $2m - 2$ residues and $2m - 1$ non-residues. Now consider $\boldsymbol{R}_{-1/s} = \left(0 \mid \sum_{i \in Q} x^{i-1/s}\right)$, the set $i - 1/s$ contains 0 (when $i = 1/s \in Q$), $2m - 2$ residues and $2m - 1$ non-residues. Let $I \subset Q$ be a set of all residues such that for all $i \in I$, $i - 1/s \in Q$. If $-1/(s+i) \in Q$ then $z - 1/s \in Q$ which means $z \in Q$ and z must belong to I. This means all $2m - 2$ residues of $-1/(s+i)$ and those of $i - 1/s$ are identical. On the contrary, if $-1/(s+i) \in N$, $z \in N$ and this means $z - 1/s \neq i - 1/s$ for all $i \in Q$, and therefore all non-residues in $-1/(s+i)$ and $i - 1/s$ are mutually disjoint. Thus, $T(\boldsymbol{R}_s) + \boldsymbol{R}_{-1/s} = \left(1 \mid 1 + \sum_{i \in N} x^i\right)$, i.e.

$$T(\boldsymbol{R}_s) = \boldsymbol{R}_{-1/s} + \boldsymbol{R}_0 + \boldsymbol{J}.$$

Similarly, $T(\boldsymbol{L}_s) = \left(0 \mid 1 + x^{-1/s}\right)$, and we can write

$$T(\boldsymbol{L}_s) = \boldsymbol{L}_{-1/s} + \boldsymbol{L}_0 + \boldsymbol{J}'.$$

For $t \in N$, we have $\boldsymbol{R}_t = \left(0 \mid \sum_{i \in Q} x^{t+i}\right)$ and $T(\sum_{i \in Q} x^{t+i}) = \sum_{i \in Q} x^{-1/(t+i)}$. Following [13, Theorem 24, Chap. 16], there are $2m - 1$ residues and $2m - 1$ non-residues in the set $t + i$ and the same distributions are contained in the set $-1/(t+i)$. Considering $\boldsymbol{R}_{-1/t} = \left(0 \mid \sum_{i \in Q} x^{i-1/t}\right)$, there are $2m - 1$ residues and $2m - 1$ non-residues in $i - 1/t$. Rewriting $-1/(t+i) = z - 1/t$, for $z = (i/t)/(t+i)$, and letting $I' \subset Q$ be a set of all residues such that for all $i \in I'$, $i - 1/t \in N$, we know that if $-1/(t+i) \in N$ then $z - 1/t \in N$ which means that $z \in Q$ and z must belong to I'. Hence, the non-residues in $i - 1/t$ and $-1/(t+i)$ are identical. If $-1/(t+i) \in Q$, however, $z \in N$ and for all $i \in Q$, $i - 1/t \neq z - 1/t$, implying that the residues in $-1/(t+i)$ and $i - 1/t$ are mutually disjoint. Thus, $T(\boldsymbol{R}_t) + \boldsymbol{R}_{-1/t} = \left(0 \mid \sum_{i \in Q} x^i\right)$, i.e.

$$T(\boldsymbol{R}_t) = \boldsymbol{R}_{-1/t} + \boldsymbol{R}_0.$$

Similarly, $T(\boldsymbol{L}_t) = \left(0 \mid 1 + x^{-1/t}\right)$, and we can write

$$T(\boldsymbol{L}_t) = \boldsymbol{L}_{-1/t} + \boldsymbol{L}_0.$$

The effect T to the circulants is summarised as follows:

T	for $p \equiv 3 \pmod 8$	for $p \equiv -3 \pmod 8$
$T(\boldsymbol{R}_0)$	$\boldsymbol{R}_0 + \boldsymbol{J}$	$\boldsymbol{R}_0$
$T(\boldsymbol{R}_s)$	$\boldsymbol{R}_{-1/s} + \boldsymbol{R}_0$	$\boldsymbol{R}_{-1/s} + \boldsymbol{J}$
$T(\boldsymbol{R}_t)$	$\boldsymbol{R}_{-1/t} + \boldsymbol{R}_0 + \boldsymbol{J}$	$\boldsymbol{R}_{-1/t} + \boldsymbol{R}_0$
$T(\boldsymbol{L}_0)$	$\boldsymbol{L}_0 + \boldsymbol{J}'$	$\boldsymbol{L}_0$
$T(\boldsymbol{L}_s)$	$\boldsymbol{L}_{-1/s} + \boldsymbol{L}_0$	$\boldsymbol{L}_{-1/s} + \boldsymbol{J}'$
$T(\boldsymbol{L}_t)$	$\boldsymbol{L}_{-1/t} + \boldsymbol{L}_0 + \boldsymbol{J}'$	$\boldsymbol{L}_{-1/t} + \boldsymbol{L}_0$

where $s \in Q$ and $t \in N$. This shows that, for $p \equiv \pm 3 \pmod 8$, the transformation T is a linear combination of at most three rows of the circulant and hence it fixes the circulant. This establishes the following theorem on $\mathrm{Aut}(\mathscr{B}_p)$ [2, 13].

Theorem 9.1 *The automorphism group of the* $(2(p+1), p+1, d)$ *binary quadratic double-circulant codes contains* $PSL_2(p)$ *applied simultaneously to both circulants.*

The knowledge of $\mathrm{Aut}(\mathscr{B}_p)$ can be exploited to deduce the modular congruence weight distributions of $\mathscr{B}_p$ as shown in Sect. 9.6.

9.5 Evaluation of the Number of Codewords of Given Weight and the Minimum Distance: A More Efficient Approach

In Chap. 5 algorithms to compute the minimum distance of a binary linear code and to count the number of codewords of a given weight are described. Assuming the code rate of the code is a half and its generator matrix contains two mutually disjoint information sets, each of rank k (the code dimension), these algorithms require enumeration of

$$\binom{k}{w/2} + 2\sum_{i=1}^{w/2-1}\binom{k}{i}$$

codewords in order to count the number of codewords of weight w. For FSD double-circulant codes with $p \equiv -3 \pmod 8$ and self-dual double-circulant codes a more efficient approach exists. This approach applies to both pure and bordered double-circulant cases.

Lemma 9.7 *Let $T_m(x)$ be a set of binary polynomials with degree at most m. Let $u_i(x), v_i(x) \in T_{k-1}(x)$ for $i = 1, 2$, and $e(x), f(x) \in T_{k-2}(x)$. The numbers of weight w codewords of the form $c_1(x) = (u_1(x)|v_1(x))$ and $c_2(x) = (v_2(x)|u_2(x))$ are equal, where*

(i) for self-dual pure double-circulant codes, $u_2(x) = u_1(x)^T$ and $v_2(x) = v_1(x)^T$;
(ii) for self-dual bordered double-circulant codes, $u_1(x) = (\varepsilon|e(x))$, $v_1(x) = (\gamma|f(x))$, $u_2(x) = (\varepsilon|e(x)^T)$ and $v_2(x) = (\gamma|f(x)^T)$, where $\gamma = \mathrm{wt}_H(e(x)) \pmod 2$;
(iii) for FSD pure double-circulant codes ($p \equiv -3 \pmod 8$), $u_2(x) = u_1(x)^2$ and $v_2(x) = v_1(x)^2$;
(iv) for FSD bordered double-circulant codes ($p \equiv -3 \pmod 8$), $u_1(x)=(\varepsilon|e(x))$, $v_1(x) = (\gamma|f(x))$, $u_2(x) = (\varepsilon|e(x)^2)$, $v_2(x) = (\gamma|f(x)^2)$, where $\gamma = \mathrm{wt}_H(e(x)) \pmod 2$.

Proof

(i) Let $\boldsymbol{G}_1 = [\boldsymbol{I}_k|\boldsymbol{R}]$ and $\boldsymbol{G}_2 = [\boldsymbol{R}^T|\boldsymbol{I}_k]$ be the two full-rank generator matrices with mutually disjoint information sets of a self-dual pure double-circulant code. Assume that $r(x)$ and $r(x)^T$ are the defining polynomials of $\boldsymbol{G}_1$ and $\boldsymbol{G}_2$, respectively. Given $u_1(x)$ as an input, we have a codeword $c_1(x) = (u_1(x)|v_1(x))$, where $v_1(x) = u_1(x)r(x)$, from $\boldsymbol{G}_1$. Another codeword $c_2(x)$ can be obtained from $\boldsymbol{G}_2$ using $u_1(x)^T$ as an input, $c_2(x) = (v_1(x)^T|u_1(x)^T)$, where $v_1(x)^T = u_1(x)^T r(x)^T = (u_1(x)r(x))^T$. Since the weight of a polynomial and that of its transpose are equal, for a given polynomial of degree at most $k-1$, there exist two distinct codewords of the same weight.

(ii) Let $\boldsymbol{G}_1$, given by (9.5b), and $\boldsymbol{G}_2$ be two full-rank generator matrices with pairwise disjoint information sets, of bordered self-dual double-circulant codes. It is assumed that the form of $\boldsymbol{G}_2$ is identical to that given by (9.21) with $\boldsymbol{R}^T = \boldsymbol{B}^T$. Let $f(x) = e(x)r(x)$, consider the following cases:

a. $\varepsilon = 0$ and $\mathrm{wt}_H(e(x))$ is odd, we have a codeword $c_1(x) = (0 \mid e(x) \mid 1 \mid f(x))$ from $\boldsymbol{G}_1$. Applying $\left(0 \mid e(x)^T\right)$ as an information vector to $\boldsymbol{G}_2$, we have another codeword $c_2(x) = \left(1 \mid e(x)^T r(x)^T \mid 0 \mid e(x)^T\right) = \left(1 \mid f(x)^T \mid 0 \mid e(x)^T\right)$.
b. $\varepsilon = 1$ and $\mathrm{wt}_H(e(x))$ is odd, $\boldsymbol{G}_1$ produces $c_1(x) = (1 \mid e(x) \mid 1 \mid f(x) + j(x))$. Applying $\left(1 \mid e(x)^T\right)$ as an information vector to $\boldsymbol{G}_2$, we have a codeword $c_2(x) = \left(1 \mid e(x)^T r(x)^T + j(x) \mid 1 \mid e(x)^T\right) = \left(1 \mid f(x)^T + j(x) \mid 1 \mid e(x)^T\right)$.
c. $\varepsilon = 0$ and $\mathrm{wt}_H(e(x))$ is even, $\boldsymbol{G}_1$ produces a codeword $c_1(x) = (0 \mid e(x) \mid 0 \mid f(x))$. Applying $\left(0 \mid e(x)^T\right)$ as an information vector to $\boldsymbol{G}_2$, we have another codeword $c_2(x) = \left(0 \mid e(x)^T r(x)^T \mid 0 \mid e(x)^T\right) = \left(0 \mid f(x)^T \mid 0 \mid e(x)^T\right)$.
d. $\varepsilon = 1$ and $\mathrm{wt}_H(e(x))$ is even, $\boldsymbol{G}_1$ produces $c_1(x) = (1 \mid e(x) \mid 0 \mid f(x) + j(x))$. Applying $\left(1 \mid e(x)^T\right)$ as an information vector to $\boldsymbol{G}_2$, we have a codeword $c_2(x) = \left(0 \mid e(x)^T r(x)^T + j(x) \mid 1 \mid e(x)^T\right) = \left(0 \mid f(x)^T + j(x) \mid 1 \mid e(x)^T\right)$.

It is clear that in all cases, $\mathrm{wt}_H(c_1(x)) = \mathrm{wt}_H(c_2(x))$ since $\mathrm{wt}_H(v(x)) = \mathrm{wt}_H(v(x)^T)$ and $\mathrm{wt}_H(v(x) + j(x)) = \mathrm{wt}_H(v(x)^T + j(x))$ for some polynomial $v(x)$. This means that given an information vector, there always exist two distinct codewords of the same weight.

(iii) Let $\boldsymbol{G}_1$, given by (9.5a) with $\boldsymbol{R} = \boldsymbol{I}_p + \boldsymbol{Q}$, and $\boldsymbol{G}_2$, given by (9.18), be two full-rank generator matrices with pairwise disjoint information sets, of pure FSD double-circulant codes for $p \equiv -3 \pmod 8$.
Given $u_1(x)$ as input, we have a codeword $c_1(x) = (u_1(x)|v_1(x))$, where $v_1(x) = u_1(x)(1 + q(x))$, from $\boldsymbol{G}_1$ and another codeword $c_2(x) = (v_2(x)|u_2(x))$, where $u_2(x) = u_1(x)^2$ and $v_2(x) = u_1(x)^2(1 + n(x)) = u_1(x)^2(1 + q(x))^2 = v_1(x)^2$, from $\boldsymbol{G}_2$. Since the weight of a polynomial and that of its square are the same over $\mathbb{F}_2$, the proof follows.

(iv) Let $\boldsymbol{G}_1$, given by (9.5b) with $\boldsymbol{B} = \boldsymbol{R}$, and $\boldsymbol{G}_2$, given by (9.22), be two full-rank generator matrices with pairwise disjoint information sets, of bordered FSD double-circulant codes for $p \equiv -3 \pmod 8$. Let $f(x) = e(x)b(x)$, consider the following cases:

a. $\varepsilon = 0$ and $\mathrm{wt}_H(e(x))$ is odd, we have a codeword $c_1(x) = (0 \mid e(x) \mid 1 \mid f(x))$ from $\boldsymbol{G}_1$. Applying $\left(0 \mid e(x)^2\right)$ as an information vector to $\boldsymbol{G}_2$, we have another codeword $c_2(x) = \left(1 \mid e(x)^2 n(x) \mid 0 \mid e(x)^2\right)$. Since $e(x)^2 n(x) = e(x)^2 b(x)^2 = f(x)^2$, the codeword $c_2 = \left(1 \mid f(x)^2 \mid 0 \mid e(x)^2\right)$.
b. $\varepsilon = 1$ and $\mathrm{wt}_H(e(x))$ is odd, $\boldsymbol{G}_1$ produces $c_1(x) = (1 \mid e(x) \mid 1 \mid f(x) + j(x))$. Applying $\left(1 \mid e(x)^2\right)$ as an information vector to $\boldsymbol{G}_2$, we have a codeword $c_2(x) = \left(1 \mid e(x)^2 n(x) + j(x) \mid 1 \mid e(x)^2\right) = \left(1 \mid f(x)^2 + j(x) \mid 1 \mid e(x)^2\right)$.
c. $\varepsilon = 0$ and $\mathrm{wt}_H(e(x))$ is even, $\boldsymbol{G}_1$ produces a codeword $c_1(x) = (0 \mid e(x) \mid 0 \mid f(x))$. Applying $\left(0 \mid e(x)^2\right)$ as an information vector to $\boldsymbol{G}_2$, we have another codeword $c_2(x) = \left(0 \mid e(x)^2 n(x) \mid 0 \mid e(x)^2\right) = \left(0 \mid f(x)^2 \mid 0 \mid e(x)^2\right)$.
d. $\varepsilon = 1$ and $\mathrm{wt}_H(e(x))$ is even, $\boldsymbol{G}_1$ produces $c_1(x) = (1 \mid e(x) \mid 0 \mid f(x) + j(x))$. Applying $\left(1 \mid e(x)^2\right)$ as an information vector to $\boldsymbol{G}_2$, we have a codeword $c_2(x) = \left(0 \mid e(x)^2 n(x) + j(x) \mid 1 \mid e(x)^2\right) = \left(0 \mid f(x)^2 + j(x) \mid 1 \mid e(x)^2\right)$.

It is clear that in all cases, $\mathrm{wt}_H(c_1(x)) = \mathrm{wt}_H(c_2(x))$ since $\mathrm{wt}_H(v(x)) = \mathrm{wt}_H(v(x)^2)$ and $\mathrm{wt}_H(v(x) + j(x)) = \mathrm{wt}_H(v(x)^2 + j(x))$ for some polynomial $v(x)$. This means that given an information vector, there always exist two distinct codewords of the same weight.

From Lemma 9.7, it follows that, in order to count the number of codewords of weight w, we only require

$$\sum_{i=1}^{w/2} \binom{k}{i}$$

codewords to be enumerated and if A_w denotes the number of codewords of weight w,

$$A_w = a_{w/2} + 2 \sum_{i=1}^{w/2-1} a_i \tag{9.26}$$

where a_i is the number of weight w codewords which have i non-zeros in the first k coordinates.

Similarly, the commonly used method to compute the minimum distance of half-rate codes with two full-rank generator matrices of mutually disjoint information sets, for example, see van Dijk et al. [18], assuming that d is the minimum distance of the code, requires as many as

$$S = 2 \sum_{i=1}^{d/2-1} \binom{n}{i}$$

codewords to be enumerated. Following Lemma 9.7, only $S/2$ codewords are required for $\mathscr{P}_p$ and $\mathscr{B}_p$ for $p \equiv -3 \pmod 8$, and self-dual double-circulant codes. Note that the bound $d/2 - 1$ may be improved for singly even and doubly even codes, but we consider the general case here.

9.6 Weight Distributions

The automorphism group of both $(p+1, \frac{1}{2}(p+1), d)$ extended QR and $(2(p+1), p+1, d)$ quadratic double-circulant codes contains the projective special linear group, $\mathrm{PSL}_2(p)$. Let $\mathscr{H}$ be a subgroup of the automorphism group of a linear code, and the number of codewords of weight i, denoted by A_i, can be categorised into two classes:

1. a class of weight i codewords which are invariant under some element of $\mathscr{H}$; and
2. a class of weight i codewords which forms an orbit of size $|\mathscr{H}|$, the order of $\mathscr{H}$. In the other words, if $\boldsymbol{c}$ is a codeword of this class, applying all elements of $\mathscr{H}$ to $\boldsymbol{c}$, $|\mathscr{H}|$ distinct codewords are obtained.

Thus, we can write A_i in terms of congruence as follows:

$$\begin{aligned} A_i &= n_i \times |\mathscr{H}| + A_i(\mathscr{H}), \\ &\equiv A_i(\mathscr{H}) \pmod{|\mathscr{H}|} \end{aligned} \tag{9.27}$$

where $A_i(\mathscr{H})$ is the number of codewords of weight i fixed by some element of $\mathscr{H}$. This was originally shown by Mykkeltveit et al. [14], where it was applied to extended QR codes for primes 97 and 103.

9.6.1 *The Number of Codewords of a Given Weight in Quadratic Double-Circulant Codes*

For $\mathscr{B}_p$, we shall choose $\mathscr{H} = \text{PSL}_2(p)$, which has order $|\mathscr{H}| = \frac{1}{2}p(p^2-1)$. Let the matrix $\left[\begin{smallmatrix} a & b \\ c & d \end{smallmatrix}\right]$ represent an element of $\text{PSL}_2(p)$, see (9.23). Since $|\mathscr{H}|$ can be factorised as $|\mathscr{H}| = \prod_j q_j^{e_j}$, where q_j is a prime and e_j is some integer, $A_i(\mathscr{H})$ (mod $|\mathscr{H}|$) can be obtained by applying the Chinese remainder theorem to $A_i(S_{q_j})$ (mod $q_j^{e_j}$) for all q_j that divides $|\mathscr{H}|$, where S_{q_j} is the Sylow-q_j-subgroup of $\mathscr{H}$. In order to compute $A_i(S_{q_j})$, a subcode of $\mathscr{B}_p$ which is invariant under S_{q_j} needs to be obtained in the first place. This invariant subcode, in general, has a considerably smaller dimension than $\mathscr{B}_p$, and hence, its weight distribution can be easily obtained.

For each odd prime q_j, S_{q_j} is a cyclic group which can be generated by some $\left[\begin{smallmatrix} a & b \\ c & d \end{smallmatrix}\right] \in \text{PSL}_2(p)$ of order q_j. Because S_{q_j} is cyclic, it is straightforward to obtain the invariant subcode, from which we can compute $A_i(S_{q_j})$.

On the other hand, the case of $q_j = 2$ is more complicated. For $q_j = 2$, S_2 is a dihedral group of order 2^{m+1}, where $m+1$ is the maximum power of 2 that divides $|\mathscr{H}|$ [?]. For $p = 8m \pm 3$, we know that

$$|\mathscr{H}| = \frac{1}{2}(8m \pm 3)\left((8m \pm 3)^2 - 1\right) = 2^2\left(64m^3 \pm 72m^2 + 26m \pm 3\right),$$

which shows that the highest power of 2 that divides $|\mathscr{H}|$ is 2^2 ($m = 1$). Following [?], there are $2^m + 1$ subgroups of order 2 in S_2, namely

$$\begin{aligned} H_2 &= \{1, P\}, \\ G_2^0 &= \{1, T\}, \text{ and} \\ G_2^1 &= \{1, PT\}, \end{aligned}$$

where $P, T \in \text{PSL}_2(p)$, $P^2 = T^2 = 1$ and $TPT^{-1} = P^{-1}$.

Let $T = \left[\begin{smallmatrix} 0 & p-1 \\ 1 & 0 \end{smallmatrix}\right]$, which has order 2. It can be shown that any order 2 permutation, $P = \left[\begin{smallmatrix} a & b \\ c & d \end{smallmatrix}\right]$, if a constraint $b = c$ is imposed, we have $a = -d$. All these subgroups, however, are conjugates in $\text{PSL}_2(p)$ [?] and therefore, the subcodes fixed by G_2^0, G_2^1 and H_2 have identical weight distributions and considering any of them, say G_2^0, is sufficient.

Apart from $2^m + 1$ subgroups of order 2, S_2 also contains a cyclic subgroup of order 4, 2^{m-1} non-cyclic subgroups of order 4, and subgroups of order 2^j for $j \geq 3$.

Following [14], only the subgroups of order 2 and the non-cyclic subgroups of order 4 make contributions towards $A_i(S_2)$. For $p \equiv \pm 3 \pmod 8$, there is only one non-cyclic subgroup of order 4, denoted by G_4, which contains, apart from an identity, three permutations of order 2 [?], i.e. a Klein 4 group,

$$G_4 = \{1, P, T, PT\}.$$

Having obtained $A_i(G_2^0)$ and $A_i(G_4)$, following the argument in [14], the number of codewords of weight i that are fixed by some element of S_2 is given by

$$A_i(S_2) \equiv 3A_i(G_2^0) - 2A_i(G_4) \pmod 4. \tag{9.28}$$

In summary, in order to deduce the modular congruence of the number of weight i codewords in $\mathscr{B}_p$, it is sufficient to do the following steps:

1. compute the number of weight i codewords in the subcodes fixed by G_2^0, G_4 and S_q, for all odd primes q that divide $|\mathscr{H}|$;
2. apply (9.28) to $A_i(G_2^0)$ and $A_i(G_4)$ to obtain $A_i(S_2)$; and then
3. apply the Chinese remainder theorem to $A_i(S_2)$ and all $A_i(S_q)$ to obtain $A_i(\mathscr{H}) \pmod{|\mathscr{H}|}$.

Given $\mathscr{B}_p$ and an element of $\mathrm{PSL}_2(p)$, how can we find the subcode consisting of the codewords fixed by this element? Assume that $Z = \left[\begin{smallmatrix} a & b \\ c & d \end{smallmatrix}\right] \in \mathrm{PSL}_2(p)$ of prime order. Let c_{l_i} (resp. c_{r_i}) and $c_{l_{i'}}$ (resp. $c_{r_{i'}}$) denote the ith coordinate and $\pi_Z(i)$th coordinate (ith coordinate with the respect to permutation π_Z), in the left (resp. right) circulant form, respectively. The invariant subcode can be obtained by solving a set of linear equations consisting of the parity-check matrix of $\mathscr{B}_p$ (denoted by $\boldsymbol{H}$), $c_{l_i} + c_{l_{i'}} = 0$ (denoted by $\boldsymbol{\pi}_Z(L)$) and $c_{r_i} + c_{r_{i'}} = 0$ (denoted by $\boldsymbol{\pi}_Z(R)$) for all $i \in \mathbb{F}_p \cup \{\infty\}$, i.e.

$$\boldsymbol{H}_{sub} = \begin{array}{|c|} \hline \\ \boldsymbol{H} \\ \\ \hline \boldsymbol{\pi}_Z(L) \\ \hline \boldsymbol{\pi}_Z(R) \\ \hline \end{array}\,.$$

The solution to $\boldsymbol{H}_{sub}$ is a matrix of rank $r > (p+1)$, which is the parity-check matrix of the $(2(p+1), 2(p+1)-r, d')$ invariant subcode. For subgroup G_4, which consists of permutations P, T and PT, we need to solve the following matrix

$$\boldsymbol{H}_{sub} = \begin{array}{|c|} \hline \\ \boldsymbol{H} \\ \\ \hline \boldsymbol{\pi}_P(L) \\ \hline \boldsymbol{\pi}_P(R) \\ \hline \boldsymbol{\pi}_T(L) \\ \hline \boldsymbol{\pi}_T(R) \\ \hline \boldsymbol{\pi}_{PT}(L) \\ \hline \boldsymbol{\pi}_{PT}(R) \\ \hline \end{array}$$

to obtain the invariant subcode. Note that the parity-check matrix of $\mathscr{B}_p$ is assumed to have the following form:

$$
\begin{array}{c}
\quad\; l_\infty\; l_0\; \ldots\; l_{p-1}\; r_\infty\; r_0\; \ldots\; r_{p-1} \\
\boldsymbol{H} = \left[\begin{array}{c|ccc|c|ccc}
0 & & & & 1 & & & \\
\vdots & & \boldsymbol{B}^T & & \vdots & & \boldsymbol{I}_p & \\
0 & & & & 1 & & & \\
\hline
1 & 1 & \ldots & 1 & 0 & 0 & \ldots & 0
\end{array}\right].
\end{array}
\qquad (9.29)
$$

One useful application of the modular congruence of the number of codewords of weight w is to verify, independently, the number of codewords of a given weight w that were computed exhaustively.

Computing the number of codewords of a given weight in small codes using a single-threaded algorithm is tractable, but for longer codes, it is necessary to use multiple computers working in parallel to produce a result within a reasonable time. Even so it can take several weeks, using hundreds of computers, to evaluate a long code. In order to do the splitting, the codeword enumeration task is distributed among all of the computers and each computer just needs to evaluate a predetermined number of codewords, finding the partial weight distributions. In the end, the results are combined to give the total number of codewords of a given weight. There is always the possibility of software bugs or mistakes to be made, particularly in any parallel computing scheme. The splitting may not be done correctly or double-counting or miscounting introduced as a result, apart from possible errors in combining the partial results. Fortunately, the modular congruence approach can also provide detection of computing errors by revealing inconsistencies in the summed results. The importance of this facet of modular congruence will be demonstrated in determining the weight distributions of extended QR codes in Sect. 9.6.2. In the following examples we work through the application of the modular congruence technique in evaluating the weight distributions of the quadratic double-circulant codes of primes 37 and 83.

Example 9.1 For prime 37, there exists an FSD $(76, 38, 12)$ quadratic double-circulant code, $\mathscr{B}_{37}$. The weight enumerator of an FSD code is given by Gleason's theorem [15]

$$
A(z) = \sum_{i=0}^{\lfloor \frac{n}{8} \rfloor} K_i (1+z^2)^{\frac{n}{2}-4i} (z^2 - 2z^4 + z^6)^i \qquad (9.30)
$$

for integers K_i. The number of codewords of any weight w is given by the coefficient of z^w of $A(z)$. In order to compute $A(z)$ of $\mathscr{B}_{37}$, we need only to compute A_{2i} for $6 \leq i \leq 9$. Using the technique described in Sect. 9.5, the number of codewords of desired weights is obtained and then substituted into (9.30). The resulting weight enumerator function giving the whole weight distribution of the $(76, 38, 12)$ code, $\mathscr{B}_{37}$ is

$$
\begin{aligned}
A(z) = &\left(1 + z^{76}\right) + 2109 \times \left(z^{12} + z^{64}\right) + \\
&86469 \times \left(z^{16} + z^{60}\right) + 961704 \times \left(z^{18} + z^{58}\right) + \\
&7489059 \times \left(z^{20} + z^{56}\right) + 53574224 \times \left(z^{22} + z^{54}\right) + \\
&275509215 \times \left(z^{24} + z^{52}\right) + 1113906312 \times \left(z^{26} + z^{50}\right) + \\
&3626095793 \times \left(z^{28} + z^{48}\right) + 9404812736 \times \left(z^{30} + z^{46}\right) + \\
&19610283420 \times \left(z^{32} + z^{44}\right) + 33067534032 \times \left(z^{34} + z^{42}\right) + \\
&45200010670 \times \left(z^{36} + z^{40}\right) + 50157375456 \times z^{38}.
\end{aligned}
\tag{9.31}
$$

Let $\mathscr{H} = \mathrm{PSL}_2(37)$, and we know that $|\mathscr{H}| = 2^2 \times 3^2 \times 19 \times 37 = 25308$. Consider the odd primes as factors q. For $q = 3$, $\left[\begin{smallmatrix} 0 & 1 \\ 36 & 1 \end{smallmatrix}\right]$ generates the following permutation of order 3:

$$
\begin{aligned}
&(\infty, 0, 1)(2, 36, 19)(3, 18, 13)(4, 12, 10)(5, 9, 23)(6, 22, 7)(8, 21, 24)\\
&(11)(14, 17, 30)(15, 29, 33)(16, 32, 31)(20, 35, 25)(26, 34, 28)(27)
\end{aligned}
$$

The corresponding invariant subcode has a generator matrix $G^{(S_3)}$ of dimension 14, which is given by

$$
\boldsymbol{G}^{(S_3)} = \begin{bmatrix}
1000000000000011101110011111101001001011001010001001110001000100000000000000 \\
0100000000000000001001100010111011000100010111001101110000001000000000100000 \\
0010000000000011101011101010110011010110101001000011110100000001000000000000 \\
0001000000000011100111111111000100001111010100110101110010000010000000000000 \\
0000100000000011101011111010101001011111111111000111110000000000000000000000 \\
0000010000000011111111111111111111111111111111111110000000000000000000000000 \\
0000001000000011101110001111111111010010101011111101111000000000000000000000 \\
0000000100000000000110011101000100011011101000110011110000000000100100000000 \\
0000000010000011101101100101110111000100011101110101110000110000000000000000 \\
0000000001000011100101110111000100011101110100110111111000000000000100000001 \\
0000000000100011100010001010011010010010111111001111110000000000001000000010 \\
0000000000010011100101100101010111000100011111111110111000000000000000001100 \\
0000000000001011101000010010101101011001101010101110111100000000000000101000 \\
0000000000000100001111101111101101010110111110111111111000000000000010000000
\end{bmatrix}
$$

and its weight enumerator function is

$$
\begin{aligned}
A^{(S_3)}(z) = &\left(1 + z^{76}\right) + 3 \times \left(z^{12} + z^{64}\right) + 24 \times \left(z^{16} + z^{60}\right) + \\
&54 \times \left(z^{18} + z^{58}\right) + 150 \times \left(z^{20} + z^{56}\right) + 176 \times \left(z^{22} + z^{54}\right) + \\
&171 \times \left(z^{24} + z^{52}\right) + 468 \times \left(z^{26} + z^{50}\right) + 788 \times \left(z^{28} + z^{48}\right) + \\
&980 \times \left(z^{30} + z^{46}\right) + 1386 \times \left(z^{32} + z^{44}\right) + 1350 \times \left(z^{34} + z^{42}\right) + \\
&1573 \times \left(z^{36} + z^{40}\right) + 2136 \times z^{38}.
\end{aligned}
\tag{9.32}
$$

For $q = 19$, $\left[\begin{smallmatrix} 0 & 1 \\ 36 & 3 \end{smallmatrix}\right]$ generates the following permutation of order 19:

$$
\begin{aligned}
&(\infty, 0, 25, 5, 18, 32, 14, 10, 21, 2, 1, 19, 30, 26, 8, 22, 35, 15, 3)\\
&(4, 36, 28, 34, 31, 33, 16, 17, 29, 27, 20, 13, 11, 23, 24, 7, 9, 6, 12).
\end{aligned}
$$

The resulting generator matrix of the invariant subcode $\boldsymbol{G}^{(S_{19})}$, which has dimension 2, is

$$\boldsymbol{G}^{(S_{19})} = \left[\begin{smallmatrix} 1011111111111111111111111111111111111110000000000000000000000000000000000000 \\ 0100000000000000000000000000000000000001111111111111111111111111111111111111 \end{smallmatrix}\right]$$

and its weight enumerator function is

$$A^{(S_{19})}(z) = 1 + 2z^{38} + z^{76}. \tag{9.33}$$

For the last odd prime, $q = 37$, a permutation of order 37

$$(\infty, 0, 18, 24, 27, 14, 30, 15, 13, 32, 25, 26, 33, 19, 7, 4, 6, 23, 34,$$
$$1, 12, 29, 31, 28, 16, 2, 9, 10, 3, 22, 20, 5, 21, 8, 11, 17, 35)(36)$$

is generated by $\left[\begin{smallmatrix} 0 & 1 \\ 36 & 35 \end{smallmatrix}\right]$ and it turns out that the corresponding invariant subcode, and hence, the weight enumerator function, are identical to those of $q = 19$.

For $q = 2$, subcodes fixed by some element of G_2^0 and G_4 are required. We have $P = \left[\begin{smallmatrix} 3 & 8 \\ 8 & 34 \end{smallmatrix}\right]$ and $T = \left[\begin{smallmatrix} 0 & 36 \\ 1 & 0 \end{smallmatrix}\right]$, and the resulting order 2 permutations generated by P, T and PT are

$$(\infty, 5)(0, 22)(1, 17)(2, 21)(3, 29)(4, 16)(6, 31)(7, 18)(8, 26)(9, 30)(10, 25)$$
$$(11, 34)(12, 14)(13, 36)(15)(19, 28)(20, 24)(23, 27)(32)(33, 35)$$

$$(\infty, 0)(1, 36)(2, 18)(3, 12)(4, 9)(5, 22)(6)(7, 21)(8, 23)(10, 11)(13, 17)$$
$$(14, 29)(15, 32)(16, 30)(19, 35)(20, 24)(25, 34)(26, 27)(28, 33)(31)$$

and

$$(\infty, 22)(0, 5)(1, 13)(2, 7)(3, 14)(4, 30)(6, 31)(8, 27)(9, 16)(10, 34)(11, 25)$$
$$(12, 29)(15, 32)(17, 36)(18, 21)(19, 33)(20)(23, 26)(24)(28, 35)$$

respectively. It follows that the corresponding generator matrices and weight enumerator functions of the invariant subcodes are

$$\boldsymbol{G}^{(G_2^0)} = \left[\begin{smallmatrix}
10000000000000000000001010101100011010101011101000000101110010000000000000000000 \\
0100000000000000000000101110000110000110000111010101101101000000010000000000000 \\
001000000000000000000001011011101100101111000100001000111101100000000000000000000 \\
000100000000000000000001100100111101010111100100110101001010000000000000001000000 \\
0000100000000000000000010000110111101101110110000000101010101000010000000000000000 \\
0000010000000000000000001011001001011101001001100111000000111010000000000000000000 \\
0000001000000000000000011111111111111111111111111111111111111111100000000000000000000 \\
0000000100000000000000000011000011110111001000000100111011111000000000000000000000100 \\
000000001000000000000000101101100110001111110001100110110100000000000000100000000 \\
0000000001000000000000001000100111111010101010111111001000100000000000100000000000 \\
000000000010000000000000100111100111100001000110111111011000100000000000000000000010 \\
0000000000001000000000001110000011100001010011111111100101000000000000000000000001000 \\
000000000000010000000000110000111010101101101010111000011000000000100000000000000000 \\
00000000000000100000001111001010101000101111111010001010101000000000001000000000 \\
0000000000000000100000011001011010111011000000110111010110100000000100000000000000 \\
000000000000000001000000001100011000101100111101101111001101000000000000000000010000 \\
000000000000000000100000101001011011010010100010011111110010000000100000000000000000 \\
0000000000000000001001110111011000000011110011001111000000000000000000000010000000 \\
0000000000000000000100111110010011010100101011001001111010100000000000000000000001 \\
00000000000000000000011001101110110000000111100110011110000000000000000000000100000
\end{smallmatrix}\right],$$

which has dimension 20, with

$$\begin{aligned}A^{(G_2^0)}(z) = {} & \left(1 + z^{76}\right) + 21 \times \left(z^{12} + z^{64}\right) + 153 \times \left(z^{16} + z^{60}\right) + \\ & 744 \times \left(z^{18} + z^{58}\right) + 1883 \times \left(z^{20} + z^{56}\right) + 4472 \times \left(z^{22} + z^{54}\right) + \\ & 10119 \times \left(z^{24} + z^{52}\right) + 21000 \times \left(z^{26} + z^{50}\right) + 36885 \times \left(z^{28} + z^{48}\right) + \\ & 58656 \times \left(z^{30} + z^{46}\right) + 85548 \times \left(z^{32} + z^{44}\right) + 108816 \times \left(z^{34} + z^{42}\right) + \\ & 127534 \times \left(z^{36} + z^{40}\right) + 136912 \times z^{38}\end{aligned} \tag{9.34}$$

and

$$\boldsymbol{G}^{(G_4)} = \left[\begin{array}{c}
1000000000000110110101101110100011011101111100011100010000100000000000010000000 \\
0100000000000000000010100111000110001000110000111010110000100000000000000000000 \\
0010000000000110110101000010100010011000000100000000000000000010000000000000010100 \\
0001000000000111111111111111111111111111111111111111110000000000000000000000000000 \\
0000100000000000000010100100000110011000100010111101000001000000000101000000000000 \\
0000010000000110110111100010110010111110110101100000000010000100000001000000000 \\
0000001000000110010110101110110000001101110101101000000000100000000000000000000 \\
00000001000000001100011000101100111101101111001101000000000000000000010000000000 \\
0000000010000000110000111101110010000001001110111110000000000000000000000100000 \\
0000000001001100001101011000100001011111110001101000000000000001100000000000001000 \\
00000000001000000100101000101100011001101111011101000000000000000000000001000011 \\
000000000000111001110001111010100100100100111001110010000000000000100010000000000
\end{array}\right],$$

which has dimension 12, with

$$\begin{aligned}A^{(G_4)}(z) = {} & \left(1 + z^{76}\right) + 3 \times \left(z^{12} + z^{64}\right) + 11 \times \left(z^{16} + z^{60}\right) + \\ & 20 \times \left(z^{18} + z^{58}\right) + 51 \times \left(z^{20} + z^{56}\right) + 56 \times \left(z^{22} + z^{54}\right) + \\ & 111 \times \left(z^{24} + z^{52}\right) + 164 \times \left(z^{26} + z^{50}\right) + 187 \times \left(z^{28} + z^{48}\right) + \\ & 224 \times \left(z^{30} + z^{46}\right) + 294 \times \left(z^{32} + z^{44}\right) + 328 \times \left(z^{34} + z^{42}\right) + \\ & 366 \times \left(z^{36} + z^{40}\right) + 464 \times z^{38}\end{aligned} \tag{9.35}$$

respectively. Consider the number of codewords of weight 12, from (9.31)–(9.35), we know that $A_{12}(G_2^0) = 21$ and $A_{12}(G_4) = 3$; applying (9.28),

$$A_{12}(S_2) \equiv 3 \times 21 - 2 \times 3 \pmod 4 \equiv 1 \pmod 4$$

and thus, we have the following set of simultaneous congruences:

$$\begin{aligned}A_{12}(S_2) &\equiv 1 \pmod{2^2}\\ A_{12}(S_3) &\equiv 3 \pmod{3^2}\\ A_{12}(S_{19}) &\equiv 0 \pmod{19}\\ A_{12}(S_{37}) &\equiv 0 \pmod{37}.\end{aligned}$$

Following the Chinese remainder theorem, a solution to the above congruences, denoted by $A_{12}(\mathscr{H})$, is congruent modulo LCM$\{2^2, 3^2, 19, 37\}$, where LCM$\{2^2, 3^2, 19, 37\}$ is the least common multiple of the moduli 2^2, 3^2, 19 and 37, which is equal to $2^2 \times 3^2 \times 19 \times 37 = 25308$ in this case. Since these moduli are pairwise coprime, by the extended Euclidean algorithm, we can write

$$
\begin{aligned}
1 &= 4 \times 1582 + \frac{25308}{4} \times (-1) \\
1 &= 9 \times 625 + \frac{25308}{9} \times (-2) \\
1 &= 19 \times 631 + \frac{25308}{19} \times (-9) \\
1 &= 37 \times 37 + \frac{25308}{37} \times (-2).
\end{aligned}
$$

A solution to the congruences above is given by

$$
\begin{aligned}
A_{12}(\mathscr{H}) = 1 \times \left[(-1)\frac{25308}{4}\right] + 3 \times \left[(-2)\frac{25308}{9}\right] + 0 \times \left[(-9)\frac{25308}{19}\right] \\
+ 0 \times \left[(-2)\frac{25308}{37}\right] \pmod{25308} \\
= -1 \times 6327 + -6 \times 2812 \pmod{25308} \\
= 2109 \pmod{25308} \\
= 25308 n_{12} + 2109.
\end{aligned}
$$

Referring to the weight enumerator function, (9.31), we can immediately see that $n_{12} = 0$, indicating that A_{12} has been accurately evaluated. Repeating the above procedures for weights larger than 12, we have Table 9.3 which shows that the weight distributions of $\mathscr{B}_{37}$ are indeed accurate. In fact, since the complete weight distrib-

Table 9.3 Modular congruence weight distributions of $\mathscr{B}_{37}$

$i/n-i$	$A_i(S_2)$ mod 2^2	$A_i(S_3)$ mod 3^2	$A_i(S_{19})$ mod 19	$A_i(S_{37})$ mod 37	$A_i(\mathscr{H})$ mod 25308	n_i in $A_i = 25308n_i + A_i(\mathscr{H})$
0/76	1	1	1	1	1	0
12/64	1	3	0	0	2109	0
16/60	1	6	0	0	10545	3
18/58	0	0	0	0	0	38
20/56	3	6	0	0	23199	295
22/54	0	5	0	0	22496	2116
24/52	3	0	0	0	6327	10886
26/50	0	0	0	0	0	44014
28/48	1	5	0	0	16169	143278
30/46	0	8	0	0	5624	371614
32/44	0	0	0	0	0	774865
34/42	0	0	0	0	0	1306604
36/40	2	7	0	0	23902	1785996
38	0	3	2	2	7032	1981878

utions can be obtained once the first few terms required by Gleason's theorem are known, verification of these few terms is sufficient.

Example 9.2 Gulliver et al. [6] have shown that the (168, 84, 24) doubly even self-dual quadratic double-circulant code $\mathscr{B}_{83}$ is not extremal since it has minimum distance less than or equal to 28. The weight enumerator of a Type-II code of length n is given by Gleason's theorem, which is expressed as [15]

$$A(z) = \sum_{i=0}^{\lfloor n/24 \rfloor} K_i(1 + 14z^4 + z^8)^{\frac{n}{8}-3i}\{z^4(1 - z^4)^4\}^i, \tag{9.36}$$

where K_i are some integers. As shown by (9.36), only the first few terms of A_i are required in order to completely determine the weight distribution of a Type-II code. For $\mathscr{B}_{83}$, only the first eight terms of A_i are required. Using the parallel version of the efficient codeword enumeration method described in Chap. 5, Sect. 9.5, we determined that all of these eight terms are 0 apart from $A_0 = 1$, $A_{24} = 571704$ and $A_{28} = 17008194$.

We need to verify independently whether or not A_{24} and A_{28} have been correctly evaluated. As in the previous example, the modular congruence method can be used for this purpose. For $p = 83$, we have $|\mathscr{H}| = 2^2 \times 3 \times 7 \times 41 \times 83 = 285852$. We will consider the odd prime cases in the first place.

For prime $q = 3$, a cyclic group of order 3, S_3 can be generated by $\left[\begin{smallmatrix} 0 & 1 \\ 82 & 1 \end{smallmatrix}\right] \in \mathrm{PSL}_2(83)$, and we found that the subcode invariant under S_3 has dimension 28 and has 63 and 0 codewords of weights 24 and 28, respectively.

For prime $q = 7$, we have $\left[\begin{smallmatrix} 0 & 1 \\ 82 & 10 \end{smallmatrix}\right]$ which generates S_7. The subcode fixed by S_7 has dimension 12 and no codewords of weight 24 or 28 are contained in this subcode.

Similarly, for prime $q = 41$, the subcode fixed by S_{41}, which is generated by $\left[\begin{smallmatrix} 0 & 1 \\ 82 & 4 \end{smallmatrix}\right]$ and has dimension 4, contains no codewords of weight 24 or 28.

Finally, for prime $q = 83$, the invariant subcode of dimension 2 contains the all-zeros, the all-ones, $\{\underbrace{0, 0, \ldots, 0, 0}_{84}, \underbrace{1, 1, \ldots, 1, 1}_{84}\}$ and $\{\underbrace{1, 1, \ldots, 1, 1}_{84}, \underbrace{0, 0, \ldots, 0, 0}_{84}\}$ codewords only. The cyclic group S_{83} is generated by $\left[\begin{smallmatrix} 0 & 1 \\ 82 & 81 \end{smallmatrix}\right]$.

For the case of $q = 2$, we have $P = \left[\begin{smallmatrix} 1 & 9 \\ 9 & 82 \end{smallmatrix}\right]$ and $T = \left[\begin{smallmatrix} 0 & 82 \\ 1 & 0 \end{smallmatrix}\right]$. The subcode fixed by S_2, which has dimension 42, contains 196 and 1050 codewords of weights 24 and 28, respectively. Meanwhile, the subcode fixed by G_4, which has dimension 22, contains 4 and 6 codewords of weights 24 and 28, respectively.

Thus, using (9.28), the numbers of codewords of weights 24 and 28 fixed by S_2 are

$$\begin{aligned} A_{24}(S_2) &= 3 \times 196 - 2 \times 4 \equiv 0 \pmod 4, \text{ and} \\ A_{28}(S_2) &= 3 \times 1050 - 2 \times 6 \equiv 2 \pmod 4 \end{aligned}$$

and by applying the Chinese remainder theorem to all $A_i(S_q)$ for $i = 24, 28$, we arrive at

$$A_{24} = n_{24} \times 285852 \tag{9.37a}$$

and

$$A_{28} = n_{28} \times 285852 + 142926\,. \tag{9.37b}$$

From (9.37) we have now verified A_{24} and A_{28}, since they have equality for non-negative integers n_{24} and n_{28} ($n_{24} = 2$ and $n_{28} = 59$). Using Gleason's theorem, i.e. (9.36), the weight enumerator function of the (168, 84, 24) code $\mathscr{B}_{83}$ is obtained and it is given by

$$\begin{aligned}
A(z) = (z^0 + z^{168}) +& \\
571704 \times (z^{24} + z^{144}) +& \\
17008194 \times (z^{28} + z^{140}) +& \\
5507510484 \times (z^{32} + z^{136}) +& \\
1252615755636 \times (z^{36} + z^{132}) +& \\
166058829151929 \times (z^{40} + z^{128}) +& \\
13047194638256310 \times (z^{44} + z^{124}) +& \\
629048483051034984 \times (z^{48} + z^{120}) +& \\
19087129808556586056 \times (z^{52} + z^{116}) +& \\
372099697089030108600 \times (z^{56} + z^{112}) +& \\
4739291490433882602066 \times (z^{60} + z^{108}) +& \\
39973673426117369814414 \times (z^{64} + z^{104}) +& \\
225696677517789500207052 \times (z^{68} + z^{100}) +& \\
860241109321000217491044 \times (z^{72} + z^{96}) +& \\
2227390682939806465038006 \times (z^{76} + z^{92}) +& \\
3935099587279668544910376 \times (z^{80} + z^{88}) +& \\
4755747411704650343205104 \times z^{84}\,.&
\end{aligned} \tag{9.38}$$

For the complete weight distributions and their congruences of the $(2(p+1), p+1, d)$ quadratic double-circulant codes, for $11 \le p \le 83$, except $p = 37$ as it has already been given in Example 9.1, refer to Appendix "Weight Distributions of Quadratic Double-Circulant Codes and their Modulo Congruence".

9.6.2 The Number of Codewords of a Given Weight in Extended Quadratic Residue Codes

We have modified the modular congruence approach of Mykkeltveit et al. [14], which was originally introduced for extended QR codes $\hat{\mathscr{L}}_p$, so that it is applicable to the quadratic double-circulant codes. Whilst $\mathscr{B}_p$ contains one non-cyclic subgroup of order 4, $\hat{\mathscr{L}}_p$ contains two distinct non-cyclic subgroups of this order, namely G_4^0 and G_4^1. As a consequence, (9.28) becomes

$$A_i(S_2) \equiv (2^m + 1)A_i(H_2) - 2^{m-1}A_i(G_4^0) - 2^{m-1}A_i(G_4^1) \pmod{2^{m+1}}, \quad (9.39)$$

where 2^{m+1} is the highest power of 2 that divides $|\mathscr{H}|$. Unlike $\mathscr{B}_p$, where there are two circulants in which each one is fixed by $\mathrm{PSL}_2(p)$, a linear group $\mathrm{PSL}_2(p)$ acts on the entire coordinates of $\hat{\mathscr{L}}_p$. In order to obtain the invariant subcode, we only need a set of linear equations containing the parity-check matrix of $\hat{\mathscr{L}}_p$, which is arranged in $(0, 1, \ldots, p-2, p-1)(\infty)$ order, and $c_i + c_{i'} = 0$ for all $i \in \mathbb{F}_p \cup \{\infty\}$. Note that c_i and $c_{i'}$ are defined in the same manner as in Sect. 9.6.1.

We demonstrate the importance of this modular congruence approach by proving that the published results for the weight distributions of $\hat{\mathscr{L}}_{151}$ and $\hat{\mathscr{L}}_{137}$ are incorrect. However, first let us derive the weight distribution of $\hat{\mathscr{L}}_{167}$.

Example 9.3 There exists an extended QR code $\hat{\mathscr{L}}_{167}$ which has identical parameters ($n = 168$, $k = 84$ and $d = 24$) as the code $\mathscr{B}_{83}$. Since $\hat{\mathscr{L}}_{167}$ can be put into double-circulant form and it is Type-II self-dual, the algorithm in Sect. 9.5 can be used to compute the number of codewords of weights 24 and 28, denoted by A'_{24} and A'_{28} for convenience, from which we can use Gleason's theorem (9.36) to derive the weight enumerator function of the code, $A'(z)$. By codeword enumeration using multiple computers we found that

$$\begin{aligned} A'_{24} &= 776216 \\ A'_{28} &= 18130188. \end{aligned} \quad (9.40)$$

In order to verify the accuracy of A'_{24} and A'_{28}, the modular congruence method is used. In this case, we have $\mathrm{Aut}(\hat{\mathscr{L}}_{167}) \supseteq \mathscr{H} = \mathrm{PSL}_2(167)$. We also know that $|\mathrm{PSL}_2(167)| = 2^3 \times 3 \times 7 \times 83 \times 167 = 2328648$. Let $P = \left[\begin{smallmatrix} 12 & 32 \\ 32 & 155 \end{smallmatrix}\right]$ and $T = \left[\begin{smallmatrix} 0 & 166 \\ 1 & 0 \end{smallmatrix}\right]$.

Let the permutations of orders 3, 7, 83 and 167 be generated by $\left[\begin{smallmatrix} 0 & 1 \\ 166 & 1 \end{smallmatrix}\right]$, $\left[\begin{smallmatrix} 0 & 1 \\ 166 & 19 \end{smallmatrix}\right]$, $\left[\begin{smallmatrix} 0 & 1 \\ 166 & 4 \end{smallmatrix}\right]$ and $\left[\begin{smallmatrix} 0 & 1 \\ 166 & 165 \end{smallmatrix}\right]$, respectively. The numbers of codewords of weights 24 and 28 in the various invariant subcodes of dimension k are

	H_2	G_4^0	G_4^1	S_3	S_7	S_{83}	S_{167}
k	42	22	21	28	12	2	1
A_{24}	252	6	4	140	0	0	0
A_{28}	1812	36	0	0	6	0	0

For $\hat{\mathscr{L}}_{167}$, equation (9.39) becomes

$$A_i(S_2) \equiv 5 \times A_i(H_2) - 2 \times A_i(G_4^0) - 2 \times A_i(G_4^1) \pmod 8 . \tag{9.41}$$

It follows that

$$\begin{aligned} A_{24}(S_2) &\equiv 0 \pmod 8 \\ A_{28}(S_2) &\equiv 4 \pmod 8 \end{aligned}$$

and thus,

$$A'_{24} = n'_{24} \times 2328648 + 776216 \tag{9.42a}$$

and

$$A'_{28} = n'_{28} \times 2328648 + 1829652 \tag{9.42b}$$

from the Chinese remainder theorem.

From (9.37a) and (9.42a), we can see that $\mathscr{B}_{83}$ and $\hat{\mathscr{L}}_{167}$ are indeed inequivalent. This is because for integers $n_{24}, n'_{24} \geq 0$, $A_{24} \neq A'_{24}$.

Comparing Eq. (9.40) with (9.42a) and (9.42b) establishes that $A'_{24} = 776216$ ($n'_{24} = 0$) and $A'_{28} = 18130188$ ($n'_{28} = 7$). The weight enumerator of $\hat{\mathscr{L}}_{167}$ is derived from (9.36) and it is given in (9.43). In comparison to (9.38), it may be seen that $\hat{\mathscr{L}}_{167}$ is a slightly inferior code than $\mathscr{B}_{83}$ having more codewords of weights 24, 28 and 32.

$$\begin{aligned} A'(z) =&(z^0 + z^{168})+ \\ & 776216 \times (z^{24} + z^{144})+ \\ & 18130188 \times (z^{28} + z^{140})+ \\ & 5550332508 \times (z^{32} + z^{136})+ \\ & 1251282702264 \times (z^{36} + z^{132})+ \\ & 166071600559137 \times (z^{40} + z^{128})+ \\ & 13047136918828740 \times (z^{44} + z^{124})+ \\ & 629048543890724216 \times (z^{48} + z^{120})+ \\ & 19087130695796615088 \times (z^{52} + z^{116})+ \\ & 372099690249351071112 \times (z^{56} + z^{112})+ \\ & 4739291519495550245228 \times (z^{60} + z^{108})+ \\ & 39973673337590380474086 \times (z^{64} + z^{104})+ \\ & 225696677727188690570184 \times (z^{68} + z^{100})+ \end{aligned}$$

$$\begin{aligned}&8602411089218607419476\,76 \times (z^{72}+z^{96})+\\&2227390683565491780127428 \times (z^{76}+z^{92})+\\&3935099586463594172460648 \times (z^{80}+z^{88})+\\&4755747412595715344169376 \times z^{84}\,.\end{aligned} \tag{9.43}$$

Example 9.4 Gaborit et al. [4] gave A_{2i}, for $22 \le 2i \le 32$, of $\hat{\mathscr{L}}_{137}$ and we will check the consistency of the published results. For $p = 137$, we have $|\mathrm{PSL}_2(137)| = 2^3 \times 3 \times 17 \times 23 \times 137 = 1285608$ and we need to compute $A_{2i}(S_q)$, where $22 \le 2i \le 32$, for all primes q dividing $|\mathrm{PSL}_2(137)|$. Let $P = \left[\begin{smallmatrix} 137 & 51 \\ 51 & 1 \end{smallmatrix}\right]$ and $T = \left[\begin{smallmatrix} 0 & 136 \\ 1 & 0 \end{smallmatrix}\right]$.

Let $\left[\begin{smallmatrix} 0 & 1 \\ 136 & 1 \end{smallmatrix}\right]$, $\left[\begin{smallmatrix} 0 & 1 \\ 136 & 6 \end{smallmatrix}\right]$ and $\left[\begin{smallmatrix} 0 & 1 \\ 136 & 11 \end{smallmatrix}\right]$ be generators of permutation of orders 3, 17 and 23, respectively. It is not necessary to find a generator of permutation of order 137 as it fixes the all-zeros and all-ones codewords only. Subcodes that are invariant under G_2^0, G_4^0, G_4^1, S_3, S_{17} and S_{23} are obtained and the number of weight i, for $22 \le 2i \le 32$, codewords in these subcodes is then computed. The results are shown as follows, where k denotes the dimension of the corresponding subcode,

	H_2	G_4^0	G_4^1	S_3	S_{17}	S_{23}	S_{137}
k	35	19	18	23	5	3	1
A_{22}	170	6	6	0	0	0	0
A_{24}	612	10	18	46	0	0	0
A_{26}	1666	36	6	0	0	0	0
A_{28}	8194	36	60	0	0	0	0
A_{30}	34816	126	22	943	0	0	0
A_{32}	114563	261	189	0	0	0	0

We have

$$A_i(S_2) \equiv 5 \times A_i(H_2) - 2 \times A_i(G_4^0) - 2 \times A_i(G_4^1) \pmod 8\,,$$

for $\hat{\mathscr{L}}_{137}$, which is identical to that for $\hat{\mathscr{L}}_{167}$ since they both have 2^3 as the highest power of 2 that divides $|\mathscr{H}|$. Using this formulation, we obtain

$$\begin{aligned}A_{22}(S_2) &= 2 \pmod 8\\A_{24}(S_2) &= 4 \pmod 8\\A_{26}(S_2) &= 6 \pmod 8\\A_{28}(S_2) &= 2 \pmod 8\\A_{30}(S_2) &= 0 \pmod 8\\A_{32}(S_2) &= 3 \pmod 8\end{aligned}$$

and combining all the results using the Chinese remainder theorem, we arrive at

$$\begin{aligned}
A_{22} &= n_{22} \times 1285608 + 321402 \\
A_{24} &= n_{24} \times 1285608 + 1071340 \\
A_{26} &= n_{26} \times 1285608 + 964206 \\
A_{28} &= n_{28} \times 1285608 + 321402 \\
A_{30} &= n_{30} \times 1285608 + 428536 \\
A_{32} &= n_{32} \times 1285608 + 1124907
\end{aligned} \tag{9.44}$$

for some non-negative integers n_i. Comparing these to the results in [4], we can immediately see that $n_{22} = 0$, $n_{24} = 1$, $n_{26} = 16$, $n_{28} = 381$, and both A_{30} and A_{32} were incorrectly reported. By codeword enumeration using multiple computers in parallel, we have determined that

$$\begin{aligned}
A_{30} &= 6648307504 \\
A_{32} &= 77865259035
\end{aligned}$$

hence, referring to (9.44) it is found that $n_{30} = 5171$ and $n_{32} = 60566$.

Example 9.5 Gaborit et al. [4] also published the weight distribution of $\hat{\mathscr{L}}_{151}$ and we will show that this has also been incorrectly reported. For $\hat{\mathscr{L}}_{151}$, $|\mathrm{PSL}_2(151)| = 2^3 \times 3 \times 5^2 \times 19 \times 151 = 1721400$ and we have $P = \left[\begin{smallmatrix} 104 & 31 \\ 31 & 47 \end{smallmatrix}\right]$ and $T = \left[\begin{smallmatrix} 0 & 150 \\ 1 & 0 \end{smallmatrix}\right]$.

Let $\left[\begin{smallmatrix} 0 & 1 \\ 150 & 1 \end{smallmatrix}\right]$, $\left[\begin{smallmatrix} 0 & 1 \\ 150 & 27 \end{smallmatrix}\right]$ and $\left[\begin{smallmatrix} 0 & 1 \\ 150 & 8 \end{smallmatrix}\right]$ be generators of permutation of orders 3, 5 and 19, respectively. The numbers of weight i codewords for $i = 20$ and 24, in the various fixed subcodes of dimension k, are

	H_2	G_4^0	G_4^1	S_3	S_5	S_{19}	S_{151}
k	38	20	19	26	16	4	1
A_{20}	38	2	0	25	15	0	0
A_{24}	266	4	4	100	0	0	0

and $A_i(S_2)$ is again the same as that for primes 167 and 137, see (9.41). Using this equation, we have $A_{20}(S_2) = A_{24}(S_2) = 2 \pmod 8$. Following the Chinese remainder theorem, we obtain

$$\begin{aligned}
A_{20} &= n_{20} \times 1721400 + 28690 \\
A_{24} &= n_{24} \times 1721400 + 717250
\end{aligned}. \tag{9.45}$$

It follows that A_{20} is correctly reported in [4], but A_{24} is incorrectly reported as 717230. Using the method in Sect. 9.5 implemented on multiple computers, we have determined that

$$\begin{aligned}
A_{20} &= 28690 \\
A_{24} &= 717250,
\end{aligned}$$

hence $n_{20} = 0$ and $n_{24} = 0$ in (9.45). Since A_{20} and A_{24} are required to derive the complete weight distribution of $\hat{\mathscr{L}}_{151}$ according to Gleason's theorem for Type-II codes (9.36), the weight distribution of $\hat{\mathscr{L}}_{151}$ given in [4] is not correct. The correct weight distribution of this code, given in terms of the weight enumerator function, is

$$\begin{aligned} A(z) = & \left(z^0 + z^{152}\right) + \\ & 28690 \times \left(z^{20} + z^{132}\right) + \\ & 717250 \times \left(z^{24} + z^{128}\right) + \\ & 164250250 \times \left(z^{28} + z^{124}\right) + \\ & 39390351505 \times \left(z^{32} + z^{120}\right) + \\ & 5498418962110 \times \left(z^{36} + z^{116}\right) + \\ & 430930711621830 \times \left(z^{40} + z^{112}\right) + \\ & 19714914846904500 \times \left(z^{44} + z^{108}\right) + \\ & 542987434093298550 \times \left(z^{48} + z^{104}\right) + \\ & 9222363801696269658 \times \left(z^{52} + z^{100}\right) + \\ & 98458872937331749615 \times \left(z^{56} + z^{96}\right) + \\ & 670740325520798111830 \times \left(z^{60} + z^{92}\right) + \\ & 2949674479653615754525 \times \left(z^{64} + z^{88}\right) + \\ & 8446025592483506824150 \times \left(z^{68} + z^{84}\right) + \\ & 15840564760239238232420 \times \left(z^{72} + z^{80}\right) + \\ & 19527364659006697265368 \times z^{76}. \end{aligned} \tag{9.46}$$

9.7 Minimum Distance Evaluation: A Probabilistic Approach

An interesting observation is that the minimum weight codewords of $\hat{\mathscr{L}}_p$, for $p \equiv \pm 1 \pmod 8$, and $\mathscr{B}_p$, for $p \equiv \pm 3 \pmod 8$ are always contained in one or more of their fixed subcodes. At least, this is true for all known cases ($n \leq 200$) and this is depicted in Table 9.4. We can see that the subcode fixed by H_2 appears in all the known cases. In Table 9.4, the column d_U denotes the minimum distance upper bound of extremal doubly even self-dual codes of a given length and the last column indicates the various subgroups whose fixed subcodes contain the minimum weight codewords. The highest n, for which the minimum distance of extended QR codes is known, is 168 [5] and we provide further results for $n = 192$, 194, and 200. We obtained the minimum distance of these extended QR codes using the parallel version of the minimum distance algorithm for cyclic codes (QR codes are cyclic) described in Chap. 5, Sect. 5.4. Note that the fact that the code is singly even ($n = 194$) or doubly

Table 9.4 The minimum distance of $\hat{\mathscr{L}}_p$ and $\mathscr{B}_p$ for $12 \leq n \leq 200$

n	p	p mod 8	d	d_U	Subgroups
12	5	−3	4		H_2, G_4
18	17	1	6		H_2, G_4^0, S_3
24	23	−1	8	8	H_2, G_4^0, G_4^1
28	13	−3	6		H_2, G_4, S_3
32	31	−1	8	8	H_2, G_4^0, S_3
40	19	3	8	8	H_2, G_4, S_3
42	41	1	10		H_2, G_4^1, S_5
48	47	−1	12	12	H_2, G_4^1, S_5
60	29	−3	12		H_2, S_3
72	71	−1	12	16	H_2, G_4^1, S_3, S_5
74	73	1	14		H_2, G_4^0, G_4^1, S_3
76	37	−3	12		H_2, G_4, S_3
80	79	−1	16	16	H_2, G_4^0, G_4^1, S_3
88	43	3	16	16	H_2, S_3, S_7
90	89	1	18		H_2, G_4^0, G_4^1, S_3
98	97	1	16		H_2, G_4^0
104	103	−1	20	20	H_2, G_4^0, S_3
108	53	−3	20		H_2, G_4
114[a]	113	1	16		H_2, G_4^1, S_7
120	59	3	20	24	H_2, G_4, S_5
124	61	−3	20		H_2, G_4, S_3, S_5
128	127	−1	20	24	H_2, S_3
136	67	3	24	24	H_2, G_4, S_3, S_{11}
138	137	1	22		H_2, G_4^0, G_4^1
152[a]	151	−1	20	28	H_2, G_4^0, S_3, S_5
168	167	−1	24	32	H_2, G_4^0, G_4^1, S_3
168	83	3	24	32	H_2, G_4, S_3
192	191	−1	28	36	H_2, G_4^1
194	193	1	28		H_2, G_4^1, S_3
200	199	−1	32	36	H_2, G_4^0, G_4^1, S_3

[a]Extended duadic code [12] has higher minimum distance

even ($n = 192, 200$) is also taken into account in order to reduce the number of codewords that need to be enumerated, see Chap. 5, Sects. 5.2.3 and 5.4. This code property is also taken into account for computing the minimum distance of $\mathscr{B}_p$ using the method described in Sect. 9.5.

Based on the above observation, a probabilistic approach to minimum distance evaluation is developed. Given $\hat{\mathscr{L}}_p$ or $\mathscr{B}_p$, the minimum distance of the code is upper bounded by

$$d \leq \min_{Z=\{G_2^0, G_4^0, G_4^1, S_{q_1}, S_{q_2}, \ldots\}} \{d(Z)\}, \tag{9.47}$$

Table 9.5 The minimum distance of $\hat{\mathscr{L}}_p$ and $\mathscr{B}_p$ for $204 \le n \le 450$

n	p	p mod 8	d	d_U	Subgroups
203	101	−3	≤ 24		H_2, G_4, S_5
216	107	3	< 24	40	H_2, G_4, S_3
220	109	−3	≤ 30		H_2, S_3
224	223	−1	≤ 32	40	H_2, G_4^0, G_4^1
234[a]	233	1	≤ 26		H_2, S_{13}
240[b]	239	−1	≤ 32	44	H_2, G_4^1
242[b]	241	1	≤ 32		H_2, G_4^1, S_3, S_5
258[b]	257	1	≤ 34		H_2, G_4^1
264[b]	263	−1	≤ 36	48	H_2, G_4^0, S_3
264[b]	131	3	≤ 40	48	H_2, G_4
272[b]	271	−1	≤ 40	48	H_2, G_4^0, G_4^1, S_3
280[b]	139	3	≤ 36	48	H_2, S_3
282[b]	281	1	≤ 36		H_2, G_4^0, G_4^1, S_3
300[b]	149	−3	≤ 36		H_2, G_4
312[b]	311	−1	≤ 36	56	H_2, G_4^0, S_3
314[b]	313	1	≤ 40		H_2, G_4^1, S_3
316[b]	157	−3	≤ 40		H_2, S_3
328[b]	163	3	≤ 44	56	H_2, G_4
338[b]	337	1	≤ 40		H_2, G_4^1, S_3
348[b]	173	−3	≤ 42		H_2, S_3
354[b]	353	1	≤ 42		H_2, G_4^1
360[b]	359	−1	≤ 40	64	H_2, G_4^0, G_4^1, Z_5
360[b]	179	3	≤ 40	64	H_2, G_4, Z_5
364[b]	181	−3	≤ 40		H_2, G_4, Z_3
368[b]	367	−1	≤ 48	64	$H_2, G_4^0, Z_3,$
384[b]	383	−1	≤ 48	68	H_2, G_4^0, Z_3
396[b]	197	−3	≤ 44		H_2, Z_{11}
402[b]	201	1	≤ 42		H_2, G_4^0, G_4^1, Z_5
410[b]	409	1	≤ 48		H_2, G_4^0, Z_3
424[b]	211	3	≤ 56	72	H_2, G_4, Z_3, Z_7
432[b]	431	−1	≤ 48	76	H_2, G_4^0, G_4^1, Z_3
434[b]	433	1	≤ 38		H_2, G_4^0, Z_3
440[b]	440	−1	≤ 48	76	H_2, G_4^0, G_4^1, Z_3
450[b]	449	1	≤ 56		H_2, G_4^1

[a]Extended duadic code [12] has higher minimum distance
[b]The minimum distance of the subcode is computed probabilistically

where $d(Z)$ is the minimum distance of the subcode fixed by $Z \in \mathrm{PSL}_2(p)$ and q runs through all odd primes that divide $|\mathrm{PSL}_2(p)|$. Note that for $\mathscr{B}_p$, $G_4^0 = G_4^1$ hence, only one is required. Using (9.47), we give an upper bound of the minimum distance of $\hat{\mathscr{L}}_p$ and $\mathscr{B}_p$ for all codes where $n \le 450$ and this is tabulated in Table 9.5. The

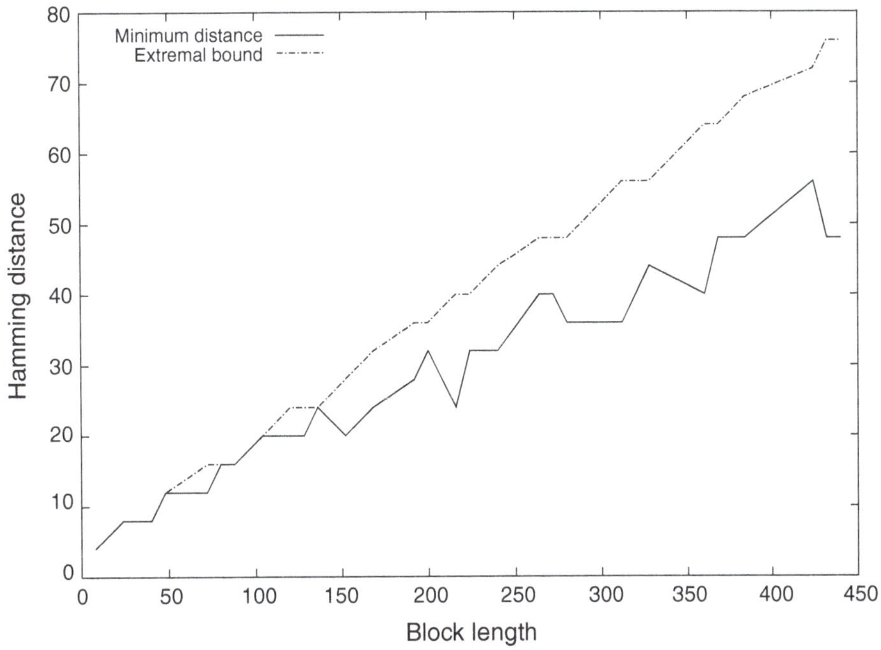

Fig. 9.1 Minimum distance and the extremal bound for distance of doubly even self-dual codes

various fixed subgroups where the minimum weight codewords are found are given in the last column of this table. As shown in Tables 9.4 and 9.5, there is no extremal extended QR or quadratic double-circulant codes for $136 < n \leq 450$ and we plot the minimum distance (or its upper bound for $n > 200$) against the extremal bound in Fig. 9.1. From this figure, it is obvious that, as the block length increases, the gap between the extremal bound and the minimum distance widens and it seems that longer block lengths will follow the same trend. Thus, we conjecture that $n = 136$ is the longest doubly even extremal self-dual double-circulant code. It is worth noting that, for extended QR codes, the results obtained using this probabilistic method are the same as those published by Leon [11].

9.8 Conclusions

Bordered double-circulant codes based on primes can be classified into two classes: $(p+1, (p+1)/2, d)$ extended QR codes, for primes $\pm 1 \pmod 8$, and $(2(p+1), p+1, d)$ quadratic double-circulant codes, for primes $\pm 3 \pmod 8$.

Whilst quadratic double-circulant codes always exist, given a prime $p \equiv \pm 3 \pmod 8$, bordered double-circulant codes may not exist given a prime $p \equiv \pm 1 \pmod 8$.

There always exist $(2p, p, d)$ pure double-circulant codes for any prime $p \equiv \pm 3 \pmod 8$.

For primes $p \equiv -1, 3 \pmod 8$, the double-circulant codes are self-dual and for other primes, the double-circulant codes are formally self-dual.

By exploiting the code structure of formally self-dual, double-circulant codes for $p \equiv -3 \pmod 8$ and also the self-dual double-circulant codes for both pure and bordered cases, we have shown that, compared to the standard method of evaluation, the number of codewords required to evaluate the minimum distance or to count the number of codewords of a given weight can be reduced by a factor of 2.

The automorphism group of the $(p+1, (p+1)/2, d)$ extended QR code contains the projective special linear group $\text{PSL}_2(p)$ acting on the coordinates $(\infty)(0, 1, \ldots, p-2, p-1)$.

The automorphism group of the $(2(p+1), p+1, d)$ quadratic double-circulant code contains $\text{PSL}_2(p)$, acting on coordinates $(\infty)(0, 1, \ldots, p-2, p-1)$, applied simultaneously to left and right circulants.

The number of codewords of weight i of prime-based double-circulant codes, denoted by A_i, can be written as $A_i = n_i \times |\text{PSL}_2(p)| + A_i(\text{PSL}_2(p)) \equiv A_i(\text{PSL}_2(p)) \pmod{|\text{PSL}_2(p)|}$ where $A_i(\text{PSL}_2(p))$ denotes the number of codewords of weight i that are fixed by some element of $\text{PSL}_2(p)$. This result was due to Mykkeltveit et al. [14] and was originally introduced for extended QR codes. We have shown in this chapter that, with some modifications, this modulo congruence method can also be applied to quadratic double-circulant codes.

The modulo congruence technique is found to be very useful in verifying the number of codewords of a given weight obtained exhaustively by computation. We have shown the usefulness of this method by providing corrections to mistakes in previously published results of the weight distributions of extended QR codes for primes 137 and 151.

The weight distribution of the (168, 84, 24) extended QR code, which was previously unknown, has been evaluated and presented above. There also exists a quadratic double-circulant code with identical parameters (n, k and d) and the weight distribution of this code has also been presented above. The (168, 84, 24) quadratic double-circulant code is a better code than the (168, 84, 24) extended QR code since it has less low-weight codewords. The usefulness of the modulo congruence method in checking weight distribution results has been demonstrated in verifying the correctness of the weight distributions of these two codes.

The weight enumerator polynomial of an extended QR code of prime p, denoted by $A_{\hat{\mathscr{L}}}(z)$, can be obtained using Gleason's theorem once the first few terms are known. Since $\text{PSL}_2(p)$ is doubly transitive [13], knowing $A_{\hat{\mathscr{L}}}(z)$ implies $A_{\mathscr{L}}(z)$, the weight enumerator polynomial of the corresponding cyclic QR code, is also known, i.e.

$$A_{\mathscr{L}}(z) = A_{\hat{\mathscr{L}}}(z) + \frac{1-z}{p+1} A'_{\hat{\mathscr{L}}}(z)$$

where $A'_{\hat{\mathscr{L}}}(z)$ is the first derivative of $A_{\hat{\mathscr{L}}}(z)$ with the respect to z [19]. As a consequence, we have been able to evaluate the weight distributions of the QR codes for primes 151 and 167. These are tabulated in Appendix "Weight Distributions of Quadratic Residues Codes for Primes 151 and 167", Tables 9.19 and 9.20, respectively.

A new probabilistic method to obtain the minimum distance of double-circulant codes based on primes has been described. This probabilistic approach is based on the observation that the minimum weight codewords are always contained in one or more subcodes fixed by some element of $\mathrm{PSL}_2(p)$. Using this approach, we conjecture that there are no extremal double-circulant self-dual codes longer than 136 and that this is the last extremal code to be found.

9.9 Summary

In this chapter, self-dual and binary double-circulant codes based on primes have been described in detail. These binary codes are some of the most powerful codes known and as such form an important class of codes due to their powerful error-correcting capabilities and their rich mathematical structure. This structure enables the entire weight distribution of a code to be determined. With these properties, this family of codes has been a subject of extensive research for many years. For these codes that are longer than around 150 bits, an accurate determination of the codeword weight distributions has been an unsolved challenge. We have shown that the code structure may be used in a new algorithm that requires less codewords to be enumerated than traditional methods. As a consequence we have presented new weight distribution results for codes of length 152, 168, 192, 194 and 200. We have shown how a modular congruence method can be used to check weight distributions and have corrected some mistakes in previously published results for codes of lengths 137 and 151. For evaluation of the minimum Hamming distance for very long codes a new probabilistic method has been presented along with results for codes up to 450 bits long. It is conjectured that the (136, 68, 24) self-dual code is the longest extremal code, meeting the upper bound for minimum Hamming distance, and no other, longer, extremal code exists.

Appendix

Circulant Analysis $p = 11$

See Tables 9.6, 9.7 and 9.8.

Table 9.6 Circulant analysis $p = 11$, $a(x) = 1 + x + x^3$, non-factors of $1 + x^p$

i	$a(x)^i$	i	$a(x)^i$	i	$a(x)^i$	i	$a(x)^i$	i	$a(x)^i$
1	0, 1, 3	2	0, 2, 6	3	0, 1, 2, 5, 6, 7, 9	4	0, 1, 4	5	0, 2, 3, 5, 7
6	0, 1, 2, 3, 4, 7, 10	7	2, 3, 4, 6, 8	8	0, 2, 8	9	1, 2, 5, 8, 9	10	0, 3, 4, 6, 10
11	1, 2, 5, 9, 10	12	0, 2, 3, 4, 6, 8, 9	13	2, 3, 8, 9, 10	14	1, 4, 5, 6, 8	15	0, 1, 2
16	0, 4, 5	17	0, 1, 3, 4, 6, 7, 8	18	2, 4, 5, 7, 10	19	0, 3, 4, 5, 6	20	0, 1, 6, 8, 9
21	1, 2, 3, 4, 6, 7, 8, 9, 10	22	2, 4, 7, 9, 10	23	0, 1, 3, 4, 8, 9, 10	24	0, 1, 4, 5, 6, 7, 8	25	2, 3, 7, 8, 10
26	4, 5, 6, 7, 9	27	1, 4, 7	28	1, 2, 5, 8, 10	29	1, 2, 3, 4, 6, 9, 10	30	0, 2, 4
31	0, 1, 2, 4, 7	32	0, 8, 10	33	0, 1, 2, 3, 8, 9, 10	34	0, 1, 2, 3, 5, 6, 8	35	3, 6, 7
36	3, 4, 8, 9, 10	37	1, 2, 3, 5, 6, 7, 8	38	0, 1, 6, 8, 10	39	0, 3, 4, 6, 7, 8, 10	40	0, 1, 2, 5, 7
41	0, 4, 6, 7, 10	42	1, 2, 3, 4, 5, 6, 7, 8, 9	43	0, 4, 5, 6, 7, 8, 9	44	3, 4, 7, 8, 9	45	0, 1, 3, 5, 6
46	0, 2, 5, 6, 7, 8, 9	47	2, 8, 9	48	0, 1, 2, 3, 5, 8, 10	49	0, 2, 3, 9, 10	50	3, 4, 5, 6, 9
51	1, 3, 6, 8, 10	52	1, 3, 7, 8, 10	53	1, 3, 6, 7, 9	54	2, 3, 8	55	0, 2, 4, 5, 6, 8, 9
56	2, 4, 5, 9, 10	57	0, 1, 3, 4, 5, 6, 7, 8, 9	58	1, 2, 4, 6, 7, 8, 9	59	0, 3, 6, 7, 9	60	0, 4, 8
61	1, 3, 4, 5, 7, 8, 9	62	0, 2, 3, 4, 8	63	1, 2, 3, 6, 7, 8, 9	64	0, 5, 9	65	0, 3, 5, 6, 8, 9, 10
66	0, 2, 4, 5, 6, 7, 9	67	0, 2, 4, 5, 7	68	0, 1, 2, 4, 5, 6, 10	69	2, 5, 8, 9, 10	70	1, 3, 6
71	1, 2, 3, 7, 9	72	5, 6, 7, 8, 9	73	0, 1, 5, 8, 9	74	1, 2, 3, 4, 5, 6, 10	75	0, 1, 2, 4, 5, 6, 8, 9, 10
76	0, 1, 2, 5, 9	77	0, 1, 4, 6, 8, 9, 10	78	0, 1, 3, 5, 6, 8, 9	79	1, 2, 5, 6, 7, 9, 10	80	0, 2, 3, 4, 10
81	1, 3, 6, 7, 10	82	0, 1, 3, 8, 9	83	1, 2, 6, 8, 10	84	1, 2, 3, 4, 5, 6, 7, 8, 10	85	1, 2, 4, 5, 6, 7, 8
86	0, 1, 3, 5, 7, 8, 10	87	0, 5, 7, 8, 9	88	3, 5, 6, 7, 8	89	0, 3, 4, 5, 6, 8, 10	90	0, 1, 2, 6, 10
91	2, 4, 5, 6, 7, 9, 10	92	0, 1, 3, 4, 5, 7, 10	93	4	94	4, 5, 7	95	4, 6, 10
96	0, 2, 4, 5, 6, 9, 10	97	4, 5, 8	98	0, 4, 6, 7, 9	99	0, 3, 4, 5, 6, 7, 8	100	1, 6, 7, 8, 10
101	1, 4, 6	102	1, 2, 5, 6, 9	103	3, 4, 7, 8, 10	104	2, 3, 5, 6, 9	105	1, 2, 4, 6, 7, 8, 10
106	1, 2, 3, 6, 7	107	1, 5, 8, 9, 10	108	4, 5, 6	109	4, 8, 9	110	0, 1, 4, 5, 7, 8, 10
111	0, 3, 6, 8, 9	112	4, 7, 8, 9, 10	113	1, 2, 4, 5, 10	114	0, 1, 2, 3, 5, 6, 7, 8, 10	115	0, 2, 3, 6, 8
116	1, 2, 3, 4, 5, 7, 8	117	0, 1, 4, 5, 8, 9, 10	118	0, 1, 3, 6, 7	119	0, 2, 8, 9, 10	120	0, 5, 8

(continued)

Table 9.6 (continued)

i	$a(x)^i$	i	$a(x)^i$	i	$a(x)^i$	i	$a(x)^i$	i	$a(x)^i$
121	1, 3, 5, 6, 9	122	2, 3, 5, 6, 7, 8, 10	123	4, 6, 8	124	0, 4, 5, 6, 8	125	1, 3, 4
126	1, 2, 3, 4, 5, 6, 7	127	1, 4, 5, 6, 7, 9, 10	128	0, 7, 10	129	1, 2, 3, 7, 8	130	0, 1, 5, 6, 7, 9, 10
131	1, 3, 4, 5, 10	132	0, 1, 3, 4, 7, 8, 10	133	0, 4, 5, 6, 9	134	0, 3, 4, 8, 10	135	0, 1, 2, 5, 6, 7, 8, 9, 10
136	0, 1, 2, 4, 8, 9, 10	137	0, 1, 2, 7, 8	138	4, 5, 7, 9, 10	139	0, 1, 2, 4, 6, 9, 10	140	1, 2, 6
141	1, 3, 4, 5, 6, 7, 9	142	2, 3, 4, 6, 7	143	2, 7, 8, 9, 10	144	1, 3, 5, 7, 10	145	0, 1, 3, 5, 7
146	0, 2, 5, 7, 10	147	1, 6, 7	148	1, 2, 4, 6, 8, 9, 10	149	2, 3, 6, 8, 9	150	0, 1, 2, 4, 5, 7, 8, 9, 10
151	0, 1, 2, 5, 6, 8, 10	152	0, 2, 4, 7, 10	153	1, 4, 8	154	0, 1, 2, 5, 7, 8, 9	155	1, 4, 6, 7, 8
156	0, 1, 2, 5, 6, 7,10	157	2, 4,9	158	1, 2, 3, 4, 7, 9, 10	159	0, 2, 4, 6, 8, 9, 10	160	0, 4, 6, 8,9
161	3, 4,5, 6,8, 9,10	162	1, 2, 3, 6, 9	163	5, 7, 10	164	0, 2,5, 6,7	165	0, 1,2, 9,10
166	1, 2,4, 5,9	167	3, 5,6, 7,8, 9,10	168	1, 2,3, 4,5, 6,8, 9,10	169	2, 4, 5, 6,9	170	1, 2, 3, 4, 5, 8,10
171	1, 2,4, 5,7, 9,10	172	0, 2,3, 5,6, 9,10	173	3, 4, 6, 7,8	174	0, 3,5, 7, 10	175	1, 2, 4, 5,7
176	1, 3,5, 6,10	177	0, 1,3, 5,6, 7,8, 9,10	178	0, 1, 5, 6,8, 9,10	179	0, 1, 3, 4, 5, 7, 9	180	0, 1,2, 4,9
181	0, 1,7, 9,10	182	1, 3,4, 7,8, 9,10	183	3, 4, 5, 6, 10	184	0, 2, 3, 6, 8, 9,10	185	0, 3, 4, 5, 7, 8,9
186	8	187	0, 8,9	188	3, 8,10	189	2, 3, 4, 6, 8, 9, 10	190	1, 8, 9
191	0, 2, 4, 8, 10	192	0, 1, 4, 7, 8, 9, 10	193	0, 1, 3, 5, 10	194	5, 8, 10	195	2, 5, 6, 9, 10
196	0, 1, 3, 7, 8	197	2, 6, 7, 9, 10	198	0, 1, 3, 5, 6, 8, 10	199	0, 5, 6, 7, 10	200	1, 2, 3, 5, 9
201	8, 9, 10	202	1, 2, 8	203	0, 1, 3, 4, 5, 8, 9	204	1, 2, 4, 7, 10	205	0, 1, 2, 3, 8
206	3, 5, 6, 8, 9	207	0, 1, 3, 4, 5, 6, 7, 9, 10	208	1, 4, 6, 7, 10	209	0, 1, 5, 6, 7, 8, 9	210	1, 2, 3, 4, 5, 8, 9
211	0, 4, 5, 7, 10	212	1, 2, 3, 4, 6	213	1, 4, 9	214	2, 5, 7, 9, 10	215	0, 1, 3, 6, 7, 9, 10
216	1, 8, 10	217	1, 4, 8, 9, 10	218	5, 7, 8	219	0, 5, 6, 7, 8, 9, 10	220	0, 2, 3, 5, 8, 9, 10
221	0, 3, 4	222	0, 1, 5, 6, 7	223	0, 2, 3, 4, 5, 9, 10	224	3, 5, 7, 8, 9	225	0, 1, 3, 4, 5, 7, 8
226	2, 4, 8, 9, 10	227	1, 3, 4, 7, 8	228	0, 1, 2, 3, 4, 5, 6, 9, 10	229	1, 2, 3, 4, 5, 6, 8	230	0, 1, 4, 5, 6
231	0, 2, 3, 8, 9	232	2, 3, 4, 5, 6, 8, 10	233	5, 6, 10	234	0, 2, 5, 7, 8, 9, 10	235	0, 6, 7, 8, 10

(continued)

Table 9.6 (continued)

i	$a(x)^i$	i	$a(x)^i$	i	$a(x)^i$	i	$a(x)^i$	i	$a(x)^i$
236	0, 1, 2, 3, 6	237	0, 3, 5, 7, 9	238	0, 4, 5, 7, 9	239	0, 3, 4, 6, 9	240	0, 5, 10
241	1, 2, 3, 5, 6, 8, 10	242	1, 2, 6, 7, 10	243	0, 1, 2, 3, 4, 5, 6, 8, 9	244	1, 3, 4, 5, 6, 9, 10	245	0, 3, 4, 6, 8
246	1, 5, 8	247	0, 1, 2, 4, 5, 6, 9	248	0, 1, 5, 8, 10	249	0, 3, 4, 5, 6, 9, 10	250	2, 6, 8
251	0, 2, 3, 5, 6, 7, 8	252	1, 2, 3, 4, 6, 8, 10	253	1, 2, 4, 8, 10	254	1, 2, 3, 7, 8, 9, 10	255	2, 5, 6, 7, 10
256	0, 3, 9	257	0, 4, 6, 9, 10	258	2, 3, 4, 5, 6	259	2, 5, 6, 8, 9	260	0, 1, 2, 3, 7, 9, 10
261	1, 2, 3, 5, 6, 7, 8, 9, 10	262	2, 6, 8, 9, 10	263	1, 3, 5, 6, 7, 8, 9	264	0, 2, 3, 5, 6, 8, 9	265	2, 3, 4, 6, 7, 9, 10
266	0, 1, 7, 8, 10	267	0, 3, 4, 7, 9	268	0, 5, 6, 8, 9	269	3, 5, 7, 9, 10	270	0, 1, 2, 3, 4, 5, 7, 9, 10
271	1, 2, 3, 4, 5, 9, 10	272	0, 2, 4, 5, 7, 8, 9	273	2, 4, 5, 6, 8	274	0, 2, 3, 4, 5	275	0, 1, 2, 3, 5, 7, 8
276	3, 7, 8, 9, 10	277	1, 2, 3, 4, 6, 7, 10	278	0, 1, 2, 4, 7, 8, 9	279	1	280	1, 2, 4
281	1, 3, 7	282	1, 2, 3, 6, 7, 8, 10	283	1, 2, 5	284	1, 3, 4, 6, 8	285	0, 1, 2, 3, 4, 5, 8
286	3, 4, 5, 7, 9	287	1, 3, 9	288	2, 3, 6, 9, 10	289	0, 1, 4, 5, 7	290	0, 2, 3, 6, 10
291	1, 3, 4, 5, 7, 9, 10	292	0, 3, 4, 9, 10	293	2, 5, 6, 7, 9	294	1, 2, 3	295	1, 5, 6
296	1, 2, 4, 5, 7, 8, 9	297	0, 3, 5, 6, 8	298	1, 4, 5, 6, 7	299	1, 2, 7, 9, 10	300	0, 2, 3, 4, 5, 7, 8, 9, 10
301	0, 3, 5, 8, 10	302	0, 1, 2, 4, 5, 9, 10	303	1, 2, 5, 6, 7, 8, 9	304	0, 3, 4, 8, 9	305	5, 6, 7, 8, 10
306	2, 5, 8	307	0, 2, 3, 6, 9	308	0, 2, 3, 4, 5, 7, 10	309	1, 3, 5	310	1, 2, 3, 5, 8
311	0, 1, 9	312	0, 1, 2, 3, 4, 9, 10	313	1, 2, 3, 4, 6, 7, 9	314	4, 7, 8	315	0, 4, 5, 9, 10
316	2, 3, 4, 6, 7, 8, 9	317	0, 1, 2, 7, 9	318	0, 1, 4, 5, 7, 8, 9	319	1, 2, 3, 6, 8	320	0, 1, 5, 7, 8
321	2, 3, 4, 5, 6, 7, 8, 9, 10	322	1, 5, 6, 7, 8, 9, 10	323	4, 5, 8, 9, 10	324	1, 2, 4, 6, 7	325	1, 3, 6, 7, 8, 9, 10
326	3, 9, 10	327	0, 1, 2, 3, 4, 6, 9	328	0, 1, 3, 4, 10	329	4, 5, 6, 7, 10	330	0, 2, 4, 7, 9
331	0, 2, 4, 8, 9	332	2, 4, 7, 8, 10	333	3, 4, 9	334	1, 3, 5, 6, 7, 9, 10	335	0, 3, 5, 6, 10
336	1, 2, 4, 5, 6, 7, 8, 9, 10	337	2, 3, 5, 7, 8, 9, 10	338	1, 4, 7, 8, 10	339	1, 5, 9	340	2, 4, 5, 6, 8, 9, 10
341	1, 3, 4, 5, 9	342	2, 3, 4, 7, 8, 9, 10	343	1, 6, 10	344	0, 1, 4, 6, 7, 9, 10	345	1, 3, 5, 6, 7, 8, 10
346	1, 3, 5, 6, 8	347	0, 1, 2, 3, 5, 6, 7	348	0, 3, 6, 9, 10	349	2, 4, 7	350	2, 3, 4, 8, 10
351	6, 7, 8, 9, 10	352	1, 2, 6, 9, 10	353	0, 2, 3, 4, 5, 6, 7	354	0, 1, 2, 3, 5, 6, 7, 9, 10	355	1, 2, 3, 6, 10

(continued)

Table 9.6 (continued)

i	$a(x)^i$	i	$a(x)^i$	i	$a(x)^i$	i	$a(x)^i$	i	$a(x)^i$
356	0, 1, 2, 5, 7, 9, 10	357	1, 2, 4, 6, 7, 9, 10	358	0, 2, 3, 6, 7, 8, 10	359	0, 1, 3, 4, 5	360	0, 2, 4, 7, 8
361	1, 2, 4, 9, 10	362	0, 2, 3, 7, 9	363	0, 2, 3, 4, 5, 6, 7, 8, 9	364	2, 3, 5, 6, 7, 8, 9	365	0, 1, 2, 4, 6, 8, 9
366	1, 6, 8, 9, 10	367	4, 6, 7, 8, 9	368	0, 1, 4, 5, 6, 7, 9	369	0, 1, 2, 3, 7	370	0, 3, 5, 6, 7, 8, 10
371	0, 1, 2, 4, 5, 6, 8	372	5	373	5, 6, 8	374	0, 5, 7	375	0, 1, 3, 5, 6, 7, 10
376	5, 6, 9	377	1, 5, 7, 8, 10	378	1, 4, 5, 6, 7, 8, 9	379	0, 2, 7, 8, 9	380	2, 5, 7
381	2, 3, 6, 7, 10	382	0, 4, 5, 8, 9	383	3, 4, 6, 7, 10	384	0, 2, 3, 5, 7, 8, 9	385	2, 3, 4, 7, 8
386	0, 2, 6, 9, 10	387	5, 6, 7	388	5, 9, 10	389	0, 1, 2, 5, 6, 8, 9	390	1, 4, 7, 9, 10
391	0, 5, 8, 9, 10	392	0, 2, 3, 5, 6	393	0, 1, 2, 3, 4, 6, 7, 8, 9	394	1, 3, 4, 7, 9	395	2, 3, 4, 5, 6, 8, 9
396	0, 1, 2, 5, 6, 9, 10	397	1, 2, 4, 7, 8	398	0, 1, 3, 9, 10	399	1, 6, 9	400	2, 4, 6, 7, 10
401	0, 3, 4, 6, 7, 8, 9	402	5, 7, 9	403	1, 5, 6, 7, 9	404	2, 4, 5	405	2, 3, 4, 5, 6, 7, 8
406	0, 2, 5, 6, 7, 8, 10	407	0, 1, 8	408	2, 3, 4, 8, 9	409	0, 1, 2, 6, 7, 8, 10	410	0, 2, 4, 5, 6
411	0, 1, 2, 4, 5, 8, 9	412	1, 5, 6, 7, 10	413	0, 1, 4, 5, 9	414	0, 1, 2, 3, 6, 7, 8, 9, 10	415	0, 1, 2, 3, 5, 9, 10
416	1, 2, 3, 8, 9	417	0, 5, 6, 8, 10	418	0, 1, 2, 3, 5, 7, 10	419	2, 3, 7	420	2, 4, 5, 6, 7, 8, 10
421	3, 4, 5, 7, 8	422	0, 3, 8, 9, 10	423	0, 2, 4, 6, 8	424	1, 2, 4, 6, 8	425	0, 1, 3, 6, 8
426	2, 7, 8	427	0, 2, 3, 5, 7, 9, 10	428	3, 4, 7, 9, 10	429	0, 1, 2, 3, 5, 6, 8, 9, 10	430	0, 1, 2, 3, 6, 7, 9
431	0, 1, 3, 5, 8	432	2, 5, 9	433	1, 2, 3, 6, 8, 9, 10	434	2, 5, 7, 8, 9	435	0, 1, 2, 3, 6, 7, 8
436	3, 5, 10	437	0, 2, 3, 4, 5, 8, 10	438	0, 1, 3, 5, 7, 9, 10	439	1, 5, 7, 9, 10	440	0, 4, 5, 6, 7, 9, 10
441	2, 3, 4, 7, 10	442	0, 6, 8	443	1, 3, 6, 7, 8	444	0, 1, 2, 3, 10	445	2, 3, 5, 6, 10
446	0, 4, 6, 7, 8, 9, 10	447	0, 2, 3, 4, 5, 6, 7, 9, 10	448	3, 5, 6, 7, 10	449	0, 2, 3, 4, 5, 6, 9	450	0, 2, 3, 5, 6, 8, 10
451	0, 1, 3, 4, 6, 7, 10	452	4, 5, 7, 8, 9	453	0, 1, 4, 6, 8	454	2, 3, 5, 6, 8	455	0, 2, 4, 6, 7
456	0, 1, 2, 4, 6, 7, 8, 9, 10	457	0, 1, 2, 6, 7, 9, 10	458	1, 2, 4, 5, 6, 8, 10	459	1, 2, 3, 5, 10	460	0, 1, 2, 8, 10
461	0, 2, 4, 5, 8, 9, 10	462	0, 4, 5, 6, 7	463	0, 1, 3, 4, 7, 9, 10	464	1, 4, 5, 6, 8, 9, 10	465	9

(continued)

Table 9.6 (continued)

i	$a(x)^i$	i	$a(x)^i$	i	$a(x)^i$	i	$a(x)^i$	i	$a(x)^i$
466	1, 9, 10	467	0, 4, 9	468	0, 3, 4, 5, 7, 9, 10	469	2, 9, 10	470	0, 1, 3, 5, 9
471	0, 1, 2, 5, 8, 9, 10	472	0, 1, 2, 4, 6	473	0, 6, 9	474	0, 3, 6, 7, 10	475	1, 2, 4, 8, 9
476	0, 3, 7, 8, 10	477	0, 1, 2, 4, 6, 7, 9	478	0, 1, 6, 7, 8	479	2, 3, 4, 6, 10	480	0, 9, 10
481	2, 3, 9	482	1, 2, 4, 5, 6, 9, 10	483	0, 2, 3, 5, 8	484	1, 2, 3, 4, 9	485	4, 6, 7, 9, 10
486	0, 1, 2, 4, 5, 6, 7, 8, 10	487	0, 2, 5, 7, 8	488	1, 2, 6, 7, 8, 9, 10	489	2, 3, 4, 5, 6, 9, 10	490	0, 1, 5, 6, 8
491	2, 3, 4, 5, 7	492	2, 5, 10	493	0, 3, 6, 8, 10	494	0, 1, 2, 4, 7, 8, 10	495	0, 2, 9
496	0, 2, 5, 9, 10	497	6, 8, 9	498	0, 1, 6, 7, 8, 9, 10	499	0, 1, 3, 4, 6, 9, 10	500	1, 4, 5
501	1, 2, 6, 7, 8	502	0, 1, 3, 4, 5, 6, 10	503	4, 6, 8, 9, 10	504	1, 2, 4, 5, 6, 8, 9	505	0, 3, 5, 9, 10
506	2, 4, 5, 8, 9	507	0, 1, 2, 3, 4, 5, 6, 7, 10	508	2, 3, 4, 5, 6, 7, 9	509	1, 2, 5, 6, 7	510	1, 3, 4, 9, 10
511	0, 3, 4, 5, 6, 7, 9	512	0, 6, 7	513	0, 1, 3, 6, 8, 9, 10	514	0, 1, 7, 8, 9	515	1, 2, 3, 4, 7
516	1, 4, 6, 8, 10	517	1, 5, 6, 8, 10	518	1, 4, 5, 7, 10	519	0, 1, 6	520	0, 2, 3, 4, 6, 7, 9
521	0, 2, 3, 7, 8	522	1, 2, 3, 4, 5, 6, 7, 9, 10	523	0, 2, 4, 5, 6, 7, 10	524	1, 4, 5, 7, 9	525	2, 6, 9
526	1, 2, 3, 5, 6, 7, 10	527	0, 1, 2, 6, 9	528	0, 1, 4, 5, 6, 7, 10	529	3, 7, 9	530	1, 3, 4, 6, 7, 8, 9
531	0, 2, 3, 4, 5, 7, 9	532	0, 2, 3, 5, 9	533	0, 2, 3, 4, 8, 9, 10	534	0, 3, 6, 7, 8	535	1, 4, 10
536	0, 1, 5, 7, 10	537	3, 4, 5, 6, 7	538	3, 6, 7, 9, 10	539	0, 1, 2, 3, 4, 8, 10	540	0, 2, 3, 4, 6, 7, 8, 9, 10
541	0, 3, 7, 9, 10	542	2, 4, 6, 7, 8, 9, 10	543	1, 3, 4, 6, 7, 9, 10	544	0, 3, 4, 5, 7, 8, 10	545	0, 1, 2, 8, 9
546	1, 4, 5, 8, 10	547	1, 6, 7, 9, 10	548	0, 4, 6, 8, 10	549	0, 1, 2, 3, 4, 5, 6, 8, 10	550	0, 2, 3, 4, 5, 6, 10
551	1, 3, 5, 6, 8, 9, 10	552	3, 5, 6, 7, 9	553	1, 3, 4, 5, 6	554	1, 2, 3, 4, 6, 8, 9	555	0, 4, 8, 9, 10
556	0, 2, 3, 4, 5, 7, 8	557	1, 2, 3, 5, 8, 9, 10	558	2	559	2, 3, 5	560	2, 4, 8
561	0, 2, 3, 4, 7, 8, 9	562	2, 3, 6	563	2, 4, 5, 7, 9	564	1, 2, 3, 4, 5, 6, 9	565	4, 5, 6, 8, 10
566	2, 4, 10	567	0, 3, 4, 7, 10	568	1, 2, 5, 6, 8	569	0, 1, 3, 4, 7	570	0, 2, 4, 5, 6, 8, 10

(continued)

Table 9.7 Circulant analysis $p = 11$, $\alpha(x) = 1 + x + x^2 + x^4$, factors of $1 + x^p$

i	$\alpha(x)^i$	i	$\alpha(x)^i$	i	$\alpha(x)^i$	i	$\alpha(x)^i$	i	$\alpha(x)^i$
1	0, 1, 2, 4	2	0, 2, 4, 8	3	0, 3, 4, 5, 9, 10	4	0, 4, 5, 8	5	0, 2, 7, 10
6	0, 6, 7, 8, 9, 10	7	1, 3, 4, 6, 8, 9	8	0, 5, 8, 10	9	1, 2, 3, 4, 5, 6, 7, 8	10	0, 3, 4, 9
11	1, 3, 4, 6, 7, 8, 9, 10	12	0, 1, 3, 5, 7, 9	13	0, 2, 5, 6, 7, 8, 9, 10	14	1, 2, 5, 6, 7, 8	15	0, 4, 6, 7, 8, 9
16	0, 5, 9, 10	17	0, 3, 4, 5, 6, 7	18	1, 2, 3, 4, 5, 6, 8, 10	19	0, 1, 4, 7, 8, 10	20	0, 6, 7, 8
21	2, 4, 6, 8	22	1, 2, 3, 5, 6, 7, 8, 9	23	2, 3, 5, 6, 8, 10	24	0, 2, 3, 6, 7, 10	25	0, 3, 4, 5, 7, 9
26	0, 1, 3, 4, 5, 7, 9, 10	27	1, 2, 3, 4, 7, 9	28	1, 2, 3, 4, 5, 10	29	0, 4, 6, 8, 9, 10	30	0, 1, 3, 5, 7, 8
31	1, 5, 6, 7, 9, 10	32	0, 7, 9, 10	33	3, 4, 7, 8	34	0, 1, 3, 6, 8, 10	35	3, 6, 9, 10
36	1, 2, 4, 5, 6, 8, 9, 10	37	1, 2, 3, 5, 8, 9	38	0, 2, 3, 5, 8, 9	39	2, 4, 8, 9	40	0, 1, 3, 5
41	0, 5, 6, 9	42	1, 4, 5, 8	43	2, 3, 4, 5, 7, 10	44	1, 2, 3, 4, 5, 6, 7, 10	45	4, 8
46	1, 4, 5, 6, 9, 10	47	4, 5, 6, 10	48	0, 1, 3, 4, 6, 9	49	2, 4, 5, 9	50	0, 3, 6, 7, 8, 10
51	0, 1, 2, 5, 6, 7, 8, 10	52	0, 2, 3, 6, 7, 8, 9, 10	53	1, 2, 3, 4, 5, 7, 8, 9	54	2, 3, 4, 6, 7, 8	55	0, 1, 2, 4, 6, 7
56	2, 4, 6, 8, 9, 10	57	5, 6, 7, 8	58	0, 1, 5, 7, 8, 9	59	0, 1, 2, 3, 4, 6	60	0, 2, 3, 5, 6, 10
61	3, 4, 6, 7, 8, 9	62	1, 2, 3, 7, 9, 10	63	0, 2, 6, 8	64	0, 3, 7, 9	65	0, 1, 3, 5, 8, 10
66	3, 5, 6, 8	67	1, 3, 4, 7	68	0, 1, 2, 5, 6, 9	69	6, 8	70	1, 6, 7, 9
71	1, 3, 5, 6	72	1, 2, 4, 5, 7, 8, 9, 10	73	0, 1, 2, 3, 5, 6, 8, 10	74	2, 4, 5, 6, 7, 10	75	1, 2, 7, 8
76	0, 4, 5, 6, 7, 10	77	0, 2, 3, 6, 7, 8	78	4, 5, 7, 8	79	0, 1, 4, 8, 9, 10	80	0, 2, 6, 10
81	7, 8	82	0, 1, 7, 10	83	0, 1, 4, 5, 7, 8, 9, 10	84	2, 5, 8, 10	85	0, 2, 4, 5, 7, 8
86	3, 4, 6, 8, 9, 10	87	2, 8	88	1, 2, 3, 4, 6, 8, 9, 10	89	1, 2, 4, 5, 6, 8	90	5, 8
91	1, 5, 6, 7, 8, 10	92	1, 2, 7, 8, 9, 10	93	0, 1, 2, 3, 4, 5, 6, 7, 9, 10	94	1, 8, 9, 10	95	1, 5, 8, 10
96	0, 1, 2, 6, 7, 8	97	1, 2, 5, 8	98	4, 7, 8, 10	99	3, 4, 5, 6, 7, 8	100	0, 1, 3, 5, 6, 9
101	2, 5, 7, 8	102	0, 1, 2, 3, 4, 5, 9, 10	103	0, 1, 6, 8	104	0, 1, 3, 4, 5, 6, 7, 9	105	0, 2, 4, 6, 8, 9
106	2, 3, 4, 5, 6, 7, 8, 10	107	2, 3, 4, 5, 9, 10	108	1, 3, 4, 5, 6, 8	109	2, 6, 7, 8	110	0, 1, 2, 3, 4, 8
111	0, 1, 2, 3, 5, 7, 9, 10	112	[illegible], 4, 5, 7, 8, 9	113	3, 4, 5, 8	114	1, 3, 5, 10	115	0, 2, 3, 4, 5, 6, 9, 10
116	0, 2, 3, 5, 7, 10	117	0, 3, 4, 7, 8, 10	118	0, 1, 2, 4, 6, 8	119	0, 1, 2, 4, 6, 7, 8, 9	120	0, 1, 4, 6, 9, 10

(continued)

Table 9.7 (continued)c

i	$\alpha(x)^i$	i	$\alpha(x)^i$	i	$\alpha(x)^i$	i	$\alpha(x)^i$	i	$\alpha(x)^i$
121	0, 1, 2, 7, 9, 10	122	1, 3, 5, 6, 7, 8	123	0, 2, 4, 5, 8, 9	124	2, 3, 4, 6, 7, 9	125	4, 6, 7, 8
126	0, 1, 4, 5	127	0, 3, 5, 7, 8, 9	128	0, 3, 6, 7	129	1, 2, 3, 5, 6, 7, 9, 10	130	0, 2, 5, 6, 9, 10
131	0, 2, 5, 6, 8, 10	132	1, 5, 6, 10	133	0, 2, 8, 9	134	2, 3, 6, 8	135	1, 2, 5, 9
136	0, 1, 2, 4, 7, 10	137	0, 1, 2, 3, 4, 7, 9, 10	138	1, 5	139	1, 2, 3, 6, 7, 9	140	1, 2, 3, 7
141	0, 1, 3, 6, 8, 9	142	1, 2, 6, 10	143	0, 3, 4, 5, 7, 8	144	2, 3, 4, 5, 7, 8, 9, 10	145	0, 3, 4, 5, 6, 7, 8, 10
146	0, 1, 2, 4, 5, 6, 9, 10	147	0, 1, 3, 4, 5, 10	148	1, 3, 4, 8, 9, 10	149	1, 3, 5, 6, 7, 10	150	2, 3, 4, 5
151	2, 4, 5, 6, 8, 9	152	0, 1, 3, 8, 9, 10	153	0, 2, 3, 7, 8, 10	154	0, 1, 3, 4, 5, 6	155	0, 4, 6, 7, 9, 10
156	3, 5, 8, 10	157	0, 4, 6, 8	158	0, 2, 5, 7, 8, 9	159	0, 2, 3, 5	160	0, 1, 4, 9
161	2, 3, 6, 8, 9, 10	162	3, 5	163	3, 4, 6, 9	164	0, 2, 3, 9	165	1, 2, 4, 5, 6, 7, 9, 10
166	0, 2, 3, 5, 7, 8, 9, 10	167	1, 2, 3, 4, 7, 10	168	4, 5, 9, 10	169	1, 2, 3, 4, 7, 8	170	0, 3, 4, 5, 8, 10
171	1, 2, 4, 5	172	1, 5, 6, 7, 8, 9	173	3, 7, 8, 10	174	4, 5	175	4, 7, 8, 9
176	1, 2, 4, 5, 6, 7, 8, 9	177	2, 5, 7, 10	178	1, 2, 4, 5, 8, 10	179	0, 1, 3, 5, 6, 7	180	5, 10
181	0, 1, 3, 5, 6, 7, 9, 10	182	1, 2, 3, 5, 9, 10	183	2, 5	184	2, 3, 4, 5, 7, 9	185	4, 5, 6, 7, 9, 10
186	0, 1, 2, 3, 4, 6, 7, 8, 9, 10	187	5, 6, 7, 9	188	2, 5, 7, 9	189	3, 4, 5, 8, 9, 10	190	2, 5, 9, 10
191	1, 4, 5, 7	192	0, 1, 2, 3, 4, 5	193	0, 2, 3, 6, 8, 9	194	2, 4, 5, 10	195	0, 1, 2, 6, 7, 8, 9, 10
196	3, 5, 8, 9	197	0, 1, 2, 3, 4, 6, 8, 9	198	1, 3, 5, 6, 8, 10	199	0, 1, 2, 3, 4, 5, 7, 10	200	0, 1, 2, 6, 7, 10
201	0, 1, 2, 3, 5, 9	202	3, 4, 5, 10	203	0, 1, 5, 8, 9, 10	204	0, 2, 4, 6, 7, 8, 9, 10	205	1, 2, 4, 5, 6, 9
206	0, 1, 2, 5	207	0, 2, 7, 9	208	0, 1, 2, 3, 6, 7, 8, 10	209	0, 2, 4, 7, 8, 10	210	0, 1, 4, 5, 7, 8
211	1, 3, 5, 8, 9, 10	212	1, 3, 4, 5, 6, 8, 9, 10	213	1, 3, 6, 7, 8, 9	214	4, 6, 7, 8, 9, 10	215	0, 2, 3, 4, 5, 9
216	1, 2, 5, 6, 8, 10	217	0, 1, 3, 4, 6, 10	218	1, 3, 4, 5	219	1, 2, 8, 9	220	0, 2, 4, 5, 6, 8
221	0, 3, 4, 8	222	0, 2, 3, 4, 6, 7, 9, 10	223	2, 3, 6, 7, 8, 10	224	2, 3, 5, 7, 8, 10	225	2, 3, 7, 9
226	5, 6, 8, 10	227	0, 3, 5, 10	228	2, 6, 9, 10	229	1, 4, 7, 8, 9, 10	230	0, 1, 4, 6, 7, 8, 9, 10
231	2, 9	232	0, 3, 4, 6, 9, 10	233	0, 4, 9, 10	234	0, 3, 5, 6, 8, 9	235	3, 7, 9, 10
236	0, 1, 2, 4, 5, 8	237	0, 1, 2, 4, 5, 6, 7, 10	238	0, 1, 2, 3, 4, 5, 7, 8	239	1, 2, 3, 6, 7, 8, 9, 10	240	0, 1, 2, 7, 8, 9

(continued)

Table 9.7 (continued)

i	$\alpha(x)^i$	i	$\alpha(x)^i$	i	$\alpha(x)^i$	i	$\alpha(x)^i$	i	$\alpha(x)^i$
241	0, 1, 5, 6, 7, 9	242	0, 2, 3, 4, 7, 9	243	0, 1, 2, 10	244	1, 2, 3, 5, 6, 10	245	0, 5, 6, 7, 8, 9
246	0, 4, 5, 7, 8, 10	247	0, 1, 2, 3, 8, 9	248	1, 3, 4, 6, 7, 8	249	0, 2, 5, 7	250	1, 3, 5, 8
251	2, 4, 5, 6, 8, 10	252	0, 2, 8, 10	253	1, 6, 8, 9	254	0, 3, 5, 6, 7, 10	255	0, 2
256	0, 1, 3, 6	257	0, 6, 8, 10	258	1, 2, 3, 4, 6, 7, 9, 10	259	0, 2, 4, 5, 6, 7, 8, 10	260	0, 1, 4, 7, 9, 10
261	1, 2, 6, 7	262	0, 1, 4, 5, 9, 10	263	0, 1, 2, 5, 7, 8	264	1, 2, 9, 10	265	2, 3, 4, 5, 6, 9
266	0, 4, 5, 7	267	1, 2	268	1, 4, 5, 6	269	1, 2, 3, 4, 5, 6, 9, 10	270	2, 4, 7, 10
271	1, 2, 5, 7, 9, 10	272	0, 2, 3, 4, 8, 9	273	2, 7	274	0, 2, 3, 4, 6, 7, 8, 9	275	0, 2, 6, 7, 9, 10
276	2, 10	277	0, 1, 2, 4, 6, 10	278	1, 2, 3, 4, 6, 7	279	0, 1, 3, 4, 5, 6, 7, 8, 9, 10	280	2, 3, 4, 6
281	2, 4, 6, 10	282	0, 1, 2, 5, 6, 7	283	2, 6, 7, 10	284	1, 2, 4, 9	285	0, 1, 2, 8, 9, 10
286	0, 3, 5, 6, 8, 10	287	1, 2, 7, 10	288	3, 4, 5, 6, 7, 8, 9, 10	289	0, 2, 5, 6	290	0, 1, 3, 5, 6, 8, 9, 10
291	0, 2, 3, 5, 7, 9	292	0, 1, 2, 4, 7, 8, 9, 10	293	3, 4, 7, 8, 9, 10	294	0, 2, 6, 8, 9, 10	295	0, 1, 2, 7
296	2, 5, 6, 7, 8, 9	297	1, 3, 4, 5, 6, 7, 8, 10	298	1, 2, 3, 6, 9, 10	299	2, 8, 9, 10	300	4, 6, 8, 10
301	0, 3, 4, 5, 7, 8, 9, 10	302	1, 4, 5, 7, 8, 10	303	1, 2, 4, 5, 8, 9	304	0, 2, 5, 6, 7, 9	305	0, 1, 2, 3, 5, 6, 7, 9
306	0, 3, 4, 5, 6, 9	307	1, 3, 4, 5, 6, 7	308	0, 1, 2, 6, 8, 10	309	2, 3, 5, 7, 9, 10	310	0, 1, 3, 7, 8, 9
311	0, 1, 2, 9	312	5, 6, 9, 10	313	1, 2, 3, 5, 8, 10	314	0, 1, 5, 8	315	0, 1, 3, 4, 6, 7, 8, 10
316	0, 3, 4, 5, 7, 10	317	0, 2, 4, 5, 7, 10	318	0, 4, 6, 10	319	2, 3, 5, 7	320	0, 2, 7, 8
321	3, 6, 7, 10	322	1, 4, 5, 6, 7, 9	323	1, 3, 4, 5, 6, 7, 8, 9	324	6, 10	325	0, 1, 3, 6, 7, 8
326	1, 6, 7, 8	327	0, 2, 3, 5, 6, 8	328	0, 4, 6, 7	329	1, 2, 5, 8, 9, 10	330	1, 2, 3, 4, 7, 8, 9, 10
331	0, 1, 2, 4, 5, 8, 9, 10	332	0, 3, 4, 5, 6, 7, 9, 10	333	4, 5, 6, 8, 9, 10	334	2, 3, 4, 6, 8, 9	335	0, 1, 4, 6, 8, 10
336	7, 8, 9, 10	337	0, 2, 3, 7, 9, 10	338	2, 3, 4, 5, 6, 8	339	1, 2, 4, 5, 7, 8	340	0, 5, 6, 8, 9, 10
341	0, 1, 3, 4, 5, 9	342	2, 4, 8, 10	343	0, 2, 5, 9	344	1, 2, 3, 5, 7, 10	345	5, 7, 8, 10
346	3, 5, 6, 9	347	0, 2, 3, 4, 7, 8	348	8, 10	349	0, 3, 8, 9	350	3, 5, 7, 8
351	0, 1, 3, 4, 6, 7, 9, 10	352	1, 2, 3, 4, 5, 7, 8, 10	353	1, 4, 6, 7, 8, 9	354	3, 4, 9, 10	355	1, 2, 6, 7, 8, 9
356	2, 4, 5, 8, 9, 10	357	6, 7, 9, 10	358	0, 1, 2, 3, 6, 10	359	1, 2, 4, 8	360	9, 10

(continued)

Table 9.7 (continued)

i	$\alpha(x)^i$	i	$\alpha(x)^i$	i	$\alpha(x)^i$	i	$\alpha(x)^i$	i	$\alpha(x)^i$
361	1, 2, 3, 9	362	0, 1, 2, 3, 6, 7, 9, 10	363	1, 4, 7, 10	364	2, 4, 6, 7, 9, 10	365	0, 1, 5, 6, 8, 10
366	4, 10	367	0, 1, 3, 4, 5, 6, 8, 10	368	3, 4, 6, 7, 8, 10	369	7, 10	370	1, 3, 7, 8, 9, 10
371	0, 1, 3, 4, 9, 10	372	0, 1, 2, 3, 4, 5, 6, 7, 8, 9	373	0, 1, 3, 10	374	1, 3, 7, 10	375	2, 3, 4, 8, 9, 10
376	3, 4, 7, 10	377	1, 6, 9, 10	378	5, 6, 7, 8, 9, 10	379	0, 2, 3, 5, 7, 8	380	4, 7, 9, 10
381	0, 1, 2, 3, 4, 5, 6, 7	382	2, 3, 8, 10	383	0, 2, 3, 5, 6, 7, 8, 9	384	0, 2, 4, 6, 8, 10	385	1, 4, 5, 6, 7, 8, 9, 10
386	0, 1, 4, 5, 6, 7	387	3, 5, 6, 7, 8, 10	388	4, 8, 9, 10	389	2, 3, 4, 5, 6, 10	390	0, 1, 2, 3, 4, 5, 7, 9
391	0, 3, 6, 7, 9, 10	392	5, 6, 7, 10	393	1, 3, 5, 7	394	0, 1, 2, 4, 5, 6, 7, 8	395	1, 2, 4, 5, 7, 9
396	1, 2, 5, 6, 9, 10	397	2, 3, 4, 6, 8, 10	398	0, 2, 3, 4, 6, 8, 9, 10	399	0, 1, 2, 3, 6, 8	400	0, 1, 2, 3, 4, 9
401	3, 5, 7, 8, 9, 10	402	0, 2, 4, 6, 7, 10	403	0, 4, 5, 6, 8, 9	404	6, 8, 9, 10	405	2, 3, 6, 7
406	0, 2, 5, 7, 9, 10	407	2, 5, 8, 9	408	0, 1, 3, 4, 5, 7, 8, 9	409	0, 1, 2, 4, 7, 8	410	1, 2, 4, 7, 8, 10
411	1, 3, 7, 8	412	0, 2, 4, 10	413	4, 5, 8, 10	414	0, 3, 4, 7	415	1, 2, 3, 4, 6, 9
416	0, 1, 2, 3, 4, 5, 6, 9	417	3, 7	418	0, 3, 4, 5, 8, 9	419	3, 4, 5, 9	420	0, 2, 3, 5, 8, 10
421	1, 3, 4, 8	422	2, 5, 6, 7, 9, 10	423	0, 1, 4, 5, 6, 7, 9, 10	424	1, 2, 5, 6, 7, 8, 9, 10	425	0, 1, 2, 3, 4, 6, 7, 8
426	1, 2, 3, 5, 6, 7	427	0, 1, 3, 5, 6, 10	428	1, 3, 5, 7, 8, 9	429	4, 5, 6, 7	430	0, 4, 6, 7, 8, 10
431	0, 1, 2, 3, 5, 10	432	1, 2, 4, 5, 9, 10	433	2, 3, 5, 6, 7, 8	434	0, 1, 2, 6, 8, 9	435	1, 5, 7, 10
436	2, 6, 8, 10	437	0, 2, 4, 7, 9, 10	438	2, 4, 5, 7	439	0, 2, 3, 6	440	0, 1, 4, 5, 8, 10
441	5, 7	442	0, 5, 6, 8	443	0, 2, 4, 5	444	0, 1, 3, 4, 6, 7, 8, 9	445	0, 1, 2, 4, 5, 7, 9, 10
446	1, 3, 4, 5, 6, 9	447	0, 1, 6, 7	448	3, 4, 5, 6, 9, 10	449	1, 2, 5, 6, 7, 10	450	3, 4, 6, 7
451	0, 3, 7, 8, 9, 10	452	1, 5, 9, 10	453	6, 7	454	0, 6, 9, 10	455	0, 3, 4, 6, 7, 8, 9, 10
456	1, 4, 7, 9	457	1, 3, 4, 6, 7, 10	458	2, 3, 5, 7, 8, 9	459	1, 7	460	0, 1, 2, 3, 5, 7, 8, 9
461	0, 1, 3, 4, 5, 7	462	4, 7	463	0, 4, 5, 6, 7, 9	464	0, 1, 6, 7, 8, 9	465	0, 1, 2, 3, 4, 5, 6, 8, 9, 10
466	0, 7, 8, 9	467	0, 4, 7, 9	468	0, 1, 5, 6, 7, 10	469	0, 1, 4, 7	470	3, 6, 7, 9
471	2, 3, 4, 5, 6, 7	472	0, 2, 4, 5, 8, 10	473	1, 4, 6, 7	474	0, 1, 2, 3, 4, 8, 9, 10	475	0, 5, 7, 10
476	0, 2, 3, 4, 5, 6, 8, 10	477	1, 3, 5, 7, 8, 10	478	1, 2, 3, 4, 5, 6, 7, 9	479	1, 2, 3, 4, 8, 9	480	0, 2, 3, 4, 5, 7

(continued)

Table 9.7 (continued)

i	$\alpha(x)^i$	i	$\alpha(x)^i$	i	$\alpha(x)^i$	i	$\alpha(x)^i$	i	$\alpha(x)^i$
481	1, 5, 6, 7	482	0, 1, 2, 3, 7, 10	483	0, 1, 2, 4, 6, 8, 9, 10	484	0, 3, 4, 6, 7, 8	485	2, 3, 4, 7
486	0, 2, 4, 9	487	1, 2, 3, 4, 5, 8, 9, 10	488	1, 2, 4, 6, 9, 10	489	2, 3, 6, 7, 9, 10	490	0, 1, 3, 5, 7, 10
491	0, 1, 3, 5, 6, 7, 8, 10	492	0, 3, 5, 8, 9, 10	493	0, 1, 6, 8, 9, 10	494	0, 2, 4, 5, 6, 7	495	1, 3, 4, 7, 8, 10
496	1, 2, 3, 5, 6, 8	497	3, 5, 6, 7	498	0, 3, 4, 10	499	2, 4, 6, 7, 8, 10	500	2, 5, 6, 10
501	0, 1, 2, 4, 5, 6, 8, 9	502	1, 4, 5, 8, 9, 10	503	1, 4, 5, 7, 9, 10	504	0, 4, 5, 9	505	1, 7, 8, 10
506	1, 2, 5, 7	507	0, 1, 4, 8	508	0, 1, 3, 6, 9, 10	509	0, 1, 2, 3, 6, 8, 9, 10	510	0, 4
511	0, 1, 2, 5, 6, 8	512	0, 1, 2, 6	513	0, 2, 5, 7, 8, 10	514	0, 1, 5, 9	515	2, 3, 4, 6, 7, 10
516	1, 2, 3, 4, 6, 7, 8, 9	517	2, 3, 4, 5, 6, 7, 9, 10	518	0, 1, 3, 4, 5, 8, 9, 10	519	0, 2, 3, 4, 9, 10	520	0, 2, 3, 7, 8, 9
521	0, 2, 4, 5, 6, 9	522	1, 2, 3, 4	523	1, 3, 4, 5, 7, 8	524	0, 2, 7, 8, 9, 10	525	1, 2, 6, 7, 9, 10
526	0, 2, 3, 4, 5, 10	527	3, 5, 6, 8, 9, 10	528	2, 4, 7, 9	529	3, 5, 7, 10	530	1, 4, 6, 7, 8, 10
531	1, 2, 4, 10	532	0, 3, 8, 10	533	1, 2, 5, 7, 8, 9	534	2, 4	535	2, 3, 5, 8
536	1, 2, 8, 10	537	0, 1, 3, 4, 5, 6, 8, 9	538	1, 2, 4, 6, 7, 8, 9, 10	539	0, 1, 2, 3, 6, 9	540	3, 4, 8, 9
541	0, 1, 2, 3, 6, 7	542	2, 3, 4, 7, 9, 10	543	0, 1, 3, 4	544	0, 4, 5, 6, 7, 8	545	2, 6, 7, 9
546	3, 4	547	3, 6, 7, 8	548	0, 1, 3, 4, 5, 6, 7, 8	549	1, 4, 6, 9	550	0, 1, 3, 4, 7, 9
551	0, 2, 4, 5, 6, 10	552	4, 9	553	0, 2, 4, 5, 6, 8, 9, 10	554	0, 1, 2, 4, 8, 9	555	1, 4
556	1, 2, 3, 4, 6, 8	557	3, 4, 5, 6, 8, 9	558	0, 1, 2, 3, 5, 6, 7, 8, 9, 10	559	4, 5, 6, 8	560	1, 4, 6, 8
561	2, 3, 4, 7, 8, 9	562	1, 4, 8, 9	563	0, 3, 4, 6	564	0, 1, 2, 3, 4, 10	565	1, 2, 5, 7, 8, 10
566	1, 3, 4, 9	567	0, 1, 5, 6, 7, 8, 9, 10	568	2, 4, 7, 8	569	0, 1, 2, 3, 5, 7, 8, 10	570	0, 2, 4, 5, 7, 9
571	0, 1, 2, 3, 4, 6, 9, 10	572	0, 1, 5, 6, 9, 10	573	0, 1, 2, 4, 8, 10	574	2, 3, 4, 9	575	0, 4, 7, 8, 9, 10
576	1, 3, 5, 6, 7, 8, 9, 10	577	0, 1, 3, 4, 5, 8	578	0, 1, 4, 10	579	1, 6, 8, 10	580	0, 1, 2, 5, 6, 7, 9, 10
581	1, 3, 6, 7, 9, 10	582	0, 3, 4, 6, 7, 10	583	0, 2, 4, 7, 8, 9	584	0, 2, 3, 4, 5, 7, 8, 9	585	0, 2, 5, 6, 7, 8
586	3, 5, 6, 7, 8, 9	587	1, 2, 3, 4, 8, 10	588	0, 1, 4, 5, 7, 9	589	0, 2, 3, 5, 9, 10	590	0, 2, 3, 4
591	0, 1, 7, 8	592	1, 3, 4, 5, 7, 10	593	2, 3, 7, 10	594	1, 2, 3, 5, 6, 8, 9, 10	595	1, 2, 5, 6, 7, 9

(continued)

Table 9.7 (continued)

i	$\alpha(x)^i$	i	$\alpha(x)^i$	i	$\alpha(x)^i$	i	$\alpha(x)^i$	i	$\alpha(x)^i$
596	1, 2, 4, 6, 7, 9	597	1, 2, 6, 8	598	4, 5, 7, 9	599	2, 4, 9, 10	600	1, 5, 8, 9
601	0, 3, 6, 7, 8, 9	602	0, 3, 5, 6, 7, 8, 9, 10	603	1, 8	604	2, 3, 5, 8, 9, 10	605	3, 8, 9, 10
606	2, 4, 5, 7, 8, 10	607	2, 6, 8, 9	608	0, 1, 3, 4, 7, 10	609	0, 1, 3, 4, 5, 6, 9, 10	610	0, 1, 2, 3, 4, 6, 7, 10
611	0, 1, 2, 5, 6, 7, 8, 9	612	0, 1, 6, 7, 8, 10	613	0, 4, 5, 6, 8, 10	614	1, 2, 3, 6, 8, 10	615	0, 1, 9, 10
616	0, 1, 2, 4, 5, 9	617	4, 5, 6, 7, 8, 10	618	3, 4, 6, 7, 9, 10	619	0, 1, 2, 7, 8, 10	620	0, 2, 3, 5, 6, 7
621	1, 4, 6, 10	622	0, 2, 4, 7	623	1, 3, 4, 5, 7, 9	624	1, 7, 9, 10	625	0, 5, 7, 8
626	2, 4, 5, 6, 9, 10	627	1, 10	628	0, 2, 5, 10	629	5, 7, 9, 10	630	0, 1, 2, 3, 5, 6, 8, 9
631	1, 3, 4, 5, 6, 7, 9, 10	632	0, 3, 6, 8, 9, 10	633	0, 1, 5, 6	634	0, 3, 4, 8, 9, 10	635	0, 1, 4, 6, 7, 10
636	0, 1, 8, 9	637	1, 2, 3, 4, 5, 8	638	3, 4, 6, 10	639	0, 1	640	0, 3, 4, 5
641	0, 1, 2, 3, 4, 5, 8, 9	642	1, 3, 6, 9	643	0, 1, 4, 6, 8, 9	644	1, 2, 3, 7, 8, 10	645	1, 6
646	1, 2, 3, 5, 6, 7, 8, 10	647	1, 5, 6, 8, 9, 10	648	1, 9	649	0, 1, 3, 5, 9, 10	650	0, 1, 2, 3, 5, 6
651	0, 2, 3, 4, 5, 6, 7, 8, 9, 10	652	1, 2, 3, 5	653	1, 3, 5, 9	654	0, 1, 4, 5, 6, 10	655	1, 5, 6, 9
656	0, 1, 3, 8	657	0, 1, 7, 8, 9, 10	658	2, 4, 5, 7, 9, 10	659	0, 1, 6, 9	660	2, 3, 4, 5, 6, 7, 8, 9
661	1, 4, 5, 10	662	0, 2, 4, 5, 7, 8, 9, 10	663	1, 2, 4, 6, 8, 10	664	0, 1, 3, 6, 7, 8, 9, 10	665	2, 3, 6, 7, 8, 9
666	1, 5, 7, 8, 9, 10	667	0, 1, 6, 10	668	1, 4, 5, 6, 7, 8	669	0, 2, 3, 4, 5, 6, 7, 9	670	0, 1, 2, 5, 8, 9
671	1, 7, 8, 9	672	3, 5, 7, 9	673	2, 3, 4, 6, 7, 8, 9, 10	674	0, 3, 4, 6, 7, 9	675	0, 1, 3, 4, 7, 8
676	1, 4, 5, 6, 8, 10	677	0, 1, 2, 4, 5, 6, 8, 10	678	2, 3, 4, 5, 8, 10	679	0, 2, 3, 4, 5, 6	680	0, 1, 5, 7, 9, 10
681	1, 2, 4, 6, 8, 9	682	0, 2, 6, 7, 8, 10	683	0, 1, 8, 10	684	4, 5, 8, 9	685	0, 1, 2, 4, 7, 9
686	0, 4, 7, 10	687	0, 2, 3, 5, 6, 7, 9, 10	688	2, 3, 4, 6, 9, 10	689	1, 3, 4, 6, 9, 10	690	3, 5, 9, 10
691	1, 2, 4, 6	692	1, 6, 7, 10	693	2, 5, 6, 9	694	0, 3, 4, 5, 6, 8	695	0, 2, 3, 4, 5, 6, 7, 8
696	5, 9	697	0, 2, 5, 6, 7, 10	698	0, 5, 6, 7	699	1, 2, 4, 5, 7, 10	700	3, 5, 6, 10
701	0, 1, 4, 7, 8, 9	702	0, 1, 2, 3, 6, 7, 8, 9	703	0, 1, 3, 4, 7, 8, 9, 10	704	2, 3, 4, 5, 6, 8, 9, 10	705	3, 4, 5, 7, 8, 9
706	1, 2, 3, 5, 7, 8	707	0, 3, 5, 7, 9, 10	708	6, 7, 8, 9	709	1, 2, 6, 8, 9, 10	710	1, 2, 3, 4, 5, 7

(continued)

Table 9.7 (continued)

i	$\alpha(x)^i$	i	$\alpha(x)^i$	i	$\alpha(x)^i$	i	$\alpha(x)^i$	i	$\alpha(x)^i$
711	0, 1, 3, 4, 6, 7	712	4, 5, 7, 8, 9, 10	713	0, 2, 3, 4, 8, 10	714	1, 3, 7, 9	715	1, 4, 8, 10
716	0, 1, 2, 4, 6, 9	717	4, 6, 7, 9	718	2, 4, 5, 8	719	1, 2, 3, 6, 7, 10	720	7, 9
721	2, 7, 8, 10	722	2, 4, 6, 7	723	0, 2, 3, 5, 6, 8, 9, 10	724	0, 1, 2, 3, 4, 6, 7, 9	725	0, 3, 5, 6, 7, 8
726	2, 3, 8, 9	727	0, 1, 5, 6, 7, 8	728	1, 3, 4, 7, 8, 9	729	5, 6, 8, 9	730	0, 1, 2, 5, 9, 10
731	0, 1, 3, 7	732	8, 9	733	0, 1, 2, 8	734	0, 1, 2, 5, 6, 8, 9, 10	735	0, 3, 6, 9
736	1, 3, 5, 6, 8, 9	737	0, 4, 5, 7, 9, 10	738	3, 9	739	0, 2, 3, 4, 5, 7, 9, 10	740	2, 3, 5, 6, 7, 9
741	6, 9	742	0, 2, 6, 7, 8, 9	743	0, 2, 3, 8, 9, 10	744	0, 1, 2, 3, 4, 5, 6, 7, 8, 10	745	0, 2, 9, 10
746	0, 2, 6, 9	747	1, 2, 3, 7, 8, 9	748	2, 3, 6, 9	749	0, 5, 8, 9	750	4, 5, 6, 7, 8, 9
751	1, 2, 4, 6, 7, 10	752	3, 6, 8, 9	753	0, 1, 2, 3, 4, 5, 6, 10	754	1, 2, 7, 9	755	1, 2, 4, 5, 6, 7, 8, 10
756	1, 3, 5, 7, 9, 10	757	0, 3, 4, 5, 6, 7, 8, 9	758	0, 3, 4, 5, 6, 10	759	2, 4, 5, 6, 7, 9	760	3, 7, 8, 9
761	1, 2, 3, 4, 5, 9	762	0, 1, 2, 3, 4, 6, 8, 10	763	2, 5, 6, 8, 9, 10	764	4, 5, 6, 9	765	0, 2, 4, 6
766	0, 1, 3, 4, 5, 6, 7, 10	767	0, 1, 3, 4, 6, 8	768	0, 1, 4, 5, 8, 9	769	1, 2, 3, 5, 7, 9	770	1, 2, 3, 5, 7, 8, 9, 10
771	0, 1, 2, 5, 7, 10	772	0, 1, 2, 3, 8, 10	773	2, 4, 6, 7, 8, 9	774	1, 3, 5, 6, 9, 10	775	3, 4, 5, 7, 8, 10
776	5, 7, 8, 9	777	1, 2, 5, 6	778	1, 4, 6, 8, 9, 10	779	1, 4, 7, 8	780	0, 2, 3, 4, 6, 7, 8, 10
781	0, 1, 3, 6, 7, 10	782	0, 1, 3, 6, 7, 9	783	0, 2, 6, 7	784	1, 3, 9, 10	785	3, 4, 7, 9
786	2, 3, 6, 10	787	0, 1, 2, 3, 5, 8	788	0, 1, 2, 3, 4, 5, 8, 10	789	2, 6	790	2, 3, 4, 7, 8, 10
791	2, 3, 4, 8	792	1, 2, 4, 7, 9, 10	793	0, 2, 3, 7	794	1, 4, 5, 6, 8, 9	795	0, 3, 4, 5, 6, 8, 9, 10
796	0, 1, 4, 5, 6, 7, 8, 9	797	0, 1, 2, 3, 5, 6, 7, 10	798	0, 1, 2, 4, 5, 6	799	0, 2, 4, 5, 9, 10	800	0, 2, 4, 6, 7, 8
801	3, 4, 5, 6	802	3, 5, 6, 7, 9, 10	803	0, 1, 2, 4, 9, 10	804	0, 1, 3, 4, 8, 9	805	1, 2, 4, 5, 6, 7
806	0, 1, 5, 7, 8, 10	807	0, 4, 6, 9	808	1, 5, 7, 9	809	1, 3, 6, 8, 9, 10	810	1, 3, 4, 6
811	1, 2, 5, 10	812	0, 3, 4, 7, 9, 10	813	4, 6	814	4, 5, 7, 10	815	1, 3, 4, 10
816	0, 2, 3, 5, 6, 7, 8, 10	817	0, 1, 3, 4, 6, 8, 9, 10	818	0, 2, 3, 4, 5, 8	819	0, 5, 6, 10	820	2, 3, 4, 5, 8, 9
821	0, 1, 4, 5, 6, 9	822	2, 3, 5, 6	823	2, 6, 7, 8, 9, 10	824	0, 4, 8, 9	825	5, 6
826	5, 8, 9, 10	827	2, 3, 5, 6, 7, 8, 9, 10	828	0, 3, 6, 8	829	0, 2, 3, 5, 6, 9	830	1, 2, 4, 6, 7, 8
831	0, 6	832	0, 1, 2, 4, 6, 7, 8, 10	833	0, 2, 3, 4, 6, 10	834	3, 6	835	3, 4, 5, 6, 8, 10

(continued)

Table 9.7 (continued)

i	$\alpha(x)^i$	i	$\alpha(x)^i$	i	$\alpha(x)^i$	i	$\alpha(x)^i$	i	$\alpha(x)^i$
836	0, 5, 6, 7, 8, 10	837	0, 1, 2, 3, 4, 5, 7, 8, 9, 10	838	6, 7, 8, 10	839	3, 6, 8, 10	840	0, 4, 5, 6, 9, 10
841	0, 3, 6, 10	842	2, 5, 6, 8	843	1, 2, 3, 4, 5, 6	844	1, 3, 4, 7, 9, 10	845	0, 3, 5, 6
846	0, 1, 2, 3, 7, 8, 9, 10	847	4, 6, 9, 10	848	1, 2, 3, 4, 5, 7, 9, 10	849	0, 2, 4, 6, 7, 9	850	0, 1, 2, 3, 4, 5, 6, 8
851	0, 1, 2, 3, 7, 8	852	1, 2, 3, 4, 6, 10	853	0, 4, 5, 6	854	0, 1, 2, 6, 9, 10	855	0, 1, 3, 5, 7, 8, 9, 10
856	2, 3, 5, 6, 7, 10	857	1, 2, 3, 6	858	1, 3, 8, 10	859	0, 1, 2, 3, 4, 7, 8, 9	860	0, 1, 3, 5, 8, 9
861	1, 2, 5, 6, 8, 9	862	0, 2, 4, 6, 9, 10	863	0, 2, 4, 5, 6, 7, 9, 10	864	2, 4, 7, 8, 9, 10	865	0, 5, 7, 8, 9, 10
866	1, 3, 4, 5, 6, 10	867	0, 2, 3, 6, 7, 9	868	0, 1, 2, 4, 5, 7	869	2, 4, 5, 6	870	2, 3, 9, 10
871	1, 3, 5, 6, 7, 9	872	1, 4, 5, 9	873	0, 1, 3, 4, 5, 7, 8, 10	874	0, 3, 4, 7, 8, 9	875	0, 3, 4, 6, 8, 9
876	3, 4, 8, 10	877	0, 6, 7, 9	878	0, 1, 4, 6	879	0, 3, 7, 10	880	0, 2, 5, 8, 9, 10
881	0, 1, 2, 5, 7, 8, 9, 10	882	3, 10	883	0, 1, 4, 5, 7, 10	884	0, 1, 5, 10	885	1, 4, 6, 7, 9, 10
886	0, 4, 8, 10	887	1, 2, 3, 5, 6, 9	888	0, 1, 2, 3, 5, 6, 7, 8	889	1, 2, 3, 4, 5, 6, 8, 9	890	0, 2, 3, 4, 7, 8, 9, 10
891	1, 2, 3, 8, 9, 10	892	1, 2, 6, 7, 8, 10	893	1, 3, 4, 5, 8, 10	894	0, 1, 2, 3	895	0, 2, 3, 4, 6, 7
896	1, 6, 7, 8, 9, 10	897	0, 1, 5, 6, 8, 9	898	1, 2, 3, 4, 9, 10	899	2, 4, 5, 7, 8, 9	900	1, 3, 6, 8
901	2, 4, 6, 9	902	0, 3, 5, 6, 7, 9	903	0, 1, 3, 9	904	2, 7, 9, 10	905	0, 1, 4, 6, 7, 8
906	1, 3	907	1, 2, 4, 7	908	0, 1, 7, 9	909	0, 2, 3, 4, 5, 7, 8, 10	910	0, 1, 3, 5, 6, 7, 8, 9
911	0, 1, 2, 5, 8, 10	912	2, 3, 7, 8	913	0, 1, 2, 5, 6, 10	914	1, 2, 3, 6, 8, 9	915	0, 2, 3, 10
916	3, 4, 5, 6, 7, 10	917	1, 5, 6, 8	918	2, 3	919	2, 5, 6, 7	920	0, 2, 3, 4, 5, 6, 7, 10
921	0, 3, 5, 8	922	0, 2, 3, 6, 8, 10	923	1, 3, 4, 5, 9, 10	924	3, 8	925	1, 3, 4, 5, 7, 8, 9, 10
926	0, 1, 3, 7, 8, 10	927	0, 3	928	0, 1, 2, 3, 5, 7	929	2, 3, 4, 5, 7, 8	930	0, 1, 2, 4, 5, 6, 7, 8, 9, 10
931	3, 4, 5, 7	932	0, 3, 5, 7	933	1, 2, 3, 6, 7, 8	934	0, 3, 7, 8	935	2, 3, 5, 10
936	0, 1, 2, 3, 9, 10	937	0, 1, 4, 6, 7, 9	938	0, 2, 3, 8	939	0, 4, 5, 6, 7, 8, 9, 10	940	1, 3, 6, 7
941	0, 1, 2, 4, 6, 7, 9, 10	942	1, 3, 4, 6, 8, 10	943	0, 1, 2, 3, 5, 8, 9, 10	944	0, 4, 5, 8, 9, 10	945	0, 1, 3, 7, 9, 10
946	1, 2, 3, 8	947	3, 6, 7, 8, 9, 10	948	0, 2, 4, 5, 6, 7, 8, 9	949	0, 2, 3, 4, 7, 10	950	0, 3, 9, 10
951	0, 5, 7, 9	952	0, 1, 4, 5, 6, 8, 9, 10	953	0, 2, 5, 6, 8, 9	954	2, 3, 5, 6, 9, 10	955	1, 3, 6, 7, 8, 10
956	1, 2, 3, 4, 6, 7, 8, 10	957	1, 4, 5, 6, 7, 10	958	2, 4, 5, 6, 7, 8	959	0, 1, 2, 3, 7, 9	960	0, 3, 4, 6, 8, 10

(continued)

Table 9.7 (continued)

i	$\alpha(x)^i$	i	$\alpha(x)^i$	i	$\alpha(x)^i$	i	$\alpha(x)^i$	i	$\alpha(x)^i$
961	1, 2, 4, 8, 9, 10	962	1, 2, 3, 10	963	0, 6, 7, 10	964	0, 2, 3, 4, 6, 9	965	1, 2, 6, 9
966	0, 1, 2, 4, 5, 7, 8, 9	967	0, 1, 4, 5, 6, 8	968	0, 1, 3, 5, 6, 8	969	0, 1, 5, 7	970	3, 4, 6, 8
971	1, 3, 8, 9	972	0, 4, 7, 8	973	2, 5, 6, 7, 8, 10	974	2, 4, 5, 6, 7, 8, 9, 10	975	0, 7
976	1, 2, 4, 7, 8, 9	977	2, 7, 8, 9	978	1, 3, 4, 6, 7, 9	979	1, 5, 7, 8	980	0, 2, 3, 6, 9, 10
981	0, 2, 3, 4, 5, 8, 9, 10	982	0, 1, 2, 3, 5, 6, 9, 10	983	0, 1, 4, 5, 6, 7, 8, 10	984	0, 5, 6, 7, 9, 10	985	3, 4, 5, 7, 9, 10
986	0, 1, 2, 5, 7, 9	987	0, 8, 9, 10	988	0, 1, 3, 4, 8, 10	989	3, 4, 5, 6, 7, 9	990	2, 3, 5, 6, 8, 9
991	0, 1, 6, 7, 9, 10	992	1, 2, 4, 5, 6, 10	993	0, 3, 5, 9	994	1, 3, 6, 10	995	0, 2, 3, 4, 6, 8
996	0, 6, 8, 9	997	4, 6, 7, 10	998	1, 3, 4, 5, 8, 9	999	0, 9	1000	1, 4, 9, 10
1001	4, 6, 8, 9	1002	0, 1, 2, 4, 5, 7, 8, 10	1003	0, 2, 3, 4, 5, 6, 8, 9	1004	2, 5, 7, 8, 9, 10	1005	0, 4, 5, 10
1006	2, 3, 7, 8, 9, 10	1007	0, 3, 5, 6, 9, 10	1008	0, 7, 8, 10	1009	0, 1, 2, 3, 4, 7	1010	2, 3, 5, 9
1011	0, 10	1012	2, 3, 4, 10	1013	0, 1, 2, 3, 4, 7, 8, 10	1014	0, 2, 5, 8	1015	0, 3, 5, 7, 8, 10
1016	0, 1, 2, 6, 7, 9	1017	0, 5	1018	0, 1, 2, 4, 5, 6, 7, 9	1019	0, 4, 5, 7, 8, 9	1020	0, 8
1021	0, 2, 4, 8, 9, 10	1022	0, 1, 2, 4, 5, 10	1023	1, 2, 3, 4, 5, 6, 7, 8, 9, 10				

Table 9.8 Circulant analysis $p = 11$, $j(x) = 1 + x + x^2 + x^3 + x^4 + x^5 + x^6 + x^7 + x^8 + x^9 + x^{10}$, factors of $1 + x^p$

i	$j(x)^i$
1	0, 1, 2, 3, 4, 5, 6, 7, 8, 9, 10

Weight Distributions of Quadratic Double-Circulant Codes and their Modulo Congruence

Primes +3 Modulo 8

Prime 11

We have $P = \left[\begin{smallmatrix} 1 & 3 \\ 3 & 10 \end{smallmatrix}\right]$ and $T = \left[\begin{smallmatrix} 0 & 10 \\ 1 & 0 \end{smallmatrix}\right]$, $P, T \in \mathrm{PSL}_2(11)$, and the permutations of order 3, 5 and 11 are generated by $\left[\begin{smallmatrix} 0 & 1 \\ 10 & 1 \end{smallmatrix}\right]$, $\left[\begin{smallmatrix} 0 & 1 \\ 10 & 3 \end{smallmatrix}\right]$ and $\left[\begin{smallmatrix} 0 & 1 \\ 10 & 9 \end{smallmatrix}\right]$, respectively. In addition,

$$\mathrm{PSL}_2(11) = 2^2 \cdot 3 \cdot 5 \cdot 11 \cdot = 660$$

and the weight enumerator polynomials of the invariant subcodes are

$$\begin{aligned}
A_{\mathscr{B}_{11}}^{G_2^0}(z) &= \left(1 + z^{24}\right) + 15 \cdot \left(z^8 + z^{16}\right) + 32 \cdot z^{12} \\
A_{\mathscr{B}_{11}}^{G_4}(z) &= \left(1 + z^{24}\right) + 3 \cdot \left(z^8 + z^{16}\right) + 8 \cdot z^{12} \\
A_{\mathscr{B}_{11}}^{S_3}(z) &= \left(1 + z^{24}\right) + 14 \cdot z^{12} \\
A_{\mathscr{B}_{11}}^{S_5}(z) &= \left(1 + z^{24}\right) + 4 \cdot \left(z^8 + z^{16}\right) + 6 \cdot z^{12} \\
A_{\mathscr{B}_{11}}^{S_{11}}(z) &= \left(1 + z^{24}\right) + 2 \cdot z^{12} .
\end{aligned}$$

The weight distributions of $\mathscr{B}_{11}$ and their modular congruence are shown in Table 9.9.

Table 9.9 Modular congruence weight distributions of $\mathscr{B}_{11}$

i	$A_i(S_2)$ mod 2^2	$A_i(S_3)$ mod 3	$A_i(S_5)$ mod 5	$A_i(S_{11})$ mod 11	$A_i(\mathscr{H})$ mod 660	n_i[a]	A_i
0	1	1	1	1	1	0	1
8	3	0	4	0	99	1	759
12	0	2	1	2	596	3	2576
16	3	0	4	0	99	1	759
24	1	1	1	1	1	0	1

[a] $n_i = \frac{A_i - A_i(\mathscr{H})}{660}$

Prime 19

We have $P = \left[\begin{smallmatrix} 1 & 6 \\ 6 & 18 \end{smallmatrix}\right]$ and $T = \left[\begin{smallmatrix} 0 & 18 \\ 1 & 0 \end{smallmatrix}\right]$, $P, T \in \mathrm{PSL}_2(19)$, and the permutations of order 3, 5 and 19 are generated by $\left[\begin{smallmatrix} 0 & 1 \\ 18 & 1 \end{smallmatrix}\right]$, $\left[\begin{smallmatrix} 0 & 1 \\ 18 & 4 \end{smallmatrix}\right]$ and $\left[\begin{smallmatrix} 0 & 1 \\ 18 & 17 \end{smallmatrix}\right]$, respectively. In addition,

$$\mathrm{PSL}_2(19) = 2^2 \cdot 3^2 \cdot 5 \cdot 19 \cdot = 3420$$

and the weight enumerator polynomials of the invariant subcodes are

$$\begin{aligned}
A_{\mathscr{B}_{19}}^{(G_2^0)}(z) &= \left(1 + z^{40}\right) + 5 \cdot \left(z^8 + z^{32}\right) + 80 \cdot \left(z^{12} + z^{28}\right) + 250 \cdot \left(z^{16} + z^{24}\right) + 352 \cdot z^{20} \\
A_{\mathscr{B}_{19}}^{(G_4)}(z) &= \left(1 + z^{40}\right) + 1 \cdot \left(z^8 + z^{32}\right) + 8 \cdot \left(z^{12} + z^{28}\right) + 14 \cdot \left(z^{16} + z^{24}\right) + 16 \cdot z^{20} \\
A_{\mathscr{B}_{19}}^{(S_3)}(z) &= \left(1 + z^{40}\right) + 6 \cdot \left(z^8 + z^{32}\right) + 22 \cdot \left(z^{12} + z^{28}\right) + 57 \cdot \left(z^{16} + z^{24}\right) + 84 \cdot z^{20} \\
A_{\mathscr{B}_{19}}^{(S_5)}(z) &= \left(1 + z^{40}\right) + 14 \cdot z^{20} \\
A_{\mathscr{B}_{19}}^{(S_{19})}(z) &= \left(1 + z^{40}\right) + 2 \cdot z^{20}.
\end{aligned}$$

The weight distributions of $\mathscr{B}_{19}$ and their modular congruence are shown in Table 9.10.

Prime 43

We have $P = \left[\begin{smallmatrix} 1 & 16 \\ 16 & 42 \end{smallmatrix}\right]$ and $T = \left[\begin{smallmatrix} 0 & 42 \\ 1 & 0 \end{smallmatrix}\right]$, $P, T \in \mathrm{PSL}_2(43)$, and the permutations of order 3, 7, 11 and 43 are generated by $\left[\begin{smallmatrix} 0 & 1 \\ 42 & 1 \end{smallmatrix}\right]$, $\left[\begin{smallmatrix} 0 & 1 \\ 42 & 8 \end{smallmatrix}\right]$, $\left[\begin{smallmatrix} 0 & 1 \\ 42 & 4 \end{smallmatrix}\right]$ and $\left[\begin{smallmatrix} 0 & 1 \\ 42 & 41 \end{smallmatrix}\right]$, respectively. In addition,

$$\mathrm{PSL}_2(43) = 2^2 \cdot 3 \cdot 7 \cdot 11 \cdot 43 \cdot = 39732$$

Table 9.10 Modular congruence weight distributions of $\mathscr{B}_{19}$

i	$A_i(S_2)$ mod 2^2	$A_i(S_3)$ mod 3^2	$A_i(S_5)$ mod 5	$A_i(S_{19})$ mod 19	$A_i(\mathscr{H})$ mod 3420	n_i[a]	A_i
0	1	1	1	1	1	0	1
8	1	6	0	0	285	0	285
12	0	4	0	0	760	6	21280
16	2	3	0	0	570	70	239970
20	0	3	4	2	2244	153	525504
24	2	3	0	0	570	70	239970
28	0	4	0	0	760	6	21280
32	1	6	0	0	285	0	285
40	1	1	1	1	1	0	1

[a] $n_i = \frac{A_i - A_i(\mathscr{H})}{3420}$

and the weight enumerator polynomials of the invariant subcodes are

$$A_{\mathscr{B}_{43}}^{(G_2^0)}(z) = \left(1+z^{88}\right) + 44\cdot\left(z^{16}+z^{72}\right) + 1232\cdot\left(z^{20}+z^{68}\right) + 10241\cdot\left(z^{24}+z^{64}\right) + 54560\cdot\left(z^{28}+z^{60}\right) + 198374\cdot\left(z^{32}+z^{56}\right) + 491568\cdot\left(z^{36}+z^{52}\right) + 839916\cdot\left(z^{40}+z^{48}\right) + 1002432\cdot z^{44}$$

$$A_{\mathscr{B}_{43}}^{(G_4)}(z) = \left(1+z^{88}\right) + 32\cdot\left(z^{20}+z^{68}\right) + 77\cdot\left(z^{24}+z^{64}\right) + 160\cdot\left(z^{28}+z^{60}\right) + 330\cdot\left(z^{32}+z^{56}\right) + 480\cdot\left(z^{36}+z^{52}\right) + 616\cdot\left(z^{40}+z^{48}\right) + 704\cdot z^{44}$$

$$A_{\mathscr{B}_{43}}^{(S_3)}(z) = \left(1+z^{88}\right) + 7\cdot\left(z^{16}+z^{72}\right) + 168\cdot\left(z^{20}+z^{68}\right) + 445\cdot\left(z^{24}+z^{64}\right) + 1960\cdot\left(z^{28}+z^{60}\right) + 4704\cdot\left(z^{32}+z^{56}\right) + 7224\cdot\left(z^{36}+z^{52}\right) + 10843\cdot\left(z^{40}+z^{48}\right) + 14832\cdot z^{44}$$

$$A_{\mathscr{B}_{43}}^{(S_7)}(z) = \left(1+z^{88}\right) + 6\cdot\left(z^{16}+z^{72}\right) + 16\cdot\left(z^{24}+z^{64}\right) + 6\cdot\left(z^{28}+z^{60}\right) + 9\cdot\left(z^{32}+z^{56}\right) + 48\cdot\left(z^{36}+z^{52}\right) + 84\cdot z^{44}$$

$$A_{\mathscr{B}_{43}}^{(S_{11})}(z) = \left(1+z^{88}\right) + 14\cdot z^{44}$$

$$A_{\mathscr{B}_{43}}^{(S_{43})}(z) = \left(1+z^{88}\right) + 2\cdot z^{44}.$$

The weight distributions of $\mathscr{B}_{43}$ and their modular congruence are shown in Table 9.11.

Prime 59

We have $P = \left[\begin{smallmatrix}1 & 23\\ 23 & 58\end{smallmatrix}\right]$ and $T = \left[\begin{smallmatrix}0 & 58\\ 1 & 0\end{smallmatrix}\right]$, $P, T \in \text{PSL}_2(59)$, and the permutations of order 3, 5, 29 and 59 are generated by $\left[\begin{smallmatrix}0 & 1\\ 58 & 1\end{smallmatrix}\right]$, $\left[\begin{smallmatrix}0 & 1\\ 58 & 25\end{smallmatrix}\right]$, $\left[\begin{smallmatrix}0 & 1\\ 58 & 3\end{smallmatrix}\right]$ and $\left[\begin{smallmatrix}0 & 1\\ 58 & 57\end{smallmatrix}\right]$, respectively. In addition,

$$\text{PSL}_2(59) = 2^2\cdot 3\cdot 5\cdot 29\cdot 59\cdot = 102660$$

and the weight enumerator polynomials of the invariant subcodes are

$$A_{\mathscr{B}_{59}}^{(G_2^0)}(z) = \left(1+z^{120}\right) + 90\cdot\left(z^{20}+z^{100}\right) + 2555\cdot\left(z^{24}+z^{96}\right) + 32700\cdot\left(z^{28}+z^{92}\right) + 278865\cdot\left(z^{32}+z^{88}\right) + 1721810\cdot\left(z^{36}+z^{84}\right) + 7807800\cdot\left(z^{40}+z^{80}\right) + 26366160\cdot\left(z^{44}+z^{76}\right) + 67152520\cdot\left(z^{48}+z^{72}\right) + 130171860\cdot\left(z^{52}+z^{68}\right) + 193193715\cdot\left(z^{56}+z^{64}\right) + 220285672\cdot z^{60}$$

Table 9.11 Modular congruence weight distributions of $\mathscr{B}_{43}$

i	$A_i(S_2) \bmod 2^2$	$A_i(S_3) \bmod 3$	$A_i(S_7) \bmod 7$	$A_i(S_{11}) \bmod 11$	$A_i(S_{43}) \bmod 43$	$A_i(\mathscr{H}) \bmod 39732$	n_i[a]	A_i
0	1	1	1	1	1	1	0	1
16	0	1	6	0	0	32164	0	32164
20	0	0	0	0	0	0	176	6992832
24	1	1	2	0	0	25069	13483	535731625
28	0	1	6	0	0	32164	418387	16623384448
32	2	0	2	0	0	8514	5673683	225426781470
36	0	0	6	0	0	5676	35376793	1405590745152
40	0	1	0	0	0	26488	104797219	4163803131796
44	0	0	0	3	2	28812	150211729	5968212445440
48	0	1	0	0	0	26488	104797219	4163803131796
52	0	0	6	0	0	5676	35376793	1405590745152
56	2	0	2	0	0	8514	5673683	225426781470
60	0	1	6	0	0	32164	418387	16623384448
64	1	1	2	0	0	25069	13483	535731625
68	0	0	0	0	0	0	176	6992832
72	0	1	6	0	0	32164	0	32164
88	1	1	1	1	1	1	0	1

[a] $n_i = \frac{A_i - A_i(\mathscr{H})}{39732}$

$$A^{(G_4)}_{\mathscr{B}_{59}}(z) = \left(1+z^{120}\right) + 6\cdot\left(z^{20}+z^{100}\right) + 19\cdot\left(z^{24}+z^{96}\right) + 132\cdot\left(z^{28}+z^{92}\right) + 393\cdot\left(z^{32}+z^{88}\right) + 878\cdot\left(z^{36}+z^{84}\right) + 1848\cdot\left(z^{40}+z^{80}\right) + 3312\cdot\left(z^{44}+z^{76}\right) + 5192\cdot\left(z^{48}+z^{72}\right) + 7308\cdot\left(z^{52}+z^{68}\right) + 8931\cdot\left(z^{56}+z^{64}\right) + 9496\cdot z^{60}$$

$$A^{(S_3)}_{\mathscr{B}_{59}}(z) = \left(1+z^{120}\right) + 285\cdot\left(z^{24}+z^{96}\right) + 21280\cdot\left(z^{36}+z^{84}\right) + 239970\cdot\left(z^{48}+z^{72}\right) + 525504\cdot z^{60}$$

$$A^{(S_5)}_{\mathscr{B}_{59}}(z) = \left(1+z^{120}\right) + 12\cdot\left(z^{20}+z^{100}\right) + 711\cdot\left(z^{40}+z^{80}\right) + 2648\cdot z^{60}$$

$$A^{(S_{29})}_{\mathscr{B}_{59}}(z) = \left(1+z^{120}\right) + 4\cdot\left(z^{32}+z^{88}\right) + 6\cdot z^{60}$$

$$A^{(S_{59})}_{\mathscr{B}_{59}}(z) = \left(1+z^{120}\right) + 2\cdot z^{60}.$$

The weight distributions of $\mathscr{B}_{59}$ and their modular congruence are shown in Table 9.12.

Prime 67

We have $P = \left[\begin{smallmatrix}1 & 20\\ 20 & 66\end{smallmatrix}\right]$ and $T = \left[\begin{smallmatrix}0 & 66\\ 1 & 0\end{smallmatrix}\right]$, $P, T \in \mathrm{PSL}_2(67)$, and the permutations of order 3, 11, 17 and 67 are generated by $\left[\begin{smallmatrix}0 & 1\\ 66 & 1\end{smallmatrix}\right]$, $\left[\begin{smallmatrix}0 & 1\\ 66 & 17\end{smallmatrix}\right]$, $\left[\begin{smallmatrix}0 & 1\\ 66 & 4\end{smallmatrix}\right]$ and $\left[\begin{smallmatrix}0 & 1\\ 66 & 65\end{smallmatrix}\right]$, respectively. In addition,

$$\mathrm{PSL}_2(67) = 2^2\cdot 3\cdot 11\cdot 17\cdot 67\cdot = 150348$$

and the weight enumerator polynomials of the invariant subcodes are

$$A^{(G_2^0)}_{\mathscr{B}_{67}}(z) = (1+z^{136}) + 578\cdot(z^{24}+z^{112}) + 14688\cdot(z^{28}+z^{108}) + 173247\cdot(z^{32}+z^{104}) + 1480768\cdot(z^{36}+z^{100}) + 9551297\cdot(z^{40}+z^{96}) + 46687712\cdot(z^{44}+z^{92}) + 175068210\cdot(z^{48}+z^{88}) + 509510400\cdot\left(z^{52}+z^{84}\right) + 1160576876\cdot\left(z^{56}+z^{80}\right) + 2081112256\cdot(z^{60}+z^{76}) + 2949597087\cdot(z^{64}+z^{72}) + 3312322944\cdot z^{68}$$

$$A^{(G_4)}_{\mathscr{B}_{67}}(z) = (1+z^{136}) + 18\cdot(z^{24}+z^{112}) + 88\cdot(z^{28}+z^{108}) + 271\cdot(z^{32}+z^{104}) + 816\cdot(z^{36}+z^{100}) + 2001\cdot(z^{40}+z^{96}) + 4344\cdot(z^{44}+z^{92}) + 8386\cdot(z^{48}+z^{88}) + 14144\cdot\left(z^{52}+z^{84}\right) + 21260\cdot\left(z^{56}+z^{80}\right) + 28336\cdot(z^{60}+z^{76}) + 33599\cdot(z^{64}+z^{72}) + 35616\cdot z^{68}$$

$$A^{(S_3)}_{\mathscr{B}_{67}}(z) = (1+z^{136}) + 66\cdot(z^{24}+z^{112}) + 682\cdot(z^{28}+z^{108}) + 3696\cdot(z^{32}+z^{104}) + 12390\cdot(z^{36}+z^{100}) + 54747\cdot(z^{40}+z^{96}) + 163680\cdot(z^{44}+z^{92}) + 318516\cdot(z^{48}+z^{88}) + 753522\cdot\left(z^{52}+z^{84}\right) + 1474704\cdot\left(z^{56}+z^{80}\right) + 1763454\cdot(z^{60}+z^{76}) + 2339502\cdot(z^{64}+z^{72}) + 3007296\cdot z^{68}$$

Table 9.12 Modular congruence weight distributions of $\mathscr{B}_{59}$

i	$A_i(S_2)$ mod 2^2	$A_i(S_3)$ mod 3	$A_i(S_5)$ mod 5	$A_i(S_{29})$ mod 29	$A_i(S_{59})$ mod 59	$A_i(\mathscr{H})$ mod 102660	n_i[a]	A_i
0	1	[illegible]	1	1	1	1	0	1
20	2	0	2	0	0	71862	0	71862
24	3	0	0	0	0	76995	372	38266515
28	0	0	0	0	0	0	59565	6114942900
32	1	0	0	4	0	32745	4632400	475562216745
36	2	[illegible]	0	0	0	17110	183370922	18824858869630
40	0	0	1	0	0	61596	3871511775	397449398883096
44	0	0	0	0	0	0	45105349212	4630515150103920
48	0	0	0	0	0	0	297404962554	30531593455793640
52	0	0	0	0	0	0	1130177151411	116023986363853260
56	3	0	0	0	0	76995	2505920073120	257257754706576195
60	0	0	3	6	2	85788	3265149944551	335200293307691448
64	3	0	0	0	0	76995	2505920073120	257257754706576195
68	0	0	0	0	0	0	1130177151411	116023986363853260
72	0	0	0	0	0	0	297404962554	30531593455793640
76	0	0	0	0	0	0	45105349212	4630515150103920
80	0	0	1	0	0	61596	3871511775	397449398883096
84	2	[illegible]	0	0	0	17110	183370922	18824858869630
88	1	0	0	4	0	32745	4632400	475562216745
92	0	0	0	0	0	0	59565	6114942900
96	3	0	0	0	0	76995	372	38266515
100	2	0	2	0	0	71862	0	71862
120	1	[illegible]	1	1	1	1	0	1

[a] $n_i = \frac{A_i - A_i(\mathscr{H})}{102660}$

$$A_{\mathscr{B}_{67}}^{(S_{11})}(z) = (1 + z^{136}) + 6 \cdot (z^{24} + z^{112}) + 16 \cdot (z^{36} + z^{100}) + 6 \cdot (z^{44} + z^{92}) + 9 \cdot (z^{48} + z^{88}) + 48 \cdot \left(z^{56} + z^{80}\right) + 84 \cdot z^{68}$$

$$A_{\mathscr{B}_{67}}^{(S_{17})}(z) = (1 + z^{136}) + 14 \cdot z^{68}$$

$$A_{\mathscr{B}_{67}}^{(S_{67})}(z) = (1 + z^{136}) + 2 \cdot z^{68}$$

The weight distributions of $\mathscr{B}_{67}$ and their modular congruence are shown in Table 9.13.

Prime 83

We have $P = \left[\begin{smallmatrix} 1 & 9 \\ 9 & 82 \end{smallmatrix}\right]$ and $T = \left[\begin{smallmatrix} 0 & 82 \\ 1 & 0 \end{smallmatrix}\right]$, $P, T \in \mathrm{PSL}_2(83)$, and the permutations of order 3, 7, 41 and 83 are generated by $\left[\begin{smallmatrix} 0 & 1 \\ 82 & 1 \end{smallmatrix}\right]$, $\left[\begin{smallmatrix} 0 & 1 \\ 82 & 10 \end{smallmatrix}\right]$, $\left[\begin{smallmatrix} 0 & 1 \\ 82 & 4 \end{smallmatrix}\right]$ and $\left[\begin{smallmatrix} 0 & 1 \\ 82 & 81 \end{smallmatrix}\right]$, respectively. In addition,

$$\mathrm{PSL}_2(83) = 2^2 \cdot 3 \cdot 7 \cdot 41 \cdot 83 \cdot = 285852$$

and the weight enumerator polynomials of the invariant subcodes are

$$\begin{aligned}
A_{\mathscr{B}_{83}}^{(G_2^0)}(z) = {} & \left(1 + z^{168}\right) + 196 \cdot \left(z^{24} + z^{144}\right) + 1050 \cdot \left(z^{28} + z^{140}\right) + \\
& 29232 \cdot \left(z^{32} + z^{136}\right) + 443156 \cdot \left(z^{36} + z^{132}\right) + \\
& 4866477 \cdot \left(z^{40} + z^{128}\right) + 42512190 \cdot \left(z^{44} + z^{124}\right) + \\
& 292033644 \cdot \left(z^{48} + z^{120}\right) + 1590338568 \cdot \left(z^{52} + z^{116}\right) + \\
& 6952198884 \cdot \left(z^{56} + z^{112}\right) + 24612232106 \cdot \left(z^{60} + z^{108}\right) + \\
& 71013075210 \cdot \left(z^{64} + z^{104}\right) + 167850453036 \cdot \left(z^{68} + z^{100}\right) + \\
& 326369180312 \cdot \left(z^{72} + z^{96}\right) + 523672883454 \cdot \left(z^{76} + z^{92}\right) + \\
& 694880243820 \cdot \left(z^{80} + z^{88}\right) + 763485528432 \cdot z^{84}
\end{aligned}$$

$$\begin{aligned}
A_{\mathscr{B}_{83}}^{(G_4)}(z) = {} & \left(1 + z^{168}\right) + 4 \cdot \left(z^{24} + z^{144}\right) + 6 \cdot \left(z^{28} + z^{140}\right) + \\
& 96 \cdot \left(z^{32} + z^{136}\right) + 532 \cdot \left(z^{36} + z^{132}\right) + 1437 \cdot \left(z^{40} + z^{128}\right) + \\
& 3810 \cdot \left(z^{44} + z^{124}\right) + 10572 \cdot \left(z^{48} + z^{120}\right) + 24456 \cdot \left(z^{52} + z^{116}\right) + \\
& 50244 \cdot \left(z^{56} + z^{112}\right) + 95030 \cdot \left(z^{60} + z^{108}\right) + 158874 \cdot \left(z^{64} + z^{104}\right) + \\
& 241452 \cdot \left(z^{68} + z^{100}\right) + 337640 \cdot \left(z^{72} + z^{96}\right) + 425442 \cdot \left(z^{76} + z^{92}\right) + \\
& 489708 \cdot \left(z^{80} + z^{88}\right) + 515696 \cdot z^{84}
\end{aligned}$$

Table 9.13 Modular congruence weight distributions of $\mathscr{B}_{67}$

i	$A_i(S_2)$ mod 2^2	$A_i(S_3)$ mod 3	$A_i(S_{11})$ mod 11	$A_i(S_{17})$ mod 17	$A_i(S_{67})$ mod 67	$A_i(\mathscr{H})$ mod 150348	n_i[a]	A_i
0	1	1	1	1	1	1	0	1
24	2	0	6	0	0	88842	26	3997890
28	0	1	0	0	0	50116	8173	1228844320
32	3	0	0	0	0	37587	1217081	182985731775
36	0	0	5	0	0	136680	95005682	14283914414016
40	1	0	0	0	0	112761	4076381478	612875802567105
44	0	0	6	0	0	13668	99752935189	14997654299809440
48	2	0	9	0	0	20502	1432445445981	215365307912371890
52	0	0	0	0	0	0	12338369112000	1855049119250976000
56	0	0	4	0	0	109344	64817708364545	9745212817192721004
60	0	0	0	0	0	0	210227711554224	31607315976754469952
64	3	0	0	0	0	37587	424499666112161	63822675800631219615
68	0	0	7	14	2	138156	536258660836183	80625417139398579840
72	3	0	0	0	0	37587	424499666112161	63822675800631219615
76	0	0	0	0	0	0	210227711554224	31607315976754469952
80	0	0	4	0	0	109344	64817708364545	9745212817192721004
84	0	0	0	0	0	0	12338369112000	1855049119250976000
88	2	0	9	0	0	20502	1432445445981	215365307912371890
92	0	0	6	0	0	13668	99752935189	14997654299809440
96	1	0	0	0	0	112761	4076381478	612875802567105
100	0	0	5	0	0	136680	95005682	14283914414016
104	3	0	0	0	0	37587	1217081	182985731775
108	0	1	0	0	0	50116	8173	1228844320
112	2	0	6	0	0	88842	26	3997890
136	1	1	1	1	1	1	0	1

[a] $n_i = \frac{A_i - A_i(\mathscr{H})}{150348}$

$$A_{\mathscr{B}_{83}}^{(S_3)}(z) = \left(1 + z^{168}\right) + 63 \cdot \left(z^{24} + z^{144}\right) + 8568 \cdot \left(z^{36} + z^{132}\right) + 617085 \cdot \left(z^{48} + z^{120}\right) + 11720352 \cdot \left(z^{60} + z^{108}\right) + 64866627 \cdot \left(z^{72} + z^{96}\right) + 114010064 \cdot z^{84}$$

$$A_{\mathscr{B}_{83}}^{(S_7)}(z) = \left(1 + z^{168}\right) + 759 \cdot \left(z^{56} + z^{112}\right) + 2576 \cdot z^{84}$$

$$A_{\mathscr{B}_{83}}^{(S_{41})}(z) = \left(1 + z^{168}\right) + 4 \cdot \left(z^{44} + z^{124}\right) + 6 \cdot z^{84}$$

$$A_{\mathscr{B}_{83}}^{(S_{83})}(z) = \left(1 + z^{168}\right) + 2 \cdot z^{84}.$$

The weight distributions of $\mathscr{B}_{83}$ and their modular congruence are shown in Table 9.14.

Primes −3 Modulo 8

Prime 13

We have $P = \left[\begin{smallmatrix} 3 & 4 \\ 4 & 10 \end{smallmatrix}\right]$ and $T = \left[\begin{smallmatrix} 0 & 12 \\ 1 & 0 \end{smallmatrix}\right]$, $P, T \in \mathrm{PSL}_2(13)$, and the permutations of order 3, 7 and 13 are generated by $\left[\begin{smallmatrix} 0 & 1 \\ 12 & 1 \end{smallmatrix}\right]$, $\left[\begin{smallmatrix} 0 & 1 \\ 12 & 3 \end{smallmatrix}\right]$ and $\left[\begin{smallmatrix} 0 & 1 \\ 12 & 11 \end{smallmatrix}\right]$, respectively. In addition,

$$\mathrm{PSL}_2(13) = 2^2 \cdot 3 \cdot 7 \cdot 13 \cdot = 1092$$

and the weight enumerator polynomials of the invariant subcodes are

$$A_{\mathscr{B}_{13}}^{G_2^0}(z) = \left(1 + z^{28}\right) + 26 \cdot \left(z^{8} + z^{20}\right) + 32 \cdot \left(z^{10} + z^{18}\right) + 37 \cdot \left(z^{12} + z^{16}\right) + 64 \cdot z^{14}$$

$$A_{\mathscr{B}_{13}}^{G_4}(z) = \left(1 + z^{28}\right) + 10 \cdot \left(z^{8} + z^{20}\right) + 8 \cdot \left(z^{10} + z^{18}\right) + 5 \cdot \left(z^{12} + z^{16}\right) + 16 \cdot z^{14}$$

$$A_{\mathscr{B}_{13}}^{S_3}(z) = \left(1 + z^{28}\right) + 6 \cdot \left(z^{8} + z^{20}\right) + 10 \cdot \left(z^{10} + z^{18}\right) + 9 \cdot \left(z^{12} + z^{16}\right) + 12 \cdot z^{14}$$

$$A_{\mathscr{B}_{13}}^{S_7}(z) = \left(1 + z^{28}\right) + 2 \cdot z^{14}$$

$$A_{\mathscr{B}_{13}}^{S_{13}}(z) = \left(1 + z^{28}\right) + 2 \cdot z^{14}.$$

The weight distributions of $\mathscr{B}_{13}$ and their modular congruence are shown in Table 9.15.

Prime 29

We have $P = \left[\begin{smallmatrix} 2 & 13 \\ 13 & 27 \end{smallmatrix}\right]$ and $T = \left[\begin{smallmatrix} 0 & 28 \\ 1 & 0 \end{smallmatrix}\right]$, $P, T \in \mathrm{PSL}_2(29)$, and the permutations of order 3, 5, 7 and 29 are generated by $\left[\begin{smallmatrix} 0 & 1 \\ 28 & 1 \end{smallmatrix}\right]$, $\left[\begin{smallmatrix} 0 & 1 \\ 28 & 5 \end{smallmatrix}\right]$, $\left[\begin{smallmatrix} 0 & 1 \\ 28 & 3 \end{smallmatrix}\right]$ and $\left[\begin{smallmatrix} 0 & 1 \\ 28 & 27 \end{smallmatrix}\right]$, respectively. In addition,

$$\mathrm{PSL}_2(29) = 2^2 \cdot 3 \cdot 5 \cdot 7 \cdot 29 \cdot = 12180$$

Table 9.14 Modular congruence weight distributions of $\mathscr{B}_{83}$

i	$A_i(S_2)$	$A_i(S_3)$	$A_i(S_7)$	$A_i(S_{41})$	$A_i(S_{83})$	$A_i(\mathscr{H})$	n_i[a]	A_i
i	mod 2^2	mod 3	mod 7	mod 41	mod 83	mod 285852		
0	1	1	1	1	0	6889	0	1
24	0	0	0	0	0	0	2	571704
28	2	0	0	0	0	142926	59	17008194
32	0	0	0	0	0	0	19267	5507510484
36	0	0	0	0	0	0	4382043	1252615755636
40	1	0	0	0	0	214389	580925895	166058829151929
44	2	0	0	4	0	156870	45643181220	13047194638256310
48	0	0	0	0	0	0	2200608997142	629048483051034984
52	0	0	0	0	0	0	66772769854878	19087129808556586056
56	0	0	3	0	0	81672	1301721510043764	372099697089030108600
60	2	0	0	0	0	142926	16579528883596695	4739291490433882602066
64	2	0	0	0	0	142926	139840453892634544	39973673426117369814414
68	0	0	0	0	0	0	789557804450518101	225696677517789500207052
72	0	0	0	0	0	0	3009393355026378047	860241109321000217491044
76	2	0	0	0	0	142926	7792111592501736790	2227390682939806465038006
80	0	0	0	0	0	0	13766213240696824038	3935099587279668544910376
84	0	2	0	6	0	211484	16637096860279621422	4755747411704650343205104
88	0	0	0	0	0	0	13766213240696824038	3935099587279668544910376
92	2	0	0	0	0	142926	7792111592501736790	2227390682939806465038006

(continued)

Table 9.14 (continued)

i i	$A_i(S_2)$ mod 2^2	$A_i(S_3)$ mod 3	$A_i(S_7)$ mod 7	$A_i(S_{41})$ mod 41	$A_i(S_{83})$ mod 83	$A_i(\mathscr{H})$ mod 285852	n_i[a]	A_i
96	0	0	0	0	0	0	3009393355026378047	860241109321000217491044
100	0	0	0	0	0	0	789557804450518101	225696677517789500207052
104	2	0	0	0	0	142926	139840453892634544	39973673426117369814414
108	2	0	0	0	0	142926	16579528883596695	4739291490433882602066
112	0	0	3	0	0	81672	1301721510043764	372099697089030108600
116	0	0	0	0	0	0	66772769854878	19087129808556586056
120	0	0	0	0	0	0	2200608997142	629048483051034984
124	2	0	0	4	0	156870	45643181220	13047194638256310
128	1	0	0	0	0	214389	580925895	166058829151929
132	0	0	0	0	0	0	4382043	1252615755636
136	0	0	0	0	0	0	19267	5507510484
140	2	0	0	0	0	142926	59	17008194
144	0	0	0	0	0	0	2	571704
168	1	1	1	1	0	6889	0	1

[a] $n_i = \frac{A_i - A_i(\mathscr{H})}{285852}$

Table 9.15 Modular congruence weight distributions of $\mathscr{B}_{13}$

i	$A_i(S_2)$ mod 2^2	$A_i(S_3)$ mod 3	$A_i(S_7)$ mod 7	$A_i(S_{13})$ mod 13	$A_i(\mathscr{H})$ mod 1092	n_i[a]	A_i
0	1	1	1	1	1	0	1
8	2	0	0	0	546	0	546
10	0	1	0	0	364	1	1456
12	1	0	0	0	273	3	3549
14	0	0	2	2	912	4	5280
16	1	0	0	0	273	3	3549
18	0	1	0	0	364	1	1456
20	2	0	0	0	546	0	546
28	1	1	1	1	1	0	1

[a] $n_i = \frac{A_i - A_i(\mathscr{H})}{1092}$

and the weight enumerator polynomials of the invariant subcodes are

$$A_{\mathscr{B}_{29}}^{(G_2^0)}(z) = \left(1+z^{60}\right) + 28\cdot\left(z^{12}+z^{48}\right) + 112\cdot\left(z^{14}+z^{46}\right) + 394\cdot\left(z^{16}+z^{44}\right) + 1024\cdot\left(z^{18}+z^{42}\right) + 1708\cdot\left(z^{20}+z^{40}\right) + 3136\cdot\left(z^{22}+z^{38}\right) + 5516\cdot\left(z^{24}+z^{36}\right) + 7168\cdot\left(z^{26}+z^{34}\right) + 8737\cdot\left(z^{28}+z^{32}\right) + 9888\cdot z^{30}$$

$$A_{\mathscr{B}_{29}}^{(G_4)}(z) = \left(1+z^{60}\right) + 12\cdot\left(z^{14}+z^{46}\right) + 30\cdot\left(z^{16}+z^{44}\right) + 32\cdot\left(z^{18}+z^{42}\right) + 60\cdot\left(z^{20}+z^{40}\right) + 48\cdot\left(z^{22}+z^{38}\right) + 60\cdot\left(z^{24}+z^{36}\right) + 96\cdot\left(z^{26}+z^{34}\right) + 105\cdot\left(z^{28}+z^{32}\right) + 136\cdot z^{30}$$

$$A_{\mathscr{B}_{29}}^{(S_3)}(z) = \left(1+z^{60}\right) + 10\cdot\left(z^{12}+z^{48}\right) + 70\cdot\left(z^{18}+z^{42}\right) + 245\cdot\left(z^{24}+z^{36}\right) + 372\cdot z^{30}$$

$$A_{\mathscr{B}_{29}}^{(S_5)}(z) = \left(1+z^{60}\right) + 15\cdot\left(z^{20}+z^{40}\right) + 32\cdot z^{30}$$

$$A_{\mathscr{B}_{29}}^{(S_7)}(z) = \left(1+z^{60}\right) + 6\cdot\left(z^{16}+z^{44}\right) + 2\cdot\left(z^{18}+z^{42}\right) + 8\cdot\left(z^{22}+z^{38}\right) + 8\cdot\left(z^{24}+z^{36}\right) + 1\cdot\left(z^{28}+z^{32}\right) + 12\cdot z^{30}$$

$$A_{\mathscr{B}_{29}}^{(S_{29})}(z) = \left(1+z^{60}\right) + 2\cdot z^{30}.$$

The weight distributions of $\mathscr{B}_{29}$ and their modular congruence are shown in Table 9.16.

Prime 53

We have $P = \left[\begin{smallmatrix} 3 & 19 \\ 19 & 50 \end{smallmatrix}\right]$ and $T = \left[\begin{smallmatrix} 0 & 52 \\ 1 & 0 \end{smallmatrix}\right]$, $P, T \in \mathrm{PSL}_2(53)$, and the permutations of order 3, 13 and 53 are generated by $\left[\begin{smallmatrix} 0 & 1 \\ 52 & 1 \end{smallmatrix}\right]$, $\left[\begin{smallmatrix} 0 & 1 \\ 52 & 8 \end{smallmatrix}\right]$ and $\left[\begin{smallmatrix} 0 & 1 \\ 52 & 51 \end{smallmatrix}\right]$, respectively. In addition,

$$\mathrm{PSL}_2(53) = 2^2\cdot 3^3\cdot 13\cdot 53\cdot = 74412$$

Table 9.16 Modular congruence weight distributions of $\mathscr{B}_{29}$

i	$A_i(S_2)$ mod 2^2	$A_i(S_3)$ mod 3	$A_i(S_5)$ mod 5	$A_i(S_7)$ mod 7	$A_i(S_{29})$ mod 29	$A_i(\mathscr{H})$ mod 12180	n_i[a]	A_i
0	1	1	1	1	1	1	0	1
12	0	1	0	0	0	4060	0	4060
14	0	0	0	0	0	0	2	24360
16	2	0	0	6	0	2610	24	294930
18	0	1	0	2	0	11020	141	1728400
20	0	0	0	0	0	0	637	7758660
22	0	0	0	1	0	3480	2162	26336640
24	0	2	0	1	0	11600	5533	67403540
26	0	0	0	0	0	0	10668	129936240
28	1	0	0	1	0	6525	15843	192974265
30	0	0	2	5	2	8412	18129	220819632
32	1	0	0	1	0	6525	15843	192974265
34	0	0	0	0	0	0	10668	129936240
36	0	2	0	1	0	11600	5533	67403540
38	0	0	0	1	0	3480	2162	26336640
40	0	0	0	0	0	0	637	7758660
42	0	1	0	2	0	11020	141	1728400
44	2	0	0	6	0	2610	24	294930
46	0	0	0	0	0	0	2	24360
48	0	1	0	0	0	4060	0	4060
60	1	1	1	1	1	1	0	1

[a] $n_i = \frac{A_i - A_i(\mathscr{H})}{12180}$

and the weight enumerator polynomials of the invariant subcodes are

$$\begin{aligned}
A_{\mathscr{B}_{53}}^{(G_2^0)}(z) = &\left(1+z^{108}\right) + 234\cdot\left(z^{20}+z^{88}\right) + 1768\cdot\left(z^{22}+z^{86}\right) + 5655\cdot\left(z^{24}+z^{84}\right) + \\
&16328\cdot\left(z^{26}+z^{82}\right) + 47335\cdot\left(z^{28}+z^{80}\right) + 127896\cdot\left(z^{30}+z^{78}\right) + \\
&316043\cdot\left(z^{32}+z^{76}\right) + 705848\cdot\left(z^{34}+z^{74}\right) + 1442883\cdot\left(z^{36}+z^{72}\right) + \\
&2728336\cdot\left(z^{38}+z^{70}\right) + 4786873\cdot\left(z^{40}+z^{68}\right) + 7768488\cdot\left(z^{42}+z^{66}\right) + \\
&11636144\cdot\left(z^{44}+z^{64}\right) + 16175848\cdot\left(z^{46}+z^{62}\right) + 20897565\cdot\left(z^{48}+z^{60}\right) + \\
&25055576\cdot\left(z^{50}+z^{58}\right) + 27976131\cdot\left(z^{52}+z^{56}\right) + 29057552\cdot z^{54}
\end{aligned}$$

$$\begin{aligned}
A_{\mathscr{B}_{53}}^{(G_4)}(z) = &\left(1+z^{108}\right) + 12\cdot\left(z^{20}+z^{88}\right) + 12\cdot\left(z^{22}+z^{86}\right) + 77\cdot\left(z^{24}+z^{84}\right) + \\
&108\cdot\left(z^{26}+z^{82}\right) + 243\cdot\left(z^{28}+z^{80}\right) + 296\cdot\left(z^{30}+z^{78}\right) + 543\cdot\left(z^{32}+z^{76}\right) + \\
&612\cdot\left(z^{34}+z^{74}\right) + 1127\cdot\left(z^{36}+z^{72}\right) + 1440\cdot\left(z^{38}+z^{70}\right) + 2037\cdot\left(z^{40}+z^{68}\right) + \\
&2636\cdot\left(z^{42}+z^{66}\right) + 3180\cdot\left(z^{44}+z^{64}\right) + 3672\cdot\left(z^{46}+z^{62}\right) + 4289\cdot\left(z^{48}+z^{60}\right) + \\
&4836\cdot\left(z^{50}+z^{58}\right) + 4875\cdot\left(z^{52}+z^{56}\right) + 5544\cdot z^{54}
\end{aligned}$$

Table 9.17 Modular congruence weight distributions of $\mathscr{B}_{53}$

i	$A_i(S_2)$ mod 2^2	$A_i(S_3)$ mod 3^3	$A_i(S_{13})$ mod 13	$A_i(S_{53})$ mod 53	$A_i(\mathscr{H})$ mod 74412	n_i[a]	A_i
0	1	1	1	1	1	0	1
20	2	0	0	0	37206	3	260442
22	0	0	0	0	0	78	5804136
24	3	18	0	0	43407	1000	74455407
26	0	0	0	0	0	10034	746650008
28	3	0	6	0	64395	91060	6776021115
30	0	18	2	0	64872	658342	48988609776
32	3	0	0	0	18603	3981207	296249593887
34	0	0	0	0	0	20237958	1505946930696
36	3	6	0	0	26871	86771673	6456853758147
38	0	0	0	0	0	315441840	23472658198080
40	1	0	8	0	67257	976699540	72678166237737
42	0	0	8	0	11448	2584166840	192293022909528
44	0	0	0	0	0	5859307669	436002802265628
46	0	0	0	0	0	11412955404	849260837522448
48	1	9	0	0	31005	19133084721	1423731100290057
50	0	0	0	0	0	27645086470	2057126174405640
52	3	0	1	0	1431	34462554487	2564427604488075
54	0	5	12	2	55652	37087868793	2759782492680368
56	3	0	1	0	1431	34462554487	2564427604488075
58	0	0	0	0	0	27645086470	2057126174405640
60	1	9	0	0	31005	19133084721	1423731100290057
62	0	0	0	0	0	11412955404	849260837522448
64	0	0	0	0	0	5859307669	436002802265628
66	0	0	8	0	11448	2584166840	192293022909528
68	1	0	8	0	67257	976699540	72678166237737
70	0	0	0	0	0	315441840	23472658198080
72	3	6	0	0	26871	86771673	6456853758147
74	0	0	0	0	0	20237958	1505946930696
76	3	0	0	0	18603	3981207	296249593887
78	0	18	2	0	64872	658342	48988609776
80	3	0	6	0	64395	91060	6776021115
82	0	0	0	0	0	10034	746650008
84	3	18	0	0	43407	1000	74455407
86	0	0	0	0	0	78	5804136
88	2	0	0	0	37206	3	260442
108	1	1	1	1	1	0	1

[a] $n_i = \frac{A_i - A_i(\mathscr{H})}{74412}$

Table 9.18 Modular congruence weight distributions of $\mathscr{B}_{61}$

i	$A_i(S_2)$ mod 2^2	$A_i(S_3)$ mod 3	$A_i(S_5)$ mod 5	$A_i(S_{31})$ mod 31	$A_i(S_{61})$ mod 61	$A_i(\mathscr{H})$ mod 113460	n_i[a]	A_i
0	1	1	1	1	1	1	0	1
20	0	0	3	0	0	90768	0	90768
22	0	1	0	0	0	75640	4	529480
24	2	2	0	0	0	94550	95	10873250
26	0	2	4	0	0	83204	1508	171180884
28	2	2	3	0	0	71858	19029	2159102198
30	0	0	1	0	0	68076	199795	22668808776
32	0	1	0	0	0	75640	1759003	199576556020
34	0	0	3	0	0	90768	13123969	1489045613508
36	2	0	3	0	0	34038	83433715	9466389337938
38	0	1	1	0	0	30256	454337550	51549138453256
40	0	2	0	0	0	37820	2128953815	241551099887720
42	0	0	3	0	0	90768	8619600220	977979841051968
44	0	0	2	0	0	22692	30259781792	3433274842143012
46	0	2	1	0	0	105896	92387524246	10482288501057056
48	0	2	0	0	0	37820	245957173186	27906300869721380
50	0	2	0	0	0	37820	572226179533	64924782329852000
52	0	2	1	0	0	105896	1165598694540	132248827882614296
54	0	2	3	0	0	15128	2081950370302	236218089014480048
56	0	2	2	0	0	60512	3264875882211	370432817595720572
58	0	2	2	0	0	60512	4499326496930	510493584341738312
60	1	2	1	0	0	20801	5452574159887	618649064180799821
62	0	2	1	2	2	102116	5813004046431	659543439108163376
64	1	2	1	0	0	20801	5452574159887	618649064180799821
66	0	2	2	0	0	60512	4499326496930	510493584341738312
68	0	2	2	0	0	60512	3264875882211	370432817595720572
70	0	2	3	0	0	15128	2081950370302	236218089014480048
72	0	2	1	0	0	105896	1165598694540	132248827882614296
74	0	2	0	0	0	37820	572226179533	64924782329852000
76	0	2	0	0	0	37820	245957173186	27906300869721380
78	0	2	1	0	0	105896	92387524246	10482288501057056
80	0	0	2	0	0	22692	30259781792	3433274842143012
82	0	0	3	0	0	90768	8619600220	977979841051968
84	0	2	0	0	0	37820	2128953815	241551099887720
86	0	1	1	0	0	30256	454337550	51549138453256

(continued)

Table 9.18 (continued)

i	$A_i(S_2)$ mod 2^2	$A_i(S_3)$ mod 3	$A_i(S_5)$ mod 5	$A_i(S_{31})$ mod 31	$A_i(S_{61})$ mod 61	$A_i(\mathscr{H})$ mod 113460	n_i[a]	A_i
88	2	0	3	0	0	34038	83433715	9466389337938
90	0	0	3	0	0	90768	13123969	1489045613508
92	0	1	0	0	0	75640	1759003	199576556020
94	0	0	1	0	0	68076	199795	22668808776
96	2	2	3	0	0	71858	19029	2159102198
98	0	2	4	0	0	83204	1508	171180884
100	2	2	0	0	0	94550	95	10873250
102	0	1	0	0	0	75640	4	529480
104	0	0	3	0	0	90768	0	90768
124	1	1	1	1	1	1	0	1

[a] $n_i = \frac{A_i - A_i(\mathscr{H})}{113460}$

$$\begin{aligned}A_{\mathscr{B}_{53}}^{(S_3)}(z) &= \left(1+z^{108}\right) + 234\cdot\left(z^{24}+z^{84}\right) + 1962\cdot\left(z^{30}+z^{78}\right) + 9672\cdot\left(z^{36}+z^{72}\right) + \\ &\quad 28728\cdot\left(z^{42}+z^{66}\right) + 55629\cdot\left(z^{48}+z^{60}\right) + 69692\cdot z^{54}\\ A_{\mathscr{B}_{53}}^{(S_{13})}(z) &= \left(1+z^{108}\right) + 6\cdot\left(z^{28}+z^{80}\right) + 2\cdot\left(z^{30}+z^{78}\right) + 8\cdot\left(z^{40}+z^{68}\right) + \\ &\quad 8\cdot\left(z^{42}+z^{66}\right) + 1\cdot\left(z^{52}+z^{56}\right) + 12\cdot z^{54}\\ A_{\mathscr{B}_{53}}^{(S_{53})}(z) &= \left(1+z^{108}\right) + 2\cdot z^{54}.\end{aligned}$$

The weight distributions of $\mathscr{B}_{53}$ and their modular congruence are shown in Table 9.17.

Prime 61

We have $P = \begin{bmatrix} 2 & 19 \\ 19 & 59 \end{bmatrix}$ and $T = \begin{bmatrix} 0 & 60 \\ 1 & 0 \end{bmatrix}$, $P, T \in \mathrm{PSL}_2(61)$, and the permutations of order 3, 5, 31 and 61 are generated by $\begin{bmatrix} 0 & 1 \\ 60 & 1 \end{bmatrix}$, $\begin{bmatrix} 0 & 1 \\ 60 & 17 \end{bmatrix}$, $\begin{bmatrix} 0 & 1 \\ 60 & 3 \end{bmatrix}$ and $\begin{bmatrix} 0 & 1 \\ 60 & 59 \end{bmatrix}$, respectively. In addition,

$$\mathrm{PSL}_2(61) = 2^2\cdot 3\cdot 5\cdot 31\cdot 61\cdot = 113460$$

and the weight enumerator polynomials of the invariant subcodes are

$$\begin{aligned}
A_{\mathscr{B}_{61}}^{(G_2^0)} = & \left(1+z^{124}\right)+208\cdot\left(z^{20}+z^{104}\right)+400\cdot\left(z^{22}+z^{102}\right)+1930\cdot\left(z^{24}+z^{100}\right)+\\
& 8180\cdot\left(z^{26}+z^{98}\right)+26430\cdot\left(z^{28}+z^{96}\right)+84936\cdot\left(z^{30}+z^{94}\right)+\\
& 253572\cdot\left(z^{32}+z^{92}\right)+696468\cdot\left(z^{34}+z^{90}\right)+1725330\cdot\left(z^{36}+z^{88}\right)+\\
& 3972240\cdot\left(z^{38}+z^{86}\right)+8585008\cdot\left(z^{40}+z^{84}\right)+17159632\cdot\left(z^{42}+z^{82}\right)+\\
& 31929532\cdot\left(z^{44}+z^{80}\right)+55569120\cdot\left(z^{46}+z^{78}\right)+90336940\cdot\left(z^{48}+z^{76}\right)+\\
& 137329552\cdot\left(z^{50}+z^{74}\right)+195328240\cdot\left(z^{52}+z^{72}\right)+260435936\cdot\left(z^{54}+z^{70}\right)+\\
& 325698420\cdot\left(z^{56}+z^{68}\right)+381677080\cdot\left(z^{58}+z^{66}\right)+419856213\cdot\left(z^{60}+z^{64}\right)+\\
& 433616560\cdot z^{62}
\end{aligned}$$

$$\begin{aligned}
A_{\mathscr{B}_{61}}^{(G_4)} = & \left(1+z^{124}\right)+12\cdot\left(z^{20}+z^{104}\right)+12\cdot\left(z^{22}+z^{102}\right)+36\cdot\left(z^{24}+z^{100}\right)+\\
& 40\cdot\left(z^{26}+z^{98}\right)+140\cdot\left(z^{28}+z^{96}\right)+176\cdot\left(z^{30}+z^{94}\right)+498\cdot\left(z^{32}+z^{92}\right)+\\
& 576\cdot\left(z^{34}+z^{90}\right)+1340\cdot\left(z^{36}+z^{88}\right)+1580\cdot\left(z^{38}+z^{86}\right)+2660\cdot\left(z^{40}+z^{84}\right)+\\
& 3432\cdot\left(z^{42}+z^{82}\right)+4932\cdot\left(z^{44}+z^{80}\right)+6368\cdot\left(z^{46}+z^{78}\right)+8820\cdot\left(z^{48}+z^{76}\right)+\\
& 10424\cdot\left(z^{50}+z^{74}\right)+12752\cdot\left(z^{52}+z^{72}\right)+14536\cdot\left(z^{54}+z^{70}\right)+15840\cdot\left(z^{56}+z^{68}\right)+\\
& 18296\cdot\left(z^{58}+z^{66}\right)+18505\cdot\left(z^{60}+z^{64}\right)+20192\cdot z^{62}
\end{aligned}$$

$$\begin{aligned}
A_{\mathscr{B}_{61}}^{(S_3)} = & \left(1+z^{124}\right)+30\cdot\left(z^{20}+z^{104}\right)+10\cdot\left(z^{22}+z^{102}\right)+50\cdot\left(z^{24}+z^{100}\right)+\\
& 200\cdot\left(z^{26}+z^{98}\right)+620\cdot\left(z^{28}+z^{96}\right)+960\cdot\left(z^{30}+z^{94}\right)+\\
& 2416\cdot\left(z^{32}+z^{92}\right)+4992\cdot\left(z^{34}+z^{90}\right)+6945\cdot\left(z^{36}+z^{88}\right)+\\
& 15340\cdot\left(z^{38}+z^{86}\right)+25085\cdot\left(z^{40}+z^{84}\right)+34920\cdot\left(z^{42}+z^{82}\right)+\\
& 68700\cdot\left(z^{44}+z^{80}\right)+87548\cdot\left(z^{46}+z^{78}\right)+104513\cdot\left(z^{48}+z^{76}\right)+\\
& 177800\cdot\left(z^{50}+z^{74}\right)+201440\cdot\left(z^{52}+z^{72}\right)+225290\cdot\left(z^{54}+z^{70}\right)+\\
& 322070\cdot\left(z^{56}+z^{68}\right)+301640\cdot\left(z^{58}+z^{66}\right)+316706\cdot\left(z^{60}+z^{64}\right)+\\
& 399752\cdot z^{62}
\end{aligned}$$

$$\begin{aligned}
A_{\mathscr{B}_{61}}^{(S_5)} = & \left(1+z^{124}\right)+3\cdot\left(z^{20}+z^{104}\right)+24\cdot\left(z^{26}+z^{98}\right)+48\cdot\left(z^{28}+z^{96}\right)+\\
& 6\cdot\left(z^{30}+z^{94}\right)+150\cdot\left(z^{32}+z^{92}\right)+8\cdot\left(z^{34}+z^{90}\right)+168\cdot\left(z^{36}+z^{88}\right)+\\
& 96\cdot\left(z^{38}+z^{86}\right)+75\cdot\left(z^{40}+z^{84}\right)+468\cdot\left(z^{42}+z^{82}\right)+132\cdot\left(z^{44}+z^{80}\right)+\\
& 656\cdot\left(z^{46}+z^{78}\right)+680\cdot\left(z^{48}+z^{76}\right)+300\cdot\left(z^{50}+z^{74}\right)+1386\cdot\left(z^{52}+z^{72}\right)+\\
& 198\cdot\left(z^{54}+z^{70}\right)+1152\cdot\left(z^{56}+z^{68}\right)+1272\cdot\left(z^{58}+z^{66}\right)+301\cdot\left(z^{60}+z^{64}\right)+\\
& 2136\cdot z^{62}
\end{aligned}$$

$$A_{\mathscr{B}_{61}}^{(S_{31})} = \left(1+z^{124}\right)+2\cdot z^{62}$$

$$A_{\mathscr{B}_{61}}^{(S_{61})} = \left(1+z^{124}\right)+2\cdot z^{62}$$

The weight distributions of $\mathscr{B}_{61}$ and their modular congruence are shown in Table 9.18.

Weight Distributions of Quadratic Residues Codes for Primes 151 and 167

See Tables 9.19 and 9.20

Table 9.19 Weight distributions of QR and extended QR codes of prime 151

i	A_i of [152, 76, 20] code	$\mathscr{A}_i$ of [151, 76, 19] code
0	1	1
19	0	3775
20	28690	24915
23	0	113250
24	717250	604000
27	0	30256625
28	164250250	133993625
31	0	8292705580
32	39390351505	31097645925
35	0	1302257122605
36	5498418962110	4196161839505
39	0	113402818847850
40	430930711621830	317527892773980
43	0	5706949034630250
44	19714914846904500	14007965812274250
47	0	171469716029462700
48	542987434093298550	371517718063835850
51	0	3155019195317144883
52	9222363801696269658	6067344606379124775
55	0	36274321608490644595
56	98458872937331749615	62184551328841105020
59	0	264765917968736096775
60	670740325520798111830	405974407552062015055
63	0	1241968201959417159800
64	2949674479653615754525	1707706277694198594725
67	0	3778485133479463579225
68	8446025592483506824150	4667540459004043244925
71	0	7503425412744902320620
72	15840564760239238232420	8337139347494335911800
75	0	9763682329503348632684
76	19527364659006697265368	9763682329503348632684

Table 9.20 Weight distributions of QR and extended QR codes of prime 167

i	A_i of [168, 84, 24] code	$\mathscr{A}_i$ of [167, 84, 23] code
0	1	1
23	0	110888
24	776216	665328
27	0	3021698
28	18130188	15108490
31	0	1057206192
32	5550332508	4493126316
35	0	268132007628
36	1251282702264	983150694636
39	0	39540857275985
40	166071600559137	126530743283152
43	0	3417107288264670
44	13047136918828740	9630029630564070
47	0	179728155397349776
48	629048543890724216	449320388493374440
51	0	5907921405841809432
52	19087130695796615088	13179209289954805656
55	0	124033230083117023704
56	372099690249351071112	248066460166234047408
59	0	1692604114105553659010
60	4739291519495550245228	3046687405389996586218
63	0	15228066033367763990128
64	39973673337590380474086	24745607304222616483958
67	0	91353417175290660468884
68	225696677727188690570184	134343260551898030101300
71	0	368674760966511746549004
72	860241108921860741947676	491566347955348995398672
75	0	1007629118755817710057646
76	2227390683565491780127428	1219761564809674070069782
79	0	1873856945935044844028880
80	3935099586463594172460648	2061242640528549328431768
83	0	2377873706297857672084688
84	4755747412595715344169376	2377873706297857672084688

References

1. Dougherty, T.G., Harada, M.: Extremal binary self-dual codes. IEEE Trans. Inf. Theory **43**(6), 2036–2047 (1997)
2. Gaborit, P.: Quadratic double circulant codes over fields. J. Comb. Theory Ser. A **97**, 85–107 (2002)
3. Gaborit, P., Otmani, A.: Tables of self-dual codes (2007). http://www.unilim.fr/pages_perso/philippe.gaborit/SD/index.html
4. Gaborit, P., Nedeloaia, C.S., Wassermann, A.: On the weight enumerators of duadic and quadratic residue codes. IEEE Trans. Inf. Theory **51**(1), 402–407 (2005)
5. Grassl, M.: On the minimum distance of some quadratic residue codes. In: Proceedings of the IEEE International Symposium on Inform. Theory, Sorento, Italy, p. 253 (2000)
6. Gulliver, T.A., Senkevitch, N.: On a class of self-dual codes derived from quadratic residue. IEEE Trans. Inf. Theory **45**(2), 701–702 (1999)
7. Huffman, W.C., Pless, V.S.: Fundamentals of Error-Correcting Codes. Cambridge University Press, Cambridge (2003) ISBN 0 521 78280 5
8. Jenson, R.: A double circulant presentation for quadratic residue codes. IEEE Trans. Inf. Theory **26**(2), 223–227 (1980)
9. Karlin, M.: New binary coding results by circulants. IEEE Trans. Inf. Theory **15**(1), 81–92 (1969)
10. Karlin, M., Bhargava, V.K., Tavares, S.E.: A note on the extended quadratic residue codes and their binary images. Inf. Control **38**, 148–153 (1978)
11. Leon, J.S.: A probabilistic algorithm for computing minimum weights of large error-correcting codes. IEEE Trans. Inf. Theory **34**(5), 1354–1359 (1988)
12. Leon, J.S., Masley, J.M., Pless, V.: Duadic codes. IEEE Trans. Inf. Theory **30**(5), 709–713 (1984)
13. MacWilliams, F.J., Sloane, N.J.A.: The Theory of Error-Correcting Codes. North-Holland, Amsterdam (1977)
14. Mykkeltveit, J., Lam, C., McEliece, R.J.: On the weight enumerators of quadratic residue codes. JPL Tech. Rep. 32-1526 **XII**, 161–166 (1972)
15. Rains, E.M., Sloane, N.J.A.: Self-dual codes. In: Pless, V.S., Huffman, W.C. (eds.) Handbook of Coding Theory. Elsevier, North Holland (1998)
16. Shannon, C.E.: Probability of error for optimal codes in a Gaussian channel. Bell. Syst. Tech. J. **38**(3), 611–656 (1959)
17. Tjhai, C.J.: A study of linear error correcting codes. Ph.D dissertation, University of Plymouth, UK (2007)
18. van Dijk, M., Egner, S., Greferath, M., Wassermann, A.: On two doubly even self-dual binary codes of length 160 and minimum weight 24. IEEE Trans. Inf. Theory **51**(1), 408–411 (2005)
19. van Lint, J.H.: Coding Theory. Lecture Notes in Mathematics vol. 201. Springer, Berlin (1970)
20. Zimmermann, K.H.: Integral hecke modules, integral generalized reed-muller codes, and linear codes. Technical Report, Technische Universität Hamburg-Harburg, Hamburg, Germany, pp. 3–96 (1996)

Chapter 10
Historical Convolutional Codes as Tail-Biting Block Codes

10.1 Introduction

In the late 1950s, a branch of error-correcting codes known as convolutional codes [1, 6, 11, 14] was explored almost independently of block codes and each discipline had their champions. For convolutional codes, sequential decoding was the norm and most of the literature on the subject was concerned with the performance of practical decoders and different decoding algorithms [2]. There were few publications on the theoretical analysis of convolutional codes. In contrast, there was a great deal of theory about linear, binary block codes and not a great deal about decoders, except for hard decision decoding of block codes. Soft decision decoding of block codes was considered to be quite impractical, except for trivial, very short codes.

With Andrew Viterbi's invention [13] of the maximum likelihood decoder in 1967, featuring a trellis based decoder, an enormous impetus was given to convolutional codes and soft decision decoding. Interestingly, the algorithm itself, for solving the travelling saleman's problem [12], had been known since 1960. Consequently, interest in hard decision decoding of convolutional codes waned in favour of soft decision decoding. Correspondingly, block codes were suddenly out of fashion except for the ubiquitous Reed–Solomon codes.

For sequential decoder applications, the convolutional codes used were systematic codes with one or more feedforward polynomials, whereas for applications using a Viterbi decoder, the convolutional codes were optimised for largest, minimum Hamming distance between codewords, d_{free}, for a given memory (the highest degree of the generator polynomials defining the code). The result is always a non-systematic code. It should be noted that in the context of convolutional codes, the minimum Hamming distance between codewords is understood to be evaluated over the constraint length, the memory of the code. This is traditionally called d_{min}. This is rather confusing when comparing the minimum Hamming distance of block codes with that of convolutional codes. A true comparison should compare the d_{free} of a convolutional code to the d_{min} of a block code, for a given code rate.

M. Tomlinson et al., *Error-Correction Coding and Decoding*,
Signals and Communication Technology,
DOI 10.1007/978-3-319-51103-0_10

Table 10.1 Best rate $\frac{1}{2}$ convolutional codes designed for Viterbi decoding

Memory	Generator Polynomial $r_1(x)$	Generator Polynomial $r_2(x)$	d_{free}
2	$1+x+x^2$	$1+x^2$	5
3	$1+x+x^2+x^3$	$1+x+x^3$	6
4	$1+x+x^2+x^4$	$1+x^3+x^4$	7
5	$1+x+x^2+x^3+x^5$	$1+x^2+x^4+x^5$	8
6	$1+x+x^2+x^3+x^6$	$1+x^2+x^3+x^5+x^6$	10
7	$1+x+x^2+x^3+x^4+x^7$	$1+x^2+x^5+x^6+x^7$	10
8	$1+x+x^2+x^3+x^5+x^7+x^8$	$1+x^2+x^3+x^4+x^8$	12

Since the early 1960s, a lot of work has been carried out on block codes and convolutional codes for applications in deep space communications, primarily because providing a high signal-to-noise ratio is so expensive. Error-correcting codes allowed the signal to noise ratio to be reduced.

The first coding arrangement implemented for space [6, 9] was part of the payload of Pioneer 9 which was launched into space in 1968. The payload featured a systematic, convolutional code designed by Lin and Lyne [7] with a d_{free} of 12 and memory of 20. The generator polynomial is

$$r(x) = 1+x+x^2+x^5+x^6+x^8+x^9+x^{12}+x^{13}+x^{14}+x^{16}+x^{17}+x^{18}+x^{19}+x^{20}.$$

This convolutional code was used with soft decision, sequential decoding featuring the Fano algorithm [2] to realise a coding gain of 3 dB. Interestingly, it was initially planned as a communications experiment and not envisaged to be used operationally to send telemetry data to Earth. However, its superior performance over the standard operational communications system which featured uncoded transmission meant that it was always used instead of the standard system.

In 1969, the Mariner'69 spacecraft was launched with a first order Reed–Muller (32, 6, 16) code [8] equivalent to the extended (32, 6, 16) cyclic code. A maximum likelihood correlation decoder was used. The coding gain was 2.2 dB [9].

By the mid 1970s, the standard for soft decision decoding on the AWGN channel notably applications for satellite communications and space communications was to use convolutional codes with Viterbi decoding, featuring the memory 7 code listed in Table 10.1. The generator polynomials are $r_1(x) = 1+x+x^2+x^3+x^6$ and $r_2(x) = 1+x^2+x^3+x^5+x^6$ convolutional code, best known, in octal representation, as the (171, 133) code. The best half rate convolutional codes designed to be used with Viterbi decoding [1, 6] are tabulated in Table 10.1.

The (171, 133) code with Viterbi soft decision decoding featured a coding gain of 5.1 dB at 10^{-5} bit error rate which was around 2 dB better than its nearest rival featuring a high memory convolutional code and hard decision, sequential decoding. The (171, 133) convolutional code is one of the recommended NASA Planetary Standard Codes [3].

However, more coding gain was achieved by concatenating the (171, 133) convolutional code with a (255, 233) Reed–Solomon (RS) code which is able to correct 16 symbol errors, each symbol being 8 bits. Quite a long interleaver needs to be used between the Viterbi decoder output and the RS decoder in order to break up the occasional error bursts which are output from the Viterbi decoder. Interleaver lengths vary from 4080 bits to 16320 bits and with the longest interleaver the coding gain of the concatenated arrangement is 7.25 dB, ($\frac{E_b}{N_0} = 2.35$ dB at 10^{-5} bit error rate), and it is a CCSDS [3] standard for space communications.

10.2 Convolutional Codes and Circulant Block Codes

It is straightforward to show that a double-circulant code is a half rate, tail-biting, feedforward convolutional code. Consider the Pioneer 9, half rate, convolutional code invented by Lin and Lyne [7] with generator polynomial

$$r(x) = 1+x+x^2+x^5+x^6+x^8+x^9+x^{12}+x^{13}+x^{14}+x^{16}+x^{17}+x^{18}+x^{19}+x^{20}$$

For a semi-infinite data sequence defined by $d(x)$, the corresponding codeword, $c(x)$, of the convolutional code consists of

$$c(x) = d(x)\|d(x)r(x) \tag{10.1}$$

where $\|$ represents interlacing of the data polynomial representing the data sequence and the parity polynomial representing the sequence of parity bits.

The same generator polynomial can be used to define a block code of length $2n$, a $(2n, n)$ double-circulant code with a codeword consisting of

$$c(x) = d(x)\|d(x)r(x) \text{ modulo } (1 + x^n) \tag{10.2}$$

(Double-circulant codewords usually consist of one circulant followed by the second but it is clear that an equivalent code is obtained by interlacing the two circulants instead.)

While comparing Eq. (10.1) with (10.2) as $n \to \infty$, it can be seen that the same codewords will be obtained. For finite n, it is apparent that the tail of the convolution of $d(x)$ and $r(x)$ will wrap around adding to the beginning as in a tail-biting convolutional code. It is also clear that if n is sufficiently long, only the Hamming weight of long convolutions, will be affected by the wrap around and these long convolution results will be of high Hamming weight anyway leading to the conclusion that if n is sufficiently long the d_{min} of the circulant code will be the same as the d_{free} of the convolutional code. Indeed, the low weight spectral terms of the two codes will be identical, as is borne out by codeword enumeration using the methods described in Chap. 5.

For the Pioneer 9 code, having a d_{free} of 12, a double-circulant code with d_{min} also equal to 12 can be obtained with n as low as 34, producing a (68, 34, 12) code. It is noteworthy that this is not a very long code, particularly by modern standards.

Codewords of the double-circulant code are given by

$$c(x) = d(x)|d(x)(1 + x + x^2 + x^5 + x^6 + x^8 + x^9 + x^{12} + x^{13} + x^{14} + x^{16} + x^{17} + x^{18} + x^{19} + x^{20}) \text{ modulo } (1 + x^{34}) \tag{10.3}$$

As a double-circulant block code, this code can be soft decision decoded, with near maximum likelihood decoding using an extended Dorsch decoder, described in Chap. 15. The results for the AWGN channel are shown plotted in Fig. 10.1. Also plotted in Fig. 10.1 are the results obtained with the same convolutional code realised as a (120, 60, 12) double-circulant code which features less wrap around effects compared to the (68, 34, 12) code.

Using the original sequential decoding with 8 level quantisation of the soft decisions realised a coding gain of 3 dB at a BER of 5×10^{-4}. Using the modified Dorsch decoder with this code can realise a coding gain of over 5 dB at a BER of 5×10^{-4} and over 6 dB at a BER of 10^{-6} as is evident from Fig. 10.1. Moreover, there is no need for termination bits with the tail-biting arrangement. However, it should be noted that the state of the art, modified Dorsch decoder with soft decision decoding

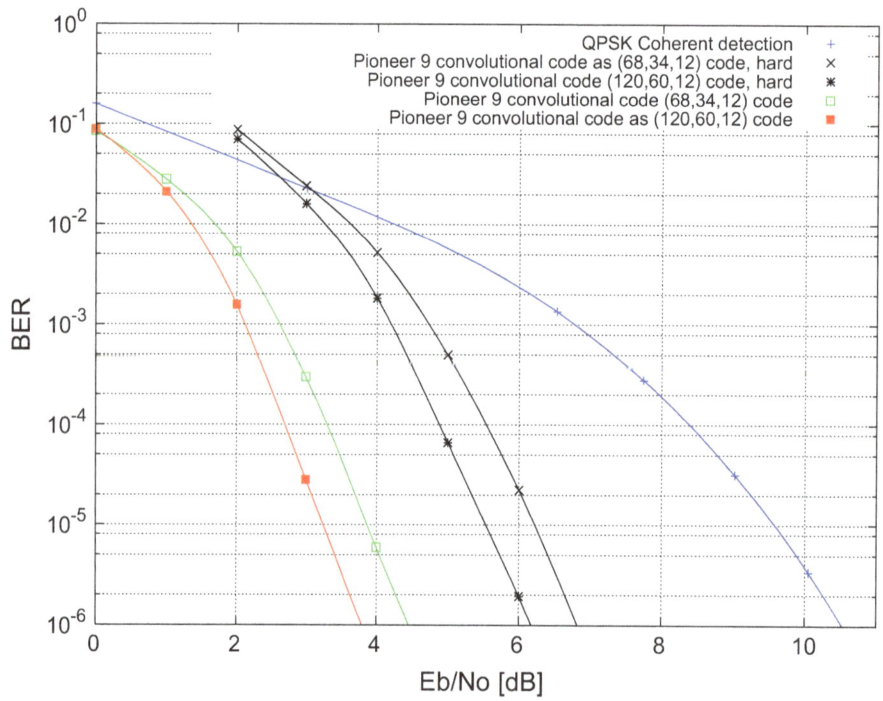

Fig. 10.1 BER performance of the Pioneer 9 convolutional code encoded as a (68, 34, 12) or (120, 60, 12) double-circulant code with soft and hard decision, extended Dorsch decoding in comparison to uncoded QPSK

needs to evaluate up to 500,000 codewords per received vector for the (68, 34, 12) double-circulant code realisation and up to 1,000,000 codewords per received vector for the (120, 60, 12) double-circulant code version in order to achieve near maximum likelihood decoding. Figure 10.1 also shows the hard decision decoding performance realised with the modified, hard decision Dorsch decoder, also described in Chap. 15. The (120, 60, 12) double-circulant code version, has a degradation of 2.3 dB at 10^{-4} BER compared to soft decision decoding, but still achieves a coding gain of 3.3 dB at 10^{-4} BER. Similarly, the (68, 34, 12) double-circulant code version, has a degradation of 2.2 dB at 10^{-4} BER compared to soft decision decoding, but still achieves a coding gain of 2.3 dB at 10^{-4} BER.

The conclusion to be drawn from Fig. 10.1 is that the Pioneer 9 coding system was limited not by the design of the code but by the design of the decoder. However to be fair, the cost of a Dorsch decoder would have been considered beyond reach back in 1967.

It is interesting to discuss the differences in performance between the (68, 34, 12) and (120, 60, 12) double-circulant code versions of the Pioneer 9 convolutional code. Both have a d_{min} of 12. However the number of weight 12 codewords, the multiplicities of weight 12 codewords of the codes' weight distributions, is higher for the (68, 34, 12) double-circulant code version due to the wrap around of the second circulant which is only of length 34. The tails of the circulants of codewords of higher weight than 12 do suffer some cancellation with the beginning of the circulants. In fact, exhaustive weight spectrum analysis, (see Chaps. 5 and 13 for description of the different methods that can be used), shows that the multiplicity of weight 12 codewords is 714 for the (68, 34, 12) code and only 183 for the (120, 60, 12) code.

Moreover, the covering radius of the (68, 34, 12) code has been evaluated and found to be 10 indicating that this code is well packed, whereas the covering radius of the (120, 60, 12) code is much higher at 16 indicating that the code is not so well packed. Indeed the code rate of the (120, 60, 12) code can be increased without degrading the minimum Hamming distance because with a covering radius of 16 at least one more information bit may be added to the code.

With maximum likelihood, hard decision decoding, which the modified Dorsch decoder is able to achieve, up to 10 hard decision errors can be corrected with the (68, 34, 12) code in comparison with up to 16 hard decision errors correctable by the (120, 60, 12) code. Note that these are considerably higher numbers of correctable errors in both cases than suggested by the d_{free} of the code (only five hard decision errors are guaranteed to be correctable). This is a recurrent theme for maximum likelihood, hard decision decoding of codes, as discussed in Chap. 3, compared to bounded distance decoding.

It is also interesting to compare the performance of other convolutional codes that have been designed for space applications and were originally intended to be used with sequential decoding. Of course now we have available the far more powerful (and more signal processing intensive) modified Dorsch decoder, which can be used with any linear code.

Massey and Costello [6, 10] constructed a rate $\frac{1}{2}$, memory 31 non-systematic code which was more powerful than any systematic code with the same memory and had the useful property that the information bits could be obtained from the two convolutionally encoded parity streams just by adding them together, modulo 2. The necessary condition for this property is that the two generator polynomials differ only in a single coefficient. The two generator polynomials, $r_0(x)$ and $r_1(x)$ may be described by the exponents of the non-zero coefficients:

$$r_0(x) \leftarrow \{0, 1, 2, 4, 5, 7, 8, 9, 11, 13, 14, 16, 17, 18, 19, 21, 22, 23, 24, 25, 27, 28, 29, 31\}$$
$$r_1(x) \leftarrow \{0, 2, 4, 5, 7, 8, 9, 11, 13, 14, 16, 17, 18, 19, 21, 22, 23, 24, 25, 27, 28, 29, 31\}$$

As can be seen the two generator polynomials differ only in the coefficient of x. This code has a d_{free} of 23 and can be realised as a double-circulant (180, 90, 23) code from the tail-biting version of the same convolutional code. This convolutional code has exceptional performance and in double-circulant form it, of course, may be decoded using the extended Dorsch decoder. The performance of the code in (180, 90, 23) form, for the soft decision and hard decision AWGN channel, is shown in Fig. 10.2. For comparison purposes, the performances of the Pioneer 9 codes are also shown in Fig. 10.2. Shorter double-circulant code constructions are possible from this convolutional code in tail-biting form, without compromising the d_{min} of the double-circulant code. The shortest version is the (166, 83, 23) double-circulant code.

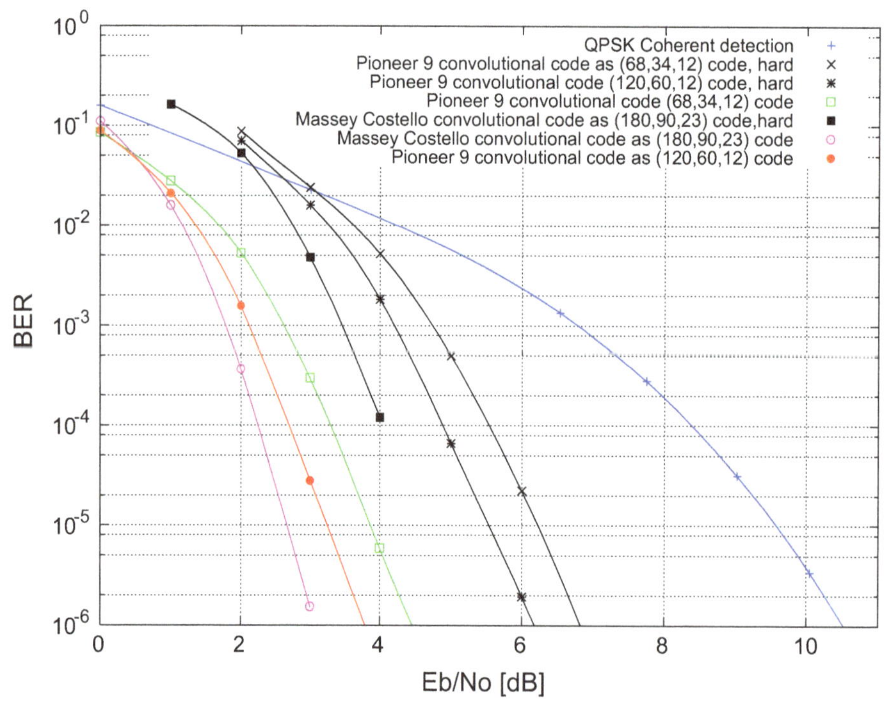

Fig. 10.2 BER performance of the Massey Costello convolutional code in (180, 90, 23) double-circulant code form for the AWGN channel, using soft and hard decisions, with extended Dorsch decoding

By truncating the generator polynomials, $r_0(x)$ and $r_1(x)$ above, a reduced memory convolutional code with memory 23 and d_{free} of 17 can be obtained as discussed by Massey and Lin [6, 10] which still has the non-systematic, quick decoding property. The generator polynomials are given by the exponents of the non-zero coefficients:

$$\hat{r}_0(x) \leftarrow \{0, 1, 2, 4, 5, 7, 8, 9, 11, 13, 14, 16, 17, 18, 19, 21, 22, 23\}$$
$$\hat{r}_1(x) \leftarrow \{0, 2, 4, 5, 7, 8, 9, 11, 13, 14, 16, 17, 18, 19, 21, 22, 23\}$$

A (160, 80, 17) double-circulant code can be obtained from the tail-biting version of this convolutional code. In fact, many double-circulant codes with high d_{min} can be obtained from tail-biting versions of convolutional codes.

It is straightforward to write a program in C++ which searches for the generator polynomials that produce the convolutional codes with the highest values of d_{free}. The only other constraint is that the the generator polynomials need to be relatively prime to each other, that is, the GCD of the generator polynomials needs to be 1 in order to avoid a catastrophic code [6]. However, it is also necessary in selecting the generator polynomials that the wrap around effects of the circulants are taken into account otherwise the d_{min} of the double-circulant code is not as high as the d_{free} of the convolutional code from which it is derived. Indeed to construct a good code in this way with high d_{free} and high d_{min}, it has to be constructed as a tail-biting convolutional code right from the start. One example of a good tail-biting convolutional code that has been found in this way has generator polynomials, $r_0(x)$ and $r_1(x)$ given by the exponents of the non-zero coefficients:

$$r_0(x) \leftarrow \{0, 2, 5, 8, 9, 10, 12, 13, 14, 15, 27\}$$
$$r_1(x) \leftarrow \{0, 1, 2, 3, 4, 5, 7, 8, 11, 12, 16, 18, 20, 23, 27\}$$

This code has a memory of 27 and a d_{free} of 26. It may be realised in double-circulant form as a (180, 90, 26) double-circulant code and weight spectrum analysis shows that this code has the same d_{min} of 26 as the best-known code with the same code parameters [4]. The two polynomials $r_0(x)$ and $r_1(x)$ factorise into polynomials with the following exponents of the non-zero coefficients:

$$r_0(x) \leftarrow \{0, 3, 5\}\{0, 2, 3, 5, 6, 7, 8, 10, 13, 14, 16, 17, 18, 20, 22\}$$
$$r_1(x) \leftarrow \{0, 3, 5, 6, 8\}\{0, 1, 3, 4, 5, 6, 8\}\{0, 2, 4, 7, 11\}$$

It can be seen that neither polynomial has a common factor and so the GCD is 1. Correspondingly, the convolutional code is not a catastrophic code.

As well as constructing double-circulant codes from convolutional codes, double-circulant codes may be used to construct good convolutional codes. The idea of generating convolutional codes from good block codes is not that new. Massey et al. in 1973 generated a convolutional code for space communications from a (89, 44, 18) quadratic residue cyclic code [5, 6]. As described in Chap. 9, prime numbers which are congruent to ± 3 modulo 8 may be used to generate double-circulant codes using the quadratic residues to construct one circulant, the other circulant being the identity circulant; the length of the circulants are equal to the prime number.

Particularly, good double-circulant codes are obtained in this way as discussed in Chap. 9. For example, the prime number 67 can be used to generate a (134, 67, 23) double-circulant code with the circulants defined by the two polynomials with the following exponents of the non-zero coefficients:

$$r_0(x) \leftarrow \{0\}$$
$$r_1(x) \leftarrow \{0, 1, 4, 6, 9, 10, 14, 15, 16, 17, 19, 21, 22, 23, 24, 25, 26, 29, 33, 35, 36, 37, 39, 40, 47, 49, 54, 55, 56, 59, 60, 62, 64, 65\}$$

Using these two polynomials as the generator polynomials for a $\frac{1}{2}$ rate convolutional code, a systematic convolutional code having a d_{free} of 30 is obtained. Interestingly, deriving another double-circulant code from the tail-biting version of this convolutional code only produces good results when the circulants are exactly equal to 67, thereby reproducing the original code. For longer circulants, the d_{min} is degraded unless the circulants are much longer. It is found that the circulants have to be as long as 110 to produce a (220, 110, 30) double-circulant code having a d_{min} equal to that of the original convolutional code. Moreover, this is a good code because the code has the same parameters as the corresponding best-known code [4].

A double-circulant code may also be used to derive a non-systematic convolutional code with much smaller memory and a d_{free} equal to the d_{min} of the double-circulant code by selecting a codeword of the double-circulant code which features low-degree polynomials in each circulant. It is necessary to check that these polynomials are relatively prime otherwise a catastrophic convolutional code is produced. In this event a new codeword is selected. The code produced is a non-systematic convolutional code with memory equal to the highest degree of the two circulant polynomials. For example, a memory 41 non-systematic convolutional code can be derived from a memory 65, systematic convolutional code based on the (134, 67, 23) double-circulant code with the following exponents of the non-zero coefficients:

$$r_0(x) \leftarrow \{0\}$$
$$r_1(x) \leftarrow \{0, 1, 4, 6, 9, 10, 14, 15, 16, 17, 19, 21, 22, 23, 24, 25, 26, 29, 33, 35\ 36, 37, 39, 40, 47, 49, 54, 55, 56, 59, 60, 62, 64, 65\}$$

Codeword analysis of the double-circulant code is carried out to find the low memory generator polynomials. The following two generator polynomials were obtained from the two circulant polynomials making up a weight 23 codeword of the (134, 67, 23) code:

$$r_0(x) \leftarrow \{0, 1, 2, 4, 5, 10, 12, 32, 34, 36, 39, 41\}$$
$$r_1(x) \leftarrow \{0, 2, 4, 13, 19, 24, 25, 26, 33, 35, 37\}$$

In another example, the outstanding (200, 100, 32) extended cyclic quadratic residue code may be put in double-circulant form using the following exponents of the non-zero coefficients:

$$r_0(x) \leftarrow \{0\}$$
$$r_1(x) \leftarrow \{0, 1, 2, 5, 6, 8, 9, 10, 11, 15, 16, 17, 18, 19, 20, 26, 27, 28, 31, 34, 35, 37, 38, 39, 42, 44, 45, 50, 51, 52, 53, 57, 58, 59, 64, 66, 67, 70, 73, 75, 76, 77, 80, 82, 85, 86, 89, 92, 93, 97, 98\}$$

Enumeration of the codewords shows that there is a weight 32 codeword that defines the generator polynomials of a memory 78, non-systematic convolutional code. The codeword consists of two circulant polynomials, the highest degree of which is 78. The generator polynomials have the following exponents of the non-zero coefficients:

$$r_0(x) \leftarrow \{0, 2, 3, 8, 25, 27, 37, 44, 50, 52, 55, 57, 65, 66, 67, 69, 74, 75, 78\}$$
$$r_1(x) \leftarrow \{0, 8, 14, 38, 49, 51, 52, 53, 62, 69, 71, 72, 73\}$$

The non-systematic convolutional code that is produced has a d_{free} of 32 equal to the d_{min} of the double-circulant code. Usually, it is hard to verify high values of d_{free} for convolutional codes, but in this particular case, as the convolutional code has been derived from the (200, 100, 32) extended quadratic residue, double-circulant code which is self-dual and also fixed by the large projective special linear group $PSL_2(199)$ the d_{min} of this code has been proven to be 32 as described in Chap. 9. Thus, the non-systematic convolutional code that is produced has to have a d_{free} of 32.

10.3 Summary

Convolutional codes have been explored from a historical and modern perspective. Their performance, as traditionally used, has been compared to the performance realised using maximum likelihood decoding featuring an extended Dorsch decoder with the convolutional codes implemented as tail-biting block codes. It has been shown that the convolutional codes designed for space applications and sequential decoding over 40 years ago were very good codes, comparable to the best codes known today. The performance realised back then was limited by the sequential decoder as shown by the presented results. An additional 2 dB of coding gain could have been realised using the modern, extended Dorsch decoder instead of the sequential decoder. However back then, this decoder had yet to be discovered and was probably too expensive for the technology available at the time.

It has also been shown that convolutional codes may be used as the basis for designing double-circulant block codes and vice versa. In particular, high, guaranteed values of d_{free} may be obtained by basing convolutional codes on outstanding double-circulant codes. A memory 78, non-systematic, half rate convolutional code with a d_{free} of 32 was presented based on the (200, 100, 32) extended quadratic residue, double-circulant code.

References

1. Clark, G.C., Cain, J.B.: Error-Correction Coding for Digital Communications. Plenum Publishing Corporation, New York (1981). ISBN 0 306 40615 2
2. Fano, R.: A heuristic discussion of probabilistic decoding. IEEE Trans. Inf. Theory **IT–9**, 64–74 (1963)
3. Gaborit, P., Otmani, A.: Tm synchronization and channel coding. CCSDS 131.0-B-1 BLUE BOOK (2003)
4. Grassl, M.: Code tables: bounds on the parameters of various types of codes (2007). http://www.codetables.de
5. Massey, J.L., Costello Jr., D.J., Justesen, J.: Polynomial weights and code constructions. IEEE Trans. Inf. Theory **IT–19**, 101–110 (1973)
6. Lin, S., Costello Jr., D.J.: Error Control Coding: Fundamentals and Applications, 2nd edn. Pearson Education Inc., Upper Saddle River (2004)
7. Lin, S., Lyne, H.: Some results on binary convolutional code generators. IEEE Trans. Inf. Theory **IT–13**, 134–139 (1967)
8. MacWilliams, F.J., Sloane, N.J.A.: The Theory of Error-Correcting Codes. North-Holland, Amsterdam (1977)
9. Massey, J.: Deep-space communications and coding: a marriage made in heaven. In: Hagenauer, J. (ed.) Advanced Methods for Satellite and Deep Space Communications. Lecture Notes in Control and Information Sciences 182. Springer, Heidelberg (1992)
10. Massey, J.L., Costello Jr., D.J.: Nonsystematic convolutional codes for sequential decoding in space applications. IEEE Trans. Commun. **COM–19**, 806–813 (1971)
11. Peterson, W.: Error-Correcting Codes. MIT Press, Cambridge (1961)
12. Pollack, M., Wiebenson, W.: Solutions of the shortest route problem - a review. Oper. Res. **8**, 224–230 (1960)
13. Viterbi, A.J.: Error bounds for convolutional codes and an asymptotically optimum decoding algorithm. IEEE Trans. Inf. Theory **IT–13**, 260–269 (1967)
14. Wozencraft, J.: Sequential decoding for reliable communications. Technical report No. 325 Research Laboratory of Electronics, MIT (1957)

Chapter 11
Analogue BCH Codes and Direct Reduced Echelon Parity Check Matrix Construction

11.1 Introduction

Analogue error-correcting codes having real and complex number coefficients were first discussed by Marshall [2]. Later on Jack Wolf [3] introduced Discrete Fourier Transform (DFT) codes having complex number coefficients and showed that an (n, k) DFT code can often correct up to $n - k - 1$ errors using a majority voting type of decoder. The codes are first defined and it is shown that (n, k) DFT codes have coordinate coefficients having complex values. These codes have a minimum Hamming distance of $n - k + 1$ and are Maximum Distance Separable (MDS) codes. The link between the Discrete Fourier Transform and the Mattson–Solomon polynomial is discussed and it is shown that the parity check algorithm used to generate DFT codes can be applied to all BCH codes including Reed–Solomon codes simply by switching from complex number arithmetic to Galois Field arithmetic. It is shown that it is straightforward to mix together quantised and non-quantised codeword coefficients which can be useful in certain applications. Several worked examples are described including that of analogue error-correction encoding and decoding being applied to stereo audio waveforms (music).

In common with standard BCH or Reed–Solomon (RS) codes, it is shown that parity check symbols may be calculated for any $n - k$ arbitrary positions in each codeword and an efficient method is described for doing this. A proof of the validity of the method is given.

11.2 Analogue BCH Codes and DFT Codes

In a similar manner to conventional BCH codes, a codeword of an analogue (n, k) BCH code is defined as

$$c(x) = c_0 + c_1x + c_2x^2 + c_3x^3 + c_4x^4 + c_5x^5 + \cdots + c_{n-1}x^{n-1}$$

M. Tomlinson et al., *Error-Correction Coding and Decoding*,
Signals and Communication Technology,
DOI 10.1007/978-3-319-51103-0_11

where

$$c(x) = g(x)d(x)$$

$g(x)$ is the generator polynomial of the code with degree $n-k$ and $d(x)$ is any data polynomial of degree less than k. Correspondingly,

$$g(x) = g_0 + g_1x + g_2x^2 + \cdots + g_{n-k}x^{n-k}$$

and

$$d(x) = d_0 + d_1x + d_2x^2 + \cdots + d_{k-1}x^{k-1}$$

The coefficients of $c(x)$ are complex numbers from the field of complex numbers. A parity check polynomial $h(x)$ is defined, where

$$h(x) = h_0 + h_1x + h_2x^2 + h_3x^3 + h_4x^4 + h_5x^5 + \cdots + h_{n-1}x^{n-1}$$

where

$$h(x)g(x) \bmod (x^n - 1) = 0$$

and accordingly,

$$h(x)c(x) \bmod (x^n - 1) = 0$$

The generator polynomial and the parity check polynomial may be defined in terms of the Discrete Fourier Transform or equivalently by the Mattson–Solomon polynomial.

Definition 11.1 (*Definition of Mattson–Solomon polynomial*) The Mattson–Solomon polynomial of any polynomial $a(x)$ is the linear transformation of $a(x)$ to $A(z)$ and is defined by [1],

$$A(z) = \text{MS}(a(x)) = \sum_{i=0}^{n-1} a(\alpha^{-i})\, z^i \tag{11.1}$$

The inverse Mattson–Solomon polynomial or inverse Fourier transform is:

$$a(x) = \text{MS}^{-1}(A(z)) = \frac{1}{n}\sum_{i=0}^{n-1} A(\alpha^i)\, x^i \tag{11.2}$$

α is a primitive root of unity with order n and for analogue BCH codes

$$\alpha = e^{j\frac{2\pi}{n}} \tag{11.3}$$

where $j = (-1)^{\frac{1}{2}}$.

In terms of a narrow sense, primitive BCH code with a generator polynomial of $g(x)$, the coefficients of G(z) are all zero from z^0 through z^{n-k-1} and the coefficients of H(z) are all zero from z^{n-k} through z^{n-1}. Consequently, it follows that the coefficient by coefficient product of $G(z)$ and $H(z)$ represented by $\odot$

$$G(z) \odot H(z) = \sum_{j=0}^{n-1} (G_j \odot H_j)\, z^j = 0 \tag{11.4}$$

The nonzero terms of $H(z)$ extend from z^0 through to z^{n-k-1} and a valid parity check matrix in the well known form is:

$$\mathbf{H} = \begin{bmatrix} 1 & 1 & 1 & \ldots & 1 \\ 1 & \alpha^1 & \alpha^2 & \ldots & \alpha^{n-1} \\ 1 & \alpha^2 & \alpha^4 & \ldots & \alpha^{2(n-1)} \\ 1 & \alpha^3 & \alpha^6 & \ldots & \alpha^{3(n-1)} \\ \ldots & \ldots & \ldots & \ldots & \ldots \\ 1 & \alpha^{n-k-1} & \alpha^{2(n-k-1)} & \ldots & \alpha^{n-k-1(n-1)} \end{bmatrix}$$

It will be noticed that each row of this matrix is simply given by the inverse Mattson–Solomon polynomial of $H(z)$, where

$$\begin{aligned} H(z) &= 1 \\ H(z) &= z \\ H(z) &= z^2 \\ H(z) &= \ldots \\ H(z) &= z^{n-k-1} \end{aligned} \tag{11.5}$$

Consider $H(z) = z - \alpha^i$, the inverse Mattson–Solomon polynomial produces a parity check equation defined by

$$1 - \alpha^i \;\; \alpha^1 - \alpha^i \;\; \alpha^2 - \alpha^i \;\ldots 0 \ldots\; \alpha^{n-1} - \alpha^i$$

Notice that this parity check equation may be derived from linear combinations of the first two rows of **H** by multiplying the first row by α^i before subtracting it from the second row of **H**. The resulting row may be conveniently represented by

$$\alpha^{a_0} \; \alpha^{a_1} \; \alpha^{a_2} \; \alpha^{a_3} \ldots 0 \ldots \alpha^{a_{n-2}} \; \alpha^{a_{n-1}}$$

It will be noticed that the ith coordinate of the codeword is multiplied by zero, and hence the parity symbol obtained by this parity check equation is independent of the value of the ith coordinate. Each one of the other coordinates is multiplied by a nonzero value. Hence any one of these $n-1$ coordinates may be solved using this parity check equation in terms of the other $n-2$ coordinates involved in the equation.

Similarly, considering $H(z)=z-\alpha^j$, the inverse Mattson–Solomon polynomial produces a parity check equation defined by

$$1-\alpha^j \ \alpha^1-\alpha^j \ \alpha^2-\alpha^j \ \ldots 0 \ldots \alpha^{n-1}-\alpha^j$$

and this may be conveniently represented by

$$\alpha^{b_0} \ \alpha^{b_1} \ \alpha^{b_2} \ \alpha^{b_3} \ \ldots 0 \ldots \alpha^{b_{n-2}} \ \alpha^{b_{n-1}}$$

Now the jth coordinate is multiplied by zero and hence the parity symbol obtained by this parity check equation is independent of the value of the jth coordinate.

Developing the argument, if we consider $H(z)=(z-\alpha^i)(z-\alpha^j)$, the inverse Mattson–Solomon polynomial produces a parity check equation defined by

$$\alpha^{a_0}\alpha^{b_0} \ \alpha^{a_1}\alpha^{b_1} \ \ldots 0 \ldots 0 \ldots \alpha^{a_{n-1}}\alpha^{b_{n-1}}$$

This parity check equation has zeros in the ith and j^{th} coordinate positions and as each one of the other coordinates is multiplied by a nonzero value, any one of these $n-2$ coordinates may be solved using this parity check equation in terms of the other $n-3$ coordinates involved in the equation.

Proceeding in this way, for $H(z)=(z-\alpha^i)(z-\alpha^j)(z-\alpha^k)$, the inverse Mattson–Solomon polynomial produces a parity check equation which is independent of the ith, j^{th} and k^{th} coordinates and these coordinate positions may be arbitrarily chosen. The parity check matrix is

$$\mathbf{H_m}=\begin{bmatrix} 1 & 1 & 1 & 1 & 1 & 1 \ldots & 1 \\ \alpha^{u_0} & \alpha^{u_1} & \alpha^{u_2} & \alpha^{u_3} & \alpha^{u_4} & 0 \ldots & \alpha^{u_{n-1}} \\ \alpha^{v_0} & 0 & \alpha^{v_2} & \alpha^{v_3} & \alpha^{v_4} & 0 \ldots & \alpha^{v_{n-1}} \\ \alpha^{w_0} & 0 & \alpha^{w_2} & 0 & \alpha^{w_4} & 0 \ldots & \alpha^{w_{n-1}} \end{bmatrix}$$

The point here is that this parity check matrix $\mathbf{H_m}$ has been obtained from linear combinations of the original parity check matrix $\mathbf{H}$ and all parity check equations from either $\mathbf{H}$ or $\mathbf{H_m}$ are satisfied by codewords of the code.

The parity check matrix $\mathbf{H_m}$ may be used to solve for 4 parity check symbols in 4 arbitrary coordinate positions defined by the ith, j^{th} and k^{th} coordinate positions plus any one of the other coordinate positions which will be denoted as the lth position. The coordinate value in the lth position is solved first using the last equation. Parity symbols in the ith, j^{th} and k^{th} positions are unknown but this does not matter as these are multiplied by zero. The third parity check equation is used next to solve for the parity symbol in the kth position. Then, the second parity check equation is used to solve for the parity symbol in the jth position and lastly, the first parity check equation is used to solve for the parity symbol in the ith position. The parity check matrix values, for $s=0$ through to $n-1$, are given by:

$$\begin{aligned}
\alpha^{u_s} &= \alpha^s - \alpha^i \\
\alpha^{v_s} &= (\alpha^s - \alpha^i)(\alpha^s - \alpha^j) = \alpha^{u_s}(\alpha^s - \alpha^j) \\
\alpha^{w_s} &= (\alpha^s - \alpha^i)(\alpha^s - \alpha^j)(\alpha^s - \alpha^k) = \alpha^{v_s}(\alpha^s - \alpha^k)
\end{aligned}$$

Codewords of the code may be produced by first deciding on the number of parity check symbols and their positions and then constructing the corresponding parity check matrix $\mathbf{H_m}$. From the information symbols, the parity check symbols are calculated by using each row of $\mathbf{H_m}$ starting with the last row as described above.

In the above, there are 4 parity check rows and hence 4 parity check symbols which can be in any positions of the code. Clearly, the method can be extended to any number of parity check symbols. Any length of code may be produced by simply assuming coordinates are always zero, eliminating these columns from the parity check matrix. The columns of the parity check matrix may also be permuted to any order but the resulting code will not be cyclic.

It follows that with the $n-k$ parity check equations constructed using the method above, codeword coordinates may be solved in any of $n-k$ arbitrary positions. In the construction of each parity check equation there is exactly one additional zero compared to the previously constructed parity check equation. Hence there are $n-k$ independent parity check equations in any of $n-k$ arbitrary positions.

Since these equations are all from the same code the minimum Hamming distance of the code is $n-k+1$ and the code is MDS. A system for the calculation of parity check symbols in arbitrary positions may be used for encoding or for the correction of erasures. A block diagram of such an encoder/erasures decoder is shown in Fig. 11.1.

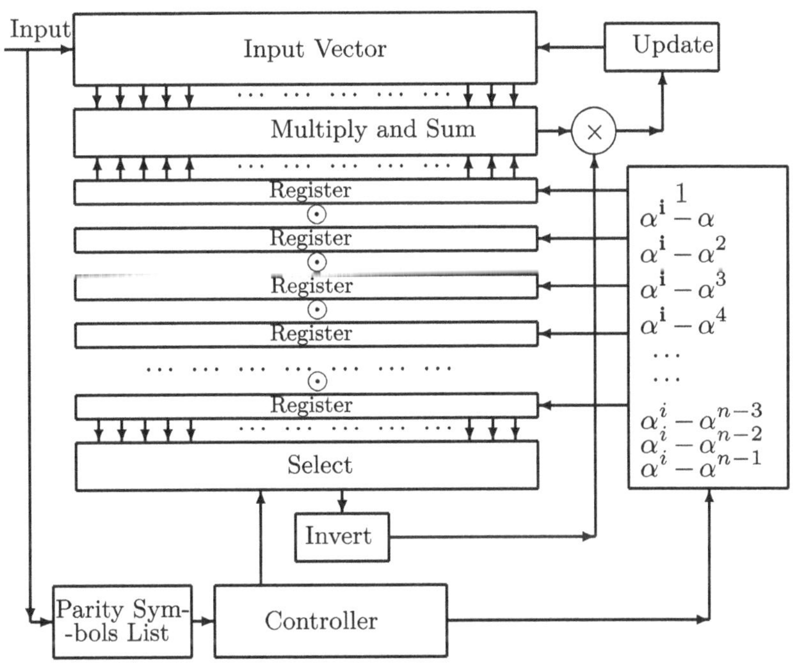

Fig. 11.1 The efficient encoder/erasures decoder realisation for BCH codes

When operating as an erasures decoder in Fig. 11.1, the List of Parity Symbols is replaced with a list of the erasures positions.

11.3 Error-Correction of Bandlimited Data

In many cases, the sampled data to be encoded with the analogue BCH code is already bandlimited or near bandlimited in which case, the higher frequency coefficients of the Mattson–Solomon polynomial, $D(z)$ of the data polynomial $d(x)$, consisting of successive PAM samples, will be zero or near zero. An important point here is that there is no need to add additional redundancy with additional parity check samples. In a sense the data, as PAM samples, already contains the parity check samples. Commonly, it is only necessary to modify a small number of samples to turn the sampled data into codewords of the analogue BCH code as illustrated in the example below. The broad sense BCH codes are used with the following parity check matrix, with $\alpha = e^{-j\frac{2\pi}{n}}$.

$$\mathbf{H_f} = \begin{bmatrix} 1 & \alpha^{\beta} & \alpha^{2\beta} & \ldots & \alpha^{(n-1)\beta} \\ 1 & \alpha^{\beta+1} & \alpha^{2(\beta+1)} & \ldots & \alpha^{((n-1)(\beta+1)} \\ 1 & \alpha^{\beta+2} & \alpha^{2(\beta+2)} & \ldots & \alpha^{(n-1)(\beta+2)} \\ 1 & \alpha^{\beta+3} & \alpha^{2(\beta+3)} & \ldots & \alpha^{(n-1)(\beta+3)} \\ \ldots & \ldots & \ldots & \ldots & \ldots \\ 1 & -1 & 1 & \ldots & 1 \end{bmatrix} \tag{11.6}$$

Using this parity check matrix will ensure that the highest $n-k$ Fourier coefficients will be zero. Several alternative procedures may be used. $n-k$ samples in each sequence of n samples may be designated as parity symbols and solved using this parity check matrix following the procedure above for constructing the reduced echelon matrix so that the values of the designated parity samples may be calculated. An alternative, more complicated procedure, is for each constructed codeword, to allow the $n-k$ parity samples to be in any of the $\frac{n!}{k!(n-k)!}$ combinations of positions and choose the combination which produces the minimum mean squared differences compared to the original $n-k$ complex samples.

11.4 Analogue BCH Codes Based on Arbitrary Field Elements

It is not necessary that the parity check matrix be based on increasing powers of α with parity check equations corresponding to the forcing of Fourier coefficients to be zero. An arbitrary ordering of complex field elements corresponding to permuted powers of α may be used. With $\alpha = e^{-j\frac{2\pi}{N}}$ where $N \geq n$, consider the parity check matrix

$$\mathbf{H_a} = \begin{bmatrix} \alpha_0 & \alpha_0 & \alpha_0 & \alpha_0 & \dots & \alpha_0 \\ \alpha_0 & \alpha_1 & \alpha_2 & \alpha_3 & \dots & \alpha_{n-1} \\ \alpha_0 & \alpha_1^2 & \alpha_2^2 & \alpha_3^2 & \dots & \alpha_{n-1}^2 \\ \alpha_0 & \alpha_1^3 & \alpha_2^3 & \alpha_3^3 & \dots & \alpha_{n-1}^3 \\ \dots & \dots & \dots & \dots & \dots & \dots \\ \alpha_0 & \alpha_1^{n-k-1} & \alpha_2^{n-k-1} & \alpha_3^{n-k-1} & \dots & \alpha_{n-1}^{n-k-1} \end{bmatrix}$$

The $\{\alpha_0, \alpha_1, \alpha_2, \alpha_3, \dots, \alpha_{n-1}\}$ complex number field elements are all distinct and arbitrary powers of α. Any combination of any $n-k$ columns, or less, of this parity check matrix are independent because the matrix transpose is a Vandermonde matrix [1]. Consequently, the code is a $(n, k, n-k+1)$ MDS code.

Following the same procedure as outlined above to produce directly a reduced echelon parity check matrix $\mathbf{H_b}$ with zeros in arbitrary columns, for example in three columns headed by, α_a, α_b and α_c.

$$\mathbf{H_b} = \begin{bmatrix} \alpha_0 & \alpha_0 & \alpha_0 & \alpha_0 & \dots & \alpha_0 \\ (\alpha_0-\alpha_a) & 0 & (\alpha_2-\alpha_a) & (\alpha_3-\alpha_a) & \dots & (\alpha_{n-1}-\alpha_a) \\ (\alpha_0-\alpha_a)(\alpha_0-\alpha_b) & 0 & 0 & (\alpha_3-\alpha_a)(\alpha_3-\alpha_b) & \dots & (\alpha_{n-1}-\alpha_a)(\alpha_{n-1}-\alpha_b) \\ (\alpha_0-\alpha_a)(\alpha_0-\alpha_b)(\alpha_0-\alpha_c) & 0 & 0 & (\alpha_3-\alpha_a)(\alpha_3-\alpha_b)(\alpha_3-\alpha_c) & \dots & 0 \end{bmatrix}$$

The parity check equation corresponding to the fourth row of this parity check matrix is

$$\sum_{i=0}^{n-1}(\alpha_i-\alpha_a)(\alpha_i-\alpha_b)(\alpha_i-\alpha_c)c_i = 0 \tag{11.7}$$

where the analogue BCH codeword consists of n complex numbers

$$\{c_0, c_1, c_2, c_3, \dots, c_{n-1}\}$$

k of these complex numbers may be arbitrary, determined by the information source and $n-k$ complex numbers are calculated from the parity check equations:

Defining

$$(\alpha_i-\alpha_a)(\alpha_i-\alpha_b)(\alpha_i-\alpha_c) = \alpha_i^3 + \beta_2\alpha_i^2 + \beta_1\alpha_i^1 + \beta_0$$

Parity check Eq. (11.7) becomes

$$\sum_{i=0}^{n-1}\alpha_i^3 c_i + \beta_2\sum_{i=0}^{n-1}\alpha_i^2 c_i + \beta_1\sum_{i=0}^{n-1}\alpha_i c_i + \beta_0\sum_{i=0}^{n-1}c_i = 0 \tag{11.8}$$

This codeword is from the same code as defined by the parity check matrix $\mathbf{H_a}$ because using parity check matrix $\mathbf{H_a}$, codewords satisfy the equations

$$\sum_{i=0}^{n-1} \alpha_i^3 c_i = 0 \quad \sum_{i=0}^{n-1} \alpha_i^2 c_i = 0 \quad \sum_{i=0}^{n-1} \alpha_i c_i = 0 \quad \sum_{i=0}^{n-1} c_i = 0$$

and consequently the codewords defined by $\mathbf{H_a}$ satisfy (11.8) as

$$0 + \beta_2 0 + \beta_1 0 + \beta_0 0 = 0$$

It is apparent that the reduced echelon matrix $\mathbf{H_b}$ consists of linear combinations of parity check matrix $\mathbf{H_a}$ and either matrix may be used to produce the same MDS, analogue BCH code.

11.5 Examples

11.5.1 *Example of Simple* (5, 3, 3) *Analogue Code*

This simple code is the extended analogue BCH code having complex sample values with $\alpha = e^{\frac{j2\pi}{4}}$ and uses the parity check matrix:

$$\mathbf{H} = \begin{bmatrix} 1 & 1 & 1 & 1 & 1 \\ 0 & 1 & j & -1 & -j \end{bmatrix}$$

This parity check matrix is used to encode 3 complex data values in the last 3 positions, viz $(0.11 + 0.98j, \quad -0.22 - 0.88j, \quad 0.33 + 0.78j)$. This produces the codeword:

$$(-0.2 - 0.22j, \quad -0.02 - 0.66j, \quad 0.11 + 0.98j, \quad -0.22 - 0.88j, \quad 0.33 + 0.78j)$$

Suppose the received vector has the last digit in error

$$(-0.2 - 0.22j, \quad -0.02 - 0.66j, \quad 0.11 + 0.98j, \quad -0.22 - 0.88j, \quad 0.4 + 0.9j)$$

Applying the first parity check equation produces $0.07 + 0.12j$. This result tells us that there is an error of $0.07 + 0.12j$ in one of the received coordinates. Applying the second parity check equation produces $0.12 - 0.07j$. Since this is the error multiplied by $-j$, this tells us that the error is in the last coordinate. Subtracting the error from the last coordinate of the received vector produces $(0.4 + 0.9j) - (0.07 + 0.12j) = 0.33 + 0.78j$ and the error has been corrected.

11.5.2 *Example of Erasures Correction Using* **(15, 10, 4)** *Binary BCH code*

This is an example demonstrating that the erasures decoder shown in Fig. 11.1 may be used to correct erasures in a binary BCH code as well as being able to correct erasures using an analogue BCH code.

The code is a binary BCH code of length $n = 15$, with binary codewords generated by the generator polynomial $g(x) = (1+x^3+x^4)(1+x)$. The Galois field is GF(2^4) generated by the primitivc root α, which is a root of the primitive polynomial $1+x+x^4$ so that $1+\alpha+\alpha^4 = 0$, and the Galois field consists of the following table of 15 field elements, plus the element, 0.

One example of a codeword from the code is

$$c(x) = x + x^3 + x^4 + x^6 + x^8 + x^9 + x^{10} + x^{11} \tag{11.9}$$

and consider that in a communication system, the codeword is received with erasures in positions in $\lambda_0 = 5$, $\lambda_1 = 0$ and $\lambda_2 = 8$, so that the received codeword is

$$\hat{c}(x) = \hat{c}_0 + x + x^3 + x^4 + \hat{c}_5 x^5 + x^6 + \hat{c}_8 x^8 + x^9 + x^{10} + x^{11} \tag{11.10}$$

To find the party check equations to solve for the erasures, referring to Fig. 11.1, the first parity check equation, $h_0(x)$, the all 1's vector is stored in the Register. The second parity check equation $h_1(x)$ has zero for the coefficient of $x^{n-\lambda_0=n-5}$ and is given by

$$h_1(x) = \sum_{j=0}^{n-1} (\alpha^j - \alpha^{10})\, x^j \tag{11.11}$$

Note that $h_i(x).\hat{c}(x) = 0$ and these polynomials are derived with the intention that the coefficient of x^0 will be evaluated. Referring to Fig. 11.1, $h_1(x)$ is stored in the corresponding Register. After substitution using Table 11.1, it is found that

$$\begin{aligned} h_1(x) = {} & \alpha^5 + \alpha^8 x + \alpha^4 x^2 + \alpha^{12} x^3 + \alpha^2 x^4 + x^5 \\ & + \alpha^7 x^6 + \alpha^6 x^7 + \alpha x^8 + \alpha^{13} x^9 + \alpha^{14} x^{11} \\ & + \alpha^3 x^{12} + \alpha^9 x^{13} + \alpha^{11} x^{14} \end{aligned} \tag{11.12}$$

Notice that although the codeword is binary, the coefficients of this equation are from the full extension field of GF(16). The third parity check equation $h_2(x)$ has zero in position $n - \lambda_1 = n - 0$ and is given by

$$h_2(x) = \sum_{j=0}^{n-1} (\alpha^j - 1)\, x^j \tag{11.13}$$

after evaluation

$$\begin{aligned} h_2(x) = {} & \alpha^4 x + \alpha^8 x^2 + \alpha^{14} x^3 + \alpha x^4 + \alpha^{10} x^5 \\ & + \alpha^{13} x^6 + \alpha^9 x^7 + \alpha^2 x^8 + \alpha^7 x^9 + \alpha^5 x^{10} \\ & + \alpha^{12} x^{11} + \alpha^{11} x^{12} + \alpha^6 x^{13} + \alpha^3 x^{14} \end{aligned} \tag{11.14}$$

Referring to Fig. 11.1, this polynomial is stored in the corresponding Register.

The parity check equation which gives the solution for coefficient $\hat{c}_8$ is $h_3(x) = h_0(x) \odot h_1(x) \odot h_2(x)$. Multiplying each of the corresponding coefficients together of the polynomials $h_0(x)$, $h_1(x)$ and $h_2(x)$ produces

$$\begin{aligned} h_3(x) = {} & \alpha^{12} x + \alpha^{12} x^2 + \alpha^{11} x^3 + \alpha^3 x^4 + \alpha^{10} x^5 \\ & + \alpha^5 x^6 + x^7 + \alpha^{10} x^8 + \alpha^5 x^9 + \alpha^{11} x^{11} \\ & + \alpha^{14} x^{12} + x^{13} + \alpha^{14} x^{14} \end{aligned} \tag{11.15}$$

Referring to Fig. 11.1, $h_3(x)$ will be input to Multiply and Sum. It should be noted that the parity check equation $h_3(x)$ has non-binary coefficients, even though the codeword is binary and the solution to the parity check equation has to be binary.

Evaluating the coefficient of x^0 of $h_3(x)\hat{c}(x)$ gives $\alpha^{14} + \alpha^{14} + \alpha^{11} + \alpha^5 + \hat{c}_8 + \alpha^5 + \alpha^{10} + \alpha^3 = 0$, which simplifies to $\alpha^{11} + \hat{c}_8 + \alpha^{10} + \alpha^3 = 0$. Using Table 11.1 gives

$$(\alpha + \alpha^2 + \alpha^3) + \hat{c}_8 + (1 + \alpha + \alpha^2) + \alpha^3 = 0$$

Table 11.1 All 15 Nonzero Galois Field elements of GF(16)

Symbol α^i	modulo $1+\alpha+\alpha^4$			
$\alpha^0 =$	1			
$\alpha^1 =$		α^1		
$\alpha^2 =$			α^2	
$\alpha^3 =$				α^3
$\alpha^4 =$	1+	α		
$\alpha^5 =$		$\alpha+$	α^2	
$\alpha^6 =$			α^2+	α^3
$\alpha^7 =$	1+	$\alpha+$		α^3
$\alpha^8 =$	1+		α^2	
$\alpha^9 =$		$\alpha+$		α^3
$\alpha^{10} =$	1+	$\alpha+$	α^2	
$\alpha^{11} =$		$\alpha+$	α^2+	α^3
$\alpha^{12} =$	1+	$\alpha+$	α^2+	α^3
$\alpha^{13} =$	1+		α^2+	α^3
$\alpha^{14} =$	1+			α^3

and $\hat{c}_8 = 1$. Referring to Fig. 11.1, Select produces from $h_3(x)$ the value of the coefficient of x^7 which is 1 and when inverted this is also equal to 1. The output of the Multiply and Add is 1, producing a product of 1, which is used by Update to update $\hat{c}_8 = 1$ in the Input Vector $\hat{c}(x)$.

The parity check equation $h_2(x)$ gives the solution for coefficient $\hat{c}_0$. Evaluating the coefficient of x^0 of $h_2(x)\hat{c}(x)$ gives

$$0 = \alpha^3 + \alpha^{11} + \alpha^{12} + \hat{c}_5\alpha^5 + \alpha^7 \\ + \alpha^9 + \alpha^{13} + \alpha^{10} + \alpha$$

Substituting using Table 11.1 gives $\hat{c}_5\alpha^5 = 0$ and $\hat{c}_5 = 0$.

Lastly the parity check equation $h_0(x)$ gives the solution for coefficient $\hat{c}_0$. Evaluating the coefficient of x^0 of $h_0(x)\hat{c}(x)$ gives

$$0 = \hat{c}_0 + 1 + 1 + 1 + 1 + 1 + 1 + 1 + 1 \tag{11.16}$$

and it is found that $\hat{c}_0 = 0$, and the updated $\hat{c}(x)$ with all three erasures solved is

$$\hat{c}(x) = x + x^3 + x^4 + x^6 + x^8 + x^9 + x^{10} + x^{11} \tag{11.17}$$

equal to the original codeword.

11.5.3 Example of (128, 112, 17) Analogue BCH Code and Error-Correction of Audio Data (Music) Subjected to Impulsive Noise

In this example, a stereo music file sampled at 44.1 kHz in complex Pulse Amplitude Modulation (PAM) format is split into sequences of 128 complex samples and encoded using an analogue (128, 112, 17) BCH code with $\alpha = e^{\frac{j2\pi}{128}}$, and reassembled into a single PAM stream. A short section of the stereo left channel waveform before encoding is shown plotted in Fig. 11.2.

The encoding parity check matrix is the $\mathbf{H_f}$ matrix for bandlimited signals given above in matrix (11.6). There are 16 parity symbols and to make these obvious they are located at the beginning of each codeword. The same section of the stereo left channel waveform as before but after encoding is shown plotted in Fig. 11.3. The parity symbols are obvious as the newly introduced spikes in the waveform.

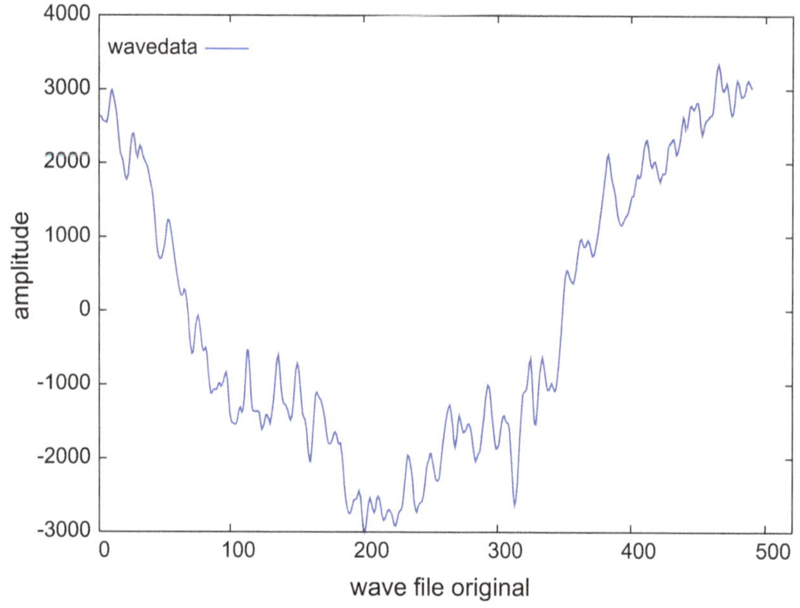

Fig. 11.2 Section of music waveform prior to encoding

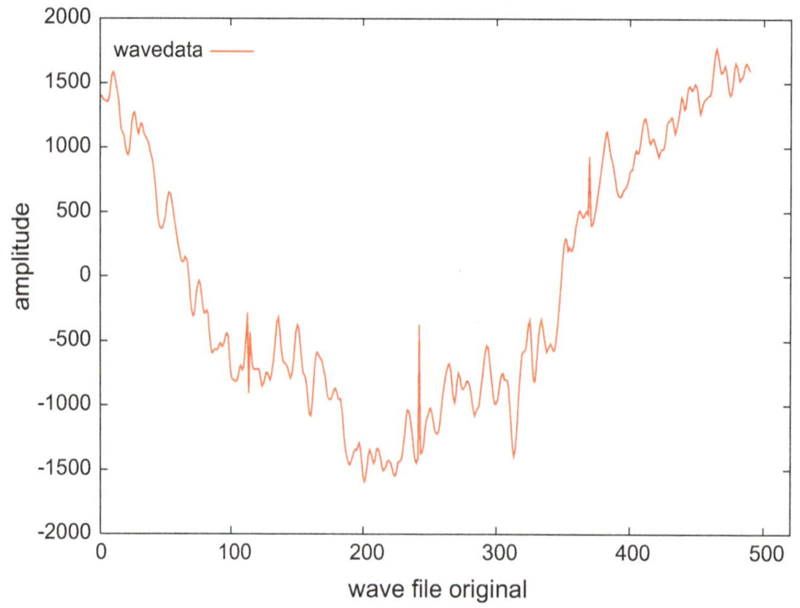

Fig. 11.3 Section of music waveform after encoding

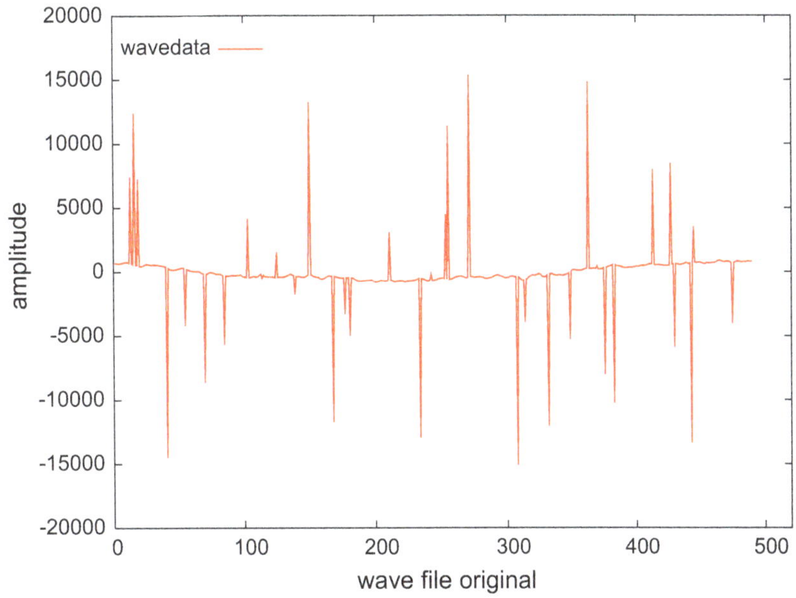

Fig. 11.4 Section of music waveform after encoding and subjected to impulse noise

The parity symbols may be calculated for any combination of 16 coordinate positions and in a more complicated encoding arrangement the positions could be selected as those that produce the minimum mean square error. However, the frequency components affected extend from 19.47 to 22.1 kHz (these components are equal to zero after encoding) and are beyond the hearing range of most people.

The encoded music waveform is subjected to randomly distributed impulse noise with a uniformly distributed amplitude in the range ±16000. The result is shown plotted in Fig. 11.4 for the same section of the waveform as before, although this is not obvious in the plot.

The decoder strategy used is that in each received codeword the 16 received PAM samples with the greatest magnitudes exceeding a dynamic threshold or with largest change relative to neighbouring samples are erased. The erasures are then solved using the parity check equations as outlined above. In several cases, correctly received PAM samples are erased, but this does not matter provided the 112 non-erased samples in each received codeword are correct. The decoded music waveform is shown in Fig. 11.5, and is apparent that waveform after decoding is the same as the encoded waveform and the impulse noise errors have been corrected.

Usually, impulse noise effects are handled by noise suppressors which produce short, zero-valued waveform sections. These audible gaps are irritating to the listener. By using analogue BCH, error-correcting codes, there are no waveform gaps following decoding.

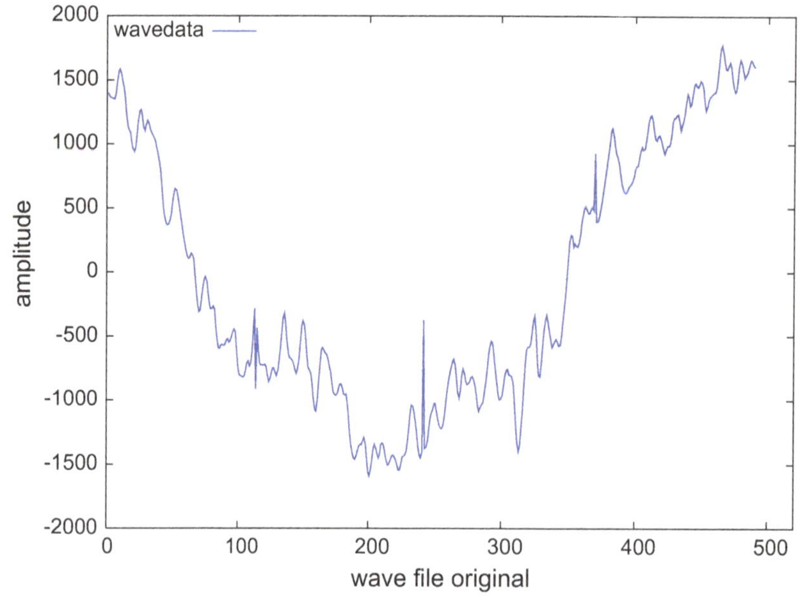

Fig. 11.5 Section of music waveform after decoding

11.6 Conclusions and Future Research

It has been demonstrated that for analogue $(n, k, n-k+1)$ BCH codes, parity check symbols having complex values may be calculated for any $n-k$ arbitrary positions in each codeword and an efficient method of calculating erased symbols for any BCH code including binary codes has been presented. Bandlimited data naturally occurs in many sources of information. In effect the source data has already been encoded with an analogue BCH code. In practice the parity check equations of the BCH code will only approximately equal zero for the PAM samples of the bandlimited source. There is scope for determining those samples which require the minimum of changes in order to satisfy the parity check equations. Similarly in decoding codewords corrupted by a noisy channel there is the opportunity to use the statistics of the noise source to design a maximum likelihood decoder for analogue BCH codes. It appears likely that the extended Dorsch decoder described in Chap. 15 may be adapted for analogue BCH codes.

There are many ad hoc noise suppression algorithms used on analogue video and audio waveforms which cause artefacts in the signal processed outputs. There appears to be an opportunity to improve on these by using analogue BCH coding since the output of the decoder is always a codeword. For high quality systems this will predominantly be the transmitted codeword and therefore the decoder output will be free of artefacts.

Whilst most communications these days is digitally based, analogue communications is usually far more bandwidth efficient, particularly in wireless applications. By using analogue BCH codes, analogue communications may be attractive once more, particularly for niche applications.

Steganography is another area in which analogue BCH codes may be utilised. Errors in parity check equations may be used to communicate data in a side channel. By virtue of the parity check equations these errors may be distributed over multiple PAM samples or pixels. Secrecy may be assured by using a combination of secret permutations of the parity check matrix columns and a secret linear matrix transformation so that the parity check equations are unknown by anyone other than the originator.

11.7 Summary

Many information sources are naturally analogue and must be digitised if they are to be transmitted digitally. The process of digitisation introduces quantisation errors and increases the bandwidth required. The use of analogue error-correcting codes eliminates the need for digitisation. It been shown that analogue BCH codes may be constructed in the same way as finite field BCH codes, including Reed–Solomon codes. The difference is that the field of complex numbers is used instead of a prime field or prime power field. It has been shown how the Mattson–Solomon polynomial or equivalently the Discrete Fourier transform may be used as the basis for the construction of analogue codes. It has also been shown that a permuted parity check matrix produces an equivalent code using a primitive root of unity to construct the code as in discrete BCH codes.

A new algorithm was presented which uses symbolwise multiplication of rows of a parity check matrix to produce directly the parity check matrix in reduced echelon form. The algorithm may be used for constructing reduced echelon parity check matrices for standard BCH and RS codes as well as analogue BCH codes. Gaussian elimination or other means of solving parallel, simultaneous equations are completely avoided by the method. It was also proven that analogue BCH codes are Maximum Distance Separable (MDS) codes. Examples have been presented of using the analogue BCH codes in providing error-correction for analogue, band-limited data including the correction of impulse noise errors in BCH encoded, analogue stereo music waveforms. It is shown that since the data is bandlimited it is already redundant and the parity check symbols replace existing values so that there is no need for bandwidth expansion as in traditional error-correcting codes. Future research areas have been outlined including an analogue, maximum likelihood, error-correcting decoder based on the extended Dorsch decoder of Chap. 15. Steganography is another future application area for analogue BCH codes.

References

1. MacWilliams, F.J., Sloane, N.J.A.: The Theory of Error-Correcting Codes. North-Holland, Amsterdam (1977)
2. Marshall, J.T.: Coding of real-number sequences for error correction: a digital signal processing problem. IEEE J. Sel. Areas Commun. **2**(2), 381–392 (1984)
3. Wolf, J.: Redundancy, the discrete fourier transform, and impulse noise cancellation. IEEE Trans. Commun. **31**(3), 458–461 (1983)

Chapter 12
LDPC Codes

12.1 Background and Notation

LDPC codes are linear block codes whose parity-check matrix—as the name implies—is sparse. These codes can be iteratively decoded using the sum product [9] or equivalently the belief propagation [24] soft decision decoder. It has been shown, for example by Chung et al. [3], that for long block lengths, the performance of LDPC codes is close to the channel capacity. The theory of LDPC codes is related to a branch of mathematics called graph theory. Some basic definitions used in graph theory are briefly introduced as follows.

Definition 12.1 (*Vertex, Edge, Adjacent and Incident*) A graph, denoted by $G(V, E)$, consists of an ordered set of vertices and edges.

- **(Vertex)** A vertex is commonly drawn as a node or a dot. The set $V(G)$ consists of vertices of $G(V, E)$ and if v is a vertex of $G(V, E)$, it is denoted as $v \in V(G)$. The number of vertices of $V(G)$ is denoted by $|V(G)|$.
- **(Edge)** An edge (u, v) *connects* two vertices $u \in V(G)$ and $v \in V(G)$ and it is drawn as a line connecting vertices u and v. The set $E(G)$ contains pairs of elements of $V(G)$, i.e. $\{(u, v) \mid u, v \in V(G)\}$.
- **(Adjacent and Incident)** If $(u, v) \in E(G)$, then $u \in V(G)$ and $v \in V(G)$ are adjacent or neighbouring vertices of $G(V, E)$. Similarly, the vertices u and v are incident with the edge (u, v).

Definition 12.2 (*Degree*) The degree of a vertex $v \in V(G)$ is the number of edges that are incident with vertex v, i.e. the number of edges that are connected to vertex v.

Definition 12.3 (*Bipartite or Tanner graph*) Bipartite or Tanner graph $G(V, E)$ consists of two disjoint sets of vertices, say $V_v(G)$ and $V_p(G)$, such that $V(G) = V_v(G) \cup V_p(G)$, and every edge $(v_i, p_j) \in E(G)$, such that $v_i \in V_v(G)$ and $p_j \in V_p(G)$ for some integers i and j.

M. Tomlinson et al., *Error-Correction Coding and Decoding*,
Signals and Communication Technology,
DOI 10.1007/978-3-319-51103-0_12

An $[n, k, d]$ LDPC code may be represented by a Tanner graph $G(V, E)$. The parity-check matrix $\boldsymbol{H}$ of the LDPC code consists of $|V_p(G)| = n - k$ rows and $|V_v(G)| = n$ columns. The set of vertices $V_v(G)$ and $V_p(G)$ are called *variable* and *parity-check* vertices, respectively. Figure 12.1 shows the parity check and the cor-

$$
\boldsymbol{H} = \begin{array}{c}
\begin{array}{cccccccccccccccc} v_0 & v_1 & v_2 & v_3 & v_4 & v_5 & v_6 & v_7 & v_8 & v_9 & v_{10} & v_{11} & v_{12} & v_{13} & v_{14} & v_{15} \end{array} \\
\left[\begin{array}{cccccccccccccccc}
1 & 1 & 1 & 1 & 0 & 0 & 0 & 0 & 0 & 0 & 0 & 0 & 0 & 0 & 0 & 0 \\
0 & 0 & 0 & 0 & 1 & 1 & 1 & 1 & 0 & 0 & 0 & 0 & 0 & 0 & 0 & 0 \\
0 & 0 & 0 & 0 & 0 & 0 & 0 & 0 & 1 & 1 & 1 & 1 & 0 & 0 & 0 & 0 \\
0 & 0 & 0 & 0 & 0 & 0 & 0 & 0 & 0 & 0 & 0 & 1 & 0 & 1 & 1 & 1 \\
0 & 0 & 0 & 1 & 1 & 0 & 0 & 0 & 0 & 0 & 0 & 0 & 0 & 1 & 1 & 0 \\
0 & 0 & 1 & 0 & 0 & 0 & 1 & 0 & 1 & 0 & 1 & 0 & 0 & 0 & 0 & 0 \\
1 & 0 & 0 & 0 & 0 & 0 & 0 & 1 & 0 & 1 & 0 & 0 & 0 & 1 & 0 & 0 \\
0 & 0 & 0 & 1 & 0 & 0 & 1 & 0 & 0 & 1 & 0 & 0 & 1 & 0 & 0 & 0 \\
0 & 0 & 1 & 0 & 0 & 0 & 0 & 0 & 1 & 0 & 0 & 0 & 1 & 0 & 0 & 1 \\
0 & 1 & 0 & 0 & 0 & 1 & 0 & 1 & 0 & 0 & 0 & 0 & 1 & 0 & 0 & 0 \\
1 & 0 & 0 & 0 & 0 & 1 & 0 & 0 & 0 & 0 & 1 & 0 & 0 & 0 & 1 & 0 \\
0 & 1 & 0 & 0 & 1 & 0 & 0 & 0 & 0 & 0 & 0 & 1 & 0 & 0 & 0 & 1
\end{array}\right]
\end{array}
\begin{array}{c} \\ p_0 \\ p_1 \\ p_2 \\ p_3 \\ p_4 \\ p_5 \\ p_6 \\ p_7 \\ p_8 \\ p_9 \\ p_{10} \\ p_{11} \end{array}
$$

(a) Parity check matrix

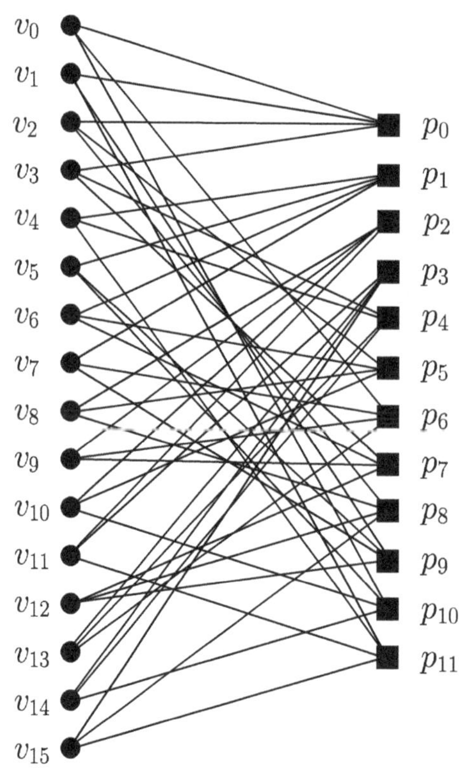

(b) Tanner graph

Fig. 12.1 Representations of a [16, 4, 4] LDPC code

responding Tanner graph of a [16, 4, 4] LDPC code. Let $V_v(G) = (v_0, v_1, \ldots, v_{n-1})$ and $V_p(G) = (p_0, p_1, \ldots, p_{n-k-1})$; we can see that for each $(v_i, p_j) \in E(G)$, the ith column and jth row of $\boldsymbol{H}$, $H_{j,i} \neq 0$, for $0 \leq i \leq n-1$ and $0 \leq j \leq n-k-1$.

Definition 12.4 (*Cycle*) A cycle in a graph $G(V, E)$ is a sequence of distinct vertices that starts and ends in the same vertex. For bipartite graph $G(V, E)$, exactly half of these distinct vertices belong to $V_v(G)$ and the remaining half belong to $V_p(G)$.

Definition 12.5 (*Girth and Local Girth*) The girth of graph $G(V, E)$ is the length of the shortest cycle in the graph $G(V, E)$. The local girth of a vertex $v \in V(G)$ is the length of shortest cycle that passes through vertex v.

The performance of a typical iteratively decodable code (e.g. an LDPC or turbo code) may be partitioned into three regions, namely erroneous, waterfall and error floor regions, see Fig. 12.2. The erroneous region occurs at low E_b/N_0 values and is indicated by the inability of the iterative decoder to correctly decode almost all of the transmitted messages. As we increase the signal power, the error rate of the iterative decoder decreases rapidly—resembling a waterfall. The E_b/N_0 value at which the waterfall region starts is commonly known as the *convergence threshold* in the literature. At higher E_b/N_0 values, the error rate starts to flatten—introducing an error floor in the frame error rate (FER) curve.

In addition to this FER curve, the offset sphere packing lower bound and the probability of error based on the union bound argument as described in Chap. 1 are also plotted in Fig. 12.2. The sphere packing lower bound represents the region of

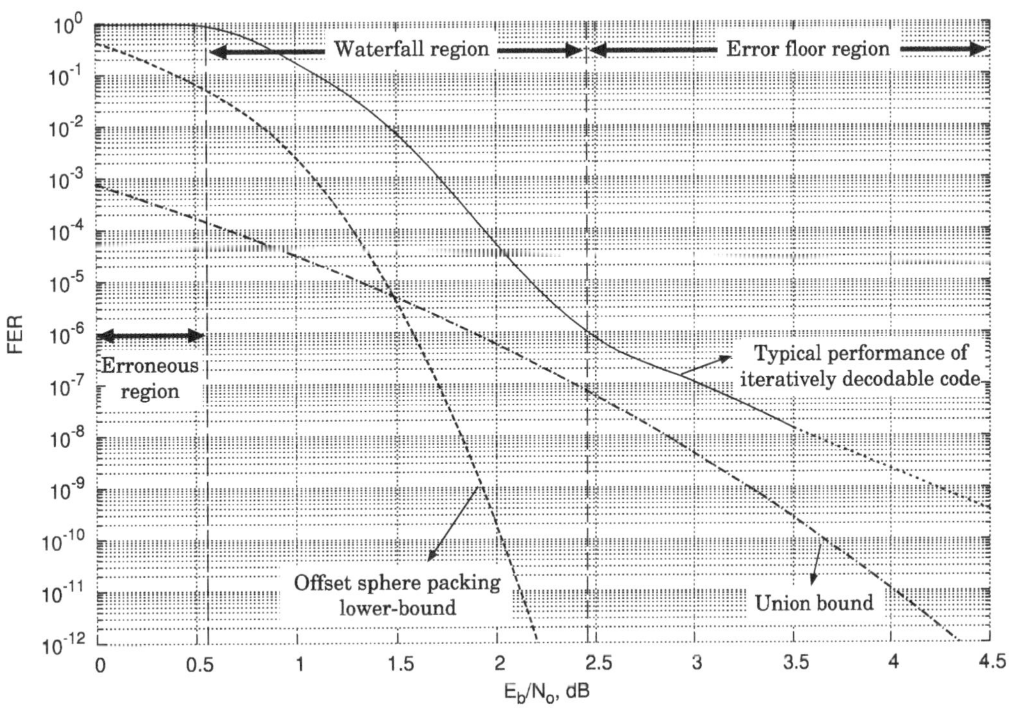

Fig. 12.2 Waterfall and error regions of a typical LDPC code for the AWGN channel

attainable performance of a coding system. The performance to the left of this lower bound is not attainable, whereas that to the right may be achieved by some coding and decoding arrangements. The other curve is the union bound of the probability of error, which is dominated by the low Hamming weight codewords and the number of codewords of these Hamming weights. The larger the minimum Hamming distance of a code, the lower the union bound typically. For iteratively decodable codes which are not designed to maximise the minimum Hamming distance, the union bound intersects with the offset sphere packing lower bound at relatively low E_b/N_0 values.

It may be seen that, with an ideal soft decision decoder, the performance of a coding system would follow the sphere packing lower bound and at higher E_b/N_0 values, the performance floors due to the limitation of the minimum Hamming weight codewords. However, as depicted in Fig. 12.2, there is a relatively wide gap between the union bound and the error floor of a typical iteratively decodable code. This is an inherent behaviour of iteratively decodable codes and it is attributed to the weakness of the iterative decoder. There are other error events, which are not caused by the minimum Hamming weight codewords, that prevent the iterative decoder from reaching the union bound.

In terms of the construction technique, we may divide LDPC codes into two categories: random and algebraic LDPC codes. We may also classify LDPC codes into two categories depending on the structure of the parity-check matrix, namely regular and irregular codes—refer to Sect. 12.1.1 for the definition. Another attractive construction method that has been shown to offer capacity-achieving performance is non-binary construction.

12.1.1 Random Constructions

Gallager [8] introduced the (n, λ, ρ) LDPC codes where n represents the block length whilst the number of non-zeros per column and the number of non-zeros per row are represented by λ and ρ, respectively.

The short notation (λ, ρ) is also commonly used to represent these LDPC codes. The coderate of the Gallager (λ, ρ) codes is given by

$$R = 1 - \frac{\lambda}{\rho}.$$

An example of the parity-check matrix of a Gallager (λ, ρ) LDPC code is shown in Fig. 12.1a. It is a [16, 4, 4] code with a λ of 3 and a ρ of 4. The parity-check matrix of the (λ, ρ) Gallager codes always have a fixed number of non-zeros per column and per row, and because of this property, this class of LDPC codes is termed regular LDPC codes. The performance of the Gallager LDPC codes in the waterfall region is not as satisfactory as that of turbo codes for the same block length and code rate. Many efforts have been devoted to improve the performance of the LDPC codes and one example that provides significant improvement is the introduction of the irregular

LDPC codes by Luby et al. [18]. The irregular LDPC codes, as the name implies, do not have a fixed number of non-zeros per column or per row and thus the level of error protection varies over a codeword. The columns of a parity check matrix that have a higher number of non-zeros provide stronger error protection than those that have a lower number of non-zeros. Given an input block in iterative decoding, errors in the coordinates of this block, whose columns of the parity-check matrix have a larger number of non-zeros, will be corrected earlier, i.e. only a small number of iterations are required. In the subsequent iterations, the corrected values in these coordinates will then be utilised to correct errors in the remaining coordinates of the block.

Definition 12.6 (*Degree Sequences*) The polynomial $\Lambda_\lambda(x) = \sum_{i\geq 1} \lambda_i x^i$ is called the symbol or variable degree sequence, where λ_i is the fraction of vertices of degree i. Similarly, $\Lambda_\rho(x) = \sum_{i\geq 1} \rho_i x^i$ is the check degree sequence, where ρ_i is the fraction of vertices of degree i.

The degree sequences given in the above definition are usually known as *vertex-oriented degree sequences*. Another representations are *edge-oriented degree sequences* which consider the fraction of edges that are connected to a vertex of certain degree. Irregular LDPC codes are defined by these degree sequences and it is assumed that the degree sequences are vertex-oriented.

Example 12.1 An irregular LDPC code with the following degree sequences

$$\Lambda_\lambda(x) = 0.5x^2 + 0.26x^3 + 0.17x^5 + 0.07x^{10}$$
$$\Lambda_\rho(x) = 0.80x^{14} + 0.20x^{15}$$

has 50, 26, 17 and 7% of the columns with 2, 3, 5 and 10 ones per column, respectively, and 80 and 20% of the rows with 14 and 15 ones per row, respectively.

Various techniques have been proposed to design good degree distributions. Richardson et al. [27] used *density evolution* to determine the convergence threshold and to optimise the degree distributions. Chung et al. [4] simplified the density evolution approach using *Gaussian approximation*. With the optimised degree distributions, Chung et al. [3] showed that the bit error rate performance of a long block length ($n = 10^7$) irregular LDPC code was within 0.04 dB away from the capacity limit for binary transmission over the AWGN channel, discussed in Chap. 1. This is within 0.18 dB of Shannon's limit [30]. The density evolution and Gaussian approximation methods, which make use of the concentration theorem [28], can only be used to design the degree distributions for infinitely long LDPC codes. The concentration theorem states that the performance of cycle-free LDPC codes can be characterised by the average performance of the ensemble. The cycle-free assumption is only valid for infinitely long LDPC codes and cycles are inevitable for finite block-length LDPC codes. As may be expected, the performance of finite block-length LDPC codes with degree distributions derived based on the concentration theorem differs considerably from the ensemble performance. There are various techniques to design good finite block-length LDPC codes, for instance see [1, 2, 10, 33]. In particular, the work of Hu et al. [10] with the introduction of the progressive edge-growth (PEG) algorithm to construct both regular and irregular LDPC codes, that of Tian et al. [33] with the introduction of *extrinsic message degree* and recently, that of Richter et al. [29] which improves the original PEG algorithm by introducing some construction constraints to avoid certain cycles involving variable vertices of degree 3, have provided significant contributions to the construction of practical LDPC codes as well as the lowering of the inherent error floor of these codes.

12.1.2 Algebraic Constructions

In general, LDPC codes constructed algebraically have a regular structure in their parity-check matrix. The algebraic LDPC codes offer many advantages over randomly generated codes. Some of these advantages are

1. The important property such as the minimum Hamming distance can be easily determined or in the worst case, lower and upper bounds may be mathematically derived. These bounds are generally more accurate than estimates for random codes.
2. The minimum Hamming distance of algebraic LDPC codes is typically higher than that of random codes. Due to the higher minimum Hamming distance, algebraic codes are not that likely to suffer from an early error floor.
3. The existence of a known structure in algebraic codes usually offers an attractive and simple encoding scheme. In the case of random codes, in order to carry out encoding, a Gaussian elimination process has to be carried out in the first place and the entire reduced echelon parity-check matrix has to be stored in the memory. Algebraically constructed codes such as cyclic or quasi-cyclic codes can be completely defined by polynomials. The encoding of cyclic or quasi-cyclic codes may be simply achieved using a linear-feedback shift-register circuit and the memory requirement is minimum. Various efficient techniques for encoding random LDPC codes have been proposed, see Ping et al. [26] for example, but none of these techniques simplifies the storage requirements. The simplicity of the encoder and decoder structure has led to many algebraically constructed LDPC codes being adopted as industry standards [5].
4. Cyclic LDPC codes have n low Hamming weight parity-check equations and therefore, compared to random codes, these cyclic LDPC codes have k extra equations for the iterative decoder to iterate with and this leads to improved performance.

One of the earliest algebraic LDPC code constructions was introduced by Margulis [21] using the Ramanujan graphs. Lucas et al. [19] showed that the well-known different set cyclic (DSC) [36] and one-step majority-logic decodable (OSMLD) [17] codes have good performance under iterative decoding. The iterative soft decision decoder offers significant improvement over the conventional hard decision majority-logic decoder. Another class of algebraic codes is the class of the Euclidean and projective geometry codes which are discussed in detail by Kou et al. [16]. Other algebraic constructions include those that use combinatorial techniques [13–15, 35].

It has been observed that in general, there is an inverse performance relationship between the minimum Hamming distance of the code and the convergence of the iterative decoder. Irregular codes converge well with iterative decoding, but the minimum Hamming distance is relatively poor. In contrast, algebraically constructed LDPC codes, which have high minimum Hamming distance, tend not to converge well with iterative decoding. Consequently, compared to the performance of irregular codes, algebraic LDPC codes may perform worse in the low SNR region and perform better in the high SNR region. This is attributed to the early error floor of the irregular codes. As will be shown later, for short block lengths ($n < 350$), cyclic algebraic LDPC codes offer some of the best performance available.

12.1.3 Non-binary Constructions

LDPC codes may be easily extended so that the symbols take values from the finite-field $\mathbb{F}_{2^m}$ and Davey et al. [6] were the pioneers in this area. Given an LDPC code over $\mathbb{F}_2$ with parity-check matrix $\boldsymbol{H}$, we may construct an LDPC code over $\mathbb{F}_{2^m}$ for $m \geq 2$ by simply replacing every non-zero element of $\boldsymbol{H}$ with any non-zero element of $\mathbb{F}_{2^m}$ in a random or structured manner. Davey et al. [6] and Hu et al. [11] have shown that the performance of LDPC codes can be improved by going beyond the binary field. The non-binary LDPC codes have better convergence behaviour under iterative decoding. Using some irregular non-binary LDPC codes, whose parity-check matrices are derived by randomly replacing the non-zeros of the PEG-constructed irregular binary LDPC codes, Hu et al. [11] demonstrated that an additional coding gain of 0.25 dB was achieved. It may be regarded that the improved performance is attributable to the improved graph structure in the non-binary arrangement. Consider a cycle of length 6 in the Tanner graph of a binary LDPC code, which is represented as the following sequence of pairs of edges $\{(v_0, p_0), (v_3, p_0), (v_3, p_2), (v_4, p_2), (v_4, p_1), (v_0, p_1)\}$. If we replace the corresponding entries in the parity-check matrix with some non-zeros over $\mathbb{F}_{2^m}$ for $m \geq 2$, provided that these six entries are not all the same, the cycle length becomes longer than 6. According to McEliece et al. [22] and Etzion et al. [7], the non-convergence of the iterative decoder is caused by the existence of cycles in the Tanner graph representation of the code. Cycles, especially those of short lengths, introduce correlations of reliability information exchanged in iterative decoding. Since cycles are inevitable for finite block length codes, it is desirable to have LDPC codes with large girth.

The non-binary LDPC codes also offer an attractive matching for higher order modulation methods. The impact of increased complexity of the symbol-based iterative decoder can be moderated as the reliability information from the component codes may be efficiently evaluated using the frequency-domain dual-code decoder based on the Fast Walsh–Hadamard transform [6].

12.2 Algebraic LDPC Codes

Based on idempotents and cyclotomic cosets, see Chap. 4, a class of cyclic codes that is suitable for iterative decoding may be constructed. This class of cyclic codes falls into the class of one-step majority-logic decodable (OSMLD) codes whose parity-check polynomial is orthogonal on each bit position—implying the absence of a girth of 4 in the underlying Tanner graph, and the corresponding parity-check matrix is sparse, and thus can be used as LDPC codes.

Definition 12.7 (*Binary Parity-Check Idempotent*) Let $\mathcal{M} \subseteq \mathcal{N}$ and let the polynomial $u(x) \in T(x)$ be defined by

$$u(x) = \sum_{s \in \mathcal{M}} e_s(x) \tag{12.1}$$

where $e_s(x)$ is an idempotent. The polynomial $u(x)$ is called a binary parity-check idempotent.

The binary parity-check idempotent $u(x)$ can be used to describe an $[n, k]$ cyclic code as discussed in Chap. 4. Since $\text{GCD}(u(x), x^n - 1) = h(x)$, the polynomial $\bar{u}(x) = x^{\deg(u(x))}u(x^{-1})$ and its n cyclic shifts $\pmod{x^n - 1}$ can be used to define the parity-check matrix of a binary cyclic code. In general, $\text{wt}_H(\bar{u}(x))$ is much lower than $\text{wt}_H(h(x))$, and therefore a low-density parity-check matrix can be derived from $\bar{u}(x)$.

Let the parity-check polynomial $\bar{u}(x) = x^{\bar{u}_0} + x^{\bar{u}_1} + \cdots + x^{\bar{u}_t}$ of weight $t + 1$. Since the code defined by $\bar{u}(x)$ is cyclic, for each non-zero coefficient $\bar{u}_i$ in $\bar{u}(x)$, there are another t parity-check polynomials of weight $t + 1$, which also have a non zero coefficient at position $\bar{u}_i$. Furthermore, consider the set of these $t + 1$ polynomials that have a non-zero coefficient at position $\bar{u}_i$, there is no more than one polynomial in the set that have a non-zero at position $\bar{u}_j$ for some integer j. In other words, if we count the number of times the positions $0, 1, \ldots, n - 1$ appear in the exponents of the aforementioned set of $t + 1$ polynomials, we shall find that all positions except $\bar{u}_i$ appear at most once. This set of $t + 1$ polynomials is said to be *orthogonal* on position $\bar{u}_i$. The mathematical expression of this orthogonality is given in the following definition and lemma.

Definition 12.8 (*Difference Enumerator Polynomial*) Let the polynomial $f(x) \in T(x)$. The difference enumerator of $f(x)$, denoted as $\mathcal{D}(f(x))$, is defined as

$$\mathcal{D}(f(x)) = f(x)\, f\left(x^{-1}\right) = d_0 + d_1 x + \cdots + d_{n-1}x^{n-1}, \tag{12.2}$$

where it is assumed that $\mathscr{D}(f(x))$ is a modulo $x^n - 1$ polynomial with coefficients taking values from $\mathbb{R}$ (real coefficients).

Lemma 12.1 *Let d_i for $0 \leq i \leq n-1$ denote the coefficients of $\mathscr{D}(\bar{u}(x))$. If $d_i \in \{0, 1\}$, for all $i \in \{1, 2, \ldots, n-1\}$, the parity-check polynomial derived from $\bar{u}(x)$ is orthogonal on each position in the n-tuple. Consequently,*

(i) the minimum distance of the resulting LDPC code is $1 + \mathrm{wt}_H(\bar{u}(x))$, and
(ii) the underlying Tanner Graph has girth of at least 6.

Proof (i) [25, Theorem 10.1] Let a codeword $c(x) = c_0 + c_1x + \cdots + c_{n-1}x^{n-1}$ and $c(x) \in T(x)$. For each non-zero bit position c_j of $c(x)$, where $j \in \{0, 1, \ldots, n-1\}$, there are $\mathrm{wt}_H(u(x))$ parity-check equations orthogonal to position c_j. Each of the parity-check equations must check another non-zero bit c_l, where $l \neq j$, so that the equation is satisfied. Clearly, $\mathrm{wt}_H(c(x))$ must equal to $1 + \mathrm{wt}_H(u(x))$ and this is the minimum weight of all codewords.
(ii) The direct consequence of having orthogonal parity-check equations is the absence of cycles of length 4 in the Tanner Graphs. Let a, b and c, where $a < b < c$, be three distinct coordinates in an n-tuple, since $d_i \in \{0, 1\}$ for $1 \leq i \leq n-1$, this implies that $b - a \neq c - b$. We know that $q(b-a) \pmod{n} \in \{1, 2, \ldots, n-1\}$ and thus, $q(b-a) \pmod{n} \equiv (c-b)$ for some integer $q \in \{1, 2, \ldots, n-1\}$. If we associate the integers a, b and c with some variable vertices in the Tanner graph, a cycle of length 6 is produced.

It can be deduced that the cyclic LDPC code with parity-check polynomial $\bar{u}(x)$ is an OSMLD code if $d_i \in \{0, 1\}$, for all $i \in \{1, 2, \ldots, n-1\}$ or a difference set cyclic (DSC) code if $d_i = 1$, for all $i \in \{1, 2, \ldots, n-1\}$, where d_i is the coefficient of $\mathscr{D}(\bar{u}(x))$.

In order to arrive at either OSMLD or DSC codes, the following design conditions are imposed on $\bar{u}(x)$ and therefore, $u(x)$:

Condition 12.1 The idempotent $u(x)$ must be chosen such that

$$\mathrm{wt}_H(u(x))\,(\mathrm{wt}_H(u(x)) - 1)) \leq n - 1.$$

Proof There are $\mathrm{wt}_H(u(x))$ polynomials of weight $\mathrm{wt}_H(u(x))$ that are orthogonal on position j for some integer j. The number of distinct positions in this set of polynomials is $\mathrm{wt}_H(u(x))\,(\mathrm{wt}_H(u(x)) - 1)$, and this number must be less than or equal to the total number of distinct integers between 1 and $n - 1$.

Condition 12.2 Following Definition 12.8, let $W = \{i \mid d_i = 1,\ 1 \leq i \leq n-1\}$, the cardinality of W must be equal to $\mathrm{wt}_H(u(x))\,(\mathrm{wt}_H(u(x)) - 1)$.

Proof The cyclic differences between the exponents of polynomial $u(x)$ are given by $\mathscr{D}(u(x)) = \sum_{i=0}^{n-1} d_i x^i$, where the coefficient d_i is the number of differences and the exponent i is the difference. The polynomial $u(x)$ and some of its cyclic shifts are orthogonal on position 0 and this means that all of the cyclic differences between the exponents of $u(x)$ (excluding zero) must be distinct, i.e. $d_i \in \{0, 1\}$ for $1 \leq i \neq n-1$. Since the weight of $u(x)$ excluding x^0 is $\text{wt}_H(u(x)) - 1$ and there are $\text{wt}_H(u(x))$ cyclic shifts of $u(x)$ that are orthogonal to x^0, the number of distinct exponents in the cyclic differences is $\text{wt}_H(u(x))(\text{wt}_H(u(x)) - 1) = W$.

Condition 12.3 The exponents of $u(x)$ must not contain a common factor of n, otherwise a degenerate code, a repetition of a shorter cyclic code, is the result.

Proof If the exponents of $u(x)$ contain a common factor of n, p with $n = pr$, then factors of $u(x)$ divide $x^r - 1$ and form a cyclic code of length r. Every codeword of the longer code is a repetition of the shorter cyclic code.

Condition 12.4 Following (12.1), unless $\text{wt}_H(e_s(x)) = 2$, the binary parity-check idempotent $e_s(x)$ must not be self-reciprocal, i.e. $e_s(x) \neq e_i\left(x^{-1}\right)$, for all $i \in \mathcal{M}$.

Proof The number of non-zero coefficients of $\mathcal{D}(e_s(x))$ is equal to

$$\text{wt}_H(e_s(x))\,(\text{wt}_H(e_s(x)) - 1)\,.$$

For a self-reciprocal case, $e_s(x)e_s\left(x^{-1}\right) = e_s^2(x) = e_s(x)$ with $\text{wt}_H(e_s(x))$ non-zero coefficients. Following Condition 12.1, the inequality

$$\text{wt}_H(e_s(x))\,(\text{wt}_H(e_s(x) - 1) \leq \text{wt}_H(e_s(x))$$

becomes equality if and only if $\text{wt}_H(e_s(x)) = 2$.

Condition 12.5 Following (12.1), $u(x)$ must not contain $e_s\left(x^{-1}\right)$, for all $i \in \mathcal{M}$, unless $e_s(x)$ is self-reciprocal.

Proof If $u(x)$ contains $e_s\left(x^{-1}\right)$ for $i \in \mathcal{M}$, then $\mathcal{D}(u(x))$ will contain both $e_s(x)e_s\left(x^{-1}\right)$ and $e_s\left(x^{-1}\right)e_s(x)$, hence, some of the coefficients of $\mathcal{D}(e_s(x))$, $d_i \neq \{0, 1\}$ for some integer i.

Although the above conditions seem overly restrictive, they turn out to be helpful in code construction. Codes may be designed in stage-by-stage by adding candidate idempotents to $u(x)$, checking the above conditions at each stage.

In order to encode the cyclic LDPC codes constructed, there is no need to determine $g(x)$. With α defined as a primitive n^{th} root of unity, it follows from Lemma 4.4 that $u(\alpha^i) \in \{0, 1\}$ for $0 \leq i \leq n-1$. Let $\mathcal{J} = (j_0, j_1, \ldots, j_{n-k-1})$ be a set of integers between 0 and $n-1$, such that $g(\alpha^j) = 0$, for all $j \in \mathcal{J}$. Because $u(x)$ does not contain α^j as its roots, it follows that $u(\alpha^j) = 1$, for all $j \in \mathcal{J}$. In $\mathbb{F}_2$, $1 + u(\alpha^j) = 0$ and the polynomial $1 + u(x) = e_g(x)$, the generating idempotent of the code may be used to generate the codewords as an alternative to $g(x)$.

The number of information symbols of the cyclic LDPC codes can be determined either from the number of roots of $u(x)$ which are also roots of unity, i.e. $n - \text{wt}_H(U(z))$, or from the degree of $(u(x), x^n - 1) = h(x)$.

Example 12.2 Consider the design of a cyclic LDPC code of length 63. The cyclotomic coset modulo 63 is given in Example 4.2. Let $u(x)$ be defined by C_9, i.e. $u(x) = e_9(x) = x^9(1 + x^9 + x^{27})$. $\mathscr{D}(\bar{u}(x))$ indicates that the parity-check matrix defined by $\bar{u}(x)$ has no cycles of length 4; however, following Condition 12.3, it is a degenerate code consisting of repetitions of codewords of length 7.

With $u(x) = e_{23}(x) = x^{23}(1 + x^6 + x^{20} + x^{23} + x^{30} + x^{35})$, the resulting cyclic code is a $[63, 31, 6]$ LDPC code which is non-degenerate and its underlying Tanner graph has girth of 6. This code can be further improved by adding $e_{21}(x)$ to $u(x)$. Despite $e_{21}(x)$ being self-reciprocal, its weight is 2 satisfying Condition 12.4. Now, $u(x) = x^{21}(1 + x^2 + x^8 + x^{21} + x^{22} + x^{25} + x^{32} + x^{37})$, and it is a $[63, 37, 9]$ cyclic LDPC code.

Based on the theory described above, an algorithm which exhaustively searches for all non-degenerate cyclic LDPC codes of length n which have orthogonal parity-check polynomials has been developed, and it is given in Algorithm 12.1.

Algorithm 12.1 CodeSearch($\mathbf{V}$, $index$)

Input:
$index \Leftarrow$ an integer that is initialised to -1
$\mathbf{V} \Leftarrow$ a vector that is initialised to $\emptyset$
$\mathscr{S} \Leftarrow \mathscr{N}$ excluding 0

Output:
CodesList contains set of cyclic codes which have orthogonal parity-check polynomial

$\mathbf{T} \Leftarrow \mathbf{V}$
for $(i{=}index{+}1;\, i \le |\mathscr{S}|;\, i{+}{+})$ **do**
  $\mathbf{T}_{\text{prev}} \Leftarrow \mathbf{T}$
  if $\left(\sum_{\forall t \in \mathbf{T}} |C_{S_t}| \le \sqrt{n},\ S_t \text{ is the } t^{\text{th}} \text{ element of } \mathscr{S}\right)$ **then**
    Append i to $\mathbf{T}$
    $u(x) = \sum_{\forall t \in \mathbf{T}} e_{S_t}(x)$
    if $(u(x)$ is non-degenerate$)$ **and** $(u(x)$ is orthogonal on each position (Lemma 12.1)$)$ **then**
      $U(z) = \mathrm{MS}\,(u(x))$
      $k = n - \mathrm{wt}_H\,(U(z))$
      $\mathscr{C} \Leftarrow$ a $[n, k, 1 + \mathrm{wt}_H(u(x))]$ cyclic code defined by $u(x)$
      if $\left(k \ge \frac{1}{4}\right)$ **and** $(\mathscr{C} \notin$ **CodeList**$)$ **then**
        Add $\mathbf{C}$ to **CodeList**
      end if
    end if
    CodeSearch($\mathbf{T}$, i)
  end if
  $\mathbf{T} \Leftarrow \mathbf{T}_{\text{prev}}$
end for

Table 12.1 lists some example of codes obtained from Algorithm 12.1. Note that all codes with code rate less than 0.25 are excluded from the table and codes of longer lengths may also be constructed. We can also see that some of the codes in Table 12.1 have the same parameters as the Euclidean and projective geometry codes, which have been shown by Jin et al. [16] to perform well under iterative decoding.

Table 12.1 Examples of 2-cyclotomic coset-based LDPC codes

$[n, k, d]$	Cyclotomic cosets
[21, 11, 6]	C_7, C_9
[63, 37, 9]	C_{21}, C_{23}
[93, 47, 8]	C_3, C_{31}
[73, 45, 10]	C_1
[105, 53, 8]	C_7, C_{15}
[219, 101, 12]	C_3, C_{73}
[255, 135, 13]	C_1, C_{119}
[255, 175, 17]	C_1, C_{27}
[273, 191, 18]	C_1, C_{91}, C_{117}
[341, 205, 16]	C_1, C_{55}
[511, 199, 19]	C_5, C_{37}
[511, 259, 13]	C_1, C_{219}
[819, 435, 13]	C_1
[819, 447, 19]	C_1, C_{351}
[1023, 661, 23]	C_1, C_{53}, C_{341}
[1023, 781, 33]	$C_1, C_{53}, C_{123}, C_{341}$
[1057, 813, 34]	C_5, C_{43}, C_{151}
[1387, 783, 28]	C_1, C_{247}
[1971, 1105, 21]	C_1, C_{657}
[2047, 1167, 23]	C_1, C_{27}
[2325, 1335, 28]	C_1, C_{57}, C_{775}
[2325, 1373, 30]	C_1, C_{525}, C_{1085}
[2359, 1347, 22]	C_1
[3741, 2229, 29]	C_1
[3813, 2087, 28]	C_1, C_{369}, C_{1271}
[4095, 2767, 49]	$C_1, C_{41}, C_{235}, C_{733}$
[4095, 3367, 65]	$C_1, C_{41}, C_{235}, C_{273}, C_{411}, C_{733}$
[4161, 2827, 39]	C_1, C_{307}, C_{1387}
[4161, 3431, 66]	$C_1, C_{285}, C_{307}, C_{357}, C_{1387}$
[4681, 2681, 31]	C_1, C_{51}
[5461, 3781, 43]	C_1, C_{77}, C_{579}

A key feature of the cyclotomic coset-based construction is the ability to increment the minimum Hamming distance of a code by adding further weight from other idempotents and so steadily decrease the sparseness of the resulting parity-check matrix. Despite the construction method producing LDPC codes with no cycles of length 4, it is important to remark that codes that have cycles of length 4 in their parity-check matrices do not necessary have bad performance under iterative decoding, and a similar finding has been demonstrated by Tang et al. [31]. It has been observed that there are many cyclotomic coset-based LDPC codes that have this property, and the constraints in Algorithm 12.1 can be easily relaxed to allow the construction of cyclic LDPC codes with girth 4.

12.2.1 Mattson–Solomon Domain Construction of Binary Cyclic LDPC Codes

The $[n, k, d]$ cyclic LDPC codes presented in Sect. 4.4 are constructed using the sum of idempotents, which are derived from the cyclotomic cosets modulo n, to define the parity-check matrix. A different insight into this construction technique may be obtained by working in the Mattson–Solomon domain.

Let n be a positive odd integer, $\mathbb{F}_{2^m}$ be a splitting field for $x^n - 1$ over $\mathbb{F}_2$, α be a generator for $\mathbb{F}_{2^m}$, and $T_m(x)$ be a polynomial with maximum degree of $n - 1$ and coefficients in $\mathbb{F}_{2^m}$. Similar to Sect. 4.4, the notation of $T(x)$ is used as an alternative to $T_1(x)$ and the variables x and z are used to distinguish the polynomials in the domain and codomain. Let the decomposition of $z^n - 1$ into irreducible polynomials over $\mathbb{F}_2$ be contained in a set $\mathscr{F} = \{f_1(z), f_2(z), \ldots, f_t(z)\}$, i.e. $\prod_{1 \le i \le t} f_i(z) = z^n - 1$. For each $f_i(z)$, there is a corresponding primitive idempotent, denoted as $\theta_i(z)$, which can be obtained by [20]

$$\theta_i(z) = \frac{z(z^n - 1) f_i'(z)}{f_i(z)} + \delta(z^n - 1) \tag{12.3}$$

where $f_i'(z) = \frac{d}{dz} f_i(z)$, $f_i'(z) \in T(z)$ and the integer δ is defined by

$$\delta = \begin{cases} 1 & \text{if } \deg(f_i(z)) \text{ is odd,} \\ 0 & \text{otherwise.} \end{cases}$$

Let the decomposition of $z^n - 1$ and its corresponding primitive idempotent be listed as follows:

$$\begin{array}{ccc} u_1(x) & \theta_1(z) & f_1(z) \\ u_2(x) & \theta_2(z) & f_2(z) \\ \vdots & \vdots & \vdots \\ u_t(x) & \theta_t(z) & f_t(z). \end{array}$$

Here $u_1(x), u_2(x), \ldots, u_t(x)$ are the binary idempotents whose Mattson–Solomon polynomials are $\theta_1(z), \theta_2(z), \ldots, \theta_t(z)$, respectively. Assume that $\mathscr{I} \subseteq \{1, 2, \ldots, t\}$, let the binary polynomials $u(x) = \sum_{\forall i \in \mathscr{I}} u_i(x)$, $f(z) = \prod_{\forall i \in \mathscr{I}} f_i(z)$, and $\theta(z) =$

$\sum_{\forall i \in \mathscr{I}} \theta_i(z)$. It is apparent that, since $u_i(x) = \text{MS}^{-1}(\theta_i(z))$, $u(x) = \text{MS}^{-1}(\theta(z))$ and $u(x)$ is an idempotent.[1]

Recall that $u(x)$ is a low-weight binary idempotent whose reciprocal polynomial can be used to define the parity-check matrix of a cyclic LDPC code. The number of distinct n^{th} roots of unity which are also roots of the idempotent $u(x)$ determines the dimension of the resulting LDPC code. In this section, the design of cyclic LDPC codes is based on several important features of a code. These features, which are listed as follows, may be easily gleaned from the Mattson–Solomon polynomial of $u(x)$ and the binary irreducible factors of $z^n - 1$.

1. **Weight of the idempotent** $u(x)$
   The weight of $u(x)$ is the number of nth roots of unity which are zeros of $f(z)$. Note that, $f(\alpha^i) = 0$ if and only if $\theta(\alpha^i) = 1$ since an idempotent takes only the values 0 and 1 over $\mathbb{F}_{2^m}$. If $u(x)$ is written as $u_0 + u_i x + \cdots + u_{n-1}x^{n-1}$, following (11.2), we have

$$u_i = \theta(\alpha^i) \pmod 2 \qquad \text{for } i = \{0, 1, \ldots, n-1\}.$$

   Therefore, $u_i = 1$ precisely when $f(\alpha^i) = 0$, giving $\text{wt}_H(u(x))$ as the degree of the polynomial $f(z)$.
2. **Number of zeros of** $u(x)$
   Following (11.1), it is apparent that the number of zeros of $u(x)$ which are roots of unity, which is also the dimension of the code k, is

$$\text{Number of zeros of } u(x) = k = n - \text{wt}_H(\theta(z)). \tag{12.4}$$

3. **Minimum Hamming distance bound**
   The lower bound of the minimum Hamming distance of a cyclic code, defined by idempotent $u(x)$, is given by its BCH bound, which is determined by the number of consecutive powers of α, taken cyclically (mod n), which are roots of the generating idempotent $e_g(x) = 1 + u(x)$. In the context of $u(x)$, it is the same as the number of consecutive powers of α, taken cyclically (mod n), which are not roots of $u(x)$. Therefore, it is the largest number of consecutive non-zero coefficients in $\theta(z)$, taken cyclically (mod n).

The method of finding $f_i(z)$ is well established and using the above information, a systematic search for idempotents of suitable weight may be developed. To be efficient, the search procedure has to start with an increasing order of $\text{wt}_H(u(x))$ and this requires rearrangement of the set $\mathscr{F}$ such that $\deg(f_i(z)) < \deg(f_i i + 1(z))$ for all i. It is worth mentioning that it is not necessary to evaluate $u(x)$ by taking the

[1] Since the Mattson–Solomon polynomial of a binary polynomial is an idempotent and vice-versa [20], the Mattson–Solomon polynomial of a binary idempotent is also a binary idempotent.

Mattson–Solomon polynomial of $\theta(z)$, for each $f(z)$ obtained. It is more efficient to obtain $u(x)$ once the desired code criteria, listed above, are met.

For an exhaustive search, the complexity is of order $\mathscr{O}\left(2^{|\mathscr{F}|}\right)$. A search algorithm, see Algorithm 12.2, has been developed and it reduces the complexity considerably by targeting the search on the following key parameters. Note that this search algorithm, which is constructed in the Mattson–Solomon domain, is not constrained to find cyclic codes that have girth at least 6.

1. **Sparseness of the parity-check matrix**
   A necessary condition for the absence of cycles of length 4 is given by the inequality $\text{wt}_H(u(x))\,(\text{wt}_H(u(x)) - 1) \le n - 1$. Since $\text{wt}_H(u(x)) = \deg(f(z))$, a reasonable bound is
$$\sum_{\forall i \in \mathscr{I}} \deg(f_i(z)) \le \sqrt{n}.$$
   In practise, this limit is extended to enable the finding of good cyclic LDPC codes which have girth of 4 in their underlying Tanner graph.
2. **Code rate**
   The code rate is directly proportional to the number of roots of $u(x)$. If R_{min} represents the minimum desired code rate, then it follows from (12.4) that we can refine the search to consider the cases where
$$\text{wt}_H(\theta(z)) \le (1 - R_{min})n\,.$$
3. **Minimum Hamming distance**
   If the idempotent $u(x)$ is orthogonal on each position, then the minimum Hamming distance of the resulting code defined by $u(x)$ is equal to $1 + \text{wt}_H(u(x))$. However, for cyclic codes with cycles of length 4, there is no direct method to determine their minimum Hamming distance and the BCH bound provides a lower bound to the minimum Hamming distance. Let d be the lowest desired minimum Hamming distance and r_θ be the largest number of consecutive non-zero coefficients, taken cyclically, of $\theta(z)$. If a cyclic code has r_θ of d, then its minimum Hamming distance is at least $1 + d$. It follows that we can further refine the search with the constraint
$$r_\theta \ge d - 1.$$

In comparison to the construction method described in Sect. 4.4, we can see that the construction given in Sect. 4.4 starts from the idempotent $u(x)$, whereas this section starts from the idempotent $\theta(z)$, which is the Mattson–Solomon polynomial of $u(x)$. Both construction methods are equivalent and the same cyclic LDPC codes are produced.

Algorithm 12.2 MSCodeSearch($\mathbf{V}$, $index$)

Input:

$\mathbf{V} \Leftarrow$ a vector initialised to $\emptyset$
$index \Leftarrow$ an integer initialised to -1
$R_{min} \Leftarrow$ minimum code rate of interest
$d \Leftarrow$ lowest expected minimum distance
$\delta \Leftarrow$ small positive integer
$\mathbf{F}(z) \Leftarrow \{f_i(z)\}\ \forall i \in I$ sorted in ascending order of the degree
$\mathbf{Q}(z) \Leftarrow \{\theta_i(z)\}\ \forall i \in I$

Output:

CodesList contains set of codes

1: $\mathbf{T} \Leftarrow \mathbf{V}$
2: **for** $\left(i{=}index{+}1;\ i \leq |\mathcal{I}|;\ i{+}{+}\right)$ **do**
3: $\quad\mathbf{T}_{\text{prev}} \Leftarrow \mathbf{T}$
4: $\quad$**if** $\left(\sum_{\forall j \in \mathbf{T}} \deg(f_j(x)) + \deg(f_i(x)) \leq \sqrt{n} + \delta\right)$ **then**
5: $\quad\quad$Append i to $\mathbf{T}$
6: $\quad\quad\theta(z) \Leftarrow \sum_{\forall j \in \mathbf{T}} \theta_j(z)$
7: $\quad\quad$**if** $\left(\mathrm{wt}_H(\theta(z)) \leq (1 - R_{min})n \textbf{ and } r_\theta > d\right)$ **then**
8: $\quad\quad\quad u(x) \Leftarrow \mathrm{MS}^{-1}\left(\theta(z)\right)$
9: $\quad\quad\quad$**if** $\left(u(x) \text{ is non-degenerate}\right)$ **then**
10: $\quad\quad\quad\quad\mathscr{C} \Leftarrow$ a cyclic code defined by $u(x)$
11: $\quad\quad\quad\quad$**if** $\left(\mathscr{C} \notin \textbf{CodeList}\right)$ **then**
12: $\quad\quad\quad\quad\quad$Add $\mathbf{C}$ to **CodeList**
13: $\quad\quad\quad\quad$**end if**
14: $\quad\quad\quad$**end if**
15: $\quad\quad$**end if**
16: $\quad\quad$MSCodeSearch($\mathbf{T}$, i)
17: $\quad$**end if**
18: $\quad\mathbf{T} \Leftarrow \mathbf{T}_{\text{prev}}$
19: **end for**

Some good cyclic LDPC codes with cycles of length 4 found using Algorithm 12.2, which may also be found using Algorithm 12.1, are tabulated in Table 12.2. A check based on Lemma 12.1 may be easily incorporated in Step 12 of Algorithm 12.2 to filter out cyclic codes whose Tanner graph has girth of 4.

Figure 12.3 demonstrates the FER performance of several cyclic LDPC codes found by Algorithm 12.2. It is assumed that binary antipodal signalling is employed and the iterative decoder uses the RVCM algorithm described by Papagiannis et al. [23]. The FER performance is compared against the sphere packing lower bound offset for binary transmission. We can see that the codes $[127, 84, 10]$ and $[127, 99, 7]$, despite having cycles of length 4, are around 0.3 dB from the offset sphere packing lower bound at 10^{-4} FER. Figure 12.3c compares two LDPC codes of block size 255 and dimension 175, an algebraic code obtained by Algorithm 12.2 and an irregular code constructed using the PEG algorithm [10]. It can be seen that, in addition to having improved minimum Hamming distance, the cyclic LDPC code is 0.4 dB superior to the irregular code, and compared to the offset sphere packing lower bound, it is within 0.25 dB away at 10^{-4} FER. The effect of the error floor is apparent in the FER performance of the $[341, 205, 6]$ irregular LDPC code, as

Table 12.2 Several good cyclic LDPC codes with girth of 4

$[n, k, d]$	$u(x)$
$[51, 26, 10]$	$1 + x^3 + x^6 + x^{12} + x^{17} + x^{24} + x^{27} + x^{34} + x^{39} + x^{45} + x^{48}$
$[63, 44, 8]$	$1+x^9+x^{11}+x^{18}+x^{21}+x^{22}+x^{25}+x^{27}+x^{36}+x^{37}+x^{42}+x^{44}+x^{45}+x^{50}+x^{54}$
$[117, 72, 12]$	$1 + x + x^2 + x^4 + x^8 + x^{11} + x^{16} + x^{22} + x^{32} + x^{44} + x^{59} + x^{64} + x^{88}$
$[127, 84, 10]$	$1+x+x^2+x^4+x^8+x^{16}+x^{32}+x^{55}+x^{59}+x^{64}+x^{91}+x^{93}+x^{109}+x^{110}+x^{118}$
$[127, 91, 10]$	$1 + x^2 + x^{10} + x^{18} + x^{29} + x^{32} + x^{33} + x^{49} + x^{50} + x^{54} + x^{58} + x^{65} + x^{74} + x^{76} + x^{78} + x^{86} + x^{87} + x^{88} + x^{92} + x^{93} + x^{95}$
$[127, 92, 7]$	$1 + x^5 + x^{10} + x^{20} + x^{29} + x^{31} + x^{33} + x^{39} + x^{40} + x^{58} + x^{62} + x^{66} + x^{78} + x^{79} + x^{80} + x^{83} + x^{103} + x^{105} + x^{115} + x^{116} + x^{121} + x^{124}$
$[127, 99, 7]$	$1+x^{13}+x^{16}+x^{18}+x^{22}+x^{26}+x^{39}+x^{42}+x^{45}+x^{46}+x^{49}+x^{57}+x^{65}+x^{68}+x^{70}+x^{78}+x^{80}+x^{90}+x^{91}+x^{92}+x^{96}+x^{97}+x^{102}+x^{103}+x^{105}+x^{108}+x^{111}$

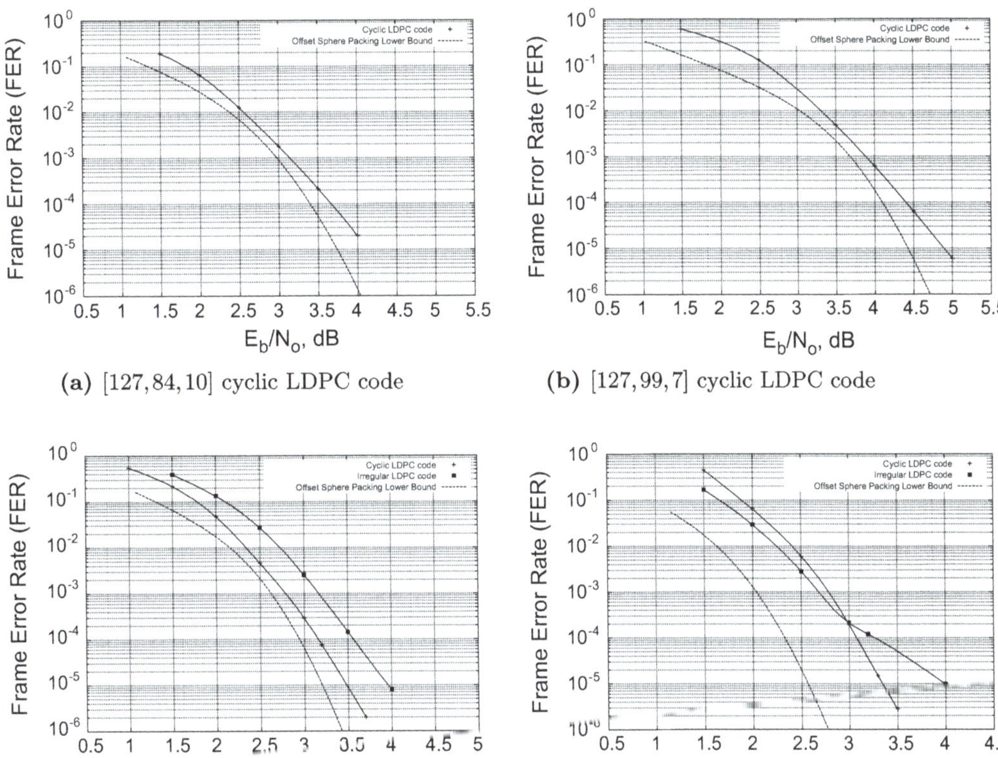

(a) [127,84,10] cyclic LDPC code

(b) [127,99,7] cyclic LDPC code

(c) [255,175,17] cyclic and [255,175,6] irregular PEG LDPC codes

(d) [341,205,16] cyclic and [341,205,6] irregular PEG LDPC codes

Fig. 12.3 FER performance of some binary cyclic LDPC codes

shown in Fig. 12.3d. The floor of this irregular code is largely attributed to minimum Hamming distance error events. Whilst this irregular code, at low SNR region, has better convergence than does the algebraic LDPC code of the same block length and dimension, the benefit of having higher minimum Hamming distance is obvious as the SNR increases. The [341, 205, 16] cyclic LDPC code is approximately 0.8 dB away from the offset sphere packing lower bound at 10^{-4} FER.

It is clear that short block length ($n \le 350$) cyclic LDPC codes have outstanding performance and the gap to the offset sphere packing lower bound is relatively close. However, as the block length increases, the algebraic LDPC codes, although these code have large minimum Hamming distance, have a convergence issue, and the threshold to the waterfall region is at larger E_b/N_0. The convergence problem arises because as the minimum Hamming distance increases, the weight of the idempotent $u(x)$, which defines the parity-check matrix, also increases. In fact, if $u(x)$ satisfies Lemma 12.1, we know that $\mathrm{wt}_H(u(x)) = d - 1$, where d is the minimum Hamming distance of the code. Large values of $\mathrm{wt}_H(u(x))$ result in a parity-check matrix that is not as sparse as that of a good irregular LDPC code of the same block length and dimension.

12.2.2 Non-Binary Extension of the Cyclotomic Coset-Based LDPC Codes

The code construction technique for the cyclotomic coset-based binary cyclic LDPC codes, which is discussed in Sect. 4.4, may be extended to non-binary fields. Similar to the binary case, the non-binary construction produces the dual-code idempotent which is used to define the parity-check matrix of the associated LDPC code.

Let m and m' be positive integers with $m \mid m'$, so that $\mathbb{F}_{2^m}$ is a subfield of $\mathbb{F}_{2^{m'}}$. Let n be a positive odd integer and $\mathbb{F}_{2^{m'}}$ be the splitting field of $x^n - 1$ over $\mathbb{F}_{2^m}$, so that $n|2^{m'} - 1$. Let $r = (2^{m'} - 1)/n$, $l = (2^{m'} - 1)/(2^m - 1)$, α be a generator for $\mathbb{F}_{2^{m'}}$ and β be a generator of $\mathbb{F}_{2^m}$, where $\beta = \alpha^l$. Let $T_a(x)$ be the set of polynomials of degree at most $n - 1$ with coefficients in $\mathbb{F}_{2^a}$. For the case of $a = 1$, we may denote $T_1(x)$ by $T(x)$ for convenience.

The Mattson–Solomon polynomial and its corresponding inverse, (11.1) and (11.2), respectively, may be redefined as

$$A(z) = \mathrm{MS}\left(a(x)\right) = \sum_{j=0}^{n-1} a(\alpha^{-rj}) z^j \tag{12.5}$$

$$a(x) = \mathrm{MS}^{-1}\left(A(z)\right) = \frac{1}{n}\sum_{i=0}^{n-1} A(\alpha^{ri}) x^i \tag{12.6}$$

where $a(x) \in T_{m'}(x)$ and $A(z) \in T_{m'}(z)$.

Recall that a polynomial $e(x) \in T_m(x)$ is termed an idempotent if the property $e(x) = e(x)^2 \pmod{x^n - 1}$ is satisfied. Note that $e(x) \neq e(x^2) \pmod{x^n - 1}$ unless $m = 1$. The following definition shows how to construct an idempotent for binary and non-binary polynomials.

Definition 12.9 (*Cyclotomic Idempotent*) Assume that $\mathscr{N}$ be a set as defined in Sect. 4.4, let $s \in \mathscr{N}$ and let $C_{s,i}$ represent the $(i+1)$th element of C_s, the cyclotomic coset of $s \pmod{n}$. Assume that the polynomial $e_s(x) \in T_m(x)$ is given by

$$e_s(x) = \sum_{0 \le i \le |C_s|-1} e_{C_{s,i}} x^{C_{s,i}}, \tag{12.7}$$

where $|C_s|$ is the number of elements in C_s. In order for $e_s(x)$ to be an idempotent, its coefficients may be chosen in the following manner:

(i) if $m = 1$, $e_{C_{s,i}} = 1$,
(ii) otherwise, $e_{C_{s,i}}$ is defined recursively as follows:

$$\begin{aligned} &\text{for } i = 0,\ e_{C_{s,i}} \in \{1, \beta, \beta^2, \ldots, \beta^{2^m-2}\}, \\ &\text{for } i > 0,\ e_{C_{s,i}} = e^2_{C_{s,i-1}}. \end{aligned}$$

We refer to the idempotent $e_s(x)$ as a cyclotomic idempotent.

Definition 12.10 (*Parity-Check Idempotent*) Let $\mathscr{M} \subseteq \mathscr{N}$ and let $u(x) \in T_m(x)$ be

$$u(x) = \sum_{s \in \mathscr{M}} e_s(x). \tag{12.8}$$

The polynomial $u(x)$ is an idempotent and it is called a parity-check idempotent.

As in Sect. 4.4, the parity-check idempotent $u(x)$ is used to define the $\mathbb{F}_{2^m}$ cyclic LDPC code over $\mathbb{F}_{2^m}$, which may be denoted by $[n, k, d]_{2^m}$. The parity-check matrix consists of n cyclic shifts of $x^{\deg(u(x))}u(x^{-1})$. For the non-binary case, the minimum Hamming distance d of the cyclic code is bounded by

$$d_0 + 1 \le d \le \min\left(\text{wt}_H(g(x)), \text{wt}_H(1 + u(x))\right),$$

where d_0 is the maximum run of consecutive ones in $U(z) = \text{MS}(u(x))$, taken cyclically mod n.

Based on the description given above, a procedure to construct a cyclic LDPC code over $\mathbb{F}_{2^m}$ is as follows.

1. For integers m and n, obtain the splitting field ($\mathbb{F}_{2^{m'}}$) of $x^n - 1$ over $\mathbb{F}_{2^m}$. Unless the condition of $m \mid m'$ is satisfied, $\mathbb{F}_{2^m}$ cyclic LDPC code of length n cannot be constructed.
2. Generate the cyclotomic cosets modulo $2^{m'} - 1$ denoted as C'.
3. Derive a polynomial $p(x)$ from C' and let $s \in \mathscr{N}$ be the smallest positive integer such that $|C'_s| = m$. The polynomial $p(x)$ is the minimal polynomial of α^s,

$$p(x) = \prod_{0 \le i < m} \left(x + \alpha^{C'_{s,i}}\right). \tag{12.9}$$

Construct all elements of $\mathbb{F}_{2^m}$ using $p(x)$ as the primitive polynomial.

4. Let C be the cyclotomic cosets modulo n and let $\mathscr{N}$ be a set containing the smallest number in each coset of C. Assume that there exists a non-empty set $\mathscr{M} \subset \mathscr{N}$ and following Definition 12.10 construct the parity-check idempotent $u(x)$. The coefficients of $u(x)$ can be assigned following Definition 12.9.
5. Generate the parity-check matrix of $\mathscr{C}$ using the n cyclic shifts of $x^{\deg(u(x))}u(x^{-1})$.
6. Compute r and l, and then take the Mattson–Solomon polynomial of $u(x)$ to produce $U(z)$. Obtain the code dimension and the lower bound of the minimum Hamming distance from $U(z)$.

Example 12.3 Consider the construction of a $n = 21$ cyclic LDPC code over $\mathbb{F}_{2^6}$. The splitting field of $x^{21} - 1$ over $\mathbb{F}_{2^6}$ is $\mathbb{F}_{2^6}$, and this implies that $m = m' = 6$, $r = 3$ and $l = 1$. Let C and C' denote the cyclotomic cosets modulo n and $2^{m'} - 1$, respectively. We know that $|C'_1| = 6$ and therefore the primitive polynomial $p(x)$ has roots of α^j, for all $j \in C'_1$, i.e. $p(x) = 1 + x + x^6$. By letting $1 + \beta + \beta^6 = 0$, all of the elements of $\mathbb{F}_{2^6}$ can be defined. If $u(x)$ is the parity-check idempotent generated by the sum of the cyclotomic idempotents defined by C_s, where $s \in \{\mathscr{M} : 5, 7, 9\}$ and $e_{C_{s,0}}$ for all $s \in \mathscr{M}$ be β^{23}, 1 and 1, respectively,

$$u(x) = \beta^{23}x^5 + x^7 + x^9 + \beta^{46}x^{10} + \beta^{43}x^{13} + x^{14} + x^{15} + \beta^{53}x^{17} + x^{18} + \beta^{58}x^{19} + \beta^{29}x^{20}$$

and its Mattson–Solomon polynomial $U(z)$ indicates that it is a $[21, 15, \geq 5]_{2^6}$ cyclic code, whose binary image is a [126, 90, 8] linear code.

The following systematic search algorithm is based on summing each possible combination of the cyclotomic idempotents to search for all possible $\mathbb{F}_{2^m}$ cyclic codes of a given length. As in Algorithm 12.2, the search algorithm targets the following key parameters:

1. **Sparseness of the resulting parity-check matrix**
   Since the parity-check matrix is directly derived from $u(x)$ which consists of the sum of the cyclotomic idempotents, only low-weight cyclotomic idempotents are of interest. Let W_{max} be the maximum $\mathrm{wt}_H(u(x))$; then the search algorithm will only choose the cyclotomic idempotents whose sum has total weight less than or equal to W_{max}.
2. **High code rate**
   The number of roots of $u(x)$ which are also roots of unity define the dimension of the resulting LDPC code. Let the integer k_{min} be defined as the minimum code dimension, and the cyclotomic idempotents that are of interest are those whose Mattson–Solomon polynomial has at least k_{min} zeros.
3. **High minimum Hamming distance**
   Let the integer d' be the smallest value of the minimum Hamming distance of the code. The sum of the cyclotomic idempotents should have at least $d' - 1$ consecutive powers of β which are roots of unity but not roots of $u(x)$.

Table 12.3 Examples of $[n, k, d]_{2^m}$ cyclic LDPC codes

q	$[n, k]$	$u(x)$	d	$d_b^\dagger$	Comment
$\mathbb{F}_{2^2}$	[51, 29]	$\beta^2x^3 + \beta x^6 + \beta^2x^{12} + \beta^2x^{17} + \beta x^{24} + \beta x^{27} + \beta x^{34} + \beta^2x^{39} + \beta x^{45} + \beta^2x^{48}$	≥5	10	$m = 2$, $m' = 8$, $r = 5$ and $l = 85$
	[255, 175]	$\beta x^7 + \beta^2x^{14} + \beta x^{28} + \beta^2x^{56} + x^{111} + \beta x^{112} + x^{123} + \beta^2x^{131} + x^{183} + x^{189} + \beta x^{193} + x^{219} + x^{222} + \beta^2x^{224} + x^{237} + x^{246}$	≥17	≤20	$m = 2$, $m' = 8$, $r = 1$ and $l = 85$
	[273, 191]	$\beta^2x^{23} + \beta x^{37} + \beta x^{46} + \beta^2x^{74} + \beta x^{91} + \beta^2x^{92} + \beta^2x^{95} + \beta^2x^{107} + x^{117} + \beta x^{148} + \beta^2x^{155} + \beta^2x^{182} + \beta x^{184} + \beta x^{190} + x^{195} + \beta x^{214} + x^{234}$	≥18	≤20	$m = 2$, $m' = 12$, $r = 15$ and $l = 1365$
$\mathbb{F}_{2^3}$	[63, 40]	$1 + \beta^5x^9 + \beta x^{13} + \beta^3x^{18} + \beta^2x^{19} + \beta^2x^{26} + \beta^6x^{36} + \beta^4x^{38} + \beta x^{41} + \beta^4x^{52}$	≥6	10	$m = 3$, $m' = 6$, $r = 1$ and $l = 9$
	[63, 43]	$\beta^2x^9 + \beta^3x^{11} + \beta^4x^{18} + x^{21} + \beta^6x^{22} + \beta^3x^{25} + x^{27} + \beta x^{36} + \beta^5x^{37} + x^{42} + \beta^5x^{44} + x^{45} + \beta^6x^{50} + x^{54}$	≥8	≤12	$m = 3$, $m' = 6$, $r = 1$ and $l = 9$
	[91, 63]	$\beta^6x + \beta^5x^2 + \beta^3x^4 + \beta^6x^8 + \beta x^{13} + \beta^5x^{16} + \beta^5x^{23} + \beta^2x^{26} + \beta^3x^{32} + \beta^5x^{37} + \beta^3x^{46} + \beta^4x^{52} + \beta^6x^{57} + \beta^6x^{64} + \beta^3x^{74}$	≥8	≤10	$m = 3$, $m' = 12$, $r = 45$ and $l = 585$
$\mathbb{F}_{2^4}$	[85, 48]	$1 + \beta^{12}x^{21} + \beta^9x^{42} + \beta^6x^{53} + \beta^3x^{69} + \beta^9x^{77} + \beta^{12}x^{81} + \beta^6x^{83} + \beta^3x^{84}$	≥7	≤12	$m = 4$, $m' = 8$, $r = 3$ and $l = 17$
$\mathbb{F}_{2^5}$	[31, 20]	$1 + \beta^{28}x^5 + \beta^7x^9 + \beta^{25}x^{10} + x^{11} + x^{13} + \beta^{14}x^{18} + \beta^{19}x^{20} + x^{21} + x^{22} + x^{26}$	≥7	12	$m = 5$, $m' = 5$, $r = l$ and $l = 1$
	[31, 21]	$\beta^{23}x^5 + \beta^{29}x^9 + \beta^{15}x^{10} + \beta x^{11} + \beta^4x^{13} + \beta^{27}x^{18} + \beta^{30}x^{20} + \beta^{16}x^{21} + \beta^2x^{22} + \beta^8x^{26}$	≥4	8	$m = 5$, $m' = 5$, $r = 1$ and $l = 1$
$\mathbb{F}_{2^6}$	[21, 15]	$\beta^{23}x^5 + x^7 + x^9 + \beta^{46}x^{10} + \beta^{43}x^{13} + x^{14} + x^{15} + \beta^{53}x^{17} + x^{18} + \beta^{58}x^{19} + \beta^{29}x^{20}$	≥5	8	$m = 6$, $m' = 6$, $r = 3$ and $l = 1$

†The minimum Hamming distance of the binary image which has been determined using the improved Zimmermann algorithm, Algorithm 5.1

Following Definition 12.10 and the Mattson–Solomon polynomial

$$U(z) = \text{MS}\left(\sum_{s \in \mathscr{M}} e_s(x)\right) = \sum_{s \in \mathscr{M}} E_s(z),$$

it is possible to maximise the run of the consecutive ones in $U(z)$ by varying the coefficients of $e_s(x)$. It is therefore important that all possible non-zero values of $e_{C_{s,0}}$ for all $s \in \mathscr{M}$ are included to guarantee that codes with the highest possible minimum Hamming distance are found.

Table 12.3 outlines some examples of $[n, k, d]_{2^m}$ cyclic LDPC codes. The non-binary algebraic LDPC codes in this table perform well under iterative decoding as shown in Fig. 12.4 assuming binary antipodal signalling and the AWGN channel. The RVCM algorithm is employed in the iterative decoder. The FER performance of these non-binary codes is compared to the offset sphere packing lower bound in Fig. 12.4.

As mentioned in Sect. 12.1.2, there is an inverse relationship between the convergence of the iterative decoder and the minimum Hamming distance of a code. The algebraic LDPC codes, which have higher minimum Hamming distances compared to irregular LDPC codes, do not converge well at long block lengths. It appears that

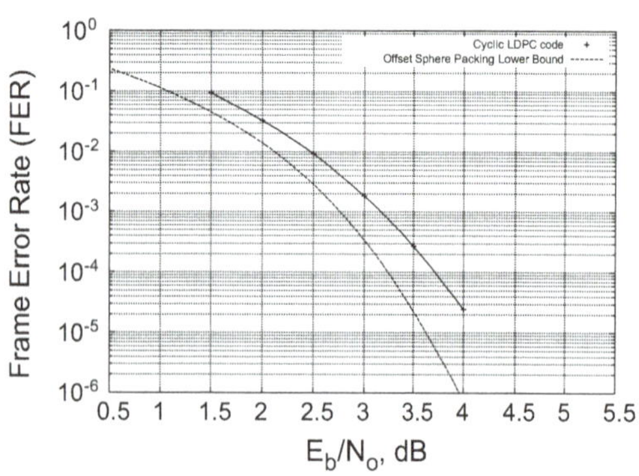

(a) $[51, 29, \geq 5]_{2^2}$ cyclic LDPC code

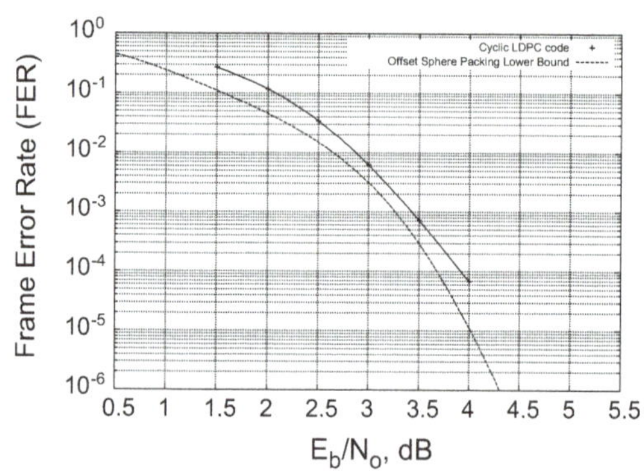

(b) $[21, 15, \geq 5]_{2^6}$ cyclic LDPC code

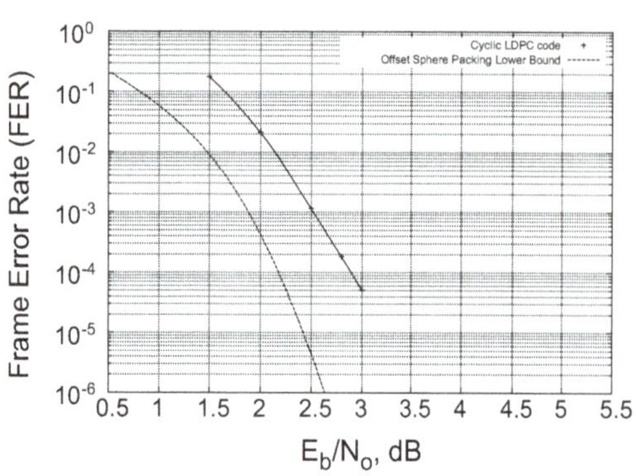

(c) $[85, 48, \geq 7]_{2^4}$ cyclic LDPC codes

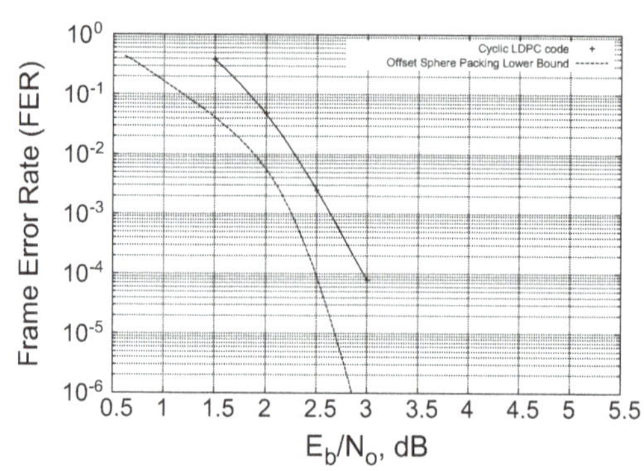

(d) $[255, 175, \geq 17]_{2^2}$ cyclic LDPC codes

Fig. 12.4 FER performance of some non-binary cyclic LDPC codes

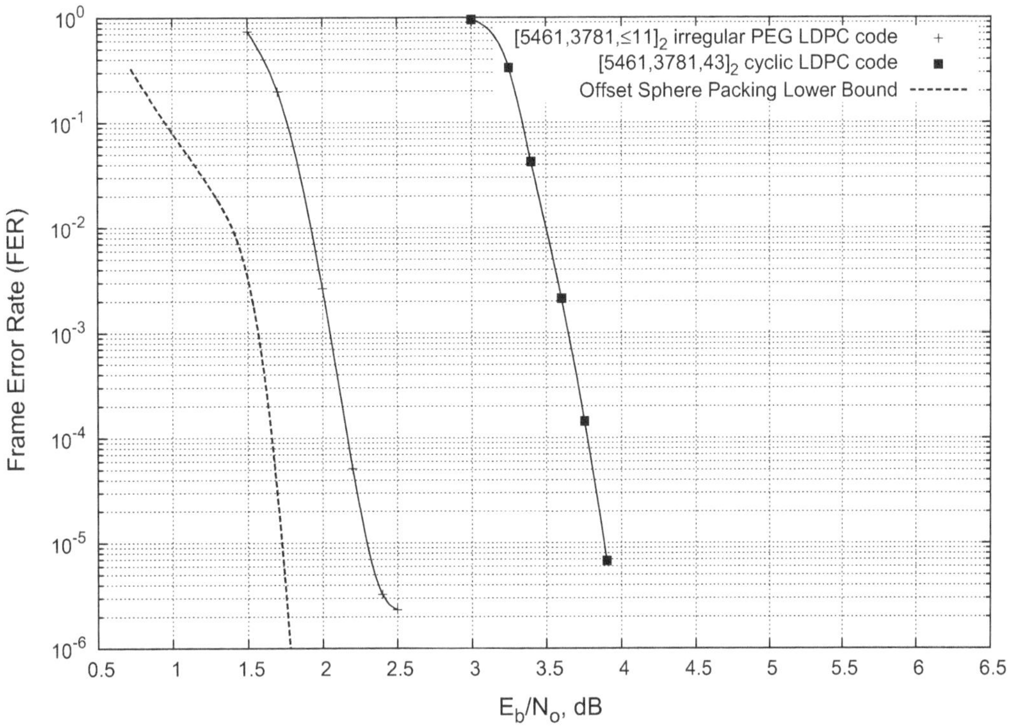

Fig. 12.5 FER performance of algebraic and irregular LDPC codes of rate 0.6924 and code length 5461 bits

the best convergence at long code lengths can only be realised by irregular LDPC codes with good degree distributions. Figure 12.5 shows the performance of two LDPC codes of block length 5461 bits and code rate 0.6924; one is an irregular code constructed using the PEG algorithm and the other one is an algebraic code of minimum Hamming distance 43 based on cyclotomic coset and idempotent construction (see Table 12.1). These results are for the AWGN channel using binary antipodal signalling with a belief propagation iterative decoder featuring 100 iterations. We can see that at 10^{-5} FER, the irregular PEG code is superior by approximately 1.6 dB compared to the algebraic cyclic LDPC code. However, for short code lengths, algebraic LDPC codes are superior. The codes have better performance and have simpler encoders than ad hoc designed LDPC codes.

12.3 Irregular LDPC Codes from Progressive Edge-Growth Construction

It is shown by Hu et al. [11] that LDPC codes obtained using the PEG construction method can perform better than other types of randomly constructed LDPC codes. The PEG algorithm adds edges to each vertex such that the local girth is maximised. The PEG algorithm considers only the variable degree sequence, and the check degree

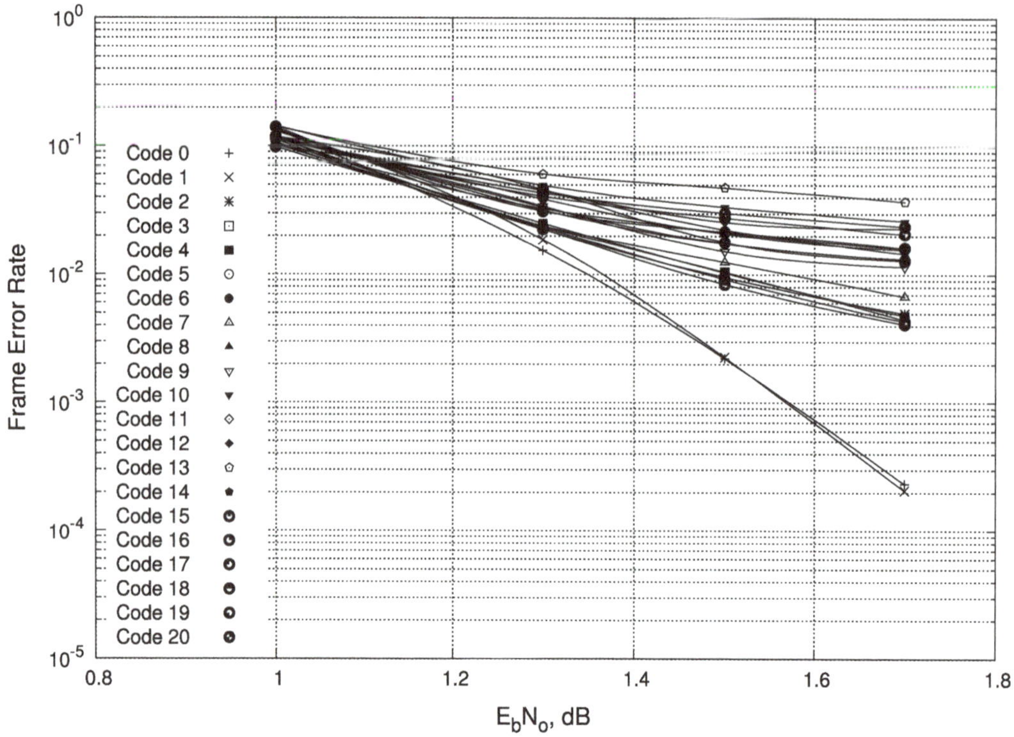

Fig. 12.6 Effect of vertex degree ordering in PEG algorithm

sequence is maintained to be as uniform as possible. In this section, the results of experimental constructions of irregular LDPC codes using the PEG algorithm are presented. Analysis on the effects of the vertex degree ordering and degree sequences have been carried out by means of computer simulations. All simulation results in this section, unless otherwise stated, were obtained using binary antipodal signalling with the belief propagation decoder using 100 iterations. Each simulation run was terminated after the decoder had produced 100 erroneous frames.

Figure 12.6 shows the FER performance of various [2048, 1024] irregular LDPC codes constructed using the PEG algorithm with different vertex degree orderings. These LDPC codes have variable degree sequence $\Lambda_\lambda(x) = 0.475x^2 + 0.280x^3 + 0.035x^4 + 0.109x^5 + 0.101x^{15}$. Let $(v_0, v_1, \ldots, v_i, \ldots, v_{n-1})$ be a set of variable vertices of an LDPC code. Code 0 and Code 1 LDPC codes were constructed with an increasing vertex degree ordering, i.e. $\deg(v_0) \le \deg(v_1) \le \cdots \le \deg(v_{n-1})$, whereas the remaining LDPC codes were constructed with random vertex degree ordering.

Figure 12.6 clearly shows that, unless the degree of the variable vertices is assigned in an increasing order, poor LDPC codes are obtained. In random degree ordering of half rate codes, it is very likely to encounter the situation where, as the construction approaches the end, there are some low-degree variable vertices that have no edge connected to them. Since almost all of the variable vertices would have already had edges connected to them, the low-degree variable vertices would not have many choice of edges to connect in order to maximise the local girth. It has been observed that, in many cases, these low-degree variable vertices are connected to each other, forming a cycle which involves all vertices, and the resulting LDPC codes often have a low minimum Hamming distance. If d variable vertices are connected to each other and a cycle of length $2d$ is formed, then the minimum Hamming distance of the resulting code is d because the sum of these d columns in the corresponding parity-check matrix $\boldsymbol{H}$ is $\mathbf{0}^T$.

In contrast, for the alternative construction which starts from an increasing degree of the variable vertices, edges are connected to the low-degree variable vertices earlier in the process. Short cycles, which involve the low-degree variable vertices and lead to low minimum Hamming distance, may be avoided by ensuring these low-degree variable vertices have edges connected to the parity-check vertices which are connected to high-degree variable vertices.

It can be expected that the PEG algorithm will almost certainly produce poor LDPC codes if the degree of the variable vertices is assigned in descending order. It is concluded that all PEG-based LDPC codes should be constructed with increasing variable vertex degree ordering.

Figure 12.7 shows the effect of low-degree variable vertices, especially λ_2 and λ_3, on the FER performance of various [512, 256] PEG-constructed irregular LDPC codes. Table 12.4 shows the variable degree sequences of the simulated irregular codes. Figure 12.7 indicates that, with the fraction of high-degree variable vertices kept constant, the low-degree variable vertices have influence over the convergence

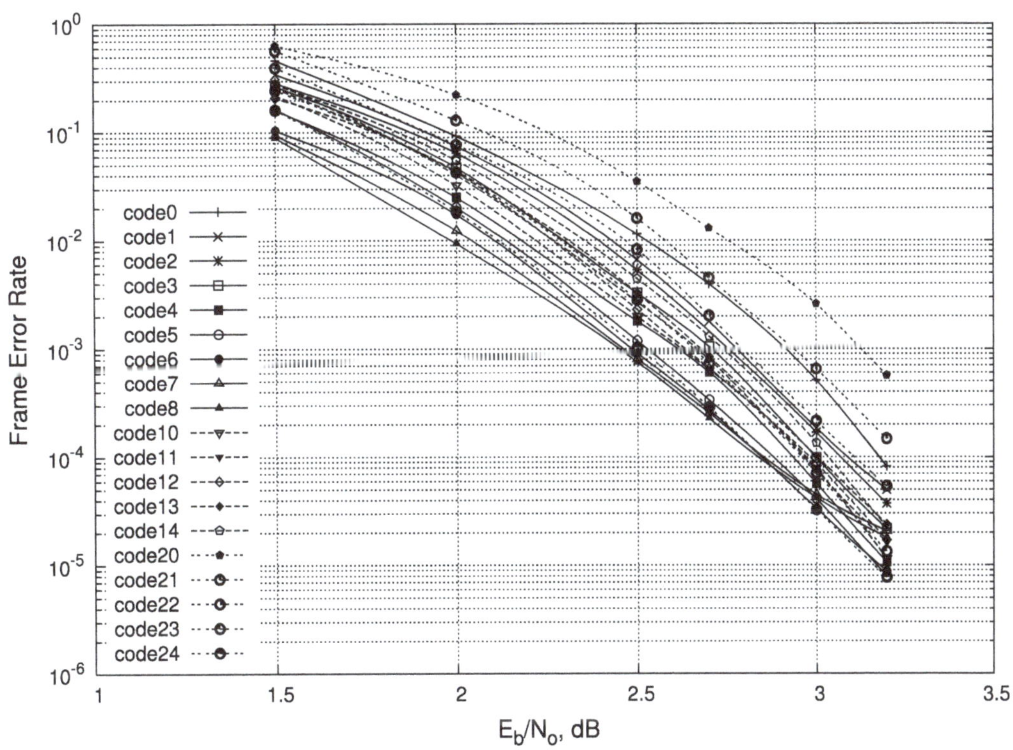

Fig. 12.7 Effect of low-degree variable vertices

Table 12.4 Variable degree sequences for codes in Fig. 12.7

Code	λ_2	λ_3	λ_4	λ_5	λ_{14}
Code 0	0.150	0.350	0.350	0.050	0.100
Code 1	0.200	0.325	0.325	0.050	0.100
Code 2	0.250	0.300	0.300	0.050	0.100
Code 3	0.300	0.275	0.275	0.050	0.100
Code 4	0.350	0.250	0.250	0.050	0.100
Code 5	0.400	0.225	0.225	0.050	0.100
Code 6	0.450	0.200	0.200	0.050	0.100
Code 7	0.500	0.175	0.175	0.050	0.100
Code 8	0.550	0.150	0.150	0.050	0.100
Code 10	0.150	0.700	0.000	0.050	0.100
Code 11	0.200	0.550	0.100	0.050	0.100
Code 12	0.250	0.400	0.200	0.050	0.100
Code 13	0.300	0.250	0.300	0.050	0.100
Code 14	0.350	0.100	0.400	0.050	0.100
Code 20	0.150	0.000	0.700	0.050	0.100
Code 21	0.200	0.100	0.550	0.050	0.100
Code 22	0.250	0.200	0.400	0.050	0.100
Code 23	0.300	0.300	0.250	0.050	0.100
Code 24	0.350	0.400	0.100	0.050	0.100

in the waterfall region. As the fraction of low-degree variable vertices is increased, the FER in the low signal-to-noise ratio (SNR) region improves. On the other hand, LDPC codes with a high fraction of low-degree variable vertices tend to have low minimum Hamming distance and as expected, these codes exhibit early error floors. This effect is clearly depicted by Code 7 and Code 8, which have the highest fraction of low-degree variable vertices among all the codes in Fig. 12.7. Of all of the codes, Code 6 and Code 24 appear to have the best performance.

Figure 12.8 demonstrates the effect of high-degree variable vertices on the FER performance. These codes are rate 3/4 irregular LDPC codes of length 1024 bits with the same degree sequences, apart from their maximum variable vertex degree. One group has maximum degree of 8 and the other group has maximum degree of 12. From Fig. 12.8, it is clear that the LDPC codes with maximum variable vertex degree of 12 converge better under iterative decoding than those codes with maximum variable vertex degree of 8.

In a similar manner to Fig. 12.7, the effect of having various low-degree variable vertices is also demonstrated in Fig. 12.9. In this case, the LDPC codes are constructed to have the advantageous linear-time encoding complexity, where the parity symbols are commonly described as having a zigzag pattern [26]. In this case, λ_1 and λ_2 of these LDPC codes are fixed and the effect of varying λ_3, λ_4 and λ_5 is investigated.

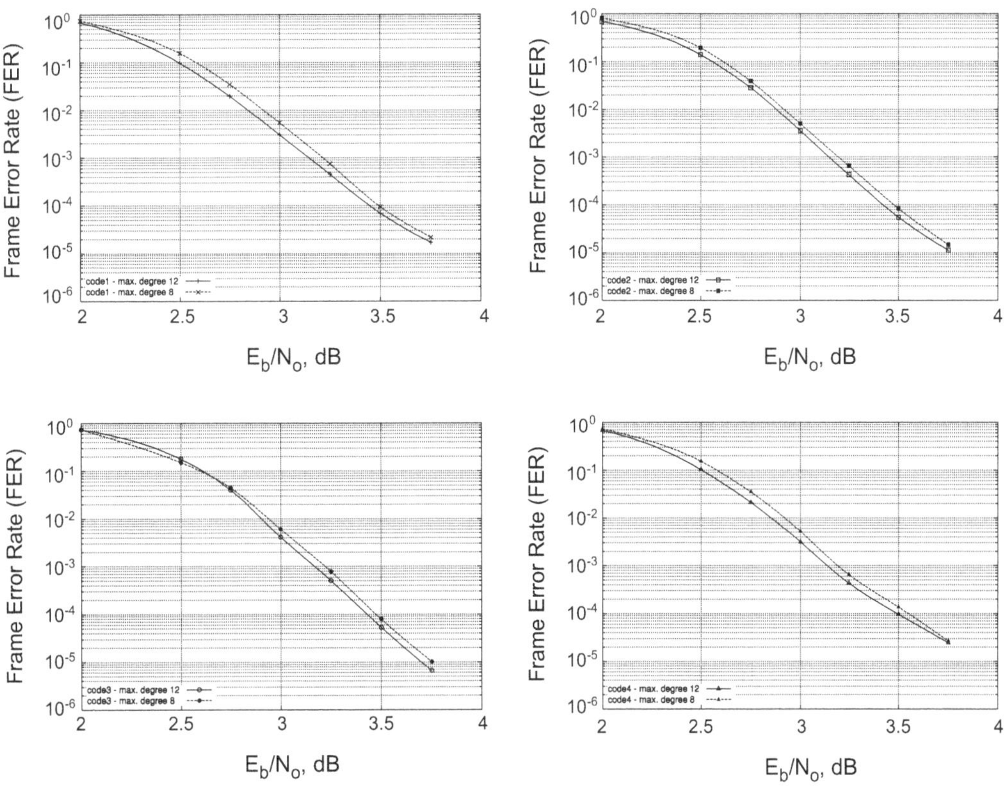

Fig. 12.8 Effect of high-degree variable vertices

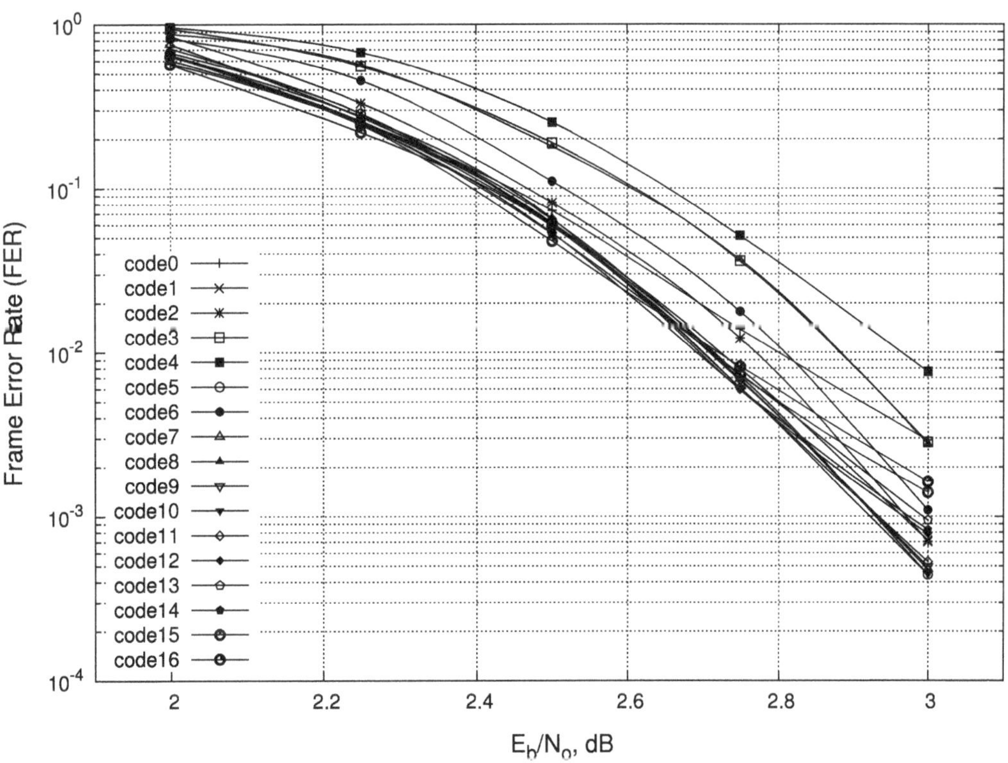

Fig. 12.9 Effect of varying low-degree variable vertices

Table 12.5 Variable degree sequences of LDPC codes in Fig. 12.9

Code	λ_1	λ_2	λ_3	λ_4	λ_5	λ_{12}
Code 0	0.000625	0.249375	0.644375			0.105625
Code 1	0.000625	0.249375	0.420000	0.224375		0.105625
Code 2	0.000625	0.249375	0.195000	0.449375		0.105625
Code 3	0.000625	0.249375		0.420000	0.224375	0.105625
Code 4	0.000625	0.249375		0.195000	0.449375	0.105625
Code 5	0.000625	0.249375	0.420000	0.111875	0.111875	0.106250
Code 6	0.000625	0.249375	0.195000	0.224375	0.224375	0.106250
Code 7	0.000625	0.249375	0.420000		0.224375	0.105625
Code 8	0.000625	0.249375	0.195000		0.449375	0.105625
Code 9	0.000625	0.249375	0.449375		0.195000	0.105625
Code 10	0.000625	0.249375	0.449375	0.097500	0.097500	0.105625
Code 11	0.000625	0.249375	0.449375	0.044375	0.150000	0.106250
Code 12	0.000625	0.249375	0.495000		0.150000	0.105000
Code 13	0.000625	0.249375	0.495000	0.075000	0.075000	0.105000
Code 14	0.000625	0.249375	0.495000	0.037500	0.111875	0.105625
Code 15	0.000625	0.249375	0.570000		0.075000	0.105000
Code 16	0.000625	0.249375	0.570000	0.037500	0.037500	0.105000

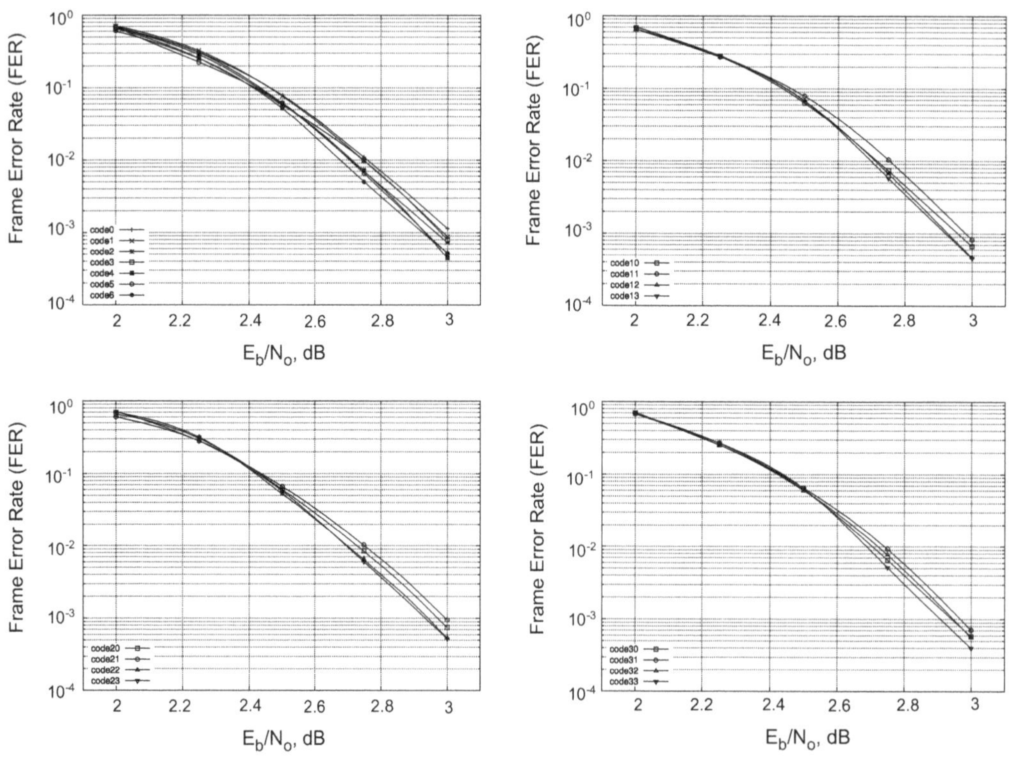

Fig. 12.10 Effect of varying high-degree variable vertices

Table 12.6 Variable degree sequences of LDPC codes in Fig. 12.10

Code	λ_3	λ_4	λ_5	λ_8	λ_9	λ_{10}	λ_{11}	λ_{12}	λ_{13}	λ_{14}
Code 0	0.420000	0.111875	0.111875	0.10625						
Code 1	0.420000	0.111875	0.111875		0.10625					
Code 2	0.420000	0.111875	0.111875			0.10625				
Code 3	0.420000	0.111875	0.111875				0.10625			
Code 4	0.420000	0.111875	0.111875					0.10625		
Code 5	0.420000	0.111875	0.111875						0.10625	
Code 6	0.420000	0.111875	0.111875							0.10625
Code 10	0.434375	0.111875	0.111875					0.091875		
Code 11	0.449375	0.111875	0.111875					0.076875		
Code 12	0.405000	0.111875	0.111875					0.121250		
Code 13	0.390000	0.111875	0.111875					0.136250		
Code 20	0.420000	0.127500	0.111875					0.090625		
Code 21	0.420000	0.142500	0.111875					0.075625		
Code 22	0.420000	0.097500	0.111875					0.120625		
Code 23	0.420000	0.082500	0.111875					0.135625		
Code 30	0.420000	0.111875	0.127500					0.090625		
Code 31	0.420000	0.111875	0.142500					0.075625		
Code 32	0.420000	0.111875	0.097500					0.120625		
Code 33	0.420000	0.111875	0.082500					0.135625		

Note that $\lambda_1 = 0.000625$ and $\lambda_2 = 0.249375$

The variable degree sequences of the LDPC codes under investigation, which are rate 3/4 codes of length 1600 bits, are depicted in Table 12.5. The results show that, as in the previous cases, these low-degree variable vertices contribute to the waterfall region of the FER curve. The contribution of λ_i is more significant than that of λ_{i+1} and this may be observed by comparing the FER curves of Code 1 with either Code 3 or Code 4, which has λ_3 of 0.0. We can also see that Code 0, which has the most variable vertices of low degree, exhibits a high error floor.

In contrast to Fig. 12.9, Fig. 12.10 shows the effect of varying high-degree variable vertices. The LDPC codes considered here all have the same code rate and code length as those in Fig. 12.9 and their variable degree sequences are shown in Table 12.6. The results show that

- The contribution of the high-degree variable vertices is in the high SNR region. Consider Code 10 to Code 33, those LDPC codes that have larger λ_{12} tend to be more resilient to errors in the high SNR region than those with smaller λ_{12}. At $E_b/N_o = 3.0$ dB, Code 10, Code 11 and Code 12 are inferior to Code 13 and similarly, Code 23 and Code 33 have the best performance in their group.
- Large values of maximum variable vertex degree may not always lead to improved FER performance. For example, Code 5 and Code 6 do not perform as well as Code 4 at $E_b/N_o = 3.0$ dB. This may be explained as follows. As the maximum variable vertex degree is increased, some of the variable vertices have many edges connected to them, in the other words the corresponding symbols are checked by many parity-check equations. This increases the chances of having unreliable information from some of these equations during iterative decoding. In addition, a larger maximum variable vertex degree also increases the number of short cycles in the underlying Tanner graph of the code. It was concluded also by McEliece et al. [22] and by Etzion et al. [7] that short cycles lead to negative contributions preventing the convergence of the iterative decoder.

12.4 Quasi-cyclic LDPC Codes and Protographs

Despite irregular LDPC codes having lower error rates than their regular counterparts, Luby et al. [18], the extra complexity of the encoder and decoder hardware structure, has made this class of LDPC codes unattractive from an industry point of view. In order to encode an irregular code which has a parity-check matrix $\boldsymbol{H}$, Gaussian elimination has to be done to transform this matrix into reduced echelon form. Irregular LDPC codes, as shown in Sect. 12.3, may also be constructed by constraining the $n - k$ low-degree variable vertices of the Tanner graph to form a zigzag pattern, as pointed out by Ping et al. [26]. Translating these $n - k$ variable vertices of the Tanner graph into matrix form, we have

$$\boldsymbol{H}_p = \begin{bmatrix} 1 & & & & \\ 1 & 1 & & & \\ & \vdots & \vdots & & \\ & & 1 & 1 & \\ & & & 1 & 1 \end{bmatrix}. \tag{12.10}$$

The matrix $\boldsymbol{H}_p$ is non-singular and the columns of this matrix may be used as the coordinates of the parity-check bits of an LDPC code.

The use of zigzag parity checks does simplify the derivation of the encoder as the Gaussian elimination process is no longer necessary and encoding, assuming that

$$\boldsymbol{H} = \left[\boldsymbol{H}_u | \boldsymbol{H}_p\right]$$

$$= \begin{array}{c} \begin{array}{ccccccccccc} v_0 & v_1 & \cdots & v_{k-2} & v_{k-1} & v_k & v_{k+1} & \cdots & v_{n-2} & v_{n-1} \end{array} \\ \left[\begin{array}{ccccc|ccccc} u_{0,0} & u_{0,1} & \cdots & u_{0,k-2} & u_{0,k-1} & 1 & & & & \\ u_{1,0} & u_{1,1} & \cdots & u_{1,k-2} & u_{1,k-1} & 1 & 1 & & & \\ \vdots & \vdots & \vdots & \vdots & \vdots & & \ddots & \ddots & & \\ u_{n-k-2,0} & u_{n-k-2,1} & \cdots & u_{n-k-2,k-2} & u_{n-k-2,k-1} & & & 1 & 1 & \\ u_{n-k-1,0} & u_{n-k-1,1} & \cdots & u_{n-k-1,k-2} & u_{n-k-1,k-1} & & & & 1 & 1 \end{array}\right] \end{array},$$

can be performed by calculating parity-check bits as follows:

$$v_k = \sum_{j=0}^{k-1} v_j u_{0,j} \pmod 2$$

$$v_i = v_{i-1} + \sum_{j=0}^{k-1} v_j u_{i-k,j} \pmod 2 \qquad \text{for } k+1 \leq i \leq n-1\,.$$

Nevertheless, zigzag parity bit checks do not lead to a significant reduction in encoder storage space as the matrix $\boldsymbol{H}_u$ still needs to be stored. It is necessary to introduce additional structure in $\boldsymbol{H}_u$, such as using a quasi-cyclic property, to reduce significantly the storage requirements of the encoder.

12.4.1 Quasi-cyclic LDPC Codes

Quasi-cyclic codes have the property that each codeword is a m-sized cyclic shift of another codeword, where m is an integer. With this property simple feedback shift registers may be used for the encoder. This type of code is known as circulant codes defined by circulant polynomials and depending on the polynomials can have significant mathematical structure as described in Chap. 9. A circulant matrix is a square matrix where each row is a cyclic shift of the previous row and the first row is the cyclic shift of the last row. In addition, each column is also a cyclic shift of the previous column and the column weight is equal to the row weight.

A circulant matrix is defined by a polynomial $r(x)$. If $r(x)$ has degree $<m$, the corresponding circulant matrix is an $m \times m$ square matrix. Let $\boldsymbol{R}$ be a circulant matrix defined by $r(x)$, then $\boldsymbol{M}$ is of the form

$$\boldsymbol{R} = \begin{bmatrix} r(x) \pmod{x^m - 1} \\ xr(x) \pmod{x^m - 1} \\ \vdots \\ x^i r(x) \pmod{x^m - 1} \\ \vdots \\ x^{m-1} r(x) \pmod{x^m - 1} \end{bmatrix} \tag{12.11}$$

where the polynomial in each row can be represented by an m-dimensional vector, which contains the coefficients of the corresponding polynomial. A quasi-cyclic code can be built from the concatenation of circulant matrices to define the generator or parity-check matrix.

Example 12.4 A quasi-cyclic code with defining polynomials $r_1(x) = 1 + x + x^3$ and $r_2(x) = 1+x^2+x^5$, where both polynomials have degree less than the maximum degree of 6, produces a parity-check matrix in the following form:

$$\boldsymbol{H} = \left[\begin{array}{ccccccc|ccccccc} 1&1&0&1&0&0&0&1&0&1&0&0&1&0 \\ 0&1&1&0&1&0&0&0&1&0&1&0&0&1 \\ 0&0&1&1&0&1&0&1&0&1&0&1&0&0 \\ 0&0&0&1&1&0&1&0&1&0&1&0&1&0 \\ 1&0&0&0&1&1&0&0&0&1&0&1&0&1 \\ 0&1&0&0&0&1&1&1&0&0&1&0&1&0 \\ 1&0&1&0&0&0&1&0&1&0&0&1&0&1 \end{array}\right].$$

Definition 12.11 (*Permutation Matrix*) A permutation matrix is a type of circulant matrix where each row or column has weight of 1. A permutation matrix, which is denoted by $\boldsymbol{P}_{m,j}$, has $r(x) = x^j \pmod{x^m - 1}$ as the defining polynomial and it satisfies the property that $\boldsymbol{P}^2_{m,j} = \boldsymbol{I}_m$, where $\boldsymbol{I}_m$ is an $m \times m$ identity matrix.

Due to the sparseness of the permutation matrix, these are usually used to construct quasi-cyclic LDPC codes. The resulting LDPC codes produce a parity-check matrix in the following form:

$$\boldsymbol{H} = \begin{bmatrix} \boldsymbol{P}_{m,O_{0,0}} & \boldsymbol{P}_{m,O_{0,1}} & \cdots & \boldsymbol{P}_{m,O_{0,t-1}} \\ \boldsymbol{P}_{m,O_{1,0}} & \boldsymbol{P}_{m,O_{1,1}} & \cdots & \boldsymbol{P}_{m,O_{1,t-1}} \\ \vdots & \vdots & \vdots & \vdots \\ \boldsymbol{P}_{m,O_{s-1,0}} & \boldsymbol{P}_{m,O_{s-1,1}} & \cdots & \boldsymbol{P}_{m,O_{s-1,t-1}} \end{bmatrix} \tag{12.12}$$

From (12.12), we can see that there exists a $s \times t$ matrix, denoted by $\boldsymbol{O}$, in $\boldsymbol{H}$. This matrix is called an *offset matrix* and it represents the exponent of $r(x)$ in each permutation matrix, i.e.

$$\boldsymbol{O} = \begin{bmatrix} O_{0,0} & O_{0,1} & \dots & O_{0,t-1} \\ O_{1,0} & O_{1,1} & \dots & O_{1,t-1} \\ \vdots & \vdots & \vdots & \vdots \\ O_{s-1,0} & O_{s-1,1} & \dots & O_{s-1,t-1} \end{bmatrix}$$

where $0 \le O_{i,j} \le m-1$, for $0 \le i \le s-1$ and $0 \le j \le t-1$. The permutation matrix $\boldsymbol{P}_{m,j}$ has m rows and m columns, and since the matrix $\boldsymbol{H}$ contains s and t of these matrices per row and column, respectively, the resulting code is a $[mt, m(t-s), d]$ quasi-cyclic LDPC code over $\mathbb{F}_2$.

In general, some of the permutation matrices $\boldsymbol{P}_{i,j}$ in (12.12) may be zero matrices. In this case, the resulting quasi-cyclic LDPC code is irregular and $O_{i,j}$ for which $\boldsymbol{P}_{i,j} = \boldsymbol{O}$ may be ignored. If none of the permutation matrices in (12.12) is a zero matrix, the quasi-cyclic LDPC code defined by (12.12) is a (s, t) regular LDPC code.

12.4.2 Construction of Quasi-cyclic Codes Using a Protograph

A protograph is a miniature prototype Tanner graph of arbitrary size, which can be used to construct a larger Tanner graph by means of replicate and permute operations as discussed by Thorpe [32]. A protograph may also be considered as an $[n', k']$ linear code $\mathscr{P}$ of small block length and dimension. A longer code may be obtained by expanding code $\mathscr{P}$ by an integer factor Q so that the resulting code has parameter $[n = n'Q, k = k'Q]$ over the same field. A simplest way to expand code $\mathscr{P}$ and also to impose structure in the resulting code is by replacing a non-zero element of the parity-check matrix of code $\mathscr{P}$ with a $Q \times Q$ permutation matrix, and a zero element with a $Q \times Q$ zero matrix. As a consequence, the resulting code has a quasi-cyclic structure. The procedure is described in detail in the following example.

Example 12.5 Consider a code $\mathscr{P} = [4, 2]$ over $\mathbb{F}_2$ as a protograph. The parity-check matrix of code $\mathscr{P}$ is given by

$$H' = \begin{array}{c|cccc|} & v_0 & v_1 & v_2 & v_3 \\ \hline c_0 & 1 & 1 & 0 & 1 \\ c_1 & 0 & 1 & 1 & 1 \\ \hline \end{array}\,. \tag{12.13}$$

Let the expansion factor $Q = 5$, the expanded code, which is a [20, 10] code, has a parity-check matrix given by

$$H = \begin{array}{c|ccccc|ccccc|ccccc|ccccc|} & v_0 & v_1 & v_2 & v_3 & v_4 & v_5 & v_6 & v_7 & v_8 & v_9 & v_{10} & v_{11} & v_{12} & v_{13} & v_{14} & v_{15} & v_{16} & v_{17} & v_{18} & v_{19} \\ \hline c_0 & \mathbf{1} & & & & & \mathbf{1} & & & & & & & & & & \mathbf{1} & & & & \\ c_1 & & 1 & & & & & 1 & & & & & & & & & & 1 & & & \\ c_2 & & & 1 & & & & & 1 & & & & & & & & & & 1 & & \\ c_3 & & & & 1 & & & & & 1 & & & & & & & & & & 1 & \\ c_4 & & & & & 1 & & & & & 1 & & & & & & & & & & 1 \\ \hline c_5 & & & & & & \mathbf{1} & & & & & \mathbf{1} & & & & & \mathbf{1} & & & & \\ c_6 & & & & & & & 1 & & & & & 1 & & & & & 1 & & & \\ c_7 & & & & & & & & 1 & & & & & 1 & & & & & 1 & & \\ c_8 & & & & & & & & & 1 & & & & & 1 & & & & & 1 & \\ c_9 & & & & & & & & & & 1 & & & & & 1 & & & & & 1 \\ \hline \end{array}, \tag{12.14}$$

where the zero elements have been omitted. This protograph construction may also be described using the Tanner graph representation as shown in Fig. 12.11.

Initially, the Tanner graph of code $\mathscr{P}$ is replicated Q times. The edges of these replicated Tanner graphs are then permuted. The edges may be permuted in many ways and in this particular example, we want the permutation to produce a code which has quasi-cyclic structure. The edges shown in bold in Fig. 12.11 or equivalently the non-zeros shown in bold in (12.14) represent the code $\mathscr{P}$.

The minimum Hamming distance of code $\mathscr{P}$ is 2 and this may be seen from its parity-check matrix, (12.13), where the summation of two column vectors, those of v_1 and v_3, produces a zero vector. Since, in the expansion, only identity matrices are

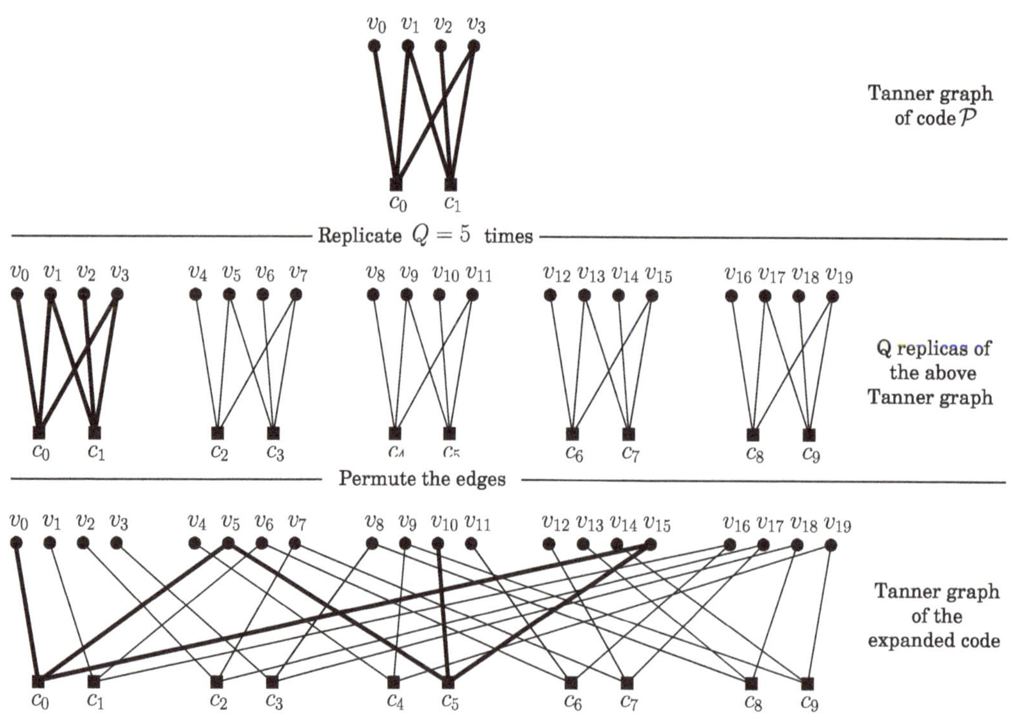

Fig. 12.11 Code construction using a protograph

employed, the expanded code will have the same minimum Hamming distance as the protograph code. This is obvious from (12.14) where the summation of two column vectors, those of v_5 and v_{15}, produces a zero vector. In order to avoid the expanded code having low minimum Hamming distance, permutation matrices may be used instead and the parity-check matrix of the expanded code is given by (12.15).

$$H = \begin{array}{c|ccccc|ccccc|ccccc|ccccc|} & v_0 & v_1 & v_2 & v_3 & v_4 & v_5 & v_6 & v_7 & v_8 & v_9 & v_{10} & v_{11} & v_{12} & v_{13} & v_{14} & v_{15} & v_{16} & v_{17} & v_{18} & v_{19} \\ \hline c_0 & & & \mathbf{1} & & & & & & & \mathbf{1} & & & & & & \mathbf{1} & & & & \\ c_1 & & & & 1 & & 1 & & & & & & & & & & & 1 & & & \\ c_2 & & & & & 1 & & 1 & & & & & & & & & & & 1 & & \\ c_3 & 1 & & & & & & & 1 & & & & & & & & & & & 1 & \\ c_4 & & 1 & & & & & & & 1 & & & & & & & & & & & 1 \\ \hline c_5 & & & & & & & \mathbf{1} & & & & & & & \mathbf{1} & & & \mathbf{1} & & & \\ c_6 & & & & & & & & 1 & & & & & & & 1 & & & 1 & & \\ c_7 & & & & & & & & & 1 & & 1 & & & & & & & & 1 & \\ c_8 & & & & & & & & & & 1 & & 1 & & & & & & & & 1 \\ c_9 & & & & & & 1 & & & & & & & 1 & & & 1 & & & & \\ \hline \end{array}, \tag{12.15}$$

The code defined by this parity-check matrix has minimum Hamming distance of 3. In addition, the cycle structure of the protograph is also preserved in the expanded code if only identity matrices are used for expansion. Since the protograph is such a small code, the variable vertex degree distribution required to design a good target code, which has much larger size than a protograph does, in general, causes many inevitable short cycles in the protograph. Using appropriate permutation matrices in the expansion, these short cycles may be avoided in the expanded code.

In the following, we describe a construction of a long quasi-cyclic LDPC code for application in satellite communications. The standard for digital video broadcasting (DVB), which is commonly known as DVB-S2, makes use of a concatenation of LDPC and BCH codes to protect the video stream. The parity-check matrices of DVB-S2 LDPC codes contain a zigzag matrix for the $n - k$ parity coordinates and quasi-cyclic matrices on the remaining k coordinates. In the literature, the code with this structure is commonly known as the irregular repeat accumulate (IRA) code [12].

The code construction described below, using a protograph and greedy PEG expansion, is aimed at improving the performance compared to the rate 3/4 DVB-S2 LDPC code of block length 64800 bits. Let the [64800, 48600] LDPC code that we will construct be denoted by $\mathscr{C}_1$. A protograph code, which has parameter [540, 405], is constructed using the PEG algorithm with a good variable vertex degree distributions obtained from Urbanke [34],

$$\begin{aligned} \Lambda_{\lambda_1}(x) = {} & \underbrace{0.00185185x + 0.248148x^2}_{\text{for zigzag matrix}} + 0.55x^3 + 0.0592593x^5 \\ & + 0.0925926x^8 + 0.00555556x^{12} + 0.00185185x^{15} + 0.0166667x^{19} \\ & + 0.00185185x^{24} + 0.00185185x^{28} + 0.0203704x^{35}. \end{aligned}$$

The constructed [540, 405] protograph code has a parity-check matrix $\boldsymbol{H}' = [\boldsymbol{H}'_u \mid \boldsymbol{H}'_p]$ where $\boldsymbol{H}'_p$ is a 135×135 zigzag matrix, see (12.10), and $\boldsymbol{H}'_u$ is an irregular matrix satisfying $\Lambda_{\lambda_1}(x)$ above. In order to construct a [64800, 48600] LDPC code $\mathscr{C}_1$, we need to expand the protograph code by a factor of $Q = 120$. In expanding the protograph code, we apply the greedy approach to construct the offset matrix $\boldsymbol{O}$ in order to obtain a Tanner graph for the [64800, 48600] LDPC code $\mathscr{C}_1$, which has local girth maximised. This greedy approach examines all offset values, from 0 to $Q - 1$, and picks an offset that results in highest girth or if there is more than one choice, one of these is randomly chosen. A 16200×48600 matrix $\boldsymbol{H}_u$ can be easily constructed by replacing a non-zero element at coordinate (i, j) in $\boldsymbol{H}'_u$ with a permutation matrix $\boldsymbol{P}_{Q,O_{i,j}}$. The resulting LDPC code $\mathscr{C}_1$ has a parity-check matrix given by $\boldsymbol{H} = [\boldsymbol{H}_u \mid \boldsymbol{H}_p]$, where, as before, $\boldsymbol{H}_p$ is given by (12.10).

In comparison, the rate 3/4 LDPC code of block length 64800 bits specified in the DVB-S2 standard takes a lower Q value, $Q = 45$. The protograph is a [1440, 1080] code which has the following variable vertex degree distributions

$$\Lambda_{\lambda_2}(x) = \underbrace{0.000694x + 0.249306x^2}_{\text{for zigzag matrix}} + 0.666667x^3 + 0.083333x^{12}.$$

For convenience, we denote the DVB-S2 LDPC code by $\mathscr{C}_2$.

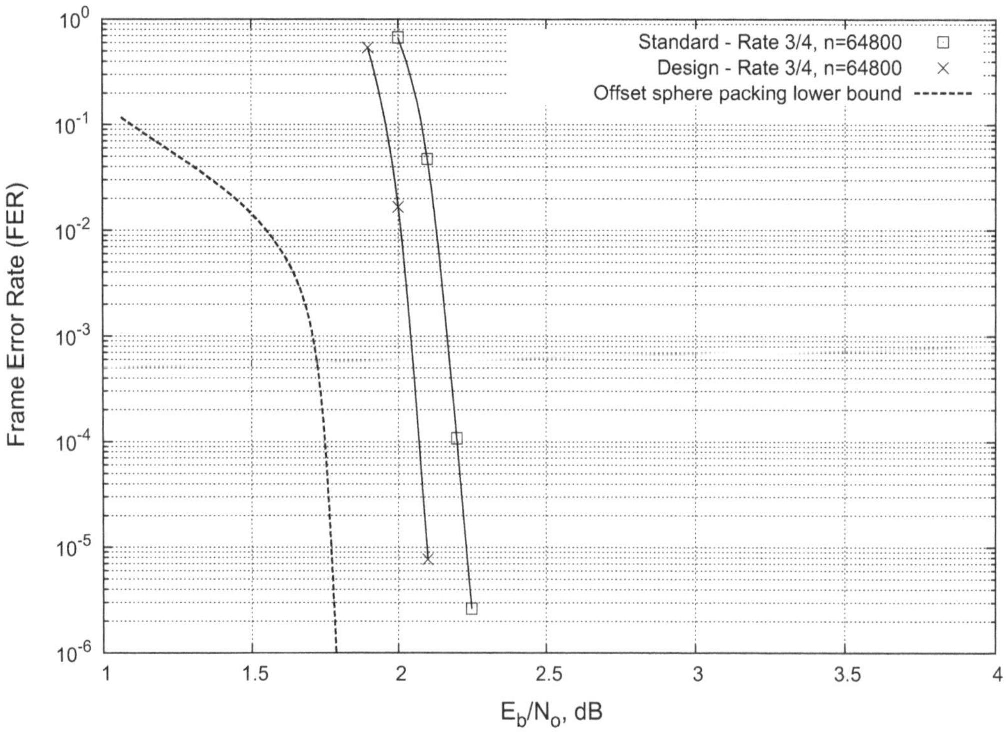

Fig. 12.12 FER performance of the DVB-S2 and the designed [64800, 48600] LDPC codes

Figure 12.12 compares the FER performance of $\mathscr{C}_1$ and $\mathscr{C}_2$ using the belief propagation decoder with 100 iterations. Binary antipodal signalling and AWGN channel are assumed. Note that, although the outer concatenation of BCH code is not used, there is still no sign of an error floor at FER as low as 10^{-6} which means that the BCH code is no longer required. It may be seen from Fig. 12.12 that the designed LDPC code, which at 10^{-5} FER performs approximately 0.35 dB away from the sphere packing lower bound offset for binary transmission loss, is 0.1 dB better than the DVB-S2 code.

12.5 Summary

The application of cyclotomic cosets, idempotents and Mattson–Solomon polynomials has been shown to produce many binary cyclic LDPC codes whose parity-check equations are orthogonal in each position. Whilst some of these excellent cyclic codes have the same parameters as the known class of finite geometry codes, other codes are new. A key feature of this construction technique is the incremental approach to the minimum Hamming distance and the sparseness of the resulting parity-check matrix of the code. Binary cyclic LDPC codes may also be constructed by considering idempotents in the Mattson–Solomon domain. This approach has provided a different insight into the cyclotomic coset-based construction. It has also been shown that, for short algebraic LDPC codes, the myths of codes which have cycles of length 4 in their Tanner graph do not converge well with iterative decoding is not necessarily true. It has been demonstrated that the cyclotomic coset-based construction can be easily extended to produce good non-binary algebraic LDPC codes.

Good irregular LDPC codes may be constructed using the progressive edge-growth algorithm. This algorithm adds edges to the variable and check vertices in a way that maximises the local girth. Many code results have been presented showing the effects of choosing different degree distributions. Guidelines are given for designing the best codes.

Methods of producing structured LDPC codes, such as those which have quasi-cyclic structure, have been described. These are of interest to industry due to the simplification of the encoder and decoder. An example of such a construction to produce a (64800, 48600) LDPC code, using a protograph, has been presented along with performance results using iterative decoding. Better results are obtained with this code than the (64800, 48600) LDPC code used in the DVB-S2 standard.

References

1. Campello, J., Modha, D.S., Rajagopalan, S.: Designing LDPC codes using bit-filling. In: Proceedings of the IEEE ICC, pp. 55–59 (2001)
2. Campello, J., Modha, D.S.: Extended bit-filling and LDPC code design. In: Proceedings of the IEEE Globecom Conference, pp. 985–989 (2001)
3. Chung, S.Y., Forney Jr., G.D., Richardson, T.J., Urbanke, R.L.: On the design of low-density parity check codes within 0.0045 db of the shannon limit. IEEE Commun. Lett. **3**(2), 58–60 (2001)
4. Chung, S.Y., Richardson, T.J., Urbanke, R.L.: Analysis of sum-product decoding of low-density parity-check codes using a gaussian approximation. IEEE Trans. Inf. Theory **47**(2), 657–670 (2001)
5. Costello, Jr. D., Forney, Jr. G.: Channel coding: the road to channel capacity (2006). Preprint available at http://arxiv.org/abs/cs/0611112
6. Davey, M.C., MacKay, D.J.C.: Low-density parity-check codes over GF(q). IEEE Commun. Lett. **2**, 165–167 (1998)
7. Etzion, T., Trachtenberg, A., Vardy, A.: Which codes have cycle-free tanner graphs? IEEE Trans. Inf. Theory **45**(6), 2173–2181 (1999)
8. Gallager, R.: Low-density parity-check codes. IRE Trans. Inf. Theory IT **8**, 21–28 (1962)
9. Gallager, R.: Low-Density Parity-Check Codes. MIT Press, Cambridge (1963)
10. Hu, X.Y., Eleftheriou, E., Arnold, D.M.: Irregular progressive edge-growth tanner graphs. In: Proceedings of IEEE International Symposium on Information Theory (ISIT), Lausanne, Switzerland (2002)
11. Hu, X.Y., Eleftheriou, E., Arnold, D.M.: Regular and irregular progressive edge-growth tanner graphs. IEEE Trans. Inf. Theory **51**(1), 386–398 (2005)
12. Jin, H., Khandekar, A., McEliece, R.J.: Irregular repeat-accumulate codes. In: Proceedings of 2nd International Symposium on Turbo Codes and Related Topics, Brest, France, pp. 1–8 (2000)
13. Johnson, S.: Low-Density Parity-Check Codes from Combinatorial Designs. Ph.D dissertation, School of Electrical Engineering and Computer Science, University of Newcastle, Callaghan, NSW 2308, Australia (2004)
14. Johnson, S.J., Weller, S.R.: Construction of low-density parity-check codes from Kirkman triple systems. In: Proceedings of IEEE Information Theory Workshop, Cairns, Australia, 2–7 Sept, pp. 90–92 (2001)
15. Johnson, S.J., Weller, S.R.: Codes for iterative decoding from partial geometries. In: Proceedings IEEE International Symposium on Information Theory, Lausanne, Switzerland, 30 June-3 July p. 310 (2002)
16. Kou, Y., Lin, S., Fossorier, M.: Low-density parity-check codes based on finite geometries: a rediscovery and new results. IEEE Trans. Inf. Theory **47**(7), 2711–2736 (2001)
17. Lin, S., Costello Jr., D.J.: Error Control Coding: Fundamentals and Applications, 2nd edn. Pearson Education, Inc, NJ (2004)
18. Luby, M.G., Shokrolloahi, M.A., Mizenmacher, M., Spielman, D.A.: Improved low-density parity-check codes using irregular graphs. IEEE Trans. Inf. Theory **47**(2), 585–598 (2001)
19. Lucas, R., Fossorier, M.P.C., Kou, Y., Lin, S.: Iterative decoding of one-step majority logic decodable codes based on belief propagation. IEEE Trans. Commun. **46**(6), 931–937 (2000)
20. MacWilliams, F.J., Sloane, N.J.A.: The Theory of Error-Correcting Codes. North-Holland, Amsterdam (1977)
21. Margulis, G.A.: Explicit constructions of graphs without short cycles and low density codes. Combinatorica **2**(1), 71–78 (1982)
22. McEliece, R.J., MacKay, D.J.C., Cheng, J.F.: Turbo decoding as an instance of pearl's "belief propagation" algorithm. IEEE J. Sel. Areas Commun. **16**, 140–152 (1998)
23. Papagiannis, E., Ambroze, M.A., Tomlinson, M.: Analysis of non convergence blocks at low and moderate SNR in SCC turbo schemes. In: SPSC 2003 8th International workshop on Signal Processing for Space Communications. Catania, Italy, pp. 121–128 (2003)

24. Pearl, J.: Probabilistic Reasoning in Intelligent Systems: Networks of Plausible Inference. Morgan Kaufmann, San Mateo (1988)
25. Peterson, W., Weldon Jr., E.J.: Error-Correcting Codes. MIT Press, Cambridge (1972)
26. Ping, L., Leung, W.K., Phamdo, N.: Low density parity check codes with semi-random parity check matrix. Electron. Lett. **35**(1), 38–39 (1999)
27. Richardson, T.J., Shokrollahi, M.A., Urbanke, R.L.: Design of capacity-approaching irregular low-density parity-check codes. IEEE Trans. Inf. Theory **47**(2), 619–637 (2001)
28. Richardson, T.J., Urbanke, R.L.: The capacity of low-density parity-check codes under message-passing decoding. IEEE Trans. Inf. Theory **47**(2), 599–618 (2001)
29. Richter, G., Hof, A.: On a construction method of irregular LDPC codes without small stopping sets. In: Proceedings of IEEE International Conference on Communications, Istanbul, Turkey, pp. 1119–1124 (2006)
30. Shannon, C.E.: A mathematical theory of communication. Bell Syst. Tech. J. **27**(3), 379–423 (1948)
31. Tang, H., Xu, J., Lin, S., Abdel-Ghaffar, K.A.S.: Codes on finite geometries. IEEE Trans. Inf. Theory **51**(2), 572–596 (2005)
32. Thorpe, J.: Low-density parity-check (LDPC) codes constructed from protographs. JPL IPN Progress Report 42–154 (2003). Available: http://tmo.jpl.nasa.gov/progress_report/42-154/154C.pdf
33. Tian, T., Jones, C., Villasenor, J., Wesel, R.: Selective avoidance of cycles in irregular LDPC code construction. IEEE Trans. Commun. **52**, 1242–1247 (2004)
34. Urbanke, R.: LdpcOpt a fast and accurate degree distribution optimizer for LDPC code ensembles (2001). Available at http://lthcwww.epfl.ch/research/ldpcopt/
35. Vasic, B., Milenkovic, M.: Combinatorial constructions of low-density parity-check codes for iterative decoding. IEEE Trans. Inf. Theory **50**(6), 1156–1176 (2004)
36. Weldon Jr., E.J.: Difference-set cyclic codes. Bell Syst. Tech. J. **45**, 1045–1055 (1966)

Part III
Analysis and Decoders

This part is about the analysis of codes in terms of their codeword and stopping set weight spectrum and various types of decoders. Decoders are described, which include hard and soft decision decoders for the AWGN channel and decoders for the erasure channel. Universal decoders are discussed, which are decoders that can be used with any linear code for hard or soft decision decoding. One such decoder is based on the Dorsch decoder and this is described in some detail together with its performance using several different code examples. Other decoders such as the iterative decoder require sparse parity-check matrices and codes specifically designed for this type of decoder. Also included in this part is a novel concatenated $|u|u+v|$ code arrangement featuring multiple near maximum likelihood decoders for an optimised matching of codes and decoders. With some outstanding codes as constituent codes, the concatenated coding arrangement is able to outperform the best LDPC and turbo coding systems with the same code parameters.

Chapter 13
An Exhaustive Tree Search for Stopping Sets of LDPC Codes

13.1 Introduction and Preliminaries

The performance of all error-correcting codes is determined by the minimum Hamming distance between codewords. For codes which are iteratively decoded such as LDPC codes and turbo codes, the performance of the codes for the erasure channel is determined by the stopping set spectrum, the weight (and number) of erasure patterns which cause the iterative decoder to fail to correct all of the erasures. Codes which perform poorly on the erasure channel do not perform well on the AWGN channel. To determine all of the stopping sets of a general (n, k) code is a prohibitive task, for example, a binary $(1000, 700)$ code would require evaluation of 2^{1000} possible stopping sets. It should be noted by the reader that all codewords are also stopping sets, but most stopping sets are not codewords. Fortunately the properties of particular types of codes may be used to reduce considerably the scale of the task, and in particular codes with sparse parity-check matrices such as LDPC codes and turbo codes are amenable to analysis in practice. As the tree search is exhaustive, the emphasis is first on focusing the search so that only low-weight stopping sets are found, up to a specified weight, and second the emphasis is on the efficiency of the algorithms involved.

In a landmark paper in 2007, Rosnes and Ytrehus [7] showed that exhaustive, low-weight stopping set analysis of codes whose parity-check matrix is sparse is feasible using a bounded tree search over the length of the code with no distinction between information and parity bits. A previous paper on the same topic of an exhaustive search of stopping sets of LDPC codes by Wang et al. [2] used a different and much less efficient algorithm. In common with this earlier research, we use similar notation in the following preliminaries.

The code $\mathscr{C}$ is defined to be binary and linear of length n and dimension k and is a k-dimensional subspace of $\{0, 1\}^n$, and may be specified as the null space of a $m \times n$ binary parity-check matrix $\mathbf{H}$ of rank $n - k$. The number of parity-check equations, m of $\mathbf{H}$ satisfies $m \geq (n - k)$, although there are, of course, only $n - k$ independent parity-check equations. It should be noted, as illustrated in the results below, that the

M. Tomlinson et al., *Error-Correction Coding and Decoding*,
Signals and Communication Technology,
DOI 10.1007/978-3-319-51103-0_13

number of parity-check equations m in excess of $n - k$ can have a dramatic effect on the stopping set weight spectrum, excluding codewords of course, as these are not affected.

As in [7], $\mathscr{S}$ is used to denote a subset of $\{0, 1\}^n$, the set of all binary vectors of length n. At any point in the tree search, a constraint set, $\mathscr{F}$ is defined consisting of bit positions p_i and the states of these bit positions $s_{p_i}, s_{p_i} \in \{0, 1\}^n$. The support set $\chi(\mathscr{F})$ of the constraint set, $\mathscr{F}$, is the set of positions where $s_{p_i} = 1$, and the Hamming weight of $\mathscr{F}$ is the number of such positions. The sub-matrix $\mathbf{H}_{\chi(\mathscr{F})}$ consists of all the columns of $\mathbf{H}$ where $s_{p_i} = 1$, and the row weight of $\mathbb{H}_{\chi(\mathscr{F})}$ is the number of $1's$ in that row. An active row of $\mathbf{H}_{\chi(\mathscr{F})}$ is a row with unity row weight. It is obvious that if all rows of $\mathbf{H}_{\chi(\mathscr{F})}$ have even row weight then $\mathscr{F}$ is a codeword, noting that for an iterative decoder codewords are also stopping sets. If at least one row has odd weight, 3 or higher and there are no active rows then $\mathscr{F}$ is a stopping set but not a codeword. If there are active rows then $\mathscr{F}$ has either to be appended with additional bit positions or one or more states s_{p_i} need to be changed to form a stopping set. With this set of basic definitions, tree search algorithms may be described which carry out an exhaustive search of $\{0, 1\}^n$ using a sequence of constraints $\mathscr{F}$ to find all stopping sets whose Hamming weight is $\leq \tau$.

13.2 An Efficient Tree Search Algorithm

At any given point in the search, the *constraint set* F is used to represent the set of searched known bits (up to this point) of a code $\mathscr{C}$, which forms a *branch* of the tree in the tree search. The set of active rows in $\mathbf{H}$ is denoted by $\{\mathfrak{h}_0, ..., \mathfrak{h}_{\phi-1}\}$, where ϕ is the total number of active rows. A constraint set F with size n is said to be *valid* if and only if there exists no active rows in $\mathbf{H}^{(F)}$. In which case the constraint set is equal to a stopping set. The pseudocode of one particularly efficient algorithm to find all the stopping sets including codeword sets with weight equal to or less than τ is given in Algorithm 13.1 below. Each time a stopping set is found, it is stored and the algorithm progresses until the entire 2^n space has been searched.

The modified iterative decoding is carried out on a n-length binary input vector containing erasures in some of the positions. Let $r_j(F)$ be the rank (ones) of row j, $j \in \{0, ..., m-1\}$ for the constrained position $\{p_i : (p_i, 1) \in F\}$ intersected by row j on $\mathbf{H}$. And let $r'_j(F)$ be the rank of row j for the unconstrained position $\{p_i : (p_i, 1) \in \{0, ..., n-1\} \backslash F\}$ intersected by row j on $\mathbf{H}$. The modified iterative decoding algorithm based on belief-propagation decoding algorithm over the binary erasure channel is shown in Algorithm 13.2. As noted in the line with marked (*), the modified iterative decoder is not invoked if the condition of $r_j \leq 1$ and $r'_j = 1$ is not met; or the branch with constraint set F has condition of $r_j = 1$ and $r'_j = 0$. This significantly speeds up the tree search. As noted in the line with marked (*), the modified iterative decoder is not necessary to call, if the condition of $r_j < 1$ and

Algorithm 13.1 Tree search based Stopping Set Enumeration (TSSE)

repeat
  Pick one untouched branch as a constraint set F.
  if $|F| = n$ and $w(F) \leq \tau$ **then**
    Constraint set F is saved, if F is valid
  else
    1). Pass F to the modified iterative decoder (*) with erasures in the unconstrained positions.
    2). Construct a new constraint set F' with new decoded positions, which is the extended branch.
    if $|F'| = n$ and $w(F') \leq \tau$ **then**
      Constraint set F' is saved, if F' is valid
    else if No contradiction is found in $\mathbf{H}^{(F')}$, and $w'(F') \leq \tau$ **then**
      a). Pick an unconstrained position p.
      b). Extending branch F' to position p to get new branch $F'' = F' \bigcup \{(p, 1)\}$ and branch $F''' = F' \bigcup \{(p, 0)\}$.
    end if
  end if
until Tree has been fully explored

Algorithm 13.2 Modified Iterative Decoding

Get rank $\mathbf{r}(F)$ and $\mathbf{r}'(F)$ for all the equation rows on $\mathbf{H}$.
repeat
  if $r_j > 1$ **then**
    Row j is flagged
  else if $r_j = 1$ and $r'_j = 0$ **then**
    Contradiction → Quit decoder
  else if $r_j \leq 1$ and $r'_j = 1$ **then**
    1). Row j is flagged
    2). The variable bit i is decoded as the **XOR** of the value of r_j.
    3). Update the value of r_j and r'_j, if $H_{ji} = 1$.
  end if
until No new unconstrained bit is decoded

$r'_j = 1$ is not met; or the branch with constraint set F can be ignored, if condition of $r_j = 1$ and $r'_j = 0$ occurs. Thus the computing complexity can be significantly reduced than calling it for every new branch with the corresponding constraint set F.

13.2.1 An Efficient Lower Bound

The tree search along the current branch may be terminated if the weight necessary for additional bits to produce a stopping set plus the weight of the current constraint set F exceeds τ. Instead of actually evaluating these bits, it is more effective to calculate a lower bound on the weight of the additional bits. The bound uses the active rows $\mathscr{I}(F) = \{I_{i_0}(F), ..., I_{i_{q-1}}(F)\}$, where $I_{i_0}(F)$ is the set of active rows with constraint set F corresponding to the i_0th column $\mathbf{h}_{i_0}$ of $\mathbf{H}$, and q is the number

Table 13.1 Low-weight stopping sets and codewords of known codes

Code Name	s_m	N_{s_m}	$N_{s_m}+1$	$N_{s_m}+2$	$N_{s_m}+3$	$N_{s_m}+4$	$N_{s_m}+5$	$N_{s_m}+6$
Tanner (155, 64) [6]	18	465 (0)	2015 (0)	9548 (1023)	23715 (0)	106175 (6200)	359290 (0)	1473585 (43865)
QC LDPC (1024, 512) [4]	15	1 (1)	1 (0)	0 (0)	1 (1)	6 (1)	6 (2)	12 (4)
PEG Reg (256, 128) [3, 5]	11	1 (0)	11 (7)	22 (12)	51 (28)	116 (46)	329 (113)	945 (239)
PEG Reg (504, 252) [3, 5]	19	2 (0)	5 (2)	8 (0)	27 (5)	78 (0)	241 (30)	()
PEG iReg (504, 252) [3, 5]	13	2 (1)	1 (1)	5 (5)	13 (11)	31 (16)	52 (28)	124 (60)
PEG iReg (1008, 504) [3, 5]	13	1 (1)	0 (0)	0 (0)	3 (3)	3 (3)	4 (4)	5 (3)
MacKay (504, 252) [5]	16	1 (0)	3 (0)	3 (0)	12 (0)	36 (2)	106 (0)	320 (22)

Table 13.2 WiMax 1/2 LDPC Codes

i	S_{min}	$N_{S_{min}}$	$N_{S_{min}+1}$	$N_{S_{min}+2}$	$N_{S_{min}+3}$	$N_{S_{min}+4}$	$N_{S_{min}+5}$	$N_{S_{min}+6}$	$N_{S_{min}+7}$	$N_{S_{min}+8}$
0	13	24(24)	0(0)	0(0)	24(24)	0(0)	24(0)	120(72)	312(96)	()
1	18	56(0)	140(56)	56(56)	308(84)	420(168)	756(224)	2296(476)	5460(1288)	()
2	18	32(0)	0(0)	96(64)	128(32)	192(96)	704(352)	992(224)	1888(672)	()
3	19	36(36)	36(36)	144(0)	324(36)	828(180)	810(162)	2304(576)	()	()
4	19	120(80)	120(40)	160(0)	280(160)	400(120)	880(120)	1760(560)	()	()
5	19	44(0)	0(0)	44(44)	132(0)	220(88)	176(44)	176(132)	()	()
6	19	48(48)	0(0)	0(0)	0(0)	0(0)	48(0)	144(144)	()	()
7	19	52(0)	0(0)	0(0)	52(52)	()	()	()	()	()
8	23	112(112)	56(0)	280(224)	560(224)	1008(280)	()	()	()	()
9	24	60(0)	60(0)	60(0)	180(60)	720(300)	()	()	()	()
10	20	64(64)	0(0)	0(0)	64(64)	64(0)	0(0)	96(96)	256(128)	()
11	27	68(68)	408(0)	()	()	()	()	()	()	()
12	21	72(72)	0(0)	0(0)	0(0)	0(0)	0(0)	216(216)	144(0)	()
13	19	76(76)	0(0)	0(0)	0(0)	0(0)	0(0)	0(0)	76(76)	76(76)
14	25	160(80)	240(80)	240(240)	400(160)	()	()	()	()	()
15	27	84(84)	84(84)	756(168)	518(182)	()	()	()	()	()
16	28	264(264)	88(0)	440(264)	()	()	()	()	()	()
17	23	92(92)	0(0)	0(0)	0(0)	0(0)	276(92)	()	()	()
18	28	96(0)	96(0)	288(0)	288(96)	624(336)	()	()	()	()

Table 13.3 WiMax $2/3A$ LDPC Codes

i	S_{min}	$N_{S_{min}}$	$N_{S_{min}+1}$	$N_{S_{min}+2}$	$N_{S_{min}+3}$	$N_{S_{min}+4}$	$N_{S_{min}+5}$	$N_{S_{min}+6}$
13	15	76(76)	228(152)	()	()	()	()	()
14	14	80(0)	80(80)	160(0)	()	()	()	()
15	15	84(84)	252(0)	()	()	()	()	()
16	15	88(88)	0(0)	()	()	()	()	()
17	15	92(92)	0(0)	92(92)	460(276)	()	()	()
18	15	96(96)	0(0)	96(96)	480(384)	()	()	()

Table 13.4 WiMax $2/3B$ LDPC Codes

i	S_{min}	$N_{S_{min}}$	$N_{S_{min}+1}$	$N_{S_{min}+2}$	$N_{S_{min}+3}$	$N_{S_{min}+4}$	$N_{S_{min}+5}$	$N_{S_{min}+6}$
6	16	96(48)	432(48)	()	()	()	()	()
7	15	52(52)	0(0)	104(104)	156(104)	728(312)	2041(533)	()
8	16	63(63)	56(56)	196(56)	560(168)	1568(196)	()	()
9	17	120(60)	()	()	()	()	()	()
10	15	64(64)	0(0)	0(0)	0(0)	128(0)	384(64)	()
11	18	204(68)	()	()	()	()	()	()
12	15	72(72)	0(0)	0(0)	72(0)	()	()	()
13	15	76(76)	0(0)	0(0)	0(0)	0(0)	76(0)	()
14	16	80(80)	80(0)	()	()	()	()	()
15	15	84(84)	0(0)	0(0)	0(0)	84(84)	294(168)	()
16	16	88(88)	88(0)	()	()	()	()	()
17	20	92(92)	92(0)	92(0)	()	()	()	()
18	15	96(96)	0(0)	0(0)	0(0)	0(0)	144(96)	()

of intersected unknown bits. Let $w(\mathbf{h}_j^{I_j(F)})$ be the weight of ones on jth column of $\mathbf{H}$, which is the number of active rows intersected with jth column. Under a worst case assumption, the $I_j(F)$ with larger column weight of ones on jth column is always with value 1, then the active rows can be compensated by $I_j(F)$ and the total number of active rows ϕ is reduced by $w(\mathbf{h}_j^{I_j(F)})$ until $\phi \leq 0$. Algorithm 13.3 shows the pseudocode of computing the smallest number of intersected unknown bits q in order to produce no active rows. The lower bound $w'(F) = w(F) + q$ is the result.

Algorithm 13.3 Simple method to find the smallest collection set of active rows

1. Arrange the set of $\mathscr{I}(F)$ in descending order, where $\mathbf{h}_{i_0'}$ is the column with the maximal column weight corresponding to constraint F.
2. q is initialised as 0.

while $\phi > 0$ **do**
  1). ϕ is subtracted by $w(\mathbf{h}_{i_0'})$.
  2). q is accumulated by 1.
end while

Table 13.5 WiMax $3/4A$ LDPC Codes

i	s_{min}	$N_{s_{min}}$	$N_{s_{min}+1}$	$N_{s_{min}+2}$	$N_{s_{min}+3}$	$N_{s_{min}+4}$	$N_{s_{min}+5}$	$N_{s_{min}+6}$
6	10	48(0)	0(0)	24(0)	240(48)	624(288)	()	()
7	12	26(0)	156(52)	260(104)	2184(416)	()	()	()
8	12	28(0)	112(0)	224(168)	952(280)	()	()	()
9	12	90(60)	60(0)	180(60)	372(192)	()	()	()
11	12	34(0)	68(68)	0(0)	0(0)	()	()	()
12	12	36(0)	0(0)	0(0)	0(0)	72(0)	504(144)	()
13	12	38(0)	76(76)	0(0)	76(76)	()	()	()
14	12	40(0)	80(0)	160(0)	240(0)	240(0)	800(160)	()
15	12	42(0)	0(0)	0(0)	0(0)	0(0)	168(84)	()
16	12	44(0)	0(0)	0(0)	88(88)	()	()	()
17	12	46(0)	0(0)	0(0)	0(0)	0(0)	0(0)	()
18	12	48(0)	0(0)	0(0)	0(0)	0(0)	0(0)	96(0)

Table 13.6 WiMax $3/4B$ LDPC Codes

i	s_{min}	$N_{s_{min}}$	$N_{s_{min}+1}$	$N_{s_{min}+2}$	$N_{s_{min}+3}$	$N_{s_{min}+4}$	$N_{s_{min}+5}$	$N_{s_{min}+6}$
7	9	52(52)	52(52)	52(52)	312(156)	988(416)	3094(1274)	11180(3952)
8	12	560(392)	616(224)	1792(616)	7784(2968)	()	()	()
9	10	60(60)	60(60)	130(10)	540(240)	2190(810)	7440(2940)	()
10	11	64(64)	128(128)	128(64)	960(640)	3648(1408)	()	()
11	13	272(204)	748(544)	2992(1564)	()	()	()	()
12	12	72(0)	576(432)	576(216)	2520(936)	()	()	()
13	12	228(228)	380(304)	988(836)	2888(836)	()	()	()
14	10	80(80)	0(0)	0(0)	0(0)	640(480)	2416(1216)	()
15	11	84(0)	84(84)	336(168)	546(294)	1260(588)	()	()
16	14	176(88)	968(792)	()	()	()	()	()
17	13	184(92)	92(92)	1012(644)	()	()	()	()
18	12	16(16)	96(96)	672(480)	()	()	()	()

13.2.2 Best Next Coordinate Position Selection

In the evaluation of the lower bound above, the selected unconstrained positions are assumed to have value 1. Correspondingly, the first position in the index list has maximal column weight and is the best choice for the coordinate to add to the constraint set F.

Table 13.7 Weight Spectra and stopping set spectra for the WiMax LDPC Codes [1]

	Code Length $N = 576 + 96i$									
i	0	1	2	3	4	5	6	7	8	9
N	576	672	768	864	960	1056	1152	1248	1344	1440
Code Rate	Minimum Codeword Weight d_m									
1/2	13	19	20	19	19	21	19	22	23	27
2/3A	10	9	8	11	13	10	14	13	14	13
2/3B	12	11	14	16	15	15	16	15	16	17
3/4A	10	10	10	12	12	13	13	13	14	12
3/4B	8	8	9	11	11	9	11	9	12	10
5/6	5	7	7	7	7	7	7	7	7	7
	Minimum Stopping Set Weight s_m									
1/2	18	18	18	21	19	19	24	19	24	24
2/3A	10	10	11	9	12	13	13	14	14	14
2/3B	10	12	13	15	14	16	16	18	18	17
3/4A	9	8	10	11	12	12	10	12	12	12
3/4B	9	10	10	10	11	11	11	12	12	12
5/6	6	6	7	7	7	7	7	9	7	8

	Code Length $N = 576 + 96i$								
i	10	11	12	13	14	15	16	17	18
N	1536	1632	1728	1824	1920	2016	2112	2208	2304
Code Rate	Minimum Codeword Weight d_m								
1/2	20	27	21	19	25	27	28	23	31
2/3A	12	13	15	15	15	15	15	15	15
2/3B	15	18	15	15	16	15	16	20	15
3/4A	14	13	17	13	17	17	15	20	19
3/4B	11	13	13	12	10	12	14	13	12
5/6	7	7	8	8	7	7	8	8	9
	Minimum Stopping Set Weight s_m								
1/2	24	28	28	28	25	29	29	28	28
2/3A	15	12	14	16	14	16	17	18	18
2/3B	19	18	18	20	17	20	17	21	20
3/4A	12	12	12	12	12	12	12	12	12
3/4B	13	13	12	13	14	11	14	13	15
5/6	8	9	7	9	7	8	9	8	10

13.3 Results

The algorithms above have been used to evaluate all of the low-weight stopping sets for some well-known LDPC codes. The results are given in Table 13.1 together with the respective references where details of the codes may be found. The total number of stopping sets are shown for a given weight with the number of codewords in parentheses. Interestingly, the Tanner code has 93 parity-check equations, 2 more than the 91 parity-check equations needed to encode the code. If only 91 parity-check equations are used by the iterative decoder there is a stopping set of weight 12 instead of 18 which will degrade the performance of the decoder. The corollary of this is that the performance of some LDPC codes may be improved by introducing additional, dependent, parity-check equations by selecting low-weight codewords of the dual code. A subsequent tree search will reveal whether there has been an improvement to the stopping sets as a result.

13.3.1 WiMax LDPC Codes

WiMax LDPC codes [1], as the IEEE 802.16e standard LDPC codes, have been fully analysed and the low-weight stopping sets for all combinations of code rates and lengths have been found. Detailed results for WiMax LDPC codes of code rates $1/2, 2/3A, 2/3B, 3/4A, 3/4B$ are given in Tables 13.2, 13.3, 13.4, 13.5, 13.6. In these tables, the code index i is linked to the code length N by the formula $N = 576+96i$. The minimum weight of non-codeword stopping sets (s_m) and codeword stopping sets (d_m) for all WiMax LDPC codes is given in Table 13.7.

13.4 Conclusions

An efficient algorithm has been presented which enables all of the low weight stopping sets to be evaluated for some common LDPC codes. Future research is planned that will explore the determination of efficient algorithms for use with multiple computers operating in parallel in order to evaluate all low weight stopping sets for commonly used LDPC codes several thousand bits long.

13.5 Summary

It has been shown that the indicative performance of an LDPC code may be determined from exhaustive analysis of the low-weight spectral terms of the code's stopping sets which by definition includes the low-weight codewords. In a breakthrough, Rosnes and Ytrehus demonstrated the feasibility of exhaustive, low-weight stopping set analysis of codes whose parity-check matrix is sparse using a bounded tree search over the length of the code, with no distinction between information and parity bits. For an (n, k) code, the potential total search space is of size 2^n but a good choice of bound dramatically reduces this search space to a practical size. Indeed, the choice of bound is critical to the success of the algorithm. It has been shown that an improved algorithm can be obtained if the bounded tree search is applied to a set of k information bits since the potential total search space is initially reduced to size 2^k. Since such a restriction will only find codewords and not all stopping sets, a class of bits is defined as unsolved parity bits, and these are also searched as appended bits in order to find all low-weight stopping sets. Weight spectrum results have been presented for commonly used WiMax LDPC codes in addition to some other well-known LDPC codes.

An interesting area of future research has been identified whose aim is to improve the performance of the iterative decoder, for a given LDPC code, by determining low-weight codewords of the dual code and using these as additional parity-check equations. The tree search may be used to determine improvements to the code's stopping sets as a result.

References

1. WiMax LDPC codes, Air interface for fixed and mobile broadband wireless access systems, IEEE Std 802.16e-2005 (2005). http://standards.ieee.org/getieee802/download/802.16e-2005.pdf
2. Wang, C.C., Kulkami. S.R., Poor, H.V.: Exhausting error-prone patterns in LDPC codes (2007). http://arxiv.org/abs/cs.IT/0609046, submitted to IEEE Transaction in Information Theory
3. Hu, X.Y., Eleftheriou, E., Arnold, D.M.: Regular and irregular progressive edge-growth tanner graphs. IEEE Trans. Inf. Theory **51**(1), 386–398 (2005)
4. Lan, L., Zeng, L., Tai, Y.Y., Chen, L., Lin, S., Abdel-Ghaffar, K.: Construction of quasi-cyclic LDPC codes for AWGN and binary erasure channels: a finite field approach. IEEE Trans. Inf. Theory **53**, 2429–2458 (2007)
5. MacKay, D.: Encyclopedia of sparse graph codes [online] (2011). http://www.inference.phy.cam.ac.uk/mackay/codes/data.html
6. Tanner, R.M., Sridhara, D., Fuja, T.: A class of group-structured LDPC codes. In: Proceedings of the International Symposium on Communications Theory and Applications (ISCTA) (2001)
7. Rosnes, E., Ytrehus, O.: An algorithm to find all small-size stopping sets of low-density parity-check matrices. In: International Symposium on Information Theory, pp. 2936–2940 (2007)

Chapter 14
Erasures and Error-Correcting Codes

14.1 Introduction

It is well known that an (n, k, d_{min}) error-correcting code $\mathscr{C}$, where n and k denote the code length and information length, can correct $d_{min} - 1$ erasures [15, 16] where d_{min} is the minimum Hamming distance of the code. However, it is not so well known that the average number of erasures correctable by most codes is significantly higher than this and almost equal to $n - k$. In this chapter, an expression is obtained for the probability density function (PDF) of the number of correctable erasures as a function of the weight enumerator function of the linear code. Analysis results are given of several common codes in comparison to maximum likelihood decoding performance for the binary erasure channel. Many codes including BCH codes, Goppa codes, double-circulant and self-dual codes have weight distributions that closely match the binomial distribution [13–15, 19]. It is shown for these codes that a lower bound of the number of correctable erasures is $n-k-2$. The decoder error rate performance for these codes is also analysed. Results are given for rate 0.9 codes and it is shown for code lengths 5000 bits or longer that there is insignificant difference in performance between these codes and the theoretical optimum maximum distance separable (MDS) codes. The results for specific codes are given including BCH codes, extended quadratic residue codes, LDPC codes designed using the progressive edge growth (PEG) technique [12] and turbo codes [1].

The erasure correcting performance of codes and associated decoders has received renewed interest in the study of network coding as a means of providing efficient computer communication protocols [18]. Furthermore, the erasure performance of LDPC codes, in particular, has been used as a measure of predicting the code performance for the additive white Gaussian noise (AWGN) channel [6, 17]. One of the first analyses of the erasure correction performance of particular linear block codes is provided in a key-note paper by Dumer and Farrell [7] who derive the erasure correcting performance of long binary BCH codes and their dual codes. Dumer and Farrell show that these codes achieve capacity for the erasure channel.

M. Tomlinson et al., *Error-Correction Coding and Decoding*,
Signals and Communication Technology,
DOI 10.1007/978-3-319-51103-0_14

14.2 Derivation of the PDF of Correctable Erasures

14.2.1 Background and Definitions

A set of s erasures is a list of erased bit positions defined as f_i where

$$0 < i < s \quad f_i \in 0 \ldots n-1$$

A codeword $\mathbf{x} = x_0, x_1 \ldots x_{n-1}$ satisfies the parity-check equations of the parity-check matrix $\mathbf{H}$

$$\mathbf{H}\,\mathbf{x}^{\mathrm{T}} = \mathbf{0}$$

A codeword with s erasures is defined as

$$\mathbf{x} = (x_{u_0}, x_{u_1} \ldots x_{u_{n-1-s}} | x_{f_0}, x_{f_1} \ldots x_{f_{s-1}})$$

where x_{u_j} are the unerased coordinates of the codeword, and the set of s erased coordinates is defined as $\mathbf{f_s}$. There are a total of $n-k$ parity check equations and provided the erased bit positions correspond to independent columns of the $\mathbf{H}$ matrix, each of the erased bits may be solved using a parity-check equation derived by the classic technique of Gaussian reduction [15–17]. For maximum distance separable (MDS) codes, [15], any set of s erasures are correctable by the code provided that

$$s \leq n-k \tag{14.1}$$

Unfortunately, the only binary MDS codes are trivial codes [15].

14.2.2 The Correspondence Between Uncorrectable Erasure Patterns and Low-Weight Codewords

Provided the code is capable of correcting the set of s erasures, then a parity-check equation may be used to solve each erasure, viz:

$$\begin{aligned}
x_{f_0} &= h_{0,0}x_{u_0} && + h_{0,1}x_{u_1} + h_{0,2}x_{u_2} && + \ldots h_{0,n-s-1}x_{u_{n-s-1}} \\
x_{f_1} &= h_{1,0}x_{u_0} && + h_{1,1}x_{u_1} + h_{1,2}x_{u_2} && + \ldots h_{1,n-s-1}x_{u_{n-s-1}} \\
x_{f_2} &= h_{2,0}x_{u_0} && + h_{2,1}x_{u_1} + h_{2,2}x_{u_2} && + \ldots h_{2,n-s-1}x_{u_{n-s-1}} \\
& \ldots && \ldots\ldots && \ldots\ldots \\
x_{f_{s-1}} &= h_{s-1,0}x_{u_0} && + h_{s-1,1}x_{u_1} + h_{s-1,2}x_{u_2} && + \ldots h_{s-1,n-s-1}x_{u_{n-s-1}}
\end{aligned}$$

where $h_{i,j}$ is the coefficient of row i and column j of $\mathbf{H}$.

As the parity-check equations are Gaussian reduced, no erased bit is a function of any other erased bits. There will also be $n - k - s$ remaining parity-check equations, which do not contain any of the erased bits' coordinates x_{f_j}:

$$h_{s,0}x_{u_0} + h_{s,1}x_{u_1} + h_{s,2}x_{u_2} + \cdots + h_{s,n-s-1}x_{u_{n-s-1}} = 0$$

$$h_{s+1,0}x_{u_0} + h_{s+1,1}x_{u_1} + h_{s+1,2}x_{u_2} + \cdots + h_{s+1,n-s-1}x_{u_{n-s-1}} = 0$$

$$h_{s+2,0}x_{u_0} + h_{s+2,1}x_{u_1} + h_{s+2,2}x_{u_2} + \cdots + h_{s+2,n-s-1}x_{u_{n-s-1}} = 0$$

$$\ldots$$

$$\ldots$$

$$h_{n-k-1,0}x_{u_0} + h_{n-k-1,1}x_{u_1} + h_{n-k-1,2}x_{u_2} + \cdots + h_{n-k-1,n-s-1}x_{u_{n-s-1}} = 0$$

Further to this, the hypothetical case is considered where there is an additional erased bit x_{f_s}. This bit coordinate is clearly one of the previously unerased bit coordinates, denoted as x_{u_p}.

$$x_{f_s} = x_{u_p}$$

Also, in this case it is considered that these $s+1$ erased coordinates do not correspond to $s+1$ independent columns of the **H** matrix, but only to $s+1$ dependent columns. This means that x_{u_p} is not contained in any of the $n-k-s$ remaining parity-check equations, and cannot be solved as the additional erased bit.

For the first s erased bits whose coordinates do correspond to s independent columns of the **H** matrix, the set of codewords is considered in which all of the unerased coordinates are equal to zero except for x_{u_p}. In this case the parity-check equations above are simplified to become:

$$\begin{aligned} x_{f_0} &= h_{0,p}x_{u_p} \\ x_{f_1} &= h_{1,p}x_{u_p} \\ x_{f_2} &= h_{2,p}x_{u_p} \\ \ldots &= \ldots \\ \ldots &= \ldots \\ x_{f_{s-1}} &= h_{s-1,p}x_{u_p} \end{aligned}$$

As there are, by definition, at least $n - s - 1$ zero coordinates contained in each codeword, the maximum weight of any of the codewords above is $s+1$. Furthermore, any erased coordinate that is zero may be considered as an unsolved coordinate, since no non-zero coordinate is a function of this coordinate. This leads to the following theorem.

Theorem 1 *The non-zero coordinates of a codeword of weight w that is not the juxtaposition of two or more lower weight codewords, provide the coordinate positions of $w-1$ erasures that can be solved and provide the coordinate positions of w erasures that cannot be solved.*

Proof The coordinates of a codeword of weight w must satisfy the equations of the parity-check matrix. With the condition that the codeword is not constructed from the juxtaposition of two or more lower weight codewords, the codeword must have $w-1$ coordinates that correspond to linearly independent columns of the **H** matrix and w coordinates that correspond to linearly dependent columns of the **H** matrix.

Corollary 1 *Given s coordinates corresponding to an erasure pattern containing s erasures, $s \leq (n-k)$, of which w coordinates are equal to the non-zero coordinates of a single codeword of weight w, the maximum number of erasures that can be corrected is $s-1$ and the minimum number that can be corrected is $w-1$.*

Corollary 2 *Given $w-1$ coordinates that correspond to linearly independent columns of the* **H** *matrix and w coordinates that correspond to linearly dependent columns of the* **H** *matrix, a codeword can be derived that has a weight less than or equal to w.*

The weight enumeration function of a code [15] is usually described as a homogeneous polynomial of degree n in x and y.

$$W_{\mathscr{C}}(x,y) = \sum_{i=0}^{n-1} A_i x^{n-i} y^i$$

The support of a codeword is defined [15] as the coordinates of the codeword that are non-zero. The probability of the successful erasure correction of s or more erasures is equal to the probability that no subset of the s erasure coordinates corresponds to the support of any codeword.

The number of possible erasure patterns of s erasures of a code of length n is $\binom{n}{s}$. For a *single* codeword of weight w, the number of erasure patterns with s coordinates that include the support of this codeword is $\binom{n-w}{s-w}$. Thus, the probability of a subset of the s coordinates coinciding with the support of a single codeword of weight w, $prob(\mathbf{x_w} \in \mathbf{f_s})$ is given by:

$$prob(\mathbf{x_w} \in \mathbf{f_s}) = \frac{\binom{n-w}{s-w}}{\binom{n}{s}}$$

and

$$prob(\mathbf{x_w} \in \mathbf{f_s}) = \frac{(n-w)!\; s!\; (n-s)!}{n!\; (s-w)!\; (n-s)!}$$

simplifying

$$prob(\mathbf{x_w} \in \mathbf{f_s}) = \frac{(n-w)!\; s!}{n!\; (s-w)!}$$

In such an event the s erasures are uncorrectable because, for these erasures, there are not s independent parity-check equations [15, 16]. However, $s-1$ erasures are correctable provided the $s-1$ erasures do not contain the support of a lower weight codeword.

The probability that s erasures will contain the support of at least one codeword of any weight, is upper and lower bounded by

$$1-\prod_{j=d_{min}}^{s} 1-A_j\frac{(n-j)!s!}{n!(s-j)!} < P_s \leq \sum_{j=d_{min}}^{s} A_j\frac{(n-j)!s!}{n!(s-j)!} \tag{14.2}$$

And given $s+1$ erasures, the probability that exactly s erasures are correctable, $Pr(s)$ is given by

$$Pr(s) = P_{s+1} - P_s \tag{14.3}$$

Given up to $n-k$ erasures the average number of erasures correctable by the code is

$$\overline{N_e} = \sum_{s=d_{min}}^{n-k} sPr(s) = \sum_{s=d_{min}}^{n-k} s\left(P_{s+1} - P_s\right). \tag{14.4}$$

Carrying out the sum in reverse order and noting that $P_{n-k+1} = 1$, the equation simplifies to become

$$\overline{N_e} = (n-k) - \sum_{s=d_{min}}^{n-k} P_s \tag{14.5}$$

An MDS code can correct $n-k$ erasures and is clearly the maximum number of correctable erasures as there are only $n-k$ independent parity-check equations. It is useful to denote an MDS shortfall

$$\text{MDS}_{\text{shortfall}} = \sum_{s=d_{min}}^{n-k} P_s \tag{14.6}$$

and

$$\overline{N_e} = (n-k) - \text{MDS}_{\text{shortfall}} \tag{14.7}$$

with

$$\sum_{s=d_{min}}^{n-k} 1-\prod_{j=d_{min}}^{s} 1-A_j\frac{(n-j)!s!}{n!(s-j)!} < \text{MDS}_{\text{shortfall}} \tag{14.8}$$

and

$$\text{MDS}_{\text{shortfall}} < \sum_{s=d_{min}}^{n-k} \sum_{j=d_{min}}^{s} A_j \frac{(n-j)!s!}{n!(s-j)!} \tag{14.9}$$

The contribution made by the high multiplicity of low-weight codewords to the shortfall in MDS performance is indicated by the probability $\hat{P}_j$ that the support of at least one codeword of weight j is contained in s erasures averaged over the number of uncorrectable erasures s, from $s = d_{min}$ to $n - k$, and is given by

$$\hat{P}_j = \sum_{s=d_{min}}^{n-k} Pr(s-1)A_j \frac{(n-j)!s!}{n!(s-j)!} \tag{14.10}$$

14.3 Probability of Decoder Error

For the erasure channel with erasure probability p, the probability of codeword decoder error, $P_d(p)$ for the code may be derived in terms of the weight spectrum of the code assuming ML decoding. It is assumed that a decoder error is declared if more than $n - k$ erasures occur and that the decoder does not resort to guessing erasures. The probability of codeword decoder error is given by the familiar function of p.

$$P_d(p) = \sum_{s=1}^{n} P_s p^s (1-p)^{(n-s)} \tag{14.11}$$

Splitting the sum into two parts

$$P_d(p) = \sum_{s=1}^{n-k} P_s p^s (1-p)^{(n-s)} + \sum_{s=n-k+1}^{n} P_s p^s (1-p)^{(n-s)} \tag{14.12}$$

The second term gives the decoder error rate performance for a hypothetical MDS code and the first term represents the degradation of the code compared to an MDS code. Using the upper bound of Eq. (14.2),

$$\begin{aligned} P_d(p) \leq & \sum_{s=1}^{n-k} \sum_{j=1}^{s} A_j \frac{(n-j)!\ s!}{n!\ (s-j)!} \frac{n!}{(n-s)!\ s!} p^s (1-p)^{(n-s)} \\ & + \sum_{s=n-k+1}^{n} \frac{n!}{(n-s)!\ s!} p^s (1-p)^{(n-s)} \end{aligned} \tag{14.13}$$

As well as determining the performance shortfall, compared to MDS codes, in terms of the number of correctable erasures it is also possible to determine the loss from capacity for the erasure channel. The capacity of the erasure channel with erasure probability p was originally determined by Elias [9] to be $1 - p$. Capacity may be approached with zero codeword error for very long codes, even using non-MDS codes such as BCH codes [7]. However, short codes and even MDS codes, will produce a non-zero frame error rate (FER). For $(n, k, n - k + 1)$ MDS codes, a codeword decoder error is deemed to occur whenever there are more than $n - k$ erasures. (It is assumed here that the decoder does not resort to guessing erasures that cannot be solved). This probability, $P_{MDS}(p)$, is given by

$$P_{MDS}(p) = 1 - \sum_{s=0}^{n-k} \frac{n!}{(n-s)!\,s!} p^s (1-p)^{(n-s)} \tag{14.14}$$

The probability of codeword decoder error for the code may be derived from the weight enumerator of the code using Eq. (14.13).

$$P_{code}(p) = \sum_{s=d_{min}}^{n-k} \sum_{j=d_{min}}^{s} \left(A_j \frac{(n-j)!\,s!}{n!\,(s-j)!} \frac{n!}{(n-s)!\,s!} p^s (1-p)^{(n-s)} \right.$$
$$\left. + \sum_{s=n-k+1}^{n} \frac{n!}{(n-s)!\,s!} p^s (1-p)^{(n-s)} \right) \tag{14.15}$$

This simplifies to become

$$P_{code}(p) = \sum_{s=d_{min}}^{n-k} \sum_{j=d_{min}}^{s} A_j \frac{(n-j)!\,(n-s)!}{(s-j)!} p^s (1-p)^{(n-s)} + P_{MDS}(p) \tag{14.16}$$

The first term in the above equation represents the loss from MDS code performance.

14.4 Codes Whose Weight Enumerator Coefficients Are Approximately Binomial

It is well known that the distance distribution for many linear, binary codes including BCH codes, Goppa codes, self-dual codes [13–15, 19] approximates to a binomial distribution. Accordingly,

$$A_j \approx \frac{n!}{(n-j)!\,j!\,2^{n-k}} \tag{14.17}$$

For these codes, for which the approximation is true, the shortfall in performance compared to an MDS code, $MDS_{shortfall}$ is obtained by substitution into Eq. (14.9)

$$MDS_{shortfall} = \sum_{s=1}^{n-k}\sum_{j=1}^{s} \frac{n!}{(n-j)!j!\,2^{n-k}}\frac{(n-j)!\,s!}{n!\,(s-j)!} \tag{14.18}$$

which simplifies to

$$MDS_{shortfall} = \sum_{s=1}^{n-k} \frac{2^s-1}{2^{n-k}} \tag{14.19}$$

which leads to the simple result

$$MDS_{shortfall} = 2 - \frac{n-k-2}{2^{n-k}} \approx 2 \tag{14.20}$$

It is apparent that for these codes the MDS shortfall is just 2 bits from correcting all $n-k$ erasures. It is shown later using the actual weight enumerator functions for codes, where these are known, that this result is slightly pessimistic since in the above analysis there is a non-zero number of codewords with distance less than d_{min}. However, the error attributable to this is quite small. Simulation results for these codes show that the actual MDS shortfall is closer to 1.6 bits due to the assumption that there is never an erasure pattern which has the support of more than one codeword.

For these codes whose weight enumerator coefficients are approximately binomial, the probability of the code being able to correct exactly s erasures, but no more, may also be simplified from (14.2) and (14.3).

$$\begin{aligned} Pr(s) = &\sum_{j=1}^{s+1} \frac{n!}{(n-j)!j!\,2^{n-k}}\frac{(n-j)!\,(s+1)!}{n!\,(s+1-j)!} \\ &- \sum_{j=1}^{s} \frac{n!}{(n-j)!j!\,2^{n-k}}\frac{(n-j)!\,s!}{n!\,(s-j)!} \end{aligned} \tag{14.21}$$

which simplifies to become

$$Pr(s) = \frac{2^s-1}{2^{n-k}} \tag{14.22}$$

for $s < n-k$ and for $s = n-k$

$$Pr(n-k) = 1 - \sum_{j=1}^{n-k} \frac{n!}{(n-j)!j!\,2^{n-k}}\frac{(n-j)!\,(n-k)!}{n!\,(n-k-j)!} \tag{14.23}$$

and

Table 14.1 PDF of number of correctable erasures for codes whose weight enumerator coefficients are binomial

Correctable erasures	Probability
$n-k$	$\frac{1}{2^{n-k}}$
$n-k-1$	$0.5-\frac{1}{2^{n-k}}$
$n-k-2$	$0.25-\frac{1}{2^{n-k}}$
$n-k-3$	$0.125-\frac{1}{2^{n-k}}$
$n-k-4$	$0.0625-\frac{1}{2^{n-k}}$
$n-k-5$	$0.03125-\frac{1}{2^{n-k}}$
$n-k-6$	$0.0150625-\frac{1}{2^{n-k}}$
$n-k-7$	$0.007503125-\frac{1}{2^{n-k}}$
$\vdots$	$\vdots$
$n-k-s$	$\frac{1}{2^s}-\frac{1}{2^{n-k}}$

$$Pr(n-k)=\frac{1}{2^{n-k}} \tag{14.24}$$

For codes whose weight enumerator coefficients are approximately binomial, the pdf of correctable erasures is given in Table 14.1.

The probability of codeword decoder error for these codes is given by substitution into (14.15),

$$P_{code}(p)=\sum_{s=0}^{n-k}\left(\frac{2^s-1}{2^{n-k}}\right)\frac{n!}{(n-s)!\,s!}p^s(1-p)^{(n-s)}+P_{MDS}(p) \tag{14.25}$$

As first shown by Dumer and Farrell [7] as n is taken to ∞, these codes achieve the erasure channel capacity. As examples, the probability of codeword decoder error for hypothetical rate 0.9 codes, having binomial weight distributions, and lengths 100 to 10,000 bits are shown plotted in Fig. 14.1 as a function of the channel erasure probability expressed in terms of relative erasure channel capacity $\frac{0.9}{1-p}$. It can be seen that at a decoder error rate of 10^{-8} the (1000, 900) code is operating at 95% of channel capacity, and the (10,000, 9,000) code is operating at 98% of channel capacity. A comparison with MDS codes is shown in Fig. 14.2. For codelengths from 500 to 50,000 bits, it can be seen that for codelengths of 5,000 bits and above, these rate 0.9 codes are optimum since their performance is indistinguishable from the performance of MDS codes with the same length and rate.

A comparison of MDS codes to codes with binomial weight enumerator coefficients is shown in Fig. 14.3 for $\frac{1}{2}$ rate codes with code lengths from 128 to 1024.

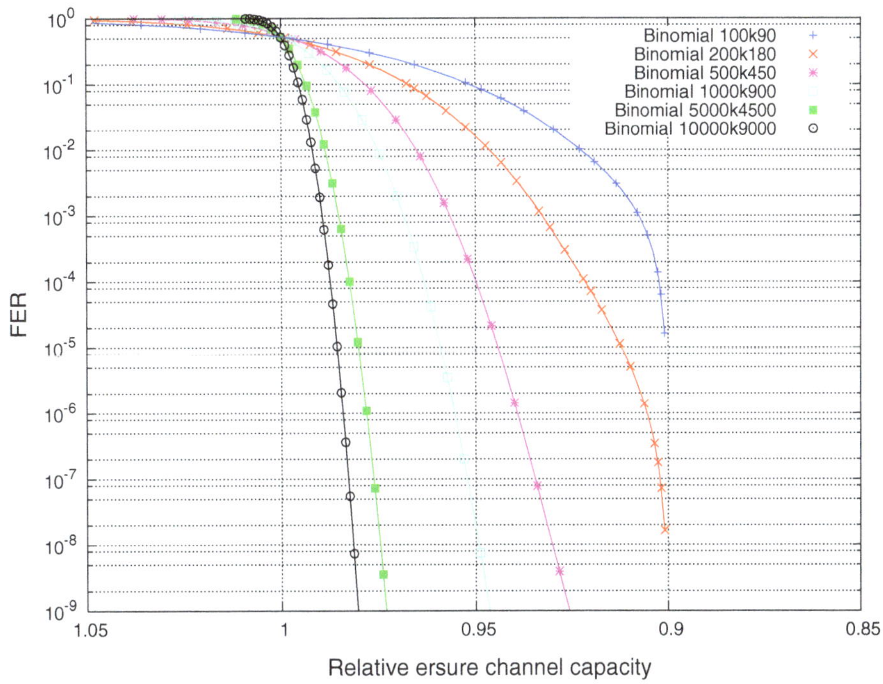

Fig. 14.1 FER performance of codes with binomial weight enumerator coefficients

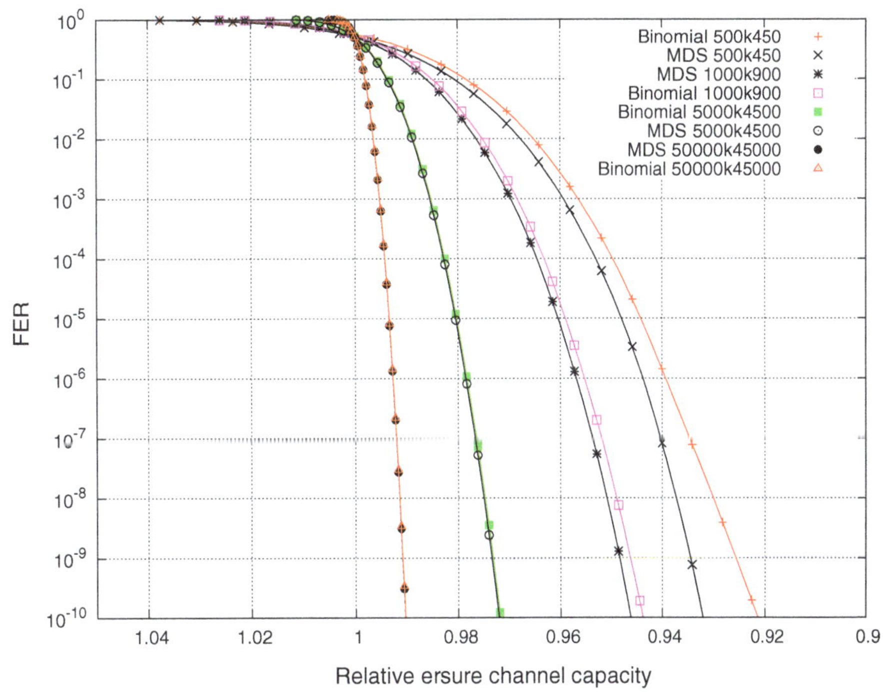

Fig. 14.2 Comparison of codes with binomial weight enumerator coefficients to MDS codes

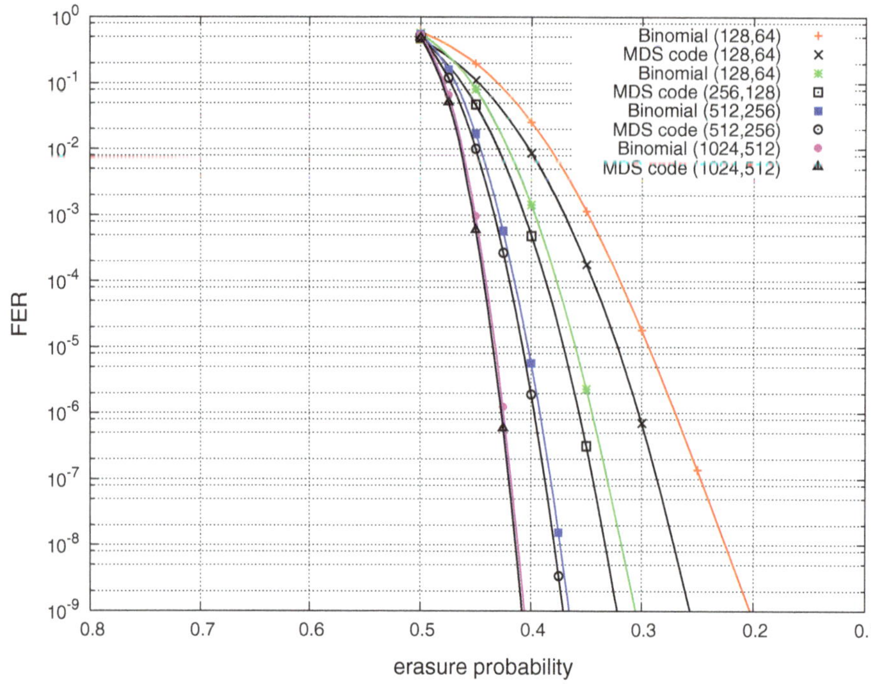

Fig. 14.3 Comparison of half rate codes having binomial weight enumerator coefficients with MDS codes as a function of erasure probability

14.5 MDS Shortfall for Examples of Algebraic, LDPC and Turbo Codes

The first example is the extended BCH code (128, 99, 10) whose coefficients up to weight 30 of the weight enumerator polynomial [5] are tabulated in Table 14.2.

Table 14.2 Low-weight spectral terms for the extended BCH (128, 99) code

Weight	A_d
0	1
10	796544
12	90180160
14	6463889536
16	347764539928
18	14127559573120
20	445754705469248
22	11149685265467776
24	224811690627712384
26	3704895377802191104
28	50486556173121673600
30	574502176730571255552

The PDF of the number of erased bits that are correctable up to the maximum of 29 erasures, derived from Eq. (14.1), is shown plotted in Fig. 14.4. Also shown plotted in Fig. 14.4 is the performance obtained numerically. It is straightforward, by computer simulation, to evaluate the erasure correcting performance of the code by generating a pattern of erasures randomly and solving these in turn using the parity-check equations. This procedure corresponds to maximum likelihood (ML) decoding [6, 17]. Moreover, the codeword responsible for any instances of non-MDS performance, (due to this erasure pattern) can be determined by back substitution into the solved parity-check equations. Except for short codes or very high rate codes, it is not possible to complete this procedure exhaustively, because there are too many combinations of erasure patterns. For example, there are 4.67×10^{28} combinations of 29 erasures in this code of length 128 bits. In contrast, there are relatively few low-weight codewords responsible for the non-MDS performance of the code. For example, each codeword of weight 10 is responsible for $\binom{118}{19} = 4.13 \times 10^{21}$ erasures patterns not being solvable.

As the d_{min} of this code is 10, the code is guaranteed to correct any erasure pattern containing up to 9 erasures. It can be seen from Fig. 14.4 that the probability of not being able to correct any pattern of 10 erasures is less than 10^{-8}. The probability of correcting 29 erasures, the maximum number, is 0.29. The average number of erasures corrected is 27.44, almost three times the d_{min}, and the average shortfall from MDS performance is 1.56 erased bits. The prediction of performance by the lower bound is pessimistic due to double codeword counting in erasure patterns featuring more than 25 bits or so. The effect of this is evident in Fig. 14.4. The lower bound average number of erasures corrected is 27.07, and the shortfall from MDS performance is 1.93 erasures, an error of 0.37 erasures. The erasure performance evaluation by simulation is complementary to the analysis using the weight distribution of the code, in that the simulation, being a sampling procedure, is inaccurate for short, uncorrectable erasure patterns, because few codewords are responsible for the performance in this region. For short, uncorrectable erasure patterns, the lower bound analysis is tight in this region because it not possible for these erasure patterns to contain more than one codeword due to codewords differing by at least d_{min}.

The distribution of the codeword weights responsible for non-MDS performance of this code is shown in Fig. 14.5.

This is in contrast to the distribution of low-weight codewords shown in Fig. 14.6. Although there are a larger number of higher weight codewords, there is less chance of an erasure pattern containing a higher weight codeword. The maximum occurrence is for weight 14 codewords as shown in Fig. 14.5.

The FER performance of the BCH (128, 107, 10) code is shown plotted in Fig. 14.7 as a function of relative capacity defined by $\frac{(1-p)n}{k}$. Also, plotted in Fig. 14.7 is the FER performance of a hypothetical (128, 99, 30) MDS code. Equations (14.15) and (14.14), respectively, were used to derive Fig. 14.7. As may be seen from Fig. 14.7, there is a significant shortfall in capacity even for the optimum MDS code. This shortfall is attributable to the relatively short length of the code. At 10^{-9} FER, the BCH (128, 99, 10) code achieves approximately 80% of the erasure channel capacity.

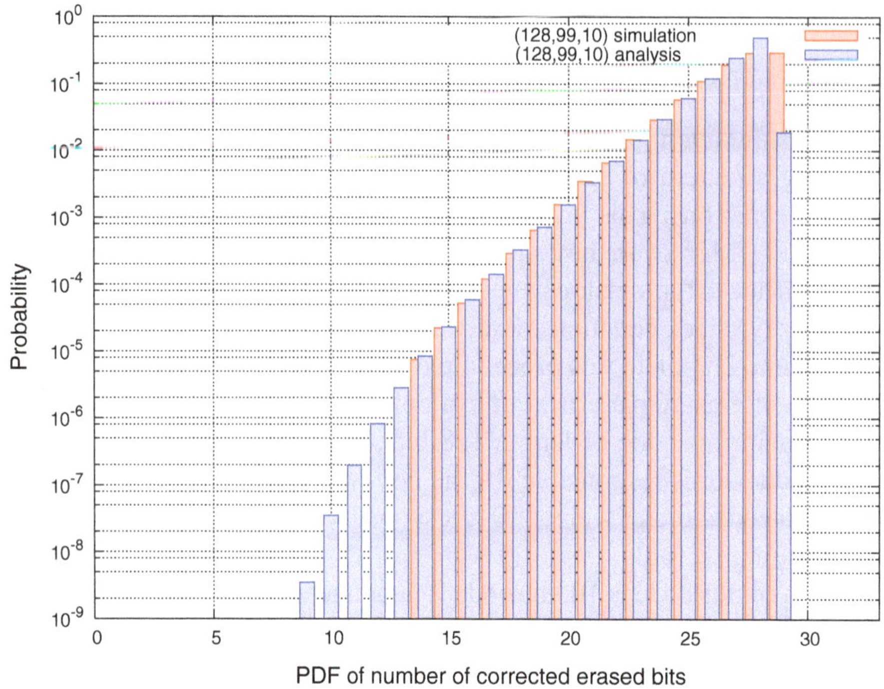

Fig. 14.4 Erasure performance for the (128, 99, 10) Extended BCH Code

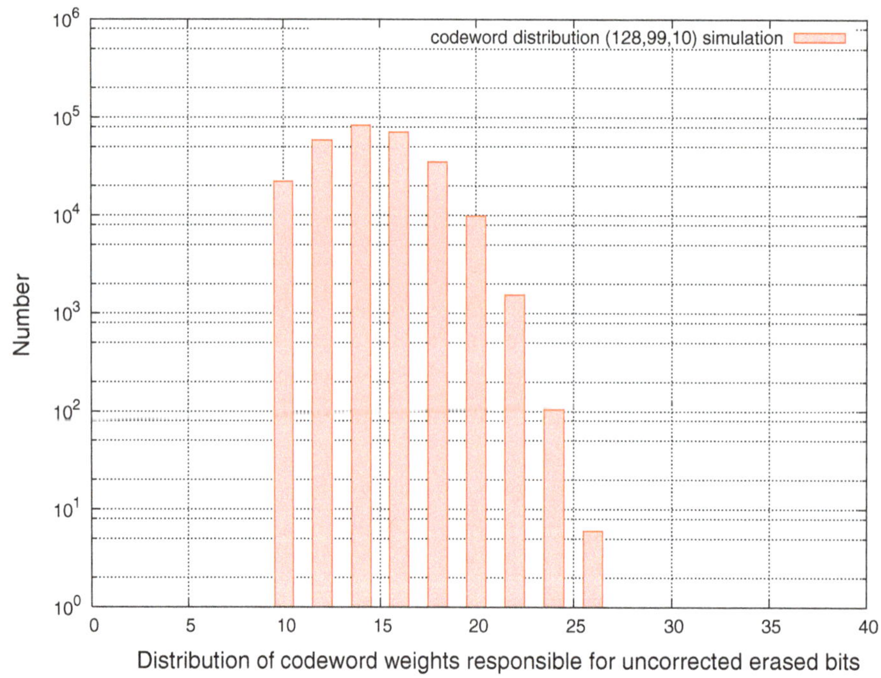

Fig. 14.5 Distribution of codeword weights responsible for non-MDS performance, of the (128, 99, 10) BCH Code

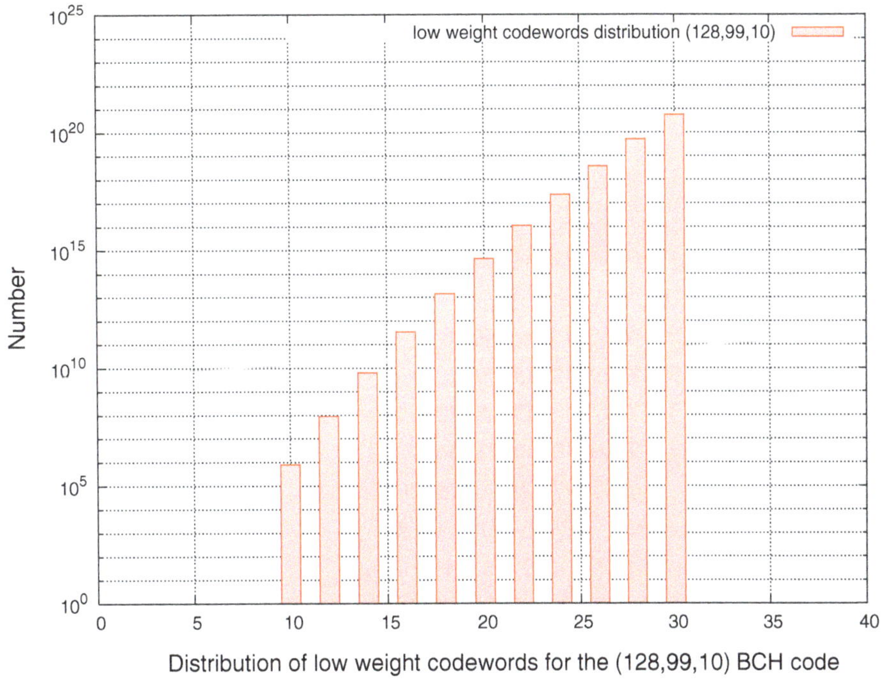

Fig. 14.6 Distribution of low-weight codewords for the (128, 99, 10) BCH code

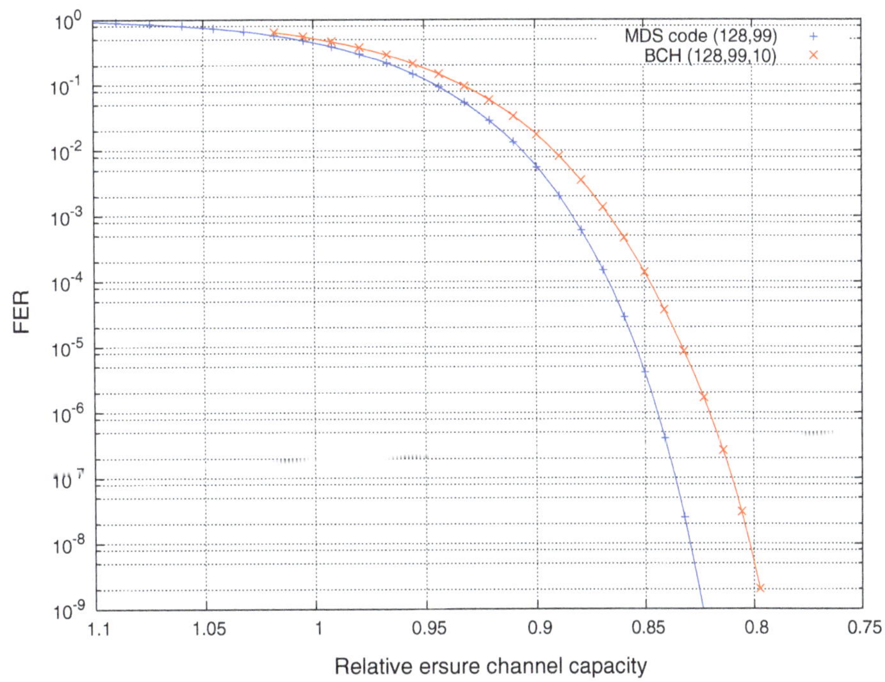

Fig. 14.7 FER performance for the (128, 99, 10) BCH code for the erasure channel

Table 14.3 Spectral terms up to weight 50 for the extended BCH (256, 207) code

Weight	A_d
0	1
14	159479040
16	36023345712
18	6713050656000
20	996444422768640
22	119599526889384960
24	11813208348266177280
26	973987499253055749120
28	67857073021007558686720
30	4036793565003066065373696
32	206926366333597318696425720
34	9212465086525810564304939520
36	358715843060045310259622139904
38	12292226836236855272009377988 0960
40	372755158433879986474102933212928
42	10052700091541303286178365979008000
44	242189310556445744774611488568535040
46	5233629101357641331155176578460897024
48	101819140628807204943892435954902207120
50	1789357109760781792970450788764603959040

The maximum capacity achievable by any (128, 99) binary code as represented by a (128, 99, 30) MDS code is approximately 82.5%.

An example of a longer code is the (256, 207, 14) extended BCH code. The coefficients up to weight 50 of the weight enumerator polynomial [10] are tabulated in Table 14.3. The evaluated erasure correcting performance of this code is shown in Fig. 14.8, and the code is able to correct up to 49 erasures. It can be seen from Fig. 14.8 that there is a close match between the lower bound analysis and the simulation results for the number of erasures between 34 and 46. Beyond 46 erasures, the lower bound becomes increasingly pessimistic due to double counting of codewords. Below 34 erasures the simulation results are erratic due to insufficient samples. It can be seen from Fig. 14.8 that the probability of correcting only 14 erasures is less than 10^{-13} (actually 5.4×10^{-14}) even though the d_{min} of the code is 14. If a significant level of erasure correcting failures is defined as 10^{-6}, then from Fig. 14.8, this code is capable of correcting up to 30 erasures even though the guaranteed number of correctable erasures is only 13. The average number of erasures correctable by the code is 47.4, an average shortfall of 1.6 erased bits. The distribution of codeword weights responsible for the non-MDS performance of this code is shown in Fig. 14.9.

The FER performance of the BCH (256, 207, 14) code is shown plotted in Fig. 14.10 as a function of relative capacity defined by $\frac{(1-p)n}{k}$. Also plotted in

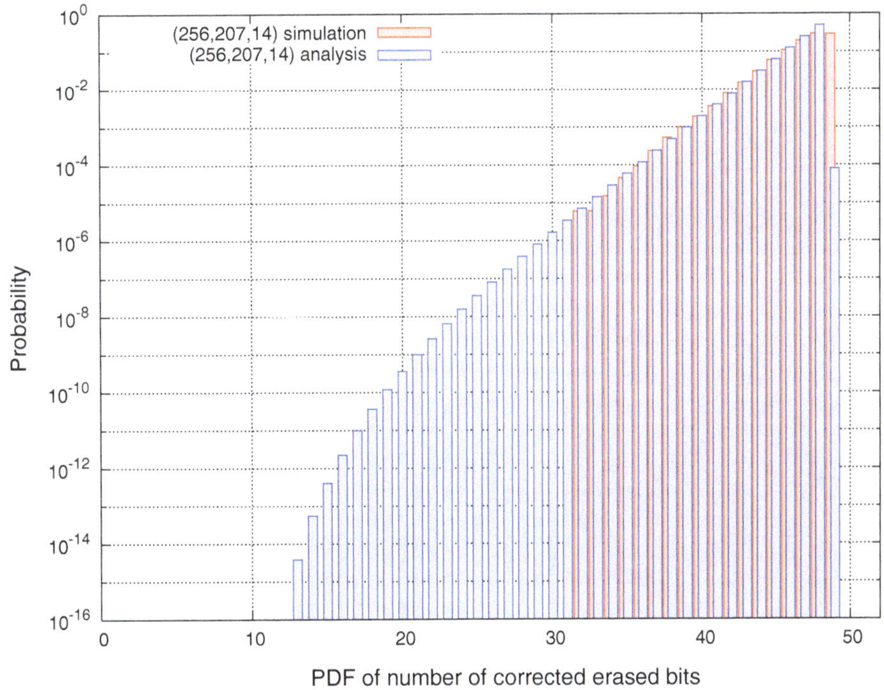

Fig. 14.8 PDF of erasure corrections for the (256, 207, 14) Extended BCH Code

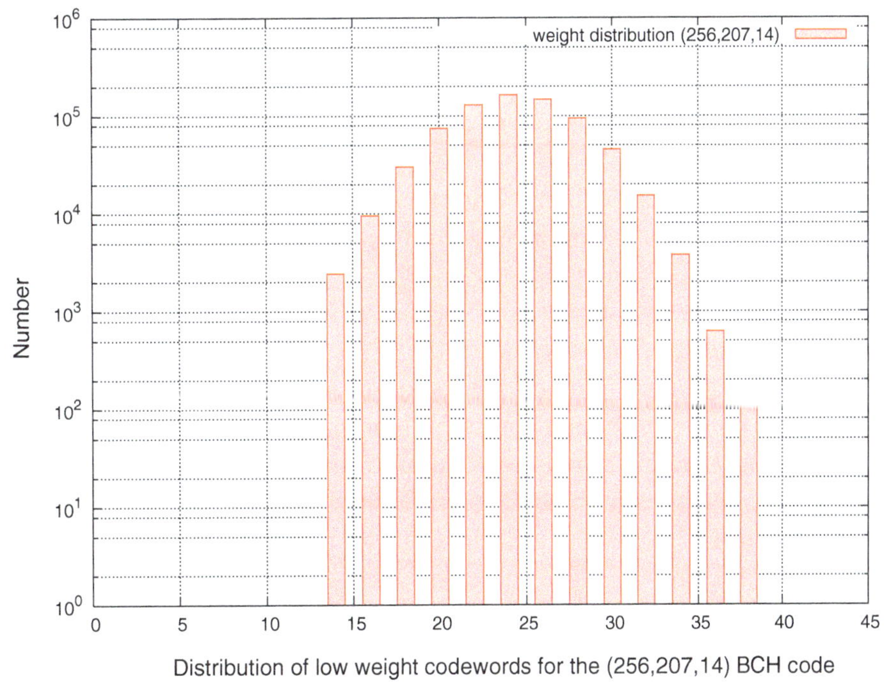

Fig. 14.9 Distribution of codeword weights responsible for non-MDS performance, for the extended (256, 207, 14) BCH Code

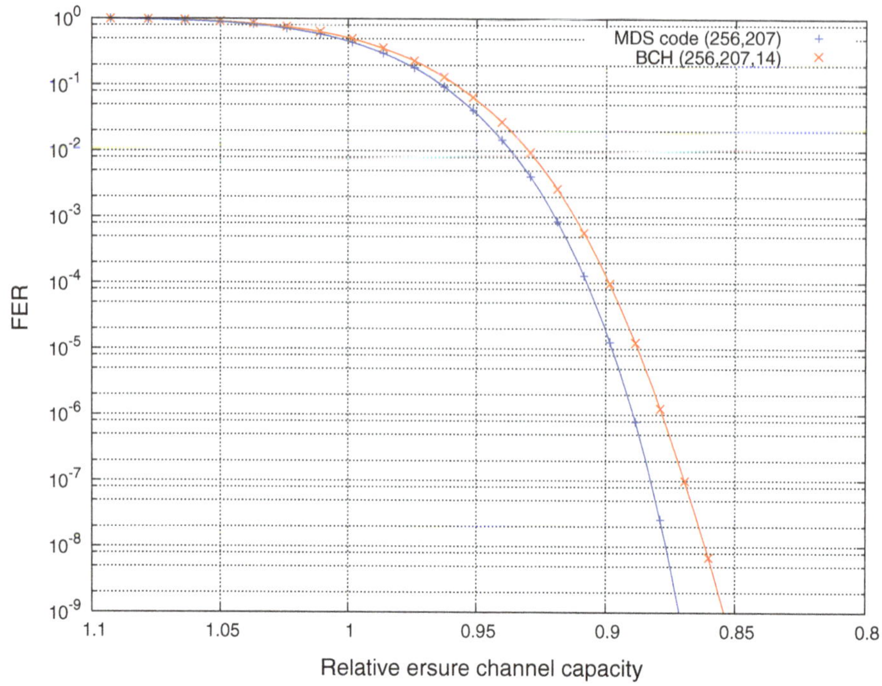

Fig. 14.10 FER performance for the (256, 207, 14) BCH Code for the erasure channel

Fig. 14.10 is the FER performance of a hypothetical (256, 207, 50) MDS code. Equations (14.15) and (14.14), respectively, were used to derive Fig. 14.10. As may be seen from Fig. 14.10, there is less of a shortfall in capacity compared to the BCH (128, 107, 10) code. At 10^{-9} FER, the BCH (256, 207, 14) code achieves approximately 85.5% of the erasure channel capacity. The maximum capacity achievable by any (256, 207) binary code as represented by the (256, 207, 50) hypothetical MDS code is approximately 87%.

The next code to be investigated is the (512, 457, 14) extended BCH code which was chosen because it is comparable to the (256, 207, 14) code in being able to correct a similar maximum number of erasures (55 *cf.* 49) and has the same d_{min} of 14. Unfortunately, the weight enumerator polynomial has yet to be determined, and only erasure simulation results may be obtained. Figure 14.11 shows the performance of this code. The average number of erasures corrected is 53.4, an average shortfall of 1.6 erased bits. The average shortfall is identical to the (256, 207, 14) extended BCH code. Also, the probability of achieving MDS code performance, i.e. being able to correct all $n - k$ erasures is also the same and equal to 0.29. The distribution of codeword weights responsible for non-MDS performance of the (512, 457, 14) code is very similar to that for the (256, 207, 14) code, as shown in Fig. 14.12.

An example of an extended cyclic quadratic residue code is the (168, 84, 24) code whose coefficients of the weight enumerator polynomial have been recently determined [20] and are tabulated up to weight 72 in Table 14.4. This code is a self-dual, doubly even code, but not extremal because its d_{min} is not 32 but 24 [3]. The FER performance of the (168, 84, 24) code is shown plotted in Fig. 14.13 as

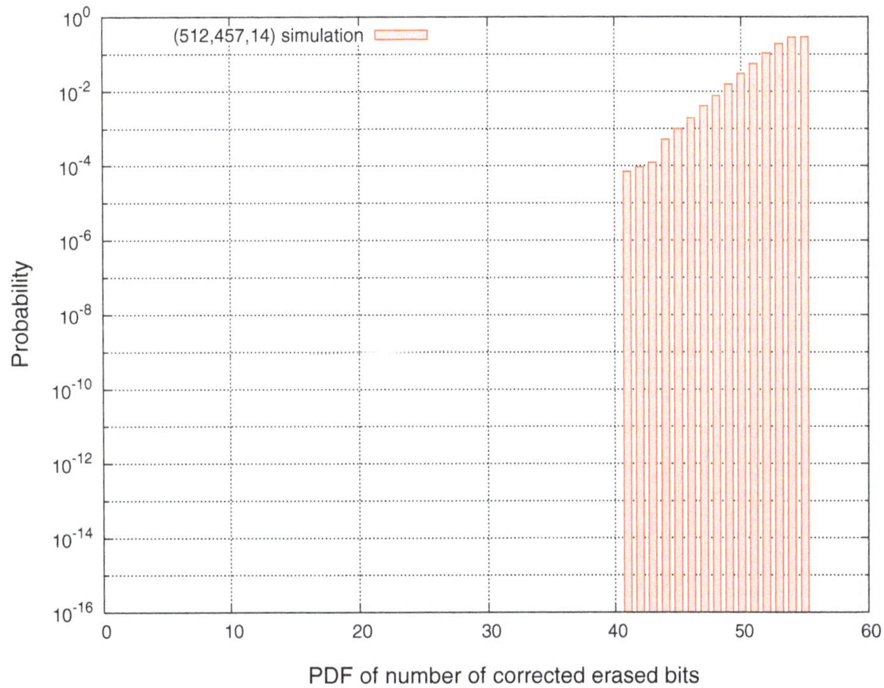

Fig. 14.11 PDF of erasure corrections for the (512, 457, 14) Extended BCH Code

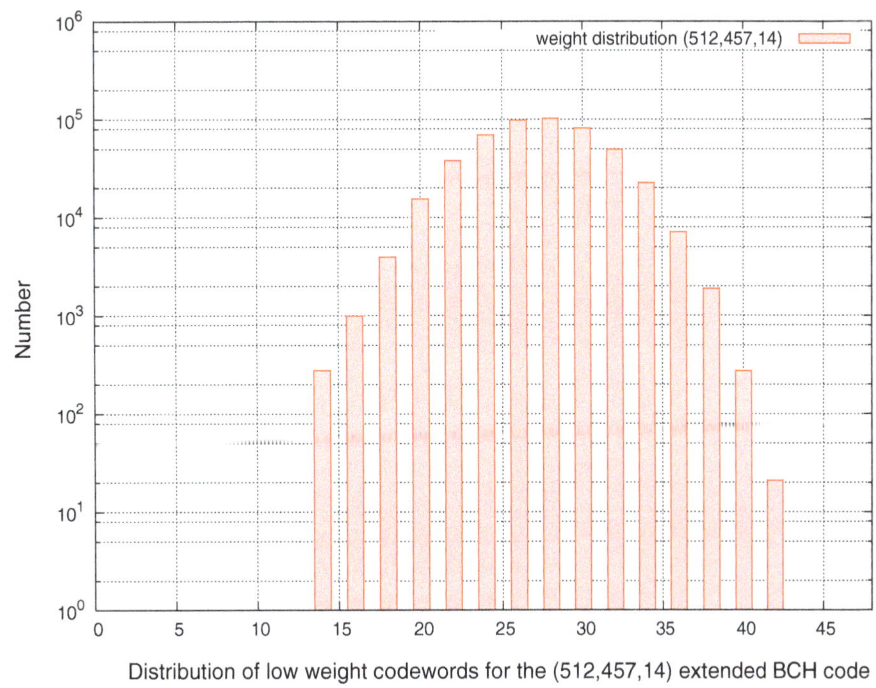

Fig. 14.12 Distribution of codeword weights responsible for non-MDS performance, for the extended (512, 457, 14) BCH Code

Table 14.4 Spectral terms up to weight 72 for the extended Quadratic Residue (168, 84) code

Weight	A_d
0	1
24	776216
28	18130188
32	5550332508
36	1251282702264
40	166071600559137
44	13047136918828740
48	629048543890724216
52	19087130695796615088
56	372099690249351071112
60	4739291519495550245228
64	39973673337590380474086
68	225696677727188690570184
72	860241108921860741947676

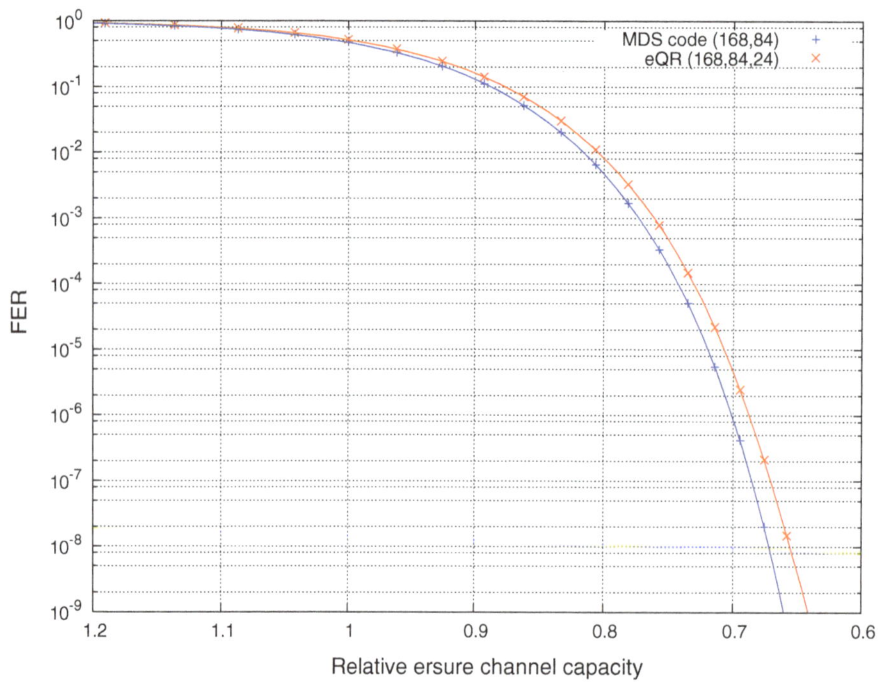

Fig. 14.13 FER performance for the (168, 84, 24) eQR Code for the erasure channel

a function of relative capacity defined by $\frac{(1-p)n}{k}$. Also plotted in Fig. 14.13 is the FER performance of a hypothetical (168, 84, 85) MDS code. Equations (14.15) and (14.14), respectively, were used to derive Fig. 14.13. The performance of the (168, 84, 24) code is close to that of the hypothetical MDS code but both codes are around 30% from capacity at 10^{-6} FER.

The erasure correcting performance of non-algebraic designed codes is quite different from algebraic designed codes as may be seen from the performance results for a (240, 120, 16) turbo code shown in Fig. 14.14. The turbo code features memory 4 constituent recursive encoders and a code matched, modified S interleaver, in order to maximise the d_{min} of the code. The average number of erasures correctable by the code is 116.5 and the average shortfall is 3.5 erased bits. The distribution of codeword weights responsible for non-MDS performance of the (240, 120, 16) code is very different from the algebraic codes and features a flat distribution as shown in Fig. 14.15.

Similarly, the erasure correcting performance of a (200, 100, 11) LDPC code designed using the Progressive Edge Growth (PEG) algorithm [12] is again quite different from the algebraic codes as shown in Fig. 14.16. As is typical of randomly generated LDPC codes, the d_{min} of the code is quite small at 11, even though the code has been optimised. For this code, the average number of correctable erasures is 93.19 and the average shortfall is 6.81 erased bits. This is markedly worse than the turbo code performance. It is the preponderance of low-weight codewords that is responsible for the inferior performance of this code compared to the other codes as shown by the codeword weight distribution in Fig. 14.17.

The relative weakness of the LDPC code and turbo code becomes clear when compared to a good algebraic code with similar parameters. There is a (200, 100, 32) extended quadratic residue code. The p.d.f. of the number of erasures corrected by this code is shown in Fig. 14.18. The difference between having a d_{min} of 32 compared to 16 for the turbo code and 10 for the LDPC code is dramatic. The average number of correctable erasures is 98.4 and the average shortfall is 1.6 erased bits. The weight enumerator polynomial of this self-dual code, is currently unknown as evaluation of the 2^{100} codewords is currently beyond the reach of today's computers. However, the distribution of codeword weights responsible for non-MDS performance of the (200, 100, 32) code which is shown in Fig. 14.19 indicates the doubly even codewords of this code and the d_{min} of 32.

14.5.1 Turbo Codes with Dithered Relative Prime (DRP) Interleavers

DRP interleavers were introduced in [4]. They have been shown to produce some of the largest minimum distances for turbo codes. However, the iterative decoding algorithm does not exploit this performance to its full on AWGN channels where the performance of these interleavers is similar to that of randomly designed interleavers having lower minimum distance. This is due to convergence problems in the error floor region. A DRP interleaver is a concatenation of 3 interleavers, the two dithers A, B and a relative prime interleaver π:

$$I(i) = B(\pi(A(i))) \tag{14.26}$$

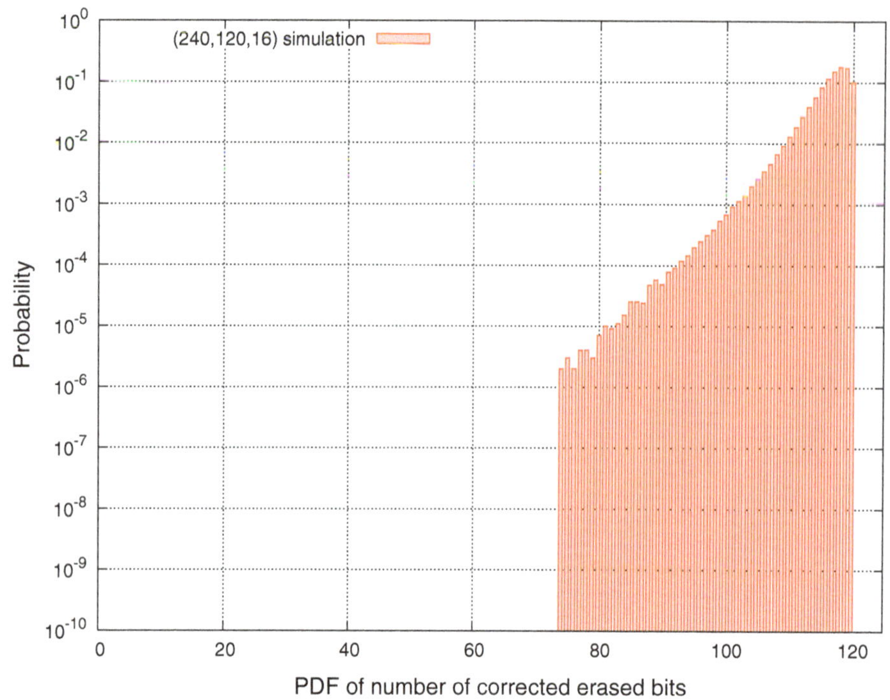

Fig. 14.14 PDF of erasure corrections for the (240, 120, 16) turbo code

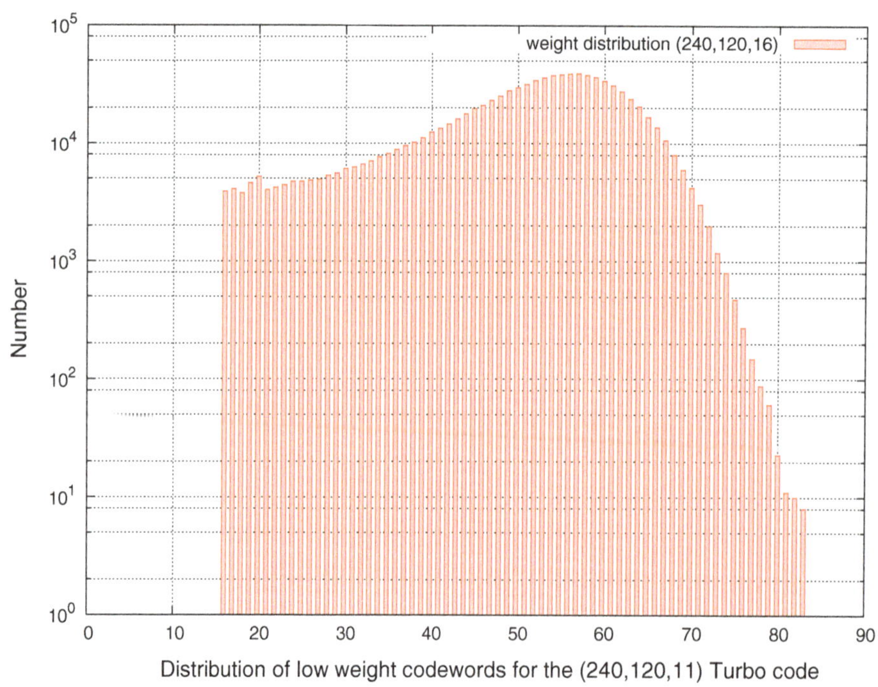

Fig. 14.15 Distribution of codeword weights responsible for non-MDS performance, for the (240, 120, 16) turbo code

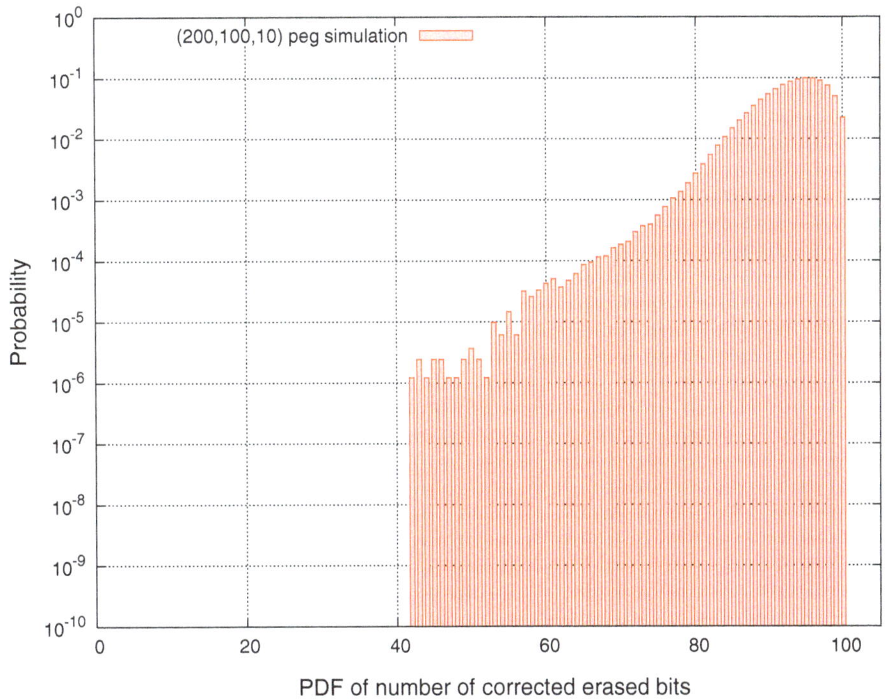

Fig. 14.16 PDF of erasure corrections for the (200, 100, 10) PEG LDPC code

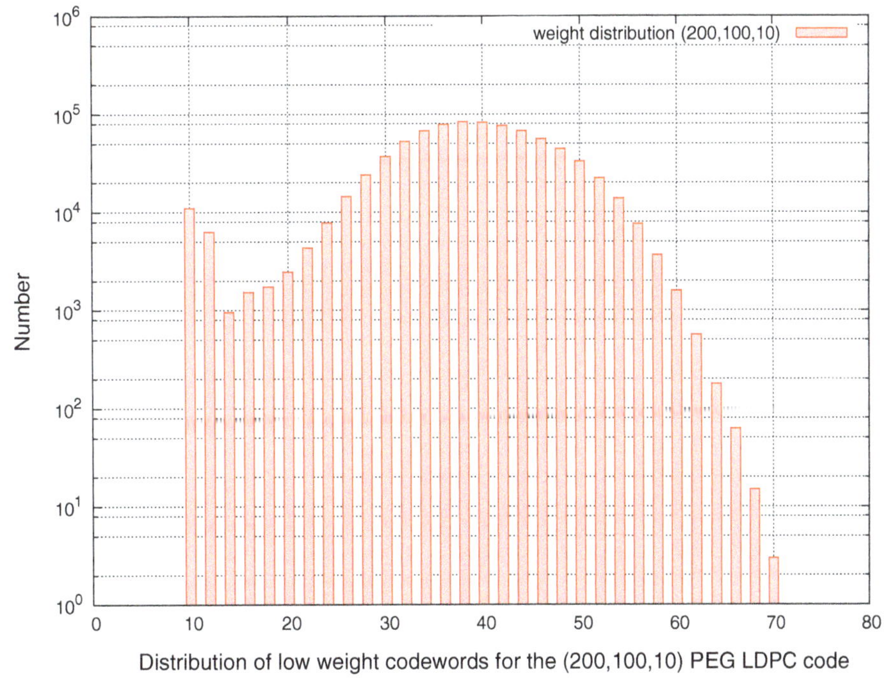

Fig. 14.17 Distribution of codeword weights responsible for non-MDS performance, for the (200, 100, 10) PEG LDPC code

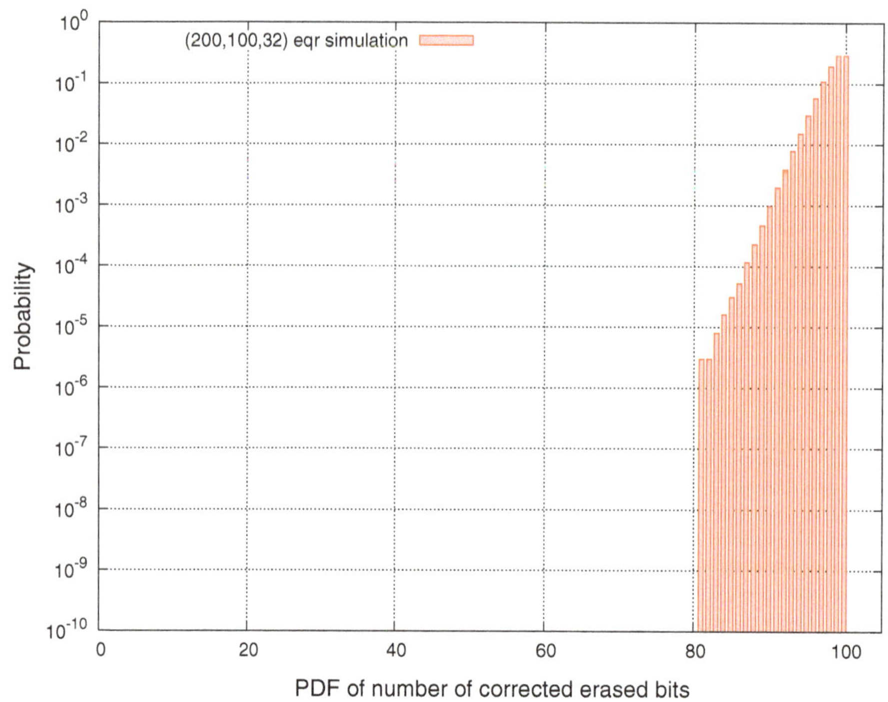

Fig. 14.18 PDF of erasure corrections for the (200, 100, 32) Extended QR Code

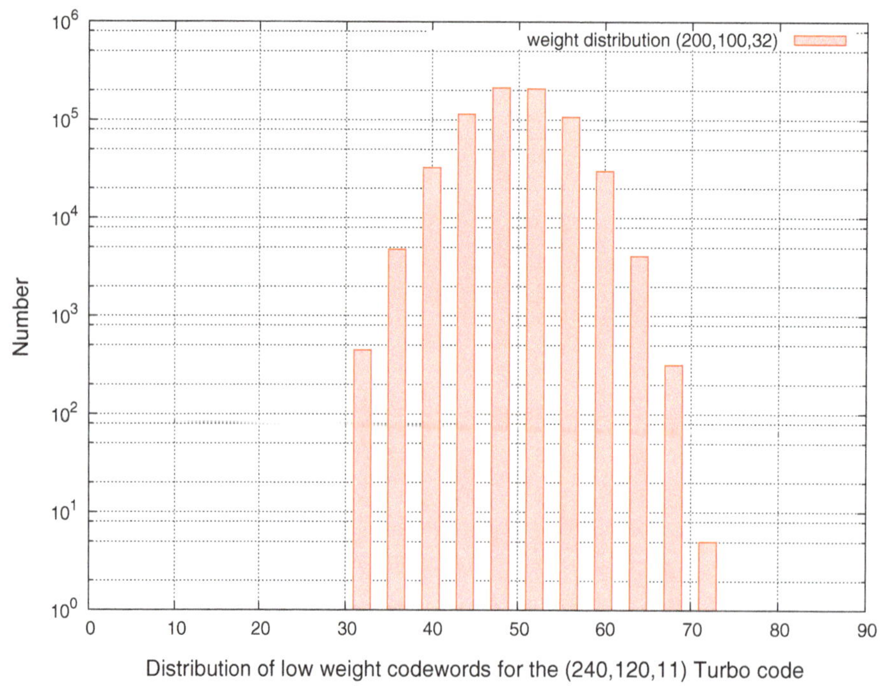

Fig. 14.19 Distribution of codeword weights responsible for non-MDS performance, for the (200, 100, 32) Extended QR Code

Table 14.5 Minimum distance of turbo codes using DRP interleavers as compared to *S*-random interleavers

k	40	200	400	1000
DRP	19	33	38	45
S-RAN	13	17	22	26

The dithers are short permutations, generally of length $m = 4, 8, 16$ depending on the length of the overall interleaver. We have

$$A(i) = m\lfloor i/m \rfloor + a_{i\%m} \tag{14.27}$$

$$B(i) = m\lfloor i/m \rfloor + b_{i\%m} \tag{14.28}$$

$$\pi(i) = (pi + q)\%m, \tag{14.29}$$

where a, b, are permutations of length m and p must be relatively prime to k. If a, b and p are properly chosen, the minimum distance of turbo codes can be drastically improved as compared to that of a turbo code using a typical $\mathscr{S}$-random interleaver. A comparison is shown in Table 14.5 for memory 3 component codes.

As an example two turbo codes are considered, one employing a DRP interleavers, having parameters (120, 40, 19) and another employing a typical $\mathscr{S}$-random interleaver and having parameters (120, 40, 13).

14.5.2 Effects of Weight Spectral Components

The weight spectrum of each of the two turbo codes has been determined exhaustively from the G matrix of each code by codeword enumeration using the revolving door algorithm. The weight spectrum of both of the two turbo codes is shown in Table 14.6. It should be noted that as the codes include the all ones codeword, $A_{n-j} = A_j$, only weights up to A_{60} are shown in Table 14.6.

Using the weight spectrum of each code the upper and lower bound cumulative distributions and corresponding density functions have been derived using Eqs. (14.2) and (14.3), respectively, and are compared in Fig. 14.20. It can be observed that the DRP interleaver produces a code with a significantly smaller probability of failing to correct a given number of erasures.

The MDS shortfall for the two codes is:

$$\text{MDS}_{\text{shortfall}}(120, 40, 19) = 2.95 \text{ bits} \tag{14.30}$$

$$\text{MDS}_{\text{shortfall}}(120, 40, 13) = 3.29 \text{ bits} \tag{14.31}$$

The distribution of the codeword weights responsible for the MDS shortfalls is shown in Fig. 14.21. For interest, also shown in Fig. 14.21 is the distribution for

Table 14.6 Weight spectrum of the (120, 40, 19) and (120, 40, 13) turbo codes. Multiplicity for weights larger than 60 satisfy $A_{60-i} = A_{60+i}$

Weight	Multiplicity	
	(120, 40, 19)	(120, 40, 13)
0	1	1
13	0	3
14	0	6
15	0	3
16	0	15
17	0	21
18	0	17
19	10	52
20	100	82
21	130	136
22	300	270
23	450	462
24	880	875
25	1860	2100
26	3200	3684
27	7510	7204
28	14715	15739
29	29080	30930
30	63469	64602
31	137130	137976
32	279815	279700
33	611030	608029
34	1313930	1309472
35	2672760	2671331
36	5747915	5745253
37	12058930	12045467
38	24137345	24112022
39	49505760	49486066
40	97403290	97408987
41	183989250	184005387
42	347799180	347810249
43	626446060	626489895
44	1086030660	1086006724
45	1855409520	1855608450
46	3021193870	3021448047
47	4744599030	4744412946
48	7286393500	7286669468
49	10691309800	10690683197
50	15157473609	15156479947

(continued)

Table 14.6 (continued)

Weight	Multiplicity	
	(120, 40, 19)	(120, 40, 13)
51	20938289040	20939153481
52	27702927865	27702635729
53	35480878330	35481273341
54	44209386960	44210370096
55	52854740864	52853468145
56	61256875090	61257409658
57	69008678970	69008947092
58	74677319465	74677092916
59	78428541430	78428875230
60	80007083570	80006086770

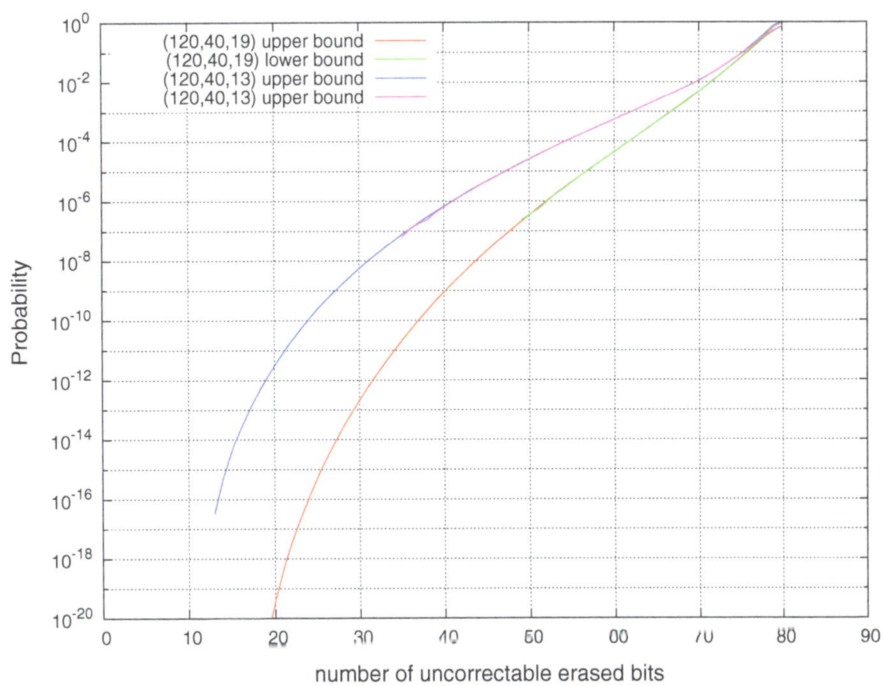

Fig. 14.20 Probability of Maximum Likelihood decoder failure

(120, 40, 28) best known linear code. This code, which is chosen to have the same block length and code rate as the turbo code, is derived by shortening a (130, 50, 28) code obtained by adding two parity checks to the (128, 50, 28) extended BCH. This linear code has an MDS shortfall of 1.62 bits and its weight spectrum consists of doubly even codewords as shown in Table 14.7. For the turbo codes the contribution made by the lower weight codewords is apparent in Fig. 14.21, and this is confirmed by the plot of the cumulative contribution made by the lower weight codewords shown in Fig. 14.22.

Table 14.7 Weight spectrum of the linear (120, 40, 28) code derived from the extended BCH (128, 50, 28) code

Weight j	Multiplicity A_j
0	1
28	5936
32	448563
36	17974376
40	379035818
44	4415788318
48	29117944212
52	110647710572
56	245341756158
60	319670621834
64	245340760447
68	110648904336
72	29117236550
76	4415980114
80	379051988
84	17949020
88	453586
92	5910
96	37

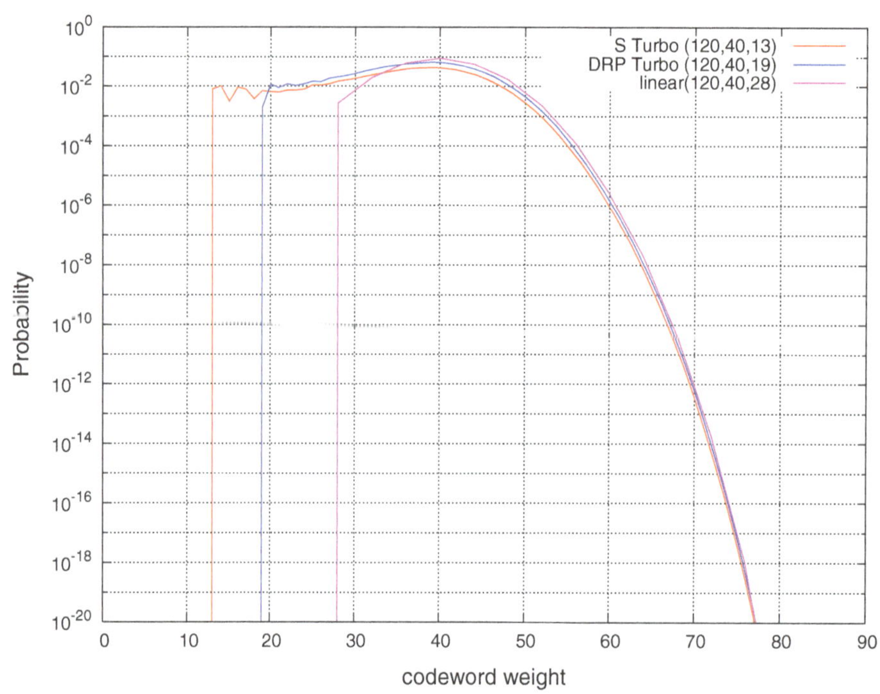

Fig. 14.21 Distribution of codeword weights responsible for non-MDS performance

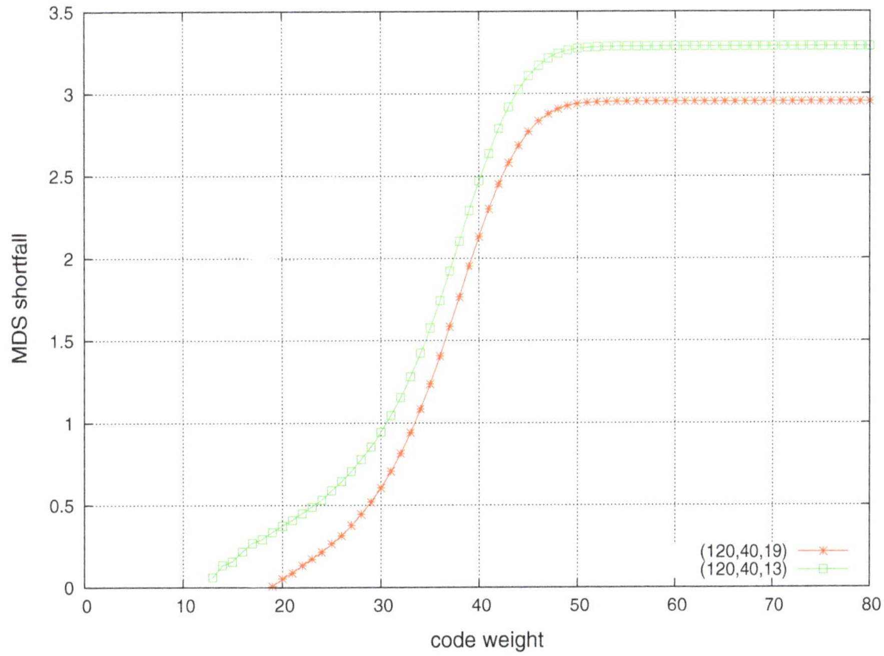

Fig. 14.22 Cumulative code weight contribution to MDS shortfall

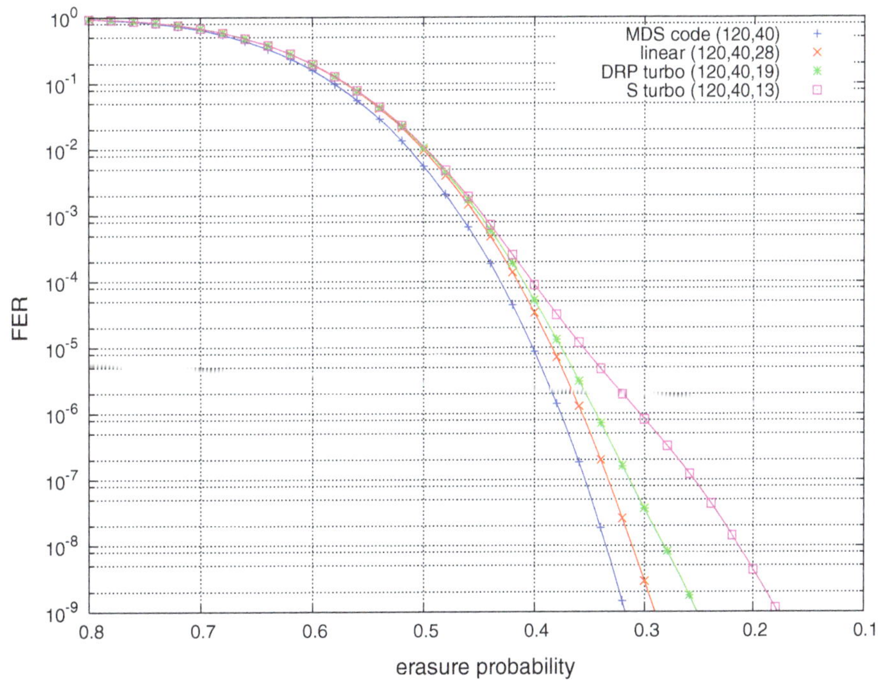

Fig. 14.23 Probability of ML decoder error for the erasure channel

For the erasure channel, the performance of the two turbo codes and the $(120, 40, 28)$ code is given by (14.15) and is shown in Fig. 14.23 assuming ML decoding. Also shown in Fig. 14.23 is the performance of a (hypothetical) binary $(120, 40, 81)$ MDS which is given by the second term of (14.15). The code derived from the shortened, extended BCH code, $(120, 40, 28)$, has the best performance and compares well to the lower bound provided by the MDS hypothetical code. The DRP interleaver turbo code also has good performance, but the $\mathscr{S}$-random interleaver turbo code shows an error floor due to the d_{min} of 13.

14.6 Determination of the d_{min} of Any Linear Code

It is well known that the determination of weights of any linear code is a Nondeterministic Polynomial time (NP) hard problem [8] and except for short codes, the best methods for determining the minimum Hamming distance, d_{min} codeword of a linear code, to date, are probabilistically based [2]. Most methods are based on the generator matrix, the G matrix of the code and tend to be biased towards searching using constrained information weight codewords. Such methods become less effective for long codes or codes with code rates around $\frac{1}{2}$ because the weights of the evaluated codewords tend to be binomially distributed with average weight $\frac{n}{2}$ [15].

Corollary 2 from Sect. 14.2 above, provides the basis of a probabilistic method to find low-weight codewords in a significantly smaller search space than the G matrix methods. Given an uncorrectable erasure pattern of $n-k$ erasures, from Corollary 2, the codeword weight is less than or equal to $n-k$. The search method suggested by this, becomes one of randomly generating erasure patterns of $n-k+1$ erasures, which of course are uncorrectable by any (n,k) code, and determining the codeword and its weight from (14.2). This time, the weights of the evaluated codewords will tend to be binomially distributed with average weight $\frac{n-k+1}{2}$. With this trend, for N_{trials} the number of codewords determined with weight d, M_d is given by

$$M_d = N_{trials} \frac{(n-k+1)!}{d!(n-k-d+1)!2^{n-k+1}} \tag{14.32}$$

As an example of this approach, the self-dual, bordered, double-circulant code (168, 84) based on the prime number 83, is described in [11] as having an unconfirmed d_{min} of 28. From (14.32) when using 18,000 trials, 10 codewords of weight 28 will be found on average. However, as the code is doubly even and only has codewords weights which are a multiple of 4, using 18,000 trials, 40 codewords are expected. In a set of trials using this method for the (168, 84) code, 61 codewords of weight 28 were found with 18,000 trials. Furthermore, 87 codewords of weight 24 were also found indicating that the d_{min} of this code is 24 and not 28 as was originally expected in [11].

The search method can be improved by biasing towards the evaluation of erasure patterns that have small numbers of erasures that cannot be solved. Recalling the analysis in Sect. 14.2, as the parity-check equations are Gaussian reduced, no erased bit is a function of any other erased bits. There will be $n - k - s$ remaining parity-check equations, which do not contain the erased bit coordinates $\mathbf{x_f}$. These remaining equations may be searched to see if there is an unerased bit coordinate, that is not present in any of the equations. If there is one such coordinate, then this coordinate in conjunction with the erased coordinates solved so far forms an uncorrectable erasure pattern involving only s erasures instead of $n - k + 1$ erasures. With this procedure, biased towards small numbers of unsolvable erasures, it was found that, for the above code, 21 distinct codewords of weight 24 and 17 distinct codewords of weight 28 were determined in 1000 trials and the search took approximately 2 s on a typical 2.8GHz, Personal Computer (PC).

In another example, the (216, 108) self dual, bordered double-circulant code is given in [11] with an unconfirmed d_{min} of 36. With 1000 trials which took 7 s on the PC, 11 distinct codewords were found with weight 24 and a longer evaluation confirmed that the d_{min} of this code is indeed 24.

14.7 Summary

Analysis of the erasure correcting performance of linear, binary codes has provided the surprising result that many codes can correct, on average, almost $n-k$ erasures and have a performance close to the optimum performance as represented by (hypothetical), binary MDS codes. It was shown that for codes having a weight distribution approximating to a binomial distribution, and this includes many common codes, such as BCH codes, Goppa codes and self-dual codes, that these codes can correct at least $n-k-2$ erasures on average, and closely match the FER performance of MDS codes as code lengths increase. The asymptotic performance achieves capacity for the erasure channel. It was also shown that codes designed for iterative decoders, the turbo and LDPC codes, are relatively weak codes for the erasure channel and compare poorly with algebraically designed codes. Turbo codes, designed for optimised d_{min}, were found to outperform LDPC codes.

For turbo codes using DRP interleavers for the erasure channel using ML decoding, the result is that these relatively short turbo codes are (on average), only about 3 erasures away from optimal MDS performance. The decoder error rate performance of the two turbo codes when using ML decoding on the erasure channel was compared to (120, 40, 28) best known linear code and a hypothetical binary MDS code. The DRP interleaver demonstrated a clear advantage over the $\mathscr{S}$-random interleaver and was not too far way from MDS performance. Analysis of the performance of longer turbo codes is rather problematic.

Determination of the erasure correcting performance of a code provides a means of determining the d_{min} of a code and an efficient search method was described. Using the method, the d_{min} results for two self-dual codes, whose d_{min} values were previously unknown were determined, and these codes were found to be (168, 84, 24) and (216, 108, 24) codes.

References

1. Berrou, C., Glavieux, A., Thitimajshima, P.: Near Shannon limit error-correcting coding: Turbo codes. In: Proceedings of the IEEE International Conference on Communications, Geneva, Switzerland, pp. 1064–1070 (1993)
2. Canteaut, A., Chabaud, F.: A new algorithm for finding minimum weight words in a linear code: application to McEliece's cryptosystem and to narrow-sense BCH codes of length 511. IEEE Trans. Inf. Theory **44**(1), 367–378 (1998)
3. Conway, J.H., Sloane, N.J.A.: A new upper bound on the minimum distance of self-dual codes. IEEE Trans. Inf. Theory **36**(6), 1319–1333 (1990)
4. Crozier, S., Guinard, P.: Distance upper bounds and true minimum distance results for Turbo codes designed with DRP interleavers. In: Proceedings of the 3rd International Symposium on Turbo Codes, pp. 169–172 (2003)
5. Desaki, Y., Fujiwara, T., Kasami, T.: The weight distribution of extended binary primitive BCH code of length 128. IEEE Trans. Inf. Theory **43**(4), 1364–1371 (1997)
6. Di, C., Proietti, D., Telatar, I.E., Richardson, T.J., Urbanke, R.L.: Finite-length analysis of low-density parity-check codes on the binary erasure channel. IEEE Trans. Inf. Theory **48**(6), 1570–1579 (2002)
7. Dumer, I., Farrell, P.: Erasure correction performance of linear block codes. In: Cohen, G., Litsyn, S., Lobstein, A., Zemor, G. (eds.) Lecture Notes in Computer Science, vol. 781, pp. 316–326. Springer, Berlin (1993)
8. Dumer, I., Micciancio, D., Sudan, M.: Hardness of approximating the minimum distance of a linear code. IEEE Trans. Inf. Theory **49**(1), 22–37 (2003)
9. Elias, P.: Coding for two noisy channels. Third London Symposium on Information Theory. Academic Press, New York (1956)
10. Fujiwara, T., Kasami, T.: The weight distribution of (256, k) extended binary primitive BCH code with $k \leq 63$, $k \geq 207$. Technical Report of IEICE IT97-46:29–33 (1997)
11. Gulliver, T.A., Senkevitch, N.: On a class of self-dual codes derived from quadratic residues. IEEE Trans. Inf. Theory **45**(2), 701–702 (1999)
12. Hu, X.Y., Eleftheriou, E., Arnold, D.M.: Irregular progressive edge-growth tanner graphs. In: Proceedings of IEEE International Symposium on Information Theory (ISIT), Lausanne, Switzerland (2002)
13. Krasikov, I., Litsyn, S.: On spectra of BCH codes. IEEE Trans. Inf. Theory **41**(3), 786–788 (1995)
14. Krasikov, I., Litsyn, S.: On the accuracy of the binomial approximation to the distance distribution of codes. IEEE Trans. Inf. Theory **41**(5), 1472–1474 (1995)
15. MacWilliams, F.J., Sloane, N.J.A.: The Theory of Error-Correcting Codes. North-Holland, Amsterdam (1977)
16. Peterson, W.: Error-Correcting Codes. MIT Press, Cambridge (1961)
17. Pishro-Nik, H., Fekri, F.: On decoding of low-density parity-check codes over the binary erasure channel. IEEE Trans. Inf. Theory **50**(3), 439–454 (2004)
18. Rizzo, L.: Effective erasure codes for reliable computer communication protocols. ACM SIGCOMM Comput. Commun. Rev. **27**(2), 24–36 (1997)
19. Roychowdhury, V.P., Vatan, F.: Bounds for the weight distribution of weakly self-dual codes. IEEE Trans. Inf. Theory **47**(1), 393–396 (2001)
20. Tjhai C., Tomlinson M., Horan R., Ahmed M., Ambroze M.: On the efficient codewords counting algorithm and the weight distribution of the binary quadratic double-circulant codes. In: Proceedings of IEEE Information Theory Workshop, Chengdu, China, 22–26 Oct. 2006, pp. 42–46 (2006)

Chapter 15
The Modified Dorsch Decoder

15.1 Introduction

In a relatively unknown paper published in 1974, Dorsch [4] described a decoder for linear binary block (n, k) codes using soft decisions quantised to J levels. The decoder is applicable to any linear block code and does not rely upon any particular features of the code, such as being a concatenated code or having a sparse parity-check matrix. In the Dorsch decoder, hard decisions are derived from the soft decisions using standard bit by bit detection, choosing the binary state closest to the received coordinate. The hard decisions are then ranked in terms of their likelihoods and candidate codewords are derived from a set of k, independent, most likely bits. This is done by producing a new parity-check matrix $\mathbf{H_I}$ obtained by reordering the columns of the original $\mathbf{H}$ matrix according to the likelihood of each coordinate, and reducing the resulting matrix to echelon canonical form by elementary row operations. After evaluation of several candidate codewords, the codeword with the minimum soft decision metric is output from the decoder. A decoder using a similar principle, but without soft decision quantisation, has been described by Fossorier [5, 6]. Other approaches, after ranking the reliability of the received bits, adopt various search strategies for finding likely codewords [11] or utilise a hard decision decoder in conjunction with a search for errors in the least likely bit positions [2, 15].

The power of the Dorsch decoder arises from the relatively unknown property that most codes, *on average*, can correct almost $n - k$ erasures [17], which is considerably more than the guaranteed number of correctable erasures of $d_{min} - 1$, or the guaranteed number of correctable hard decision errors of $\frac{d_{min}-1}{2}$, where d_{min} is the minimum Hamming distance of the code. In its operation, the Dorsch decoder needs to correct any combination of $n - k$ erasures which is impossible unless the code is an MDS code [12]. Dorsch did not discuss this problem, or potential solutions, in his original paper [4], although at least one solution is implied by the results he presented.

M. Tomlinson et al., *Error-Correction Coding and Decoding*,
Signals and Communication Technology,
DOI 10.1007/978-3-319-51103-0_15

In this chapter, a solution to the erasure correcting problem of being able to solve $n-k$ erasures for a non-MDS code is described. It is based on using alternative columns of the parity-check matrix without the need for column permutations. It is also shown that it is not necessary to keep recalculating each candidate codeword and its associated soft decision metric in order to find the most likely codeword. Instead, an incremental correlation approach is adopted which features low information weight codewords and a correlation function involving only a small number of coordinates of the received vector [17]. It is proven that maximum likelihood decoding is realised provided all codewords are evaluated up to a bounded information weight. This means that maximum likelihood decoding may be achieved for a high percentage of received vectors. The decoder lends itself to a low complexity, parallel implementation involving a concatenation of hard and soft decision decoding. It produces near maximum likelihood decoding for codes that can be as long as 1000 bits, provided the code rate is high enough. When implementing the decoder, it is shown that complexity may be traded-off against performance in a flexible manner. Decoding results, achieved by the decoder, are presented for some of the most powerful binary codes known and compared to Shannon's sphere packing bound [14].

The extension to non-binary codes is straightforward and this is described in Sect. 15.5.

15.2 The Incremental Correlation Dorsch Decoder

Codewords with binary coordinates having state 0 or 1, are denoted as:

$$\mathbf{x} = (x_0, x_1, x_2, \ldots, x_{n-1})$$

For transmission, bipolar transmission is used with coordinates having binary state 0 mapped to $+1$ and having state 1 mapped to -1. Transmitted codewords are denoted as

$$\mathbf{c} = (c_0, c_1, c_2, \ldots, c_{n-1})$$

The received vector $\mathbf{r}$ consists of n coordinates $(r_0, r_1, r_2, \ldots, r_{n-1})$ equal to the transmitted codeword plus Additive White Gaussian Noise with variance σ^2. The received vector processed by the decoder is assumed to have been matched filtered and free from distortion so that $\frac{1}{\sigma^2} = \frac{2E_b}{N_o}$, where E_b is the energy per information bit and N_o is the single sided noise power spectral density. Accordingly,

$$\sigma^2 = \frac{N_o}{2E_b}$$

The basic principle that is used is that the k most reliable bits of the received vector are initially taken as correct and the $n-k$ least reliable bits are treated as erasures. The parity-check equations of the code, as represented by $\mathbf{H}$, are used to solve for these erased bits and a codeword $\hat{\mathbf{x}}$ is obtained. This codeword is either equal to the transmitted codeword or needs only small changes to produce a codeword equal to the transmitted codeword. One difficulty is that, depending on the code, the $n - k$ least reliable bits usually cannot all be solved as erasures. This depends on the positions of the erased coordinates and the power of the code. Only Maximum Distance Separable (MDS) codes [12] are capable of solving $n - k$ erasures regardless of the positions of the erasures in the received codeword. Unfortunately, there are no binary MDS codes apart from trivial examples. However, a set of $n - k$ erasures can always be solved from $n - k + s$ least reliable bit positions, and, depending on the code, s is usually a small integer. In order to obtain best performance it is important that the very least reliable bit positions are solved first, since the corollary that the $n - k$ least reliable bits usually cannot all be solved as erasures is that the k most reliable bits, used to derive codeword $\hat{\mathbf{x}}$, must include a small number of least reliable bits. However, for most received vectors, the difference in reliability between ranked bit k and ranked bit $k + s$ is usually small. For any received coordinate, the a priori log likelihood ratio of the bit being correct is proportional to $|r_i|$. The received vector $\mathbf{r}$ with coordinates ranked in order of most likely to be correct is defined as $(r_{\mu_0}, r_{\mu_1}, r_{\mu_2}, \ldots, r_{\mu_{n-1}})$, where $|r_{\mu_0}| > |r_{\mu_1}| > |r_{\mu_2}| > \cdots > |r_{\mu_{n-1}}|$.

The decoder is most straightforward for a binary MDS code. The codeword coordinates $(x_{\mu_0}, x_{\mu_1}, x_{\mu_2}, \ldots, x_{\mu_{k-1}})$ are formed directly from the received vector $\mathbf{r}$ using the bitwise decision rule $x_{\mu_i} = 1$ if $r_{\mu_i} < 0$ else $x_{\mu_i} = 0$. The $n - k$ coordinates $(x_{\mu_k}, x_{\mu_{k+1}}, x_{\mu_{k+2}}, \ldots, x_{\mu_{n-1}})$ are considered to be erased and derived from the k most reliable codeword coordinates $(x_{\mu_0}, x_{\mu_1}, x_{\mu_2}, \ldots, x_{\mu_{k-1}})$ using the parity-check equations.

For a non-MDS code, the $n - k$ coordinates cannot always be solved from the parity-check equations because the parity-check matrix is not a Cauchy or Vandermonde matrix [12]. To get around this problem a slightly different order is defined $(x_{\eta_0}, x_{\eta_1}, x_{\eta_2}, \ldots, x_{\eta_{n-1}})$.

The label of the last coordinate η_{n-1} is set equal to μ_{n-1} and $x_{\eta_{n-1}}$ solved first by flagging the first parity-check equation that contains $x_{\eta_{n-1}}$, and then subtracting this equation from all other parity-check equations containing $x_{\eta_{n-1}}$. Consequently, $x_{\eta_{n-1}}$ is now only contained in one equation, the first flagged equation.

The label of the next coordinate η_{n-2} is set equal to μ_{n-2} and an attempt is made to solve $x_{\eta_{n-2}}$ by finding an unflagged parity-check equation containing $x_{\eta_{n-2}}$. In the event that there is not an unflagged equation containing $x_{\eta_{n-2}}$, η_{n-2} is set equal to μ_{n-3} the label of the next most reliable bit, $x_{\mu_{n-3}}$ and the procedure repeated until an unflagged equation contains $x_{\eta_{n-2}}$. As before, this equation is flagged that it will be used to solve for $x_{\eta_{n-2}}$ and is subtracted from all other unflagged equations containing $x_{\eta_{n-2}}$. The procedure continues until all of the $n - k$ codeword coordinates $x_{\eta_{n-1}}, x_{\eta_{n-2}}, x_{\eta_{n-3}}, \ldots, x_{\eta_k}$ have been solved and all $n - k$ equations have been flagged. In effect, the least reliable coordinates are skipped if they cannot be solved. The remaining k ranked received coordinates are set equal to $(r_{\eta_0}, r_{\eta_1}, r_{\eta_2}, \ldots, r_{\eta_{k-1}})$ in most reliable order, where $|r_{\eta_0}| > |r_{\eta_1}| > |r_{\eta_2}| > \cdots > |r_{\eta_{n-1}}|$ and $(x_{\eta_0}, x_{\eta_1}, x_{\eta_2}, \ldots, x_{\eta_{k-1}})$ determined using the bit decision rule $x_{\eta_i} = 1$ if $r_{\eta_i} < 0$ else $x_{\eta_i} = 0$. The flagged parity-check equations are in upper triangular form and have to be solved in reverse order starting with the last flagged equation. This equation gives the solution to x_{η_k} which is back substituted into the other equations and $x_{\eta_{k+1}}$ is solved next, back substituted, and so on, with coordinate $x_{\eta_{n-1}}$ solved last.

This codeword is denoted as $\hat{\mathbf{x}}$ and the mapped version of the codeword is denoted as $\hat{\mathbf{c}}$.

As is well-known [13], the codeword most likely to be transmitted is the codeword, denoted as $\breve{\mathbf{x}}$, which has the smallest squared Euclidean distance, $D(\breve{\mathbf{x}})$, between the mapped codeword, $\breve{\mathbf{c}}$, and the received vector.

$$D(\breve{\mathbf{x}}) = \sum_{j=0}^{n-1} (r_j - \breve{c}_j)^2$$

$D(\breve{\mathbf{x}}) < D(\mathbf{x})$ for all other codewords $\mathbf{x}$.

Equivalently $\breve{\mathbf{x}}$ is the codeword, after mapping, which has the highest cross correlation

$$Y(\breve{\mathbf{x}}) = \sum_{j=0}^{n-1} r_j \times \breve{c}_j \tag{15.1}$$

$Y(\breve{\mathbf{x}}) > Y(\mathbf{x})$ for all other codewords $\mathbf{x}$.

The decoder may be simplified if the cross correlation function is used to compare candidate codewords. The cross correlation is firstly determined for the codeword $\hat{\mathbf{x}}$

$$Y(\hat{\mathbf{x}}) = \sum_{j=0}^{n-1} r_j \times \hat{c}_j \tag{15.2}$$

It is interesting to make some observations about $Y(\hat{\mathbf{x}})$. Since the summation can be carried out in any order

$$Y(\hat{\mathbf{x}}) = \sum_{j=0}^{n-1} r_{\eta_j} \times \hat{c}_{\eta_j} \tag{15.3}$$

and

$$Y(\hat{\mathbf{x}}) = \sum_{j=0}^{k-1} r_{\eta_j} \times \hat{c}_{\eta_j} + \sum_{j=k}^{n-1} r_{\eta_j} \times \hat{c}_{\eta_j} \tag{15.4}$$

Considering the first term

$$\sum_{j=0}^{k-1} r_{\eta_j} \times \hat{c}_{\eta_j} = \sum_{j=0}^{k-1} |r_{\eta_j}| \tag{15.5}$$

This is because the sign of $\hat{c}_{\eta_j}$ equals the sign of $\hat{c}_{\eta_j}$ for $j < k$. Thus, this term is independent of the code and Eq. (15.4) becomes

$$Y(\hat{\mathbf{x}}) = \sum_{j=0}^{k-1} |r_{\eta_j}| + \sum_{j=k}^{n-1} r_{\eta_j} \times \hat{c}_{\eta_j} \tag{15.6}$$

Almost all of the k largest received coordinates (all of the k largest terms for an MDS code) are contained in the first term of Eq. (15.6) and this ensures that the codeword $\hat{\mathbf{x}}$, after mapping, has a high correlation with $\mathbf{r}$.

A binary, (hard decision), received vector $\mathbf{b}$ may be derived from the received vector $\mathbf{r}$ using the bitwise decision rule $b_j = 1$ if $r_j < 0$, else $b_j = 0$ for $j = 0$ to $n - 1$. It should be noted that in general the binary vector $\mathbf{b}$ is not a codeword.

It is useful to define a binary vector $\hat{\mathbf{z}}$ as

$$\hat{\mathbf{z}} = \mathbf{b} \oplus \hat{\mathbf{x}} \tag{15.7}$$

The maximum attainable correlation Y_{max} is given by

$$Y_{max} = \sum_{j=0}^{n-1} |r_{\eta_j}| \tag{15.8}$$

This correlation value occurs when there are no bit errors in transmission and provides an upper bound to the maximum achievable correlation for $\check{\mathbf{x}}$. The correlation $Y(\hat{\mathbf{x}})$ may be expressed in terms of Y_{max} and $\hat{\mathbf{x}}$ for

$$Y(\hat{\mathbf{x}}) = Y_{max} - 2\sum_{j=0}^{n-1} \hat{z}_{\eta_j} \times |r_{\eta_j}| \tag{15.9}$$

equivalently,

$$Y(\hat{\mathbf{x}}) = Y_{max} - Y_{\Delta}(\hat{\mathbf{x}}), \tag{15.10}$$

where $Y_{\Delta}(\hat{\mathbf{x}})$ is the shortfall from the maximum achievable correlation for the codeword $\hat{\mathbf{x}}$ and is evidently

$$Y_{\Delta}(\hat{\mathbf{x}}) = 2\sum_{j=0}^{n-1} \hat{z}_{\eta_j} \times |r_{\eta_j}| \tag{15.11}$$

Some observations may be made about the binary vector $\hat{\mathbf{z}}$. The coordinates $\hat{z}_{\eta_j}$ for $j = 0$ to $(k - 1)$ are always equal to zero. The maximum possible weight of $\hat{\mathbf{z}}$ is thus $n - k$ and the average weight is $\frac{n-k}{2}$ at low $\frac{E_b}{N_o}$ values. At high $\frac{E_b}{N_o}$ values, the average weight of $\hat{\mathbf{z}}$ is small because there is a high chance that $\hat{\mathbf{x}}$ is equal to the transmitted codeword. It may be seen from Eq. (15.11) that, in general, the lower the weight of $\hat{\mathbf{z}}$ the smaller will be $Y_\Delta(\hat{\mathbf{x}})$ and the larger will be the correlation value $Y(\hat{\mathbf{x}})$.

Since there is no guarantee that the codeword $\hat{\mathbf{x}}$ is the transmitted codeword, the decoder has to evaluate additional codewords since one or more of these may produce a correlation higher than $\hat{\mathbf{x}}$. There are $2^k - 1$ other codewords which may be derived by considering all other $2^k - 1$ sign combinations of c_{η_j} for $j = 0$ to $k - 1$. For any of these codewords denoted as $\mathbf{c_i}$ the first term of the correlation given in Eq. (15.6) is bound to be smaller since

$$\sum_{j=0}^{k-1} r_{\eta_j} \times c_{i,\eta_j} < \sum_{j=0}^{k-1} |r_{\eta_j}| \tag{15.12}$$

This is because there has to be, by definition, at least one sign change of c_{i,η_j} compared to $\hat{c}_{\eta_j}$ for $j = 0$ to $k - 1$. In order for $Y(\mathbf{x_i})$ to be larger than $Y(\hat{\mathbf{x}})$ the second term of the correlation $\sum_{j=k}^{n-1} r_{\eta_j} \times c_{i,\eta_j}$ which uses the bits from the solved parity-check equations must be larger than $\sum_{j=k}^{n-1} r_{\eta_j} \times \hat{c}_{\eta_j}$ plus the negative contribution from the first term.

However, the first term has higher received magnitudes than the second term because the received coordinates are ordered. It follows that codewords likely to have a higher correlation than $\hat{\mathbf{x}}$ will have small number of differences in the coordinates x_{η_j} for $j = 0$ to $k - 1$. As the code is linear these differences will correspond to a codeword and codewords may be generated that have low weight in coordinates x_{η_j} for $j = 0$ to $k - 1$. These codewords are represented as $\tilde{\mathbf{x}}_\mathbf{i}$ and referred to as low information weight codewords since coordinates x_{η_j} for $j = 0$ to $k - 1$ form an information set. Thus, codewords $\mathbf{x_i}$ are given by

$$\mathbf{x_i} = \hat{\mathbf{x}} \oplus \tilde{\mathbf{x}}_\mathbf{i} \tag{15.13}$$

and $\tilde{\mathbf{x}}_\mathbf{i}$ are codewords chosen to have increasing weight in coordinates x_{η_j} for $j = 0$ to $k - 1$ as i is incremented. This means that for increasing i it will become less likely that a codeword will be found that has higher correlation than the correlation of a codeword already found.

The difference in the correlation value $Y_\Delta(\mathbf{x_i})$ as a function of $\tilde{\mathbf{x}}_\mathbf{i}$ may be derived. Firstly, the binary vector $\mathbf{z_i}$ is given by

$$\mathbf{z_i} = \mathbf{b} \oplus \hat{\mathbf{x}} \oplus \tilde{\mathbf{x}}_\mathbf{i} \tag{15.14}$$

which may be simplified to

$$\mathbf{z_i} = \hat{\mathbf{z}} \oplus \tilde{\mathbf{x}}_\mathbf{i} \tag{15.15}$$

The cross correlation $Y(\mathbf{x_i})$ is given by

$$Y(\mathbf{x_i}) = Y_{max} - 2\sum_{j=0}^{n-1} z_{i,\eta_j} \times |r_{\eta_j}| \tag{15.16}$$

equivalently

$$Y(\mathbf{x_i}) = Y_{max} - Y_{\Delta}(\mathbf{x_i}) \tag{15.17}$$

The shortfall from maximum correlation, $Y_{\Delta}(\mathbf{x_i})$, is evidently

$$Y_{\Delta}(\mathbf{x_i}) = 2\sum_{j=0}^{n-1} z_{i,\eta_j} \times |r_{\eta_j}| \tag{15.18}$$

Substituting for $\mathbf{z_i}$ gives $Y_{\Delta}(\mathbf{x_i})$ as a function of $\tilde{\mathbf{x}}_\mathbf{i}$.

$$Y_{\Delta}(\mathbf{x_i}) = 2\sum_{j=0}^{n-1} (\hat{z}_j \oplus \tilde{x}_{i\eta_j}) \times |r_{\eta_j}| \tag{15.19}$$

It is apparent that instead of the decoder determining $Y(\mathbf{x_i})$ for each codeword, $\mathbf{x_i}$, it is sufficient for the decoder to determine $Y_{\Delta}(\mathbf{x_i})$ for each codeword $\tilde{\mathbf{x}}_\mathbf{i}$ and compare the value with the smallest value obtained so far, denoted as $Y_{\Delta}(\mathbf{x_{min}})$, starting with $Y_{\Delta}(\hat{\mathbf{x}})$:

$$Y_{\Delta}(\mathbf{x_{min}}) = min\big(Y_{\Delta}(\mathbf{x})\big) \tag{15.20}$$

Thus it is more efficient for the decoder to compute the correlation (partial sum) of the $\tilde{\mathbf{x}}_\mathbf{i}$ instead of deriving $(\hat{\mathbf{x}} \oplus \tilde{\mathbf{x}}_\mathbf{i})$ by solving $\mathbf{H_I}$ and computing the squared Euclidean distance. Since codewords $\tilde{\mathbf{x}}_\mathbf{i}$ produce low weight in $\mathbf{z_i}$, the number of non-zero terms that need to be evaluated in Eq. (15.18) is typically $\frac{n-k}{2}$ rather than the $\frac{n}{2}$ terms of Eq. (15.1) which makes for an efficient, fast decoder. Before Eq. (15.19) is evaluated, the Hamming weight of $\mathbf{z_i}$ may be compared to a threshold and the correlation stage bypassed if the Hamming weight of $\mathbf{z_i}$ is high. There is an associated performance loss and results are presented in Sect. 15.4.

The maximum information weight $w_{inf\,max}$ necessary to achieve maximum likelihood decoding may be upper bounded from $Y_{\Delta}(\hat{\mathbf{x}})$ and $|r_{\eta_j}|$ initially, updated by $Y_{\Delta}(\mathbf{x_{min}})$ as decoding progresses, since

$$Y_{\Delta}(\mathbf{x_i}) \geq \sum_{m=0}^{w_{inf}} |r_{\eta_{k-m-1}}| \tag{15.21}$$

This is reasonably tight since there is a possibility of at least one codeword with information weight $w_{inf\,max}$, for which all of the coordinates of the binary vector $\mathbf{z_i}$ corresponding to the parity bits of $\tilde{\mathbf{x}}_\mathbf{i}$ are zero. Correspondingly, $w_{inf\,max}$ is the smallest integer such that

$$\sum_{m=0}^{w_{inf\,max}} |r_{\eta_{k-m-1}}| \geq Y_{\Delta}(\hat{\mathbf{x}}) \tag{15.22}$$

The codewords $\tilde{\mathbf{x}}_\mathbf{i}$ may be most efficiently derived from the $\mathbf{G}$ matrix corresponding to the solved $\mathbf{H}$ matrix because the maximum information weight given by Eq. (15.22) turns out to be small. Each row, i, of the solved $\mathbf{G}$ matrix is derived by setting $x_{\eta_j} = 0$ for $j = 0$ to $k-1, j \neq i$, and using the solved parity-check equations to determine x_{η_j} for $j = k$ to $n-1$. The maximum number of rows of the $\mathbf{G}$ matrix that need to be combined to produce $\tilde{\mathbf{x}}_\mathbf{i}$ is $w_{inf\,max}$.

15.3 Number of Codewords that Need to Be Evaluated to Achieve Maximum Likelihood Decoding

For each received vector the decoder needs to evaluate the correlation shortfall for the codewords $\tilde{\mathbf{x}}_\mathbf{i}$ for information weights up to the maximum information weight of $w_{inf\,max}$ in order to achieve maximum likelihood decoding. The number of codewords that need to be evaluated is a function of the received vector. Not all of the codewords having information weight less than or equal to $w_{inf\,max}$ need be evaluated because lower bounds may be derived for $Y_{\Delta}(\mathbf{x_i})$ in terms of the coordinates of the information bits, their total weight and the magnitudes of selected coordinates of the received vector. For an information weight of w_{inf}, $Y_{\Delta}(\mathbf{x_i})$ is lower bounded by

$$Y_{\Delta}(\mathbf{x_i}) \geq |r_{\eta_j}| + \sum_{m=0}^{w_{inf}-1} |r_{\eta_{k-m-1}}| \quad 0 \leq j < k-m \tag{15.23}$$

and

$$|r_{\eta_{j_{min}(w_{inf})}}| \geq Y_{\Delta}(\mathbf{x_i}) - \sum_{m=0}^{w_{inf}-1} |r_{\eta_{k-m-1}}| \quad 0 \leq j < k-m \tag{15.24}$$

where $j_{min}(w_{inf})$ is defined as the lower limit for j to satisfy Eq. (15.24). The minimum number of codewords that need to be evaluated as a function of the received vector $N(\mathbf{r})$ is given by the total number of combinations

$$N(\mathbf{r}) = \sum_{m=0}^{w_{inf}} \binom{k - j_{min}(m) - 1}{m} \tag{15.25}$$

For many short codes the minimum number of codewords that need to be evaluated is surprisingly small in comparison to the total number of codewords.

15.4 Results for Some Powerful Binary Codes

The decoder can be used with any linear code and best results are obtained for codes which have the highest known d_{min} for a given codelength n and number of information symbols k. The best binary codes are tabulated up to length 257 in Marcus Grassl's on line data base [7]. Non-binary codes, for example, ternary codes of length up to 243 symbols and GF(4) codes of length up to 256 symbols are also tabulated.

A particularly good class of codes are the binary self-dual, double-circulant codes first highlighted in a classic paper by Karlin [8]. For example the (24, 12, 8) extended Golay code is included since it may be put in double-circulant form. There is also the (48, 24, 12) bordered double-circulant code, based on quadratic residues of the prime 47 and the (136, 68, 24) bordered double-circulant code based on quadratic residues of the prime 67. These codes are extremal [3] and are doubly even, only having codeword weights that are a multiple of 4, and in these cases it is necessary that the codelengths are a multiple of 8 [3]. For higher code rates of length greater than 256, the best codes are tabulated in [12], and some of these include cyclic codes and Goppa codes.

15.4.1 The (136, 68, 24) Double-Circulant Code

This code is a bordered double-circulant code based on the identity matrix and a matrix whose rows consist of all cyclic combinations, modulo $1 + x^{67}$, of the polynomial $b(x)$ defined by

$$\begin{aligned} b(x) = 1 &+ x + x^4 + x^6 + x^9 + x^{10} + x^{14} + x^{15} + x^{16} + x^{17} + x^{19} + x^{21} + x^{22} + x^{23} + x^{24} + x^{25} + x^{26} + x^{29} \\ &+ x^{33} + x^{35} + x^{36} + x^{37} + x^{39} + x^{40} + x^{47} + x^{49} + x^{54} + x^{55} + x^{56} + x^{59} + x^{60} + x^{62} + x^{64} + x^{65} \end{aligned} \tag{15.26}$$

The Frame Error Rate (FER) of this code using the extended Dorsch decoder with a maximum number of codewords limited to 3×10^6 is shown in Fig. 15.1. Also, shown in Fig. 15.1 is Shannon's [14] sphere packing bound offset by the loss for binary transmission [1], which is 0.19 dB for a code rate of $\frac{1}{2}$.

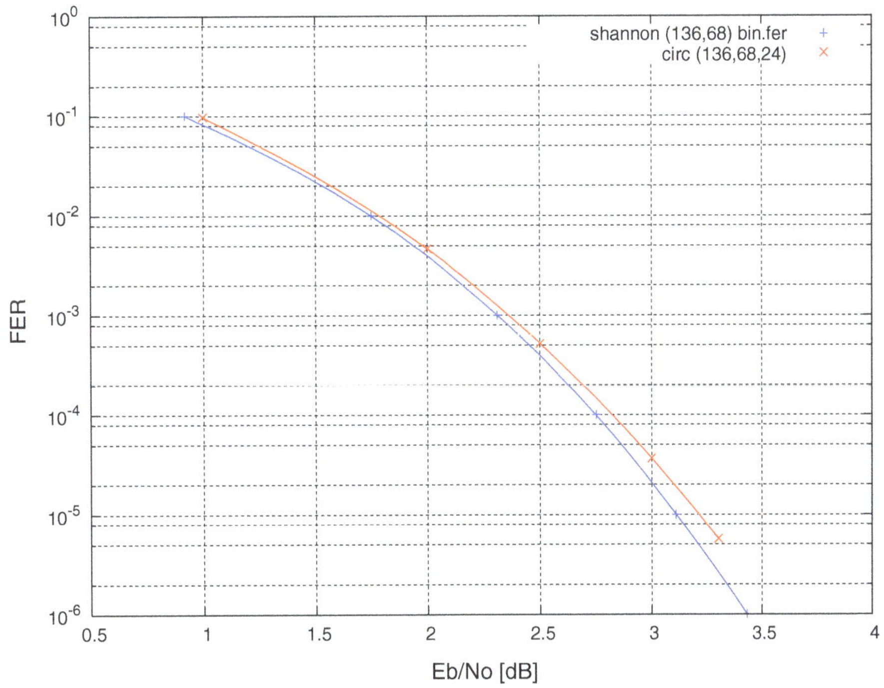

Fig. 15.1 FER as a function of $\frac{E_b}{N_o}$ for the double-circulant (136, 68, 24) code using incremental correlation decoding compared to the sphere packing bound, offset for binary transmission

It may be seen from Fig. 15.1 that the performance of the decoder in conjunction with the double-circulant code is within 0.2 dB of the best achievable performance for any (136, 68) code at 10^{-5} FER. Interestingly, there is a significant number of maximum likelihood codeword errors which have a Hamming distance of 36 or 40 from the transmitted codeword. This indicates that a bounded distance decoder would not perform very well for this code. At the typical practical operating point of $\frac{E_b}{N_o}$ equal to 3.5 dB, the probability of the decoder processing each received vector as a maximum likelihood decoder is shown plotted in Fig. 15.2 as a function of the number of codewords evaluated.

Of course to guarantee maximum likelihood decoding, all $2^{68} = 2.95 \times 10^{20}$ codewords need to be evaluated by the decoder. Equation (15.21) has been evaluated for the double-circulant (136, 68, 24) code in computer simulations, at an $\frac{E_b}{N_o}$ of 3.5 dB, for each received vector and the cumulative distribution derived. Figure 15.2 shows that by evaluating 10^7 codewords per received vector, 65% of received vectors are guaranteed to be maximum likelihood decoded. For the remaining 35% of received vectors, although maximum likelihood decoding is not guaranteed, the probability is very small that the codeword with the highest correlation is not the transmitted codeword or a codeword closer to the received vector than the transmitted codeword. This last point is illustrated by Fig. 15.3 which shows the FER performance of the decoder as a function of the maximum number of evaluated codewords.

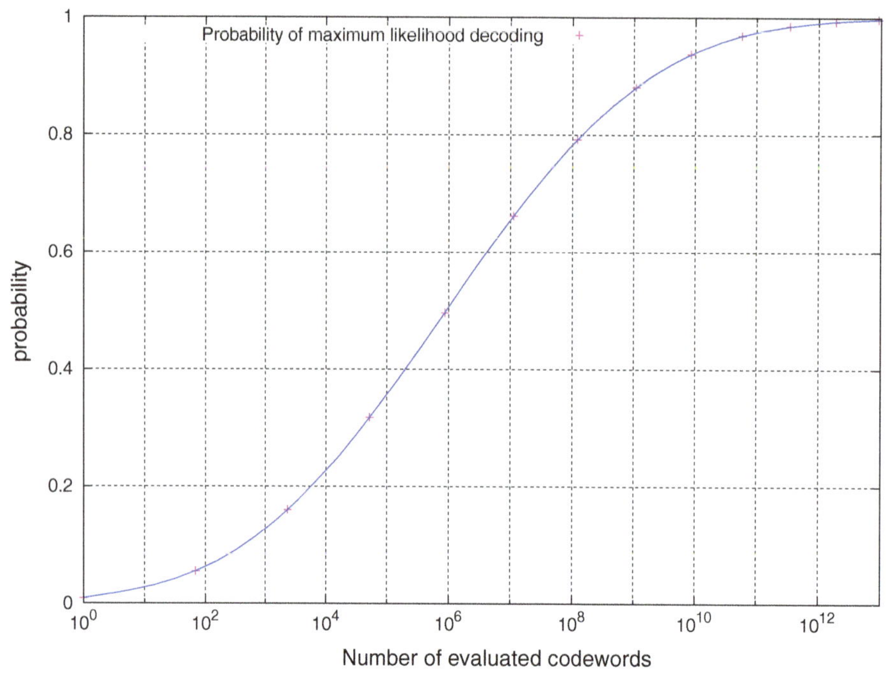

Fig. 15.2 Probability of a received vector being maximum likelihood decoded as a function of number of evaluated codewords for the (136, 68, 24) code at $\frac{E_b}{N_o} = 3.5$ dB

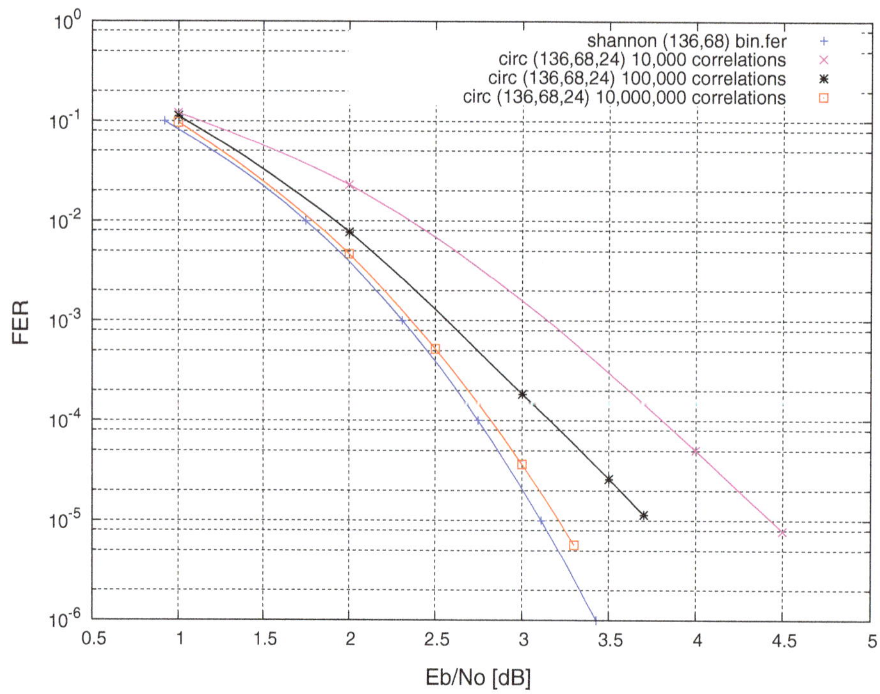

Fig. 15.3 FER performance of the (136, 68, 24) code as a function of number of evaluated codewords

Although this EG code performs well with iterative decoding it is apparent that the incremental correlation decoder is able to improve the performance of the code for the AWGN channel by 0.45 dB at 10^{-3} FER.

15.4.3 The (513, 467, 12) Extended Binary Goppa Code

Goppa codes are frequently better than the corresponding BCH codes because there is an additional information bit and the Goppa code is only one bit longer than the BCH code. For example, the (512, 467, 11) binary Goppa Code has one more information bit than the (511, 466, 11) BCH code and may be generated by the irreducible Goppa polynomial $1 + x^2 + x^5$, whose roots have order 31 which is relatively prime to 511. The d_{min} of the binary Goppa code [12] is equal to twice the degree of the irreducible polynomial plus 1 and is the same as the (511, 466, 11) BCH code. The Goppa code may be extended by adding an overall parity check, increasing the d_{min} to 12.

The FER performance of the extended Goppa code is shown in Fig. 15.8 and was obtained using the incremental correlation decoder. Also shown in Fig. 15.8 is the sphere packing bound offset by the binary transmission loss. It can be seen that the realised performance of the decoder is within 0.3 dB at 10^{-4}.

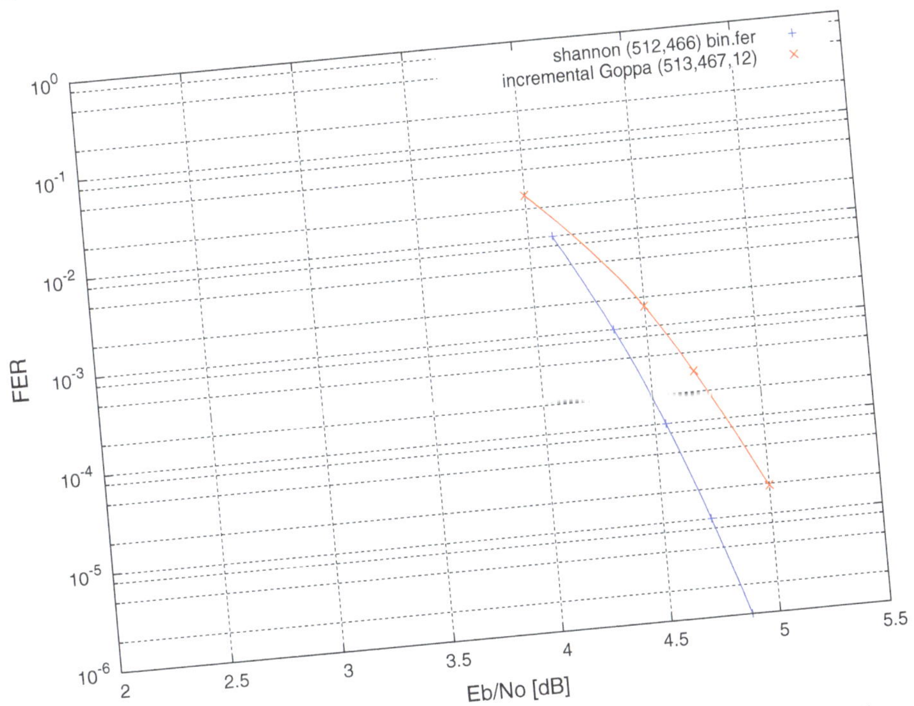

Fig. 15.8 FER performance of the (513, 467, 12) binary Goppa code using incremental correlation decoding

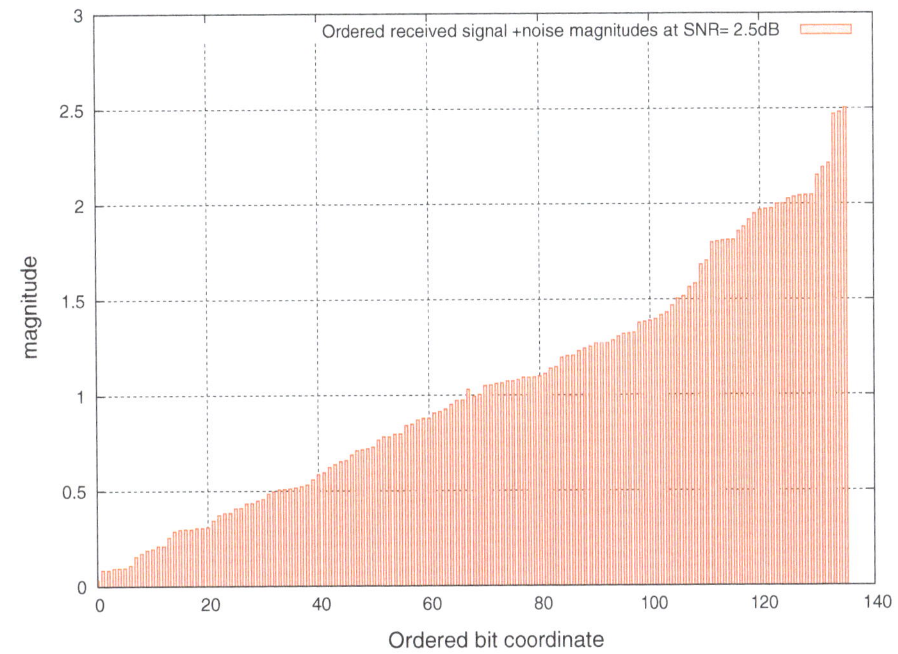

Fig. 15.4 An example of received coordinate magnitudes in their solved order for the (136, 68, 24) code at $\frac{E_b}{N_o} = 2.5$ dB for a single received vector

The detailed operation of the decoder may be seen by considering an example of a received vector at $\frac{E_b}{N_o}$ of 2.5 dB. The magnitudes of the received coordinates, ordered in their solved order, is shown in Fig. 15.4. In this particular example, it is not possible to solve for ordered coordinates 67 and 68 (in their order prior to solving of the parity-check matrix) and so these coordinates are skipped and become coordinates 68 and 69, respectively, in the solved order. The transmitted bits are normalised with magnitudes 1 and the σ of the noise is $\approx$1.07. The shift in position of coordinate 69 (in original position) to 67 (in solved order) is evident in Fig. 15.4. The positions of the bits received in error in the same solved order is shown in Fig. 15.5. It may be noted that the received bit errors are concentrated in the least reliable bit positions. There are a total of 16 received bit errors and only two of these errors correspond to the (data) bit coordinates 11 and 34 of the solved **G** matrix. Evaluation of 10^7 codewords indicates that the minimum value of $Y_\Delta(\mathbf{x_{min}})$ is $\approx$13.8, and this occurs for the 640th codeword producing a maximum correlation of $\approx$126.2 with $Y_{max} \approx 140$. The weight of $\mathbf{z_{min}}$ is 16 corresponding to the 16 received bit errors.

In practice, it is not necessary for $Y_\Delta(\mathbf{x_i})$ given by the partial sum equation (15.18) to be evaluated for each codeword. In most cases, the weight of the binary vector $\mathbf{z_i}$ is sufficiently high to indicate that this codeword is not the most likely codeword. Shown in Fig. 15.6 are the cumulative probability distributions for the weight of $\mathbf{z_i}$ for the case where $\mathbf{x_i}$ is equal to the transmitted codeword, and the case where it is not equal to the transmitted codeword. Two operating values for $\frac{E_b}{N_o}$ are shown: 3.5 dB and 4 dB. Considering the decoding rule that a weight 29 or more for $\mathbf{z_i}$ is unlikely to be produced by the transmitted codeword means that 95.4% of candidate codewords

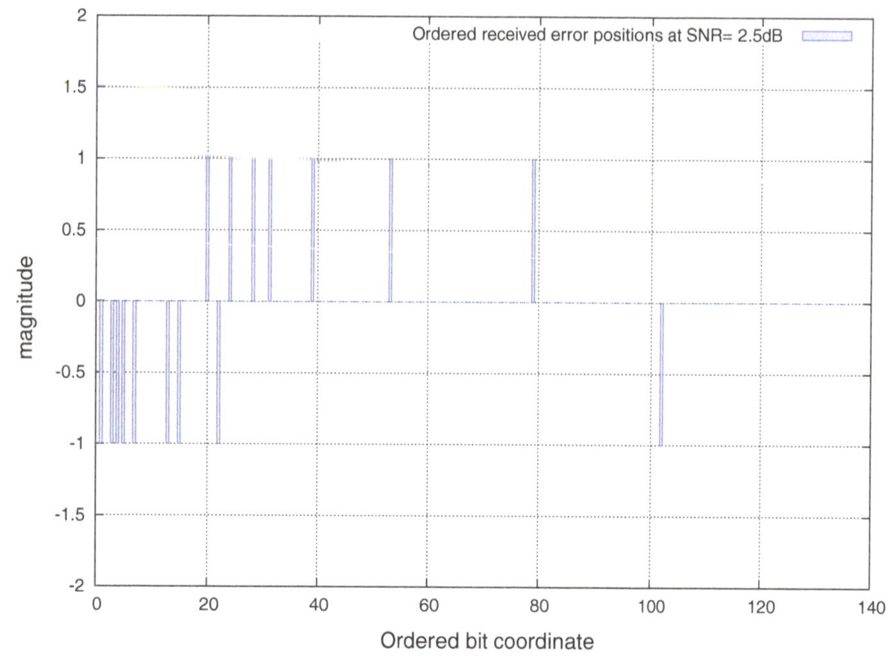

Fig. 15.5 Received bits showing bit error positions for the same received vector and same order as that shown in Fig. 15.4

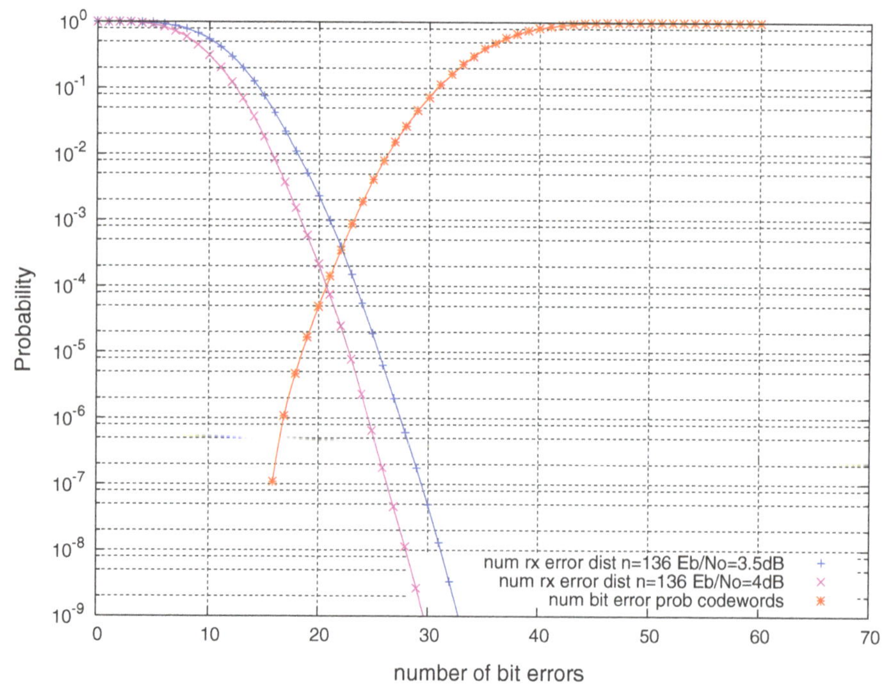

Fig. 15.6 Cumulative probability distributions for the number of bit errors for the transmitted codeword and non-transmitted, evaluated codewords for the (136, 68, 24) code

may be rejected at this point, and that the partial sum equation (15.18) need only be evaluated for 4.6% of the candidate codewords. In reducing the decoder complexity in this way, the degradation to the FER performance as a result of rejection of a transmitted codeword corresponds to ≈3% increase in the FER and is not significant.

15.4.2 The (255, 175, 17) Euclidean Geometry (EG) Code

This code is an EG code originally used in hard decision, one-step majority-logic decoding by Lin and Costello, Jr. [10]. Finite geometry codes also have applications as LDPC codes using iterative decoding with the belief propagation algorithm [9]. The (255, 175, 17) code is a cyclic code and its parity-check polynomial $p(x)$ may conveniently be generated from the cyclotomic idempotents as described in Chap. 12. The parity-check polynomial is

$$p(x) = 1 + x + x^3 + x^7 + x^{15} + x^{26} + x^{31} + x^{53} + x^{63} + x^{98} \tag{15.27}$$
$$+ x^{107} + x^{127} + x^{140} + x^{176} + x^{197} + x^{215} \tag{15.28}$$

The FER performance of the code is shown in Fig. 15.7 and was obtained using the incremental correlation decoder and is shown in comparison to using the iterative decoder. Also shown in Fig. 15.7 is the sphere packing bound offset by the binary transmission loss.

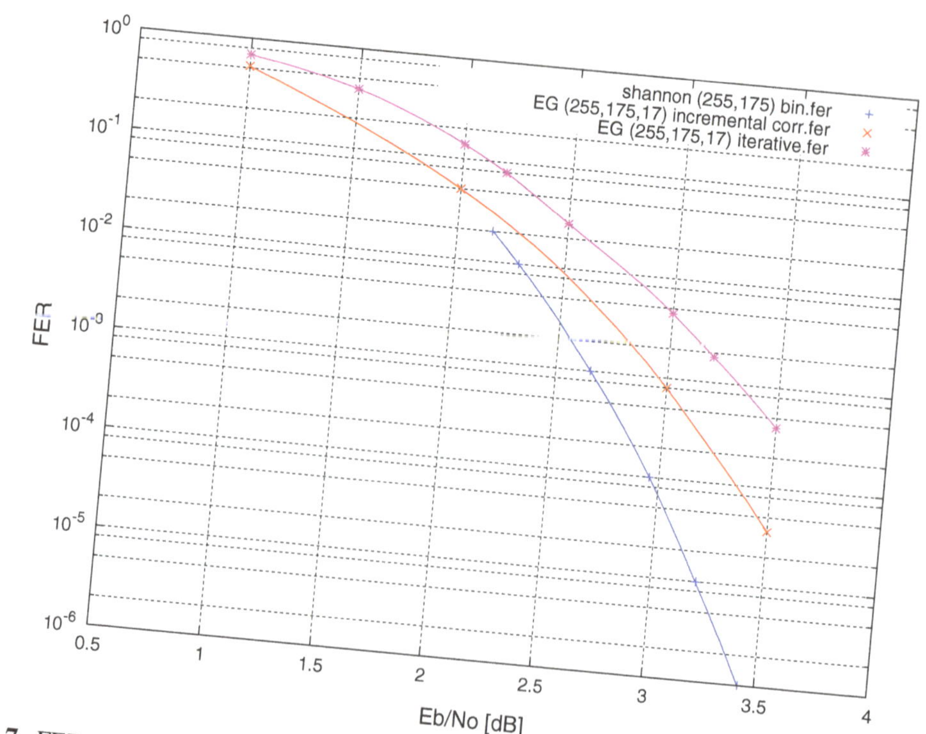

Fig. 15.7 FER performance of the (255, 175, 17) EG code using belief propagation, iterative decoding, compared to incremental correlation decoding

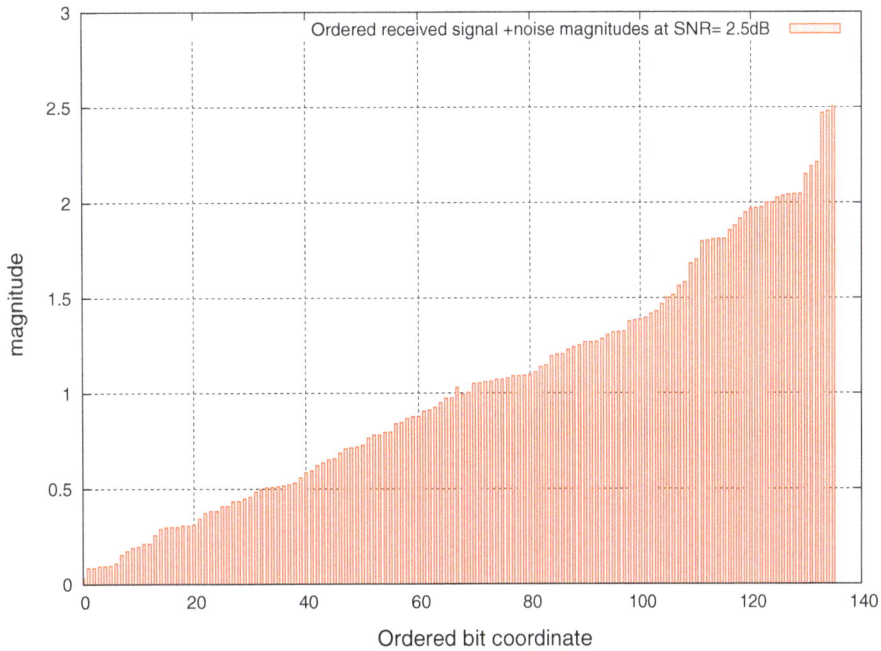

Fig. 15.4 An example of received coordinate magnitudes in their solved order for the (136, 68, 24) code at $\frac{E_b}{N_o} = 2.5$ dB for a single received vector

The detailed operation of the decoder may be seen by considering an example of a received vector at $\frac{E_b}{N_o}$ of 2.5 dB. The magnitudes of the received coordinates, ordered in their solved order, is shown in Fig. 15.4. In this particular example, it is not possible to solve for ordered coordinates 67 and 68 (in their order prior to solving of the parity-check matrix) and so these coordinates are skipped and become coordinates 68 and 69, respectively, in the solved order. The transmitted bits are normalised with magnitudes 1 and the σ of the noise is $\approx$1.07. The shift in position of coordinate 69 (in original position) to 67 (in solved order) is evident in Fig. 15.4. The positions of the bits received in error in the same solved order is shown in Fig. 15.5. It may be noted that the received bit errors are concentrated in the least reliable bit positions. There are a total of 16 received bit errors and only two of these errors correspond to the (data) bit coordinates 11 and 34 of the solved **G** matrix. Evaluation of 10^7 codewords indicates that the minimum value of $Y_\Delta(\mathbf{x_{min}})$ is $\approx$13.8, and this occurs for the 640th codeword producing a maximum correlation of $\approx$126.2 with $Y_{max} \approx 140$. The weight of $\mathbf{z_{min}}$ is 16 corresponding to the 16 received bit errors.

In practice, it is not necessary for $Y_\Delta(\mathbf{x_i})$ given by the partial sum equation (15.18) to be evaluated for each codeword. In most cases, the weight of the binary vector $\mathbf{z_i}$ is sufficiently high to indicate that this codeword is not the most likely codeword. Shown in Fig. 15.6 are the cumulative probability distributions for the weight of $\mathbf{z_i}$ for the case where $\mathbf{x_i}$ is equal to the transmitted codeword, and the case where it is not equal to the transmitted codeword. Two operating values for $\frac{E_b}{N_o}$ are shown: 3.5 dB and 4 dB. Considering the decoding rule that a weight 29 or more for $\mathbf{z_i}$ is unlikely to be produced by the transmitted codeword means that 95.4% of candidate codewords

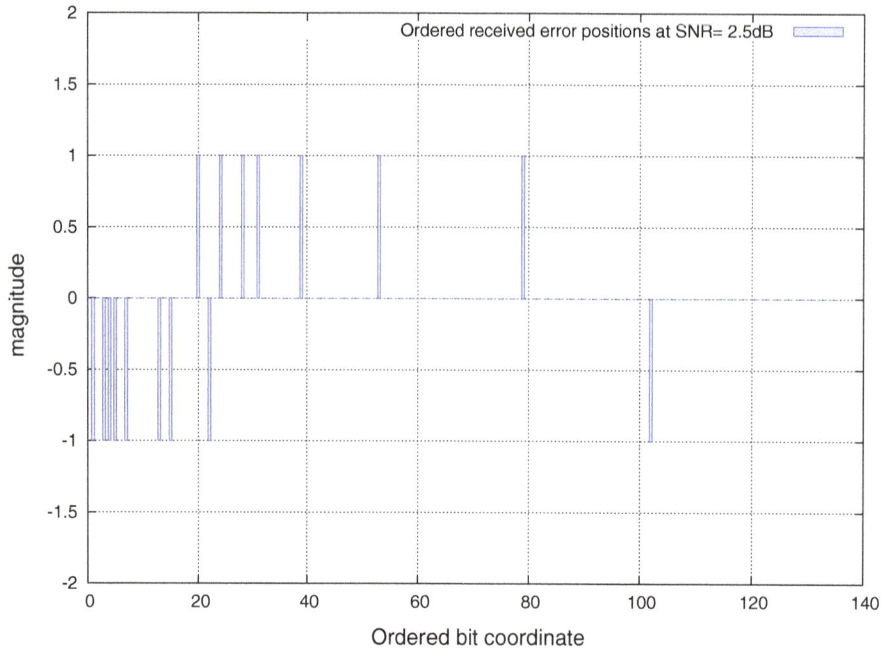

Fig. 15.5 Received bits showing bit error positions for the same received vector and same order as that shown in Fig. 15.4

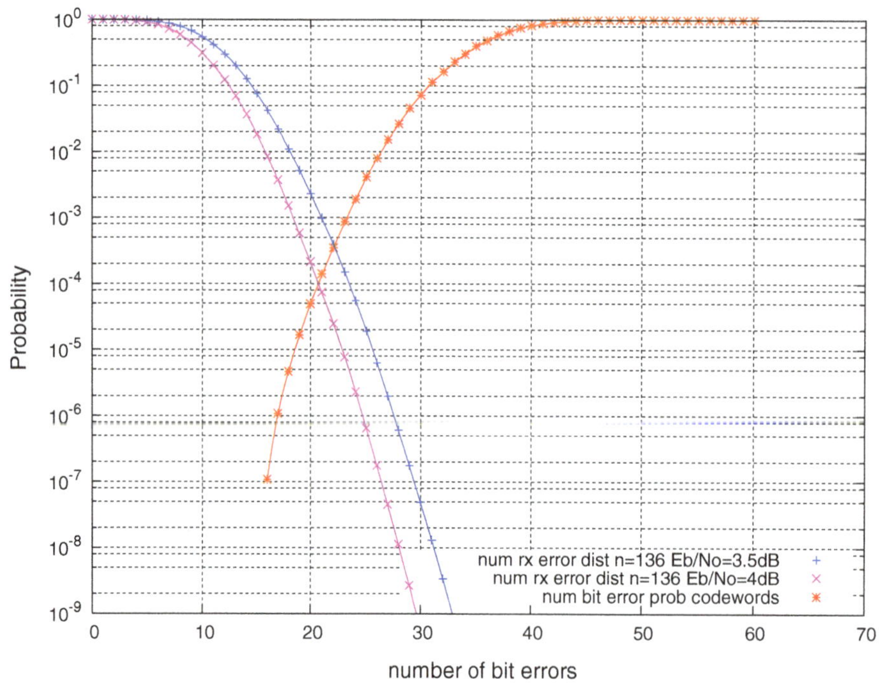

Fig. 15.6 Cumulative probability distributions for the number of bit errors for the transmitted codeword and non-transmitted, evaluated codewords for the (136, 68, 24) code

may be rejected at this point, and that the partial sum equation (15.18) need only be evaluated for 4.6% of the candidate codewords. In reducing the decoder complexity in this way, the degradation to the FER performance as a result of rejection of a transmitted codeword corresponds to ≈3% increase in the FER and is not significant.

15.4.2 The (255, 175, 17) Euclidean Geometry (EG) Code

This code is an EG code originally used in hard decision, one-step majority-logic decoding by Lin and Costello, Jr. [10]. Finite geometry codes also have applications as LDPC codes using iterative decoding with the belief propagation algorithm [9]. The (255, 175, 17) code is a cyclic code and its parity-check polynomial $p(x)$ may conveniently be generated from the cyclotomic idempotents as described in Chap. 12. The parity-check polynomial is

$$p(x) = 1 + x + x^3 + x^7 + x^{15} + x^{26} + x^{31} + x^{53} + x^{63} + x^{98} \quad (15.27)$$

$$+ x^{107} + x^{127} + x^{140} + x^{176} + x^{197} + x^{215} \quad (15.28)$$

The FER performance of the code is shown in Fig. 15.7 and was obtained using the incremental correlation decoder and is shown in comparison to using the iterative decoder. Also shown in Fig. 15.7 is the sphere packing bound offset by the binary transmission loss.

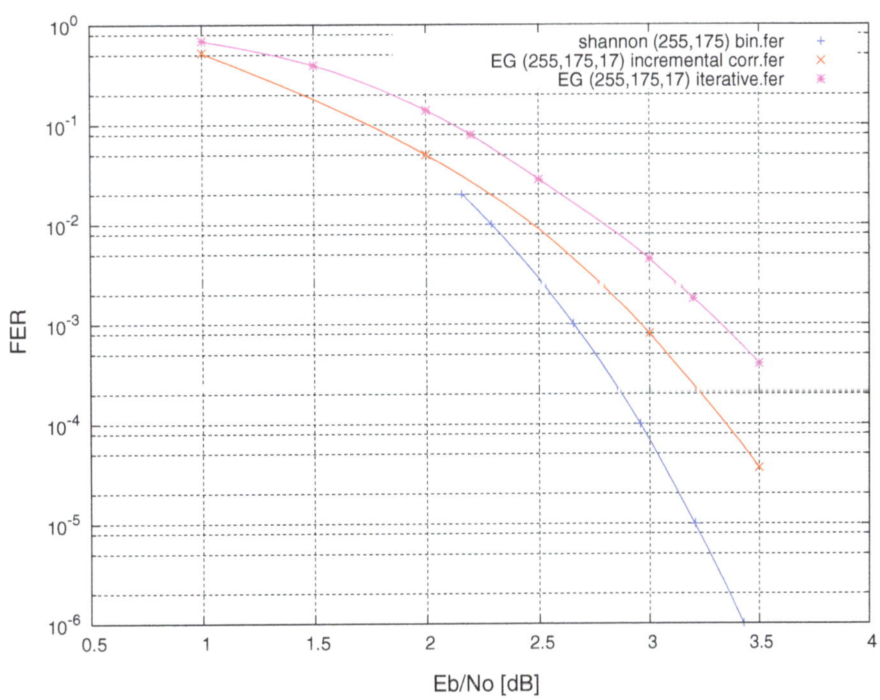

Fig. 15.7 FER performance of the (255, 175, 17) EG code using belief propagation, iterative decoding, compared to incremental correlation decoding

Although this EG code performs well with iterative decoding it is apparent that the incremental correlation decoder is able to improve the performance of the code for the AWGN channel by 0.45 dB at 10^{-3} FER.

15.4.3 The (513, 467, 12) Extended Binary Goppa Code

Goppa codes are frequently better than the corresponding BCH codes because there is an additional information bit and the Goppa code is only one bit longer than the BCH code. For example, the (512, 467, 11) binary Goppa Code has one more information bit than the (511, 466, 11) BCH code and may be generated by the irreducible Goppa polynomial $1 + x^2 + x^5$, whose roots have order 31 which is relatively prime to 511. The d_{min} of the binary Goppa code [12] is equal to twice the degree of the irreducible polynomial plus 1 and is the same as the (511, 466, 11) BCH code. The Goppa code may be extended by adding an overall parity check, increasing the d_{min} to 12.

The FER performance of the extended Goppa code is shown in Fig. 15.8 and was obtained using the incremental correlation decoder. Also shown in Fig. 15.8 is the sphere packing bound offset by the binary transmission loss. It can be seen that the realised performance of the decoder is within 0.3 dB at 10^{-4}.

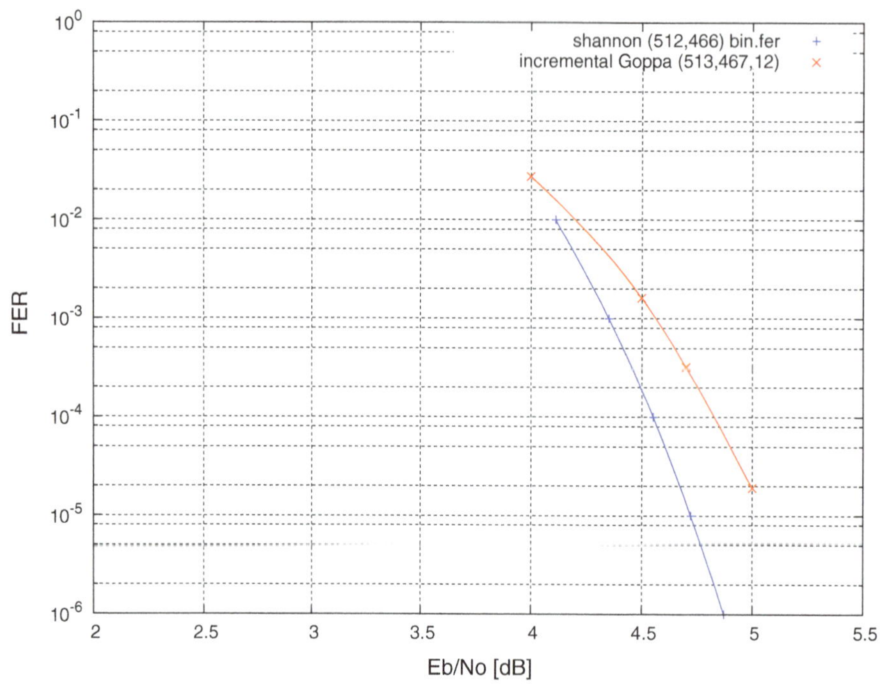

Fig. 15.8 FER performance of the (513, 467, 12) binary Goppa code using incremental correlation decoding

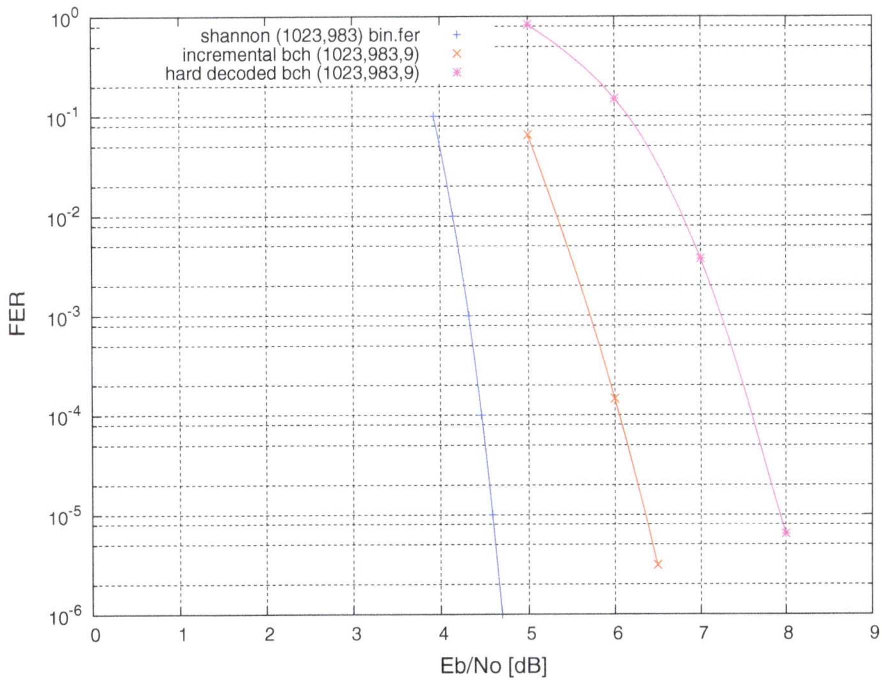

Fig. 15.9 FER performance of the (1023, 983, 9) binary BCH code using incremental correlation decoding compared to hard decision decoding

15.4.4 The (1023, 983, 9) BCH Code

This code is a standard BCH code that may be found in reference text book tables such as by Lin and Costello, Jr. [10]. This example is considered here in order to show that the decoder can produce near maximum likelihood performance for relatively long codes. The performance obtained is shown in Fig. 15.9 with evaluation of candidate codewords limited to 10^6 codewords. At 10^{-5} FER, the degradation from the sphere packing bound, offset for binary transmission, is 1.8 dB. Although this may seem excessive, the degradation of hard decision decoding is 3.6 dB as may also be seen from Fig. 15.9.

15.5 Extension to Non-binary Codes

The extension of the decoder to non-binary codes is relatively straightforward, and for simplicity binary transmission of the components of each non-binary symbol is assumed. Codewords are denoted as before by $\mathbf{x_i}$ but redefined with coefficients, $\gamma_{j\,i}$ from $GF(2^m)$

$$\mathbf{x_i} = (\gamma_{0\,i}x_0, \gamma_{1\,i}x_1, \gamma_{2\,i}x_2, \ldots, \gamma_{n-1\,i}x_{n-1}) \tag{15.29}$$

The received vector **r** with coordinates ranked in order of those most likely to be correct is redefined as

$$\mathbf{r} = \sum_{l=0}^{m-1} (r_{l\,\mu_0}, r_{l\,\mu_1}, r_{l\,\mu_2}, \ldots, r_{l\,\mu_{n-1}}) \tag{15.30}$$

so that the received vector consists of n symbols, each with m values. The maximum attainable correlation Y_{max} is straightforward and is given by

$$Y_{max} = \sum_{j=0}^{n-1} \sum_{l=0}^{m-1} |r_{l\,j}| \tag{15.31}$$

The hard decided received vector **r**, is redefined as

$$\mathbf{b} = \sum_{j=0}^{n-1} \theta_j \, x^j \tag{15.32}$$

where θ_j is the $GF(2^m)$ symbol corresponding to $sign(r_{l\,j})$ for $l = 0$ to $m - 1$.

Decoding follows in a similar manner to the binary case. The received symbols are ordered in terms of their symbol magnitudes $|r_{\mu_j}|_S$ where each symbol magnitude is defined as

$$|r_{\eta_j}|_S = \sum_{l=0}^{m-1} |r_{l\,\eta_j}| \tag{15.33}$$

The codeword $\hat{\mathbf{x}}$ is derived from the k coordinates x_{η_j} whose coefficients ν_{η_j} are the $GF(2^m)$ symbols corresponding to $sign(r_{l\,\eta_j})$ for $l = 0$ to $m - 1$; for $j = 0$ to $k - 1$ and then using the solved parity-check equations for the remaining $n - k$ coordinates.

The vector $\mathbf{z_i}$ is given by

$$\mathbf{z_i} = \mathbf{b} \oplus \hat{\mathbf{x}} \oplus \tilde{\mathbf{x}}_\mathbf{i} \mod GF(2^m) \tag{15.34}$$

which may be simplified as before to

$$\mathbf{z_i} = \hat{\mathbf{z}} \oplus \tilde{\mathbf{x}}_\mathbf{i} \mod GF(2^m) \tag{15.35}$$

Denoting the n binary vectors $\rho_{i\,l\,j}$ corresponding to the n $GF(2^m)$ coefficients of $\mathbf{z_i}$

$$Y(\mathbf{x_i}) = Y_{max} - Y_\Delta(\mathbf{x_i}) \tag{15.36}$$

where $Y_\Delta(\mathbf{x_i})$, the shortfall from maximum correlation is given by

$$Y_\Delta(\mathbf{x_i}) = 2\sum_{j=0}^{n-1}\sum_{l=0}^{m-1} \rho_{ilj} \times |r_{lj}| \tag{15.37}$$

In the implementation of the decoder, as in the binary case, the Hamming weight of the vector $\mathbf{z_i}$ may be used to decide whether it is necessary to evaluate the soft decision metric given by Eq. (15.37) for each candidate codeword.

15.5.1 Results for the (63, 36, 13) $GF(4)$ BCH Code

This is a non-binary BCH code with the generator polynomial $g(x)$ defined by roots

$$\{\alpha^1, \alpha^4, \alpha^{16}, \alpha^2, \alpha^8, \alpha^{32}, \alpha^3, \alpha^{12}, \alpha^{48}, \alpha^5, \alpha^{20}, \alpha^{17}, \alpha^6, \alpha^{24}, \alpha^{33}, \alpha^7, \alpha^{28}, \alpha^{29}, \alpha^9, \alpha^{36}, \alpha^{18}, \alpha^{10}, \alpha^{40}, \alpha^{34}, \alpha^{11}, \alpha^{44}, \alpha^{50}\}$$

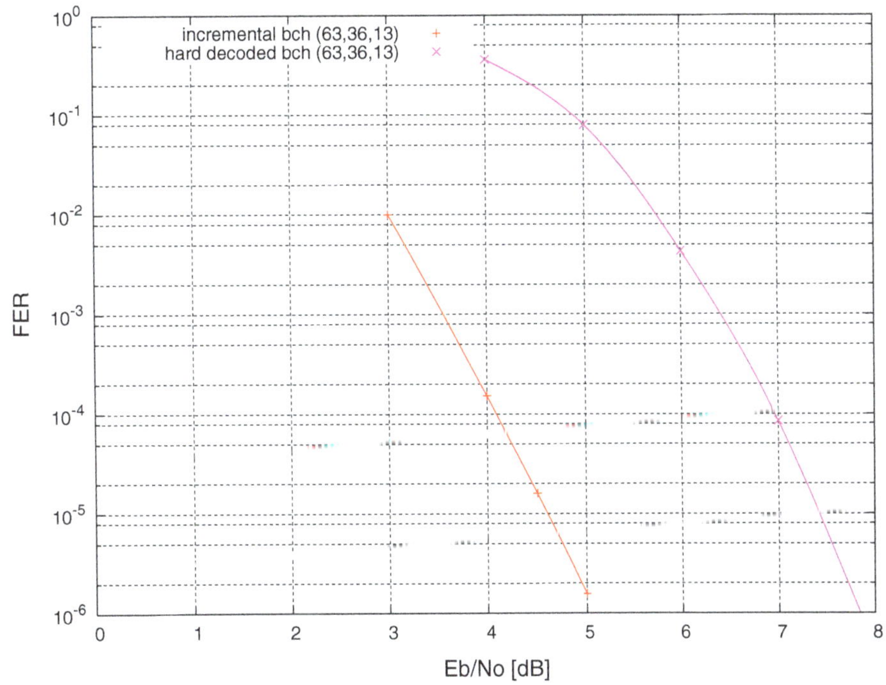

Fig. 15.10 FER performance of the (63, 36, 13) GF(4) BCH code using incremental correlation decoding compared to hard decision decoding

The benefit of having $GF(4)$ coefficients is that $g(x)$ does not need to contain the roots

$$\{\alpha^{14}, \alpha^{56}, \alpha^{35}, \alpha^{22}, \alpha^{25}, \alpha^{37}\}$$

which are necessary to constrain $g(x)$ to binary coefficients [12]. Correspondingly, the binary version of this BCH code is the lower rate (63, 30, 13) code with 6 less information symbols (bits).

The performance of the (63, 36, 13) $GF(4)$ BCH Code is shown in Fig. 15.10 for the AWGN channel using Quadrature Amplitude Modulation (QAM). Also shown in Fig. 15.10 is the performance of the code with hard decision decoding. It may be seen that at 10^{-4} FER the performance of the incremental correlation decoder is 2.9 dB better than the performance of the hard decision decoder.

15.6 Conclusions

It has been shown that the extended Dorsch decoder may approach maximum likelihood decoding by an incremental correlation approach in which for each received vector a partial summation metric is evaluated as a function of low information weight codewords. Furthermore, the number of information weight codewords that need to be evaluated to achieve maximum likelihood decoding may be calculated as an upper bound for each received vector. Consequently, for each received vector it is known whether the decoder has achieved maximum likelihood decoding. An efficient decoder structure consisting of a combination of hard decision threshold decoding followed by partial sum correlation was also described, which enables practical decoders to trade-off performance against complexity.

The decoder for non-binary codes was shown to be straightforward for the AWGN channel and an example was described for a GF(4) (63, 36, 13) BCH code using QAM to transmit each GF(4) symbol. It is readily possible to extend the decoder to other modulation formats by extensions to the incremental correlation of Eq. (15.37) although this inevitably involves an increase in complexity. It is hoped that there will sufficient interest from the coding community to address this research area.

Another interesting conclusion is just how well some codes in Brouwer's table perform with maximum likelihood decoding. In particular, the (136, 68, 24) double-circulant, extremal, self-dual code is shown to be an outstanding code.

It seems that the implementation of this type of decoder coupled with the availability of powerful processors will eventually herald a new era in the application of error control coding with the re-establishment of the importance of the optimality of codes rather than the ease of decoding. Certainly, this type of decoder is more complex than an iterative decoder, but the demonstrable performance, which is achievable for short codes, can approach theoretical limits for error-correction coding performance such as the sphere packing bound.

15.7 Summary

The current day, unobtainable goal of a practical realisation of the maximum likelihood decoder that can be used with any error-correcting code has been partially addressed with the description of the modified Dorsch decoder presented in this chapter. A decoder based on enhancements to the original Dorsch decoder has been described which achieves near maximum likelihood performance for all codes whose codelength is not too long. It is a practical decoder for half rate codes having a codelength less than about 180 bits or so using current digital processors. The performance achieved by the decoder when using different examples of outstanding binary codes has been evaluated and the results presented in this chapter. A description of the decoder suitable for use with non-binary codes has also been given. An example showing the results obtained by the decoder using a (63, 36, 13) GF(4) non-binary code for the AWGN channel has also been presented.

References

1. Butman, S., McEliece, R.J.: The ultimate limits of binary coding for a wideband Gaussian channel. JPL Deep Space Netw. Prog. Rep. **42–22**, 78–80 (1974)
2. Chase, D.: A class of algorithms for decoding block codes with channel measurement information. IEEE Trans. Inf. Theory IT **18**, 170–182 (1972)
3. Conway, J.H., Sloane, N.J.A.: A new upper bound on the minimum distance of self-dual codes. IEEE Trans. Inf. Theory **36**(6), 1319–1333 (1990)
4. Dorsch, B.G.: A decoding algorithm for binary block codes and J-ary output channels. IEEE Trans. Inf. Theory **20**, 391–394 (1974)
5. Fossorier, M., Lin, S.: Soft-decision decoding of linear block codes based on ordered statistics. IEEE Trans. Inf. Theory **41**(5), 1379–1396 (1995)
6. Fossorier, M., Lin, S.: Computationally efficient soft-decision decoding of linear block codes based upon ordered statistics. IEEE Trans. Inf. Theory **42**, 738–750 (1996)
7. Grassl, M.: Code Tables: Bounds on the parameters of various types of codes. http://www.codetables.de (2007)
8. Karlin, M.: New binary coding results by circulants. IEEE Trans. Inf. Theory **15**(1), 81–92 (1969)
9. Kou, Y., Lin, S., Fossorier, M.: Low-density parity-check codes based on finite geometries: a rediscovery and new results. IEEE Trans. Inf. Theory **47**(7), 2711–2736 (2001)
10. Lin, S., Costello Jr., D.J.: Error Control Coding: Fundamentals and Applications, 2nd edn. Pearson Education, Inc, Englewood Cliffs (2004)
11. Lous, N.J.C., Bours, P.A.H., van Tilborg, H.C.A.: On maximum likelihood soft-decision decoding of binary linear codes. IEEE Trans. Inf. Theory **39**, 197–203 (1993)
12. MacWilliams, F.J., Sloane, N.J.A.: The Theory of Error-Correcting Codes. North-Holland, Amsterdam (1977)
13. Proakis, J.: Digital Communications, 4th edn. McGraw-Hill (2001)
14. Shannon, C.E.: Probability of error for optimal codes in a Gaussian channel. Bell Syst. Tech. J. **38**(3), 611–656 (1959)
15. Snyders, J.: Reduced lists of error patterns for maximum likelihood soft decision decoding. IEEE Trans. Inf. Theory **37**, 1194–1200 (1991)

16. Tjhai, C.J., Tomlinson, M., Ambroze, M., Ahmed, M.: Cyclotomic idempotent-based binary cyclic codes. Electron. Lett. **41**(3), 341–343 (2005)
17. Tomlinson, M., Tjhai, C.J., Ambroze, M., Ahmed, M.: Improved error correction decoder using ordered symbol reliabilities. UK Patent Application GB0637114.3 (2005)

Chapter 16
A Concatenated Error-Correction System Using the $|u|u + v|$ Code Construction

16.1 Introduction

There is a classical error-correcting code construction method where two good codes are combined together to form a new, longer code. It is a method first pioneered by Plotkin [1]. The Plotkin sum, also known as the $|u|u + v|$ construction method [3], consists of one or more codes having replicated codewords to which are added codewords from one or more other codes to form a concatenated code. This code construction may be exploited in the receiver with a decoder that first decodes one or more individual codewords prior to the Plotkin sum from a received vector. The detected codewords from this first decoding are used to undo the code concatenation within the received vector to allow the replicated codewords to be decoded. The output from the overall decoder of the concatenated code consists of the information symbols from the first decoder followed by the information symbols from the second stage decoder. Multiple codewords may be replicated and added to the codewords from other codes so that the concatenated code consists of several shorter codewords which are decoded first and the decoded codewords used to decode the remaining codewords. It is possible to utilise a recurrent construction whereby the replicated codewords are themselves concatenated codewords. It follows that the receiver has to use more than two stages of decoding.

With suitable modifications, any type of error-correction decoder may be utilised including iterative decoders, Viterbi decoders, list decoders, and ordered reliability decoders, and of particular importance the modified Dorsch decoder described in Chap. 15. It is well known that for a given code rate longer codes have better performance than shorter codes, but implementation of a maximum likelihood decoder is much more difficult for longer codes. The Plotkin sum code construction method provides a means whereby several decoders for short codes may be used together to implement a near maximum likelihood decoder for a long code.

M. Tomlinson et al., *Error-Correction Coding and Decoding*,
Signals and Communication Technology,
DOI 10.1007/978-3-319-51103-0_16

16.2 Description of the System

Figure 16.1 shows the generic structure of the transmitted signal in which the codeword of length n_1 from code u, denoted as $\mathscr{C}_u$ is followed by a codeword comprising the sum of the same codeword and another codeword from code v, denoted as $\mathscr{C}_v$ to form a codeword denoted as $\mathscr{C}_{cat}$ of length $2n_1$. This code construction is well known as the $|u|u+v|$ code construction [3]. The addition is carried out symbol by symbol using the arithmetic rules of the Galois Field being used, namely $GF(q)$. If code u is an (n_1, k_1, d_1) code with k_1 information symbols and Hamming distance d_1 and code v is an (n_1, k_2, d_2) code with k_2 information symbols and Hamming distance d_2, the concatenated code $\mathscr{C}_{cat}$ is an $(2n_1, k_1 + k_2, d_3)$ code with Hamming distance d_3 equal to the smaller of $2 \times d_1$ and d_2.

Prior to transmission, symbols from the concatenated codeword are mapped to signal constellation points in order to maximise the Euclidean distance between transmitted symbols in keeping with current best transmission practice. For example see the text book by Professor J. Proakis [4]. The mapped concatenated codeword is denoted as $\mathscr{X}_{cat}$ and is given by

$$\mathscr{X}_{cat} = |\mathscr{X}_u|\mathscr{X}_{u+v}| = |\mathscr{X}_u|\mathscr{X}_w|, \tag{16.1}$$

where $\mathscr{X}_w$ is used to represent $\mathscr{X}_{u+v}$.

$\mathscr{X}_{cat}$ consists of $2 \times n_1$ symbols and the first n_1 symbols are the n_1 symbols of $\mathscr{X}_u$ and the second n_1 symbols are the n_1 symbols resulting from mapping of the symbols resulting from the summation, symbol by symbol, of the n_1 symbols of $\mathscr{C}_u$, and the n_1 symbols of codeword $\mathscr{C}_v$.

The encoding system to produce the concatenated codeword format shown in Fig. 16.1 is shown in Fig. 16.2. For each concatenated codeword, k_1 information symbols are input to the encoder for the (n_1, k_1, d_1) code and n_1 symbols are produced at the output of the encoder and are stored in the codeword buffer A as shown in Fig. 16.2. Additionally, for each concatenated codeword, k_2 information symbols are input to the encoder for the (n_1, k_2, d_2) code and n_1 symbols are produced at the output and are stored in the codeword buffer B as shown in Fig. 16.2. The encoded symbols

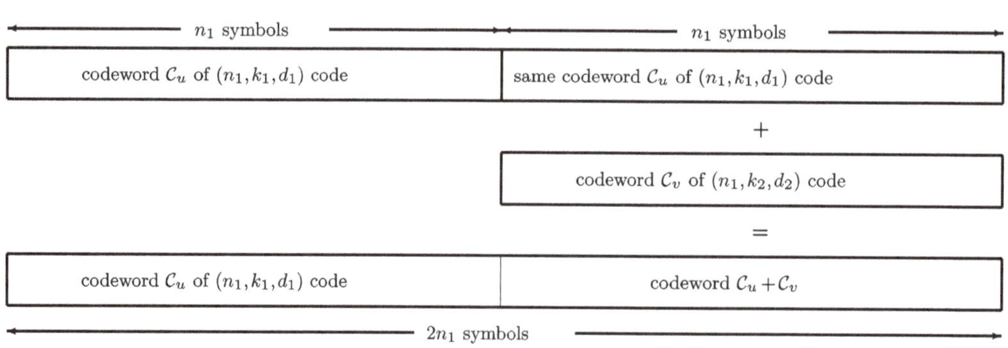

Fig. 16.1 Format of transmitted codeword consisting of two shorter codewords

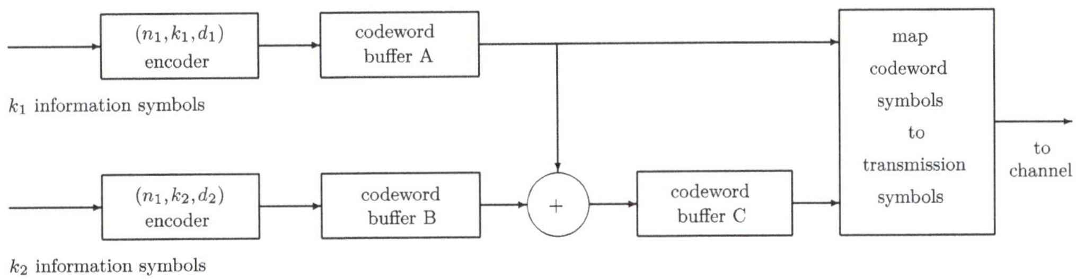

Fig. 16.2 Concatenated code encoder and mapping for transmission

output from the codeword buffer A are added symbol by symbol to the encoded symbols output from the codeword buffer B and the results are stored in codeword buffer C. The codeword stored in codeword buffer A is $\mathscr{C}_u$ as depicted in Fig. 16.1 and the codeword stored in codeword buffer C is $\mathscr{C}_u + \mathscr{C}_v$ as also depicted in Fig. 16.1. The encoded symbols output from the codeword buffer A are mapped to transmission symbols and transmitted to the channel, and these are followed sequentially by the symbols output from the codeword buffer C which are also mapped to transmission symbols and transmitted to the channel as shown in Fig. 16.2.

After transmission through the communications medium each concatenated mapped codeword is received as the received vector, denoted as $\mathscr{R}_{cat}$ and given by

$$\mathscr{R}_{cat} = |\mathscr{R}_u|\mathscr{R}_{u+v}| = |\mathscr{R}_u|\mathscr{R}_w|. \tag{16.2}$$

Codeword $\mathscr{C}_v$ is decoded first as shown in Fig. 16.3. It is possible by comparing the received samples $\mathscr{R}_u$ with the received samples $\mathscr{R}_{u+v}$ that the a priori log likelihoods of the symbols of $\mathscr{R}_v$ may be determined, since it is clear that the difference between the respective samples, in the absence of noise and distortion, is attributable to $\mathscr{C}_v$. This is done by the soft decision metric calculator shown in Fig. 16.3.

Binary codeword symbols are considered with values which are either 0 or 1. The ith transmitted sample, $X_{u_i} = (-1)^{C_{u_i}}$ and the n_1 + ith transmitted sample, $X_{u_i+v_i} = (-1)^{C_{u_i}} \times (-1)^{C_{v_i}}$. It is apparent that X_{v_i} and C_{v_i} may be derived from X_{u_i} and $X_{u_i+v_i}$.

An estimate of X_{v_i} and C_{v_i} may be derived from R_{u_i} and $R_{u_i+v_i}$. First:

$$X_{v_i} = X_{u_i} \times X_{u_i+v_i} = (-1)^{C_{u_i}} \times (-1)^{C_{u_i}} \times (-1)^{C_{v_i}} = (-1)^{C_{v_i}} \tag{16.3}$$

Second, in the absence of distortion and with Gaussian distributed additive noise with standard deviation σ, and normalised signal power, the log likelihood that $C_{v_i} = 0$, $L_{log}(C_{v_i} = 0)$ is given by

$$L_{\log}(C_{v_i} = 0) = \log\left[\cosh\left(\frac{R_{u_i} + R_{u_i+v_i}}{\sigma^2}\right)\right] - \log\left[\cosh\left(\frac{R_{u_i} - R_{u_i+v_i}}{\sigma^2}\right)\right]. \tag{16.4}$$

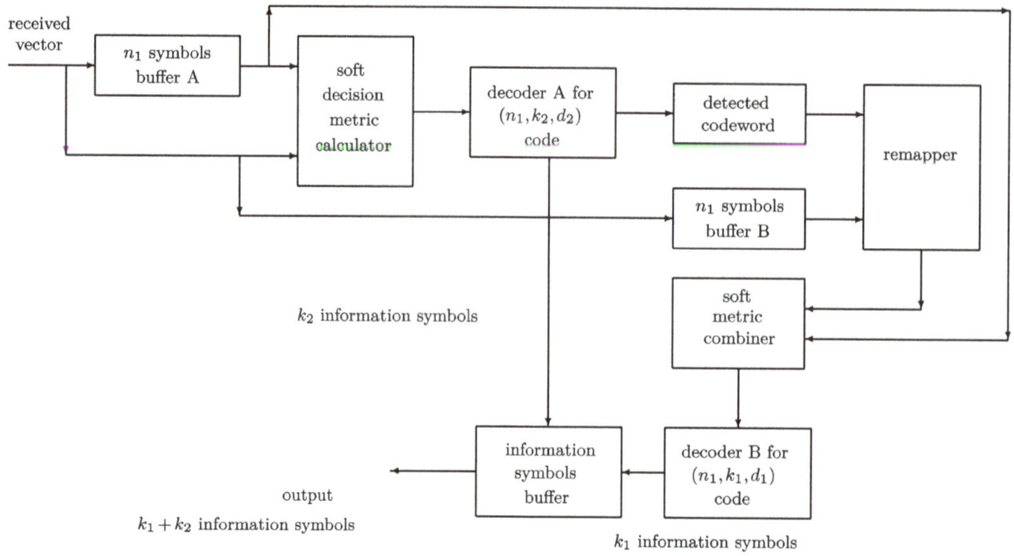

Fig. 16.3 Decoder for the concatenated code with the codeword format shown in Fig. 16.1

The soft decision metric calculator, shown in Fig. 16.3, calculates these log likelihoods according to Eq. (16.4) and these are input to the decoder A shown in Fig. 16.3. The decoder A determines the most likely codeword $\mathscr{C}_{\hat{v}}$ of the (n_1, k_2, d_2) code. With the knowledge of the detected codeword, $\mathscr{C}_{\hat{v}}$, the received samples $\mathscr{R}_{u+v}$, which are stored in the n_1 symbols buffer B, are remapped to form $\mathscr{R}_{\hat{u}}$ by multiplying $\mathscr{R}_{u+v}$ by $\mathscr{X}_{\hat{v}}$.

$$\mathscr{R}_{\hat{u}} = \mathscr{R}_{u+v} \times \mathscr{X}_{\hat{v}}. \tag{16.5}$$

This remapping function is provided by the remapper shown in Fig. 16.3. The output of the remapper is $\mathscr{R}_{\hat{u}}$. If the decoder's output is correct, $\mathscr{C}_{\hat{v}} = \mathscr{C}_v$ and there are now two independent received versions of the transmitted, mapped codeword $\mathscr{C}_u$, $\mathscr{R}_{\hat{u}}$ and the original received $\mathscr{R}_u$. Both of these are input to the soft metric combiner shown in Fig. 16.3, $\mathscr{R}_{\hat{u}}$ from the output of the remapper and $\mathscr{R}_u$ from the output of the n_1 symbols buffer A.

The soft metric combiner calculates the log likelihood of each bit of $\mathscr{C}_u$, C_{u_i} from the sum of the individual log likelihoods:

$$L_{\log}(C_{u_i} = 0) = \frac{2R_{u_i}}{\sigma^2} + \frac{2R_{\hat{u}_i}}{\sigma^2}. \tag{16.6}$$

These log likelihood values, $L_{\log}(C_{u_i} = 0)$, output from the soft metric combiner shown in Fig. 16.3 are input to the decoder B. The output of Decoder B is the k_1 information bits of the detected codeword $\mathscr{C}_{\hat{u}}$ of the (n_1, k_1, d_1) code, and these are input to the information symbols buffer shown in Fig. 16.3. The other input to the information symbols buffer is the k_2 information bits of the detected codeword

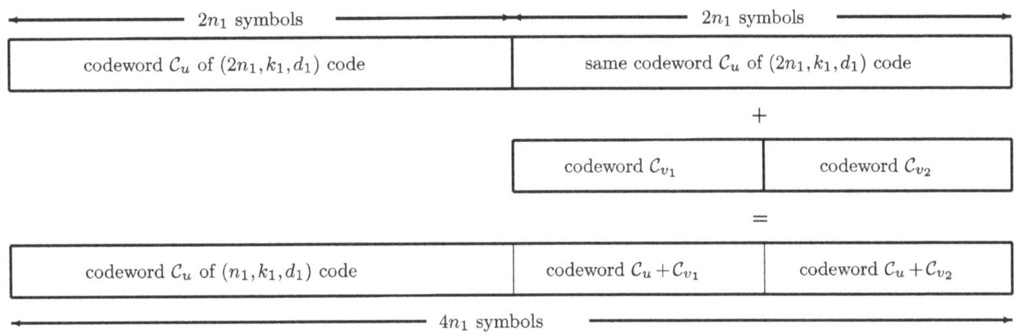

Fig. 16.4 Format of transmitted codeword consisting of three shorter codewords

$\mathscr{C}_{\hat{v}}$ of the (n_1, k_2, d_2) code, provided at the output of decoder A. The output of the information symbols buffer, for each received vector, is the $k_1 + k_2$ information bits which were originally encoded, provided both decoders' outputs, A and B, are correct.

In similar fashion to previous constructions, Fig. 16.4 shows the format of a concatenated codeword of length $4 \times n_1$ symbols consisting of three shorter codewords. The codeword of length $2 \times n_1$ from a $(2n_1, k_1, d_1)$, code u, denoted as $\mathscr{C}_u$ is replicated as shown in Fig. 16.4. The first half of the replicated codeword, $\mathscr{C}_u$, is added to the codeword $\mathscr{C}_{v1}$ and the second half of the replicated codeword, $\mathscr{C}_u$, is added to the codeword $\mathscr{C}_{v2}$, as shown in Fig. 16.4. Each codeword $\mathscr{C}_{v1}$ and $\mathscr{C}_{v2}$ is the result of encoding k_2 information symbols using code v, a (n_1, k_2, d_2) code. The concatenated codeword that results, $\mathscr{C}_{cat}$, is from a $(4n_1, k_1 + 2k_2, d_3)$ concatenated code where d_3 is the smaller of $2d_1$ or d_2.

The decoder for the concatenated code with codeword format shown in Fig. 16.4 is similar to the decoder shown in Fig. 16.3 except that following soft decision metric calculation each of the two codewords $\mathscr{C}_{v_1}$ and $\mathscr{C}_{v_2}$ are decoded independently. With the knowledge of the detected codewords, $\mathscr{C}_{\hat{v}_1}$ and $\mathscr{C}_{\hat{v}_2}$, the received samples $\mathscr{R}_{u+v_1}$, which are buffered, are remapped to form the first n_1 symbols of $\mathscr{R}_{\hat{u}}$ by multiplying $\mathscr{R}_{u+v_1}$ by $\mathscr{X}_{\hat{v}_1}$ and the second n_1 symbols of $\mathscr{R}_{\hat{u}}$ are obtained by multiplying $\mathscr{R}_{u+v_2}$ by $\mathscr{X}_{\hat{v}_2}$.The two independent received versions of the transmitted, mapped codeword $\mathscr{C}_u$, $\mathscr{R}_{\hat{u}}$ and the original received $\mathscr{R}_u$ are input to a soft metric combiner prior to decoding the codeword $\mathscr{C}_{\hat{u}}$.

In another code arrangement, Fig. 16.5 shows the format of a concatenated codeword of length $3 \times n_1$ symbols. The concatenated codeword is the result of three layers of concatenation. A codeword of length n_1 from a (n_1, k_1, d_1), code u, denoted as $\mathscr{C}_u$ is replicated twice, as shown in Fig. 16.5. A second codeword of length n_1 from a (n_1, k_2, d_2), code v, denoted as $\mathscr{C}_v$ is replicated and each of these two codewords is added to the two replicated codewords $\mathscr{C}_u$, as shown in Fig. 16.5. A third codeword of length n_1 from a (n_1, k_3, d_3), code w, denoted as $\mathscr{C}_w$ is added to the codeword summation $\mathscr{C}_u + \mathscr{C}_v$, as shown in Fig. 16.5. The concatenated codeword that results, $\mathscr{C}_{cat}$, is from a $(3n_1, k_1 + k_2 + k_3, d_4)$ concatenated code where d_4 is the smallest of $3d_1$ or $2d_2$ or d_3.

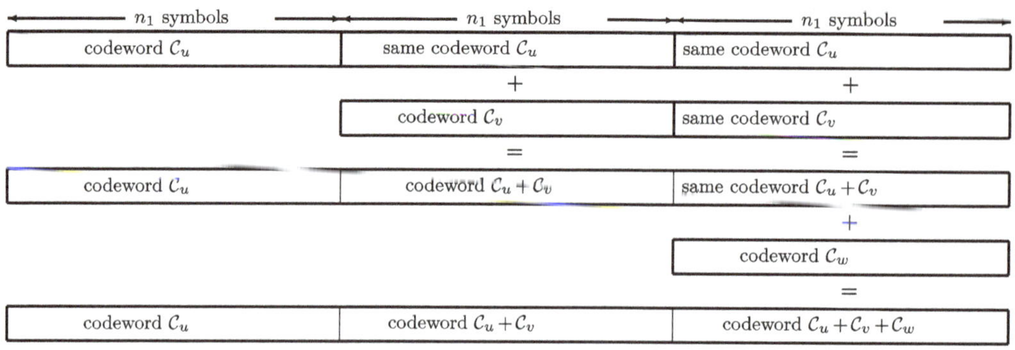

Fig. 16.5 Format of transmitted codeword consisting of two levels of concatenation and three shorter codewords

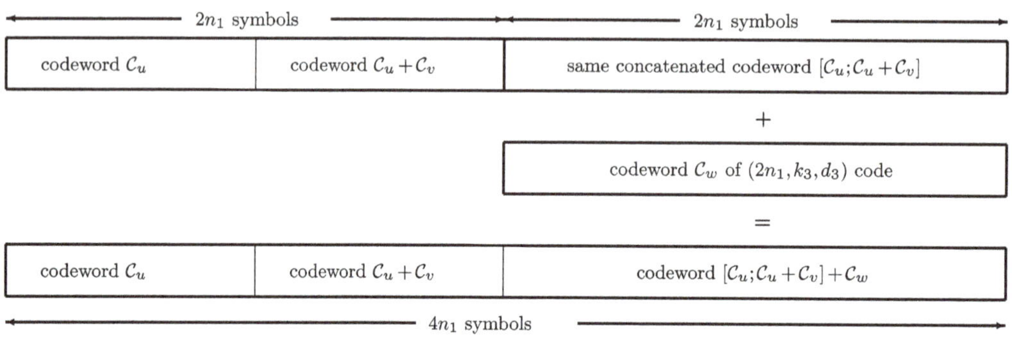

Fig. 16.6 Format of transmitted codeword after two stages of concatenation

The decoder for the three layered concatenated code with codeword format shown in Fig. 16.5 uses similar signal processing to the decoder shown in Fig. 16.3 with changes corresponding to the three layers of concatenation. The codeword $\mathscr{C}_{\hat{w}}$ is decoded first following soft decision metric calculation using the$\mathscr{R}_{u+v}$ and $\mathscr{R}_{u+v+w}$ sections of the received vector. The detected codeword $\mathscr{C}_{\hat{w}}$ is used to obtain two independent received versions of the transmitted, mapped result of the two codewords summation $\mathscr{C}_{u+v}$, $\mathscr{R}_{u\hat{+}v}$ and the original received $\mathscr{R}_{u+v}$. These are input to a soft metric combiner and the output is input to the soft decision metric calculation together with $\mathscr{R}_u$, prior to decoding of codeword $\mathscr{C}_{\hat{v}}$. With the knowledge of codeword $\mathscr{C}_{\hat{v}}$, remapping and soft metric combining is carried out prior to the decoding of codeword $\mathscr{C}_{\hat{u}}$.

Figure 16.6 shows the format of a concatenated codeword of length $4 \times n_1$ symbols. The concatenated codeword is the result of three layers of concatenation. A concatenated codeword with the format shown in Fig. 16.1 is replicated and added to a codeword, $\mathscr{C}_{\hat{w}}$, of length $2n_1$ symbols from a $(2n_1, k_3, d_3)$ code to form a codeword of an overall concatenated code having parameters $(4n_1, k_1 + k_2 + k_3, d_4)$, where d_4 is equal to the smallest of $4d_1$, $2d_2$ or d_3.

The decoder for the three layered concatenated code with codeword format shown in Fig. 16.6 is similar to the decoder described above. Codeword $\mathscr{C}_{\hat{w}}$ is detected first following soft decision metric calculation using $\mathscr{R}_u$ and $\mathscr{R}_{u+v}$ sections of the received vector as one input and the $\mathcal{R}_{u+w}$ and $\mathcal{R}_{u+v+w}$ sections of the received vector as the other input. The detected codeword $\mathcal{C}_{\hat{w}}$ is used to obtain two independent received versions of the concatenated codeword of length $2n_1$ symbols with format equal to that of Fig. 16.1. Accordingly, following soft metric combining of the two independent received versions of the concatenated codeword of length $2n_1$ symbols, a vector of length equal to $2n_1$ symbols is obtained which may be input to the concatenated code decoder shown in Fig. 16.3. This decoder provides at its output the $k_1 + k_2$ detected information symbols which together with the k_3 information symbols already detected provide the complete detected output of the overall three layer concatenated code.

Any type of code, binary or non-binary, LDPC, Turbo or algebraically constructed code, may be used. Any corresponding type of decoder, for example an iterative decoder or a list decoder may be used. As an illustration of this, Decoder A and Decoder B, shown in Fig. 16.3, do not have to be the same type of decoder.

There are particular advantages in using the modified Dorsch decoder, described in Chap. 15, because the Dorsch decoder may realise close to maximum likelihood decoding, with reasonable complexity of the decoder. The complexity increases exponentially with codelength. Using modified Dorsch decoders. Both decoder A and decoder B shown in Fig. 16.3 operate on n_1 received samples and may realise close to maximum likelihood decoding with reasonable complexity even though the concatenated codelength is $2 \times n_1$ symbols and the total number of received samples is $2 \times n_1$ samples. Using a single modified Dorsch decoder to decode the $2 \times n_1$ samples of the concatenated code directly will usually result in non-maximum likelihood performance unless the list of codewords evaluated for each received vector is very long. For example, a modified Dorsch decoder with moderate complexity, typically will process 100,000 codewords for each received vector and realise near maximum likelihood performance. Doubling the codelength will require typically in excess of 100,000,000 codewords to be processed for each received vector if near maximum likelihood performance is to be maintained.

An example of the performance that may be achieved is shown in Fig. 16.7 for the concatenated codeword format shown in Fig. 16.1. The encoder used is the same as that shown in Fig. 16.2 and the concatenated code decoder is the same as that shown in Fig. 16.3. The results were obtained by computer simulation using Quaternary Phase Shift Keying (QPSK) modulation and featuring the Additive White Gaussian Noise (AWGN) channel. The decoder error rate, the ratio of the number of incorrect codewords output by the decoder to the total number of codewords output by the decoder, is denoted by the Frame Error Rate (FER) and this is plotted against $\frac{E_b}{N_o}$, the ratio of the energy per information bit to the noise power spectral density. Binary codes are used and the length of the concatenated code is 256 bits. For best results, it is important to use outstanding codes for the constituent codes, particularly for code v which is decoded first. In this example, code u is the (128,92,12) extended Bose Chaudhuri Hocquenghem (BCH) code. Code v is the (128,36,36) extended cyclic code, an optimum code described in [5] by D. Schoemaker and M. Wirtz. The (128,36,36) extended cyclic code is not an extended BCH code as it has roots {1, 3, 5, 7, 9, 11, 13, 19, 21, 27, 43, 47, 63}. The minimum Hamming distance

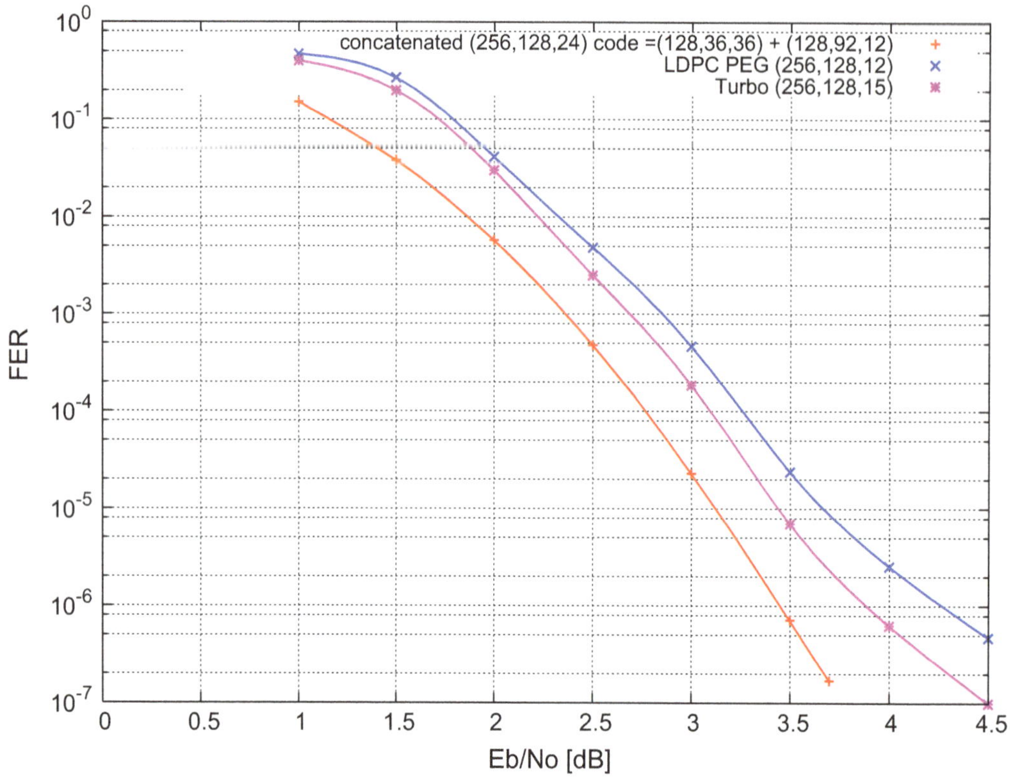

Fig. 16.7 The error rate performance for a (256,128,24) concatenated code compared to iterative decoding of a (256,128,15) Turbo code and a (256,128,12) LDPC code

of the concatenated code is $2d_1 = 24$. Both decoder A and decoder B, as shown in Fig. 16.3, are a modified Dorsch decoder and for both code u and code v, near maximum likelihood performance is obtained with moderate decoder complexity. For each point plotted in Fig. 16.7, the number of codewords transmitted was chosen such that were at least 100 codewords decoded in error.

Also shown in Fig. 16.7 is the performance of codes and decoders designed according to the currently known state of the art in error-correction coding that is Low Density Parity Check (LDPC) codes using Belief Propagation (BP) iterative decoding, and Turbo codes with BCJR iterative decoding. Featured in Fig. 16.7 is the performance of an optimised Low Density Parity Check (LDPC) (256,128,12) code using BP, iterative decoding and an optimised (256,128,15) Turbo code with iterative decoding. As shown in Fig. 16.7 both the (256,128,15) Turbo code and the (256,128,12) LDPC code suffer from an error floor for $\frac{E_b}{N_o}$ values higher than 3.5dB whilst the concatenated code features a FER performance with no error floor. This is attributable to the significantly higher minimum Hamming distance of the concatenated code which is equal to 24 in comparison to 15 for the Turbo code and 12 for the LDPC code. Throughout the entire range of $\frac{E_b}{N_o}$ values the concatenated code can be seen to outperform the other codes and decoders.

For (512,256) codes, using the concatenated code arrangement, the performance achievable is shown in Fig. 16.8. The concatenated code arrangement uses the

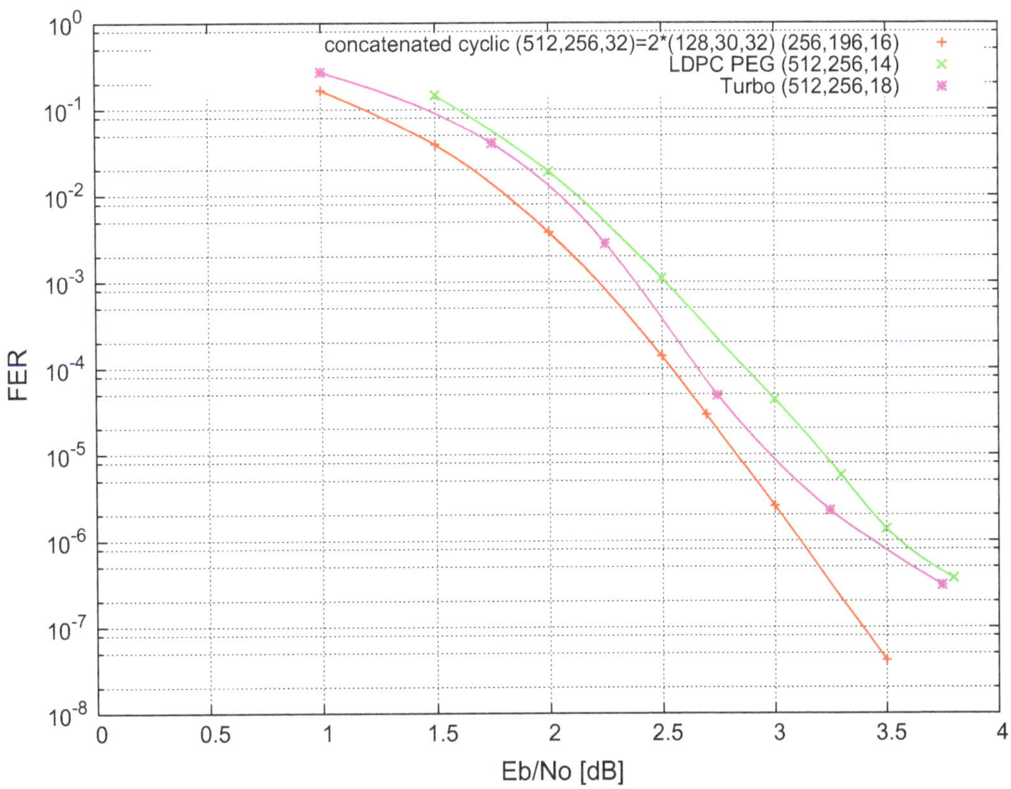

Fig. 16.8 Comparison of the error rate performance for a (512,256,32) concatenated code compared to iterative decoding of a (512,256,18) Turbo code and a (512,256,14) LDPC code

concatenated codeword format which is shown in Fig. 16.4. As before, the FER results were obtained by computer simulation using QPSK modulation and the AWGN channel. Both codes v_1 and v_2 are the same and equal to the outstanding (128,30,38) best-known code [6]. Code u is equal to a (256,196,16) extended cyclic code. Featured in Fig. 16.8 is the performance of an optimised Low Density Parity Check (LDPC) (512,256,14) code using BP iterative decoding and an optimised (512,256,18) Turbo code with iterative decoding. For each point plotted in Fig. 16.8, the number of codewords transmitted was chosen such that were at least 100 codewords decoded in error. As shown in Fig. 16.8 both the (512,256,18) Turbo code and the (512,256,14) LDPC code suffer from an error floor for $\frac{E_b}{N_o}$ values higher than 3.4 dB whilst the concatenated code features a FER performance with no error floor. As before this is attributable to the significantly higher minimum Hamming distance of the concatenated code which is equal to 32 in comparison to 18 for the Turbo code and 14 for the LDPC code. Throughout the entire range of $\frac{E_b}{N_o}$ values, the concatenated code system can be seen to outperform the other coding arrangements for (512,256) codes.

16.3 Concatenated Coding and Modulation Formats

With the $|u|u+v|$ code construction and binary transmission, the received vector for the codeword of code v suffers the full interference from the codeword of code u because it is transmitted as $u+v$. The interference is removed by differential detection using the first version of the codeword of code u. However, although the effects of code u are removed, differential detection introduces additional noise power due to noise times noise components. One possible solution to reduce this effect is to use multi-level modulation such as 8-PSK. Code u is transmitted as 4-PSK and code v modulates the 4-PSK constellation by ± 22.5 degrees. Now there is less direct interference between code u and code v. Initial investigations show that this approach is promising, particularly for higher rate systems.

16.4 Summary

Concatenation of good codes is a classic method of constructing longer codes which are good. As codes are increased in length, it becomes progressively harder to realise a near maximum likelihood decoder. This chapter presented a novel concatenated code arrangement featuring multiple near maximum likelihood decoders for an optimised matching of codes and decoders. It was demonstrated that by using some outstanding codes as constituent codes, the concatenated coding arrangement is able to outperform the best LDPC and Turbo coding systems with the same code parameters. The performance of a net (256,128) code achieved with the concatenated arrangement is compared to a best (256,128) LDPC code and a best (256,128) Turbo code. Similarly, the performance of a (512,256) net concatenated code is compared to a best (512,256) LDPC code and a best (512,256) Turbo code. In both cases, the new system was shown to outperform the LDPC and Turbo systems. To date, for the AWGN channel and net, half rate codes no other codes or coding arrangement is known that will outperform the system presented in this chapter for codes of lengths 256 and 512 bits.

References

1. Plotkin, M.: Binary codes with specified minimum distances. IEEE Trans. Inf. Theory **6**, 445–450 (1960)
2. Tomlinson, M., Tjhai, C.J., Ambroze, M.: Extending the Dorsch decoder towards achieving maximum-likelihood decoding for linear codes. IET Proc. Commun. **1**(3), 479–488 (2007)
3. MacWilliams, F.J., Sloane, N.J.A.: The Theory of Error-Correcting Codes. North-Holland (1977)
4. Proakis, J.: Digital Communications, 4th edn. McGraw-Hill, New York (2001)

5. Schomaker, D., Wirtz, M.: On binary cyclic codes of odd lengths from 101 to 127. IEEE Trans. Inf. Theory **38**(2), 516–518 (1992)
6. Grassl, M.: Code Tables: Bounds on the Parameters of Various Types of Codes. http://www.codetables.de

Part IV
Applications

This part is concerned with a wide variety of applications using error-correcting codes. Analysis is presented of combined error-detecting and error-correcting codes which enhance the reliability of digital communications by using the parity check bits for error detection as well as using the parity check bits for error-correction. A worked example of code construction for a (251, 113, 20) incremental redundancy error-correcting code is described. The idea is that additional sequences of parity bits may be transmitted in stages until the decoded codeword satisfies a Cyclic Redundancy Check (CRC). A soft decision scheme for measuring codeword reliability is also described which does not require a CRC to be transmitted. The relative performance of the undetected error rate and throughput of the different systems is presented.

In this part it is also shown that error-correcting codes may be used for the automatic correction of small errors in password authentication systems or in submitting personal identification information. An adaptive mapping of $GF(q)$ symbols is used to convert a high percentage of passwords into Reed–Solomon codewords without the need for additional parity check symbols. It is shown that a BCH decoder may be used for error-correction or error detection. Worked examples of codes and passwords are included.

Goppa codes are used as the basis of a public key cryptosystem invented by Professor Robert McEliece. The way in which Goppa codes are designed into the cryptosystem is illustrated with step by step worked examples showing how a ciphertext is constructed and subsequently decrypted. The cryptosystem is described in considerable detail together with some proposed system variations designed to reduce the ciphertext length with no loss in security. An example is presented in which the system realises 256 bits of security, normally requiring a ciphertext length of 8192 bits, that uses a ciphertext length of 1912 bits.

Different attacks, designed to break the McEliece cryptosystem are described including the information set decoding attack. Analysis is provided showing the security level achieved by the cryptosystem as a function of Goppa code length. Vulnerabilities of the standard McEliece to chosen plaintext and chosen ciphertext attacks are described, together with system modifications that defeat these attacks. Some commercial applications are described that are based on using a smartphone for secure messaging and cloud based, encrypted information access.

The use of error-correcting codes in impressing watermarks on different media by using dirty paper coding is included in this part. The method described, is based on firstly decoding the media or white noise with a cross correlating decoder so as to find sets of codewords from a given code that will cause minimum change to the media, but still be detectable. The best codewords are added to the media as a watermark so as to convey additional information as in steganography. Some examples are included using a binary (47, 24, 11) quadratic residue code.

Chapter 17
Combined Error Detection and Error-Correction

17.1 Analysis of Undetected Error Probability

Let the space of vectors over a field with q elements $\mathbb{F}_q$ of length n be denoted by $\mathbb{F}_q^n$. Let $[n, k, d]_q$ denote a linear code over $\mathbb{F}_q$ of length n symbols, dimension k symbols and minimum Hamming distance d. We know that a code with minimum Hamming distance d can correct $t = \lfloor (d-1)/2 \rfloor$ errors. It is possible for an $[n, k, d = 2t+1]_q$ linear code, which has q^{n-k} syndromes, to use a subset of these syndromes to correct $\tau < t$ errors and then to use the remaining syndromes for error detection. For convenience, let $\mathscr{C}$ denote an $[n, k, d]_q$ linear code with cardinality $|\mathscr{C}|$, and let a codeword of $\mathscr{C}$ be denoted by $\boldsymbol{c}_l = (c_{l,0}, c_{l,1}, \ldots, c_{l,n-1})$, where $0 \leq l < |\mathscr{C}|$.

Consider a codeword $\boldsymbol{c}_i$, for some integer i, which is transmitted over a q-ary symmetric channel with symbol transition probability $p/(q-1)$. At the receiver, a length n vector $\boldsymbol{y}$ is received. This vector $\boldsymbol{y}$ is not necessarily the same as $\boldsymbol{c}_i$ and, denoting $\mathrm{d}_H(\boldsymbol{a}, \boldsymbol{b})$ as the Hamming distance between vectors $\boldsymbol{a}$ and $\boldsymbol{b}$, the following possibilities may occur assuming that nearest neighbour decoding algorithm is employed:

1. (no error) $\mathrm{d}_H(\boldsymbol{y}, \boldsymbol{c}_i) \leq \tau$ and $\boldsymbol{y}$ is decoded as $\boldsymbol{c}_i$;
2. (error) $\mathrm{d}_H(\boldsymbol{y}, \boldsymbol{c}_j) > \tau$ for $0 \leq j < |\mathscr{C}|$; and
3. (undetected error) $\mathrm{d}_H(\boldsymbol{y}, \boldsymbol{c}_j) \leq \tau$ for $j \neq i$ and $\boldsymbol{y}$ is decoded as $\boldsymbol{c}_j$

Definition 17.1 A sphere of radius t centered at a vector $\boldsymbol{v} \in \mathbb{F}_q^n$, denoted by $S_q^t(\boldsymbol{v})$, is defined as

$$S_q^t(\boldsymbol{v}) = \{\boldsymbol{w} \mid wt_H(\boldsymbol{v} - \boldsymbol{w}) \leq t \text{ for all } \boldsymbol{w} \in \mathbb{F}_q^n\}. \tag{17.1}$$

It can be seen that, in an error-detection-after-correction case, $S_q^\tau(\boldsymbol{c})$ may be drawn around all $|\mathscr{C}|$ codewords of the code $\mathscr{C}$. For any vector falling within $S_q^\tau(\boldsymbol{c})$, the decoder returns $\boldsymbol{c}$ the corresponding codeword which is the center of the sphere. It is worth noting that all these $|\mathscr{C}|$ spheres are pairwise disjoint, i.e.

M. Tomlinson et al., *Error-Correction Coding and Decoding*,
Signals and Communication Technology,
DOI 10.1007/978-3-319-51103-0_17

$$\bigcup_{\substack{0\le i,j<|\mathscr{C}|\\ i\neq j}} S_q^{\tau}(\boldsymbol{c}_i)\cap S_q^{\tau}(\boldsymbol{c}_j)=\emptyset.$$

In a pure error-detection scenario, the radius of these spheres is zero and the probability of an undetected error is minimised. When the code is used to correct a given number of errors, the radius increases and so does the probability of undetected error.

Lemma 17.1 *The number of length n vectors over $\mathbb{F}_q$ of weight j within a sphere of radius τ centered at a length n vector of weight i, denoted by $N_q^{\tau}(n,i,j)$, is equal to*

$$N_q^{\tau}(n,i,j)=\sum_{e=e_L}^{e_U}\sum_{\delta=\delta_L}^{\delta_U}\binom{i}{e}\binom{e}{\delta}\binom{n-i}{j-i+\delta}(q-1)^{j-i+\delta}(q-2)^{e-\delta} \tag{17.2}$$

where $e_L=\max(0,i-j)$, $e_U=\min(\tau,\tau+i-j)$, $\delta_L=\max(0,i-j)$ *and* $\delta_U=\min(e,\tau+i-j-e,n-j)$.

Proof Let $\boldsymbol{u}$ be a vector of weight i and let $\sup(\boldsymbol{u})$ and $\overline{\sup}(\boldsymbol{u})$ denote the support of $\boldsymbol{u}$, and the non-support of $\boldsymbol{u}$, respectively, that is

$$\begin{aligned}\sup(\boldsymbol{u})&=\{i\mid u_i\neq 0,\ \text{for } 0\le i\le n-1\}\\ \overline{\sup}(\boldsymbol{u})&=\{0,1,\ldots,n-1\}\setminus\sup(\boldsymbol{u}).\end{aligned}$$

A vector of weight j, denoted by $\boldsymbol{v}$, may be obtained by adding a vector $\boldsymbol{w}$, which has e coordinates which are the elements of $\sup(\boldsymbol{u})$ and f coordinates which are the elements of $\overline{\sup}(\boldsymbol{u})$. In the case where $q>2$, considering the coordinates in $\sup(\boldsymbol{u})$, it is obvious that vector $\boldsymbol{v}=\boldsymbol{u}+\boldsymbol{w}$ can have more than $i-e$ non-zeros in these coordinates. Let δ, where $0\le\delta\le e$, denote the number of coordinates for which $v_i=0$ among $\sup(\boldsymbol{u})$ of $\boldsymbol{v}$, i.e.

$$\delta=|\sup(\boldsymbol{u})\setminus(\sup(\boldsymbol{u})\cap\sup(\boldsymbol{v}))|$$

Given an integer e, there are $\binom{i}{e}$ ways to generate e coordinates for which $w_i\neq 0$ where $i\in\sup(\boldsymbol{u})$. For each way, there are $\binom{e}{e-\delta}(q-2)^{e-\delta}$ ways to generate $e-\delta$ non-zeros in the coordinates $\sup(\boldsymbol{u})\cap\sup(\boldsymbol{w})$ such that $v_i\neq 0$. It follows that $f=j-(i-e)-(e-\delta)=j-i+\delta$ and there are $\binom{n-i}{j-i+\delta}(q-1)^{j-i+\delta}$ ways to generate f non-zero coordinates such that $v_i\neq 0$ where $i\in\overline{\sup}(\boldsymbol{u})$. Therefore, for given integers e and δ, we have

$$\binom{i}{e}\binom{e}{\delta}\binom{n-i}{j-i+\delta}(q-1)^{j-i+\delta}(q-2)^{e-\delta} \tag{17.3}$$

vectors $\boldsymbol{w}$ that produce $wt_H(\boldsymbol{v})=j$. Note that $\binom{e}{e-\delta}=\binom{e}{\delta}$.

It is obvious that $0 \le e, f \le \tau$ and $e + f \le \tau$. In the case of $j \le i$, the integer e may not take the entire range of values from 0 to τ, it is not possible to have $e < i - j$. On the other hand, for $j \ge i$, the integer $e \ge 0$ and thus, the lower limit on the value of e is $e_L = \max(0, i - j)$. The upper limit of e, denoted by e_U, is dictated by the condition $e + f = \tau$. For $j \le i$, $e_U = \tau$ since for any value of e, δ may be adjusted such that $wt_H(\boldsymbol{v}) = j$. For the case $j \ge i$, $f \ge 0$ and for any value of e, there exists at least one vector for which $\delta = 0$, implying $e_U = \tau - f = \tau + i - j$. It follows that $e_U = \min(\tau, \tau + i - j)$.

For a given value of e, δ takes certain values in the range between 0 and e such that $wt_H(\boldsymbol{v}) = j$. The lower limit of δ is obvious $\delta_L = e_L$. The upper limit of δ for $j \ge i$ case is also obvious, $\delta_U = e$, since $f \ge 0$. For the case $j \le i$, we have $e + f = e + (j - i + \delta_U) \le \tau$, implying $\delta_U \le \tau - e + i - j$. In addition, $n - i \ge j - i + \delta_U$ and thus, we have $\delta_U = \min(e, \tau - e + i - j, n - j)$.

Corollary 17.1 *For $q = 2$, we have*

$$N_2^{\tau}(n, i, j) = \sum_{e=\max(0,i-j)}^{\lfloor(\tau+i-j)/2\rfloor} \binom{i}{e}\binom{n-i}{j-i+e} \tag{17.4}$$

Proof For $q = 2$, it is obvious that $\delta = e$ and $0^0 = 1$. Since $e + f \le \tau$ and $f = j - i + e$, the upper limit of e, e_L, becomes $e_L \le \lfloor(\tau + i - j)/2\rfloor$.

Theorem 17.1 *For an $[n, k, d = 2t + 1]_q$ linear code $\mathscr{C}$, the probability of undetected error after correcting at most τ errors, where $\tau \le t$, in a q-ary symmetric channel with transition probability $p/(q - 1)$, is given by*

$$P_{ue}^{(\tau)}(\mathscr{C}, p) = \sum_{i=d}^{n} A_i \sum_{j=i-\tau}^{i+\tau} N_q^{\tau}(n, i, j) \left(\frac{p}{q-1}\right)^j (1-p)^{n-j} \tag{17.5}$$

where A_i is the number of codewords of weight i in $\mathscr{C}$ and $N_q^{\tau}(n, i, j)$ is given in Lemma 17.1.

Proof An undetected error occurs if the received vector falls within a sphere of radius τ centered at any codeword $\mathscr{C}$ except the transmitted codeword. Without loss of generality, as the code is linear, the transmission of the all zeros codeword may be assumed. Consider $\boldsymbol{c}_i$ a codeword of weight $i > 0$, all vectors within $S_q^{\tau}(\boldsymbol{c}_i)$ have weights ranging from $i - \tau$ to $i + \tau$ with respect to the transmitted all zeros codeword. For each weight j in the range, there are $N_q^{\tau}(n, i, j)$ such vectors in the sphere.

Following [2], if B_j denotes the number of codewords of weight j in $\mathscr{C}^{\perp}$, the dual code of $\mathscr{C}$, A_j may be written as

$$A_m = \frac{1}{|\mathscr{C}^{\perp}|} \sum_{i=0}^{n} B_i P_q(n, m, i) \tag{17.6}$$

where

$$P_q(n, m, i) = \sum_{j=0}^{m} (-1)^j q^{m-j} \binom{n-m+j}{j} \binom{n-i}{m-j} \quad (17.7)$$

is a Krawtchouk polynomial. Using (17.6) and (17.7), the probability of undetected error after error-correction (17.5) may be rewritten in terms of the weight of the codewords in the dual code.

17.2 Incremental-Redundancy Coding System

17.2.1 Description of the System

The main area of applications is two-way digital communication systems with particular importance to wireless communication systems which feature packet digital communications using a two-way communications medium. In wireless communications, each received packet is subject to multipath effects and noise plus interference causing errors in some of the received symbols. Typically forward error-correction (FEC) is provided using convolutional codes, turbo codes, LDPC codes, or algebraic block codes and at the receiver a forward error-correction decoder is used to correct any transmission errors. Any residual errors are detected using a cyclic redundancy check (CRC) which is included in each transmitted codeword. The CRC is calculated for each codeword that is decoded from the corresponding received symbols and if the CRC is not satisfied, then the codeword is declared to be in error. If such an error is detected, the receiver requests the transmitter by means of a automatic repeat request (ARQ) either to retransmit the codeword or to transmit additional redundant symbols. Since this is a hybrid form of error-correction coupled with error-detection feedback through the ARQ mechanism, it is commonly referred to as a hybrid automatic repeat request (HARQ) system.

The two known forms of HARQ are Chase combining and incremental redundancy (IR). Chase combining is a simplified form of HARQ, wherein the receiver simply requests retransmission of the original codeword and the received symbols corresponding to the codeword are combined together prior to repeated decoding and detection. IR provides for a transmission of additional parity symbols extending the length of the codeword and increasing the minimum Hamming distance, d_{min} between codewords. This results in a lower error rate following decoding of the extended codeword. The average throughput of such a system is higher than a fixed code rate system which always transmits codewords of maximum length and redundancy. In HARQ systems, it is a prerequisite that a reliable means be provided to detect errors in each decoded codeword. A system is described below which is able to provide an improvement to current HARQ systems by providing a more reliable means of error detection using the CRC and also provides for an improvement in

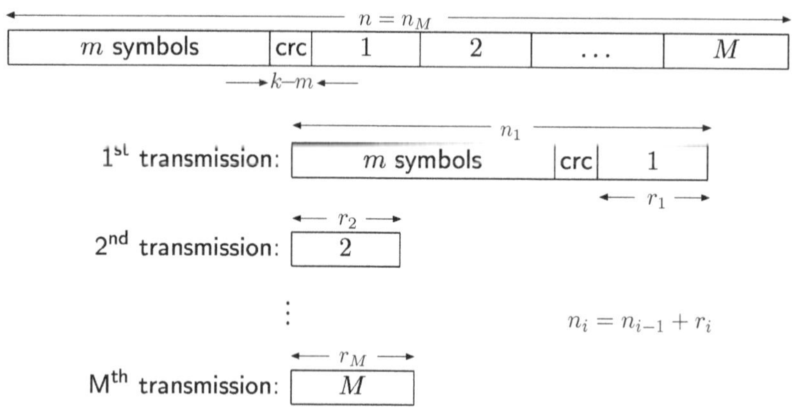

Fig. 17.1 Codeword format for conventional incremental-redundancy ARQ schemes

throughput by basing the error detection on the reliability of the detected codeword without the need to transmit the CRC.

Figure 17.1 shows the generic structure of the transmitted signal for a punctured codeword system. The transmitted signal comprises the initial codeword followed by additional parity symbols which are transmitted following each ARQ request up to a total of M transmissions for each codeword. All of the different types of codes used in HARQ systems: convolutional codes, turbo codes, LDPC codes, and algebraic codes can be constructed to fit into this generic codeword structure. As shown in Fig. 17.1, the maximum length of each codeword is n_M symbols transmitted in a total of M transmissions resulting from the reception of $M - 1$ negative ACK's (NACK's). The first transmission consists of m information symbols encoded into a total of n_1 symbols. There are r_1 parity symbols in addition to the CRC symbols. This is equivalent to puncturing the maximum length codeword in the last $n_M - n_1$ symbols. If this codeword is not decoded correctly, a NACK is received by the transmitter, (indicated either by the absence of an ACK being received or by a NACK signal being received), and r_2 parity symbols are transmitted as shown in Fig. 17.1.

The detection of an incorrect codeword is derived from the CRC in conventional HARQ systems. After the decoding of the received codeword, the CRC is recalculated and compared to the CRC symbols contained in the decoded codeword. If there is no match, then an incorrect codeword is declared and a NACK is conveyed to the transmitter. Following the second transmission, the decoder has a received codeword consisting of $n_1 + r_2$ symbols which are decoded. The CRC is recalculated and compared to the decoded CRC symbols. If there is still no match, a NACK is conveyed to the transmitter and the third transmission consists of the r_3 parity symbols and the net codeword consisting of $n_1 + r_2 + r_3$ symbols is decoded, and so on. The IR procedure ends either when an ACK is received by the transmitter or when a codeword of total length n_M symbols has been transmitted in a total of M transmissions.

Most conventional HARQ systems first encode the m information symbols plus CRC symbols into a codeword of length n_M symbols, where $\mathscr{C}_M = [n_M, k, d_M]$ denotes this code. The code $\mathscr{C}_M$ is then punctured by removing the last $n_M - n_{M-1}$

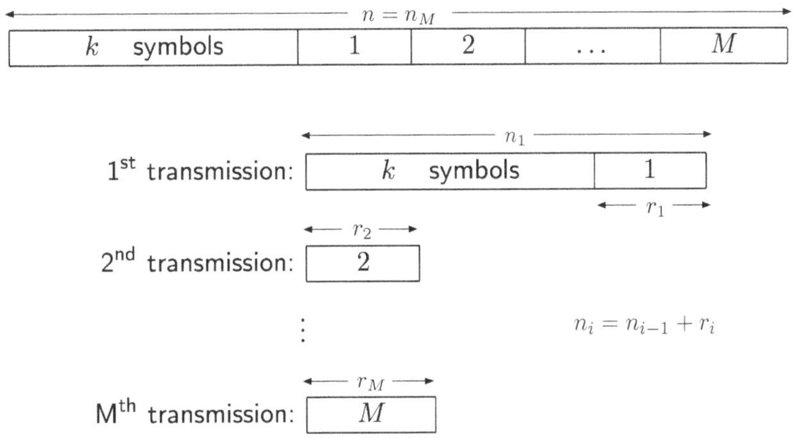

Fig. 17.2 Codeword format for the incremental-redundancy ARQ scheme without a CRC

symbols to produce a code $\mathscr{C}_{M-1} = [n_{M-1}, k, d_{M-1}]$, the code $\mathscr{C}_{M-1}$ is then punctured by removing the last $n_{M-1} - n_{M-2}$ symbols to produce a code $\mathscr{C}_{M-2}$, and so forth until a code $\mathscr{C}_1 = [n_1, k, d_1]$ is obtained. In this way, a sequence of codes $\mathscr{C}_1 = [n_1, k, d_1]$, $\mathscr{C}_2 = [n_2, k, d_2]$, ..., $\mathscr{C}_M = [n_M, k, d_M]$ is obtained. In the first transmission stage, a codeword $\mathscr{C}_1$ is transmitted, in the second transmission stage, the punctured parity symbols of $\mathscr{C}_2$ is transmitted and so on as shown in Fig. 17.1.

An alternative IR code construction method is to produce a sequence of codes using a generator matrix formed from a juxtaposition of the generator matrices of a nested block code. In this way, no puncturing is required.

Figure 17.2 shows the structure of the transmitted signal. The transmitted signal format is the same as Fig. 17.1 except that no CRC symbols are transmitted. The initial codeword consists only of the m information symbols plus the r_1 parity symbols. Additional parity symbols are transmitted following each ARQ request up to a total of M transmissions for each codeword. All of the different types of codes used in HARQ systems: convolutional codes, turbo codes, LDPC codes, and algebraic codes may be used in this format including the sequence of codes based on a nested block code construction.

Figure 17.3 shows a variation of the system where the k information symbols, denoted by vector $\boldsymbol{u}$, are encoded with the forward error-correction (FEC) encoder into n_M symbols denoted as $\boldsymbol{c}_M$ which are stored in the transmission controller. In the first transmission, n_1 symbols are transmitted. At the end of the ith stage, a codeword of total length n_i symbols has been transmitted. This corresponds to a codeword of length n_M symbols punctured in the last $n_M - n_i$ symbols. In Fig. 17.3, the codeword of length n_i is represented as a vector $\boldsymbol{v}$, which is then passed through the channel to produce $\boldsymbol{y}'$ and buffered in the Received buffer as $\boldsymbol{y}$ which is forward error-correction (FEC) decoded in the FEC decoder which produces the most likely codeword $\boldsymbol{c}_1$ and the next most likely codeword $\boldsymbol{c}_2$.

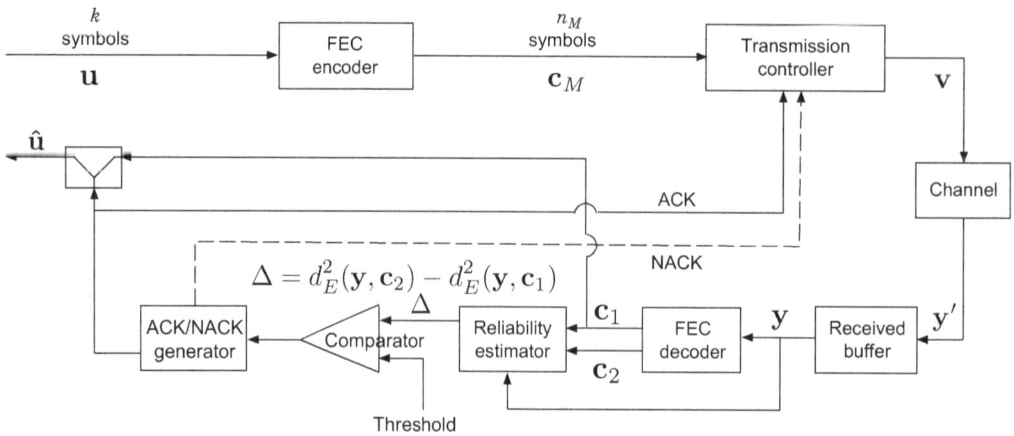

Fig. 17.3 The incremental-redundancy ARQ scheme with adjustable reliability without using a CRC

Let us consider that the IR system has had i transmissions so that a total of n_i symbols have been received and the total length of the transmitted codeword is n_i symbols.

$\boldsymbol{c}_1$ is denoted as

$$\boldsymbol{c}_1 = c_{10} + c_{11}x + c_{12}x^2 + c_{13}x^3 + c_{14}x^4 + \cdots + c_{1(n_i-1)}x^{n_i-1} \tag{17.8}$$

and $\boldsymbol{c}_2$ is denoted as

$$\boldsymbol{c}_2 = c_{20} + c_{21}x + c_{22}x^2 + c_{23}x^3 + c_{24}x^4 + \cdots + c_{2(n_i-1)}x^{n_i-1} \tag{17.9}$$

and the received symbols $\boldsymbol{y}$ are denoted as

$$\boldsymbol{y} = y_0 + y_1x + y_2x^2 + y_3x^3 + y_4x^4 + \cdots + y_{(n_i-1)2}x^{n_i-1} \tag{17.10}$$

For each decoded codeword, $\boldsymbol{c}_1$ and $\boldsymbol{c}_2$, the squared Euclidean distances $d_E^2(\boldsymbol{y}, \boldsymbol{c}_1)$ and $d_E^2(\boldsymbol{y}, \boldsymbol{c}_2)$ respectively are calculated between the codewords and the received symbols $\boldsymbol{y}$ stored in the Received buffer.

$d_E^2(\boldsymbol{y}, \boldsymbol{c}_1)$ is given by

$$d_E^2(\boldsymbol{y}, \boldsymbol{c}_1) = \sum_{j=0}^{n_i-1}(y_j - c_{1j})^2 \tag{17.11}$$

$d_E^2(\boldsymbol{y}, \boldsymbol{c}_2)$ is given by

$$d_E^2(\boldsymbol{y}, \boldsymbol{c}_2) = \sum_{j=0}^{n_i-1}(y_j - c_{2j})^2 \tag{17.12}$$

The function of the Reliability estimator shown in Fig. 17.3 is to determine how much smaller is $d_E^2(\boldsymbol{y}, \boldsymbol{c}_1)$ compared to $d_E^2(\boldsymbol{y}, \boldsymbol{c}_2)$ in order to estimate the likelihood that the codeword $\boldsymbol{c}_1$ is correct. The Reliability estimator calculates the squared Euclidean distances $d_E^2(\boldsymbol{y}, \boldsymbol{c}_1)$ and $d_E^2(\boldsymbol{y}, \boldsymbol{c}_2)$, and determines the difference Δ given by

$$\Delta = d_E^2(\boldsymbol{y}, \boldsymbol{c}_2) - d_E^2(\boldsymbol{y}, \boldsymbol{c}_1) \tag{17.13}$$

Δ is compared to a threshold which is calculated from the minimum Hamming distance of the first code in the sequence of codes, the absolute noise power, and a multiplicative constant, termed κ. As shown in Fig. 17.3, Δ is compared to the threshold by the Comparator. If Δ is not greater than the threshold, $\boldsymbol{c}_1$ is considered to be insufficiently reliable, and the output of the comparator causes the ACK/NACK generator to convey a NACK to the transmitter for more parity symbols to be transmitted. If Δ is greater than or equal to the threshold then $\boldsymbol{c}_1$ is considered to be correct, the output of the comparator causes the ACK/NACK generator to convey an ACK to the transmitter and in turn, the ACK/NACK generator causes the switch to close and $\boldsymbol{c}_1$ is switched to the output $\hat{\boldsymbol{u}}$. The ACK causes the entire IR procedure to begin again with a new vector $\boldsymbol{u}$. The way that Δ works as an indication of whether the codeword $\boldsymbol{c}_1$ is correct or not. If $\boldsymbol{c}_1$ is correct, then $d_E^2(\boldsymbol{y}, \boldsymbol{c}_1)$ is a summation of squared noise samples only because the signal terms cancel out. The codeword $\boldsymbol{c}_2$ differs from $\boldsymbol{c}_1$ in a number of symbol positions equal to at least the minimum Hamming distance of the current code, d_{min}. With the minimum squared Euclidean distance between symbols defined as d_S^2, Δ will be greater or equal to $d_{min} \times d_S^2$ plus a noise term dependent on the signal to noise ratio. If $\boldsymbol{c}_1$ is not correct $d_E^2(\boldsymbol{y}, \boldsymbol{c}_1)$ and $d_E^2(\boldsymbol{y}, \boldsymbol{c}_2)$ will be similar and Δ will be small.

If more parity symbols are transmitted because Δ is less than the threshold, the d_{min} of the code increases with each increase of codeword length and provided $\boldsymbol{c}_1$ is correct, Δ will increase accordingly.

The Reliability measure shown in Fig. 17.3 uses the squared Euclidean distance but it is apparent that equivalent soft decision metrics including cross-correlation and log likelihood may be used to the same effect.

In the system shown in Fig. 17.4 a CRC is transmitted in the first transmitted codeword. The m information symbols, shown as vector $\boldsymbol{u}$ in Fig. 17.4 are encoded with the CRC encoder to form a total of k symbols, shown as vector $\boldsymbol{x}$. The k symbols are encoded by the FEC encoder into n_M symbols denoted as $\boldsymbol{c}_M$ which are stored in the transmission controller. In the first transmission, n_1 symbols are transmitted. At the end of the ith stage, a codeword of total length n_i symbols has been transmitted. This corresponds to a codeword of length n_M symbols punctured in the last $n_M - n_i$ symbols. In Fig. 17.4, the codeword of length n_i is represented as a vector $\boldsymbol{v}$, which is then passed through the channel to produce $\boldsymbol{y}'$ and buffered in the Received buffer as $\boldsymbol{y}$, which is forward error-correction (FEC) decoded in the FEC decoder. The FEC decoder produces L codewords with decreasing reliability as measured by the squared Euclidean distance between each codeword and the received symbols or as measured by an equivalent soft decision metric such as cross-

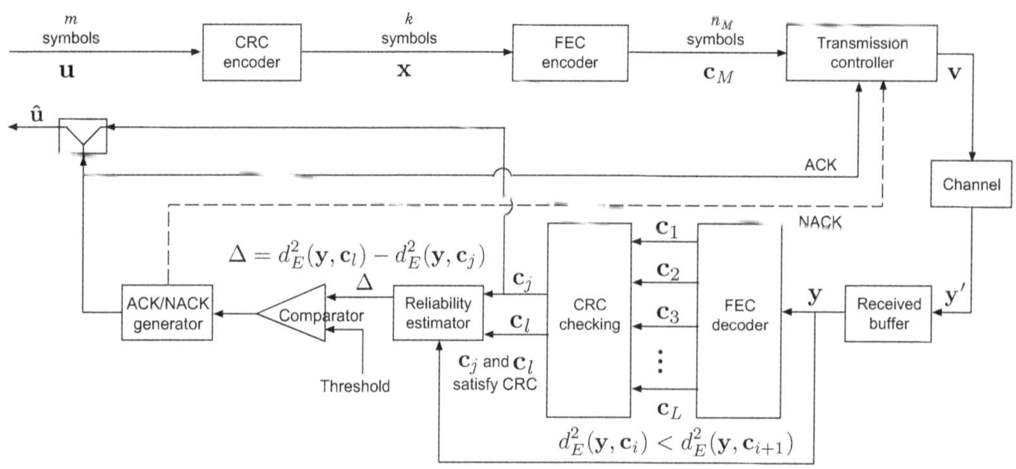

Fig. 17.4 The incremental-redundancy ARQ scheme with adjustable reliability using a CRC

correlation between each codeword and the received symbols. The L codewords are input to CRC checking which determines the most reliable codeword, $\boldsymbol{c}_j$, which satisfies the CRC and the next most reliable codeword, $\boldsymbol{c}_l$, which satisfies the CRC. The Reliability estimator shown in Fig. 17.4 determines the difference, Δ, of the squared Euclidean distances between codewords $\boldsymbol{c}_j$ and $\boldsymbol{c}_l$ and the corresponding received symbols.

Δ is given by

$$\Delta = d_E^2(\boldsymbol{y}, \boldsymbol{c}_l) - d_E^2(\boldsymbol{y}, \boldsymbol{c}_j) \tag{17.14}$$

Δ is compared to a threshold which is calculated from the minimum Hamming distance of the first code in the sequence of codes, the absolute noise power, and a multiplicative constant termed κ. As shown in Fig. 17.4, Δ is compared to the threshold by the comparator. If Δ is not greater than the threshold, $\boldsymbol{c}_j$ is considered to be insufficiently reliable, and the output of the comparator causes the ACK/NACK generator to convey a NACK to the transmitter for more parity symbols to be transmitted. If Δ is greater than or equal to the threshold then $\boldsymbol{c}_j$ is considered to be correct, the output of the comparator causes the ACK/NACK generator to convey an ACK to the transmitter and in turn, the ACK/NACK generator causes the switch to close and $\boldsymbol{c}_j$ is switched to the output $\hat{\boldsymbol{u}}$. The ACK causes the entire IR procedure to begin again with a new vector $\boldsymbol{u}$.

The Reliability measure shown in Fig. 17.4 uses the squared Euclidean distance but it is apparent that equivalent soft decision metrics including cross correlation and log likelihood ratios may be used to the same effect.

17.2.1.1 Code Generation Using Nested Block Codes

If $\mathscr{C}$ is a cyclic code, then there exists a generator polynomial $g(x) \in \mathbb{F}_2[x]$ and a parity-check polynomial $h(x) \in \mathbb{F}_2[x]$ such that $g(x)h(x) = x^{n_1} - 1$. Two cyclic codes, $\mathscr{C}_1$ with $g_1(x)$ as the generator polynomial and $\mathscr{C}_2$ with $g_2(x)$ as the generator polynomial, are said to be chained or nested, if $g_1(x)|g_2(x)$, and we denote them by $\mathscr{C}_1 \supset \mathscr{C}_2$. With reference to this definition, it is clear that narrow-sense BCH codes of the same length form a chain of cyclic codes. Given a chain of two codes, using a code construction method known as Construction X, a construction method first described by Sloane et al. [5], the code with larger dimension can be lengthened to produce a code with increased length and minimum distance.

A generalised form of Construction X involves more than two codes. Let $\mathscr{B}_i$ be an $[n_1, k_i, d_i]$ code, given a chain of M codes, $\mathscr{B}_1 \supset \mathscr{B}_2 \supset \cdots \supset \mathscr{B}_M$, and a set of auxiliary codes $\mathscr{A}_i = [n'_i, k'_i, d'_i]$, for $1 \leq i \leq M-1$, where $k'_i = k_1 - k_i$, a code $\mathscr{C}_X = [n_1 + \sum_{i=1}^{M-1} n'_i, k_1, d]$ can be constructed, where $d = \min\{d_M, d_{M-1} + d'_{M-1}, d_{M-2} + d'_{M-2} + d'_{M-1}, \ldots, d_1 + \sum_{i=1}^{M-1} d'_i\}$.

Denoting z as a vector of length n_1 formed by the first n_1 coordinates of a codeword of $\mathscr{C}_X$. A codeword of $\mathscr{C}_X$ is a juxtaposition of codewords of $\mathscr{B}_i$ and $\mathscr{A}_i$, where

$$\begin{array}{lll}
(\; \boldsymbol{b}_M \mid \boldsymbol{0} \mid \boldsymbol{0} \mid \ldots \mid \boldsymbol{0} \mid \boldsymbol{0} \;) & \text{if } z \in \mathscr{B}_M, \\
(\; \boldsymbol{b}_{M-1} \mid \boldsymbol{0} \mid \boldsymbol{0} \mid \ldots \mid \boldsymbol{0} \mid \boldsymbol{a}_{M-1} \;) & \text{if } z \in \mathscr{B}_{M-1}, \\
(\; \boldsymbol{b}_{M-2} \mid \boldsymbol{0} \mid \boldsymbol{0} \mid \ldots \mid \boldsymbol{a}_{M-2} \mid \boldsymbol{a}_{M-1} \;) & \text{if } z \in \mathscr{B}_{M-2}, \\
\qquad \vdots & \quad \vdots \\
(\; \boldsymbol{b}_2 \mid \boldsymbol{0} \mid \boldsymbol{a}_2 \mid \ldots \mid \boldsymbol{a}_{M-2} \mid \boldsymbol{a}_{M-1} \;) & \text{if } z \in \mathscr{B}_2, \\
(\; \boldsymbol{b}_1 \mid \boldsymbol{a}_1 \mid \boldsymbol{a}_2 \mid \ldots \mid \boldsymbol{a}_{M-2} \mid \boldsymbol{a}_{M-1} \;) & \text{if } z \in \mathscr{B}_1,
\end{array}$$

where $\boldsymbol{b}_i \in \mathscr{B}_i$ and $\boldsymbol{a}_i \in \mathscr{A}_i$.

17.2.1.2 Example of Code Generation Using Nested Block Codes

There exists a chain of extended BCH codes of length 128 bits,

$$\mathscr{B}_1 = [128, 113, 6] \supset \mathscr{B}_2 = [128, 92, 12] \supset \mathscr{B}_3 = [128, 78, 16] \supset \mathscr{B}_4 = [128, 71, 20].$$

Applying Construction X to $[128, 113, 6] \supset [128, 92, 12]$ with an $[32, 21, 6]$ extended BCH code as auxiliary code, a $[160, 113, 12]$ code is obtained, giving

$$[160, 113, 12] \supset [160, 92, 12] \supset [160, 78, 16] \supset [160, 71, 20].$$

Additionally, using a $[42, 35, 4]$ shortened extended Hamming code as the auxiliary code in applying Construction X to $[160, 113, 12] \supset [160, 78, 16]$, giving

$$[202, 113, 16] \supset [202, 92, 16] \supset [202, 78, 16] \supset [202, 71, 20].$$

Finally, applying Construction X to $[202, 113, 16] \supset [202, 71, 20]$ with the shortened extended Hamming code $[49, 42, 4]$ as the auxiliary code, giving

$$[251, 113, 20] \supset [251, 92, 20] \supset [251, 78, 20] \supset [251, 71, 20].$$

The resulting sequence of codes which are used in this example are [128, 113, 6], [160, 113, 12], [202, 113, 16] and [251, 113, 20].

The generator matrix of the last code, the [251, 113, 20] code is given by

$\mathbf{G} =$

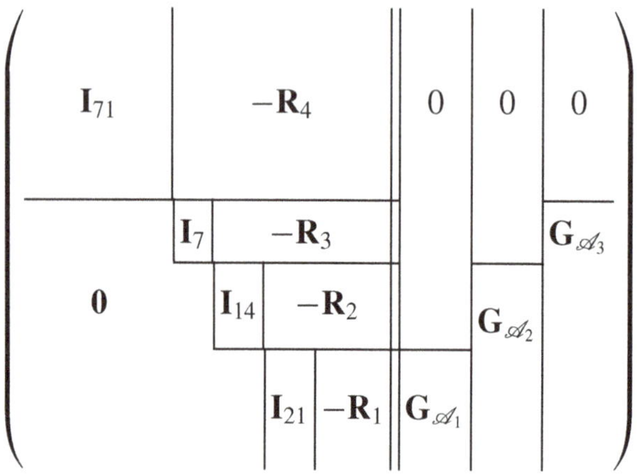

. (17.15)

On the left hand side of the double bar, the generator matrix of the code $\mathscr{B}_1$ is decomposed along the chain $\mathscr{B}_1 \supset \mathscr{B}_2 \supset \mathscr{B}_3 \supset \mathscr{B}_4$. The matrices $\mathbf{G}_{\mathscr{A}_i}$, for $1 \leq i \leq 3$ are the generator matrices of the auxiliary codes $\mathscr{A}_i$.

This generator matrix is used to generate each entire codeword of length $n_M = 251$ bits, but these bits are not transmitted unless requested. The first 128 bits of each entire codeword are selected to form the codeword of the code [128, 113, 6] and are transmitted first, bit 0 through to bit 127. The next transmission (if requested by the IR system) consists of 32 parity bits. These are bit 128 through to bit 159 of the entire codeword. These 32 parity bits plus the original 128 bits form a codeword of the [160, 113, 12] code. The next transmission (if requested by the IR system) consists of 42 parity bits. These are bit 160 through to bit 201 of the entire codeword. These 42 parity bits plus the previously transmitted 160 bits form a codeword from the [202, 113, 16] code. The last transmission (if requested by the IR system) consists of 49 parity bits. These are the last 49 bits, bit 202 through to bit 250, of the entire codeword. These 49 parity bits plus the previously transmitted 202 bits form a codeword from the [251, 113, 20] code. The sequence of increasing length codewords with each transmission (if requested by the IR system) has a minimum Hamming distance which starts with 6, increases from 6 to 12, then to 16 and finally, to 20. In turn this will produce an increasing reliability given by Eq. (17.13) or (17.14) depending on the type of system.

A completely different method of generating nested codes is to use the external parity checks, augmentation method first suggested by Goppa in which independent columns are added incrementally to the parity-check matrix. The method is described in detail in Chap. 6 and can be applied to any Goppa or BCH code.

In order to be used in the HARQ systems, a FEC decoder is needed that will decode these nested block codes. One such universal decoder is the modified Dorsch decoder described in Chap. 15 and results using this decoder are presented below.

17.2.1.3 List Decoder for Turbo and LDPC Codes

If LDPC or turbo codes are to be used, the HARQ system needs a decoder that provides several codewords at its output in order that the difference between the squared Euclidean distances (or an equivalent soft decision metric) of the most likely transmitted codeword and the next most likely transmitted codeword may be determined and compared to the threshold. For turbo codes, the conventional decoder is not a list decoder but Narayanan and Stuber [3] show how a list decoder may be provided for turbo codes. Similarly for LDPC codes, Kristensen [1] shows how a list decoder may be provided for LDPC codes.

17.2.1.4 Performance Results Using the Nested Codes

Computer simulations using the nested codes constructed above have been carried out featuring all three HARQ systems. These systems include the traditional HARQ system using hard decision checks of the CRC and the two new systems featuring the soft decision, decoded codeword/received vector check, with or without a CRC. All of the simulations of the three systems have been carried out using a modified Dorsch decoder as described in Chap. 15. The modified Dorsch decoder can be easily configured as a list decoder with hard and soft decision outputs.

For each one of the nested codes, the decoder exhibits almost optimum maximum likelihood performance by virtue of its delta correlation algorithm corresponding to a total of 10^6 codewords, that are closest to the received vector, being evaluated each time there is a new received vector to input. Since the decoder knows which of the nested codes it is decoding, it is possible to optimise the settings of the decoder for each code.

For the CRC cases, an 8 bit CRC polynomial $(1 + x)(1 + x^2 + x^5 + x^6 + x^7)$ was used, the 8 CRC bits being included in each codeword. It should be noted that in calculating the throughput these CRC bits are not counted as information bits. In the CRC cases, there are 105 information bits per transmitted codeword. In the computer simulations, an ACK is transmitted if Δ is greater than threshold or there have been M IR transmissions, otherwise a NACK is transmitted.

The traditional HARQ system using a CRC is compared to the new system not using a CRC in Figs. 17.5 and 17.6. The comparative frame error rate (FER) performance is shown in Fig. 17.5 and the throughput is shown in Fig. 17.6 as a function

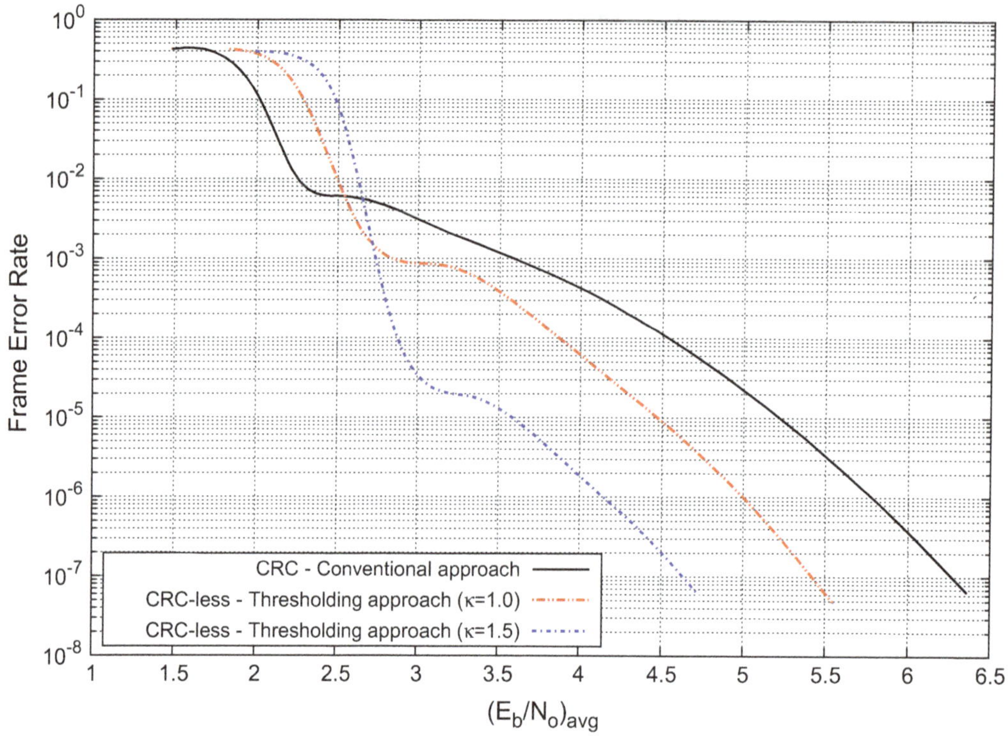

Fig. 17.5 The error rate performance in comparison to the classical HARQ scheme using a CRC

of the average $\frac{E_b}{N_o}$ ratio. The traditional CRC approach shows good throughput, but exhibits an early error-floor of the FER, which is caused by undetected error events. The FER performance shows the benefit of having increased reliability of error detection compared to the traditional CRC approach. Two threshold settings are provided using the multiplicative constant κ and the effects of these are shown in Figs. 17.5 and 17.6. It is apparent from the graphs that the threshold setting may be used to trade-off throughput against reduced FER. The improvements in both throughput and FER provided by the new HARQ systems compared to the conventional HARQ system, featuring a hard decision CRC check, are evident from Figs. 17.5 and 17.6.

The comparative FER performance and throughput with a CRC compared to not using a CRC is shown in Figs. 17.7 and 17.8 for the new system where the threshold is fixed by $\kappa = 1$. The new system using a CRC shows an improvement in FER, Fig. 17.7, over the entire range of average $\frac{E_b}{N_o}$ and an improvement in throughput, Fig. 17.8, also over the entire range of average $\frac{E_b}{N_o}$ compared to the traditional HARQ approach using a CRC.

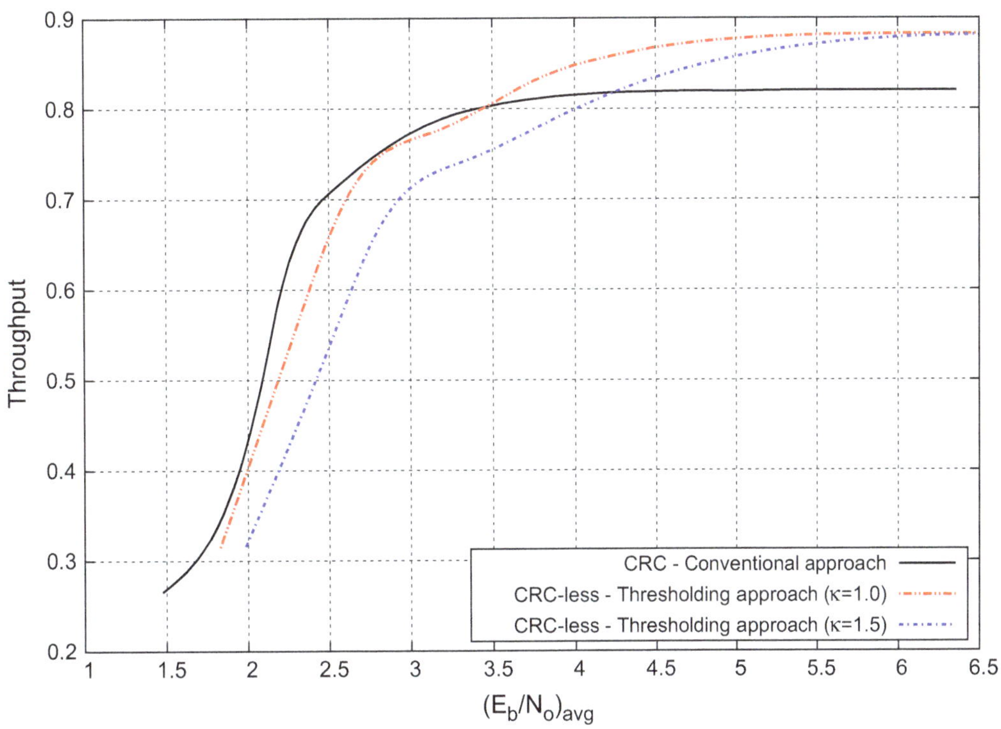

Fig. 17.6 The throughput performance without using a CRC in comparison to the classical HARQ scheme using a CRC

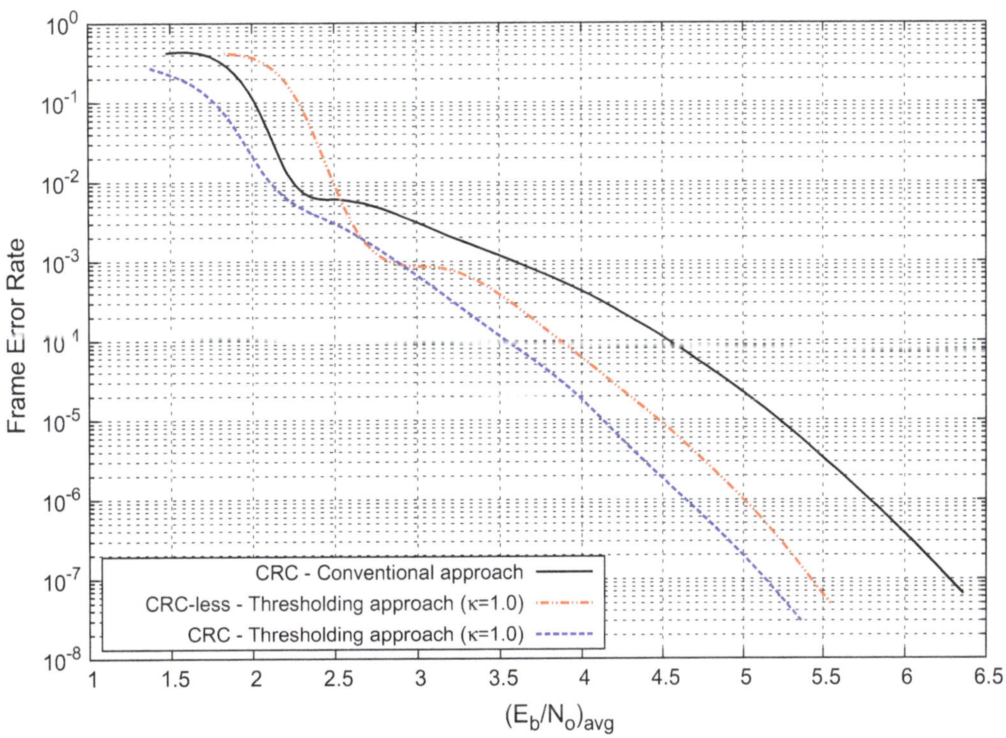

Fig. 17.7 The error rate performance using a CRC in comparison to the classical HARQ scheme using a CRC

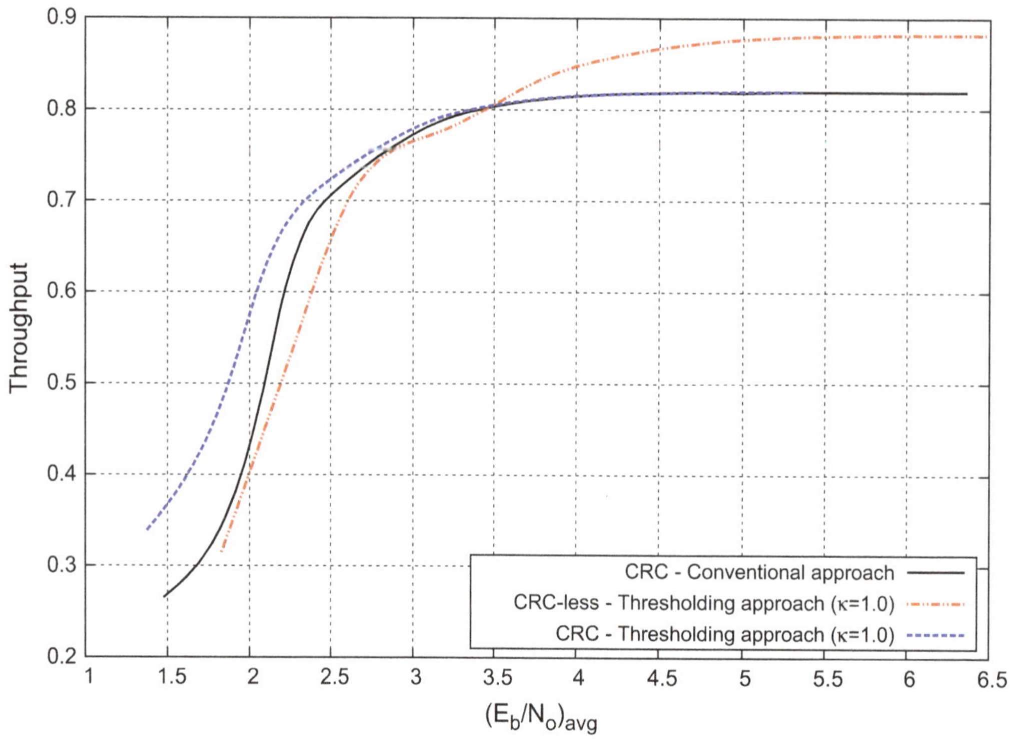

Fig. 17.8 The throughput performance with a CRC in comparison to the classical HARQ scheme using a CRC

17.3 Summary

This chapter has discussed the design of codes and systems for combined error detection and correction, primarily aimed at applications featuring retransmission of data packets which have not been decoded correctly. Several such Hybrid Automatic ReQuest, HARQ, systems have been described including a novel system variation which uses a retransmission metric based on a soft decision; the Euclidean distance between the decoded codeword and the received vector. It has been shown that a cyclic redundancy check, CRC, is not essential for this system and need not be transmitted.

It has also been shown how to construct the generator matrix of a nested set of block codes of length 251 bits by applying Construction X three times in succession starting with an extended BCH (128, 113, 6) code. The resulting nested codes have been used as the basis for an incremental-redundancy system whereby the first 128 bits transmitted is a codeword from the BCH code, followed by the transmission of a further 32 bits, if requested, producing a codeword from a (160, 113, 12) code. Further requests for additional transmitted bits finally result in a codeword from a (251, 113, 20) code, each time increasing the chance of correct decoding by increasing the minimum Hamming distance of the net received codeword. Performance graphs have been presented showing the comparative error rate performances and throughputs of the new HARQ systems compared to the standard HARQ system. The advantages of lower error floors and increased throughputs are evident from the presented graphs.

References

1. Kristensen J.T.: List Decoding of LDPC Codes, Masters thesis 2007-02, Technical University of Denmark
2. MacWilliams, F.J., Sloane, N.J.A.: The Theory of Error-Correcting Codes. North-Holland, Amsterdam (1977)
3. Narayanan, K.R., Stuber, G.L.: List decoding of Turbo codes. IEEE Trans. Commun. **46**(6), 754–762 (1998)
4. Peterson, W.W., Weldon Jr., E.J.: Error-Correcting Codes. MIT Press, Cambridge (1972)
5. Sloane, N.J., Reddy, S.M., Chen, C.L.: New binary codes. IEEE Trans. Inf. Theory **18**(4), 503–510 (1972)

Chapter 18
Password Correction and Confidential Information Access System

18.1 Introduction and Background

Following the trend of an increasing need for security and protection of confidential information, personal access codes and passwords are increasing in length with the result that they are becoming more difficult to remember correctly. The system described in this chapter provides a solution to this problem by correcting small errors in an entered password without compromising the security of the access system. Moreover, additional levels of security are provided by the system by associating passwords with codewords of an error-correcting code and using a dynamic, user-specific, mapping of Galois field symbols. This defeats password attacking systems consisting of Rainbow tables because each user transmits what appears to be a random byte stream as a password. A description of this system was first published by the authors as a UK patent application in 2007 [1].

The system is a method for the encoding and decoding of passwords and the encoding and decoding of confidential information which is accessed by use of these passwords. Passwords may be composed of numbers and alphanumeric characters and easily remembered names, phrases or notable words are the most convenient from the point of view of users of the system.

Passwords are associated with the codewords of an error-correcting code and consequently any small number of errors of an entered password may be automatically corrected. Several additional parity symbols may be calculated by the system to extend the password length prior to hashing so as to overwhelm any attacks based on Rainbow tables. Dynamic mapping of code symbols is used to ensure a password when first registered by the user is a codeword of the error-correcting code. In this process sometimes a password, due to symbol contradictions, cannot be a codeword and an alternative word or phrase, which is a codeword, is offered to the user by the system. Alternatively, the user may elect to register a different password.

Feedback can be provided to users by the system of the number of errors corrected for each user, re-entered password. Valid passwords are associated with a subset of

M. Tomlinson et al., *Error-Correction Coding and Decoding*,
Signals and Communication Technology,
DOI 10.1007/978-3-319-51103-0_18

the totality of all the codewords of the error-correcting code and an entered password, may be confirmed to the user, as a valid password, or not.

Confidential information, for example Personal Identification Numbers (PIN)s, bank account numbers, safe combinations, or more general confidential messages, are encrypted at source and stored as a sequence of encrypted messages. Each encrypted message is uniquely associated with a cryptographic hash of a valid codeword. Retrieval of the confidential information is achieved by the user entering a password which is equal to the corresponding valid password or differs from a valid password in a small number of character positions. Any small number of errors are corrected automatically and feedback is provided to the user that a valid password has been decoded. The valid password is mapped to a single codeword from a very large number of codewords that comprise the error-correcting code.

On receiving a valid hash, the cloud sends the stored encrypted message that corresponds to the valid codeword. The encryption key may be derived from the reconstituted password in conjunction of other user entered credentials, such as a fingerprint. The retrieved encrypted message is decrypted and the confidential information displayed to the user.

Security is provided by the system at a number of different levels. Codewords of the error-correcting code are composed of a sequence of symbols with each symbol taken from a set of Galois Field (GF) elements. Any size of Galois Field may be used provided the number of GF elements is greater or equal to the alphabet of the language used to construct passwords. The mapping of alphabet characters to GF elements may be defined uniquely by each user and consequently there are at least $q!$ possible mappings.

The number of possible codewords of the error-correcting code is extremely large and typically there can be 10^{500} possible codewords. The number of valid codewords in the subset of codewords is typically less than 10^2 and so the brute force chance of a randomly selected codeword being a valid codeword is $\frac{1}{10^{500}}$. Even if an attacker, or an eavesdropper enters a valid codeword, the information that is obtained is encrypted and the confidential information cannot be retrieved without the encryption key, which requires the user's credentials.

One possible application of the system is as an information retrieval app on a smartphone with encrypted information stored in the cloud. For each registered user a cloud-based message server has stored a list of cryptographic hashes of valid codewords of the error-correcting code and an associated list of encrypted messages or files. The mapping of password characters to codeword GF symbols is carried out within the user's smartphone and is not able to be easily accessed by an eavesdropper or an attacker unless the smartphone is stolen along with user login credentials. Additionally, the decryption of received encrypted messages is also carried out within the user's mobile phone. To access a long, hard to remember PIN or a long sequence of cryptic characters, the user can enter the password, which is mapped to a GF symbol stream, which is automatically corrected by the smartphone before cryptographic hashing. The hash is encrypted, using a random session key exchanged using public key cryptography, before being transmitted by the smartphone to the cloud. This is to prevent replay attacks. If the codeword hash is correct, the cloud transmits the corresponding encrypted message or file, together with an acknowledgement. The user's smartphone receives the cipher text and decrypts it into the requested information.

18.2 Details of the Password System

A block diagram of the system showing how user defined passwords are mapped to sequences of GF symbols, encoded into codewords of an error-correcting code and associated with encrypted confidential information is shown in Fig. 18.1.

We consider as an example, a system using passwords consisting of sequences of up to 256 characters long with characters taken from the ANSI (American National Standards Institute) single byte character set and an error-correcting code which is a Reed–Solomon (RS) error-correcting code [2] described in Chaps. 7 and 11. RS codes are MDS codes, constructed from GF(q) field elements. For the finite field case, q is a prime or a power of a prime. In this example $q = 2^8$ and RS codewords are constructed as sequences of GF(256) symbols. Codewords can be designed to be any length up to $q + 1$ symbols long if the doubly extended version of the RS code is utilised.

In general, any character set may be used in the system and any RS code may be used provided the sequence of characters is less than or equal to the length of the error-correcting code and each symbol of the error-correcting code is from an alphabet size equal or greater than the alphabet size of the character set used to define passwords. For maximum security, the mapping is chosen by a cryptographic random number generator, with a seed provided by the user so that there is a high probability that the resulting mapping table is unique to each user of the information retrieval system.

It is convenient to use a binary base field and the Galois Field [3], GF(256), that is used is an extension field consisting of 8 binary GF(2) field elements, generated by residues of α^n, $n = 0$ to 255 modulo $1 + x^2 + x^3 + x^4 + x^8$, where $1 + x^2 + x^3 + x^4 + x^8$ is an example of a primitive polynomial, plus the zero symbol $GF(0)$.

As an example, the registered password "silver" is considered, whose corresponding sequence of ANSI numbers is

$$115 \quad 105 \quad 108 \quad 118 \quad 101 \quad 114$$

As shown in Fig. 18.1, a mapping table is used to map these numbers to GF(256) symbols. In this example, the error-correcting code that is used is the (256, 254, 3) extended RS code which is capable of correcting either two erased symbols or one erroneous symbol, and the code has 254 information symbols and 2 parity-check symbols. The first two symbols are chosen as parity-check symbols and denoted as p_1 and p_2, respectively. Putting the parity symbols first is convenient because short codewords can easily by accommodated by assuming any unused information symbols have value zero and therefore do not affect the parity symbols. A general

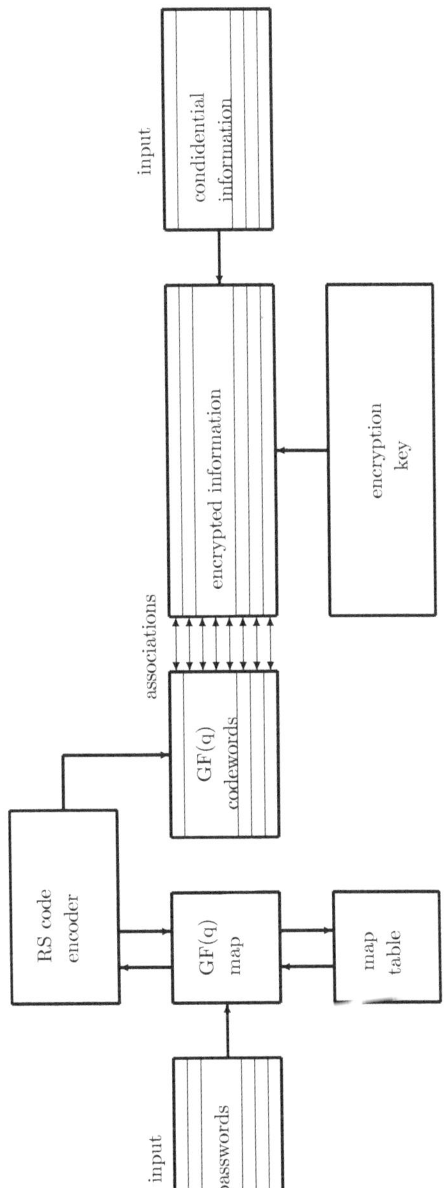

Fig. 18.1 Encoding of input (defined) passwords and association with encrypted input confidential information

codeword of this code as an extended GF(256) RS code is

$$p_1 \quad p_2 \quad x_1 \quad x_2 \quad x_3 \quad x_4 \quad \ldots \quad x_{254}$$

The general parity-check matrix of an extended RS code with $n-k$ parity-check symbols is

$$\mathbf{H} = \begin{bmatrix} 1 & 1 & 1 & \ldots & 1 & 1 \\ 1 & \alpha^1 & \alpha^2 & \ldots & \alpha^{n-1} & 0 \\ 1 & \alpha^2 & \alpha^4 & \ldots & \alpha^{2(n-1)} & 0 \\ 1 & \alpha^3 & \alpha^6 & \ldots & \alpha^{3(n-1)} & 0 \\ \ldots & \ldots & \ldots & \ldots & \ldots & \\ 1 & \alpha^{n-k-1} & \alpha^{2(n-k-1)} & \ldots & \alpha^{n-k-1(n-1)} & 0 \end{bmatrix}$$

To provide more flexibility in symbol mapping, as described below, the generalised extended RS code may be used with parity-check matrix $\mathbf{H}_\eta$.

$$\mathbf{H}_\eta = \begin{bmatrix} \eta_0 & \eta_1 & \eta_2 & \ldots \eta_{n-1} & \eta_n \\ \eta_0 & \eta_1\alpha^1 & \eta_2\alpha^2 & \ldots \eta_{n-1}\alpha^{n-1} & 0 \\ \eta_0 & \eta_1\alpha^2 & \eta_2\alpha^4 & \ldots \eta_{n-1}\alpha^{2(n-1)} & 0 \\ \eta_0 & \eta_1\alpha^3 & \eta_2\alpha^6 & \ldots \eta_{n-1}\alpha^{3(n-1)} & 0 \\ \ldots & \ldots & \ldots & \ldots \quad \ldots & \ldots \\ \eta_0 & \eta_1\alpha^{n-k-1} & \eta_2\alpha^{2(n-k-1)} & \ldots \eta_{n-1}\alpha^{n-k-1(n-1)} & 0 \end{bmatrix}$$

The constants $\eta_1, \eta_2, \eta_3, \ldots \eta_n$ may be arbitrarily chosen provided they are non-zero symbols of $GF(q)$.

With two parity-check symbols, only the first two rows of $\mathbf{H}$ are needed and we may conveniently place the last column first to obtain the reduced echelon parity-check matrix $\mathbf{H_2}$

$$\mathbf{H_2} = \begin{bmatrix} 1 & 1 & 1 & 1 & \ldots & 1 \\ 0 & 1 & \alpha^1 & \alpha^2 & \ldots & \alpha^{n-1} \end{bmatrix}$$

Any pseudo random, one to one, mapping of ANSI numbers to GF(256) symbols may be used. It is convenient to map always the null character, ANSI number = 32, to the field element $GF(0)$ otherwise each password would consist of 256 characters and 256 password characters would have to be entered for each password. With the null character mapping, a shortened RS codeword is equal to the full length codeword since any of the $GF(0)$ symbols may be deleted without affecting the parity-check symbols. Consequently, short passwords may be accommodated very easily.

It is possible to choose a fixed one to one mapping of ANSI numbers to GF(256) symbols and make this equal for all users but in this case many passwords on first registering would fail as non-valid passwords, unless arbitrarily assigned characters are allowed in the parity symbol positions. However, this is an unnecessary constraint on the system since codewords and passwords of different users are processed independently from each other. Moreover, security is enhanced if each user uses a different mapping.

In the following example, dynamic mapping is used and the mapping chosen is such that the information symbols of the RS codeword corresponding to "silver" are equal to a primitive root α raised to the power corresponding to the ANSI number of the character of the password in the same respective position as the codeword, except for the null character which is set to $GF(0)$. As the codeword has parity symbols in the first two positions, and these symbols are a function of the other symbols in the codeword, (the information symbols), the mapping of the first two characters needs to be different. Accordingly, the codeword is

$$p_1 \quad p_2 \quad \alpha^{108} \quad \alpha^{118} \quad \alpha^{101} \quad \alpha^{114} \quad 0 \quad \ldots \quad 0$$

From the parity-check matrix $\mathbf{H_2}$, the parity-check symbols are given by

$$p_2 = \sum_{i=1}^{254} \alpha^i x_i \tag{18.1}$$

$$p_1 = \sum_{i=1}^{254} x_i + p_2 \tag{18.2}$$

After substituting into Eq. (18.2) and then Eq. (18.1), it is found that $p_1 = \alpha^{220}$ and $p_2 = \alpha^{57}$ and the complete RS codeword corresponding to the defined password "silver" is

$$\alpha^{220} \quad \alpha^{57} \quad \alpha^{108} \quad \alpha^{118} \quad \alpha^{101} \quad \alpha^{114} \quad 0 \quad \ldots \quad 0$$

The RS codeword encoder is shown in Fig. 18.1 and uses the mapping of defined password characters to GF(256) symbols as input and outputs to the mapping table, as shown in Fig. 18.1, the mapping of the parity symbols. Accordingly, the mapping of the first two characters of the password is that ANSI number 115 is mapped to α^{220} and ANSI number 220 is mapped to α^{115} and the mapping table is updated accordingly, as indicated in Fig. 18.1 by the two directional vectors. Of course in order for these mappings to be valid, neither ANSI number 115, nor ANSI number 220, nor GF(256) symbols α^{220}, nor α^{115} must be already mapped otherwise the mapping of prior defined passwords will be affected.

As new passwords are defined and new valid codewords calculated, it is relatively straightforward to amend the list of assigned and unassigned mappings of the mapping table in a dynamic manner. This dynamic mapping assignment is a feature in that it not only increases the range of possible passwords but has a secondary advantage. This secondary advantage arises from the mapping of entered passwords and the subsequent error-correction decoding. Any entered password character not having an assigned ANSI number cannot be part of a valid password and accordingly the corresponding GF(256) symbol is marked as an erased symbol. Since on average, twice as many erased characters can be corrected by an error-correcting code compared to the number of correctable erroneous characters, a distinct advantage arises.

The confidential information corresponding to "silver" is, for example: "The safe combination is 29 53 77 22" and as shown in Fig. 18.1 confidential information input to the system is encrypted using an encryption key. The encryption key is usually chosen from a master encryption key, unique to each user. Once input, the confidential information is only retained in encrypted form. The encrypted confidential information associated with "silver" forms the encrypted text:

```
AjelMHjq+iw&nd^fh)y!"16f@h:G#)P7=3Mq|2=0+YX?z/+6sGs+2|Z1
-GWp<)g/,HDZ)H4D7F/j+gFAqY1FcXZPMY6$3"/
```

As shown in Fig. 18.2, in order to retrieve this confidential information, the user re-enters their password. However, this entered password is allowed to contain errors. For example the password "solver" may be entered and has the corresponding ANSI number sequence:

$$115 \quad 111 \quad 108 \quad 118 \quad 101 \quad 114 \quad \ldots \quad 32 \quad 32 \quad 32$$

Following input of the password, as shown in Fig. 18.2, the mapping table is used to map the entered password into the GF(256) sequence

$$\alpha^{220} \quad \alpha^{88} \quad \alpha^{108} \quad \alpha^{118} \quad \alpha^{101} \quad \alpha^{114} \quad 0 \ldots \quad 0$$

The decoder for the RS code, as shown in Fig. 18.2, decodes the sequence of GF(256) symbols, resulting from the mapping using the mapping table, into a codeword of the error-correcting code. In order to carry this out, two syndromes, s_1 and s_2, are calculated from the two parity-check equations for the extended (256, 254, 3) RS code:

$$s_1 = \sum_{i=1}^{254} x_i + p_2 + p_1 \tag{18.3}$$

$$s_2 = \sum_{i=1}^{254} \alpha^i x_i + p_2 \tag{18.4}$$

The two syndromes, in this case, are found both to be equal to α^{43} indicating there is an error in the second information symbol position of the codeword and this error is

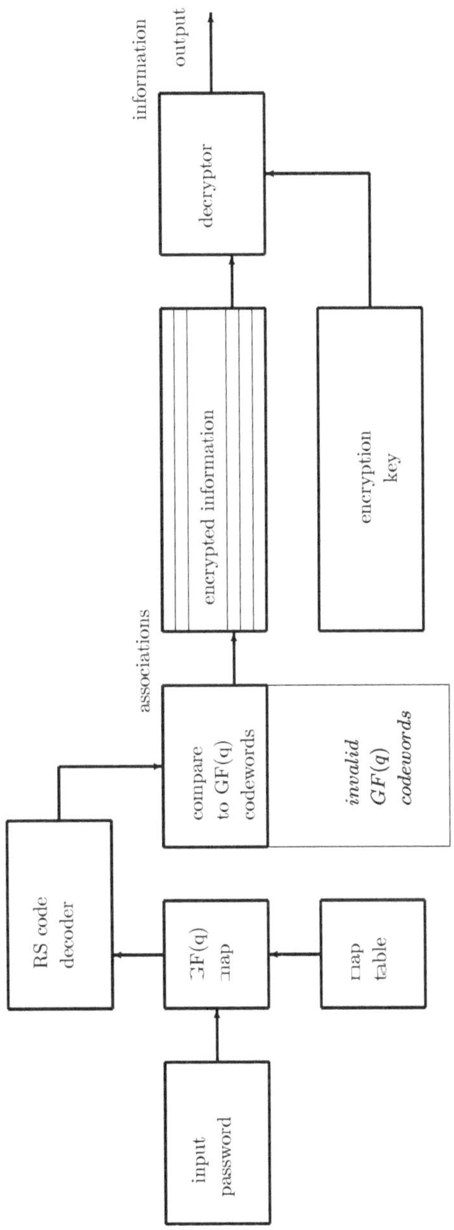

Fig. 18.2 Decoding of entered password, codeword validation and decryption of confidential information (valid password case)

α^{43}. If the same error had been in the third symbol position, say, the two syndromes would have been equal to α^{43} and α^{44}.

Subtracting the error, α^{43}, from the entered symbol (after mapping) of α^{88} produces the correct GF(256) symbol α^{57}. The codeword of the RS code is thus corrected to be

$$\alpha^{220} \quad \alpha^{88} \quad \alpha^{108} \quad \alpha^{118} \quad \alpha^{101} \quad \alpha^{114} \quad 0\ldots \quad 0$$

Applying the inverse mapping to each GF(256) symbol produces the ANSI number sequence 115 105 108 118 101 114 32 32 32 32…32

$$115 \quad 105 \quad 108 \quad 118 \quad 101 \quad 114 \quad 32 \quad 32 \quad 32 \quad 32\ldots \quad 32$$

This corresponds to "silver", the corrected entered password.

As shown in Fig. 18.2, the decoded codeword is compared to the list of valid codewords of the error-correcting code. The codewords of the error-correcting code are split into two groups, the valid codewords and rest of the codewords, invalid codewords. The codeword

$$\alpha^{220} \quad \alpha^{88} \quad \alpha^{108} \quad \alpha^{118} \quad \alpha^{101} \quad \alpha^{114} \quad 0\ldots \quad 0$$

is verified as a valid codeword associated with the encrypted confidential information:

```
AjelMHjq+iw&nd^fh)y!"16f@h:G#)P7=3Mq|2=0+YX?z/+6sGs+2|Z1-GW
p<)g/,HDZ)H4D7F/j+gFAqY1FcXZPMY6$3"/
```

As shown in Fig. 18.2, this is decrypted, using the encryption key and the confidential information is output: "The safe combination is 29 53 77 22"

In a further extension of the system, as shown in Fig. 18.3, the encoded RS codeword, denoted as $\mathbf{c_x}$ which results from the mapped, defined password is convolved with a fixed RS codeword denoted as

$$\mathbf{y_x} = \alpha^{y_0} + \alpha^{y_1}x + \alpha^{y_2}x^2 + \alpha^{y_3}x^3 + \cdots \alpha^{y_{255}}x^{254}$$

Note that the standard polynomial-based RS codes of length $q-1$ are used in this system variation. The fixed RS codeword is the result of encoding a random set of GF(256) information symbols. The reason for doing this is to ensure that the resulting codeword after convolution, $\mathbf{r_x}$

$$\mathbf{r_x} = \mathbf{c_x}.\mathbf{y_x} \text{ modulo } 1 + x^{255} \tag{18.5}$$

does not have a long sequence of $GF(0)$ symbols which may compromise the security of the information retrieval system. Correspondingly, it is the codeword $\mathbf{r_x}$ which is associated with the encrypted message. As shown in Fig. 18.4, retrieval of encrypted

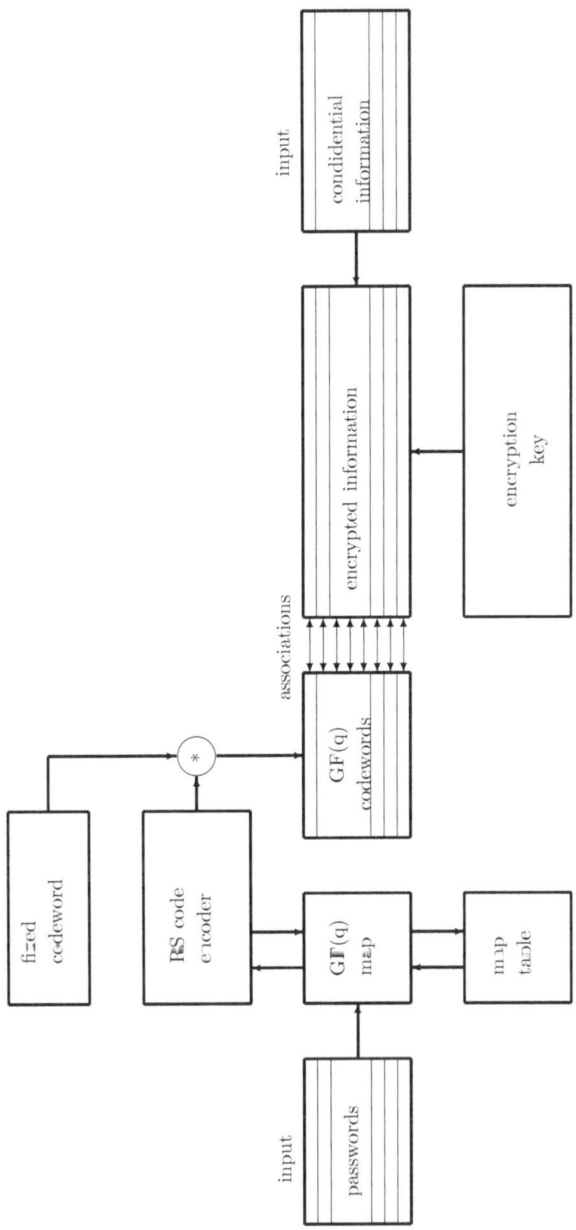

Fig. 18.3 Encoding of input (defined passwords, addition of fixed codeword and association with encrypted confidential information

information is carried out by entering a password. After the decoding of the sequence of GF(256) symbols resulting from the mapping of the entered password, using the mapping table, into a codeword of the error-correcting code, this codeword is convolved with the fixed codeword as shown in Fig. 18.4. The resulting codeword is compared to the list of valid codewords of the error-correcting code.

One feature, particularly with long passwords hard to remember, is that a partially known password may be entered, deliberately using characters known not to be contained in the mapping table, in order for the system to fill in the missing parts of the password. Characters may be reserved for this purpose. As a simple example, the password may be entered "si**er" where it is known that the character * will not be contained in the mapping table, because the character * had been previously defined as a reserved character. The corresponding codeword is

$$\alpha^{220} \quad \alpha^{88} \quad erase_1 \quad erase_2 \quad \alpha^{101} \quad \alpha^{114} \quad 0\ldots \quad 0$$

where $erase_1$ and $erase_2$ represent erased (unknown) GF(256) symbols. The decoder for the RS (256,254,3) error-correcting code may be used to solve straightforwardly for these erased symbols. The first step is to produce a reduced echelon parity-check matrix with zeros in the columns corresponding to the positions of the erased symbols, bar one. The procedure is described in detail in Chap. 11.

For two erasures the procedure is trivial and the reduced echelon parity-check matrix $\mathbf{H_e}$ is

$$\mathbf{H_e} = \begin{bmatrix} 1 & 1 & 1 & 1 & \ldots & 1 \\ \alpha^1 & (1+\alpha^1) & 0 & (\alpha^2+\alpha^1) & \ldots & (\alpha^{n-1}+\alpha^1) \end{bmatrix}$$

Now, the erased symbol $erase_2$ may be solved directly using the second row of $\mathbf{H_e}$

$$\alpha^{220}\alpha^1 + \alpha^{88}(1+\alpha^1) + erase_2(\alpha^2+\alpha^1) + \alpha^{101}(\alpha^3+\alpha^1) + \alpha^{114}(\alpha^3+\alpha^1) = 0$$

and

$$erase_2 = (\alpha^2+\alpha^1)^{q-2}\alpha^{221} + \alpha^{88} + \alpha^{89} + \alpha^{104} + \alpha^{102} + \alpha^{117} + \alpha^{115} = \alpha^{118}$$

Using the first row of $\mathbf{H_e}$, $erase_1$ can now be solved

$$\alpha^{220} + \alpha^{88} + \alpha^{118} + erase_1 + \alpha^{101} + \alpha^{114} = 0$$

and

$$erase_1 = \alpha^{220} + \alpha^{88} + \alpha^{118} + \alpha^{101} + \alpha^{114} = \alpha^{108}$$

With the reverse mapping the complete password is reconstituted allowing the encrypted information to be retrieved as before.

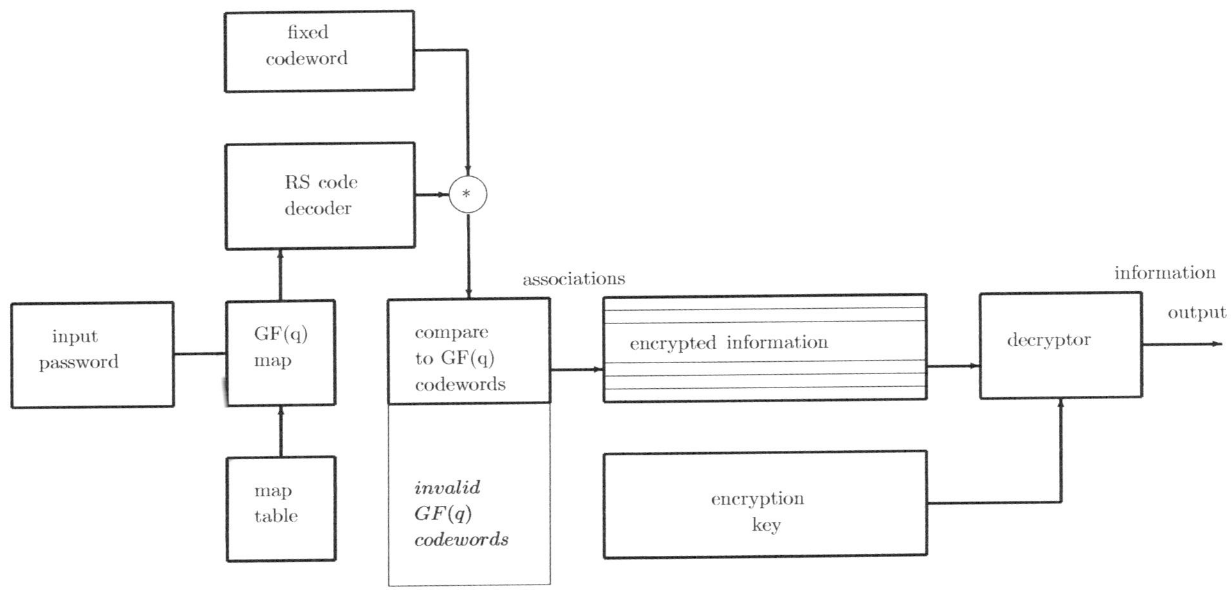

Fig. 18.4 Decoding of entered password, addition of fixed codeword, codeword validation and decryption of confidential information (valid password case)

The advantage of defining erasures is that each unknown symbol may be solved for each parity symbol in the RS codeword. Having a relatively large number of parity symbols allows several parts of the entered password to be filled in automatically. Obviously security is compromised if this procedure is used to extreme.

18.3 Summary

This chapter has described the use of Reed–Solomon codes to correct user mistakes or missing parts of long entered passwords. The system is ideally suited to a smartphone-based encrypted, information retrieval system or a password-based authentication system. Dynamic, user-specific mapping of Galois field elements is used to ensure that passwords, arbitrarily chosen by the user, are valid codewords. A typical system is described based on GF(256) and the ANSI character set with worked examples given. Security is also enhanced by having, user-specific, Galois field symbol mapping because, with long passwords, this defeats Rainbow tables.

References

1. Tomlinson, M., Tjhai, C.J., Ambroze, M., Ahmed M.Z.: Password Correction and Confidential Information Access System, UK Patent Application GB0724723.3 (2007)
2. MacWilliams, F.J., Sloane, N.J.A.: The Theory of Error-Correcting Codes. North-Holland, Amsterdam (1977)
3. Lin, S., Costello, Jr D.J.: Error Control Coding: Fundamentals and Applications, 2nd edn. Pearson Education, Inc., NJ (2004)

Chapter 19
Variations on the McEliece Public Key Cryptoystem

19.1 Introduction and Background

In 1978, the distinguished mathematician Robert McEliece invented a public key encryption system [8] based upon encoding the plaintext as codewords of an error-correcting code from the family of Goppa [6] codes. In this system, the ciphertext, sometimes termed the cryptogram, is formed by adding a randomly chosen error pattern containing up to t bits to each codeword. One or more such corrupted codewords make up the ciphertext. On reception, the associated private key is used to invoke an error-correcting decoder based upon the underlying Goppa code to correct the errored bits in each codeword, prior to retrieval of the plaintext from all of the information bits present in the decoded codewords.

Since the original invention there have been a number of proposed improvements. For example, in US Patent 5054066, Riek and McFarland improved the security of the system by complementing the error patterns so as to increase the number of errors contained in the cryptogram [14] and cited other variations of the original system.

This chapter is concerned with a detailed description of the original system plus some refinements which enhance the bandwidth efficiency and security of the original arrangement. The security strength of the system is discussed and analysed.

19.1.1 Outline of Different Variations of the Encryption System

In the originally proposed system [8] a codeword is generated from plaintext message bits by using a permuted, scrambled generator matrix of a Goppa code [6] of length n symbols, capable of correcting t errors. This matrix is the public key. The digital cryptogram is formed from codewords corrupted by exactly t randomly, or t pseudorandomly, chosen bit errors. The security is provided by the fact that it is

M. Tomlinson et al., *Error-Correction Coding and Decoding*,
Signals and Communication Technology,
DOI 10.1007/978-3-319-51103-0_19

impossible to remove the unknown bit errors unless the original unpermuted Goppa code, the private key, is known in which case the errors can be removed by correcting them and then descrambling the information bits in the codeword to recover the original message. Any attempt to descramble the information bits without removing the errors first just results in a scrambled mess. In the original paper by McEliece [8], the Goppa codeword length n is 1024 and t is 50. The number of possible error combinations is 3.19×10^{85} equivalent to a secret key of length 284 bits given a brute force attack. (There are more sophisticated attacks which reduce the equivalent secret key length and these are discussed later in this chapter.)

In a variation of the original theme, after first partitioning the message into message vectors of length k bits each and encoding these message vectors into codewords, the codewords are corrupted by a combination of bit errors and bit deletions to form the cryptogram. The number of bit errors in each corrupted codeword is not fixed, but is an integer s, which is randomly chosen, with the constraint that, $s \leq t$. This increases the number of possible error combinations, thereby increasing the security of the system. As a consequence $2(t - s)$ bits may be deleted from each codeword in random positions adding to the security of the cryptogram as well as reducing its size, without shortening the message. In the case of the original example, above, with $\frac{t}{2} \leq s \leq t$ the number of possible error combinations is increased to 3.36×10^{85} and the average codeword in the cryptogram is reduced to 999 bits from 1024 bits.

Most encryption systems are deterministic in which there is a one-to-one correspondence between the message and the cryptogram with no random variations. Security can be improved through the use of a truly, random integer generator, not a pseudorandom generator to form the cryptogram. Consequently, the cryptogram is not predictable or deterministic. Even with the same message and public key, the cryptogram produced will be different each time and without knowledge of the random errors and bit deletions, which may be determined only by using the structure of the Goppa code, recovery of the original message is practically impossible.

The basic McEliece encryption system has little resistance to chosen-plaintext (message) attacks. For example, if the same message is encrypted twice and the two cryptograms are added modulo 2, the codeword of the permuted Goppa code cancels out and the result is the sum of the two error patterns. Clearly the encryption method does not provide indistinguishability under chosen-plaintext attack (IND-CPA), a quality measure used by the cryptographic community.

However, an additional technique may be used which does provide (IND-CPA) and results in semantically secure cryptograms. The technique is to scramble the message twice by using a second scrambler. With scrambling the message using the fixed non-singular matrix contained in the public key as well, a different scrambler is used to scramble each message in addition. The scrambling function of this second scrambler is derived from the random error vector which is added to the codeword to produce the corrupted codeword after encoding using the permuted, scrambled generator matrix of a Goppa code. As the constructed digital cryptogram is a function of truly randomly chosen vectors, not pseudorandomly chosen vectors, or a fixed vector, the security of this public key encryption system is enhanced compared to the standard system. Even with an identical message and using exactly the same public key, the resulting cryptograms will have no similarity at all to any previously generated cryptograms. This is not true for the standard McEliece public key system as each codeword will only differ in a maximum of $2t$ bit positions. Providing this semantic security eliminates the risk from known plaintext attacks and is useful in several applications such as in RFID, and these are discussed later in the chapter.

An alternative to using a second scrambler is to use a cryptographic hash function such as SHA-256 [11] or SHA-3 [12] to calculate the hash of each t bit error pattern and add, modulo 2, the first k bits of the hash values to the message prior to encoding. Effectively the message is encrypted with a stream cipher prior to encoding.

Having provided additional message scrambling, it now becomes safe to represent the generator matrix in reduced echelon form, i.e. a $k \times k$ identity matrix followed by a $(n-k) \times k$ matrix for the parity bits. Consequently, the public key may be reduced in size from a $n \times k$ matrix to a $(n-k) \times k$ matrix corresponding typically to a reduction in size of around 65%. This is useful because one of the criticisms of the McEliece system is the relatively large size of the public keys.

Most attacks on the McEliece system are blind attacks and rely on the assumption that there are exactly t errors in each corrupted codeword. If there are more than t errors these attacks fail. Consequently, to enhance the security of the system, additional errors known only to intended recipients may be inserted into the digital cryptogram so that each corrupted codeword contains more than t errors. A sophisticated method of introducing the additional errors is not necessary since provided there are sufficient additional errors to defeat decryption based on guessing the positions of the additional errors the message is theoretically unrecoverable from the corrupted digital cryptogram even with knowledge of the private key. This feature may find applications where a message needs to be distributed to several recipients using the same or different public/private keys at the same time, possibly in a commercial, competitive environment. The corrupted digital cryptograms may be sent to each recipient arriving asynchronously, due to variable network delays and only a relatively short secret key containing information of the additional error positions needs to be sent at the same time to all recipients.

In another arrangement designed to enhance the security of the system, additional errors are inserted into each codeword in positions defined by a position vector, which is derived from a cryptographic hash of the previous message vector. Standard hash functions may be used such as SHA-256 [11] or SHA-3 [12]. The first message vector can use a position vector derived from a hash or message already known by the recipient of the cryptogram.

These arrangements may be used in a wide number of different applications such as active and passive RFID, secure barcodes, secure ticketing, magnetic cards, message services, email applications, digital broadcasting, digital communications, video communications and digital storage. Encryption and decryption is amenable to high speed implementation operating at speeds beyond 1 Gbit/s.

19.2 Details of the Encryption System

The security strength of the McEliece public key encryption system stems from the fact that a truly random binary error pattern is added to the encoded message as part of the digital cryptogram. Even with the same message and the same public key a different digital cryptogram is produced each time. Each message is encoded with a scrambled, binary mapped, permuted, version of a $GF(2^m)$ Goppa code. Without the knowledge of the particular Goppa code that is used, the error pattern cannot be corrected and the message cannot be recovered. It is not possible to deduce which particular Goppa code is being used from the public key, which is the matrix used for encoding, because this matrix is a scrambled, permuted version of the original encoding matrix of the Goppa code, plus the fact that for a given m there are an extremely large number of Goppa codes [8].

The message information to be sent, if not in digital form, is digitally encoded into binary form comprising a sequence of information bits. The method of encryption is shown in Fig. 19.1. The message comprising a sequence of information bits is formatted by appending dummy bits as necessary into an integral number m of binary message vectors of length k bits each. This is carried out by *format into message vectors* shown in Fig. 19.1. Each message vector is scrambled and encoded into a codeword, n bits long, defined by an error-correcting code which is derived from a binary Goppa code and a scrambling matrix. The binary Goppa code is derived from a non-binary Goppa code and the procedure is described below for a specific example.

The *encode using public key* shown in Fig. 19.1 carries out the scrambling and codeword encoding for each message vector by selecting rows of the codeword generator matrix according to the message bits contained in the message vector. This operation is described in more detail below for a specific example. The codeword generator matrix to be used for encoding is defined by the public key which is stored in a buffer memory, *public key* shown in Fig. 19.1. As shown in Fig. 19.1, a random number generator generates a number s internally constrained to be less than or equal to t and this is carried out by *generate number of random errors* (s). The parameter t is the number of bit errors that the Goppa code can correct.

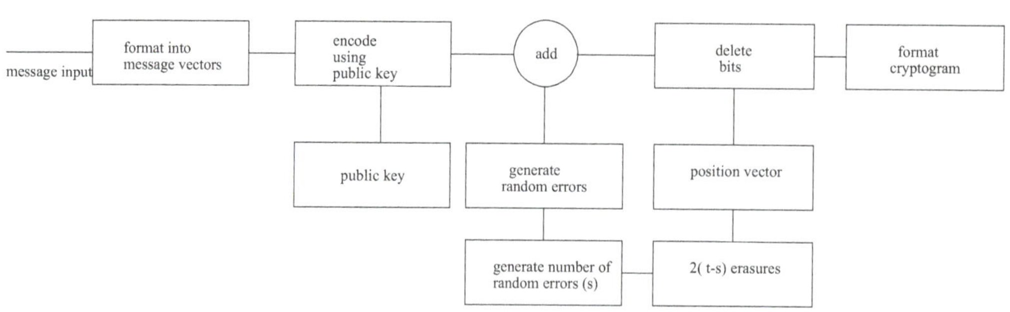

Fig. 19.1 Public key encryption system with s random bit errors and $2(t-s)$ bit deletions

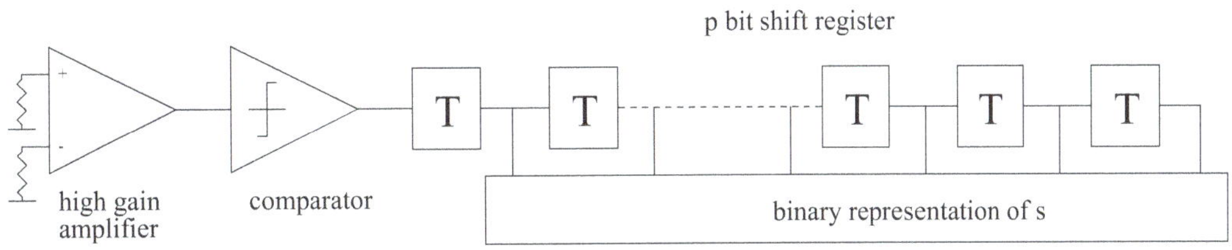

Fig. 19.2 Random integer generator of the number of added, random bit errors, s

The number of random errors s is input to *generate random errors* which for each codeword, initialises an n bit buffer memory with zeros, and uses a random number generator to generate s 1's in s random positions of the buffer memory. The contents of the n bit buffer are added to the codeword of n bits by *add* shown in Fig. 19.1. The 1's are added modulo 2 which inverts the codeword bits in these positions so that these bits are in error. In Fig. 19.1, $t - s$ *erasures* takes the input s, calculates $2(t-s)$ and outputs this value to *position vector* which comprises a buffer memory of n bits containing a sequence of integers corresponding to a position vector described below. The first $2(t - s)$ integers are input to *delete bits* which deletes the bits in the corresponding positions of the codeword so that $2(t - s)$ bits of the codeword are deleted. The procedure is carried out for each codeword so that each codeword is randomly shortened due to deleted bits and corrupted with a random number of bit errors in random positions. In Fig. 19.1, *format cryptogram* has the sequence of shortened corrupted codewords as input and appends these together, together with formatting information to produce the cryptogram.

The highest level of security is provided when the block *generate number of random errors (s)* of Fig. 19.1 is replaced by a truly random number generator and not a pseudorandom generator. An example of a random number generator is shown in Fig. 19.2.

The differential amplifier with high gain amplifies the thermal noise generated by the resistor terminated inputs. The output of the amplifier is the amplified random noise which is input to a comparator which carries out binary quantisation. The comparator output is 1 if the amplifier output is a positive voltage and 0 otherwise. This produces 1's and 0's with equal probability at the output of the comparator. The output of the comparator is clocked into a shift register having p shift register stages, each of delay T. The clock rate is $\frac{1}{T}$. After p clock cycles, the contents of the shift register represent a number in binary which is the random number s having a uniform probability distribution between 0 and $2^p - 1$.

One or more of the bits output from the shift register may be permanently set to 1 to provide a lower limit to the random number of errors s. As an example, if the 4th bit (counting from the least significant bits) is permanently set to 1 then s has a uniform probability distribution between $2^3 = 8$ and $2^p - 1$.

Similarly, the highest level of security is provided if the positions of the errors generated by *generate random errors* of Fig. 19.1 is a truly random number generator and not a pseudorandom generator. An example of an arrangement which generates

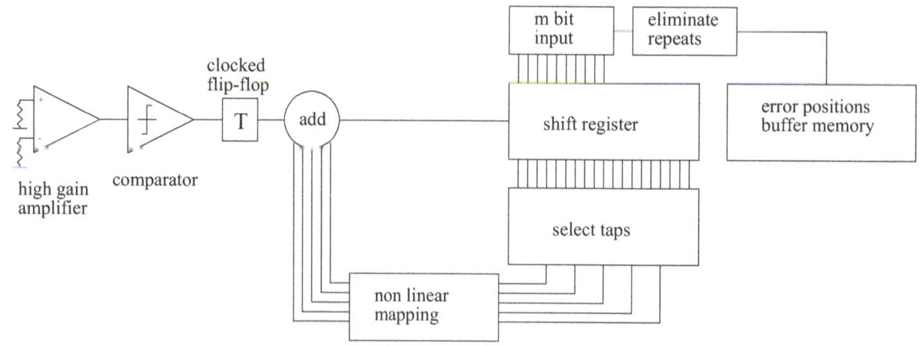

Fig. 19.3 Random integer generator of error positions

truly random positions in the range of 0 to $2^m - 1$ corresponding to the codeword length is shown in Fig. 19.3.

As shown in Fig. 19.3, the differential amplifier, with high gain amplifies the thermal noise generated by the resistor terminated inputs. The output of the amplifier is the amplified random noise which is input to a comparator which outputs a 1 if the amplifier output is a positive voltage and a 0 otherwise. This produces 1's and 0's with equal probability at the output of the comparator. The output of the comparator is clocked into a flip-flop clocked at $\frac{1}{T}$, with the same clock source as the shift register shown in Fig. 19.3, *shift register*. The output of the flip-flop is a clocked output of truly random 1's and 0's which is input to a nonlinear feedback shift register arrangement.

The output of the flip-flop is input to a modulo 2, adder *add* which is added to the outputs of a nonlinear mapping of u selected outputs of the shift register. Which outputs are to be selected correspond to the key which is being used. The parameter u is a design parameter, typically equal to 8.

The nonlinear mapping *nonlinear mapping* shown in Fig. 19.3 has a pseudorandom one-to-one correspondence between each of the 2^u input states to each of the 2^u output states. An example of such a one to one correspondence, for $u = 4$ is given in Table 19.1. For example, the first entry, 0000, value 0 is mapped to 0011, value 3.

The shift register typically has a relatively large number of stages, 64 is a typical number of stages and a number of tapped outputs, typically 8. The relationship between the input of the shift register a_{in} and the tapped outputs is usually represented by the delay operator D. Defining the tap positions as w_i, for $i = 0$ to i_{max}, the input to the nonlinear mapping *nonlinear mapping* shown in Fig. 19.3, defined as x_i for $i = 0$ to i_{max}, is

$$x_i = a_{in} D^{w_i} \tag{19.1}$$

and the output y_j after the mapping function, depicted as M is

$$y_j = M[x_i] = M[a_{in} D^{w_i}] \tag{19.2}$$

The input to the shift register is the output of the adder given by the sum of the random input R_{nd} and the summed output of the mapped outputs. Accordingly,

Table 19.1 Example of nonlinear mapping for $u = 4$

Input		Output
0000	→	0011
0001	→	1011
0010	→	0111
0011	→	0110
0100	→	1111
0101	→	0001
0110	→	1001
0111	→	1100
1000	→	1010
1001	→	0000
1010	→	1000
1011	→	0010
1100	→	0101
1101	→	1110
1110	→	0100
1111	→	1101

$$a_{in} = R_{nd} + \sum_{j=0}^{i_{max}} y_j = R_{nd} + \sum_{j=0}^{i_{max}} M[x_i] = R_{nd} + \sum_{j=0}^{i_{max}} M[a_{in}D^{w_i}] \tag{19.3}$$

It can be seen that the shift register input a_{in} is a nonlinear function of delayed outputs of itself added to the random input R_{nd}, and so will be a random binary function.

The positions of the errors are given by the output of *m-bit input* shown in Fig. 19.3, an m bit memory register and defined as e_{pos}. Consider that the first m outputs of the shift register are used as the input to *m-bit input*. The output of *m-bit input* is a binary representation of a number given by

$$e_{pos} = \sum_{j=0}^{m-1} 2^j \times a_{in}D^j \tag{19.4}$$

Since a_{in} is a random binary function, e_{pos} will be an integer between 0 and $2^m - 1$ randomly distributed with a uniform distribution. As shown in Fig. 19.3, these randomly generated integers are stored in memory in *error positions buffer memory* after *eliminate repeats* has eliminated any repeated numbers, since repeated integers will occur from time to time in any independently distributed random integer generator.

The random bit errors and bit deletions can only be corrected with the knowledge of the particular non-binary Goppa code, the private key, which is used in deriving the codeword generator matrix. Reviewing the background on Goppa codes: Goppa defined a family of codes [6] where the coordinates of each codeword

$\{c_0, c_1, c_2, \ldots c_{2^m-1}\}$ with $\{c_0 = x_0, c_1 = x_1, c_2 = x_2, \ldots c_{2^m-1} = x_{2^m-1}\}$ satisfy the congruence $p(z)$ modulo $g(z) = 0$ where $g(z)$ is now known as the Goppa polynomial and $p(z)$ is the Lagrange interpolation polynomial.

Goppa codes have coefficients from $GF(2^m)$ and provided $g(z)$ has no roots which are elements of $GF(2^m)$ (which is straightforward to achieve) the Goppa codes have parameters $(2^m, k, 2^m - k + 1)$. Goppa codes can be converted into binary codes. Provided that $g(z)$ has no roots which are elements of $GF(2^m)$ and has no repeated roots, the binary code parameters are $(2^m, 2^m - mt, d_{min})$ where $d_{min} \geq 2t + 1$, the Goppa code bound on minimum Hamming distance. Most binary Goppa codes have equality for the bound and t is the number of correctable errors.

For a Goppa polynomial of degree r, there are r parity check equations defined from the congruence. Denoting $g(z)$ by

$$g(z) = g_r z^r + g_{r-1} z^{r-1} + g_{r-2} z^{r-2} + \cdots + g_1 z + g_0 \tag{19.5}$$

$$\sum_{i=0}^{2^m-1} \frac{c_i}{z - \alpha_i} = 0 \quad \text{modulo}\, g(z) \tag{19.6}$$

Since Eq. (19.6) is modulo $g(z)$ then $g(z)$ is equivalent to 0, and we can add $g(z)$ to the numerator. Dividing each term $z - \alpha_i$ into $1 + g(z)$ produces the following

$$\frac{g(z) + 1}{z - \alpha_i} = q_i(z) + \frac{r_m + 1}{z - \alpha_i} \tag{19.7}$$

where r_m is the remainder, an element of $GF(2^m)$ after dividing $g(z)$ by $z - \alpha_i$.

As r_m is a scalar, $g(z)$ may simply be pre-multiplied by $\frac{1}{r_m}$ so that the remainder cancels with the other numerator term which is 1.

$$\frac{\frac{g(z)}{r_m} + 1}{z - \alpha_i} = \frac{q_i(z)}{r_m} + \frac{\frac{r_m}{r_m} + 1}{z - \alpha_i} = \frac{q(z)}{r_m} \tag{19.8}$$

As

$$g(z) = (z - \alpha_i) q_i(z) + r_m \tag{19.9}$$

When $z = \alpha_i$, $r_m = g(\alpha_i)$.

Substituting for r_m in Eq. (19.8) produces

$$\frac{\frac{g(z)}{g(\alpha_i)} + 1}{z - \alpha_i} = \frac{q_i(z)}{g(\alpha_i)} \tag{19.10}$$

Since $\frac{g(z)}{g(\alpha_i)}$ modulo $g(z) = 0$

$$\frac{1}{z - \alpha_i} = \frac{q_i(z)}{g(\alpha_i)} \tag{19.11}$$

The quotient polynomial $q_i(z)$ is a polynomial of degree $r-1$ with coefficients which are a function of α_i and the Goppa polynomial coefficients. Denoting $q_i(z)$ as

$$q_i(z) = q_{i,0} + q_{i,1}z + q_{i,2}z^2 + q_{i,3}z^3 + \cdots + q_{i,(r-1)}z^{r-1} \tag{19.12}$$

Since the coefficients of each power of z sum to zero the r parity check equations are given by

$$\sum_{i=0}^{2^m-1} \frac{c_i q_{i,j}}{g(\alpha_i)} = 0 \quad \text{for} \quad j = 0 \quad \text{to} \quad r-1 \tag{19.13}$$

If the Goppa polynomial has any roots which are elements of $GF(2^m)$, say α_j, then the codeword coordinate c_j has to be permanently set to zero in order to satisfy the parity check equations. Effectively the codelength is shortened by the number of roots of $g(z)$ which are elements of $GF(2^m)$. Usually the Goppa polynomial is chosen to have distinct roots which are not in $GF(2^m)$.

The security depends upon the number of bit errors added and in practical examples to provide sufficient security, it is necessary to use long Goppa codes of length 2048 bits, 4096 bits or longer. For brevity, the procedure will be described using an example of a binary Goppa code of length 32 bits capable of correcting 4 bit errors. It is important to note that all binary Goppa codes are derived from non-binary Goppa codes which are designed first.

In this example, the non-binary Goppa code consists of 32 symbols from the Galois field $GF(2^5)$ and each symbol takes on 32 possible values with the code capable of correcting two errors. There are 28 information symbols and 4 parity check symbols. (It should be noted that when the Goppa code is used with information symbols restricted to binary values as in a binary Goppa code, twice as many errors can be corrected). The non-binary Goppa code has parameters of a (32, 28, 5) code. There are 4 parity check symbols defined by the 4 parity check equations and the Goppa polynomial has degree 4. Choosing arbitrarily as the Goppa polynomial, the polynomial $1 + z + z^4$ which has roots only in $GF(16)$ and none in $GF(32)$, we determine $q_i(z)$ by dividing by $z - \alpha_i$.

$$q_i(z) = z^3 + \alpha_i z^2 + \alpha_i^2 z + (1 + \alpha_i^3) \tag{19.14}$$

The 4 parity check equations are

$$\sum_{i=0}^{31} \frac{c_i}{g(\alpha_i)} = 0 \tag{19.15}$$

$$\sum_{i=0}^{31} \frac{c_i \alpha_i}{g(\alpha_i)} = 0 \tag{19.16}$$

$$\sum_{i=0}^{31} \frac{c_i \alpha_i^2}{g(\alpha_i)} = 0 \tag{19.17}$$

$$\sum_{i=0}^{31} \frac{c_i (1 + \alpha_i^3)}{g(\alpha_i)} = 0 \tag{19.18}$$

Using the $GF(2^5)$ Table 19.2 to evaluate the different terms for $GF(2^5)$, the parity check matrix is

$$\mathbf{H}_{(32,28,5)} = \begin{bmatrix} 1 & 1 & \alpha^{14} & \alpha^{28} & \alpha^{20} & \alpha^{25} & \ldots & \alpha^{10} \\ 0 & 1 & \alpha^{15} & \alpha^{30} & \alpha^{23} & \alpha^{29} & \ldots & \alpha^{9} \\ 0 & 1 & \alpha^{16} & \alpha^{1} & \alpha^{26} & \alpha^{2} & \ldots & \alpha^{8} \\ 1 & 0 & \alpha^{12} & \alpha^{24} & \alpha^{5} & \alpha^{17} & \ldots & \alpha^{5} \end{bmatrix} \tag{19.19}$$

To implement the Goppa code as a binary code, the symbols in the parity check matrix are replaced with their m-bit binary column representations of each respective $GF(2^m)$ symbol. For the (32, 28, 5) Goppa code above, each of the 4 parity symbols will be represented as a 5-bit symbol from Table 19.2. The parity check matrix will now have 20 rows for the binary code. The minimum Hamming distance of the binary Goppa code is improved from $r + 1$ to $2r + 1$. Correspondingly, the example binary Goppa code becomes a (32, 12, 9) code with parity check matrix:

$$\mathbf{H}_{(32,12,9)} = \begin{bmatrix} 1\,1\,1\,0\,0\,1 \ldots 1 \\ 0\,0\,0\,1\,0\,0 \ldots 0 \\ 0\,0\,1\,1\,1\,0 \ldots 0 \\ 0\,0\,1\,0\,1\,1 \ldots 0 \\ 0\,0\,1\,1\,0\,1 \ldots 1 \\ 0\,1\,1\,0\,1\,1 \ldots 0 \\ 0\,0\,1\,1\,1\,0 \ldots 1 \\ 0\,0\,1\,0\,1\,0 \ldots 0 \\ 0\,0\,1\,0\,1\,1 \ldots 1 \\ 0\,0\,1\,1\,0\,0 \ldots 1 \\ 0\,1\,1\,1\,1\,0 \ldots 1 \\ 0\,0\,1\,0\,1\,0 \ldots 0 \\ 0\,0\,0\,0\,1\,1 \ldots 1 \\ 0\,0\,1\,0\,0\,0 \ldots 1 \\ 0\,0\,1\,0\,1\,0 \ldots 0 \\ 1\,0\,0\,0\,1\,1 \ldots 1 \\ 0\,0\,1\,1\,0\,1 \ldots 0 \\ 0\,0\,1\,1\,1\,0 \ldots 1 \\ 0\,0\,1\,1\,0\,0 \ldots 0 \\ 0\,0\,0\,1\,0\,1 \ldots 0 \end{bmatrix} \tag{19.20}$$

Table 19.2 $GF(32)$ non-zero extension field elements defined by $1+\alpha^2+\alpha^5=0$

$\alpha^0 = 1$
$\alpha^1 = \alpha$
$\alpha^2 = \alpha^2$
$\alpha^3 = \alpha^3$
$\alpha^4 = \alpha^4$
$\alpha^5 = 1+\alpha^2$
$\alpha^6 = \alpha+\alpha^3$
$\alpha^7 = \alpha^2+\alpha^4$
$\alpha^8 = 1+\alpha^2+\alpha^3$
$\alpha^9 = \alpha+\alpha^3+\alpha^4$
$\alpha^{10} = 1+\alpha^4$
$\alpha^{11} = 1+\alpha+\alpha^2$
$\alpha^{12} = \alpha+\alpha^2+\alpha^3$
$\alpha^{13} = \alpha^2+\alpha^3+\alpha^4$
$\alpha^{14} = 1+\alpha^2+\alpha^3+\alpha^4$
$\alpha^{15} = 1+\alpha+\alpha^2+\alpha^3+\alpha^4$
$\alpha^{16} = 1+\alpha+\alpha^3+\alpha^4$
$\alpha^{17} = 1+\alpha+\alpha^4$
$\alpha^{18} = 1+\alpha$
$\alpha^{19} = \alpha+\alpha^2$
$\alpha^{20} = \alpha^2+\alpha^3$
$\alpha^{21} = \alpha^3+\alpha^4$
$\alpha^{22} = 1+\alpha^2+\alpha^4$
$\alpha^{23} = 1+\alpha+\alpha^2+\alpha^3$
$\alpha^{24} = \alpha+\alpha^2+\alpha^3+\alpha^4$
$\alpha^{25} = 1+\alpha^3+\alpha^4$
$\alpha^{26} = 1+\alpha+\alpha^2+\alpha^4$
$\alpha^{27} = 1+\alpha+\alpha^3$
$\alpha^{28} = \alpha+\alpha^2+\alpha^4$
$\alpha^{29} = 1+\alpha^3$
$\alpha^{30} = \alpha+\alpha^4$

The next step is to turn the parity check matrix into reduced echelon form by using elementary matrix row and column operations so that there are 20 rows representing 20 independent parity check equations for each parity bit. From the reduced echelon parity check matrix, the generator matrix can be obtained straightforwardly as it is the transpose of the reduced echelon parity check matrix. The resulting generator matrix is:

$$
\mathbf{G}_{(32,12,9)} = \begin{bmatrix}
1&0&0&0&0&0&0&0&0&0&0&0&1&1&0&0&0&1&1&1&1&0&1&0&0&1&1&1&0&1&1&0\\
0&1&0&0&0&0&0&0&0&0&0&0&1&1&0&1&0&1&1&0&0&0&0&1&1&0&1&0&1&1&0&0\\
0&0&1&0&0&0&0&0&0&0&0&0&0&0&1&1&0&0&1&1&0&0&0&0&1&1&0&1&0&0&1&0\\
0&0&0&1&0&0&0&0&0&0&0&0&1&1&1&1&1&1&0&1&0&1&0&1&1&0&0&0&1&0&1&0\\
0&0&0&0&1&0&0&0&0&0&0&0&1&1&0&0&0&0&1&1&0&1&0&0&1&1&1&0&0&0&0&1\\
0&0&0&0&0&1&0&0&0&0&0&0&1&1&1&1&1&0&0&1&1&0&0&0&1&0&1&0&1&0&0&1\\
0&0&0&0&0&0&1&0&0&0&0&0&0&0&0&0&1&0&1&1&1&1&1&1&0&0&0&1&1&1&0&0\\
0&0&0&0&0&0&0&1&0&0&0&0&0&0&1&1&1&0&0&1&1&0&0&1&1&1&0&0&0&1&0&0\\
0&0&0&0&0&0&0&0&1&0&0&0&1&1&1&1&0&1&1&1&1&1&1&1&0&1&1&1&1&1&0&1\\
0&0&0&0&0&0&0&0&0&1&0&0&1&1&1&0&1&1&0&1&1&1&1&0&0&1&0&0&1&0&1&1\\
0&0&0&0&0&0&0&0&0&0&1&0&0&0&0&1&1&1&0&0&1&1&1&0&1&1&0&0&1&1&0&0\\
0&0&0&0&0&0&0&0&0&0&0&1&1&1&1&0&1&1&0&0&1&0&0&1&1&0&1&1&0&1&1&1
\end{bmatrix} \tag{19.21}
$$

It will be noticed that the generator matrix is in reduced echelon form and has 12 rows, one row for each information bit. Each row is the codeword resulting from that information bit equal to a 1, all other information bits equal to 0.

The next step is to scramble the information bits by multiplying by a $k \times k$ non-singular matrix, that is one that is invertible. As a simple example, the following 12×12 matrix is invertible.

$$
\mathbf{NS}_{12\times 12} = \begin{bmatrix}
0&1&1&1&0&1&0&0&1&1&1&0\\
0&0&1&1&1&0&1&0&0&1&1&1\\
1&0&0&1&1&1&0&1&0&0&1&1\\
1&1&0&0&1&1&1&0&1&0&0&1\\
1&1&1&0&0&1&1&1&0&1&0&0\\
0&1&1&1&0&0&1&1&1&0&1&0\\
0&0&1&1&1&0&0&1&1&1&0&1\\
1&0&0&1&1&1&0&0&1&1&1&0\\
0&1&0&0&1&1&1&0&0&1&1&1\\
1&0&1&0&0&1&1&1&0&0&1&1\\
1&1&0&1&0&0&1&1&1&0&0&1\\
1&1&1&0&1&0&0&1&1&1&0&0
\end{bmatrix} \tag{19.22}
$$

The above is invertible using the following matrix:

$$
\mathbf{NS}^{-1}_{12\times 12} = \begin{bmatrix}
1&1&0&1&0&0&0&0&0&0&0&0\\
0&1&1&0&1&0&0&0&0&0&0&0\\
0&0&1&1&0&1&0&0&0&0&0&0\\
0&0&0&1&1&0&1&0&0&0&0&0\\
0&0&0&0&1&1&0&1&0&0&0&0\\
0&0&0&0&0&1&1&0&1&0&0&0\\
0&0&0&0&0&0&1&1&0&1&0&0\\
0&0&0&0&0&0&0&1&1&0&1&0\\
0&0&0&0&0&0&0&0&1&1&0&1\\
1&0&0&0&0&0&0&0&0&1&1&0\\
0&1&0&0&0&0&0&0&0&0&1&1\\
1&0&1&0&0&0&0&0&0&0&0&1
\end{bmatrix} \tag{19.23}
$$

The next step is to scramble the generator matrix with the non-singular matrix to produce the scrambled generator matrix given below. The code produced with this generator matrix has the same codewords as the generator matrix given by matrix (19.21) and can correct the same number of errors but there is a different mapping to codewords from a given information bit pattern.

$$\mathbf{SG}_{(32,12,9)} = \begin{bmatrix}
0&1&1&1&0&1&0&0&1&1&1&0&1&1&1&0&0&1&0&1&1&1&1&0&0&0&0&0&0&1&1&1\\
0&0&1&1&1&0&1&0&0&1&1&1&0&0&0&1&1&0&0&1&1&0&1&0&0&0&1&1&0&1&0&1\\
1&0&0&1&1&1&0&1&0&0&1&1&1&1&0&0&1&0&0&1&1&1&0&1&0&0&0&0&1&0&1&1\\
1&1&0&0&1&1&1&0&1&0&0&1&0&0&1&1&1&0&1&1&1&1&1&0&0&1&0&0&0&1&0&0\\
1&1&1&0&0&1&1&1&0&1&0&0&0&0&0&0&0&1&1&0&0&1&1&0&1&0&0&1&0&0&1&0\\
0&1&1&1&0&0&1&1&1&0&1&0&1&1&0&0&0&0&1&1&1&1&1&0&0&0&1&1&1&1&0&1\\
0&0&1&1&1&0&0&1&1&1&0&1&1&1&0&0&0&0&0&0&1&1&0&1&0&1&0&1&1&1&0&0\\
1&0&0&1&1&1&0&0&1&1&1&0&0&0&0&0&0&1&1&0&1&1&0&0&0&1&0&0&1&1&1&0\\
0&1&0&0&1&1&1&0&0&1&1&1&1&1&1&1&1&0&1&0&1&0&1&1&1&1&0&0&1&0&0&0\\
1&0&1&0&0&1&1&1&0&0&1&1&1&1&0&0&1&1&0&1&1&1&1&0&0&0&0&0&1&1&1&0\\
1&1&0&1&0&0&1&1&1&0&0&1&1&1&0&0&0&1&0&1&1&1&1&0&0&1&0&0&0&0&1&0\\
1&1&1&0&1&0&0&1&1&1&0&0&1&1&0&0&0&0&0&0&1&0&1&0&1&0&1&1&1&0&1&1
\end{bmatrix} \tag{19.24}$$

It may be seen that, for example, the first row of this matrix is the modulo 2 sum of rows 1, 2, 3, 5, 8, 9 and 10 of matrix (19.21) in accordance with the non-singular matrix (19.22).

The final step in producing the public key generator matrix for the codewords from the message vectors is to permute the columns of the matrix above. Any permutation may be randomly chosen. For example we may use the following permutation:

$$\begin{array}{cccccccccccccccccccccccccccccccc}
27&15&4&2&19&21&17&14&7&16&20&1&29&8&11&12&25&5&30&24&6&18&13&3&0&26&23&28&22&31&9&10\\
0&1&2&3&4&5&6&7&8&9&10&11&12&13&14&15&16&17&18&19&20&21&22&23&24&25&26&27&28&29&30&31
\end{array} \tag{19.25}$$

so that for example column 0 of matrix (19.24) becomes column 24 of the permuted generator matrix and column 31 of matrix (19.24) becomes column 29 of the permuted generator matrix. The resulting, permuted generator matrix is given below.

$$\mathbf{PSG}_{(32,12,9)} = \begin{bmatrix}
0&0&0&1&1&1&1&1&0&0&1&1&1&1&0&1&0&1&1&0&0&0&1&1&0&0&0&0&1&1&1&1\\
1&1&1&1&1&0&0&0&0&1&1&0&1&0&1&0&0&0&0&0&1&0&0&1&0&1&0&0&1&1&1&1\\
0&0&1&0&1&1&0&0&1&1&1&0&0&0&1&1&0&1&1&0&0&0&1&1&1&0&1&1&0&1&0&1\\
0&1&1&0&1&1&0&1&0&1&1&1&1&1&1&0&1&1&0&0&1&1&0&0&1&0&0&0&1&0&0&0\\
1&0&0&1&0&1&1&0&1&0&0&1&0&0&0&0&0&1&1&1&1&1&0&0&1&0&0&0&1&0&1&0\\
1&0&0&1&1&1&0&0&1&0&1&1&1&1&0&1&0&0&0&0&1&1&1&1&0&1&0&1&1&1&0&1\\
1&0&1&1&0&1&0&0&1&0&1&0&1&1&1&1&1&0&0&0&0&0&1&1&0&0&1&1&0&0&1&0\\
0&0&1&0&0&1&1&0&0&0&1&0&1&1&0&0&1&1&1&0&0&1&0&1&1&0&0&1&0&0&1&1\\
0&1&1&0&0&0&0&1&0&1&1&1&0&0&1&1&1&1&0&1&1&1&1&0&0&0&1&1&1&0&1&1\\
0&0&0&1&1&1&1&0&1&1&1&0&1&0&1&1&0&1&1&0&1&0&1&0&1&0&0&1&1&0&0&1\\
0&0&0&0&1&1&1&0&1&0&1&1&0&1&1&1&1&0&1&0&1&0&1&1&1&0&0&0&1&0&0&0\\
1&0&1&1&0&0&0&0&1&0&1&1&0&1&0&1&0&0&1&1&0&0&1&0&1&1&0&1&1&1&1&0
\end{bmatrix} \tag{19.26}$$

With this particular example of the Goppa code, the message needs to be split into message vectors of length 12 bits, adding padding bits as necessary so that there is an integral number of message vectors. As a simple example of a plaintext message, consider that the message consists of a single message vector with the information bit pattern:

$$\{0, 1, 0, 1, 1, 1, 0, 0, 0, 0, 0, 1\}$$

Starting with an all 0's vector, where the information bit pattern is 1, the corresponding row from the permuted, scrambled matrix, matrix (19.26) with the same position is added modulo 2 to the result so far to produce the codeword which will form the digital cryptogram plus added random errors. In this example, this codeword is generated from adding modulo 2, rows 2, 4, 5, 6 and 12 from the permuted, scrambled matrix, matrix (19.26) to produce:

$$\begin{array}{c}
\text{0 0 0 1 1 1 1 1 0 0 1 1 1 1 0 1 0 1 1 0 0 0 1 1 0 0 0 0 1 1 1 1} \\
\text{+ +} \\
\text{0 0 1 0 1 1 0 0 1 1 1 0 0 0 1 1 0 1 1 0 0 0 1 1 1 0 1 1 0 1 0 1} \\
\text{+ +} \\
\text{0 1 1 0 1 1 0 1 0 1 1 1 1 1 1 0 1 1 0 0 1 1 0 0 1 0 0 0 1 0 0 0} \\
\text{+ +} \\
\text{1 0 0 1 0 1 1 0 1 0 0 1 0 0 0 0 0 1 1 1 1 1 0 0 1 0 0 0 1 0 1 0} \\
\text{+ +} \\
\text{1 0 1 1 0 0 0 0 1 0 1 1 0 1 0 1 0 0 1 1 0 0 1 0 1 1 0 1 1 1 1 0} \\
\text{= =} \\
\text{0 1 1 1 1 0 0 0 1 0 0 0 0 1 0 1 1 0 0 0 0 0 1 0 0 1 1 0 0 1 1 0}
\end{array} \qquad (19.27)$$

The resulting codeword is:

$$\{0 1 1 1 1 0 0 0 1 0 0 0 0 1 0 1 1 0 0 0 0 0 1 0 0 1 1 0 0 1 1 0\}$$

This Goppa code can correct up to 4 errors, ($t = 4$), so a random number is chosen for the number of bits to be in error (s) and $2(t-s)$ bits are deleted from the codeword in pre-determined positions. The pre-determined positions may be given by a secret key, a position vector, known only to the originator and intended recipient of the cryptogram. It may be included as part of the public key, or may be contained in a previous cryptogram sent to the recipient. An example of a position vector, which defines the bit positions to be deleted is:

$$\{19, 3, 27, 17, 8, 30, 11, 15, 2, 5, 19, \ldots, 25\}.$$

The notation being, for example, that if there are 2 bits to be deleted, the bit positions to be deleted are the first 2 bit positions in the position vector, 19 and 3. As well as the secret key, the position vector, the recipient needs to know the number of bits deleted, preferably with the information provided in a secure way. One method is for the message vector to contain, as part of the message, a number indicating the number of errors to be deleted in the next codeword, the following codeword (not the current codeword); the first codeword having a known, fixed number of deleted bits.

The number of bit errors and the bit error positions are randomly chosen to be in error. A truly random source such as a thermal noise source as described above produces the most secure results, but a pseudorandom generator can be used instead, particularly, if seeded from the time of day with a fine time resolution such as 1 ms. If the number of random errors chosen is too few, the security of the digital cryptogram will be compromised. Correspondingly, the minimum number of errors chosen is a design parameter depending upon the length of the Goppa code and t, the number of correctable errors. A suitable choice for the minimum number of errors chosen in practice lies between $\frac{t}{2}$ and $\frac{3t}{4}$.

For the example above, consider that the number of bit errors is 2 and these are randomly chosen to be in positions 7 and 23 (starting the position index from 0). The bits in these positions in the codeword are inverted to produce the result:

$$\{0\,1\,1\,1\,1\,0\,0\,1\,1\,0\,0\,0\,0\,1\,0\,1\,1\,0\,0\,0\,0\,0\,1\,1\,0\,1\,1\,0\,0\,1\,1\,0\}.$$

As there are 2 bits in error, 4 bits $(2(t-s) = 2(4-2))$ may be deleted. Using the position vector example above, the deleted bits are in positions {19, 3, 27 and 17} resulting in 28 bits,

$$\{0\,1\,1\,1\,0\,0\,1\,1\,0\,0\,0\,0\,1\,0\,1\,1\,0\,0\,0\,1\,1\,0\,1\,1\,0\,1\,1\,0\}.$$

This vector forms the digital cryptogram which is transmitted or stored depending upon the application.

The intended recipient of this cryptogram retrieves the message in a series of steps. Figure 19.4 shows the decryption system. The retrieved cryptogram is formatted into corrupted codewords by *format into corrupted codewords* shown in Fig. 19.4. In the formatting process, the number of deleted bits in each codeword is determined from the retrieved length of each codeword. The next step is to insert 0's in the deleted bit positions so that each corrupted codeword is of the correct length. This is carried out using *fill erased positions with 0's* as input, the position vector stored in a buffer memory as *position vector* in Fig. 19.4 and the number of deleted (erased) bits from *format into corrupted codewords*. For the example above, the recipient first receives or otherwise retrieves the cryptogram {0 1 1 1 0 0 1 1 0 0 0 0 1 0 1 1 0 0 0 1 1 0 1 1 0 1 1 0}. Knowing the number of deleted bits and their positions, the recipient inserts 0's in positions {19, 3, 27 and 17} to produce:

$$\{0\,1\,1\,0\,1\,0\,0\,1\,1\,0\,0\,0\,0\,1\,0\,1\,1\,0\,0\,0\,0\,0\,1\,1\,0\,1\,1\,0\,0\,1\,1\,0\}n$$

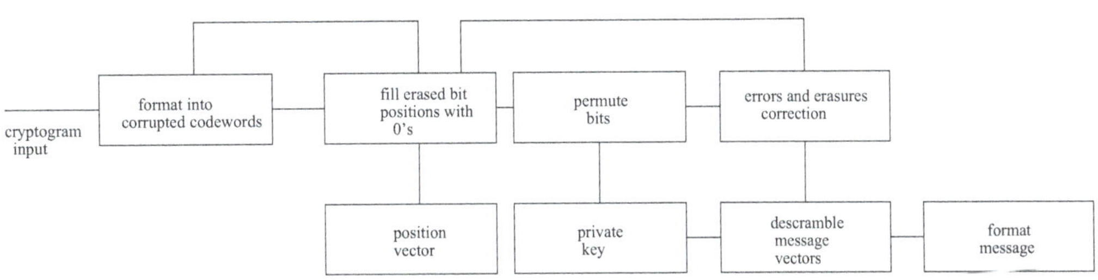

Fig. 19.4 Private key decryption system with s random bit errors and $2(t-s)$ bit deletions

The private key contains the information of which Goppa code was used, the inverse of the non-singular matrix used to scramble the data and the permutation applied to codeword symbols in constructing the public key generator matrix. This information is stored in *private key* in Fig. 19.4.

For the example, the private key is used to undo the permutation applied to codeword symbols by applying the following permutation:

$$\begin{array}{cccccccccccccccccccccccccccccccc} 0 & 1 & 2 & 3 & 4 & 5 & 6 & 7 & 8 & 9 & 10 & 11 & 12 & 13 & 14 & 15 & 16 & 17 & 18 & 19 & 20 & 21 & 22 & 23 & 24 & 25 & 26 & 27 & 28 & 29 & 30 & 31 \\ 27 & 15 & 4 & 2 & 19 & 21 & 17 & 14 & 7 & 16 & 20 & 1 & 29 & 8 & 11 & 12 & 25 & 5 & 30 & 24 & 6 & 18 & 13 & 3 & 0 & 26 & 23 & 28 & 22 & 31 & 9 & 10 \end{array} \tag{19.28}$$

so that, for example, bit 24 becomes bit 0 after permutation and bit 27 becomes bit 31 after permutation. The resulting, corrupted codeword is:

$$\{0 0 0 1 1 0 0 1 1 1 0 0 1 1 1 1 0 0 0 1 0 0 0 1 0 1 1 0 0 0 0 1\}$$

The permutation is carried out by *permute bits* shown in Fig. 19.4.

The next step is to treat the bits in the corrupted codeword as $GF(2^5)$ symbols and use the parity check matrix, matrix (19.19), from the private key to calculate the syndrome value for each row of the parity check matrix to produce α^{28}, α^{7}, α^{13}, and α^{19}. This is carried out by an errors and erasures decoder as a first step in correcting the errors and erasures. The errors and erasures are corrected by *errors and erasures correction*, which knows the positions of the erased bits from *fill erased positions with 0's* shown in Fig. 19.4.

In the example, the errors and erasures are corrected using the syndrome values to produce the uncorrupted codeword. There are several published algorithms for errors and erasures decoding [1, 13, 16]. Using, for example, the method described by Sugiyama [16], the uncorrupted codeword is obtained:

$$\{1 0 0 0 1 0 0 1 1 1 0 0 0 0 1 0 0 1 1 1 0 1 1 0 0 1 1 0 0 1 0 1\}$$

The scrambled information data is the first 12 bits of this codeword:

$$\{1 0 0 0 1 0 0 1 1 1 0 0\}$$

The last step is to unscramble the scrambled data using matrix (19.23) to produce the original message after formatting the unscrambled data:

$$\{0, 1, 0, 1, 1, 1, 0, 0, 0, 0, 0, 1\}$$

In Fig. 19.4, *descramble message vectors* take as input the matrix which is the inverse of the non-singular matrix stored in *private key* and output the descramble message vectors to *format message*.

In practice, much longer codes of length n would be used than described above. Typically n is set equal to 1024, 2048, 4096 bits or longer. Longer codes are more secure but the public key is larger and encryption and decryption take longer time.

Consider an example with $n = 1024$, correcting $t = 60$ bit errors with a randomly chosen irreducible Goppa polynomial of degree 60, say, $g(z) = 1+z+z^2+z^{23}+z^{60}$.

Setting the number of inserted bit errors s as a randomly chosen number from 40 to 60, the number of deleted bits correspondingly, is $2(t - s)$, ranging from 40 to 0 and the average codeword length is 994 bits. There are 9.12×10^{96} different bit error combinations providing security, against naive brute force decoding, equivalent to a random key of length 325 bits. The message vector length is 424 bits per codeword of which 6 bits may be assigned to indicate the number of deleted bits in the following codeword. It should be noted that there are more effective attacks than brute force decoding as discussed in Sect. 19.5.

As another example with $n = 2048$ and correcting $t = 80$ bit errors with a randomly chosen irreducible Goppa polynomial of degree 80, an example being $g(z) = 1 + z + z^3 + z^{17} + z^{80}$.

Setting the number of inserted bit errors s as a randomly chosen number from 40 to 80, the number of deleted bits correspondingly, is $2(t-s)$, ranging from 80 to 0 and the average codeword length is 2008 bits. There are 2.45×10^{144} different bit error combinations providing security, against naive brute force decoding, equivalent to a random key of length 482 bits. The message vector length is 1168 bits per codeword of which 7 bits may be assigned to indicate the number of deleted bits in the following codeword.

In a hybrid arrangement where the sender and recipient share secret information, additional bits in error may be deliberately added to the cryptogram using a secret key, the position vector to determine the positions of the additional error bits. The number of additional bits in error is randomly chosen between 0 and $n - 1$. The recipient needs to know the number of additional bits in error (as well as the position vector), preferably with this information provided in a secure way. One method is for the message vector to contain, as part of the message, the number of additional bits in error in the next codeword that is the following codeword (not the current codeword). It is arranged that the first codeword has a known, fixed number of additional bits in error.

As each corrupted codeword contains more than t bits in error, it is theoretically impossible, even with the knowledge of the private key to recover the original codewords free from errors and to determine the unknown bits in the deleted bit positions. It should be noted that this arrangement defeats attacks based on information set decoding, which is discussed later. The system is depicted in Fig. 19.5.

This encryption arrangement is as shown in Fig. 19.1 except that the system accommodates additional errors added by *generate additional errors* shown in Fig. 19.5 using a random integer generator between 0 and $n-1$ generated by *generate random number of additional errors*. Any suitable random integer generator may be used. For example, the random integer generator design shown in Fig. 19.2 may be used with the number of shift register stages p now set equal to m, where $n = 2^m$.

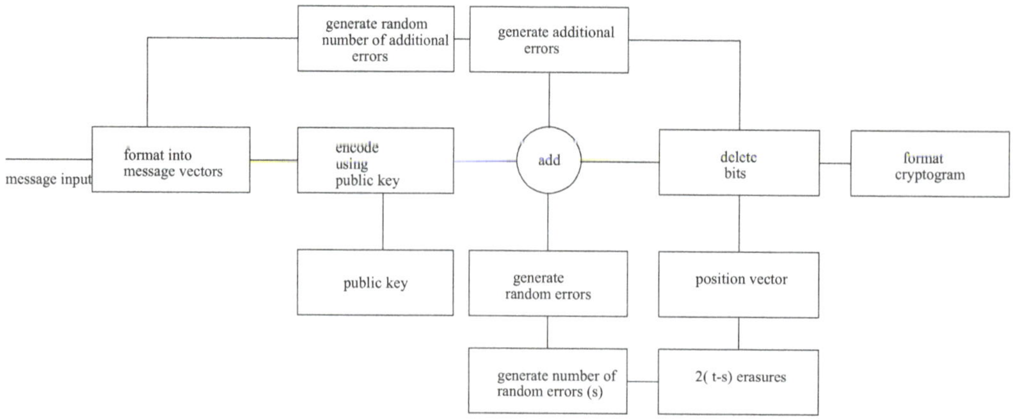

Fig. 19.5 Public key encryption system with s random bit errors, $2(t-s)$ bit deletions and a random number of additional errors

Additional errors may be added in the same positions as random errors, as this provides for a simpler implementation or may take account of the positions of the random errors. However, there is no point in adding additional bit errors to bits which will be subsequently deleted.

As shown in Fig. 19.5, the number of additional errors is communicated to the recipient as part of the message vector in the preceding codeword with the information included with the message. This is carried out by *format into message vectors* shown in Fig. 19.5. In this case, usually 1 or 2 more message vectors in total will be required to convey the information regarding numbers of additional errors and the position vector (if this has not been already communicated to the recipient). Clearly, there are alternative arrangements to communicate the numbers of additional errors to the recipient such as using a previously agreed sequence of numbers or substituting a pseudorandom number generator for the truly random number generator (*generate random number of additional errors* shown in Fig. 19.5) with a known seed.

Using the previous example above, with the position vector:

$$\{19,\ 3,\ 27,\ 17,\ 8,\ 30,\ 11,\ 15,\ 2,\ 5,\ 19,\ \ldots,\ 25\}$$

The errored bits are in positions 7 and 23 (starting the position index from 0) and the deleted bits are in positions {19, 3, 27 and 17}. The encoded codeword prior to corruption is:

$$\{0\,1\,1\,1\,1\,0\,0\,0\,1\,0\,0\,0\,0\,1\,0\,1\,1\,0\,0\,0\,0\,0\,1\,0\,0\,1\,1\,0\,0\,1\,1\,0\}$$

The number of additional bits in error is randomly chosen to be 5, say. As the first 4 positions (index 0–3) in the position vector are to be deleted bits, starting from index 4, the bits in codeword positions {8, 30, 11, 15, and 2} are inverted in addition to the errored bits in positions 7 and 23. The 32 bit corrupted codeword is produced: {0 1 0 1 1 0 0 1 0 0 0 1 0 1 0 0 1 0 0 0 0 0 1 1 0 1 1 0 0 1 0 0}.

The bits in positions {19, 3, 27 and 17} are deleted to produce the 28 bit corrupted codeword:

{0 1 0 1 0 0 1 0 0 0 1 0 1 0 0 1 0 0 0 1 1 0 1 1 0 1 0 0}

The additional bits in error are removed by the recipient of the cryptogram prior to errors and erasures correction as shown in Fig. 19.6. The number of additional bits in error in the following codewords is retrieved from the descrambled message vectors by *format message* shown in Fig. 19.6 and input to *number of additional errors* which outputs this number to *generate additional errors* which is the same as in Fig. 19.5. The position vector is stored in a buffer memory in *position vector* and outputs this to *generate additional errors*. Each additional error is corrected by the adder *add*, shown in Fig. 19.6, which adds, modulo 2, a 1 which is output from *generate additional errors* in the same position of each additional error. Retrieval of the message from this point follows correction of the errors and erasures, descrambling and formatting as described for Fig. 19.5.

Using the number of deleted bits and the position vector, 0's are inserted in the deleted bit positions to form the 32 bit corrupted codeword:

{0 1 0 0 1 0 0 1 0 0 0 1 0 1 0 0 1 0 0 0 0 0 1 1 0 1 1 0 0 1 0 0}

After the addition of the output from *generate additional errors* the bits in positions {8, 30, 11, 15, and 2} are inverted, thereby correcting the 5 additional errors to form the less corrupted codeword:

{0 1 1 0 1 0 0 1 1 0 0 0 0 1 0 1 1 0 0 0 0 0 1 1 0 1 1 0 0 1 1 0}

As in the first approach, this corrupted codeword is permuted, the syndromes calculated and the errors plus erasures corrected to retrieve the original message:

$$\{0, 1, 0, 1, 1, 1, 0, 0, 0, 0, 0, 1\}$$

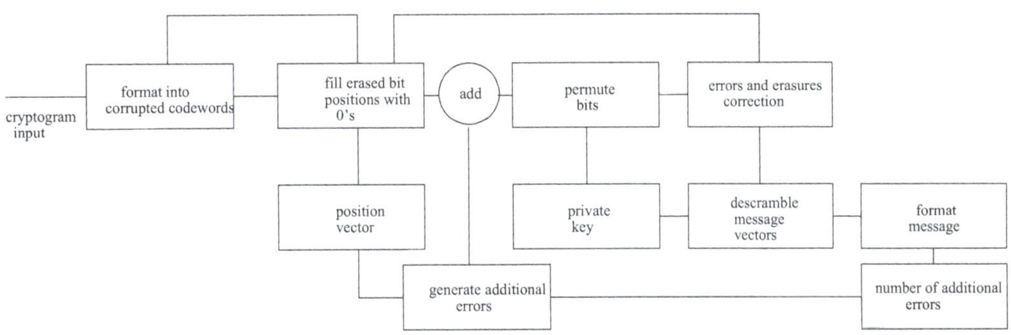

Fig. 19.6 Private key decryption system with s random bit errors, $2(t - s)$ bit deletions and a random number of additional errors

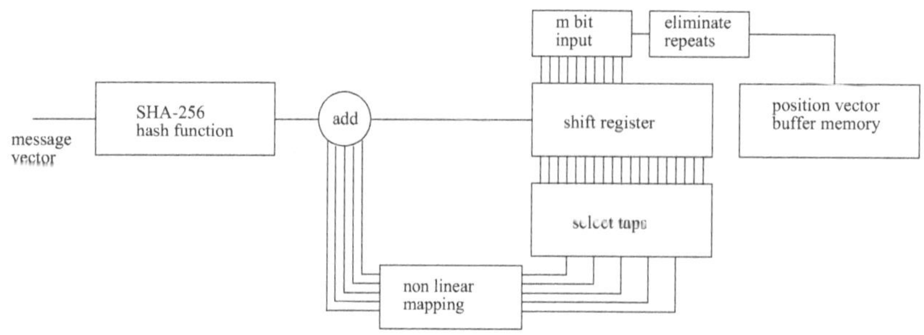

Fig. 19.7 Position vector generated by hash of message vector and nonlinear feedback shift register

In a further option, the position vector, instead of being a static vector, may be derived from a cryptographic hash of a previous message vector. Any standard cryptographic hash function may be used such as SHA-256 [11] or SHA-3 [12] as shown in Fig. 19.7. The message vector of length k bits is hashed using SHA-256 or SHA-3 to produce a binary hash vector of length 256 bits.

For example, the binary hash vector may be input to a nonlinear feedback shift register consisting of *shift register* having p stages, typically 64 stages with outputs determined by *select taps* enabling different scrambling keys to be used by selecting different outputs. The nonlinear feedback shift register arrangement to produce a position vector in *error positions buffer memory* is the same as that of Fig. 19.3 whose operation is described above.

As the hash vector is clocked into the nonlinear feedback shift register of Fig. 19.7, a derived position vector is stored in *error positions buffer memory*, and used for encrypting the message vector as described above. The current message vector is encrypted using a position vector derived from the hash of the previous message vector. As the recipient of the cryptogram has decrypted the previous message vector, the recipient of the cryptogram can use the same hash function and nonlinear feedback shift register to derive the position vector in order to decrypt the current corrupted codeword. There are a number of arrangements that may be used for the first codeword. For example, a static position vector, known only to the sender and recipient of the cryptogram could be used or alternatively a position vector derived from a fixed hash vector known only to the sender and recipient of the cryptogram or the hash of a fixed message known only to the sender and recipient of the cryptogram. A simpler arrangement may be used where the shift register has no feedback so that the position vector is derived directly from the hash vector. In this case the hash function needs to produce a hash vector $\geq n$, the length of the codeword.

As discussed earlier, the original McEliece system is vulnerable to chosen-plaintext attacks. If the same message is encrypted twice, the difference between the two cryptograms is just $2t$ bits or less, the sum of the two error patterns. This vulnerability is completely solved by encrypting or scrambling the plaintext prior to the McEliece system, using the error pattern as the key. To do this, the random error pattern needs to be generated first before the codeword is constructed by encoding with the scrambled generator matrix.

This scrambler which is derived from the error vector, for each message vector, may be implemented in a number of ways. The message vector may be scrambled by multiplying by a $k \times k$ non-singular matrix derived from the error vector.

Alternatively, the message vector may be scrambled by treating the message vector as a polynomial $m_1(x)$ of degree $k-1$ and multiplying it by a circulant polynomial $p_1(x)$ modulo $1+x^k$ which has an inverse [7]. The circulant polynomial $p_1(x)$ is derived from the error vector. Denoting the inverse of the circulant polynomial $p_1(x)$ as $q_1(x)$ then

$$p_1(x)q_1(x) = 1 \text{ modulo } 1 + x^k \tag{19.29}$$

Accordingly the scrambled message vector is $m_1(x)p_1(x)$ which is encoded into a codeword using the scrambled generator matrix. Each message vector is scrambled in a different way as the error patterns are random and different from corrupted codeword to corrupted codeword. The corrupted codewords form the cryptogram.

On decoding of each codeword, the corresponding error vector is obtained with retrieval of the scrambled message vector. Considering the above example, the circulant polynomial $p_1(x)$ is derived from the error vector and the inverse $q_1(x)$ is calculated using Euclid's method [7] from $p_1(x)$. The original message vector is obtained by multiplying the retrieved scrambled message vector $m_1(x)p_1(x)$ by $p_1(x)$ because

$$m_1(x)p_1(x)q_1(x) = m_1(x) \text{ modulo } 1 + x^k \tag{19.30}$$

Another method of scrambling each message vector using a scrambler derived from the error vector is to use two nonlinear feedback shift registers as shown in Fig. 19.8. The first operation is for the error vector, which is represented as a s-bit sequence is input to a modulo 2 adder *add* whose output is input to *shift register A* as shown in Fig. 19.8. The nonlinear feedback shift registers are the same as in Fig. 19.3 with operation as described above but *select taps* will usually have a different setting and *nonlinear mapping* also will usually have a different mapping, but this is not essential. After clocking the s-bit error sequence into the nonlinear feedback shift register, *shift register A* shown in Fig. 19.8 will essentially contain a random binary vector. This vector is used by *define taps* to define which outputs of *shift register B* are to be input to *nonlinear mapping B* whose outputs are added modulo 2 to the message vector input to form the input to *shift register B* shown in Fig. 19.8. The scrambling of the message vector is carried out by a nonlinear feedback shift register whose feedback connections are determined by a random binary vector derived from the error vector, the s-bit error sequence.

The corresponding descrambler is shown in Fig. 19.9. Following decoding of each corrupted codeword, having correcting the random errors and bit erasures, the scrambled message vector is obtained and the error vector is in the form of the s-bit error sequence. As in the scrambler, the s bit error sequence is input to a modulo 2 adder *add* whose output is input to *shift register A* as shown in Fig. 19.9. After clocking the s-bit error sequence into the nonlinear feedback shift register, *shift register A* shown in Fig. 19.9 will contain exactly the same binary vector as *shift register A* of Fig. 19.8. Consequently, exactly the same outputs of *shift register B* to

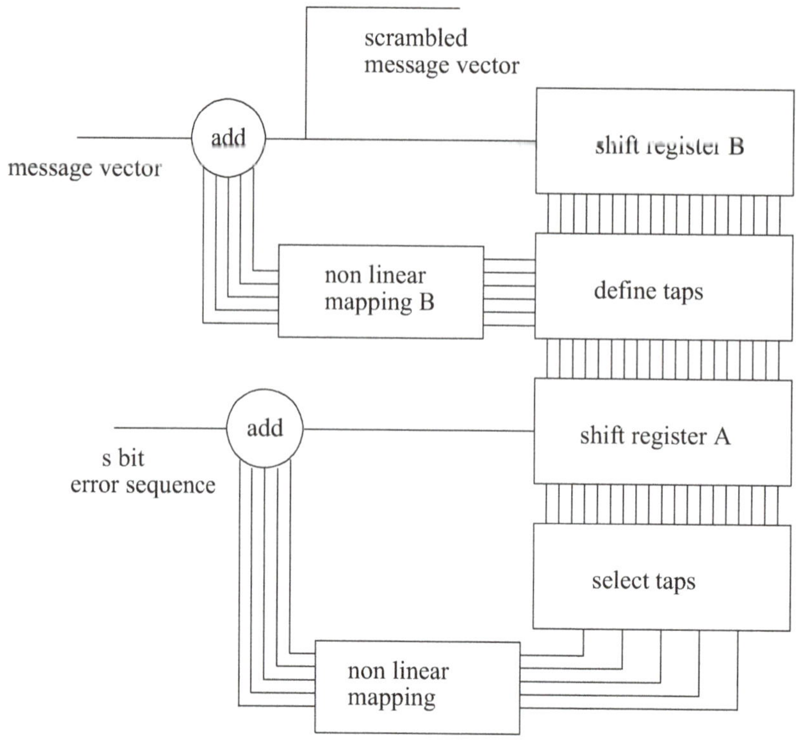

Fig. 19.8 Message vector scrambling by nonlinear feedback shift register with taps defined by s-bit error pattern

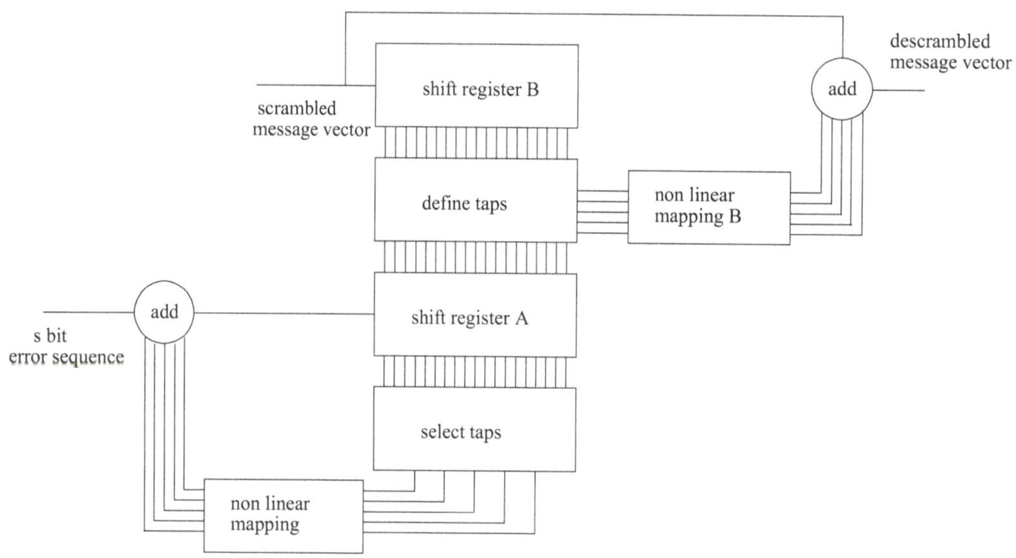

Fig. 19.9 Descrambling independently each scrambled message vector by nonlinear feedback shift register with taps defined by s-bit error pattern

be input to *non linear mapping B* will be defined by *define taps*. Moreover, comparing the input of *shift register B* of the scrambler Fig. 19.8 to the input of *shift register B* of the descrambler Fig. 19.9 it will be seen that the contents are identical and equal to the scrambled message vector.

Consequently, the same selected shift register outputs will be identical and with the same nonlinear mapping *nonlinear mapping B* the outputs of *nonlinear mapping B* in Fig. 19.9 will be identical to those that were the outputs of *nonlinear mapping B* in Fig. 19.8. The result of the addition of these outputs modulo 2 with the scrambled message vector is to produce the original message vector at the output of *add* in Fig. 19.9.

This is carried out for each scrambled message vector and associated error vector to recover the original message.

In some applications, a reduced size cryptogram is essential perhaps due to limited communications or storage capacity. For these applications, a simplification may be used in which the cryptogram consists of only one corrupted codeword containing random errors, the first codeword. The following codewords are corrupted by only deleting bits. The number of deleted bits is $2t$ bits per codeword using a position vector as described above.

For example, with $n = 1024$, and the Goppa code correcting $t = 60$ bit errors, there are $2t$ bits deleted per codeword so that apart from the first corrupted codeword, each corrupted codeword is only 904 bits long and conveys 624 message vector bits per corrupted codeword.

A similar approach is to hash the error vector of the first corrupted codeword and use this hash value as the key of a symmetric encryption system such as the Advanced Encryption Standard (AES) [10] and encrypt any following information this way. Effectively, this is AES encryption operating with a random session key since the error pattern is chosen randomly as in the classic, hybrid encryption system.

19.3 Reducing the Public Key Size

In the original McEliece system, the public key is the $k \times n$ generator matrix which can be quite large. For example with $n = 2048$ and $k = 1148$, the generator matrix needs to be represented by $1148 \times 2048 = 2.35 \times 10^6$ bits. Representing the generator matrix in reduced echelon form reduces the generator matrix to $k \times (n-k) = 1.14 \times 10^6$ bits. In the $n = 32$ example above, the generator matrix is the 12×32 matrix, $\mathbf{PSG}_{(32,12,9)}$, given by Eq. (19.26). Rows of this matrix may be added together using modulo 2 arithmetic so as to produce a matrix with k independent columns. This matrix is a reduced echelon matrix, possibly permuted to obtain k independent columns, and may be straightforwardly derived by using the Gauss–Jordan variable elimination procedure. With permutations, there are a large number of possible solutions which may be derived and candidate column positions may be selected, either initially in consecutive order to determine a solution, or optionally, selected in random order to arrive at other solutions.

Consider as an example the first option of selecting candidate column positions in consecutive order. For the $\mathbf{PSG}_{(32,12,9)}$ matrix (19.26), the following permuted reduced echelon generator matrix is produced:

$$
\mathbf{PSGR}_{(32,12,9)} = \begin{bmatrix}
1&0&0&0&0&0&0&0&0&0&0&0&0&1&0&0&1&1&0&1&0&1&0&1&0&0&0&1&1&1&1&1\\
0&1&0&0&0&0&0&0&0&0&1&0&0&0&0&0&1&0&0&1&0&0&0&1&0&0&1&0&1&1&1&0\\
0&0&1&0&0&0&0&0&0&0&0&0&0&1&0&1&0&0&1&0&1&1&1&1&0&0&1&1&0&0&0&0\\
0&0&0&1&0&0&0&0&0&0&1&1&0&1&0&0&1&0&1&0&0&0&0&1&0&1&0&0&0&0&1&0\\
0&0&0&0&1&0&0&0&0&0&0&1&0&1&0&0&1&1&1&1&0&1&0&0&1&0&0&0&1&0&0&1\\
0&0&0&0&0&1&0&0&0&0&0&1&0&0&0&0&0&1&0&0&1&1&0&1&1&1&1&1&1&1&0&1\\
0&0&0&0&0&0&1&0&0&0&1&1&0&0&0&0&0&0&1&0&1&1&0&0&1&0&0&1&1&0&0&1\\
0&0&0&0&0&0&0&1&0&0&1&1&0&1&0&0&1&1&1&1&1&1&0&0&0&1&1&0&0&1&1&1\\
0&0&0&0&0&0&0&0&1&0&0&0&0&0&0&0&0&1&1&0&1&0&0&1&1&0&1&1&0&0&1&1\\
0&0&0&0&0&0&0&0&0&1&1&0&0&0&0&1&1&1&0&0&1&0&1&1&1&0&0&1&0&1&0&0\\
0&0&0&0&0&0&0&0&0&0&0&0&1&0&0&1&1&0&1&0&1&0&1&1&1&1&0&0&0&1&1&1\\
0&0&0&0&0&0&0&0&0&0&0&0&0&0&1&1&0&1&0&1&0&1&1&1&1&1&0&1&0&1&1&0
\end{bmatrix} \tag{19.31}
$$

The permutation defined by the following input and output bit position sequences is used to rearrange the columns of the permuted, reduced echelon generator matrix.

$$
\begin{matrix}
0\ 1\ 2\ 3\ 4\ 5\ 6\ 7\ 8\ 9\ 12\ 14\ 10\ 11\ 13\ 15\ 16\ 17\ 18\ 19\ 20\ 21\ 22\ 23\ 24\ 25\ 26\ 27\ 28\ 29\ 30\ 31\\
0\ 1\ 2\ 3\ 4\ 5\ 6\ 7\ 8\ 9\ 10\ 11\ 12\ 13\ 14\ 15\ 16\ 17\ 18\ 19\ 20\ 21\ 22\ 23\ 24\ 25\ 26\ 27\ 28\ 29\ 30\ 31
\end{matrix} \tag{19.32}
$$

This permutation produces a classical reduced echelon generator matrix [7], denoted as $\mathbf{Q}_{(32,12,9)}$:

$$
\mathbf{Q}_{(32,12,9)} = \begin{bmatrix}
1&0&0&0&0&0&0&0&0&0&0&0&0&0&1&0&1&1&0&1&0&1&0&1&0&0&0&1&1&1&1&1\\
0&1&0&0&0&0&0&0&0&0&0&0&1&0&0&0&1&0&0&1&0&0&0&1&0&0&1&0&1&1&1&0\\
0&0&1&0&0&0&0&0&0&0&0&0&0&0&1&1&0&0&1&0&1&1&1&1&0&0&1&1&0&0&0&0\\
0&0&0&1&0&0&0&0&0&0&0&0&1&1&1&0&1&0&1&0&0&0&1&0&1&0&0&0&0&0&1&0\\
0&0&0&0&1&0&0&0&0&0&0&0&0&1&1&0&1&1&1&1&0&1&0&0&1&0&0&0&1&0&0&1\\
0&0&0&0&0&1&0&0&0&0&0&0&0&1&0&0&0&1&0&0&1&1&0&1&1&1&1&1&1&1&0&1\\
0&0&0&0&0&0&1&0&0&0&0&0&1&1&0&0&0&0&1&0&1&1&0&0&1&0&0&1&1&0&0&1\\
0&0&0&0&0&0&0&1&0&0&0&0&1&1&1&0&1&1&1&1&1&1&0&0&0&1&1&0&0&1&1&1\\
0&0&0&0&0&0&0&0&1&0&0&0&0&0&0&0&0&1&1&0&1&0&0&1&1&0&1&1&0&0&1&1\\
0&0&0&0&0&0&0&0&0&1&0&0&1&0&0&1&1&1&0&0&1&0&1&1&1&0&0&1&0&1&0&0\\
0&0&0&0&0&0&0&0&0&0&1&0&0&0&0&1&1&0&1&0&1&0&1&1&1&1&0&0&0&1&1&1\\
0&0&0&0&0&0&0&0&0&0&0&1&0&0&0&1&0&1&0&1&0&1&1&1&1&1&0&1&0&1&1&0
\end{bmatrix} \tag{19.33}
$$

Codewords generated by this matrix are from a systematic code [7] with the first 12 bits being information bits and the last 20 bits being parity bits. Correspondingly, the matrix above, $\mathbf{Q}_{(32,12,9)}$ consists of an identity matrix followed by a matrix denoted as $\mathbf{QT}_{(32,12,9)}$ which defines the parity bits part of the generator matrix. The transpose of this matrix is the parity check matrix of the code [7]. As shown in Fig. 19.10, the public key consists of the parity check matrix, less the identity sub-matrix, and a sequence of n numbers representing a permutation of the codeword bits after encoding. By permuting the codewords with the inverse permutation, the resulting permuted codewords will be identical to codewords produced by $\mathbf{PSG}_{(32,12,9)}$, the public key of the original McEliece public key system [8]. However, whilst the codewords are identical, the information bits will not correspond.

The permutation is defined by the following input and output bit position sequences.

$$
\begin{matrix}
0\ 1\ 2\ 3\ 4\ 5\ 6\ 7\ 8\ 9\ 12\ 14\ 10\ 11\ 13\ 15\ 16\ 17\ 18\ 19\ 20\ 21\ 22\ 23\ 24\ 25\ 26\ 27\ 28\ 29\ 30\ 31\\
0\ 1\ 2\ 3\ 4\ 5\ 6\ 7\ 8\ 9\ 10\ 11\ 12\ 13\ 14\ 15\ 16\ 17\ 18\ 19\ 20\ 21\ 22\ 23\ 24\ 25\ 26\ 27\ 28\ 29\ 30\ 31
\end{matrix} \tag{19.34}
$$

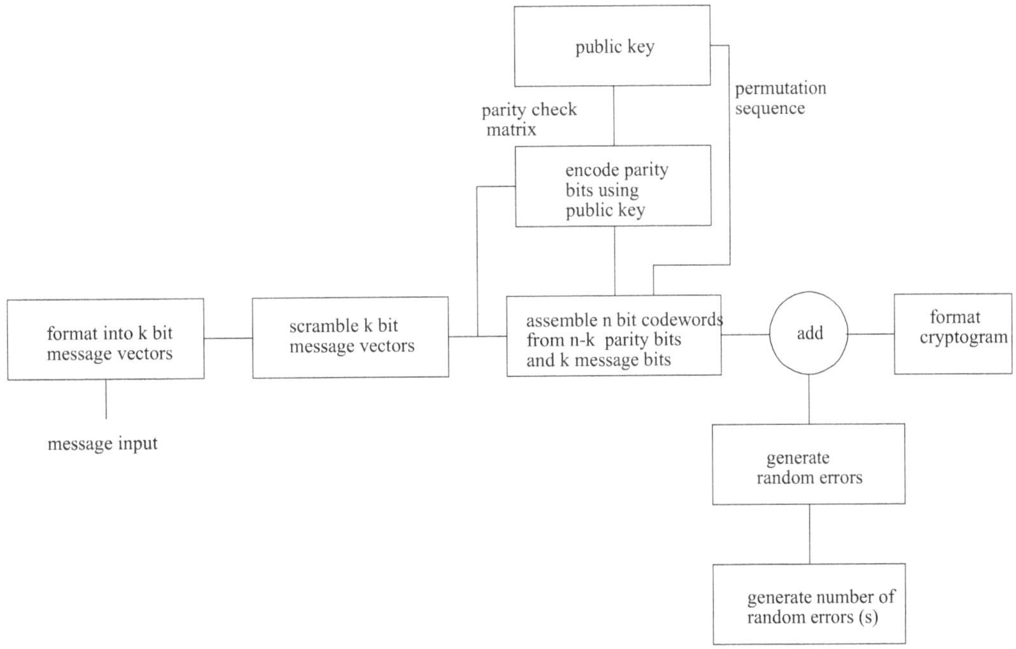

Fig. 19.10 Reduced size public key encryption system

As the output bit position sequence is just a sequence of bits in natural order, the permutation may be defined only by the input bit position sequence.

In this case, the public key consists of an n position permutation sequence and in this example the sequence chosen is:

$$0\ 1\ 2\ 3\ 4\ 5\ 6\ 7\ 8\ 9\ 12\ 14\ 10\ 11\ 13\ 15\ 16\ 17\ 18\ 19\ 20\ 21\ 22\ 23\ 24\ 25\ 26\ 27\ 28\ 29\ 30\ 31 \tag{19.35}$$

and the $k \times (n-k)$ matrix, $\mathbf{QT}_{(32,12,9)}$, which in this example is the 12×20 matrix:

$$\mathbf{QT}_{(32,12,9)} = \begin{bmatrix} 0&0&1&0&1&1&0&1&0&1&0&1&0&0&0&1&1&1&1&1 \\ 1&0&0&0&1&0&0&1&0&0&0&1&0&0&1&0&1&1&1&0 \\ 0&0&1&1&0&0&1&0&1&1&1&1&0&0&1&1&0&0&0&0 \\ 1&1&1&0&1&0&1&0&0&0&0&1&0&1&0&0&0&0&1&0 \\ 0&1&1&0&1&1&1&1&0&1&0&0&1&0&0&0&1&0&0&1 \\ 0&1&0&0&0&1&0&0&1&1&0&1&1&1&1&1&1&1&0&1 \\ 1&1&0&0&0&0&1&0&1&1&0&0&1&0&0&1&1&0&0&1 \\ 1&1&1&0&1&1&1&1&1&1&0&0&0&1&1&0&0&1&1&1 \\ 0&0&0&0&0&1&1&0&1&0&0&1&1&0&1&1&0&0&1&1 \\ 1&0&0&1&1&1&0&0&1&0&1&1&1&0&0&1&0&1&0&0 \\ 0&0&0&1&1&0&1&0&1&0&1&1&1&1&0&0&0&1&1&1 \\ 0&0&0&1&0&1&0&1&0&1&1&1&1&1&0&1&0&1&1&0 \end{bmatrix} \tag{19.36}$$

The public key of this system is much smaller than the public key of the original McEliece public key system, since as discussed below, there is no need to include the permutation sequence in the public key.

The message is split into message vectors of length 12 bits adding padding bits as necessary so that there is an integral number of message vectors. Each message vector, after scrambling, is encoded as a systematic codeword using $\mathbf{QT}_{(32,\,12,\,9)}$, part of the public key. Each systematic codeword that is obtained is permuted using the permutation (19.35), the other part of the public key. The resulting codewords are identical to codewords generated using the generator matrix $\mathbf{PSG}_{(32,\,12,\,9)}$ (19.26), the corresponding public key of the original McEliece public key system, but generated by different messages.

It should be noted that it is not necessary to use the exact permutation sequence that produces codewords identical to that produced by the original McEliece public key system for the same Goppa code and input parameters. As every permutation sequence has an inverse permutation sequence, any arbitrary permutation sequence, randomly generated or otherwise, may be used for the permutation sequence part of the public key. The permutation sequence that is the inverse of this arbitrary permutation sequence is absorbed into the permutation sequence used in decryption and forms part of the private key. The security of the system is enhanced by allowing arbitrary permutation sequences to be used and permutation sequences do not need to be part of the public key.

The purpose of scrambling each message vector using the fixed scrambler shown in Fig. 19.10 is to provide a one-to-one mapping between the 2^k possible message vectors and the 2^k scrambled message vectors such that the reverse mapping, which is provided by the descrambler, used in decryption, produces error multiplication if there are any errors present. For many messages, some information can be gained even if the message contains errors. The scrambler and corresponding descrambler prevents information being gained this way from the cryptogram itself or by means of some error guessing strategy for decryption by an attacker. The descrambler is designed to have the property that it produces descrambled message vectors that are likely to have a large Hamming distance between vectors for input scrambled message vectors which differ in a small number of bit positions.

There are a number of different techniques of realising such a scrambler and descrambler. One method is to use symmetric key encryption such as the Advanced Encryption Standard (AES) [10] with a fixed key.

An alternative means is provided by the scrambler arrangement shown in Fig. 19.11. The same arrangement may be used for descrambling but with different shift register taps and is shown in Fig. 19.12. Denoting each k bit message vector as a polynomial $m(x)$ of degree $k-1$:

$$m(x) = m_0 + m_1x + m_2x^2 + m_3x^3 \cdots + m_{k-1}x^{k-1} \tag{19.37}$$

and denoting the tap positions determined by *define taps* of Fig. 19.11 by $\mu(x)$ where

$$\mu(x) = \mu_0 + \mu_1x + \mu_2x^2 + \mu_3x^3 \cdots + \mu_{k-1}x^{k-1} \tag{19.38}$$

where the coefficients μ_0 through to μ_{k-1} have binary values of 1 or 0.

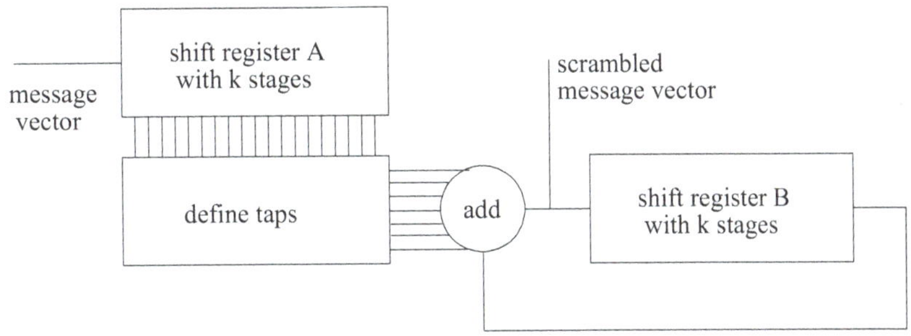

Fig. 19.11 Scrambler arrangement

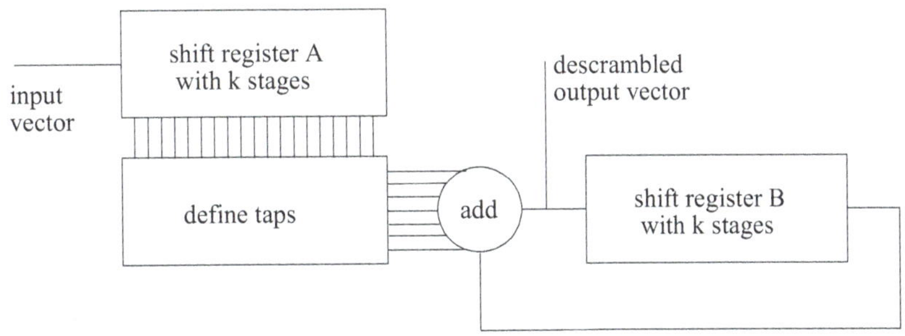

Fig. 19.12 Descrambler arrangement

The output of the scrambler, denoted by the polynomial, $scram(x)$, is the scrambled message vector given by the polynomial multiplication

$$scram(x) = m(x).\mu(x) \text{ modulo } (1 + x^k) \tag{19.39}$$

The scrambled message vector is produced by the arrangement shown in Fig. 19.11 after *shift register A with k stages* and *shift register B with k stages* have been clocked $2k$ times and is present at the input of *shift register B with k stages* whose last stage output is connected to the adder, *adder* input. The input of *shift register B with k stages* corresponds to the scrambled message vector for the next additional k clock cycles, with these bits defining the binary coefficients of $scram(x)$. The descrambler arrangement is shown in Fig. 19.12 and is an identical circuit to that of the scrambler but with different tap settings. The descrambler is used in decryption.

For $k = 12$ an example of a good scrambler polynomial, $\mu(x)$ is

$$\mu(x) = 1 + x + x^4 + x^5 + x^8 + x^9 + x^{11} \tag{19.40}$$

For brevity, the binary coefficients may be represented as a binary vector. In this example, $\mu(x)$ is represented as $\{1\,1\,0\,0\,1\,1\,0\,0\,1\,1\,0\,1\}$. This is a good scrambler polynomial because it has a relatively large number of taps (seven taps) and its inverse, the descrambler polynomial also has a relatively large number of taps (seven taps). The corresponding descrambler polynomial, $\theta(x)$ is

$$\theta(x) = 1 + x + x^3 + x^4 + x^7 + x^8 + x^{11} \tag{19.41}$$

which may be represented by the binary vector {1 1 0 1 1 0 0 1 1 0 0 1}. It is straightforward to verify that

$$\begin{aligned} \mu(x) \times \theta(x) = & \quad 1 + x^2 + x^3 + x^4 + x^5 + x^6 + x^8 + x^{10} \\ & + x^{14} + x^{15} + x^{16} + x^{17} + x^{18} + x^{20} + x^{22} \\ = & \quad 1 \text{ modulo } (1 + x^k) \end{aligned} \tag{19.42}$$

and so

$$scram(x) \times \theta(x) = m(x) \text{ modulo } (1 + x^k) \tag{19.43}$$

As a simple example of a message, consider that the message consists of a single message vector with the information bit pattern {0, 1, 0, 1, 0, 0, 0, 0, 0, 0, 0, 1} and so:

$$m(x) = x + x^3 + x^{11} \tag{19.44}$$

This is input to the scrambling arrangement shown in Fig. 19.11. The scrambled message output is $scram(x) = m(x) \times \mu(x)$ given by

$$\begin{aligned} scram(x) = & \ (1 + x + x^4 + x^5 + x^8 + x^9 + x^{11}).(x + x^3 + x^5 + x^{11}) \\ = & \quad x + x^2 + x^5 + x^6 + x^9 + x^{10} + x^{12} \\ & + x^3 + x^4 + x^7 + x^8 + x^{11} + x^{12} + x^{14} \\ & + x^{11} + x^{12} + x^{15} + x^{16} + x^{19} + x^{20} + x^{22} \\ & \quad \text{modulo } (1 + x^{12}) \\ = & \quad 1 + x + x^5 + x^6 + x^9 \end{aligned} \tag{19.45}$$

and the scrambling arrangement shown in Fig. 19.11 produces the scrambled message bit pattern {1, 1, 0, 0, 0, 1, 1, 0, 0, 1, 0, 0}.

Referring to Fig. 19.10, the next stage is to use the parity check matrix part of the *public key* to calculate the parity bits from the information bits. Starting with an all 0's vector, where the information bit pattern is a 1, the corresponding row from $\mathbf{QT}_{(32,12,9)}$ (19.36) with the same position is added modulo 2 to the result so far to produce the parity bits which with the information bits will form the digital cryptogram plus added random errors after permuting the order of the bits. In this example, this codeword is generated from adding modulo 2, rows 1, 2, 6, 7 and 10 of $\mathbf{QT}_{(32,12,9)}$ to produce:

$$
\begin{array}{cccccccccccccccccccc}
0 & 0 & 1 & 0 & 1 & 1 & 0 & 1 & 0 & 1 & 0 & 1 & 0 & 0 & 0 & 1 & 1 & 1 & 1 & 1 \\
+ & + & + & + & + & + & + & + & + & + & + & + & + & + & + & + & + & + & + & + \\
1 & 0 & 0 & 0 & 1 & 0 & 0 & 1 & 0 & 0 & 0 & 1 & 0 & 0 & 1 & 0 & 1 & 1 & 1 & 0 \\
+ & + & + & + & + & + & + & + & + & + & + & + & + & + & + & + & + & + & + & + \\
0 & 1 & 0 & 0 & 0 & 1 & 0 & 0 & 1 & 1 & 0 & 1 & 1 & 1 & 1 & 1 & 1 & 1 & 0 & 1 \\
+ & + & + & + & + & + & + & + & + & + & + & + & + & + & + & + & + & + & + & + \\
1 & 1 & 0 & 0 & 0 & 0 & 1 & 0 & 1 & 1 & 0 & 0 & 1 & 0 & 0 & 1 & 1 & 0 & 0 & 1 \\
+ & + & + & + & + & + & + & + & + & + & + & + & + & + & + & + & + & + & + & + \\
1 & 0 & 0 & 1 & 1 & 1 & 0 & 0 & 1 & 0 & 1 & 1 & 1 & 0 & 0 & 1 & 0 & 1 & 0 & 0 \\
\shortparallel & \shortparallel & \shortparallel & \shortparallel & \shortparallel & \shortparallel & \shortparallel & \shortparallel & \shortparallel & \shortparallel & \shortparallel & \shortparallel & \shortparallel & \shortparallel & \shortparallel & \shortparallel & \shortparallel & \shortparallel & \shortparallel & \shortparallel \\
1 & 0 & 1 & 1 & 1 & 1 & 1 & 0 & 1 & 1 & 1 & 0 & 1 & 1 & 0 & 0 & 0 & 0 & 0 & 1
\end{array}
\tag{19.46}
$$

The resulting systematic code, codeword is:

$$\{1\,1\,0\,0\,0\,1\,1\,0\,0\,1\,0\,0\,1\,0\,1\,1\,1\,1\,1\,0\,1\,1\,1\,0\,1\,1\,0\,0\,0\,0\,0\,1\}$$

The last step in constructing the final codeword which will be used to construct the cryptogram is to apply an arbitrary preset permutation sequence. Referring to Fig. 19.10, the operation *assemble n bit codewords from n-k parity bits and k message bits* simply takes each codeword encoded as a systematic codeword and applies the preset permutation sequence.

In this example, the permutation sequence that is used is not chosen arbitrarily but is the permutation sequence that will produce the same codewords as the original McEliece public key system for the same Goppa code and input parameters. The permutation sequence is:

$$0\ 1\ 2\ 3\ 4\ 5\ 6\ 7\ 8\ 9\ 12\ 14\ 10\ 11\ 13\ 15\ 16\ 17\ 18\ 19\ 20\ 21\ 22\ 23\ 24\ 25\ 26\ 27\ 28\ 29\ 30\ 31 \tag{19.47}$$

The notation is that the 10th bit should move to the 12th position, the 11th bit should move to the 14th position, the 12th bit should move to the 10th position, the 13th bit should move to the 11th position, the 14th bit should move to the 13th position and all other bits remain in their same positions.

Accordingly, the permuted codeword becomes:

$$\{1\,1\,0\,0\,0\,1\,1\,0\,0\,1\,1\,0\,0\,1\,0\,1\,1\,1\,1\,0\,1\,1\,1\,0\,1\,1\,0\,0\,0\,0\,0\,1\}$$

and this will be the input to the adder, *add* of Fig. 19.1.

The Goppa code used in this example can correct up to 4 errors, ($t = 4$), and a random number is chosen for the number of bits to be in error, (s) with $s \leq 4$.

A truly random source such as a thermal noise source as described above produces the most secure results, but a pseudorandom generator can be used instead, particularly if seeded from the time of day with fine time resolution such as 1mS. If the number of random errors chosen is too few, the security of the digital cryptogram will be compromised. Correspondingly, the minimum number of errors chosen is a design parameter depending upon the length of the Goppa code and t, the number of correctable errors. A suitable choice for the minimum number of errors chosen in practice lies between $\frac{t}{2}$ and t. If the cryptogram is likely to be subject to additional errors due to transmission over a noisy or interference prone medium such as wireless, or stored and read using an imperfect reader such as in barcode applications, then these additional errors can be corrected as well as the deliberately introduced errors provided the total number of errors is no more than t errors.

For such applications typically the number of deliberate errors is constrained to be between $\frac{t}{3}$ and $\frac{2t}{3}$.

For the example above, consider that the number of bit errors is 3 and these are randomly chosen to be in positions 4, 11 and 27 (starting the position index from 0). The bits in these positions in the codeword are inverted to produce the result

$$\{1\,1\,0\,0\,1\,1\,1\,0\,0\,1\,1\,1\,0\,1\,0\,1\,1\,1\,1\,0\,1\,1\,1\,0\,1\,1\,0\,1\,0\,0\,0\,1\}$$

The dcryptogram is this corrupted codeword, which is transmitted or stored depending upon the application.

The intended recipient of this cryptogram retrieves the message in a series of steps. Figure 19.13 shows the system used for decryption. The retrieved cryptogram is formatted into corrupted codewords by *format into corrupted codewords* shown in Fig. 19.4. For the example above, the recipient first receives or otherwise retrieves the cryptogram, which may contain additional errors.

$$\{1\,1\,0\,0\,1\,1\,1\,0\,0\,1\,1\,1\,0\,1\,0\,1\,1\,1\,1\,0\,1\,1\,1\,0\,1\,1\,0\,1\,0\,0\,0\,1\}.$$

It is assumed in this example that no additional errors have occurred although with this particular example one additional error can be accommodated.

The private key contains the information of which Goppa code was used and a first permutation sequence which when applied to the retrieved, corrupted codewords which make up the cryptogram produces corrupted codewords of the Goppa code with the bits in the correct order. Usually, the private key also contains a second

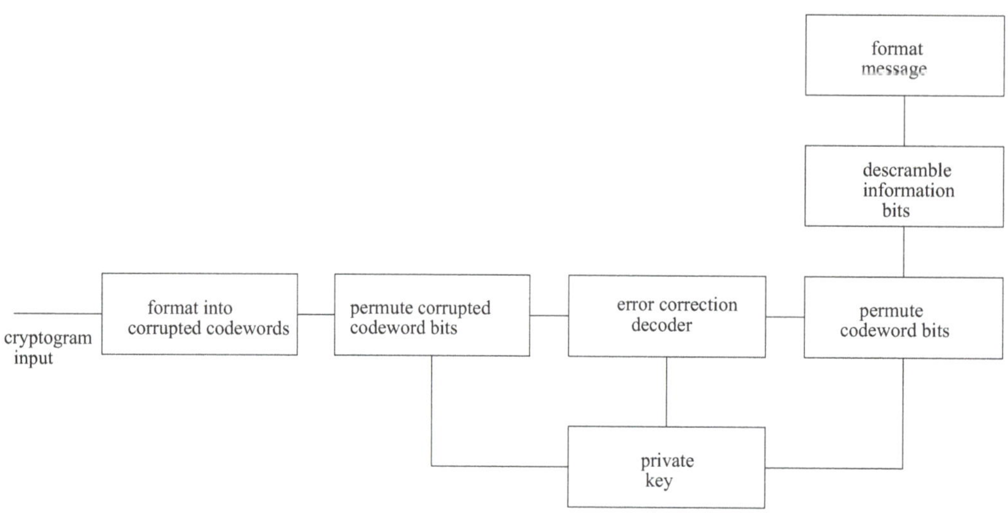

Fig. 19.13 Private key decryption system

permutation sequence which when applied to the error-corrected Goppa codewords puts the scrambled information bits in natural order. Sometimes the private key also contains a third permutation sequence which when applied to the error vectors found in decoding the corrupted corrected Goppa codewords puts the bit errors in the same order that they were when inserted during encryption. All of this information is stored in *private key* in Fig. 19.13. Other information necessary to decrypt the cryptogram, such as the descrambler required may also be stored in the private key or be implicit.

There are two permutation sequences stored as part of the private key and the decryption arrangement is shown in Fig. 19.13. The corrupted codewords retrieved from the received or read cryptogram are permuted with a first permutation sequence which will put the bits in each corrupted codeword in the same order as the Goppa codewords. In this example, the first permutation sequence stored as part of the private key is:

$$24\ 11\ 3\ 23\ 2\ 17\ 20\ 8\ 13\ 30\ 31\ 14\ 15\ 22\ 7\ 1\ 9\ 6\ 21\ 4\ 10\ 5\ 28\ 26\ 19\ 16\ 25\ 0\ 27\ 12\ 18\ 29 \tag{19.48}$$

This defines the following permutation input and output sequences:

$$\begin{matrix} 0 & 1 & 2 & 3 & 4 & 5 & 6 & 7 & 8 & 9 & 10 & 11 & 12 & 13 & 14 & 15 & 16 & 17 & 18 & 19 & 20 & 21 & 22 & 23 & 24 & 25 & 26 & 27 & 28 & 29 & 30 & 31 \\ 27 & 15 & 4 & 2 & 19 & 21 & 17 & 14 & 7 & 16 & 20 & 1 & 29 & 8 & 11 & 12 & 25 & 5 & 30 & 24 & 6 & 18 & 13 & 3 & 0 & 26 & 23 & 28 & 22 & 31 & 9 & 10 \end{matrix} \tag{19.49}$$

so that for example bit 23 becomes bit 3 after permutation and bit 30 becomes bit 9 after permutation. The resulting, permuted corrupted codeword is:

$$\{1\,1\,0\,0\,0\,1\,1\,0\,1\,0\,1\,0\,1\,1\,0\,1\,1\,1\,1\,1\,1\,1\,0\,0\,0\,1\,1\,1\,1\,0\,1\,0\}$$

The permutation is carried out by *permute corrupted codeword bits* shown in Fig. 19.13 with the first permutation sequence input from *private key*.

Following the permutation of each corrupted codeword, the codeword bits are in the correct order to satisfy the parity check matrix, matrix (19.19) if there were no codeword bit errors. (In this case all of the syndrome values would be equal to 0). The next step is to treat each bit in each permuted corrupted codeword as a $GF(2^5)$ symbol with a 1 equal to α^0 and a 0 equal to 0 and use the parity check matrix, matrix (19.19), stored as part of *private key* to calculate the syndrome values for each row of the parity check matrix. The syndrome values produced in this example, are respectively α^{30}, α^{27}, α^4, and α^2. In Fig. 19.13 *error-correction decoder* calculates the syndromes as a first step in correcting the bit errors.

The bit errors are corrected using the syndrome values to produce an error free codeword from the Goppa code for each permuted corrupted codeword. There are many published algorithms for correcting bit errors for Goppa codes, but the most straightforward is to use a BCH decoder as described by Retter [13] because Berlekamp–Massey may then be used to solve the key equation. After decoding, the error free permuted codeword is obtained:

{1 0 0 0 0 1 1 0 1 0 1 0 1 1 0 1 1 1 1 0 1 1 0 0 0 1 1 1 0 0 1 0}

and the error pattern, defined as a 1 in each error position is

{0 1 0 0 0 0 0 0 0 0 0 0 0 0 0 0 0 0 1 0 0 0 0 0 0 0 0 1 0 0 0}.

As shown in Fig. 19.13 *permute codeword bits* takes the output of *error-correction decoder* and applies the second permutation sequence stored as part of the private key to each corrected codeword.

Working through the example, consider that the following permutation input and output sequences is applied to the error free permuted codeword (the decoded codeword of the Goppa code).

27 15 4 2 19 21 17 14 7 16 20 1 29 8 11 12 25 5 30 24 6 18 13 3 0 26 23 28 22 31 9 10
0 1 2 3 4 5 6 7 8 9 10 11 12 13 14 15 16 17 18 19 20 21 22 23 24 25 26 27 28 29 30 31 (19.50)

The result is that the scrambled message bits correspond to bit positions:

{0 1 2 3 4 5 6 7 8 9 12 14}

from the encryption procedure described above. The scrambled message bits may be repositioned in bit positions:

{0 1 2 3 4 5 6 7 8 9 10 11}

by absorbing the required additional permutations into a permutation sequence defined by the following permutation input and output sequences:

27 15 4 2 19 21 17 14 7 16 29 11 20 8 1 12 25 5 30 24 6 18 13 3 0 26 23 28 22 31 9 10
0 1 2 3 4 5 6 7 8 9 10 11 12 13 14 15 16 17 18 19 20 21 22 23 24 25 26 27 28 29 30 31 (19.51)

The second permutation sequence which corresponds to this net permutation and which is stored as part of the private key, *private key* shown in Fig. 19.13 is:

24 14 3 23 2 17 20 8 13 30 31 11 15 22 7 1 9 6 21 4 12 5 28 26 19 16 25 0 27 10 18 29 (19.52)

The second permutation sequence is applied by *permute codeword bits*. Since the encryption and decryption permutation sequences are all derived at the same time in forming the public key and private key from the chosen Goppa code, it is straightforward to calculate and store the net relevant permutation sequences as part of the private key.

Continuing working through the example, applying the second permutation sequence to the error free permuted codeword produces the output of *permute codeword bits*. The first 12 bits of the result will be the binary vector, {1 1 0 0 0 1 1 0 0 1 0 0} and it can be seen that this is identical to the scrambled message vector produced from the encryption operation. Represented as a polynomial the binary vector is $1 + x + x^5 + x^6 + x^9$.

As shown in Fig. 19.13, the next step is for the k information bits of each permuted error free codeword to be descrambled by *descramble information bits*. In this example, *descramble information bits* is carried out by the descrambler arrangement shown in Fig. 19.12 with *define taps* corresponding to polynomial $1 + x + x^3 + x^4 + x^7 + x^8 + x^{11}$.

The output of the descrambler in polynomial form is $(1 + x + x^5 + x^6 + x^9).(1 + x + x^3 + x^4 + x^7 + x^8 + x^{11})$ modulo $1 + x^{12}$. After polynomial multiplication, the result is $(x + x^3 + x^{11})$ corresponding to the message

{0 1 0 1 0 0 0 0 0 0 0 1}

It is apparent that this is the same as the original plaintext message prior to encryption.

With each cryptogram restricted to contain s errors, the cryptosystem as well as providing security, is able automatically to correct $t - s$ errors occurring in the communication of the cryptogram as shown in Fig. 19.14. It makes no difference to the decryption arrangement of Fig. 19.13, whether the bit errors were introduced deliberately during encryption or were introduced due to errors in transmitting the cryptogram. A correct message is output after decryption provided the total number of bit errors is less than or equal to t, the error-correcting capability of the Goppa code used to construct the public and private keys.

As an illustration, a (512, 287, 51) Goppa code of length 512 bits with message vectors of length 287 bits can correct up to 25 bit errors, ($t = 25$). With $s = 15$, 15 bit errors are added to each codeword during encryption. Up to 10 additional bit errors can occur in transmission of each corrupted codeword and the message will be still recovered correctly from the received cryptogram.

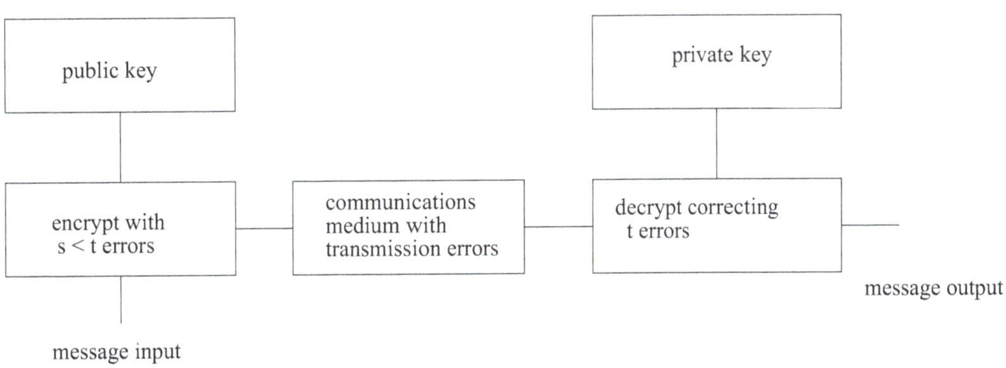

Fig. 19.14 Public key encryption system correcting communication transmission errors

The system will also correct errors in the reading of cryptograms stored in data media. As an example a medium to long range ISO 18000 6B RFID system operating in the 860–930 MHz with 2048 bits of user data can be read back from a tag. A (2048, 1388, 121) Goppa code of length 2048 bits with message vectors of length 1388 bits can correct 60 errors, ($t = 60$). With $s = 25$, 25 bit errors are added to the codeword during encryption and this is written to each passive tag as a cryptogram, stored in non-volatile memory. As well as providing confidentiality of the tag contents, up to 35 additional bit errors can be tolerated in reading each passive tag, thereby extending the operational range. The plaintext message, the encrypted tag payload information of 1388 bits will be recovered more reliably with each scanning of the tag.

19.4 Reducing the Cryptogram Length Without Loss of Security

In many applications a key encapsulation system is used. This is a hybrid encryption system in which a public key cryptosystem is used to send a random session key to the recipient and a symmetric key encryption system, such as AES [10] is used to encrypt the following data. Typically a session key is 256 bits long. To provide the same 256 bit security level a code length of 8192 bits needs to be used, with a code rate of 0.86. Security analysis of the McEliece system is provided in Sect. 19.5. The Goppa code is the (8192, 7048, 177) code. There are 7048 information bits available in each codeword, but only 256 bits are needed to communicate the session key. The code could be shortened in the traditional manner by truncating the generator matrix but this will leave less room to insert the t errors thereby reducing the security. The obvious question is can the codeword be shortened without reducing the security?

Niederreiter [9] solved this problem by transmitting only the $n-k$ bits of the syndrome calculated from the error pattern. Niederreiter originally proposed in his paper a system using Generalised Reed–Solomon codes but this scheme was subsequently broken with an attack by Sidelnikov and Shestakov [15]. However their attack fails if binary Goppa codes are used instead of Generalised Reed–Solomon codes and the Niederreiter system is now associated with the transmission of the $n - k$ parity bits as syndromes of the McEliece system.

It is unfortunate that only a tiny fraction of the 2^{n-k} syndromes correspond to correctable error patterns. For the (8192, 7048, 177) Goppa code, it turns out that there are 2^{697} correctable syndromes out of the total of 2^{1144} syndromes. The probability of an arbitrary syndrome being decoded is 2^{-447}, around 3×10^{-135}. This is the limitation of the Niederreiter system. The plaintext message has to be mapped into an error pattern consisting of t bits uniformly distributed over n bits. Any deterministic method of doing this will be vulnerable to a chosen-plaintext attack. Of course the Niederreiter system can be used to send random messages, first generated as a random error pattern, as in a random session key. However, additional information really needs to be sent as well, such as a MAC, timestamp, sender ID, digital signature or other supplementary information.

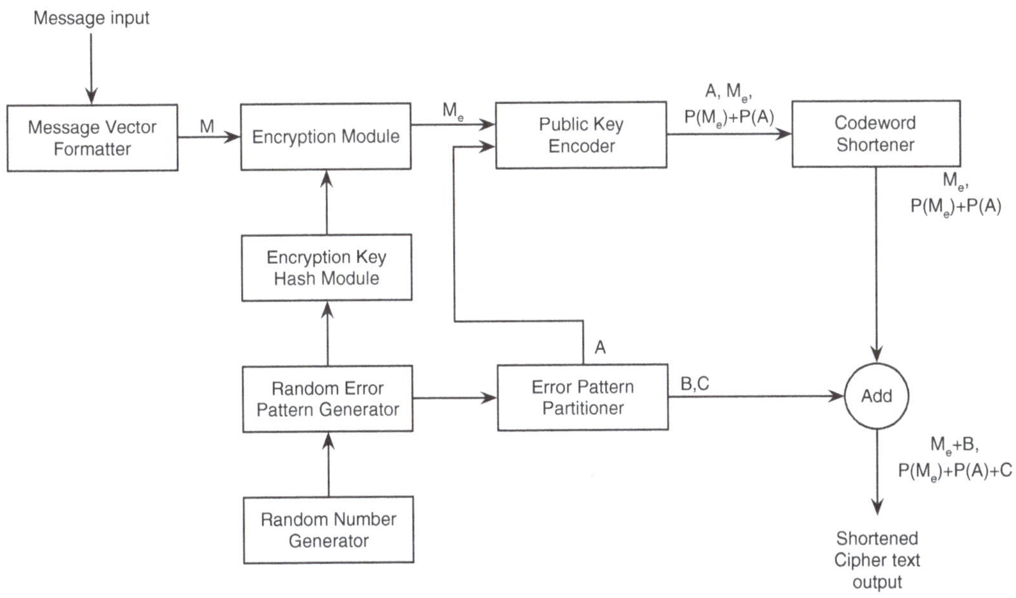

Fig. 19.15 Public key encryption system with shortened codewords

One solution is to use the system shown in Fig. 19.15. The plaintext message, M, consisting of a 256 bit random session key concatenated with 512 bits of supplementary information such as a MAC, time stamp and sender ID is encrypted in the encryption module with a key that is a cryptographic hash, such as SHA-3 [12], of the error pattern. The encrypted message is M_e. The error pattern consists of t bit errors randomly distributed over n bits. This bit pattern is partitioned into three parts, A, B and C as shown in Fig. 19.16. Using the (8192, 7048, 177) Goppa code, part A covers the first 6280 bits, part B covers the next 768 bits and part C covers the last 1144 bits, the parity bits.

The public key encoder consists of the public key generator matrix in reduced echelon form which is used to encode information bits consisting of part A, concatenated with M_e as shown in Fig. 19.15. After encoding, the $n - k$ parity bits of the codeword are $P(A) + P(M_e)$. The codeword is then shortened by removing the first 6280 bits of the codeword. The error pattern parts B and C are added to the shortened codeword of length 1912 bits to form the ciphertext of length 1912 bits. The format of the various parts of the error pattern, the hash derivation, encryption and codeword are shown in Fig. 19.16.

The principle that this system uses, is that the syndrome of any codeword is zero and that the cryptosystem is linear. The codeword resulting from the encoding of A is $\{A\ 0\ \ldots\ 0\ P(A)\}$, where $P(A)$ are the parity bits. The sum of syndromes from sections, of this codeword must be zero. Hence:

$$\text{Syndrome}(A) + 0\ \ldots\ 0 + P(A) = 0$$

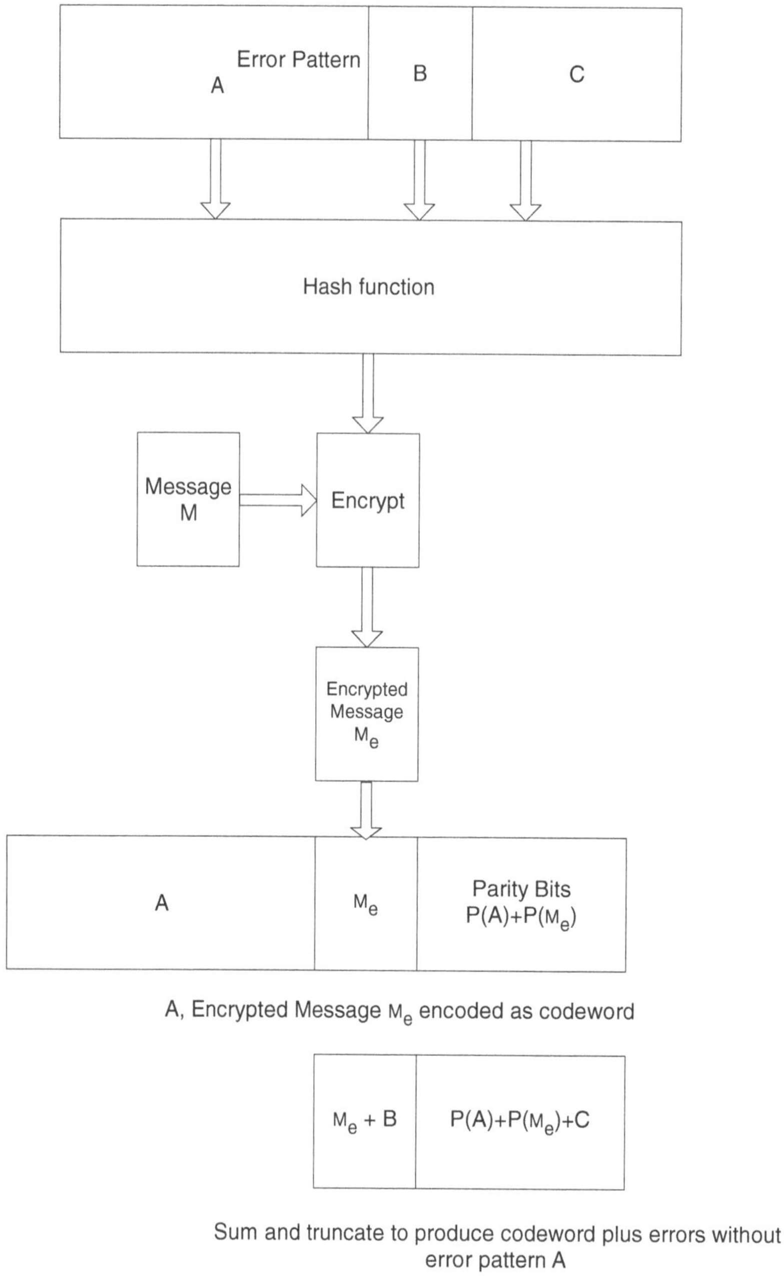

Fig. 19.16 Format of the shortened codeword and error pattern

As the base field is 2,

$$\text{Syndrome}(A) = P(A)$$

Consequently, by including $P(A)$ instead of the error bits, part A in the ciphertext results in the same syndrome being calculated in the decoder, namely $P(A) + P(B) +$

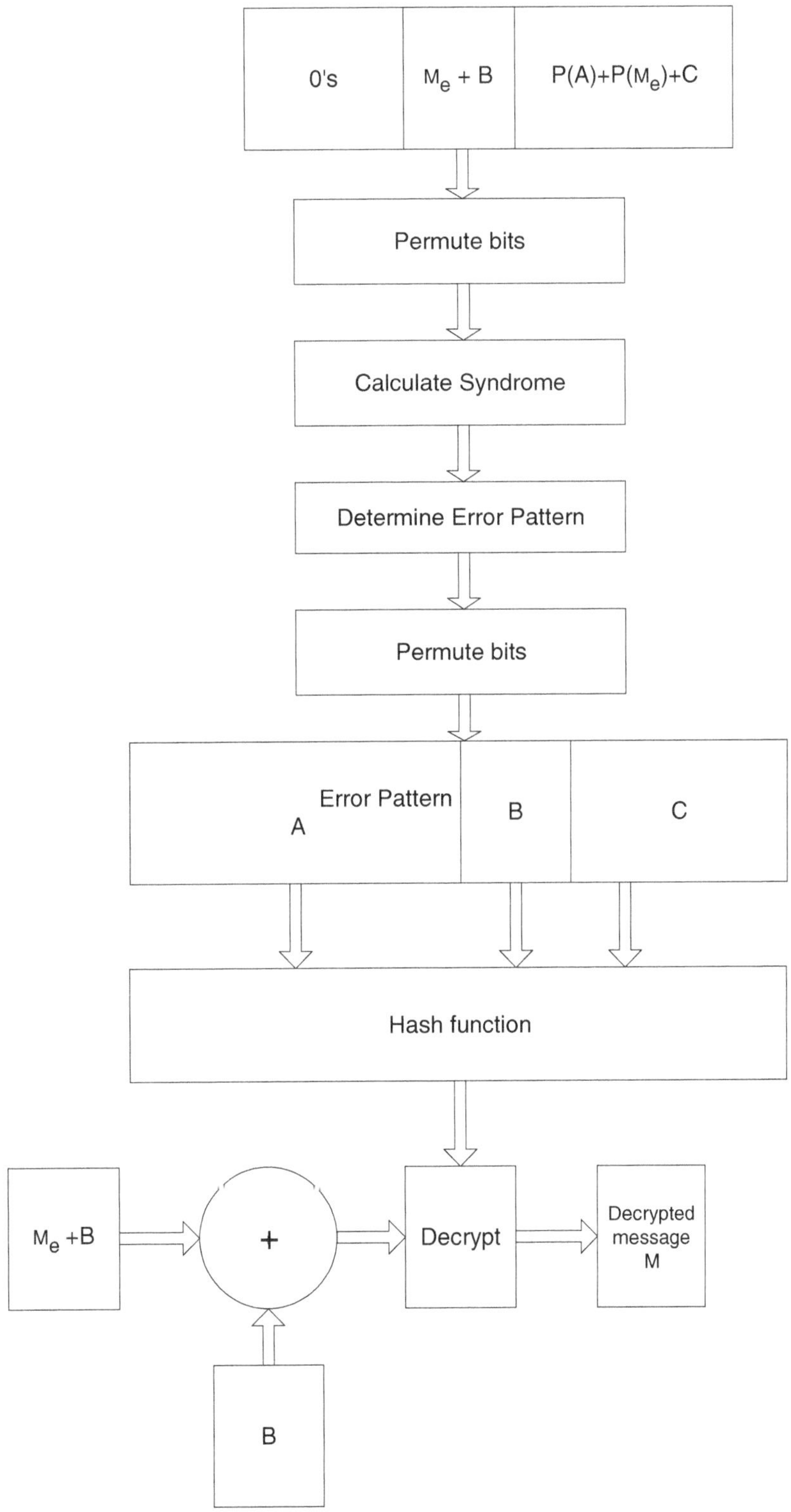

Fig. 19.17 Decryption method for the shortened ciphertext

$P(C)$. Removing error pattern, part A shortens the ciphertext whilst including $P(A)$ instead requires no additional bits in the ciphertext. Since it is necessary to derive the complete error pattern of length n bits in order to decrypt the ciphertext, there is no loss of security from shortening the ciphertext.

The method used to decrypt the ciphertext by using the private key is shown in Fig. 19.17. The received ciphertext is padded with leading zeros to restore its length to 8192 bits. This is then permuted to be in the same order as the Goppa code and the parity check matrix of the Goppa code is used to calculate the syndrome. A Goppa code error-correcting decoder is then used to find the permuted error pattern A, B and C from this syndrome. The most straightforward error-correcting decoder to use is based on Retter's decoding method [13]. This involves calculating a syndrome having $2(n-k)$ parity bits from the parity check matrix of $g^2(z)$ where $g(z)$ is the Goppa polynomial of the code, then using the Berlekamp–Massey method to solve the key equation as in a standard BCH decoder to find the error bit positions. It is because the codewords are binary codewords and the base field is 2, that the Goppa code codewords satisfy the parity checks of $g^2(z)$ as well as the parity checks of $g(z)$, since $1^2 = 1$.

As shown in Fig. 19.17, once the error pattern is determined, it is inverse permuted to produce A, B and C which is hashed to produce the decryption key needed to decrypt M_e back into the plaintext message M. Part B of the derived error pattern is added to the $M_e + B$ contained in the received ciphertext to produce M_e as shown in Fig. 19.17.

19.5 Security of the Cryptosystem

If we consider the parameters that Professor McEliece originally chose, a code of length 1024 bits correcting 50 errors and 524 information bits, then a brute force attack may be based on guessing the error pattern, adding this to the cryptogram and checking if the result is a valid codeword. Checking if the result is a codeword is easy. By using elementary matrix operations on the public key, the generator matrix, we can turn it into a reduced echelon matrix whose transpose is the parity check matrix. We simply use this parity check matrix to calculate the syndrome of the n bit input vector. If the syndrome is equal to zero then the input vector is a codeword.

The maximum number of syndromes that need to be calculated is equal to the number of different error patterns which is $\binom{n}{t} = \binom{1024}{50} = 3.19^{85} \approx 2^{284}$. This may be described as being equivalent to a symmetric key encryption system with a key length of 284 bits.

However, there are much more efficient ways of determining the error pattern in the cryptogram. An attack called information set decoding [2] as described by Professor McEliece in his original paper [8], may be used. For any (n, k, d) code, k columns of the generator matrix may be randomly selected and using Gauss–Jordan elimination of the rows, there is a probability that a permuted, reduced echelon generator matrix will be obtained which generates the same codeword as the original code. The $k \times k$ sub-matrix resulting from the k selected columns needs to be full rank and the probability of this depends on the particular code. For Goppa codes the probability turns out to be the same as the probability of a randomly chosen $k \times k$ binary matrix being full rank. This probability is 0.2887 as described below in Sect. 19.5.1.

Given a cryptogram containing t errors an attacker can select k bits randomly, construct the corresponding permuted, reduced echelon generator matrix with a chance of 0.29. The attacker then uses the matrix to generate a codeword and finds the Hamming distance between this codeword and the cryptogram. If the Hamming distance is exactly t then the cryptogram has been cracked.

For this to happen all of the k selected bits from the cryptogram need to be error free. The probability of this is:

$$\prod_{i=0}^{k-1} \frac{n-t-i}{n-i} = \frac{(n-t)!(n-k)!}{(n-t-k)!n!}$$

Including the chance of 0.29 that the selected matrix has rank k, the average number of selections of k bits from the cryptogram before the cryptogram is cracked, N_{ck} is given by

$$N_{ck} = \frac{(n-t-k)!n!}{0.29(n-t)!(n-k)!} \tag{19.53}$$

For the original code parameters (1024, 524, 101), $N_{ck} = 4.78 \times 10^{16} \approx 2^{55}$.

This is equivalent to a symmetric key encryption system with a key length of 55 bits, a lot less than 284 bits, the base 2 logarithm of the number of error combinations.

Using a longer code offers much more security. For example using code parameters (2048, 1300, 137), $N_{ck} = 1.45 \times 10^{31} \approx 2^{103}$, equivalent to a symmetric key length of 103 bits.

For code parameters (8192, 5124, 473), with a Goppa code which corrects 236 errors, it turns out that $N_{ck} = 5.60 \times 10^{103} \approx 2^{344}$, equivalent to a symmetric key length of 344 bits.

The success of this attack depends upon the code rate. The effect of the code rate, R and the security as expressed in the equivalent symmetric key length in bits is shown in Fig. 19.18 for a code length of 2048 bits. The code rate, R that maximises N_{ck} for a given n is tabulated in Table 19.3, together with t the number of correctable errors and the equivalent symmetric key length in bits.

19.5.1 Probability of a k × k Random Matrix Being Full Rank

The probability of a randomly chosen $k \times k$ binary matrix being full rank is a classical problem related to the erasure correcting capability of random binary codes [4, 5].

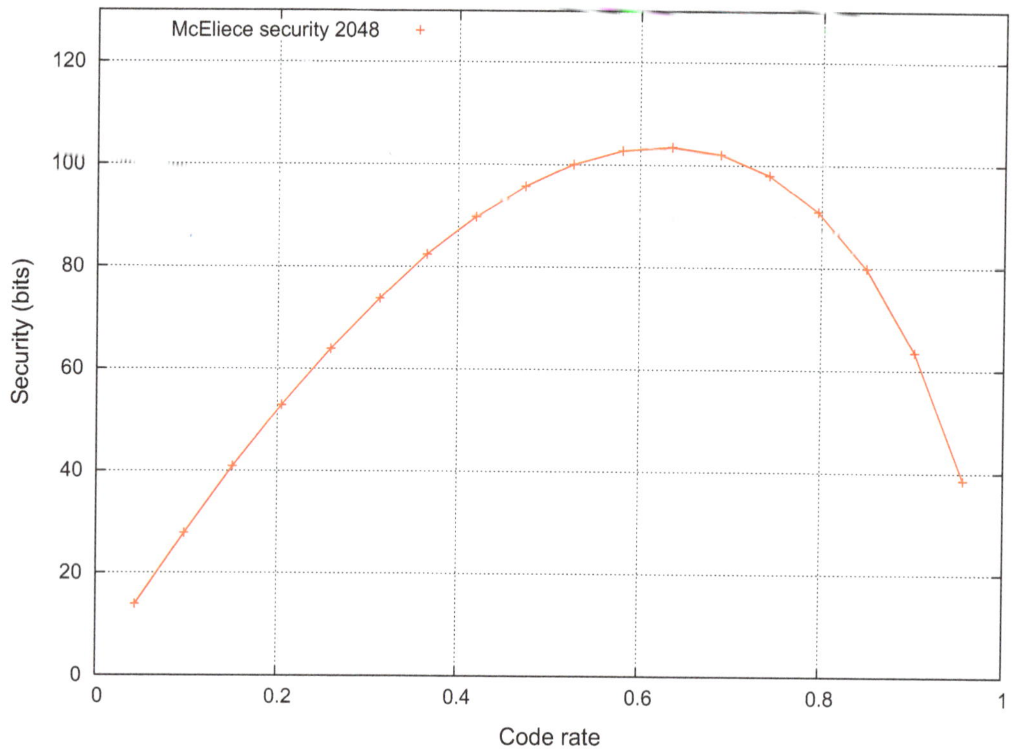

Fig. 19.18 Effect of code rate on security for a code length of 2048 bits

Table 19.3 Optimum code rate giving maximum security as a function of code length

n	R	t	Security (bits)
512	0.6309	21	33.0
1024	0.6289	38	57.9
2048	0.6294	69	103.5
4096	0.6279	127	187.9
8192	0.6287	234	344.6
16384	0.6292	434	637.4

For the binary case it is straightforward to derive the probability, P_k of a $k \times k$ randomly chosen matrix being full rank by considering the process of Gauss–Jordan elimination. Starting with the first column of the matrix, the probability of finding a 1 in at least one of the rows is $(1 - \frac{1}{2})^k$.

Selecting one of these non-zero bit rows, the bit in the second column will be arbitrary and considering the first two bits there are 2^1 linear combinations. As there are 2^2 combinations of two bits, the chances of not finding an independent 2-bit combination in the remaining $k - 1$ rows are $\frac{1}{2^{k-1}}$. Assuming an independent row is found, we next consider the third column and the first three bits. There are 2^2 linear combinations of 3 bits from the two previously found independent rows and there are a total possible 2^3 combinations of 3 bits. The probability of not finding an independent 3-bit pattern in any of the remaining $k - 2$ rows is $(\frac{2^2}{2^3})^{k-2} = \frac{1}{2^{k-2}}$.

Table 19.4 Probability of a random binary $k \times k$ matrix having full rank

k	P_k
5	0.298004
10	0.289070
15	0.288797
20	0.288788
50	0.288788

Proceeding in this way to k rows, it is apparent that P_k is given by

$$P_k = \prod_{i=0}^{k-1} 1 - \left(1 - \frac{1}{2}\right)^{k-i} = \prod_{i=0}^{k-1} 1 - \frac{1}{2^{k-i}} \tag{19.54}$$

The probability of P_k as a function of k is tabulated in Table 19.4. The asymptote of 0.288788 is reached for k exceeding 18.

19.5.2 Practical Attack Algorithms

Practical attack algorithms of course need to factor in the processing cost of Gauss–Jordan elimination compared to the problem of constructing different generator matrices. A completely different set of k coordinates does not need to be selected each time to generate a different generator matrix as discussed in [2]. Also, even if the k selected columns of the generator matrix do not have full rank, usually discarding and adding one or two columns will produce a full rank matrix. It can be shown that on average only 1.6 additional columns are necessary to achieve full rank. Canteaut and Chabaud [3] showed that by including the cryptogram as an additional row in the generator matrix of the $(n, k, 2t + 1)$ code, a code is produced with parameters $(n, k+1, t)$ for which there is only a single codeword with weight t, the original error pattern in the cryptogram. In this case algorithms for finding low-weight codewords may be deployed to break the cryptogram. However these low-weight codeword search algorithms are all very similar to the original algorithm aimed at searching for a codeword of the $(n, k, 2t + 1)$ code with Hamming distance t from the cryptogram. The conclusions of the literature are that information set decoding is an efficient method of attacking the McEliece system but that from a practical viewpoint the system is unbreakable provided the code is long enough. Bernstein et al. [2] give recommended code lengths and their corresponding security, providing similar results to that of Table 19.3.

The standard McEliece system is vulnerable to chosen-plaintext attacks. The encoder is the public key, usually publically available, and the attacker can simply guess the plaintext, construct the corresponding ciphertext and compare this to the target ciphertext. In addition, if the same plaintext message is encrypted twice the sum of the two ciphertexts is a n bit vector of $2t$ bits or less.

The standard McEliece system is also vulnerable to chosen-ciphertext attack. Assuming a decryption oracle is available, the attacker inverts two bits randomly in the ciphertext and sends the result to the decryption oracle. With probability $\frac{t(n-t)}{n(n-1)}$, a different ciphertext will be produced containing exactly t errors and the decryption oracle will output the plaintext, breaking the system.

Encrypting the plaintext using a key derived from the error pattern, as described above, defeats all of these attacks.

19.6 Applications

Public key encryption is attractive in a wide range of different applications, particularly those involving communications because the public keys may be exchanged initially using clear text messages followed by information encrypted using the public keys. The private keys remain private because they do not need to be communicated and the public keys are of no help to an eavesdropper.

An example of an application for the iPhone and iPad using the McEliece public key encryption system is the S2S app pictured in Fig. 19.19. In this app, files are encrypted with users' public keys and stored in the cloud so that they may be shared. Sharing is by means of links that index the encrypted files on the cloud and each user uses their private key to decrypt the shared files.

If the same type of application was implemented using symmetric key encryption, it would be necessary for users to share passwords with all of the associated risks that entails. Using public key encryption avoids these risks. Another application example is the secure Instant Messaging (IM) system, PQChat for the iPhone, iPad and Android devices. There is an option button which shows messages in their received encrypted format, as shown in Fig. 19.20. The application is called PQChat and the name stands for Post-Quantum Chat as the McEliece cryptosystem is relatively immune to attack by a quantum computer, unlike the public key encryption systems in common use today, such as Rivest Shamir Adleman, (RSA) and Elgamal.

As with other public key methods, the system may be used for mutual authentication. Party X sends a randomly chosen nonce x_1, together with a timestamp to Party Y using Party Y's public key. Party Y returns a randomly chosen nonce y_1, timestamp and $hash(hash(x_1, y_1))$ to Party X using Party X's public key. Party X replies with an encrypted timestamp and acknowlegement, using symmetric key cryptography with encryption key $hash(x_1, y_1)$, the preimage of $hash(hash(x_1, y_1))$, to Party Y. The session key $hash(x_1, y_1)$ is used for further exchanges of information, for the duration of the session. A cryptographic hash function is used such as SHA-3 [12] which also has good, second preimage resistance.

It is assumed that the private keys have been kept secret and the association of IDs with public keys has been independently verified. In this case, Party X knows Party Y holds the private key of Y and is the only one able to learn x_1. Party Y knows Party

Fig. 19.19 S2S application for sharing encrypted files

X holds the private key of X and is the only one able to learn y_1. Consequently Party X and Party Y are the only ones with knowledge of x_1 and y_1. Using the preimage of *hash*(*hash*(x_1, y_1)) as the session key provides added assurance as both x_1 and y_1 need to be known in order to generate the key, *hash*(x_1, y_1). The timestamps prevent replay attacks being used.

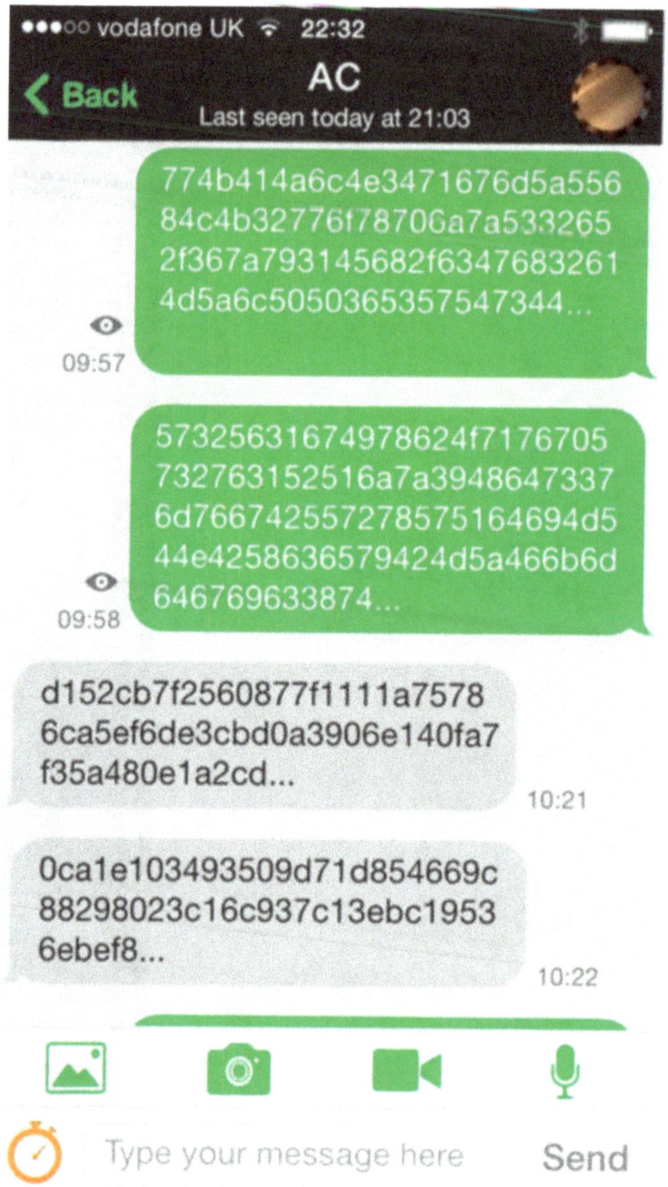

Fig. 19.20 PQChat secure instant messaging app featuring McEliece cryptosystem (with view as received, enabled)

19.7 Summary

A completely novel type of public key cryptosystem was invented by Professor Robert McEliece in 1978 and this is based on error-correcting codes using Goppa codes. Other well established public key cryptosystems are based on the difficulty of determining logarithms in finite fields which, in theory, can be broken by quantum computers. Despite numerous attempts by the crypto community, the McEliece system remains unbroken to this day and is one of the few systems predicted to survive attacks by powerful computers in the future. In this chapter, some variations to the McEliece system have been described including a method which destroys the deterministic link between plaintext messages and ciphertexts, thereby providing semantic security. Consequently, this method nullifies the chosen-plaintext attack, of which the classic McEliece is vulnerable. It is shown that the public key size can be reduced and by encrypting the plaintext with a key derived from the ciphertext random error pattern, the security of the system is improved since an attacker has to determine the exact same error pattern used to produce the ciphertext. This defeats a chosen-ciphertext attack in which two random bits of the ciphertext are inverted. The standard McEliece system is vulnerable to this attack. The security of the McEliece system has been analysed and a shortened ciphertext system has been proposed which does not suffer from any consequent loss of security due to shortening. This is important because to achieve 256 bits of security, the security analysis has shown that the system needs to be based on Goppa codes of length 8192 bits. Key encapsulation and short plaintext applications need short ciphertexts in order to be efficient. It is shown that the ciphertext may be shortened to 1912 bits, provide 256 bits of security and an information payload of 768 bits. Some examples of interesting applications that have been implemented on a smartphone in commercial products, such as a secure messaging app and secure cloud storage app, have been described in this chapter.

References

1. Berlekamp, E.R.: Algebraic Coding Theory, Revised edn. Aegean Park Press, Laguna Hills (1984). ISBN 0 894 12063 8
2. Bernstein, D., Lange, T., Peters, C.: Attacking and defending the McEliece cryptosystem. In: Buchmann, J., Ding, J. (eds.) PQCrypto, pp. 31–46 (2008)
3. Canteaut, A., Chabaud, F.: A new algorithm for finding minimum weight words in a linear code: Application to McEliece's cryptosystem and to narrow-sense BCH codes of length 511. IEEE Trans. Inf. Theory **44**(1), 367–378 (1998)
4. Cooper, C.: On the distribution of rank of a random matrix over a finite field. Random Struct. Algorithms **17**, 197–212 (2000)
5. Dumer, I., Farrell, P.: Erasure correction performance of linear block codes. In: Cohen, G., Litsyn, S., Lobstein, A., Zemor, G. (eds.) Lecture Notes in Computer Science, vol. 781, pp. 316–326. Springer, Berlin (1993)
6. Goppa, V.D.: A new class of linear error-correcting codes. Probl. Inf. Transm. **6**, 24–30 (1970)
7. MacWilliams, F.J., Sloane, N.J.A.: The Theory of Error-Correcting Codes. North-Holland, Amsterdam (1977)
8. McEliece, R.J.: A public-key cryptosystem based on algebraic coding theory. DSN Prog. Rep. **42–44**, 114–116 (1978)
9. Niederreiter, H.: Knapsack-type cryptosystems and algebraic coding theory. Probl. Control Inf. Theory **15**, 159–166 (1986)
10. Publications FIPS. Advanced Encryption Standard (AES). FIPS PUB 197 (2001)
11. Publications FIPS. Secure Hash Standard (SHS). FIPS PUB 180-3 (2008)
12. Publications FIPS. SHA-3 Standard Permutation Based Hash and Extendable Output Functions. FIPS PUB 202 (2015)
13. Retter, C.T.: Decoding Goppa codes with a BCH decoder. IEEE Trans. Inf. Theory IT **21**, 112–112 (1975)
14. Riek, J., McFarland, G.: Error correcting public key cryptographic method and program. US Patent 5054066 (1988)

15. Sidelnikov, V., Shestakov, S.: On insecurity of cryptosystems based on generalized Reed-Solomon codes. Discrete Math. Appl. **2**(4), 439–444 (1992)
16. Sugiyama, Y., Kasahara, M., Namekawa, T.: An erasures-and-errors decoding algorithm for Goppa codes. IEEE Trans. Inf. Theory IT **22**, 238–241 (1976)

Chapter 20
Error-Correcting Codes and Dirty Paper Coding

20.1 Introduction and Background

In the following we are concerned with impressing information on an independent signal, such as an image or an audio stream with the aim of the additional energy used consistent with reliable detection of the information. Information can even be impressed on background noise with no apparent signal present. A secondary aim is that in impressing the information, the independent signal should suffer a minimal amount of degradation or distortion to the point that in some circumstances the difference is virtually undetectable.

20.2 Description of the System

The following, for simplicity, is first described in terms of using binary codes and binary information. It is shown later that the method may be generalised to non-binary codes and non-binary information. The independent signal or noise is denoted by the waveform $v(t)$ and the information carrying signal to be impressed on the waveform $v(t)$ is denoted by $s(t)$. The resulting waveform $w(t)$ is simply given by the sum:

$$w(t) = v(t) + s(t) \tag{20.1}$$

The decoder which is used to determine $s(t)$ from the received waveform will usually be faced with additional noise, interference and sometimes distortion due to the receiving equipment or the transmission. With no distortion, the input to the decoder is denoted by $r(t)$ and given by:

$$r(t) = v(t) + s(t) + n(t) \tag{20.2}$$

M. Tomlinson et al., *Error-Correction Coding and Decoding*,
Signals and Communication Technology,
DOI 10.1007/978-3-319-51103-0_20

In its simplest form $s(t)$ carries only one bit of information and

$$s(t) = k_0 s_0(t) - k_1 s_1(t) \tag{20.3}$$

to convey data 0, and

$$s(t) = k_0 s_1(t) - k_1 s_0(t) \tag{20.4}$$

to convey data 1.

The multiplicative constants, k_0 and k_1 are chosen to adjust the energy of the information carrying signal and k_1 is used to reduce the correlation of the alternative information carrying signal that could cause an error in the decoder. The multiplicative constants, k_0 and k_1 are normally chosen as a function of $v(t)$, the main component of interference in the decoder, which is attempting to decode $r(t)$.

In conventional communications, $s_0(t)$ (or $s_1(t)$) is transmitted or stored and $s_0(t)$ (or $s_1(t)$) is decoded despite the presence of interference or noise. $s(t)$ is added to $v(t)$ and $s_0(t)$ (or $s_1(t)$) is decoded from the composite waveform $v(t) + s(t)$ despite the presence of additional interference or noise.

Noting that the transmitter has no control over the independent signal or noise $v(t)$, a good strategy is to choose $s_0(t)$ and $s_1(t)$ from a large number of possible waveforms in order to produce waveforms which have a large correlation with respect to $v(t)$. Each possible waveform is constrained to be a codeword from an (n, k, d_{min}) error-correcting code. In one approach using binary codes, the 2^k codewords are partitioned into two disjoint classes, codewords having even parity and codewords having odd parity. The codeword $s_0(t)$ is the even parity codeword with highest correlation out of all even parity codewords and the codeword $s_1(t)$ is the odd parity codeword with highest correlation out of all odd parity codewords. The idea is that $w(t)$ should have maximum correlation with $s_0(t)$ if the information data is 0 compared to any of the other $2^k - 1$ codewords. Conversely if the information data is 1, $w(t)$ should have maximum correlation with $s_1(t)$ compared to any of the other $2^k - 1$ codewords. As there is a minimum Hamming distance of d_{min}, between codewords, this prevents small levels of additional noise or interference causing an error in detecting the data in the decoder.

As an example, consider a typical sequence of 47 Gaussian noise samples $v(t)$ as shown in Fig. 20.1. A binary quadratic residue [4] code, described in Chap. 4, the (47, 24, 11) code is used and the highest correlation codeword having even parity is determined using a near maximum likelihood decoder, the modified Dorsch decoder described in Chap. 15. The waveform of Fig. 20.1 is input to the decoder. The highest correlation codeword, which has a correlation value of 20.96 is the codeword:

{010010011001100100100001000010000100 10000000000}

The highest correlation, odd parity codeword, is then determined. This codeword, which has a correlation value of 22.65, is the codeword:

{ 111010010110000110011001000100000100 00100000000 }

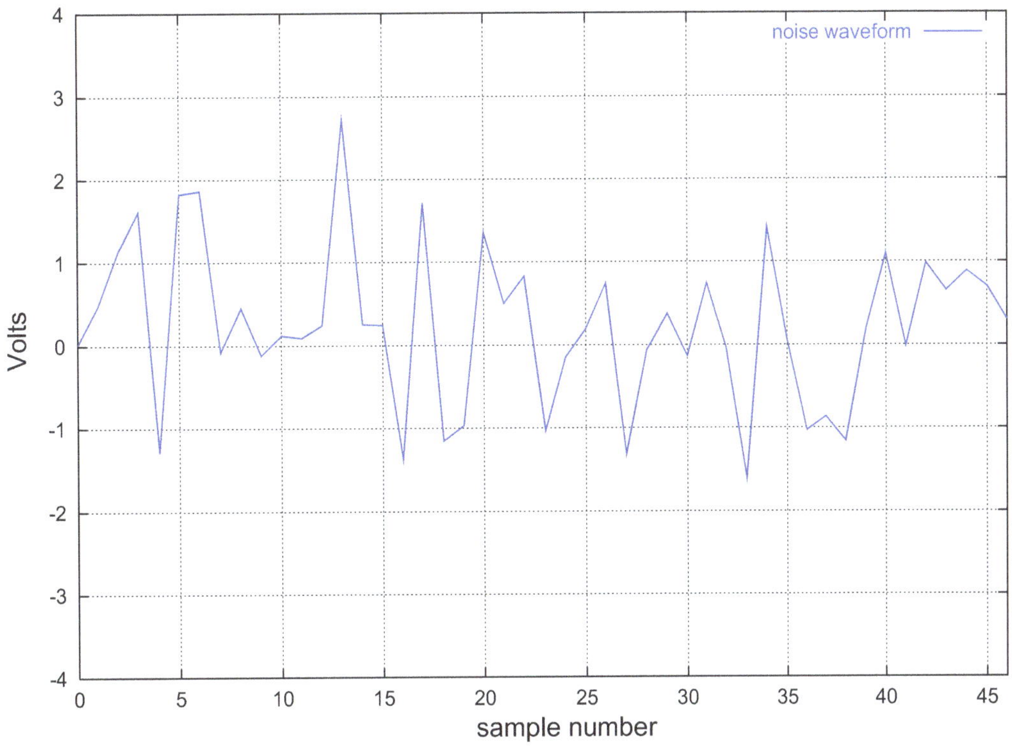

Fig. 20.1 Noise waveform to be impressed with data

It should be noted that in carrying out the correlations, codeword 1's are mapped to -1's and codeword 0's are mapped to $+1$'s.

The information bit to be impressed on the noise waveform is say, data 0, in which case the watermarked waveform $w(t)$ needs to produce a maximum correlation with an even parity codeword. Correspondingly, the value given to k_0 is 0.156 and to k_1 is 0.02 in order to make sure that the codeword which produces the maximum correlation with the marked waveform is the previously found even parity codeword:

{0 1 0 0 1 0 0 1 1 0 0 1 1 0 0 1 0 0 1 0 0 0 0 1 0 0 0 0 1 0 0 0 0 1 0 0 1 0 0 0 0 0 0
0 0 0 0}

The marked waveform $w(t)$ is as shown in Fig. 20.2. It may be observed that the difference between the marked wavefordm and the original waveform is small. In the decoder it is found that the codeword with highest correlation with the marked waveform $w(t)$ is indeed the even parity codeword:

{01001001100110010010000100001000010010000000000 }

and this codeword has a correlation of 28.31.

One advantage of this watermarking system over conventional communications is that the watermarked waveform may be tested using the decoder. If there is insufficient margin, adjustments may be made to the variables k_0 and k_1 and a new

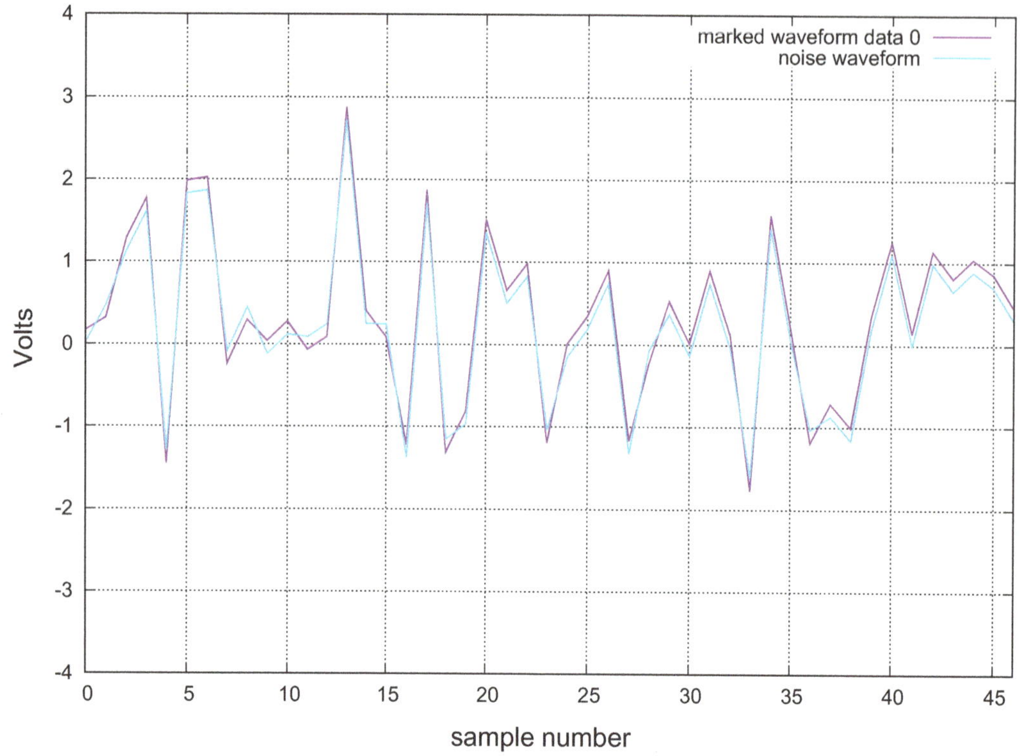

Fig. 20.2 Noise waveform impressed with data 0

watermarked waveform produced. Conversely, if there is more than adequate margin, adjustments may be made to the variables k_0 and k_1, so that there is less degradation to the original waveform $v(t)$.

The highest correlation, odd parity codeword with correlation 25.31 is the codeword:

{ 1110100101100001100110010001000000100 00100000000 }

It should be noted that this odd parity codeword is the same odd parity codeword as determined in the encoder, but this is not always the case depending upon the choice of values for k_0 and k_1.

For the case where the information bit is a 1, the marked waveform $w(t)$ needs to produce a maximum correlation with an odd parity codeword. In this case, the value of k_0 is 0.043 and the value of k_1 is 0.02 and $s(t) = k_0 s_1(t) - k_1 s_0(t)$. The marked waveform $w(t)$ is as shown in Fig. 20.3. This time in the decoder it is found that the codeword with highest correlation with $w(t)$ is indeed the odd parity codeword:

{ 111,0100101100001100110010001000000100001000000000 }

and this codeword has a correlation of 24.70. The highest correlation,even parity, codeword has a correlation of 22.02.

In the encoding and decoding procedure above, the maximum correlation codeword needs to be determined. For short codes a maximum likelihood decoder [6]

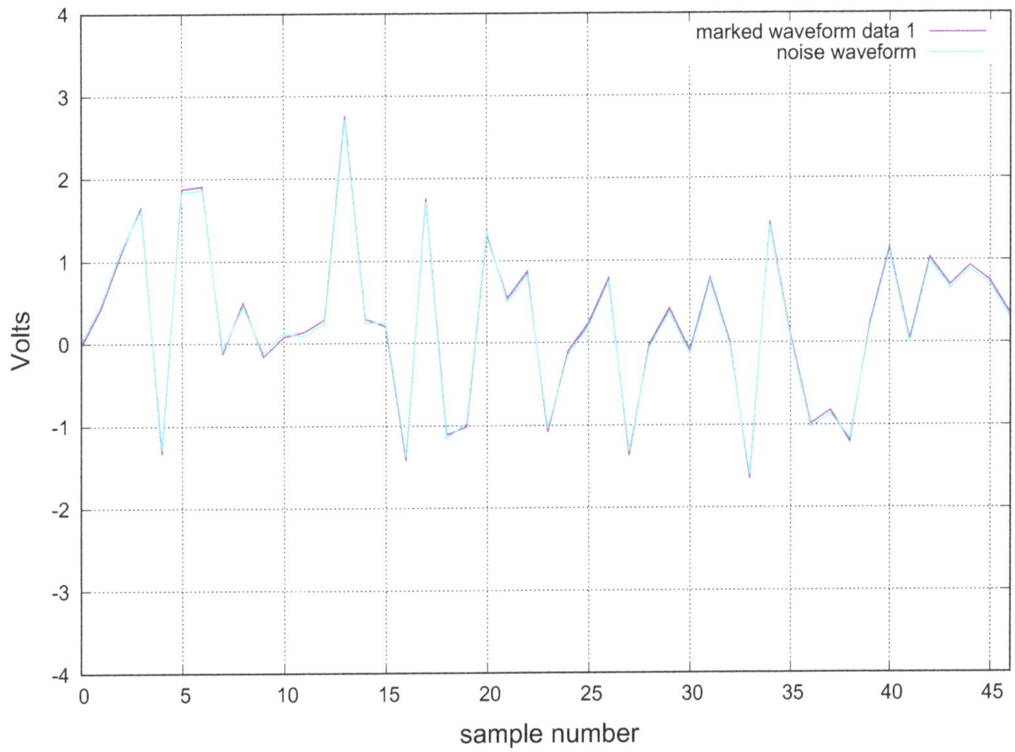

Fig. 20.3 Noise waveform impressed with data 1

may be used. For medium length codes, up to 200 bits long, the near maximum likelihood decoder, the modified Dorsch decoder of Chap. 15 is the best choice. For longer codes, decoders such as an LDPC decoder [3], turbo code decoder [1], or turbo product code decoder [7] may be used in conjunction with the appropriate iterative decoder. An example of a decoder for LDPC codes is given in by Chen [2].

Once the maximum correlation codeword has been found, codewords with similar, high correlation values, may be found from the set of codewords having small Hamming distance from the highest correlation codeword. Linear codes are the most useful codes because the codewords with high correlations with the target waveform are given by the sum of the highest correlation codeword and the low-weight codewords of the code, modulo q, (where $GF(q)$ is the base field [4] of the code). The low-weight codewords of the code are fixed and may be derived directly as described in Chaps. 9 and 13, or determined from the weight distribution of the dual code [4].

For practical implementations, the most straightforward approach is to restrict the codes to binary codes less than 200 bits long and determine the high correlation codewords by means of the modified Dorsch decoder. This conveniently, can output a ranked list of the high cross correlation codewords is together with their correlation values. It is straightforward to modify the decoder so as to provide the output codewords in odd and even parity classes, with the maximum correlation codeword for each class. The results for the example above were determined in this way.

Additional information may be impressed upon the independent signal or noise by partitioning the 2^k codewords into more disjoint classes (other than binary). For example four disjoint classes may be obtained by partitioning the codewords according to odd and even parity for the odd numbered codeword bits and odd and even parity for the even numbered codeword bits. Namely, if the codewords are represented as:

$$c(x) = c_0 + c_1x + c_2x^2 + c_3x^3 + c_4x^4 + \cdots + c_{k-1}x^{k-1} \tag{20.5}$$

then the codewords are partitioned according to the values of p_0 and p_1 given by

$$p_0 = c_0 + c_2 + c_4 + c_6 \cdots + c_{k-1} \text{ modulo } 2 = 0$$
$$p_1 = c_1 + c_3 + c_5 + c_7 \cdots + c_{k-2} \text{ modulo } 2 = 0$$

or with the result

$$p_0 = c_0 + c_2 + c_4 + c_6 \cdots + c_{k-1} \text{ modulo } 2 = 1$$
$$p_1 = c_1 + c_3 + c_5 + c_7 \cdots + c_{k-2} \text{ modulo } 2 = 1$$

Clearly the procedure may be extended to m parity bits by partitioning the 2^k codewords into 2^m disjoint classes. In this case, following encoding, m bits of information will be conveyed by the marked waveform and determined from the codeword which has the highest correlation with the marked waveform. This is by virtue of which of the 2^m classes this codeword resides.

An alternative to this procedure is to use non-binary codes [5] with a base field of $GF(q)$ as described in Chap. 7. For convenience a base field of $GF(2^m)$ may be used so that each symbol of a codeword is represented by m bits. In this case codewords are partitioned into 2^m classes according to the value of the overall parity sum:

$$p_0 = c_0 + c_1 + c_2 + c_3 + c_4 + \cdots + c_{k-1} \text{ modulo } 2^m \tag{20.6}$$

The n non-binary symbols of each codeword may be mapped into n Pulse Amplitude Modulation (PAM) symbols [6] or into $n.m$ binary symbols or a similar hybrid combination before correlation with $v(t)$.

Rather than maximum correlation with the waveform to be marked, codewords may be chosen that have near zero correlation with the waveform to be marked. Information is conveyed by the watermarked marked waveform by the addition of a codeword to $v(t)$, which is orthogonal or near orthogonal to the codeword which has maximum correlation to the independent signal or noise waveform $v(t)$. In this case, the codeword with maximum correlation to $v(t)$ is denoted as $s_{max}(t)$. Codewords that are orthogonal or near orthogonal to $s_{max}(t)$ are denoted as $s_{max,i}(t)$ for $i = 1$ to 2^m. The signal impressed upon $v(t)$ is:

$$s(t) = k_0 s_{max}(t) + k_1 s_{max,\eta}(t) \tag{20.7}$$

where η determines which one of the 2^m orthogonal codewords is impressed on the waveform to convey the m bits of information data. The addition of the maximum correlation codeword $k_0 s_{max}(t)$ to $v(t)$ is to make sure that $s_{max}(t)$ is still the codeword with maximum correlation after the waveform has been marked. Although the codewords $s_{max,i}(t)$ for $i = 1$ to 2^m are orthogonal to $k_0 s_{max}(t)$ they are not necessarily orthogonal to $v(t)$. In this case, the signal impressed upon $v(t)$ needs to be:

$$s(t) = k_0 s_{max}(t) + \sum_{i=1}^{2^m} k_i s_{max,i}(t) \tag{20.8}$$

The coefficients k_i will usually be small in order to produce near zero correlation of the codewords $s_{max,i}$ with $w(t)$ except for the coefficient k_j in order to produce a strong correlation with the codeword $s_{max,j}$.

The choice of the multiplicative constants, k_0 and k_1 or the multiplicative constants k_i for the general case (these adjust the energy of the components of the information signal), depends upon the expected levels of additional noise or interference and acceptable levels of decoder error probability. If the marked signal to noise ratio is represented as SNR_z, the marked signal energy as E_z, and the difference in highest correlation to next highest correlation of the codewords is Δ_c, then the probability of decoder error $p(e)$ is lower bounded by:

$$p(e) \leqslant \frac{1}{2} erfc\left(\frac{\Delta_c^2 . SNR_z}{8.E_z}\right)^{0.5} \tag{20.9}$$

This is under the assumption that there is only one codeword close in Euclidean distance to the maximum correlation codeword.

The multiplicative constants may be selected “open loop” or “closed loop”. In “closed loop”, which is a further variation of the system, the encoding is followed by a testing phase. After encoding, the information is decoded from the marked waveform and the margin for error determined. Different levels of noise or interference may be artificially added to the marked waveform, prior to decoding, in order to assist in determining the margin for error. If the margin for error is found to be insufficient, then the multiplicative constants may be adjusted and a new marked waveform $w(t)$ produced and tested.

In the decoder, once the maximum correlation codeword has been detected from the marked signal or noise waveform, candidate orthogonal, or near orthogonal codewords, are generated from the maximum correlation codeword and these codewords are cross correlated with the marked signal or noise waveform in order to determine which weighted orthogonal, or near orthogonal, codewords have been added to the marked signal or noise waveform. In turn the detected orthogonal, or near orthogonal, codewords from the cross correlation coefficients are used to determine the additional information which was impressed on the marked signal or noise waveform.

In order to clarify the description, Fig. 20.4 shows a block diagram of the encoder for the example of a system conveying two information bits. The independent signal

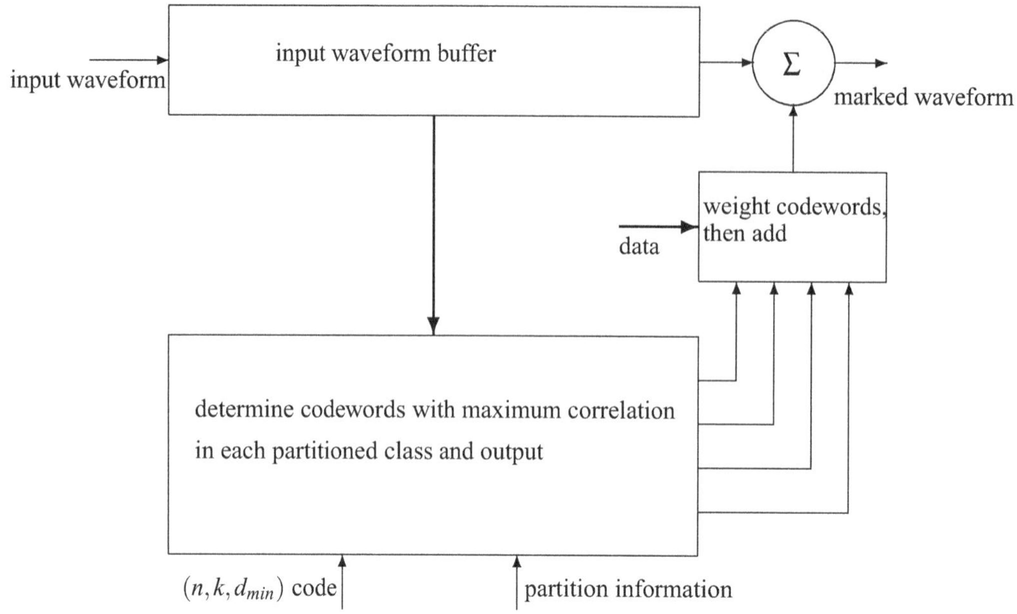

Fig. 20.4 Encoder for two information bits using near orthogonal codewords

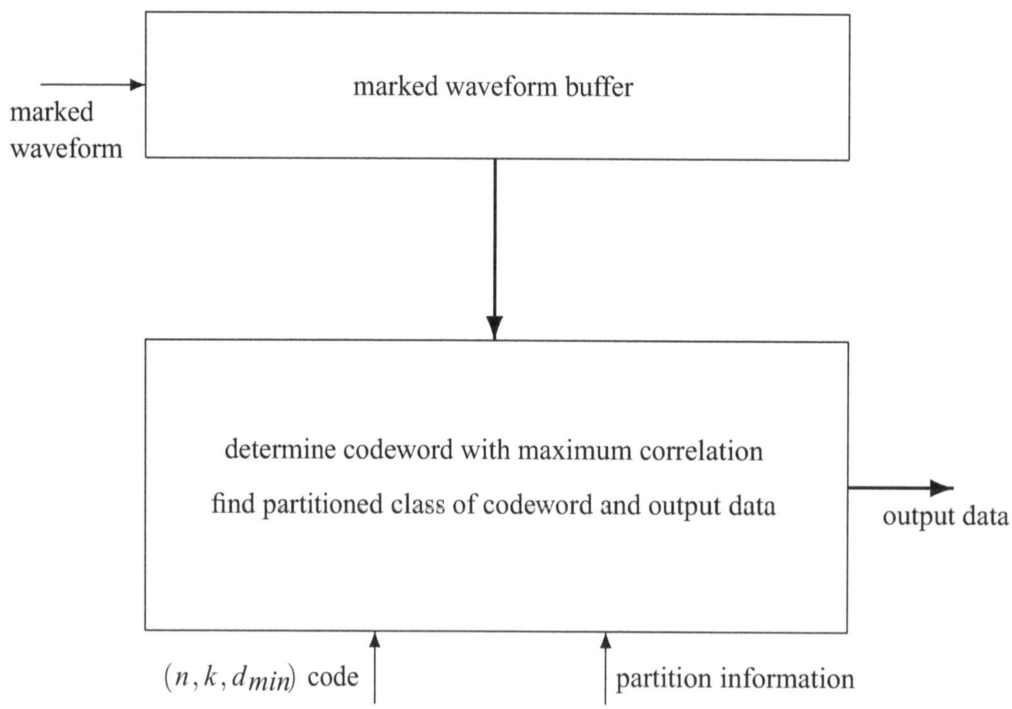

Fig. 20.5 Decoder for marked waveform containing orthogonal codewords

or noise is input to a buffer memory which feeds a maximum correlation decoder, which usually will be a modified Dorsch decoder. The maximum correlation decoder has as input the error-correcting code parameters (n, k, d_{min}) and the code partition information. In this case the partition information is used to partition the codewords into four classes. The codewords, in each class, having highest correlation, and their correlation values are output as shown in Fig. 20.4. From the input data and these correlation values, the multiplicative constants are determined. The coefficients of each codeword are weighted by these constants, and added to the stored independent signal or noise to produce the marked waveform, which is output from the encoder.

Figure 20.5 shows a block diagram of the corresponding decoder. The marked waveform is input to the buffer memory which feeds a maximum correlation decoder. The error-correcting code parameters of the same (n, k, d_{min}) code and the code partition information are also input to the maximum correlation decoder. The codeword with the highest correlation is determined. The class in which the codeword resides is found and the two bits of data identifying this class are output from the decoder.

In a further approach, additional information may be conveyed by adding weighted codewords to the marked signal or noise waveform such that these codewords are orthogonal, or near orthogonal, to the codeword having maximum correlation with the marked signal or noise waveform.

20.3 Summary

This chapter has described how error-correcting codes can be used to impress additional information onto waveforms with a minimal level of distortion. Applications include watermarking and steganography. A method has been described in which the modified Dorsch decoder of Chap. 15 is used to find codewords from partitioned classes of codewords, whose waveforms may be used as a watermark which is almost invisible, and still be reliably detected.

References

1. Berrou, C., Thitimajshima, P., Glavieux, A.: Near Shannon limit error correcting coding and decoding: turbo codes. In: Proceedings of IEEE International Conference on Communications, pp. 1064-1070. Geneva, Switzerland (1993)
2. Chen, J., Fossorier, M.P.C.: Near optimum universal belief propagation based decoding of low-density parity check codes. IEEE Trans. Comm **50**(3), 406–414 (2002). March
3. Gallager, R.G.: Low-Density Parity Check Codes. M.I.T. Press, Cambridge (1963)
4. MacWilliams, F.J., Sloane, N.J.A.: The Theory of Error-Correcting Codes. North-Holland (1977)
5. Peterson, W., Weldon, E.J,Jr: Error-Correcting-Codes, 2nd edn. MIT Press, Cambridge (1972)
6. Proakis, J.G.: Digital Communications. McGraw-Hill, New York (1995)
7. Pyndiah, R.M.: Near-optimum decoding of product codes: block Turbo codes. IEEE Trans. Commun. **46**, 1003–1010 (1998). Aug

Index

M. Tomlinson et al., *Error-Correction Coding and Decoding*,
Signals and Communication Technology,
DOI 10.1007/978-3-319-51103-0